10th Annual International Conference on Science and Engineering (10th AIC 2020)

IOP Conference Series: Materials Science and Engineering
Volume 1087

Banda Aceh, Indonesia
15-16 October 2020

Editors:

Sugiarto
Fitri Arnia
Saiful
Bambang Setiawan
Benazir

ISBN: 978-1-7138-4588-1
ISSN: 1757-8981

Printed from e-media with permission by:

Curran Associates, Inc.
57 Morehouse Lane
Red Hook, NY 12571

Institute of Physics
Dirac House, Temple Back
Bristol BS1 6BE UK

Phone: 44 1 17 929 7481
Fax: 44 1 17 920 0979

techtracking@iop.org

Additional copies of this publication are available from:

Curran Associates, Inc.
57 Morehouse Lane
Red Hook, NY 12571 USA
Phone: 845-758-0400
Fax: 845-758-2633
Email: curran@proceedings.com
Web: www.proceedings.com

TABLE OF CONTENTS

doi:10.1088/1757-899X/1087/1/011001

The 10th Annual International Conference 2020 on Sciences and Engineering

Universitas Syiah Kuala

"Connecting the World through Inovation and Sustainable Development"

Banda Aceh, Aceh, Indonesia
October 15-16, 2020

Preface
Welcome Speech from Rector of Universitas Syiah Kuala

Assalamu'alaikum warahmatullahi wabarakatuh,

In the name of Allah, the Most Beneficent and the Most Merciful. May peace, mercy, and blessings of Allah be upon you. Sholawat and salam may everlastingly be upon our adoration, Great Prophet Muhammad SAW, who had guided and led us from darkness into the shining bright path.

Dear colleagues, professors, lecturers, researchers, ladies and gentlemen. On behalf of Universitas Syiah Kuala, I would like to express my sincere gratitude and welcome you to the 10th Annual International Conference (AIC) 2020. It is an annual conference, started in 2011 and organized by Universitas Syiah Kuala, the largest and oldest public university in Aceh. This year, the conference is conducted virtually. As usual, the conference is divided into three sub-conferences and one of them is the 10th AIC on Sciences and Engineering.

I would like to thanks to the keynote and invited speakers, namely Prof. Dr. Wouter Dewulf from The University of Antwerp, Belgium; and Prof. Hamzah Fansuri, Ph.D from Institut Teknologi Sepuluh Nopember, Indonesia. who have allocated their time to fulfil the invitation, and to share knowledge and experiences from their respective academic institutions.

There has always been a gap between peoples and communities who can make effective use of technology and those who cannot. For that matter, all academicians, researchers, and practitioners should consider humanity to make a peaceful world. With humanity, we could change to a better environment and a better life. We believe in knowledge, research, and technology for all people, minorities, people with disabilities, and helping anyone. No matter how much we can accomplish by ourselves, it is never sufficient in this world of knowledge. Therefore, this conference's focal drive is to exchange ideas, and by participating in this exchange, it is hoped that all parties who may benefit from the conference can use it in their work areas. The conference's agenda covers a wide range of exciting topics related to all theoretical and practical aspects, particularly in Sciences and Engineering.

I hope that all findings disseminated in the conference will be published in international proceedings, and all parties involved can benefit and apply the new ideas in the future. I also hope that this conference can bridge and enhance further cooperation and network among the committee, attendees, and keynote speakers in research, publication, or other future academic activities.

Finally, my deepest gratitude goes to the advisory board, organizing committee, international scientific committee, institutions, companies, and volunteers who have contributed to this seminar's success. The committee has organized a vibrant scientific program and presented a highly respected and internationally notorious speakers. Please accept our sincere apologies should some inconveniences occur before, during, or after the event.

Finally, by Allah's grace and permission, **the 10th Annual International Conference Universitas Syiah Kuala 2020** is officially opened. I wish you a very productive, successful, and enjoyable conference.

Wassalamu'alaikum warahmatullahi wabarakatuh,

Prof. Dr. Ir. Samsul Rizal, M.Eng.
Rector of Universitas Syiah Kuala

IOP Conf. Series: Materials Science and Engineering 1087 (2021) 011001 doi:10.1088/1757-899X/1087/1/011001

Welcome Speech from Chairperson

Assalamu'alaikum warahmatullahi wabarakatuhu,

On behalf of the organizing committees, I would like to welcome all of you to Banda Aceh, Indonesia, for the 10th Annual International Conference (AIC) 2020 on Science and Engineering (SE).

Under the theme of "*Connecting the world through innovation and sustainable development*", the keynote and invited speakers in this 10th AIC 2020 on SE will be Prof. Dr. Wouter Dewulf from The University of Antwerp, Belgium; and Prof. Hamzah Fansuri, Ph.D from Institut Teknologi Sepuluh Nopember, Indonesia.

The conference has received 87 full papers for oral and poster presentation and will be published on IOP Conference Series: Material and Science Engineering indexed by Scopus in the first quarter of 2021. To all members of the organizing committees, the international scientific committees, the reviewers, the editors, the moderators and the collaboration partners, we would like to thank all of them for their tremendous efforts to organize this virtual conference successfully. My sincere gratitude goes to Dr. Eng. Sugiarto for his tremendous efforts to direct the conference.

We look forward to having a successful conference, and we hope that all attendees enjoy and benefit from this conference. Also, let's hope this Covid-19 global pandemic will be over as soon as possible so this virtual conference will be the first and the last.

Best regards,

Dr. Dewi Yunita
Chairperson

IOP Conf. Series: Materials Science and Engineering 1087 (2021) 011001 doi:10.1088/1757-899X/1087/1/011001

Organizing Committee

General Chairperson
Dr. Dewi Yunita, Universitas Syiah Kuala, Indonesia

General Secretary
Dr. Suhartono, Universitas Syiah Kuala, Indonesia

Treasurer and Finance Chair
Dr. Hartati Oktarina, Universitas Syiah Kuala, Indonesia

Conference chairperson & Editor in Chief
Dr. Eng. Sugiarto, Universitas Syiah Kuala, Indonesia

Website Developer
Muhammad Chandra Gunawan, Universitas Syiah Kuala, Indonesia

OCS Personnel
Wahyu Rinaldi, ST., MT, Universitas Syiah Kuala, Indonesia
Ridwan Saputra, S.Pt, Universitas Syiah Kuala, Indonesia

IOP Conf. Series: Materials Science and Engineering 1087 (2021) 011001 doi:10.1088/1757-899X/1087/1/011001

Scientific Committee

IOP Conf. Series: Materials Science and Engineering 1087 (2021) 011001 doi:10.1088/1757-899X/1087/1/011001

Keynote & Invited Speakers

Prof. Dr. Wouter Dewulf

University of Antwerp and C-MAT (Belgium)

He is a professor at the University of Antwerp and C-MAT (Belgium) and is academic director of C-MAT. The C-MAT (Centre for Maritime & Air Transport Management) is one of the world's premier suppliers of highly specialized academic as well as practice-based education and research in transport economics, with a focus on maritime transport, air transport and transport research. His main areas of research are in the fields of air cargo, air transport strategy, Ecommerce and urban logistics. He is a visiting professor at University of Hasselt (Belgium), University of Chongqing (China) and Trisakti Institute of Transportation and Logistics (Jakarta, Indonesia).

Prof. Hamzah Fansuri, Ph.D

Institut Teknologi Sepuluh Nopember (ITS) (Indonesia)

Hamzah Fansuri is a Professor at the Department of Chemistry, Institut Teknologi Sepuluh Nopember (ITS) in Surabaya, Indonesia. Hamzah's research interest is in Inorganic materials, particularly Geopolymer, Zeolite, Heterogeneous Catalyst and Inorganic membranes.

IOP Conf. Series: Materials Science and Engineering 1087 (2021) 011002 doi:10.1088/1757-899X/1087/1/011002

Peer review declaration

All papers published in this volume of *IOP Conference Series: Materials Science and Engineering* have been peer reviewed through processes administered by the Editors. Reviews were conducted by expert referees to the professional and scientific standards expected of a proceedings journal published by IOP Publishing.

- **Type of peer review: Single-blind**
- **Describe criteria used by Reviewers when accepting/declining papers. Was there the opportunity to resubmit articles after revisions?: the organizing committee received submitted manuscript and distributed the manuscript to reviewers through open conference system (OCS) of Universitas Syiah Kuala. Submitted manuscript was assigned to reviewers (two reviewers for each of paper). The papers were reviewed based on the Clarity, Originality, Appropriateness of Material and method, Clarity of Findings. If reviewers accepted the papers or accepted with revision then the committee sent the manuscript back to authors, and the authors could address all comments from the reviewers. The authors sent the final manuscripts back to the committee through the OCS. Finally, the 10th AIC-SE editors finalized the final manuscripts by proofreading and checking the format of manuscripts based on IOP Publishing guidelines.**

- **Conference submission management system: Open Conference Systems (OCS) of Universitas Syiah Kuala (http://conference.unsyiah.ac.id/AIC-SE/AIC2020-SE/index)**

- **Number of submissions received: 103 manuscripts.**

- **Number of submissions sent for review: 98 manuscripts.**

- **Number of submissions accepted: 87 manuscripts.**

- **Acceptance Rate (Number of Submissions Accepted / Number of Submissions Received X 100): 84.47%**

- **Average number of reviews per paper: 2 reviewers.**

- **Total number of reviewers involved: 45 reviewers.**

- **Any additional info on review process: The papers were evaluated based on the technical criteria, quality criteria and presentation criteria. The referees contended detailed their reviews and provided an overall evaluation (accepted, accepted with revision, or rejected). For each manuscript, the similarity index was determined by the organisers using Turnitin software at Universitas Syiah Kuala.**

- **Contact person for queries: Assoc. Prof. Dr. Eng. Sugiarto; sugiarto@unsyiah.ac.id**

Stress-strain and absorption of lightweight concrete with artificial lightweight aggregate from diatomaceous earth

M Hasan*, R Putra, T Saidi, M Jamil, N Husna, M Syukranlillah
Department of Civil Engineering, Universitas Syiah Kuala, Banda Aceh, Indonesia

*E-mail: muttaqin@unsyiah.ac.id

Abstract. Lightweight concrete can be produced by using artificial lightweight aggregate (ALWA). Diatomaceous earth has extremely low density therefore can be used as a raw material to make ALWA. It is important to understand a stress-strain relationship of materials before they can be used as structural members. This paper presents the stress-strain of lightweight concrete with ALWA from diatomaceous earth based on the experimental observation. In addition, the relationship between the absorption of the lightweight concrete and the immersion time is also presented. Two types of ALWA were produced which are the ALWA made of diatomaceous earth only and the ALWA with the addition of 5% sawdust. A polynomial model of the stress-strain curve was proposed for lightweight concrete containing a different type of ALWA. The absorption of lightweight concrete with ALWA containing sawdust was higher than that of diatomaceous earth only.

1. Introduction

A deposit of 40,353,700 tons of weight diatomaceous earth in Aceh Besar District, Indonesia has been reported [1]. Some studies on the utilizing of diatomaceous earth from Aceh Besar District on supplementary cementing material have been conducted [2-6]. Due to the low density of diatomaceous earth, there is possible to make artificial lightweight aggregate (ALWA) from diatomaceous earth [7-9] and consequently lightweight concrete. The advantage of using lightweight concrete for both structural and non-structural elements is the reduction of the dead load-carrying by the members which causes the low earthquake load working on the structures as well as the low forces transfer to the soil by the foundation.

In this study lightweight concrete with ALWA from diatomaceous earth originated from Aceh Besar District was developed. For the stress analysis of existing structures as well as the design of new structures, the understanding of the stress-strain relationship of concrete is essential [10-15]. Therefore, as a new material, developing a stress-strain model of this lightweight concrete is necessary. Since diatomaceous earth is a porous material then the understanding of absorption of the developed lightweight concrete is also necessary.

The purpose of this study is to develop the stress-strain model of lightweight concrete with artificial lightweight coarse aggregate based on the experimental data. This model can be used in the designing of reinforced concrete structures using lightweight concrete as a structural material. The other aim of this study is to propose the relationship between the absorption of lightweight concrete with the immersion time.

2. Experimental program

The lightweight concrete in this study was produced by mixing cement, fine aggregate, coarse aggregate, and water in concrete mixer and mixed for about 5 minutes. The concrete specimens were cast in the steel mould. The specimens used for the stress-strain test were cylinder with a diameter of 100 mm and a height of 200 mm while those for absorption test were cube with a size of 50 mm. After 24 hours of casting, the molds were removed and the specimens were cured in water. The stress-strain and absorption tests were conducted when the specimens reached the age of 28 days.

Ordinary Portland Cement (OPC) with a specific gravity of 3.16 was used as cement material. River sand and artificial lightweight aggregate (ALWA) were used as fine and coarse aggregate, respectively. ALWA was made from diatomaceous earth, which is originated from Beureunut Village, Aceh Besar District. Two types of ALWA were made. For the first type, only diatomaceous earth was used as the raw material. For the second type, 5% of sawdust was added on the diatomaceous earth. The diatomaceous earth was ground and sieved by sieve #200. The diatomaceous earth powder was mixed with 5% sawdust the produce the second type of ALWA. The water was added on the diatomaceous earth powder little by little and pellets with the size ranging from 5 mm to 19 mm were made by hand. Later the pellets were roasted at the temperature of 100°C for 24 hours and burned at the temperature of 650°C for 5 hours in the brick burning furnace. The produced ALWA is shown in Figure 1.

(a) (b)

Figure 1. ALWA: (a) made of diatomaceous earth only; (b) made of diatomaceous earth and 5% sawdust.

Table 1. Physical properties of aggregate.

Physical Properties	River Sand	ALWA Type 1	ALWA Type 2
Bulk density (kg/litre)	1.684	0.724	0.623
Specific gravity (oven dry)	2.575	1.461	1.360
Specific gravity (saturated surface dry)	2.518	1.645	1.516
Absorption (%)	2.272	11.499	13.253
Fineness modulus	2.879	6.665	6.874

The bulk density, specific gravity, absorption, and fineness modulus of aggregate were tested according to ASTM C29/C29M-17a, ASTM C127-15, ASTM C128-15, and ASTM C136/C136M-14, respectively [16-19] and the results are shown in Table 1. The concrete mix proportion was designed based on ACI 211.2-98 [20] and shown in Table 2.

Table 2. Mix proportion of concrete for 1 m^3 volume.

Materials	ALWA Type 1	ALWA Type 2
Cement (kg)	448.271	448.271
Sand (kg)	718.961	778.309
ALWA (kg)	623.163	536.230
Water (kg)	201.722	201.722
w/c	0.45	0.45

To obtain the stress-strain curve of lightweight concrete, a compression load was applied on the cylinder specimens according to ASTM C39/39M-18 [21] through an actuator until the specimens failed as shown in Figure 2. The deformation of the specimens was recorded by placing 2 transducers on the frame which was connected to the specimens while the load increment was recorded by the load cell. The data of load and deformation was transferred to a data logger and the stress and strain were then calculated.

Figure 2. Stress-strain test on concrete specimens.

The absorption of lightweight concrete was tested on the 50 mm cube specimens. The specimens were first roasted at the temperature of 100°C for 24 hours and the weight of the specimens was measured. The specimens were again roasted at the temperature of 100°C for 24 hours and the weight of the specimens was again measured. This process was repeated until the weight of the specimens gets constant. The specimens then were immersed in the water and the weight was measured. At the first 30 minutes of the immersion, the weight of the specimens was measured every 5 minutes. In the second 30 minutes of the immersion, the weight of the specimens was measured every 10 minutes. A the third 30 minutes of the immersion, the weight of the specimens was measured every 15 minutes. In this stage, the weight of the specimen became relatively constant. The immersion was continued until the weight of the specimens gets constant.

3. Results and discussion

3.1. Stress-strain relationship

Stress-strain relationships of 5 specimens of lightweight concrete with ALWA type 1 and ALWA type 2 are plotted in Figure 3 and 4, respectively. Those figures show that the stress-strain curve of lightweight concrete is nonlinear even at the beginning of the application of loading. This nonlinear of stress-strain curve is due to the slip of aggregate from the cement paste and the close of void in the concrete when the loading is applied. The stiffness of the concrete decreases with the increase of applied stress due to the initiation and propagation of a crack in the concrete. Therefore, in this study, the non-linear stress-strain models for lightweight concrete with ALWA made from diatomaceous earth were developed.

To develop the stress-strain model, the stress data in Figure 3 and 4 was normalized with maximum stress while the strain data was normalized with strain at peak stress. By using the polynomial regression analysis and considering the initial condition of the material before loading (strain is zero at zero stress), then the stress-strain models of lightweight concrete with ALWA made from diatomaceous earth can be written as follows:

For lightweight concrete with ALWA type 1:

$$\sigma_c' = f_c'\left[2.2\left(\frac{\varepsilon_c'}{\varepsilon_{co}'}\right) - 1.2\left(\frac{\varepsilon_c'}{\varepsilon_{co}'}\right)^2\right] \quad (1)$$

$R^2 = 0.96$

For lightweight concrete with ALWA type 2:

$$\sigma_c' = f_c'\left[1.8\left(\frac{\varepsilon_c'}{\varepsilon_{co}'}\right) - 0.8\left(\frac{\varepsilon_c'}{\varepsilon_{co}'}\right)^2\right] \quad (2)$$

$R^2 = 0.98$

where: σ_c' = concrete stress; ε_c' = concrete strain; f_c' = concrete compressive strength; and ε_{co}' = concrete strain at peak stress. Since the concrete compressive strength and concrete strain at peak stress are material properties, the stress-strain curve of concrete can be drawn once those two parameters are known. The stress-strain curves of lightweight concrete tested in this study calculated by equations (1) and (2) are also plotted in Figure 3 and 4. A good correlation between presented models and experimental results proves the applicability of the models.

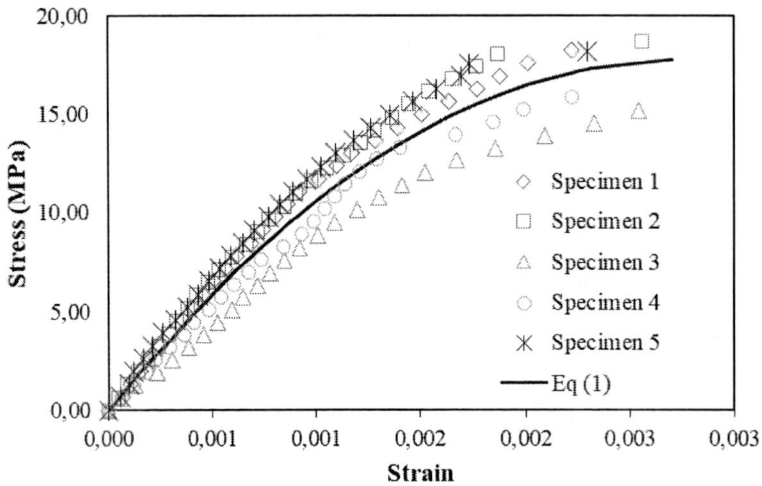

Figure 3. Stress-strain relationship of lightweight concrete with ALWA type 1.

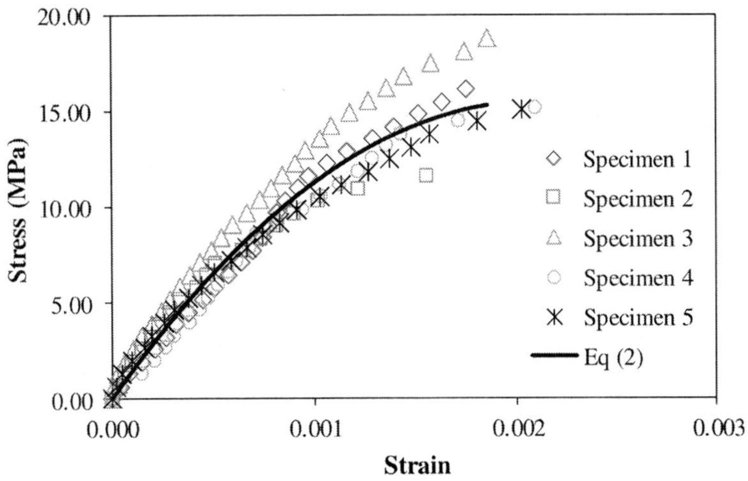

Figure 4. Stress-strain relationship of lightweight concrete with ALWA type 2.

3.2 Absorption

The absorption of 5 lightweight concrete cube specimens with ALWA type 1 and ALWA type 2 as a function of immersion time is shown in Figure 5. Since after 105 minutes of immersion the absorption became constant, only the absorption of lightweight concrete until 105 minutes of immersion is shown in Figure 5. This figure shows that the absorption of lightweight concrete with ALWA type 2 had higher absorption. This is because of the presence of sawdust in the ALWA type 2 so that can absorb more water. As shown in Table 1 the ALWA type 2 had higher absorption compared to ALWA type 1 resulted in higher absorption of lightweight concrete with ALWA type 2. Using the regression analysis, the relation between the absorption of lightweight concrete with ALWA from diatomaceous earth can be written as follows:

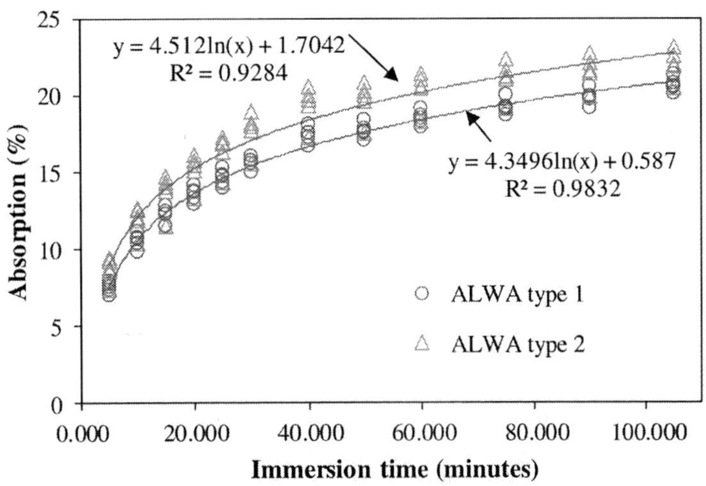

Figure 5. Absorption of lightweight concrete with ALWA from diatomaceous earth.

For lightweight concrete with ALWA type 1:

$$y = 4.350 \ln(x) + 0.587 \quad (3)$$

$R^2 = 0.98$

For lightweight concrete with ALWA type 2:

$$y = 4.512 \ln(x) + 1.704 \quad (4)$$

$R^2 = 0.93$

where y = absorption (%) and x = immersion time (minutes).

4. Conclusions

Based on the results in this study the following conclusions can be drawn:

1. A two-degree polynomial stress-strain model of lightweight concrete with artificial lightweight coarse aggregate (ALWA) was developed by using regression analysis. Two types of ALWA were produced which are ALWA made from diatomaceous earth only and ALWA made from diatomaceous earth with the addition of 5% sawdust. The model for concrete with the two different types of ALWA was the same as the difference is only in the regression constants.
2. The absorption of lightweight concrete with artificial lightweight coarse aggregate (ALWA) increased with the immersion time until 105 minutes. After 105 minutes of immersion in the water, the absorption became constant. The absorption of concrete with ALWA containing sawdust had higher absorption.

5. Acknowledgement

This research was supported by a grant provided by Universitas Syiah Kuala Research and Community Service Center (LPPM) under contract number of 85/UN11.2.1/PT.01.03/PNBP/2020.

References

[1] Pemerintah Kabuaten Aceh Besar 2006 *Laporan RTRW Kabupaten Aceh Besar.* Pemerintah Kabuoaten Aceh Besar.

[2] Hasan M, Saidi T 2020 *IOP Conf. Ser. Mater. Sci. Eng.,* **796** 012034.

[3] Hasan M, Saidi T, Muyasir A, Alkhali Y R, Muslimsyah. 2020 *IOP Conf. Ser. Mater. Sci. Eng. (in press).*

[4] Saidi T, Hasan M, Riski A D D, Ayunizar R R, Mubarak A 2020 *IOP Conf. Ser. Mater. Sci. Eng. (in press).*

[5] Hasan M, Muyasir A, Saidi T, Husaini, Azzahra R 2020 *Defect and Diffusion Forum,* **402** 7-13.

[6] Hasan M, Riski A D D, Saidi T, Husaini, Rahman P N 2020 *Defect and Diffusion Forum,* **402** 50-55.

[7] Fragoulis D, Stamatakis M G, Chaniotakis E, Columbus G 2004 *Materials Characterization,* **53** 307-316.

[8] Posi P, Lertnimoolchai S, Sata V, Chindaprasirt P 2013 *Construction and Building Materials,* **47** 896-901.

[9] Pimraksa K, Chindaprasirt P 2009 *Ceramics International,* **35** 471-478.

[10] Hasan M, Okuyama H, Sato Y, Ueda T 2004 *Journal of Advanced Concrete Technology*, **2(1)** 89-99.

[11] Hasan M, Ueda T, Sato Y 2008 *Journal of Materials in Civil Engineering,* **20** 37-45.

[12] Hasan M, Nagai K, Sato Y, Ueda T 2002 *Proc. JCI,* **24** 109-114.

[13] Hasan M, Husin S, Nursaniah C 2016 *Key Engineering Materials,* **711** 302-309.

[14] Hasan M, Husaini, Abdullah N M 2019 *IOP Conf. Ser. Mater. Sci. Eng.,* **523** 012034.

[15] Hamzani, Munirwansyah, Hasan M, Sugiarto S 2019 *IOP Conf. Ser. Mater. Sci. Eng.,* **523** 012037.
[16] ASTM C29/29M-17a 2017 *Standard Test Method for Bulk Density (Unit Weight) and Voids Aggregate.*
[17] ASTM C127-15 2015 *Standard Test Method for Relative Density (Specific Gravity) and Absorption of Coarse Aggregate.*
[18] ASTM C128-15 2015 *Standard Test Method for Relative Density (Specific Gravity) and Absorption of Fine Aggregate.*
[19] ASTM C136/C136M-14 2014 *Standard Test Method for Sieve Analysis of Fine and Coarse Aggregate.*
[20] ACI 211.2-98 2004 *Standard Practice for Selecting Proportions for Structural Lightweight Concrete.*
[21] ASTM C39/C39M-18 2018 *Standard Test Method for Compressive Strength of Cylindrical Concrete Specimens.*

Impact of dedicated mini roundabout on the capacity and level of services of the unsignalized intersection

K Rahmat[1], S Sugiarto[2,*], S M Saleh[2]
[1]Civil Engineering Postgraduate Program, Department of Civil Engineering, Universitas Syiah Kuala, Banda Aceh, Indonesia
[2]Department of Civil Engineering, Universitas Syiah Kuala, Banda Aceh, Indonesia

*E-mail: sugiarto@unsyiah.ac.id

Abstract. This paper is aiming at investigating the impact of dedicated mini-roundabout on the capacity and level of services at a four-legged unsignalized intersection. The assessment was carried out based on two scenarios namely before the installation mini-roundabout (existing) and after installation (improvement). Datasets were collected at Ujung Stasiun KA unsignalized intersection, which is in the center of Bireuen City, Aceh province of Indonesia. A video-cameras are installed to observed traffic patterns including hourly volume, movement trajectories, and vehicle composition during three days on a morning, afternoon, and evening peak hour traffic. The Indonesian highway capacity manual (IHCM, 1997) is used to evaluate the capacity and level of service in existing conditions while the simulation approach using VISSIM is then performed to assess the scenario of virtual implementation dedicated mini roundabout. The result of analysis demonstrates that the capacity of unsignalized slightly increases by 5% while the delay and level of service are significantly reduced to 42% (31.02 sec to 18 sec), and improved from E to C, for delay and LOS, respectively. It can be concluded that the improvement of the existing condition by implemented a dedicated mini roundabout could improve the capacity and LOS in the study area.

1. Introduction

Unsignalized intersections are one of the major traffic facilities implemented in emerging cities including in Bireuen City, Aceh province of Indonesia. Such traffic facility has significantly affected the quality of services of urban road network especially traffic condition in the area highly consists of motorcycles and cars. Moreover, drivers' behaviors in developing countries are extremely complicated, for instances they do not follow lane discipline, easy to maneuver (lane changing) and ignoring safety gab during driving. This driving condition has significantly degraded the capacity and level of services of the unsignalized intersection. Several previous studies have noticed that the quality of services of urban corridors are noticeably dependent on how traffic mechanisms occur within the systematic [1] and virtual bottlenecks such as interference on median opening [2], disturbance on the on-street segment [3], driving maneuver at roundabout [4]. These findings decided that the U-turn and on-street parking behaviors are drastically influenced by the traffic mechanism on its segments.

A roundabout is one of the types of an intersection that forms at least three-legs and plays a vital role in delivering traffic [4]. The roundabout is mostly installed at the center of the intersection and normally has an elevated level compared to the normal road surface. This could avoid drivers passing through the roundabout. The roundabout is also one of an unsignalized intersection especially deal with moderate traffic volume. Troutbeck and Brilon [5] says that the roundabout may be less potential for traffic accidents since traffic flow merges and diverges at small angles and lower speeds. Additionally, the roundabout has been widely implemented as a substitute for unsignalized intersection due to its benefit in decreasing delay and delivering a safer vehicle's movement compared with ordinary unsignalized intersection. Lastly, the roundabout is also provided better capacity as contrasted to a two-way-stop controlled or an all-way-stopped control intersection, under low traffic demand [6].

The Indonesian Capacity Highway Capacity Manual (IHCM) 1997 [7] is a guide for engineers and planners for planning, designing and operating traffic facilities in Indonesia. This traffic code consists of urban street facilities, roundabout, highways and including signalized and unsignalized intersections. As regard to unsignalized intersection operation, the dynamic vehicles and driver's behavior sometimes create irregularity movements within the conflict area of unsignalized intersection. The drivers mostly always rejected the and the role of priority movement at unsignalized intersection they ignored. This situation is repeatable and occurs most of the time particularly during peak hour traffic. It should be noted that traffic breakdown in emerging cities could be induced either by high traffic demand or the irregularity movement created by drivers within the conflict area. In this case, managing movements of the road users within the conflict area could improve the regularity traffic movements. Since roundabout able to circulate the movement, it should be reduced potential conflict by the driver and enhance the quality of service of the intersection.

To sum, therefore, this paper is aiming at investigating the impact of dedicated mini-roundabout on the capacity and level of services at a four-legged unsignalized intersection. The assessment was carried out based on two scenarios namely before the installation mini-roundabout (ex-ante) and after installation (ex-post). Datasets were collected at Ujung Stasion KA unsignalized intersection, which is in the center of Bireuen City, Aceh province of Indonesia. A video-cameras are installed to observed traffic patterns including hourly volume, movement trajectories, and vehicle composition during three days on a morning, afternoon, and evening peak hour traffic. IHCM 1997 is used to evaluate the capacity and level of service in existing conditions while the simulation approach using VISSIM is then performed to assess the scenario of virtual implementation dedicated mini roundabout. The remainder of this paper is organized as follows. The following section describes the area study, data collection, and processing. The next section explains the analysis method using IHCM and the setting of simulation using VISSIM. Lastly, discussions and conclusions are presented at the end of the paper.

2. Area of Study, data Collection and processing

Figure 1 shows the area of study in which the targeted location is one of the most congestion and irregular traffic movement within the city of the Bireuen. The intersection named Ujung Stasiun KA in which this intersection has four-legged with no priority roles are operated. It is observed that the traffic movements at the targeted intersection mostly consists of passenger car and motorcycle, and a few of them are heavy vehicles such as bus or truck.

Regarding the data collection method, the geometric condition including approach width, lane width, the dimension of intersection and conflict area was measured directly onsite by surveyors. As for geometric was observed during off-peak hour traffic to avoid obstacles due to high traffic. These geometric data were measured directly and filled to the form of the survey by surveyors. In addition to this, the side friction data has also collected directly on site. It is observed that the targeted intersection was considerably had medium to high side friction. This partially due to the location of intersection is located in the dense urban road which pedestrian, entry-exit vehicle at parking lots, the slow-moving vehicle is quite significant even if non-motorized transport is less observed. Furthermore, Traffic stream data such as trajectory movements, signs, and typical potential traffic conflict were observed directly at the location. Moreover, traffic stream data such as volume, vehicle composition, and type of vehicle movements were recorded using a video camera for each approach. The cameras were mounted on the high-rise building at the location of the survey. Traffic stream characteristic was recorded during peak hour traffic within the morning peak (07:15-08:50), afternoon (12:00-14:00) and evening (16:30-18:00).

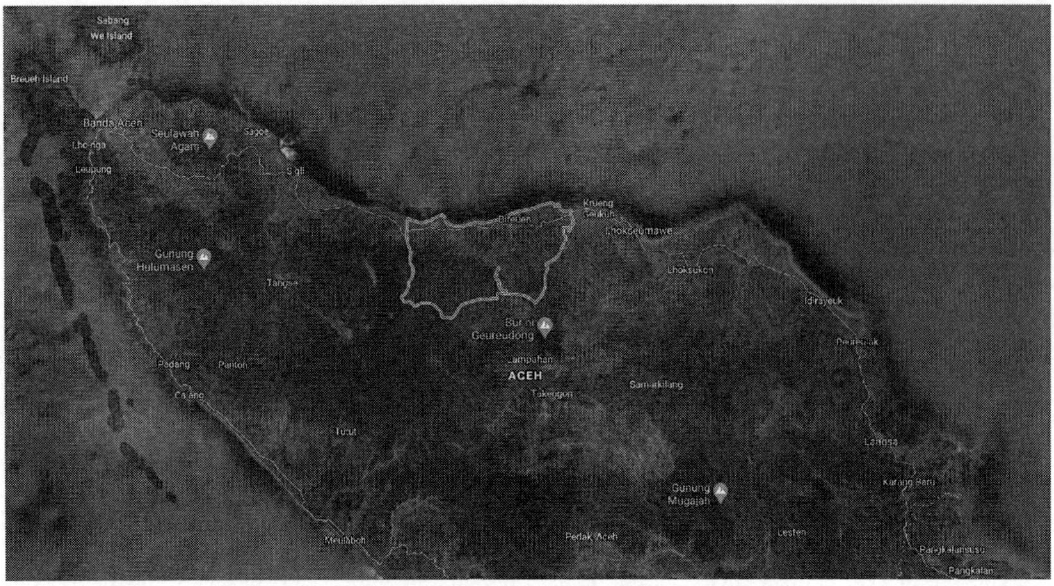

Figure 1. Map of Bireuen city (Google Maps 2020).

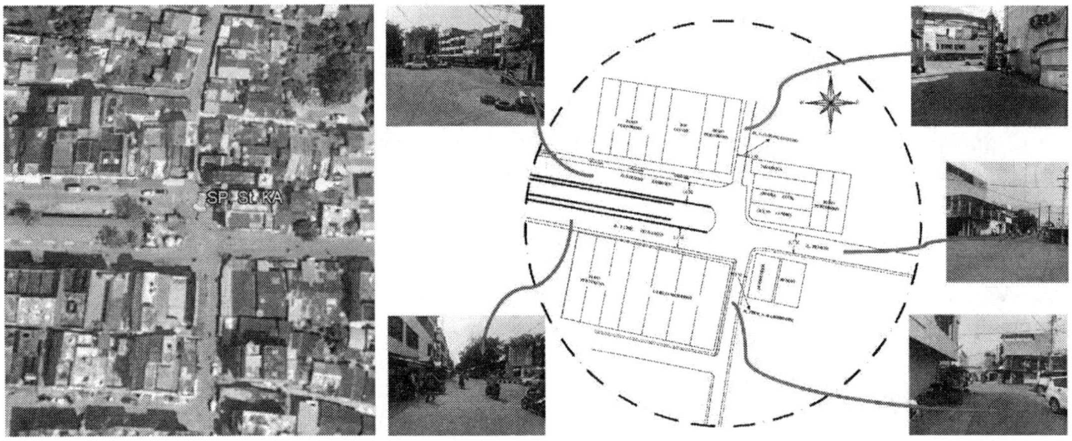

(a) (b)

Figure 2. Map of (a) Intersection of study (Google Maps 2020), (b) Layout of the intersection.

The traffic data was extracted in the traffic system laboratory from the recorded video camera. At first, recorded video data was extracted to get traffic volume and composition for each approach of the intersection. Then, the trajectories of vehicle movements have also observed by replying recorded videos in traffic laboratories. Finally, all processed data related to traffic stream characteristics within the intersection were tabulated in Ms. Excel before used in the analysis. As mentioned before, the analysis in this study is focused on capacity and level of services considering two scenarios, that is the existing condition (unsignalized intersection) and implemented dedicated mini roundabout (improvement of an existing. The method used for the analysis will be described in the following section.

IOP Conf. Series: Materials Science and Engineering 1087 (2021) 012002 doi:10.1088/1757-899X/1087/1/012002

3. Analysis method

In this study, as mentioned above, the data analysis consists of two scenarios namely (1) assessing the capacity and level of service considering unsignalized intersection (existing scenario), and (2) determining new capacity and level of services when the dedicated mini roundabout implemented. As for evaluating the existing scenario, the IHCM 1997 is used for evaluating the capacity of the unsignalized intersection. The level of service is then determined using the Ministry of Transportation Decree [8]. The level of service (LOS) for unsignalized intersection is based on the average delay (second/PCU) and classified them as LOS A to F (see Table 1). It should be noted that according to the Ministry of Public Work Decree [9] the upper limit of degree of saturation (DS) is 0.85, and 0.9 for arterial/collector, and local road, respectively. As regards to the second scenario, the evaluation of a dedicated mini roundabout is assessed using the microsimulation approach implemented in VISSIM. The radius and diameter of the mini roundabout were modelled using VISSIM and then observed traffic demand is tested into the VISSIM simulation model.

Table 1. The Level of Service (LOS) Criteria.

Criteria	Average Delay (Second/PCU)
A	< 5
B	5-10
C	11-20
D	21-30
E	31-45
F	> 45

Source: The Ministry of Transportation Decree [8]

The IHCM 1997 defines the capacity of unsignalized intersection consists of a multiplicative model among base capacity (C_0) with adjustment factors such as adjustments factor for approach width (F_W), type of median (F_M), city size (F_{CS}), the ratio of unmotorized (F_{RSU}), the ratio left turn vehicle (F_{LT}), the ration right turns vehicle (F_{RT}) and the ratio of flow from the minor road (F_{MI}), as it is written in Equation (1).

$$C = C_o \times F_W \times F_M \times F_{CS} \times F_{RSU} \times F_{LT} \times F_{RT} \times F_{MI} \quad (1)$$

Where C represents the capacity (PCU/hr), and C_0 is base capacity defined using default value in IHCM 1997 chapter 2 (unsignalized intersection). As for the delay, the IHCM has consisted of two delay namely delay due to traffic stream (DT) and delay due to geometric condition (DG). The DT must be determined using the delay curve versus the degree of saturation Figure C-2:1 and Figure C-2:2 (IHCM, Chapter 2, page 3:40-41). The DG can be calculated using equation (2) for DS < 1, and equation (3) for DS ≥ 1.

$$DG = (1-DS) \times (PT \times 6 + (1-P_T) \times 3) + DS \times 4 \text{ (second/PCU)} \quad (2)$$

$$DG = 4 \quad (3)$$

The simulation method is employed to analyze the implementation of the mini roundabout. Traffic simulation is distinguished into three categories mainly microscopic, mesoscopic, and macroscopic. Microscopic models predict the state of individual vehicles and primarily focus on individual vehicle speeds and locations. Macroscopic models aggregate the description of traffic flow, and the measures of effectiveness, which are speed, flow, and density [10].

The VISSIM has been widely used to simulate traffic performance based on the improvement scenario, for example testing roundabout improvement [11-13]. The VISSIM is a microscopic, time step and a behavior-based simulation model developed to model urban traffic and public transit operations. The VISSIM stands for "Verkehr Stadten–SIMulationsmodell" which is developed by Planung Transport Verkehr AG (PTV-AG) in Karlsruhe Germany. The program can analyze traffic and transit operations under constraints such as lane configuration, traffic composition, traffic signals, transit stops, and it makes uses tool for the evaluation of various alternatives based on the transportation engineering and planning measures of effectiveness.

In this study, the simulation model using VISSIM is involved in several steps such as (1) proposed a new type of traffic operation, (2) coding network/calibrated geometric condition, (3) establish the measurement of effectiveness (MOEs), (4) generating several numbers of seeds for multiple runs, usually 5 to 10 times, (5) established a value of simulation time, the default value of 3600 seconds for VISSIM, (6) running simulations to obtain the MOEs. These steps practically can be done by several instructions namely (1) input background, (2) network coding, (3) building link-connector, (4) vehicle/traffic input, (5) inputting desire speed, (6) inputting vehicle compositions, (7) assigning vehicle route, (8) calibrating driving behaviour's, (9) setting simulation runs and random seeds, (10) simulation runs, (11) establishing MOEs, (12) evaluating assessment of new proposed traffic operation.

4. Analysis of results and discussion

After conducting field observation, the aggregation of traffic flow data has shown maximum flow obtained during evening peak hours (16:30-18:00). Therefore, the analysis explained in this section is based on the evening peak hours data set observation. Figure 3 depicts traffic flow during evening observation in PCU/h for each of the approaches. The composition of the vehicle is classified into three types of common vehicles namely motorcycle (MC), light vehicle (LV) and heavy vehicle (HV). To convert the mixed traffic condition into the homogenous one, the term of passenger equivalent unit (PCU) is used. The PCU is a unit to measure various vehicles into a unit of LV by multiplying them with passenger car equivalent (PCE). The PCEs used in this study are based on the value in IHCM 1997.

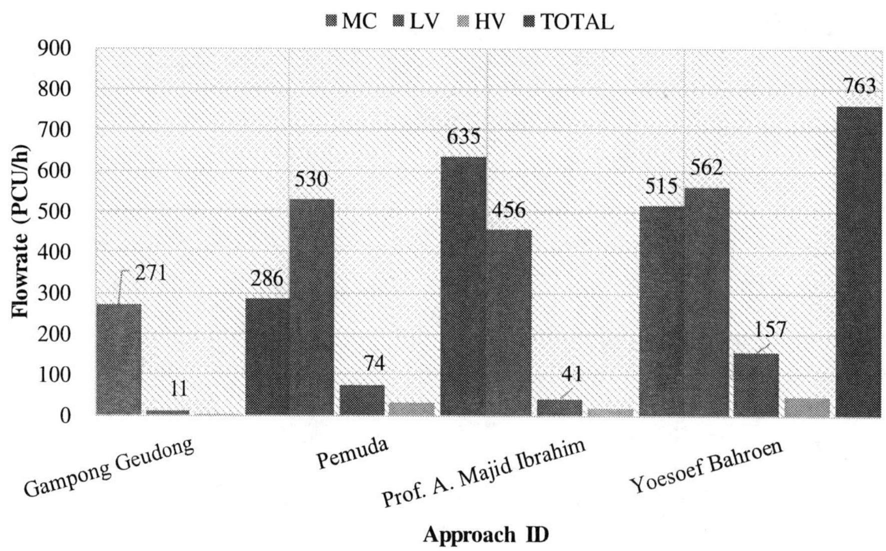

Figure 3. Traffic flow at location of the intersection.

Figure 3 shows that the traffic flow has shown about 763, 635, 515, and 286 PCU/h for Yoesoef Bahroen, Pemuda, Prof. A. Majid Ibrahim, and Gampong Geudong, respectively. The traffic flow has considerably dominated at Pemuda and Yoesoef Bahroen approaches. Looking at vehicle composition, Figure 4 has presented the distribution of vehicle composition among MC, LV, and HC. Notably, the

MC has dominated the traffic accounting for 95, 83, 89, and 74% for Gampong Geudung, Pemuda, Prof. A. Majid Ibrahim, and Yoesoef Bahroen approaches. Following by LV approximately accounted for an average between 10-21% for all approaches, and less than 5% of them are HV observed in study.

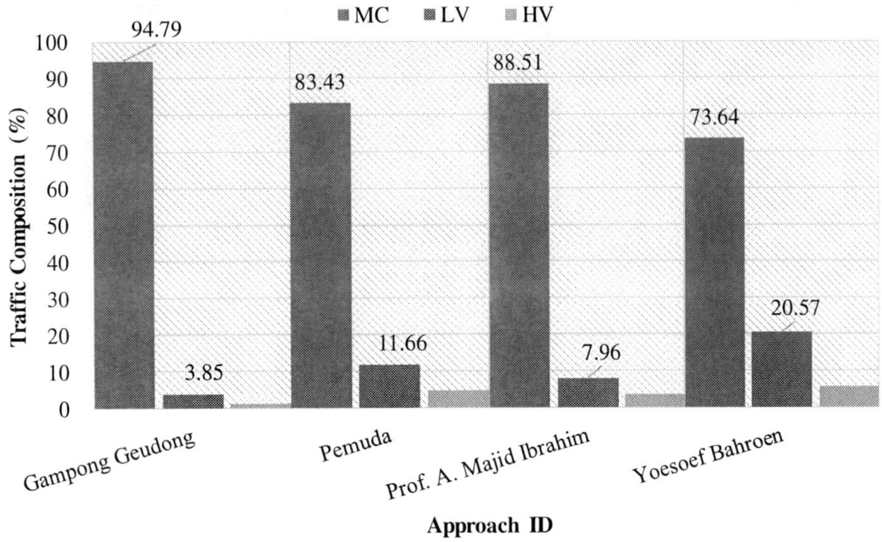

Figure 4. Distribution of traffic composition.

Table 2. The Capacity of Unsignalized Intersection (Existing).

Performance Measures Descriptions*	Average Approach Width (m)	Base capacity C_0 (PCU/h)	Calculated Capacity	Hourly Volume (PCU/h)	Calculated Total Delay (sec)	*Degree Of Saturation (DS)	**Level of Services (LOS)
Value of Performance Measures	4.275	2,900	3,834	4382	31.02	1.14	E

**traffic performance calculated using [7] and LOS using [8]*

As mentioned in section 3, the capacity was calculated utilizing the IHCM formula [7] as it is shown in equation (1) and delay calculation using equation (2-3). The calculated delay is utilized to determine the level of service using the Ministry of Transportation Decree [8]. Table 2 summarizes the calculation of capacity, delay, and level of service in existing conditions before simulating using dedicated mini roundabout. Considering an existing condition with an unsignalized intersection the capacity can be obtained as much as 3,834 PCU/h and the total delay has shown 31.02 sec. The DS value can be computed by dividing the observed hourly volume to capacity, the DS value is about 1.14. Using the Ministry of Transportation Decree [8], then the LOS criteria of E is determined. The LOS of E has meant that the traffic performance using unsignalized intersection has been considerably oversaturated and certain improvement must be proposed.

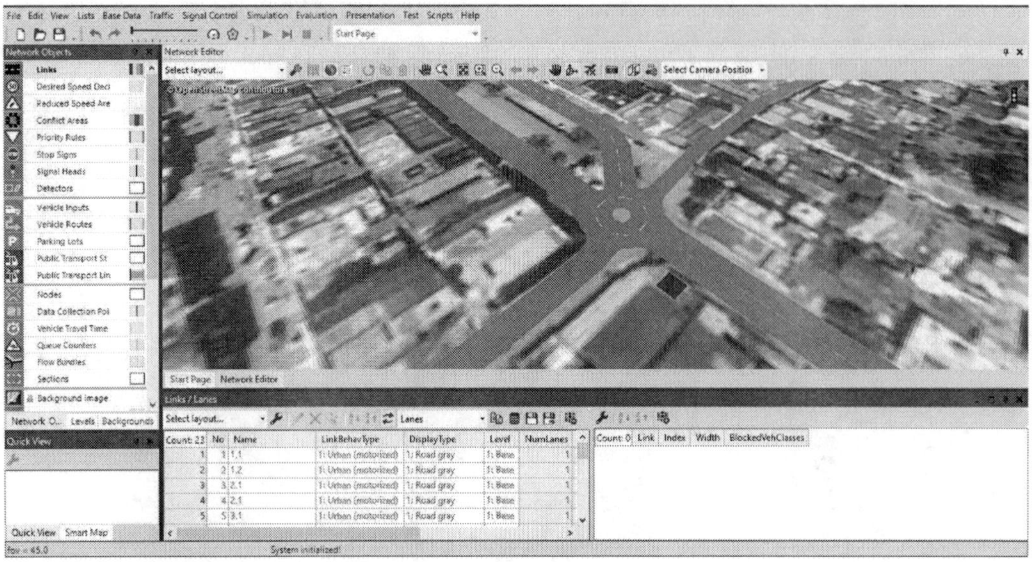

Figure 5. Simulation dedicated roundabout using PTV Vissim (Academic licensed to Universitas Syiah Kuala).

Table 3. The capacity dedicated mini roundabout.

Simulation Run/Seed Number	Simulation Interval (sec)	Queue Length (QlenMax)	Dynamic capacity (PCU/h)	Total Delay (sec)	Level of Services (LOS) Value	Level of Services (LOS)
1	900-4500	47.36	4,012	18.06	3	C
2	900-4500	50.89	4,028	18.65	3	C
3	900-4500	42.05	4,058	17.31	3	C
4	900-4500	28.94	3,946	17.88	3	C
5	900-4500	40.48	4,040	18.10	3	C
Average		41.95	4,017	18.00	3	C
Standard Deviation		8.37	43	0.48	-	-
Maximum		50.89	4,058	18.65	3	C

As regards the improvement for an existing condition, the dedicated mini roundabout with 4 m diameter is testing and simulating to obtain related traffic performance measures. Figure 5 illustrates the simulation setting and process using PTV VISSIM. The procedures of simulation using VISSIM has been explained in section 3. To ensure the consistency of stochastic nature, the simulation has considered generating 5 random seed as recommended by [2-3], [14]. Table 3 summarizes the measurement of effectiveness (MOE) obtained from the simulation. The MOEs consists of a maximum of queue length, the capacity, and obtained LOS. The average value of MOEs is 41.95 sec, 4,017 PCU/h, and C for a maximum of queue length, the capacity, and obtained LOS, respectively.

Comparing the performance measures among existing (unsignalized intersection) with the improvement (implemented roundabout) has considerably improved the performance in terms of capacity and LOS. The capacity seems to increase from 3,834 PCU/h to 4,017 PCU/h or improve about 5% compared to the existing condition. While the delay has dropped for about 42% (31.02 sec to 18 sec). This enhances the value of delay leads to an increase in the LOS from E to C. The LOS of C is considerably excellent services for the urban street facility. Therefore, it can be concluded that the improvement of the existing condition by implemented a dedicated mini roundabout could improve the capacity and LOS in the study area.

5. Conclusion

This paper investigates the impact of dedicated mini-roundabout on the capacity and level of services at a four-legged unsignalized intersection. The Indonesian highway capacity manual (IHCM, 1997) is used to evaluate the capacity and level of service in existing conditions while the simulation approach using VISSIM is then performed to assess the implementation dedicated mini roundabout. The result of analysis demonstrates that the capacity of unsignalized slightly increases by 5% while the delay and level of service are significantly reduced to 42% (31.02 sec to 18 sec), and improved from E to C, for delay and LOS, respectively. It can be concluded that the improvement of the existing condition by implemented a dedicated mini roundabout could improve the capacity and LOS in the study area.

6. Acknowledgement

The authors would like to express their honest gratitude to the staff and students of Transportation System Laboratory, Civil Engineering Department of Syiah Kuala University for their substantial help with the data collection and processing.

References

[1] Saleh S M, Sugiarto S, Hilal A, Ariansyah D 2017 *AIP Conference Proceedings*, **1903(1)** 06005.
[2] Sugiarto S, Limanond T, Nakatsuji, T 2012 *Aceh International Journal of Science and Technology,* **3** 86-92.
[3] Sugiarto S, Limanond T 2013 *Aceh International Journal of Science and Technology,* **2(1)** 63-69.
[4] Sugiarto S, Apriandy F, Faisal R, Saleh S M 2018 *Aceh International Journal of Science and Technology,* **7(2)** 77-84.
[5] Troutbeck R, Brilon W 2001 Unsignalised Intersection Theory Chapter 8, *Traffic Flow Theory, A State-of-the-Art Report, Committee on Traffic Flow Theory and Characteristics.*
[6] Sonu M, Dhamaniya A, Arkatkar S, Joshi G 2016 *Transportation Letters: the International Journal of Transportation Research (Online First).*
[7] Directorate General Highways, Ministry of Public Works 199. *Indonesian Highway Capacity Manual (IHCM) 1997.* Jakarta: Directorate General Highways, Ministry of Public Work.
[8] The Ministry of Transportatio. 2006 *The Decree No. 14 2006* Jakarta: The Ministry of Transportation.
[9] The Ministry of Transportation 2011 *The Decree No. 19 2011* Jakarta: The Ministry of Public Work.
[10] Boxil S A, Yu L 2000 *An Evaluation of Traffic Simulation Model for Supporting ITS Development.* Center of Transportation Training and Research. Texas Southern University, USA.
[11] Sugiarto S, Ruhdi F, Muhammed R 2019 *Teras Jurnal,* **8(2)** 416-425.
[12] Defry B, Sugiarto S, Renni A. 2017 *Jurnal Teknik Sipil,* **1(1)** 17-28.
[13] Rosadi D T, Sugiarto S, Anggraini R 2019 *Teras Jurnal,* **9(2)** 125-132.
[14] Faisal R, Sugiarto S, Syara A 2017 *Jurnal Teknik Sipil,* **6(2)** 183-194.

Disaster preparedness collaboration between Banda Aceh and Japan

M Fuady[1,*], Buraida[2], M R F Fuady[1]
[1]Department of Architecture and Planning, Faculty of Engineering, Universitas Syiah Kuala, Indonesia
[2]Department of Civil Engineering, Faculty of Engineering, Universitas Syiah Kuala, Indonesia

*E-mail: mirzafuady@unsyiah.ac.id

Abstract. As a disaster-prone area, the city of Banda Aceh was one of the locations that experienced an earthquake and tsunami that destroyed parts of the city at the end of 2004. An earthquake with a magnitude of 9.3 SR accompanied by a tsunami resulted in almost half of the settlement area of Banda Aceh being severely damaged. After destroyed by the biggest earthquake in Indonesia in the past few decades, the city of Banda Aceh has now recovered and is preparing itself as a city that is alert to disasters. The city recovery process has proceeded relatively quickly, and this has influenced the enthusiasm of citizens to be involved in working with various domestic and foreign institutions that helped build Aceh post-disaster. One of the countries that have collaborated with the Banda Aceh city government in disaster preparedness is the Japanese government. The collaboration was manifested in the form of sharing experiences, training, and discussions on disaster mitigation in the construction of the city of Banda Aceh as a sustainable city that is disaster prepared. With the openness of the city government to accept input and support from outside, many disaster preparedness-based activities are carried out to educate the public to be alert for disasters. For this reason, the study seeks to evaluate the disaster preparedness of the city of Banda Aceh, especially the coastal settlements of the city of Banda Aceh that are prone to the tsunami disaster. This aspect of the review in disaster preparedness-based research includes post-disaster plans to develop coastal urban settlements and rescue routes. The study began with a discussion of the initial mitigation plan a year after the 2004 disaster, followed by the conformity of the implementation of mitigation plans for several earthquakes that occurred in Banda Aceh between 2005 and 2018. In comparison, disasters and mitigation plans in Japan were also discussed. The results of this study are expected to be a reference for a better urban disaster mitigation planning not only for Banda Aceh but also can be used in other cities in Indonesia.

1. Introduction

The major disaster that struck Aceh province on December 26, 2004, began with an earthquake measuring 9.3 on the Richter scale cantered on the Indian Ocean, west of the island of Sumatra, Indonesia. This earthquake caused tsunami with a height of 30 meters and resulted in 230,000 deaths in 14 countries, 130,000 of which occurred in Banda Aceh, Indonesia [1]. After that day, many smaller aftershocks have continued to occur in the Aceh region to date.

Banda Aceh City area which is on the active fault line of Sumatera Semangko Fault is one of the areas prone to earthquakes and landslides. For this reason, according to existing regulations for disaster-prone areas, the Indonesian government has obligations and responsibilities in anticipating disasters before or after a disaster, namely with gradual efforts from disaster mitigation, emergency response and reconstruction rehabilitation [2]. Of the three stages, mitigation is interpreted as an effort to minimize the impact caused by the disaster. While the second, the emergency response period is defined as the speed in aiding when a disaster occurs in a region. And the third, the rehabilitation-reconstruction phase is to rebuild the area damaged by the disaster by paying attention to disaster mitigation-based spatial planning.

Readiness to deal with disasters is a necessity. Disasters that have occurred in Indonesia, especially in Aceh, have increasingly opened the eyes to the importance of responding quickly to overcoming the effects of disasters. In disaster preparedness, several important things that need to be dealt with quickly include handling victims and improving the environment in the disaster area, improving public infrastructure and facilities, building community houses, trauma counselling, socio-economic and cultural recovery, restoring security and order also restoring government functions and public services. In addition to aiding, evaluation and research actions on post-disaster handling steps need to be continuously improved. Therefore, research on the sustainability of post-disaster handling also needs to be strengthened. The research can be used as a reference for relevant officials and institutions in carrying out their duties in the field.

In connection with the existence of disaster-prone areas of Aceh, local governments and communities are also demanded awareness to be better prepared to face the possibility of future disasters. Various efforts to increase awareness and disaster preparedness training have been carried out by the government, both at the central and regional levels. Likewise, the policy of the Banda Aceh city government which continued to conduct socialization and disaster preparedness training to realize the city of Banda Aceh as a resilient city.

To improve disaster preparedness, the Banda Aceh City Government continues to strengthen the collaboration with various parties. One of them is the Japanese government which is a country with potential disaster risks like Indonesia, namely the earthquake and tsunami. As a city just emerging from the devastating earthquake and tsunami in 2004, Banda Aceh still has a lot of homework to be improved, starting from the implementation of the concept of disaster risk reduction to the realization of a resilient and sustainable city [3].

For this reason, the city government needs a lot of support both from within and outside the country. In addition to disaster preparedness, Banda Aceh also needs support in the development of the economic, education and health sectors. At present Japan has implemented a multi-system in disaster risk reduction efforts, starting from an early warning system, building 400 kilometres of sea walls, coastal forests, water channel canals, and determining safety zones for residents. This multi-system implementation has proven to be effective where the death toll can be significantly reduced when the tsunami hit the Higashimatsushima city area in 2011. Of the potential fatalities at that time predicted in the disaster scenario around 200 thousand can be reduced to 20 thousand people.

From the experience of the Japanese government, ideally the development of the economic sector and the development of disaster risk reduction system in a region must be balanced. For the concept of disaster mitigation, Japan has invested a long time after the defeat of the war in 1945. While in addition to infrastructure and technology investment, disaster education including research continues to be carried out starting from disaster education at the school level and training and disaster simulation for the community. Given that Banda Aceh and several cities in Japan have the same potential for disaster, the process of mutual learning and sharing of best practices in disaster mitigation will benefit each other, including sharing the experience of Aceh's local wisdom in the form of preserving the term *Smong* to educate the public.

As a follow-up to the collaboration that has been fostered since the post-disaster rehabilitation and reconstruction period of 2004, the Banda Aceh city government continues to foster relations with the Japanese government in cooperation with sharing disaster mitigation experiences including the HOPE project (Higashimatsushima Organization for Progress and Economy, Education and Energy) and CoMu (Community based Mutual Reconstruction Acceleration Program by Utilization of Local

Resources in Banda Aceh City and Higashimatsushima City). For this reason, the study seeks to evaluate the disaster preparedness of the city of Banda Aceh, especially the coastal settlements of the city of Banda Aceh that are prone to the tsunami disaster. This aspect of the review in disaster preparedness-based research includes post-disaster plans to develop coastal urban settlements and rescue routes. For comparison disaster mitigation plans in Japan also discussed. The results of this study are expected to be a reference for a better urban disaster mitigation planning not only for Banda Aceh but also can be used in other cities in Indonesia.

2. Literature review

As part of the disaster preparedness plan, the concept of disaster mitigation is defined as an effort to reduce the risk of events or series of events that threaten and disrupt the lives and livelihoods of the community. Disaster mitigation efforts can be carried out by increasing the ability to face the threat of disaster. The disaster is defined as an event that threatens, disrupts life and can lead to human casualties, environmental damage, property losses, and psychological impacts that can be caused by natural factors or human factors.

The stages of disaster mitigation are part of disaster management as an effort of all components of the government, society, and the private sector to reduce casualties and property. In the disaster management cycle, there are three components carried out, namely pre-disaster activities, emergency response during disasters and post-disaster activities. If disaster management has focused more on the aspects of emergency response and post-disaster management which have many weaknesses, disaster management focuses more on pre-disaster activities, namely disaster mitigation activities within the framework of reducing the risk and impact of disasters. Disaster events do not escape the study of public policy because it involves actions that must be carried out or not carried out by the government [4].

Based on Indonesia Law number 24 of 2007 concerning Disaster Management, disaster mitigation is defined as an effort to reduce disaster risk, both through physical development and awareness and capacity building in the face of disaster threats. Based on the understanding of the above provisions, disaster mitigation is divided into 2 (two) patterns: (a) Structural mitigation is an effort to minimize disasters carried out through the construction of various physical infrastructures and using technological approaches, such as making special canals for flood prevention, detection devices of volcanic activity, earthquake resistant buildings, or the Early Warning System used to predict tsunami waves; and (b) Non-structural mitigation, namely efforts to reduce the impact of disasters, apart from physical efforts as existing in structural mitigation.

Non-structural mitigation can be done by making urban spatial planning, community capacity building, legislation, regional planning, and insurance. Mitigation policies, both structural and non-structural, must be integrated with each other. The use of technology to predict, anticipate and reduce the risk of occurrence of a disaster must be balanced with the creation and enforcement of an adequate set of regulations that are supported by an appropriate spatial plan.

Disaster mitigation is a demand for areas that have a low level of disaster vulnerability to a high level of vulnerability. In general, disaster mitigation has not become a necessity as a development priority. Such illiteracy can be seen through the vision and mission of development from areas that have the potential to be disaster prone. Considering aspects of disaster mitigation means disaster mitigation as well as an evaluative policy process that causes policy reformulation. Although theoretically disaster management has its own stage, namely, pre-disaster emergency-post-disaster. This concept is quite clear in evaluating actions that should be carried out but does not rule out the possibility of unclear initial and final policy processes.

At the same time, disaster mitigation can be reviewed, controlled, modified, and even stopped. Disaster mitigation can be continuously formulated, implemented, evaluated, and adjusted. But this process does not occur and develops in a clear sequence, but it is a stage that continues to mix and be bound in a continuous process.

3. Methodology

This research is a type of qualitative research conducted by evaluating the disaster preparedness of the city of Banda Aceh, especially the coastal settlements of the city of Banda Aceh that are prone to tsunami disaster. The aspects that analysed in disaster preparedness-based research include post-disaster plans to develop urban coastal settlements and rescue routes. The analysis begins with a discussion of the initial mitigation plan a year after the 2004 disaster, followed by the conformity of the implementation of mitigation plans for several small earthquakes that occurred in Banda Aceh between 2005 and 2018. The mitigation plans in Japan and their experience in facing disasters are also discussed for comparison.

As supporting material in this study, secondary data information was collected from several government offices such as the 2009-2029 City Spatial Planning (RTRW) document from Bappeda office of Banda Aceh City, Banda Aceh in figures in 2011-2018 documents from BPS office of Banda Aceh. The data is a supporter in reviewing the disaster preparedness of the city of Banda Aceh. In addition, direct observation of the use of disaster evacuation lanes and spaces in the field is equipped with photo recording as primary data as well as collection of other secondary data from various literature relating to disaster mitigation research in several cities in Indonesia.

4. Discussion

4.1 Earthquake and tsunami disasters in Aceh

As a disaster-prone area, the city of Banda Aceh was one of the locations that experienced massive earthquake and tsunami that destroyed parts of the city at the end of 2004. Earthquake with a magnitude of 9.3 SR accompanied by tsunami resulted in almost half of the settlement area of Banda Aceh being severely damaged (see Figure 1).

Figure 1. Satellite imagery of the Banda Aceh City area before and after the 2004 tsunami disaster [3].

Banda Aceh is the city with the most population in Aceh Province, where before the tsunami disaster in 2004, there were around 239,000 people spread across 9 sub-districts. There are four sub-districts in the coastal city of Banda Aceh directly adjacent to the Andaman Sea and the Indian Ocean. Among these sub-districts, Meuraxa and Kutaradja Subdistrict, are two sub-districts which were severely damaged by the tsunami disaster in 2004. These two regions are also sub-districts that have a high population residing in the coastal area.

After struck by the biggest earthquake in Indonesia in 2004, the city of Banda Aceh has now recovered and is preparing itself as a city that is prepared to disasters. The city recovery process has proceeded relatively quickly, and this has influenced the enthusiasm of citizens to be involved in working with various domestic and foreign institutions that helped build Aceh post-disaster.

Figure 2. Satellite imagery of the Banda Aceh city area that was damaged in 2004 and after recovering in 2012 [3].

4.1.1 Development of population in the city of Banda Aceh after 2004 disaster.
As the administrative center of Aceh Province, Banda Aceh City also acts as a center for regional trade and services and at the same time as the center of city-scale settlements. Administratively, the city of Banda Aceh consists of 9 sub-districts with an area of 6,135.9 ha. In one year after tsunami disaster in 2004, developments in the population of the city showed growth from the number of 177,881 in 2005 increased to 259,931 in 2017 (see Table 1).

Table 1. Urban population development in Banda Aceh in 2005 and 2017.

No	District	Area (Ha)	Year 2005		Year 2017	
			Population	Density (p/Ha)	Population	Density (p/Ha)
1	Meuraxa	725,80	2.221	3	19.770	27
2	Jaya Baru	378,00	12.340	33	25.503	67
3	Banda Raya	478,90	24.257	51	23.919	50
4	Baiturrahman	453,90	33.582	74	36.721	81
5	Lueng Bata	534,10	19.284	36	25.607	48
6	Kuta Alam	1.004,70	35.033	35	51.614	51
7	Kuta Raja	521,10	2.978	6	13.365	26
8	Syiah Kuala	1424,40	25.418	18	37.193	26
9	Ulee Kareng	615,00	22.768	37	26.221	43
	Total	**6.135,90**	**177.881**		**259.931**	

Based on city area of 6,135.9 ha with a population of 259,931 people, the average population density of Banda Aceh city in 2017 is 42 people/ha [5]. This population density figure shows an increase from the previous 29 people / ha in 2005. The district that has the highest population density is Baiturrahman Sub-district located in the city center with a population density of 81 people/ha. While the lowest subdistricts were Kuta Raja and Syiah Kuala sub-districts with a population density of 26 people/ha.

Geographically, Kuta Raja Subdistrict, Kuta Alam, Syiah Kuala and Meuraxa are coastal areas of the city of Banda Aceh which are directly adjacent to the coast. Coastal settlements in the area are prone to disasters and have experienced severe damage during the tsunami disaster, namely most residential areas in Kuta Raja and Meuraxa Districts and some settlements in the Syiah Kuala and Kuta Alam areas.

4.1.2 Roads and rescue development in settlements

Meanwhile the plan to develop the Banda Aceh city road network was directed at a concentric radial network pattern with 2 (two) centers of cities to be served, namely in Peunayong (P1) and Lamdom/ Batoh (P2) as new centers combined with cross-town route patterns. In this concentric radial concept that functions as a ring road is the South Ring road and the North Ring road [6]. Whereas roads that function as crossroad will pass or intersect with the road network that connects the two planned centers.

On a smaller scale of services in a residential environment, the environmental road network must also be clearly connected to the road network system above. However, the environmental road has not been fully planned in the City Spatial Plan like RTRW and RDTRK, therefore it is recommended that environmental road links with larger road networks be formed to form adequate circulation relationships [7]. Related to the rescue line and the building for evacuation, 3 escape buildings have been built around coastal settlements in Meuraxa sub-district [8].

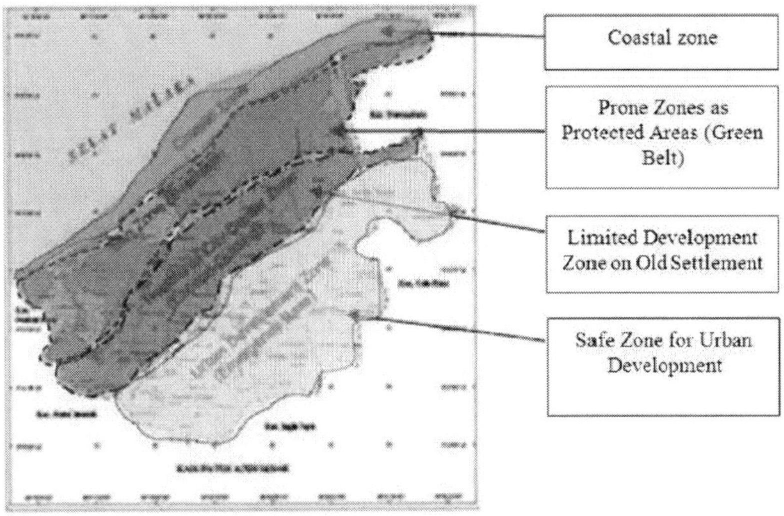

Figure 3. Post-Disaster Banda Aceh City Space Development Plan [8].

4.1.3 Development of Banda Aceh post-disaster settlements

In the Banda Aceh City Spatial Plan for 2009-2029, the settlement of Banda Aceh after tsunami disaster planned with consideration of population distribution and character development of urban space and consideration of the carrying capacity and capacity of space [9]. Plans for developing residential areas are classified into 3 (three), namely:

a. Low density residential areas, directed at relatively tsunami-prone locations in the coastal area north of the tsunami-affected city, are spread in Ulee Pata Village, Cot Lamkuwueh, Blang Village, Asoe Nanggroe, Lamjame, Lamjabat, Lampoh Daya, Lambung, Ulee Lheue, Deah Geulumpang, Alue Deah Teungoh, Deah Baro, Gampong Baro, Blang Oi, Aceh Lampaseh, Pelanggahan, Gampong Pande, Gampong Jawa, Tibang, Lamdingin, Deah Raya, and Alue Naga.
b. Medium density residential areas, directed at a relatively safe location, namely in the central, eastern and southern regions, which are spread in Jaya Baru District, Baiturrahman, Banda Raya, Ulee Kareng, Lueng Bata, and Syiah Kuala.
c. High density residential areas, directed at a safe location around the service center of Kampung Baru/Peunayong, Lampaseh Kota, Keudah, Merduati, Sukaramai, Peuniti, Neusu Jaya, Sukadamai, Seutui, Lamteumen, Keuramat, Kuta Alam, Laksana and Mulia.

4.2 Banda Aceh city disaster preparedness program

Strengthening disaster preparedness has been carried out so that the people of Banda Aceh become a disaster resilient. This is because the Banda Aceh community had experienced a devastating disaster, namely a strong earthquake followed by tsunami on 26 December 2004. One of them was the activity of the Disaster Preparedness School by the Tsunami and Disaster Mitigation Research Center (TDMRC). This activity was carried out by teaching children about disasters at the School SDN No. 2 Banda Aceh. This disaster preparedness program was realized with the collaboration between Unesco, LIPI and TDMRC, with the hope that in the future there will be a disaster preparedness school in Aceh [10].

Likewise, the Regional Disaster Management Agency of the Banda Aceh City and Government of Aceh Province, continues to strengthen disaster mitigation to the community, so that the Banda Aceh City community is resilient to disasters. The city of Banda Aceh is a disaster-prone area, not only earthquakes and tsunamis, but also extreme weather, tornadoes and strong winds. With the strength of disaster mitigation, the community in the capital city of Aceh Province was able to reduce the impact of a disaster and be able to rise from disaster [11]. But the Banda Aceh City Government cannot build a resilient community of disasters by itself. The government needs support and participation from all parties, including the business group.

Corporate social responsibility programs are encouraged to be able to support disaster mitigation training so that communities become resilient in facing disasters. With the support of the business community on disaster mitigation, many people will receive training carried out in a sustainable manner. Disasters cannot be predicted when they occur, but people must be prepared to face them.

The Banda Aceh Regional Disaster Management Agency has made community resilience training in the program named of Resilient Village or Desa Tangguh Bencana (Destana). The activity which was attended by 40 participants was held as a follow-up of a series of activities that have been carried out in 2 Gampong (Villages), namely the Gampong Deah Glumpang and Gampong Jawa, which are included in Destana. Destana's community resilience program has been running since 2016. Several activities have been produced from the Destana program, namely the establishment of the Destana Working Group in two Gampong (Deah Glumpang and Gampong Jawa), the formation of the Village Volunteer team in Meuraxa and Kuta Raja Subdistricts. Disaster Management Plans (DMP) and Community Action Plans (CAP). The integration of DMP and CAP into the Gampong development plan, the establishment of disaster risk map documents, contingency plans and evacuation routes, has also led to the formation of a Disaster Risk Reduction forum.

Desa Tangguh Bencana program has independent ability to adapt and deal with the potential threat of disasters and make it easier for themselves to avoid the adverse effects of disasters. Some of the objectives of the Destana program are to protect people living in disaster-prone areas, increasing community participation especially vulnerable groups in managing resources to reduce disaster risk. The program can also increase institutional capacity in managing resources and maintaining local wisdom for disaster risk reduction. They also increase the capacity of the Government in providing resource and technical support in disaster risk reduction. The Destana program can also enhance collaboration between stakeholders in disaster risk reduction such as Local Governments, the private sector, Universities, NGOs, CSOs and other concerned parties.

4.3 Disaster mitigation plan in Japan

To restore the area damaged by tsunami disaster in 2011, the Japanese government prepared a special disaster response plan document such for the 2011 Tohoku Earthquake. In this document, there is a recovery plan until 2013. In this document the disaster response plan is divided into 3 stages: (a) Emergency response; (b) Rehabilitation; (c) Recovery plan.

In these stages each aspect of the planning approach is discussed as follows:

(1) Transportation; Sanriku Coastal Road is a road that was built outside the tsunami prone zone but at that time the condition was severely damaged while this route had an important role in emergency transportation routes. This road is a route for decomposing goods for victims of natural disasters. In

addition, public transportation in the form of the Shinkansen fast train was also repaired. Until March 2012 even expressways in the Tohoku region were opened free for the public. The recovery of the public vehicle lane aims to accommodate victims to meet their needs outside Tohoku and to promote local tourism [12].

(2) Residence Security; The earthquake and tsunami disaster in Tohoku destroyed more than 120,000 houses, and damaged more than 240,000 houses. For this reason, the government has created a housing reconstruction program specifically for lower-middle-income housing such as public housing so that victims who do not have a place to live can temporarily rent a place to live there.

(3) Plans for rebuilding; Steps to rebuilding in disaster affected areas in the Tohoku region, are as follows: (a) Simplify ownership licensing procedures to follow a zoning plan; (b) To accommodate new development integrated with built up areas and agriculture; (c) Immediate construction of residence.

In restoring urban areas affected by disasters, there are additional separate plans, namely (a) developing urban areas; (b) move coastal settlements to safe areas; (c) developing emergency transportation routes; (d) develop rainwater reservoir drainage; (e) building parks to protect areas from disasters; (f) development of rivers and dikes; (g) construction of artificial forests to prevent disaster risks.

In addition, the planning of the disaster mitigation concept also divides the area threatened by the tsunami into two parts: (a) The first part is the level 1 tsunami area which is a potential area affected by the tsunami with a return period of 50-150 years. Tsunamis generated by this type of earthquake have an average height of 7 to 10 meters; (b) The second part is a level 2 tsunami area which is a tsunami affected area with a return period of over 500 years to thousands of years. This type of tsunami can reach a height of 20-30 meters [13].

4.4 Collaboration between Banda Aceh city and Higashimatsushima city

Relations between Banda Aceh City and Higashimatsushima City began to intertwine after the earthquake and tsunami that hit Japan in 2011. Higashimatsushima City is one of the cities in Japan that was affected by the 2011 disaster. Meanwhile, Banda Aceh City has also experienced similar things in 2004. This collaboration gave birth to the HOPE (Higashimatsushima Organization for Progress and Economy, Education and Energy) project which was intended as the Organization for the Advancement of Economics, Education and Energy of Higashimatsushima [14].

HOPE project activities aim to increase citizen awareness and understanding the importance of disaster mitigation, especially earthquakes and tsunamis. The proposed activities in this program are: (a) Family Escape Plan; (b) Emergency Food Preparation; (c) Making a disaster pamphlet and poster; (d) Establishing a disaster information center in Escape Building; and (e) Escape Building Renovation.

As a continuation to strengthen the cooperation relations between the two cities, a program has been designed called the CoMU Project facilitated by JICA. The aim of the CoMU Project is to encourage the creation of mutually beneficial cooperation in sustainable development between Higashimatsushima City and Banda Aceh City. The CoMU Project (Community-based Mutual Reconstruction Acceleration Program by Utilization of Local Resources in Banda Aceh City and Higashimatsushima City) is a Community-based Reconstruction Acceleration Program utilizing local resources in Banda Aceh City and Higashimatsushima City.

This project seeks to involve as many participants as possible, through seminars, training and workshops, in order to increase participants' understanding of sustainable development and disaster management. This project has four themes in the implementation of cooperation in the field of development between Higashimatsushima City and Banda Aceh City, namely: (a) Sustainable city development; (b) Regional disaster management; (c) Community businesses that utilize local resources; and (d) Increasing the capacity of the city government.

One of the activities carried out by Higashimatsushima's team while in the city of Banda Aceh was holding a Workshop facilitated by one of the staff of the Disaster Prevention Section of Higashimatsushima City, Japan. The purpose of this activity is to map the capacity, vulnerability, and risk of the village and to be mapped together with the Gampong Alue Deah Teungoh community. The locations of the project activities are in three gampong areas that have Japanese-assisted Tsunami Escape Building: Lambung Village, Deah Glumpang Village and Alue Deah Teungoh Village. The three gampongs are in Meuraxa District. The community involved in the project consisted of more than 800 people.

At the beginning of 2018, CoMU Project in collaboration with the Disaster Risk Reduction Forum (DRRF) conducted a workshop related to Risk Map Mapping, Vulnerability and Capacity owned by Gampong Alue Deah Teungoh and socialization of Disaster Alert Bags for the community Gampong Alue Deah Teungoh. In addition, the workshop session was also filled with an update on the Vulnerability, Risk and Capacity (VCA) Map owned by the Village. The map is also submitted to the Head of the Gampong Alue Deah Teungoh Village, with the aim that the map can be socialized to the village apparatus to be updated annually if there is insufficient information.

5. Conclusion

In general, aspects of disaster mitigation have been considered in the preparation of the Banda Aceh City Spatial Plan. The spatial pattern of the city of Banda Aceh has been planned based on consideration of the condition of the pattern of space utilization before the disaster, the tendency of urban development after a disaster, optimization and efficiency of space utilization, environmental sustainability, and disaster mitigation. For disaster mitigation purposes, it is planned to have mangrove forests along the coast and be determined by the criteria of a green belt along the coast.

However, there are also deviations from the plan, namely the failure of the plan which limits development in the disaster-prone zone. Residential areas developed into coastal areas which are in the north of Banda Aceh City. Likewise, the population density in the coastal areas of Banda Aceh shows an upward trend over time. Likewise, places and evacuation routes have not been re-evaluated to adjust to current population developments, including by planning disaster scenarios and estimating the shortest arrival time of tsunami waves.

Considering the Banda Aceh community had experienced the earthquake and tsunami disaster in 2004. Then strengthening disaster mitigation must continue to be carried out, so that the people of Banda Aceh City become resilient to the disaster. Various community debriefing activities, the implementation of preventive disaster policies, and increasing knowledge of the community about local natural characteristics that provide an indication of the threat of disaster, must be carried out continuously. Likewise strengthening the coordination of the education office on disaster preparedness curricula in the implementation of disaster preparedness for school students needs joint support between the government and the community.

The following are some important notes regarding the differences in disaster mitigation-based planning in Banda Aceh City with Japan, among others: (a) The city of Banda Aceh does not yet have a clear mitigation concept in its coastal areas, while in Japan there are clear mitigation zones and their designation; (b) There are no strict controls for spatial planning in Indonesia based on disaster mitigation, although spatial patterns have been planned in Banda Aceh, but still the population in coastal areas is still growing rapidly; (c) The disaster recovery scenario in Banda Aceh is unclear, while Japan has a target for the next 5 years, the recovery is more detailed and directed.

References

[1] Badan Rekonstruksi dan Rehabilitasi (BRR) NAD-Nias 2009 *Membentang Atap Berpilar Asa.* Banda Aceh: BRR.
[2] Tsunami and Disaster Mitigation Research Center (TDMRC) 2010 *Perencanaan Pembangunan Berbasis Mitigasi Bencana (Aceh Disaster Risk Map).* Banda Aceh: TDMRC-Unsyiah.
[3] Humas Aceh 2019 Available at https://www.humas.acehprov.go.id.
[4] Badan Koordinasi Nasional (Bakornas) 2002 *Arahan Kebijakan Mitigasi Bencana Perkotaan di Indonesia.* Jakarta: Sekretariat Badan Koordinasi Nasional Penanggulangan Bencana dan Penanganan Pengungsi (BAKORNAS PBP).

[5] Badan Pusat Statistik (BPS) 2018 *Banda Aceh dalam Angka 2018*. Banda Aceh: Badan Pusat Statistik Kota Banda Aceh.
[6] Badan Perencanaan Daerah (Bappeda). 2009. *RTRW Kota Banda Aceh 2009-2029*. Banda Aceh: BAPPEDA Kota Banda Aceh.
[7] Fuady M. 2016. *ICESReD 2016, Proceedings of International Conference on Engineering and Science for Research and Development*, **1** 109-116.
[8] Fuady M 2015 *Journal of Architecture and Built Environment*, **42(2)** 51-58.
[9] Wijaya I K, Roychansyah M S 2014 *Evaluasi Rencana Tata Ruang Wilayah Kota Banda Aceh Berbasis Mitigasi Bencana Tsunami*. Dissertation, Universitas Gadjah Mada, Indonesia.
[10] Oktari R S, Shiwaku K, Munadi K, Shaw R 2015 *International Journal of Disaster Risk Reduction,* **12** 300-310.
[11] Meilianda E, Munadi K, Direzkia Y, Oktari R S 2017 *IOP Conf. Ser. Earth Environ. Sci.,* **56(1)** 012019.
[12] Larasati, Aisharya, Medha 2016 *Jurnal Perencanaan Wilayah dan Kota ITS (in Bahasa Indonesia).*
[13] Supardiono 2014 *Jurnal Fisika,* **3**.
[14] Antaranews. 2018. Aceh tingkatkan kesiagaan dengan simulasi tsunami berkala. Available at https://www.antaranews.com/berita/773555/aceh-tingkatkan-kesiagaan-dengan-simulasi-tsunami-berkala.

Utilization of wood processing dust as a substitute for a part of cement in concrete

B Bunyamin[1,*], **R P Munirwan**[2], **M Ridha**[3], **N Hendrifa**[2]
[1]Department of Civil Engineering, Faculty of Engineering, Universitas Iskandar Muda, Banda Aceh, Indonesia, 23234
[2]Department of Civil Engineering, Faculty of Engineering, Universitas Syiah Kuala, Banda Aceh, Indonesia, 23111
[3]Department of Civil Engineering, Faculty of Engineering, Universitas Abulyatama, Banda Aceh, Indonesia, 24415

*E-mail: bunyamin@unida-aceh.ac.id

Abstract. The types of wood in Banda Aceh used was *seumantok*, *meranti*, and *sembarang* wood. The process of wood leaves sawdust. The sawdust becames rubbish and could pollute the environment. There was five quality of woods, namely quality I - quality V. The purpose of this study is to optimize the sawdust of *seumantok* (quality I), meranti (quality II), and *sembarang* wood (quality III) to be processed into wood ash as a substitute for a part of cement in concrete with its percentage were 5% and 10%. The research method was experimentally in a laboratory-based on ASTM (American Society for Testing and Materials) and ACI (American Concrete Institute). The maximum diameter of the aggregate was 25.4 mm, and the compressive strength of concrete was 14.5 MPa with 21 specimens of cylindrical concrete. Concrete compressive strength results at 28 days with 5% wood ash variation and 10% respectively for quality I were 15.83 MPa and 14.20 MPa, quality II was 13.56 MPa and 12.75 MPa, and quality III were 13.79 MPa and 10.75 MPa. While the normal compressive strength of concrete was 15.86 MPa.

1. Introduction

Since the aftermath of the 2004 earthquake and tsunami, housing construction in Banda Aceh has increased rapidly. The use of wood as a building material has also increased. Many substitute building materials to replace wood which is cheap and environmentally friendly, such as aluminium and light steel. However, the people of Aceh continue to use wood as building material for housing construction.

The wood that is used as a frame, stances or other needs is ordered in wood processing. The woods which commonly available are *seumantok, meranti, and sembarang* wood. The result of the sawing of this type of wood is called sawdust waste, which if burned, can cause air pollution to the surrounding environment. This type of sawdust is very abundant and it will produce pozzolan that can be used as a substitute for cement in concrete if it is burned. Adding 5% of wood ash into cement will improve the quality of concrete [1]. However, the clear information of the difference in the quality of wood ash for each quality of wood and on how it affects the quality of concrete compressive strength has not been obtained yet. Therefore, this study utilizes wood species that are often used in Banda Aceh, namely, *seumantok, meranti and sembarang* wood. In this research, the wood ash of *seumantok, meranti and sembarang* are used as a substitute for a part of cement. *Seumantok* wood belongs to Quality I, *meranti* wood belongs to Quality II, and *sembarang* wood belongs to Quality III [2].

Pozzolan is an additional material derived from nature and rocks, which mostly contain silica (Si) and alumina (Al) compounds, which pozzolan react to Ca(OH)2 compounds resulting from the hydration reaction between cement and water [3]. Pozzolan can be divided into two parts, namely:

a. Natural pozzolan
Natural material is sedimentation from ash or mountain larvae containing active silica (volcanic ash).
b. Artificial Pozzolan
This artificial pozzolan is the combustion residue from the furnace, as well as the utilization of waste which is processed into ash consisting of reactive silica by combustion processes, such as fly ash, silica fume, and others.

Wood is a building material that is often used and is a natural resource that can be renewed if replanting is always carried out on trees that are cut down. To use wood as a building material, the shrinkage factor is important to note because it is related to the water content which is contained in it. Naturally, wood has its durability for each different type of wood [4]. The quality of wood is divided into 5 quality as shown in the following Table 1.

Table 1. Quality of wood [4].

No.	Quality of wood	Specific weight
1	I (Very good)	$\geq$ 0,90
2	II (good)	0,60 – 0,90
3	III (Good enough)	0,40 – 0,60
4	IV (Not good)	0,30 – 0,40
5	V (bad)	$\leq$ 0,30

The waste of wood ash can be used as a substitute for cement in high-quality concrete if the addition is 10% [5]. The research of compressive sand tensile strength of concrete against the addition of wood ash by 10%, 15%, and 20% into the cement was carried out at the age of 7, 14, and 28 days. The results show that concrete with wood ash has lower workability than concrete without wood ash. In terms of strength, concrete with wood ash has a compressive and tensile strength which is still below the strength of normal concrete but has a better strength improvement than normal concrete [3].

The sawdust obtained from the sawing is collected and burned with a certain temperature to the size of the grain like the size of a powder of cement and used in concrete as a substitute for a portion of cement. From the researcher, the parameters measured are compressive strength, tensile strength, and flexural strength of beams with the composition of wood ash into cement are 5%, 10%, 15%, 18%, and 20% with W/C (Cement Water Ratio) are 0.4 and 0.45, where the ratio of cement and water is still controlled as planned. The results show that there was a slight decrease in the compressive strength of the concrete against the addition of wood ash to the cement, but the strength increased with a longer maintenance period. Wood ash contains Silica which can be used as a substitute for cement and does not affect the strength of concrete [1].

The addition of 5 grams of sawdust into a 15 cm x 15 cm x 15 cm concrete cube causes the concrete compressive strength to increase by 1.08% when compared to normal concrete. The properties of fresh concrete such as workability were decreased by 50% than normal concrete [6].

The physical properties of wood ash vary, where the specific gravity (density) of wood ash is 2.26 - 2.60, and the bulk density (volume weight) is 162 kg / m3 - 1376 kg / m^3 [7]. The physical and chemical properties are shown in Table 2 and Table 3.

IOP Conf. Series: Materials Science and Engineering 1087 (2021) 012004 doi:10.1088/1757-899X/1087/1/012004

Tabel 2. The content of chemical in 100 grams of wood [7].

No.	Chemical Properties	Volume (%)
1	SiO_2	65,3
2	AL_2O_3	4,25
3	Fe_2O_3	2,24
4	CaO	9,98
5	MgO	5,32
6	Na_2O	2,60
7	K_2O	1,90
8	*Loss on Ignition*	4,67

Table 3. Physical properties of wood ash [7].

No.	Physical Properties	Volume
1	Specific gravity	2,16
2	Grain size	170 μm
3	Volume weight	720 Kg/m^3

The chemical composition contained in wood ash is different from wood. Wood ash particles are coarser when compared to cement because they are porous and irregular. The reaction occurs when wood ash is added to cement [8], as follows:

a. Adding wood ash to concrete will reduce the value of the slump plan.
b. An increase in water absorption by adding wood ash to the concrete.
c. There is a decrease in compressive strength and tensile strength against the addition of wood ash, but it increases with age due to increased pozzolanic reactions.
d. The addition of wood ash up to 10% can be used as a good additive for cement.

The research was carried out experimentally in the laboratory and by collecting the wood processing dust of quality I, II, and III in Banda Aceh. The objectives of this research are:

1. To reduce sawdust waste in Banda Aceh's wood processing;
2. To determine the difference in the quality of sawdust quality I (*seumantok*), quality II (*meranti*), and quality III (*sembarang*) added to the cement; and
3. To find out the concrete workability of adding wood ash of quality I, II, and III to cement by 0%, 5%, and 10%.

This research was conducted using 21 specimen of cylindrical concrete with the percentage of quality I, II, and III wood ash added to cement by 5% and 10%. The scope of the study is: an examination of physical properties of fine and coarse aggregates, slump test examination, and concrete compressive strength test.

Concrete compressive strength testing is carried out if the concrete treatment has reached 28 days using a compressive test machine. Dimensions and weight of the test specimen are measured before testing. Loading on the test specimen will produce fine cracks and if the load is increased continuously, the test specimen will experience destruction [9]. Concrete compressive strength (f'c) is calculated using equation (1) as follows:

$$f'c = \frac{P}{A} \tag{1}$$

Where:
f'c = Concrete compressive strength (Mpa)
P = vertical load (N)
A = Cross-sectional area (mm^2)

2. Experimental method

The study was conducted experimentally in the Laboratory of Department Civil Engineering of Universitas Iskandar Muda based on the ASTM (American Society for Testing of Materials) and ACI (American Concrete Institute) regulations.

2.1 Wood ash and aggregate preparation

Preparation of wood ash begins with the collection of sawdust from the sawing in a wood processing. The sawdust was divided into 3 parts, where the sawdust of *seumantok* was put into a quality I bag, *meranti* into a quality II bag, and *sembarang* wood into a quality III bag. The sawdust collected was brought to the Laboratory and weighed until it reaches about 40 kg each. The sawdust was burned to ashes and ground using a blender. Then it was sieved that passes the number of 200 sieve using an ASTM sieve. Fine and coarse aggregates were prepared and physical aggregate was examined [10]. The result of the physical properties of the aggregate bye the regulations required, then it can be used as a concrete mixing material.

2.2 Concrete mix design

A concrete mix design was carried out based on the ASTM methods [10] and the ACI 211.1-91 [11]. The cement used was type 1 with a slump of 75-100 mm, fine aggregate used was that passed the sieve size of 4.76 mm and 9.52 mm and the maximum diameter of the coarse aggregate was 25.4 mm [12]. The addition of wood ash to cement were 0%, 5%, and 10% for each with the number of test specimen per wood quality as many as 3 (three) specimens. The total test specimen in this study was 21 (twenty-one) pieces. A slump test examination was done by looking at the workability of adding wood ash by 5% and 10%.

2.3 Equipment used

The equipments used were compressive test machine, concrete mixer, cylindrical concrete specimen, oven, iron diameter 16 mm with length 60 cm, ruler, aggregate filler spoon, cylindrical steel container, soaking bucket, sand cone mold with 15 mm diameter iron pounder with 25 cm long, glass (jar) with a glass plate lid, basin, brass brush, digital scales of 5 kg, a set of ASTM standard filters, and Abram's cone and a 45 cm x 45 cm x 45 cm steel plate.

2.4 Testing of concrete compressive strength

Concrete consists of cement, fine aggregate, coarse aggregate, water and with or added material. The physical properties of aggregates are very influential in determining the strength of concrete [13]. Concrete compressive strength is the maximum load arising divided by the cross-sectional area of cylindrical concrete specimen.

3. Result and discussion

The study was carried out experimentally in a laboratory by collecting the sawdust from the sawing in Wood Processing in the city of Banda Aceh. The results of the concrete mix design f'c 14.5 MPa with a Cement Water Ratio (W/C) of 0.66 were shown in the following Table 4.

Table 4. Concrete mix design (Kg/m^3).

No.	Mix Design for 1 m^3 of concrete	Material Weight (kg)
1	Water	193.00
2	Cement	292.64
3	Coarse Aggregate	1138.57
4	Coarse sand	517.82
5	Fine sand	237.97
	Total	2380.00

Test specimen were cylindrical with 21 test specimens. Addition of variations in wood ash quality by 5% and 10% combined with cement weight in kg/m^3. The results of the concrete mix design for variations in wood ash quality were shown in Table 5.

Table 5. Concrete mix design (Kg/volume of 3 cylinders).

No	Wood Ash (%)	Concrete Type	Water (kg)	Portland Cement (kg)	Wood Ash (kg)	Coarse Aggregate (kg)	Fine Sand (kg)	Coarse Sand (kg)
1	0	NC-01 to NC-03	3,529	5,347	0	20,831	4,426	9,404
2		WAC.1-01 to WAC.1-03	3,529	5,080	0,267	20,831	4,426	9,404
3	5	WAC.2-01 to WAC.2-03	3,529	5,080	0,267	20,831	4,426	9,404
4		WAC.3-01 to WAC.3-03	3,529	5,080	0,267	20,831	4,426	9,404
5		WAC.1-04 to WAC.1-06	3,529	4,812	0,534	20,831	4,426	9,404
6	10	WAC.2-04 to WAC.2-06	3,529	4,812	0,534	20,831	4,426	9,404
7		WAC.3-04 to WAC.3-06	3,529	4,812	0,534	20,831	4,426	9,404

Where:
NC-01 to NC-03 = Normal concrete of specimen 1 to 3
WAC.1-01 to WAC.1-03 = *Seumantok* wood ash concrete 5 % of specimen 1 to 3
WAC.2-01 to WAC.2-03 = *Meranti* wood ash concrete 5 % of specimen 1 to 3
WAC.3-01 to WAC.3-03 = *Sembarang* wood ash concrete 5 % of specimen 1 to 3
WAC.1-04 to WAC.1-06 = *Seumantok* wood ash concrete10 % of specimen 1 to 3
WAC.2-04 to WAC.2-06 = *Meranti* wood ash concrete 10 % of specimen 1 to 3
WAC.3-04 to WAC.3-06 = *Sembarang* wood ash concrete 10 % of specimen 1 to 3

The results of the slump test as planned were 75-100 mm with a workability of 5% for the variation of wood ash 10%. It happened because of high levels of water absorption for the quality of wood types I and II. For more details, see Table 6.

Table 6. Slump value for variation in wood ash.

No.	Type of test object	Wood Ash Content (%)	Slump Value (mm)	Workability (± Water Quantity)	High Slump Value Planned
1	NC-01 to NC-03	0	82	-	
2	WAC.1-01 to WAC.1-03	5	75	-	
3	WAC.1-04 to WAC.1-06	10	97	+ 5%	
4	WAC.2-01 to WAC.2-03	5	92	-	75 - 100 mm
5	WAC.2-04 to WAC.2-06	10	80	+ 5%	
6	WAC.3-01 to WAC.3-03	5	75	-	
7	WAC.3-04 to WAC.3-06	10	75	-	

The results of compressive strength tests using cylindrical test specimens measuring 15 cm x 30 cm when the test specimen at 28 days old were shown in Figure 1.

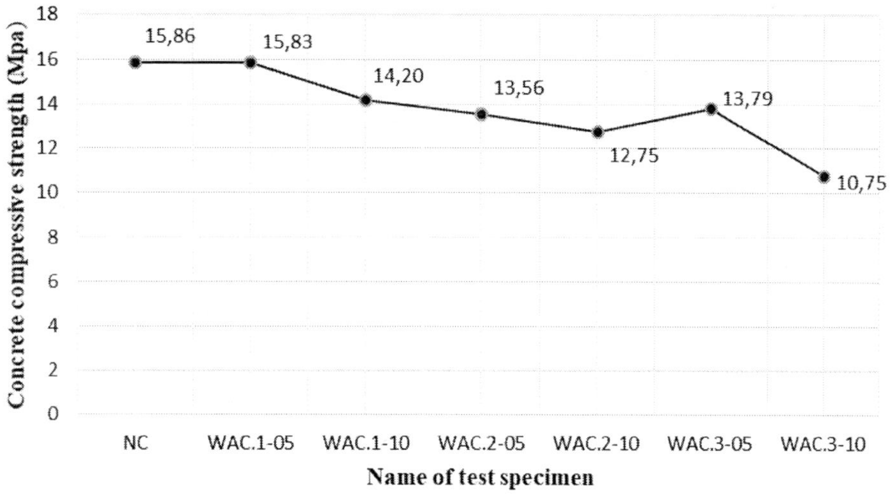

Figure 1. The result of concrete compressive strength.

Where:

NC	= Average of normal concrete of specimen 1 to 3
WAC.1-05	= Average of *seumantok* wood ash concrete 5 % of specimen 1 to 3
WAC.1-10	= Average of *seumantok* wood ash concrete 10 % of specimen 1 to 3
WAC.2-05	= Average of *meranti* wood ash concrete 5 % of specimen 1 to 3
WAC.2-10	= Average of *meranti* wood ash concrete 10 % of specimen 1 to 3
WAC.3-05	= Average of *sembarang* wood ash concrete 5 % of specimen 1 to 3
WAC.3-10	= Average of *sembarang* wood ash concrete 10 % of specimen 1 to 3

Figure 1 describes that the higher quality of wood, the higher the quality of concrete produced. Burning sawdust into wood ash greatly affects the quality of concrete. The greatest average compressive strength of concrete was found in normal concrete, which was 15.86 MPa. The compressive strength of *seumantok* wood ash concrete with a 5% mixed percentage that was equal to 15.83 MPa. The compressive strength of concrete obtained for the addition of 5% wood ash to cement was still as planned, which was above 14.5 MPa. The compressive strength value of concrete is almost the same as normal concrete compressive strength, so the percentage of the addition of *seumantok* wood ash concrete of 5% into the cement is still effective.

4. Conclusion

The conclusions from the results of the study by utilizing the remaining sawdust in Banda Aceh wood processing as follows:

1. The replacement of cement with wood ash quality I, II and III of 5%, and 10% are very influential in increasing the compressive strength of concrete.
2. Concrete compressive strength results at the age of 28 days with 5% and 10% variation of quality I wood ash were 15.83 MPa and 14.20 MPa, quality II wood ash were 13.56 MPa and 12.75 MPa, and quality III wood ash were 13.79 MPa and 10.75 MPa, respectively. While the normal compressive strength of concrete at 28 days was 15.86 MPa.
3. The most effective addition of wood ash into cement is 5 % seumantok wood ash and the quality of concrete produced by mix design concrete which is above 14.5 MPa.
4. The properties of fresh concrete for wood ash quality I, II, and III are very different when viewed from workability. At the time of the slump test, there was an additional 5% water for wood ash of quality I and quality II.

As for some suggestions to improve the results of research on variations in the quality of wood ash to concrete as follows:

1. It is necessary to check the physical properties of each wood ash to determine what percentage of water absorption.
2. It is necessary to check the sawdust combustion temperature to get an accurate quality.
3. It is necessary to examine the properties of other fresh concrete, namely the volume weight of the concrete to see the difference in weight of the concrete if added variations in the quality of wood ash.

References

[1] Chowdhury S, Maniar A, Suganya O M 2015 *Journal of advance research.* **6(6)** 907–913.
[2] Departemen Pekerjaan Umum 1961 *Peraturan Konstruksi Kayu NI-5 PPKI 1961* Bandung: Yayasan Dana Norm Indones.
[3] Indra M J, Tjondro C, Sugiharto H 2013 *Jurnal Dimensi Pratama Teknik Sipil,* **2(2)**.
[4] Kawet R S S I 2018 *Konstruksi Bangunan* Yogyakarta: Deepublish.
[5] Ghorpade V G 2012 *Environment Pollution Technology,* **11(1)** 121–124.
[6] Saifuddin M I, Edison B, Fahmi K 2014 *Jurnal Mahasiswa Teknik,* **1(1)**.
[7] Naik T R, Kraus R N, Siddique R 2003 *Material Jornal,* **100(3)** 208–215.
[8] Chowdhury S, Mishra M, Suganya O M 2015 *Ain Shams Engineering Journal,* **6(2)** 429–437.
[9] Bunyamin B 2020 *Jurnal Serambi Engineering,* **5(2)**.
[10] ASTM 2004 Annual Book of ASTM Standard 2004 Section 4, **04(2)**, Concrete and Aggregates. International Standard.
[11] ACI 2005 ACI Manual of Concrete Practice 2005, Part I, Report: ACI 104-71 (97) to ACI 223-98, Selecting Proportions For Mass Concrete (ACI 211.1-91). Detroit: Am Concr Institute.
[12] Bunyamin 2019 *AIP Conference Proceedings*, 20039.
[13] Mulyono T 2004 *Teknologi Beton* Yogyakarta: Andi Offset.

IOP Conf. Series: Materials Science and Engineering 1087 (2021) 012005 doi:10.1088/1757-899X/1087/1/012005

Potentials of plastic oil as Buton asphalt solvent

D Hariadi[1,*], S M Saleh[2], R A Yamin[3], S Aprilia[4]
[1]Doctoral Program, School of Engineering, Universitas Syiah Kuala, Banda Aceh, Indonesia
[2]Department of Civil Engineering, Universitas Syiah Kuala, Banda Aceh, Indonesia.
[3]Institute of Road Engineering, Bandung, Indonesia.
[4]Department of Chemical Engineering, Universitas Syiah Kuala, Banda Aceh, Indonesia.

*E-mail: dedy_har@yahoo.com

Abstract. Plastic is a polymer that is difficult to be decomposed, one way to decompose it through pyrolysis. Buton Asphalt (Asbuton) has a high hardness (low penetration) so Asbuton cannot be used directly. Therefore, its use must be mixed with other asphalt which is softer, lighter oil, or other solvents. This research looks the effect of quality Plastic-type Low-Density Polyethylene (LDPE) to the potential of plastic oil resulting from the pyrolysis of plastic waste as asphalt solvent in Asbuton. Based on the quality of the LDPE plastic used in this study be distinguished; LDPE-White, Mix-plastic LDPE, and Blend-Plastic LDPE. From this case study, it can be concluded that based on its chemical compound, all plastic oils produced in this study can be categorized as carbon compounds rather than hydrocarbon compounds and only oils from LDPE blend-plastics contain carbonyl groups. Therefore, the oil is more suitable to be used as a solvent or solvent.

1. Introduction

There are two types of pavement system used in Indonesia, namely: flexible pavement and rigid pavement. Flexible Pavement is the most widely used type of pavement because it is considered more economical in terms of development and maintenance [1]. Bending material is usually known as asphalt. According to the American Society for Testing and Materials, Asphalt is a material that has a dark brown to black, solid, or semi-solid consisting of natural bits or obtained from petroleum residues.

The American Association of State Highway and Transportation Office [2], the type of asphalt can be determined based on penetration rates on the asphalt used. The greater the penetration value of the asphalt, the lower the level of asphalt hardness. Each level of penetration in asphalt can be used in a mixture of asphalt aggregates such as 40/60, 60/70, 80/100 penetration. In Indonesia, the type of asphalt used is asphalt penetration 60/70 and asphalt penetration 80/100. The Buton asphalt (Asbuton) is a natural asphalt with deposits located in the area of Buton, Southeast Sulawesi. The content in asbuton consists of asphalt and minerals which have been mixed into one unit with bitumen content of 15% - 30% and minerals 70% - 85% [3]. Penetrations of less than 30 with this nature of asbuton can be used as a binder asphalt mixture. However, because asbuton has a high hardness (low penetration), asbuton cannot be used directly. Therefore, in its use, it must be mixed with other asphalt which is softer, lighter oil, or other solvents.

Previous studies have been shown that the use of solvents that are most often used in asbuton includes gasoline/Naptha, kerosene, and diesel. In this study, the solvent used in the asbuton mixture is a light oil derived from the pyrolysis of plastic waste. Besides being used as a solvent, the utilization of the pyrolysis results of this garbage will also reduce the quantity of plastic waste that continues to grow every day. Therefore, this study aims to determine the potential of plastic oil as a solvent (solvent) asphalt. In the next section gives a brief description on the existing relevance studies and method used in this research. The finding, discussion, and conclusion of this work are presented at the end of this paper.

2. Materials

2.1 Solvents in Asbuton

Asbuton is a natural asphalt found from the island of Buton, Southeast Sulawesi. Generally, solid asbuton is formed naturally due to geological processes. The process of the formation of asbuton derived from petroleum which is pushed to the surface to infiltrate between the pivot rocks [4]. There are two main types of elements in Asbuton, namely asphalt (bitumen) and minerals. The use of this element in asphalting work will affect the planned asphalt pavement performance. Asbuton extraction which still has minerals between 50% to 60% has a relatively low level of workability so that asbuton can be used it requires solvents or solvents. Solvents that are often used in asbuton are usually derived from naphtha and kerosene. Solvents (solvents) can be added to asbuton to increase emulsification, reduce settlement, increase curring rates at low temperatures, and provide proper binder viscosity after curring. The maximum amount of solvents in asbuton is usually predetermined or contains up to 25% solvents to provide proper workability performance characteristics.

Based on a study conducted by [5], solvents used to extract pure asbuton bitumen are Trichlor Ethylene (TCE), Tetra Hydro Furan (THF), ethyl acetate, acetone, furfural, toluene, limonene, bromopropan, and turpentine. From the results of the research conducted, all data show that, even though the classification of pure Asbuton is the same as the classification of asphalt pen 60 oils, but after being mixed with aggregates in hot asphalt mixtures, pure Asbuton has a better tendency than oil asphalt. Solvents (solvents) that have been done in previous studies are the result of extraction between pure asbuton bitumen and chemicals which are still relatively difficult to obtain sources. A source of solvents that can be used through easily available materials is plastic. LDPE plastic can be used as an economical asbuton solvent by pyrolysis of plastic to produce plastic oil. This oil will be a solvent that can be used in Asbuton.

2.2 Pyrolysis of plastic waste into plastic oil

The production of plastic waste in Indonesia is very large because Indonesia's total waste production reaches 189-kilo tons/day far greater than the countries in Southeast Asia [6]. As of all the plastic waste, LDPE plastic is the most dominant type of plastic (56%) found and 75% comes from household use. This is due to the fact that plastic bags which are actually LDPE plastic are very widely used in daily life. One of the uses of plastic waste in improving the quality of asbuton is by conducting plastic pyrolysis into the oil. The process of breaking down plastic waste into liquid quality fuel products is called pyrolysis.

The pyrolysis process is the process of fractionating material by temperature. The pyrolysis process is also known as a thermal cracking process, namely the breaking of the polymer structure by heating the polymer material without oxygen or a little oxygen. The pyrolysis process starts at a temperature of around 200°C when components that are thermally unstable and volatile matter in the waste will break and evaporate together with other components. This pyrolysis process is a thermal decomposition process of organic material at high temperatures without oxygen. One of the factors that influence the pyrolysis process is temperature, which temperature has an important role in the thermal cracking reaction that occurs during the pyrolysis process.

Plastics that undergo pyrolysis process will be decomposed into materials in the liquid phase in the form of fuel oil, gas phase in the form of a mixture of gas that can be condensed or cannot be condensed and solid phase in the form of residues or tar [7]. Compared to biofuels such as biodiesel and bioethanol, oil from plastic pyrolysis has several advantages. Pyrolysis oil does not contain water, so the caloric value is greater. In addition, pyrolysis oil does not contain oxygen so it does not cause corrosion [8]. Furthermore, Pyrolysis oils are flammable, have a pungent odor, and release soot. Although it has unsaturated properties, this plastic oil can be reprocessed until it has a saturated and stable nature [9]. The utilization of LDPE plastic waste into oil by the pyrolysis process was also carried out by [10]. The results of his research show that the pyrolysis process is needed within 2 hours at a temperature of 250°C with a pressure of 2 bar to produce 525mL of oil. However, they found that it has limitations regarding the inability of the tool to determine the value of the flashpoint.

2.3 Solvent (solvent) asphalt

The solvent is a diluent that is mixed into the asphalt cement to become somewhat liquid so that it can be used as a road construction material. Solvent usually divided by chemical structure or physical characteristics. Solvent classification based on its chemical structure, including:

1) Hydrocarbons, this group is further divided into three subgroups, namely: aliphatic, aromatic, and halogenated hydrocarbons. While the aliphatic subgroup is subdivided into saturated and unsaturated aliphatic. Hydrocarbon group solvents are almost entirely derived from petroleum distillation which is a mixture of several sub-groups (not pure compounds), so that the boiling point is in the range from minimum to maximum, not a single boiling point.
2) Oxygenated solvent, the Oxygenated solvent or solvents with oxygen atoms are solvents whose chemical structure contains oxygen atoms. Included in this category are esters, ketones, and alcohol esters.
   - Solubility Solvent parameter: hydrocarbon solvents have a proportional relationship to the price of Kauri Butanol (KB); the greater the price of the KB, the greater the solubility parameter or in other words the greater the solubility of the solvent. For some aliphatic hydrocarbon solvents range from 28 - 40, while for aromatic hydrocarbons greater than 70.
   - Hydrogen Bonding Index is a measure of the strength of the bond between hydrogen atoms (relatively positive) and negative atoms such as oxygen in the solvent, the price ranges from -15 to +18.

2.4 Solvent selection factor

According to the previous study conducted by [11] the solvent selection is generally influenced by the following factors, including (1) Selectivity, the solvent must have high selectivity, the solubility of the substance to be separated in the solvent must be large, while the solubility of the impurity solid is small or neglected. Solubility Able to dissolve large solutes or according to the needs of the solvent; (2) Unmixed, Solvents must not dissolve in extraction materials; (3) Density, it has a large difference in density between the solvent and the dissolved material; (4) Reactivity, Solvents must not cause chemical changes in the components of the solute; (5) Boiling point, the boiling points of the two ingredients (solvent and dissolved) must not be too close and may not form azeotropes, and (6) Other criteria, Solvents are available in large quantities, non-toxic, non-flammable, non-explosive when mixed with air, non-corrosive.

2.5 Effect of solvent on asphalt properties

Solvent or solvent has a function as a material that reduces the viscosity of asphalt, especially Asbuton, making it easier to do work using the Asbuton. The right solvent composition will also have an optimal effect on the evaporation of existing solvents. Analysis and testing conducted by [12] on the type of solvent for asphalt extraction in asphalt concrete wearing course (AC-WC) strength explained that a solvent that has a high octane will dissolve the asphalt more completely. That is because asphalt contains Asphaltenes and maltenes asphaltenes which dissolve easily in heptane mixed with octane.

3. Methods

This study describes LDPE plastic by pyrolysis non-catalysis in the laboratory so that the results obtained from the potential of plastic oil from plastic waste as solvent (Asbuton) asphalt. The reactor used, as shown in Figure 1, is a retort equipped with a thermocouple and regulator to maintain a stable operating temperature. Retorts from these devices are 60 x 60 x 30 cm. The reactor has three liquid outlets; 1, 2, and 3. The plastic used as input has three variations in the quality of LDPE plastics, namely White LDPE plastics (P-plastics): plastics derived from white-colored plastic bags; Mix-plastic LDPE (plastic-M) is plastic that comes from a mixture of plastic bags of various colors; and LDPE-plastic Blend (plastic-B), as shown in Figure 2.

The result of pyrolysis yield in the form of plastic oil is known by weighing the liquid (oil: M-1, M-2, and M-3) produced at any given time interval. At the end of pyrolysis, the solid (residue) left in the retort is weighed to determine its weight. The amount of gas wasted from the pyrolysis process is known from the difference between the input weight and the total weight of the oil yield and residue. To see the potential of plastic oil on Asbuton, the Fourier Transforms Infrared (FTIR) analysis was performed. This analysis is used to determine the functional groups contained in the sample variations in the quality of the plastic. The test was carried out in the Chemical Engineering laboratory, Bandung Institute of Technology (ITB), Bandung of Indonesia.

Figure 1. Pyrolysis reactor equipment.

(a) (b) (c)

Figure 2. Typical counts and quality of LDPE plastics: (a) white LDPE; (b) Mix-plastic LDPE; (c) Blend-plastic LPDE.

IOP Conf. Series: Materials Science and Engineering 1087 (2021) 012005 doi:10.1088/1757-899X/1087/1/012005

4. Results and discussions

4.1 White LDPE plastics (plastics-P)

Figure 3, 4 and 5 illustrates that the oil produced by white LDPE plastic pyrolysis, namely P-1, P-2 oil, and P-3 oil has a relatively equal (infrared) X-ray absorption distribution. It has 10 absorption peaks. Also, these oils contain types of chemical chain compounds or groups, namely: single bonding alkene (CH), aromatic ring (CH), alkane (CH), double bonding alkene (C = H), hydrogen bonding alcohol and phenol (OH), amines and amides (NH). Thus, it can be said that the oil obtained from the pyrolysis of white LDPE plastic not only contains hydrocarbon compounds but also other compounds. Therefore, it is not classified as a hydrocarbon compound but classified as a carbon compound.

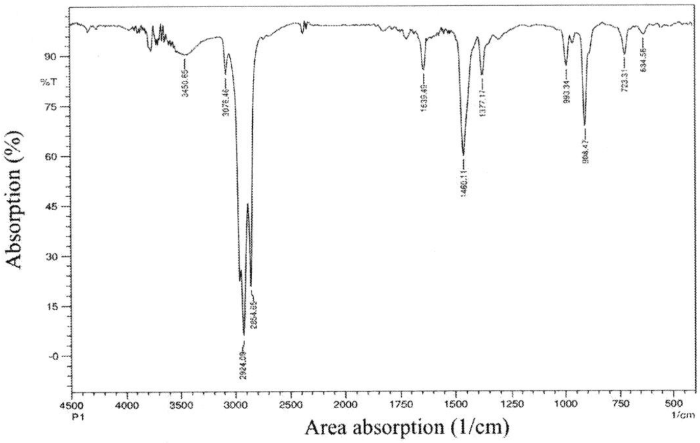

Figure 3. White plastic oil spectrophotometry: P-1 oil spectrophotometry LDPE.

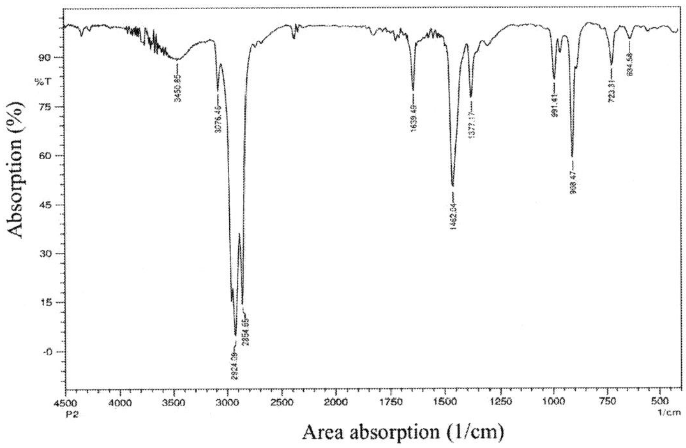

Figure 4. White plastic oil spectrophotometry: P-2 oil spectrophotometry.

4.2 LDPE mix-plastic pyrolysis (Plastic-M)

Figure 6, 7 and 8 reveals that the FTIR test results are shown on the LDPE mix-plastic pyrolysis oil, which is M-1 oil; M-2 and M-3 oil. It seems that the distribution of infrared rays from M-2 oil and M-3 oil is relatively the same and different from the infrared distribution from M-1 oil. The infrared distribution of M-2 oil and M-3 oil has 11 absorption peaks, while M-1 oil has only 10 absorption peaks.

Even though they have different infrared peak peaks, however, these oils contain relatively similar types of compounds or major chemical chain groups.

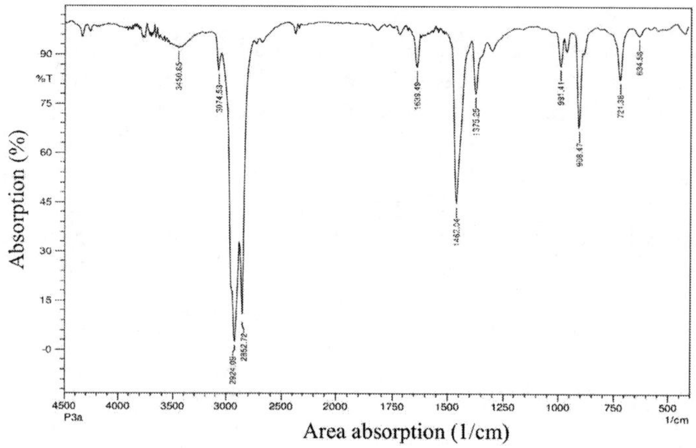

Figure 5. White plastic oil spectrophotometry: P-3 oil spectrophotometry.

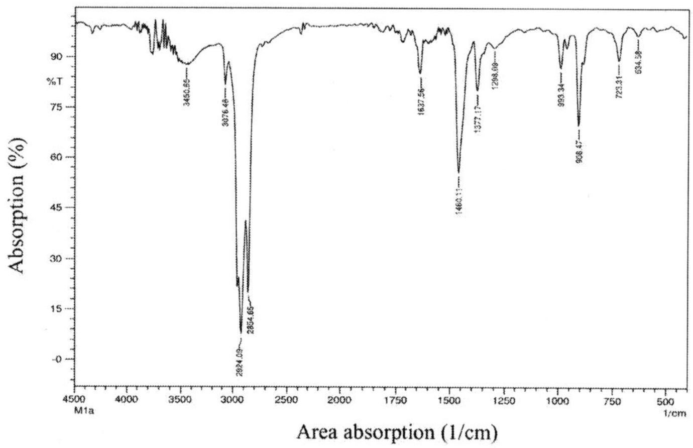

Figure 6. Plastic-mix oil spectrophotometry: M-1 oil spectrophotometry.

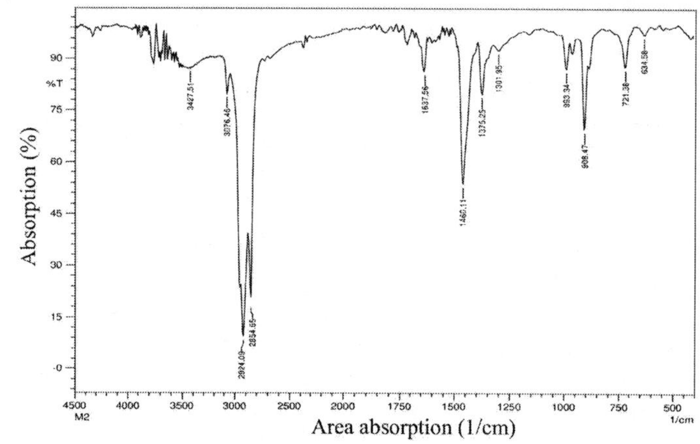

Figure 7. Plastic-mix oil spectrophotometry: M-2 oil spectrophotometry.

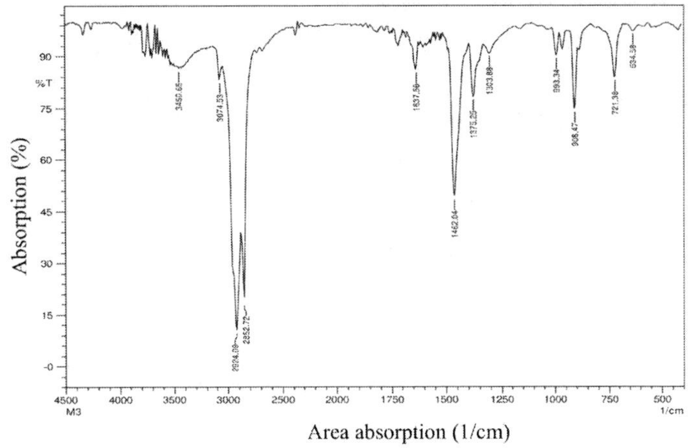

Figure 8. Plastic-mix oil spectrophotometry: M-3 oil spectrophotometry.

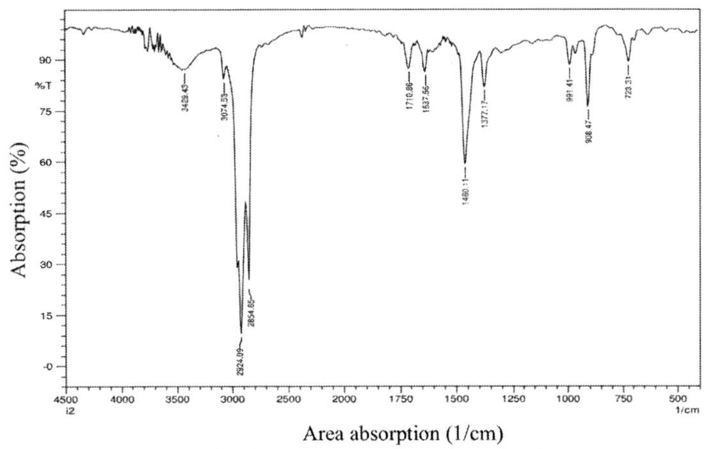

Figure 9. Plastic-blend oil spectrophotometry: B-1 oil spectrophotometry.

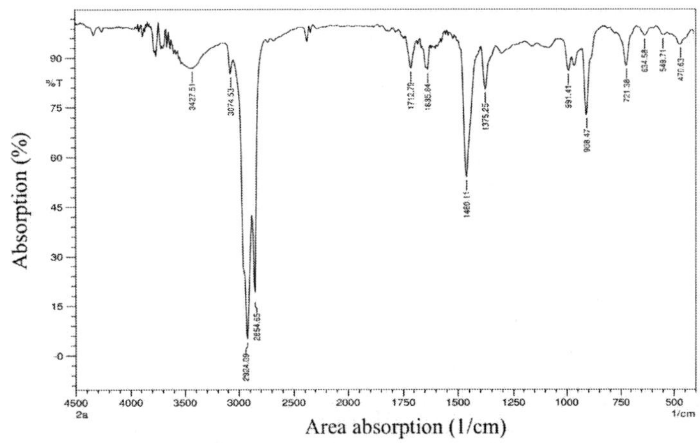

Figure 10. Plastic-blend oil spectrophotometry: B-2 oil spectrophotometry.

4.3 Blend-plastic pyrolysis LDPE (Plastic-B)

Figure 9, 10 and 11 have shown that the oil produced by LDPE blend-plastic pyrolysis, namely B-1 and B-2 oils has a relatively similar distribution of infrared light absorption and differs from the spread of infrared absorbing of B-3 oil. The distribution of infrared light absorption of B-1 and B-2 oils has 11 absorption peaks while B-3 oil has 13 peaks. The B-1 and B-2 oils contain the same types of compounds or chemical chain groups, namely: single bonding alkene (CH), aromatic ring (CH), alkane (CH), double bonding alkene (C = H), aldehydes, Kenton, carboxylic acids, and esters (C = O), hydrogen and phenol (OH) bonding alcohols, amines, and amides (NH). While the B-3 oil contains a single bonded alkene (CH), aromatic ring (CH), amines and amides (NH), amines and amides (CN), nitro compounds (NO2), alkanes (CH), double bonding alkenes (C = H), aldehydes, Kenton, carboxylic acids, and esters (C = O), hydrogen and phenol (OH) bonding alcohols, amines, and amides (NH). Thus, it can be said that the oil obtained from the LDPE blend-plastic pyrolysis does not belong to the group of hydrocarbon compounds but can be classified as carbon compounds.

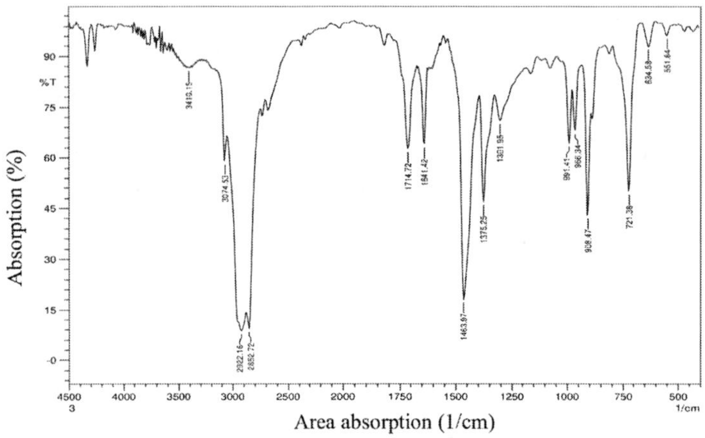

Figure 11. Plastic-blend oil spectrophotometry: B-3 oil spectrophotometry

4.4 Oil categories

LDPE type plastic pyrolysis process carried out in this study produces three outputs in the form of plastic oil. all pyrolysis oils are known that plastic oil from white LDPE plastic is liquid and bright yellow to light brown. Plastic oils from the LDPE plastic mix are also liquid and brown to black. Whereas oil from the LDPE blend-palette is black and only B-1 and B-2 oils are liquid. The B-3 oil hardens (freezes) at ambient temperatures but is liquid at relatively high temperatures.

The boiling point, the oil produced by pyrolysis of white LDPE plastic, LDPE plastic mix, and LDPE blend-plastic can all be classified as gasoline, heavy gasoline (naphtha) and kerosene (kerosene). However, although the pyrolysis process of each LDPE plastic produces oil in the same category, the quality and color of the oil are different. This difference is believed to be due to differences in color, uniformity of the LDPE plastic used. All the plastic oils produced in this study, only B-3 oil is the closest to the characteristics and chemical groups of asphalt and is best used as an asphalt modifier.

5. Conclusions

There are several conclusions can be addressed according to this study:

a. Based on its chemical compounds, all plastic oils produced in this study can be categorized as carbon compounds rather than hydrocarbon compounds.
b. Based on its chemical group, only oil from LDPE blend-plastic contains carbonyl groups. Therefore, the oil is more suitable to be used as a solvent or a hydrocarbon solvent.

c. B-3 oil is more suitable to be used as a solvent than B-1 and B2 oils because it has a CSR value of 9.5 and 13 times the CSR value of B-1 and B-2 oils and is a type of hydrocarbon solvent.
d. Based on its boiling point, plastic oil produced from LDPE plastic pyrolysis carried out in this study produced 3 types of oil, namely gasoline (P1, M-1 and B-1 oils), naphtha (P-2, M-2, and B oils -2), and kerosene (oils P3, M-3, and B-3).

In this study and research, LDPE plastic oil produced by chemical blend-plastic pyrolysis has the potential to be used as an asphalt solvent, but further studies are still needed to see the effect of oil addition on asphalt rheological properties, especially asphalt with a high level of hardness such as Asbuton.

References

[1] Rahman H 2010 *Evaluasi Model Modulus Bitumen Asbuton dan Model Modulus Campuran yang Mengandung Bitumen Asbuton*. Doctoral Dissertation. Bandung Institute of Technology, Indonesia.
[2] AASHTO 1982 *Guide for Design of Pavement Structures*. Washington, DC: AASHTO.
[3] Sidiq M, Rachmadani S, Altway A, Nurkhamidah S 2013 *Jurnal Teknik ITS.* **2(2)** 303-305.
[4] Pedoman Konstruksi dan Bangunan 2006 *Pemanfaatan Asbuton buku 1*. Jakarta: Ministry of Public Works Directorate General of Highways.
[5] Kurniadji 2014 *Jurnal Jalan-Jembatan*, **31(1)** 12-23.
[6] Kholidah N, Faizal M, Said M 2018 *Science & Technology Indonesia*, **3** 1-6.
[7] Hamidi N, Tebyanian F, Massoudi R, Whitesides L 2013 *British Journal of Applied Science & Technology,* **3(3)** 417-439.
[8] Hidayah N, Syafrudin 2018 *Proceeding of the 2nd International Conference on Energy, Environmental and Information System (ICENIS)*. Diponegoro University, Semarang, Indonesia.
[9] Pareira B C 2009 *Recycling of Plastic Waste*. Available at: http://www.ecoreccycle.vic.gov.au.
[10] Iswadi D, Nurisa F, Liastuti E 2017 *Jurnal Ilmiah Teknik Kimia UNPAM.* **1(2)**.
[11] Handojo L 1995 *Teknologi kimia*. Jakarta: PT Pradya Paramita.
[12] Musa M I, Slamet W 2019 *Jurnal Mahasiswa Teknik Sipil Universitas Tanjungpura.* **6(3)**.

Number of layer effect of abaca fiber as natural FRP material for shear-strengthened reinforced concrete beam

T Saidi[1], Z Amalia[1,*], M Hasan[1], I Hasanuddin[2], K Setiawan[1]
[1]Department of Civil Engineering, Universitas Syiah Kuala, Banda Aceh, Indonesia
[2]Department of Mechanical Engineering, Universitas Syiah Kuala, Banda Aceh, Indonesia

*E-mail: zahra.amalia@unsyiah.ac.id

Abstract. In this paper, an experimental study was carried out to evaluate the number of layer effect of abaca fiber composite as natural reinforced polymer (NFRP) material for shear-strengthened reinforced concrete (RC) beams. Two externally bonded shear-strengthened beams with a different number of NFRP layers and a sound beam as a control beam were prepared. The test was conducted by applying two axial loads on the beam. The results showed that the use of one-layer and two-layer of abaca fiber composite as NFRP material for the shear-strengthened beam increased 9.78% and 9.92% of maximum load compared to sound beam respectively. Abaca fiber composite NFRP material contributed 11% and 18.57% of the total maximum shear load for one-layer and two-layer laminates respectively. In addition, externally bonded shear-strengthened beams affected the crack pattern and deflection value. However, debonding failure of NFRP laminates occurred in two-layer of the NFRP shear strengthened beam. It caused the beam did not work optimally.

1. Introduction

Strengthening of RC structures are one of the major parts of the construction industry nowadays. Environmental factors such as earthquakes and unexpected loading caused by usage changing of structure can affect and reduce the performance of existing RC structures. In [1], shear and flexural failure of RC concrete structures were still majority found due to earthquake in Pidie, Aceh 2016. Therefore, strengthening of existing RC structures come into a major need to improve and extend its service life. This becomes a demand for developing strengthening technology and method to meet environmentally friendly and economical needs.

Strengthening and repairing methods have been developed for shear and flexural performance of the RC structures. One of the methods is using fiber-reinforced polymer (FRP). The use of FRP composite materials externally or internally bond becomes an alternative which has been commonly used in the construction industry for strengthening and repairing RC structures [2-4]. It caused by the advantages of FRP materials, such as high tensile strength, high stiffness, and resistance to environmental damage such as corrosion [5-6]. There have been many studies about the usage and influence of FRP materials on the performance of RC structures. In [7], carbon fiber reinforced polymer (CFRP) can obstruct the initial crack and increase the ultimate load of the RC beam. However, the usage of synthetic materials may affect the environment. Synthetic material is hard to be recycled or reused and may affect human health when production and usage process [8]. Furthermore, synthetic FRP material has a relatively high cost which can be a problem in developing countries, such as Indonesia. It may lead to explore a possible alternative of green and bio-based composite natural fiber material which are eco-friendly and relatively low cost.

The usage of natural fiber as a composite material can be an alternative. Natural fiber has high tensile strength, abundant in nature, renewable, light, non-toxicity, and relatively low cost at production [9]. Various natural fibers have been explored and studied to produce a composite material [10-12]. Tensile strength of sisal, kenaf, jute, and bamboo were found to be approximately between 140 MPa to 800 MPa. A few studies have been developed related to the usage of natural composite FRP material for shear and flexural strengthened RC structures [12-15]. Based on these studies, natural fiber is a possible alternative for FRP material.

Although many studies related to the usage of natural composite FRP material for strengthening and repairing RC structures have been conducted, the researches are still at the infancy stage. Moreover, the strength of natural fiber depends on its composition and where it grows. Therefore, this study aims to evaluate the contribution of abaca fiber composite NFRP material for shear-strengthened RC beams.

2. Experimental Programme

2.1. Specimen design and material properties of concrete

The test specimen had a cross-section area of 150 x 300 mm^2 and length 2200 mm. Figure 1 shows the specimen geometry. The reinforcements used in this study had a yield strength of 448.54 MPa, 457.23 MPa, 487.63 MPa, and 494.56 MPa for the rebar diameter of ø6, D13, D16, and D19 respectively. Specimens were designed to be failed in the shear mode. A total of three specimens were prepared to evaluate the number of layer contribution of abaca fiber composite as NFRP material for shear-strengthened beams. It was included of two shear strengthened beams and a sound beam as a control beam. A detailed study parameter can be seen in Table 1.

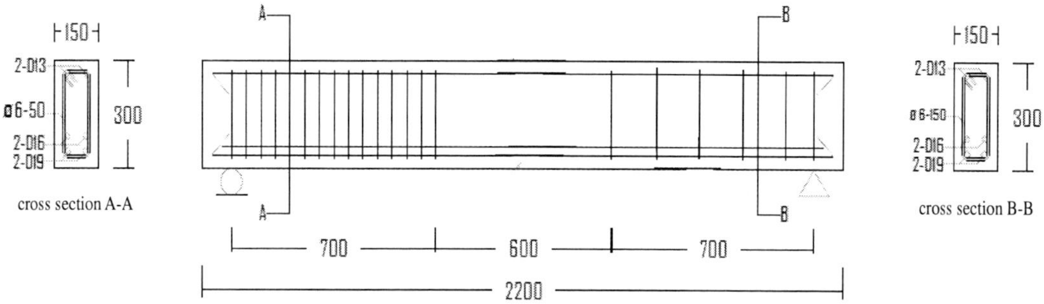

Figure 1. Specimen geometry.

Table 1. Parameter of the study.

No	Specimen	Number of abaca NFRP layer
1	BN	-
2	BSTV1	1
3	BSTV2	2

The concrete was made using ordinary Portland cement. The mix proportion is shown in Table 2. The maximum used aggregate had a diameter of 19.1mm. The specimens were cured by sealing all its surface using wet fabric and it was watered periodically to keep the humid condition for 28 days. A compressive test of the concrete cylinder was conducted before the loading test of the beam to controlled the concrete quality. The compressive strength of the concrete for BN, BSTV1, and BSTV2 were 44.1 MPa, 38.69 MPa, and 42.79 MPa, respectively.

Table 2. Concrete mixture proportion.

W/C (%)	Unit (kg/m^3)				
	Water	Cement	Sand	Aggregate	Superplasticizer
43.0	203	474.5	920.9	753.5	2.7

2.2. Fabrication and tensile strength of abaca NFRP laminates

Abaca NFRP laminates were fabricated by hand lay-up method using glass mold. Polyester type Yukalac 157 was used in this study as an adhesive which had resin and hardener. The adhesive ratio between resin and hardener was 1:1. The fiber was pressed gently during the fabrication process to ensure there is no air void inside the laminate. The drying process was 24 hours. Figure 2 shows the fabricated of abaca NFRP laminates.

Figure 2. Fabricated of abaca NFRP laminates.

Before fixing the laminates on the beam, tensile strength was conducted. A total of three tensile test specimens were prepared based on [16]. The proportion of an abaca NFRP tensile specimen consisted of 1.2 grams of fibers and 4.2 grams of mixed resin and hardener. The results showed that the average of abaca fiber tensile strength and elastic modulus were 54.58 MPa and 3474.34 MPa respectively. The average elongation of abaca NFRP material was 1.58%.

2.3. Shear strengthening of beam specimen

Abaca NFRP were fixed on both side of the beam as shown in Figure 3. One-layer abaca NFRP has a width of 50 mm and length 350mm. The proportion of one-layer abaca NFRP laminate consisted of 3.8 grams of fibers and 13.5 grams of mixed resin and hardener. Abaca NFRP laminates were placed with 150 mm spacing.

Before fixing the abaca NFRP, loose particles at the bonding surface of the concrete beam were removed by using a grinder. Abaca NFRP was directly fixed on the concrete surface by using resin and hardener with a 1:1 ratio. The abaca NFRP layer was exceeded 25 mm for each side of the upper and lower beam, thus the length of abaca NFRP was 300 mm as shear span. The abaca NFRP fiber was pressed to avoid the trapped air inside the laminates. The laminates were set for 24 hours. The test was conducted by applying two axial loads on the beam as can be seen in the Figure 3.

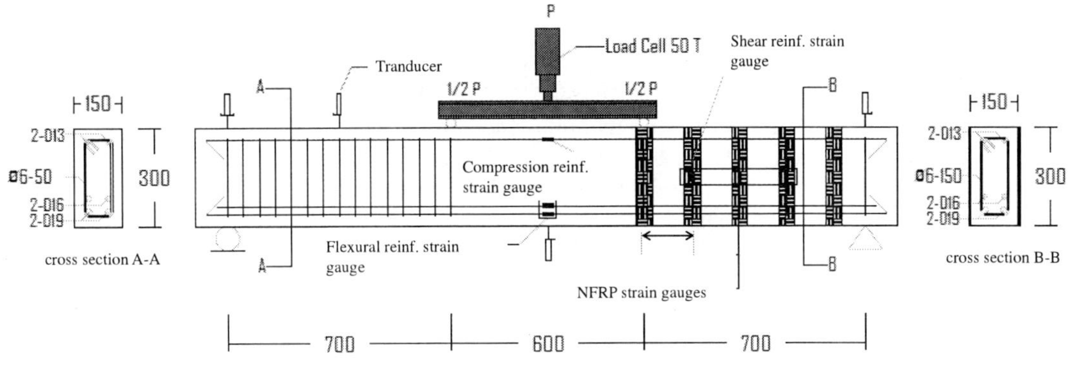

Figure 3. Shear-strengthened beam specimen and setting test.

3. Experimental Results and Discussion

3.1. Structural behavior of the shear-strengthened beam

Figure 4 shows the relationship between load and deflection of the beams. It can be seen that the maximum load for BN, BSTV1, and BSTV2 were 26.79 tf, 29.41 tf and 29.45 tf, respectively. Comparing to the sound beam, the maximum load of BSTV1 increased as 9.78% and BSTV2 as 9.92%. The use of abaca fiber composite as NFRP material slightly increases the maximum load. All specimens had almost the same load-deflection behavior. The deflection at the mid-span of the beams also presents in Figure 4. Strengthened beams had higher deflection compare to the sound beam. It is found to be approximately 19.87% and 19.67% higher than the sound beam for BSTV1 and BSTV2, respectively. Higher inclined angle of shear crack results in a high value of deflection [17]. The number of abaca NFRP layer has less effect to the load-deflection behavior. It can be happened due to the debonding of some NFRP laminates of BSTV2, thus the performance of abaca NFRP laminates did not optimal.

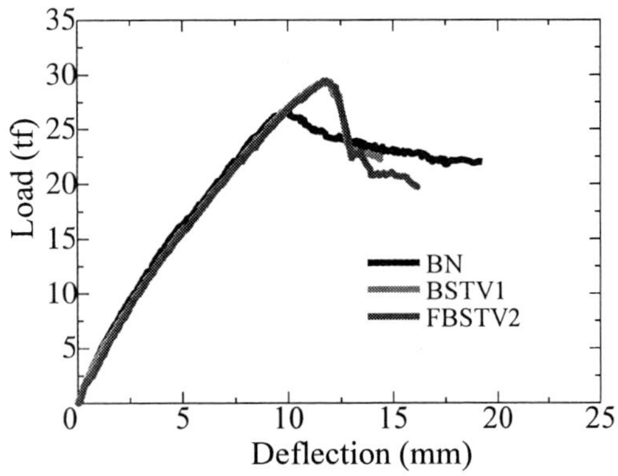

Figure 4. Load-deflection behavior of the beams.

Figure 5 presents the stress-strain behavior of flexural reinforcement at the bottom part of the mid-span of the beam. Monitored strain gages is shown as a red dot in the figure. All beams had failed after yielding the flexural reinforcement and had almost the same trend of flexural strain development. The maximum stresses of the flexural reinforcements were 499.08 MPa, for BN and 504.02 MPa for BSTV1 and BSTV2. The yield loads of flexural reinforcement in BN, BSTV1, and BSTV2 were 13.86 tf, 10.03 tf, and 8.31 tf, respectively. Before reached the yield load, an initial flexural crack occurred in the mid-span of the beam. Initial flexural cracks occurred at 6.2 tf, 5 tf and 4 tf for BN, BSTV1, and BSTV2, respectively. Concrete strength might contribute to the crack initiation development which BN beam had the highest concrete strength.

Similar stress-strain behavior was also found for all beams of shear reinforcement as can be seen in Figure 6. Red dot shows the location of the monitored strain. The maximum stresses of the shear reinforcements were 440.01 MPa, 451.32 MPa, 453.57 MPa for BN, BSTV1, and BSTV2 respectively. It shows that strengthened beams had slightly higher maximum stress. Results showed that initial shear cracks occurred at the loading of 7 tf, 6.4 tf and 6.3 tf. The yield loads of strain reinforcement in BN, BSTV1, and BSTV2 were 15.83 tf, 19.27 tf, and 19.83 tf, respectively. Strengthened beams showed a delay in shear reinforcement yielding. It happened due to the contribution of abaca NFRP. High number of abaca NFRP layers slightly increases the yield load of shear reinforcement.

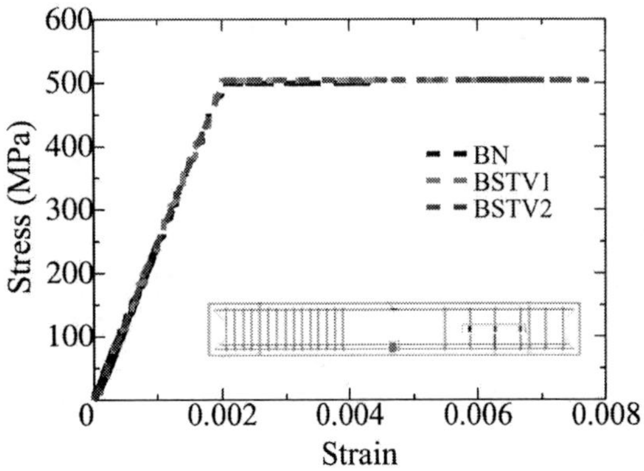

Figure 5. Strain behavior of flexural reinforcement.

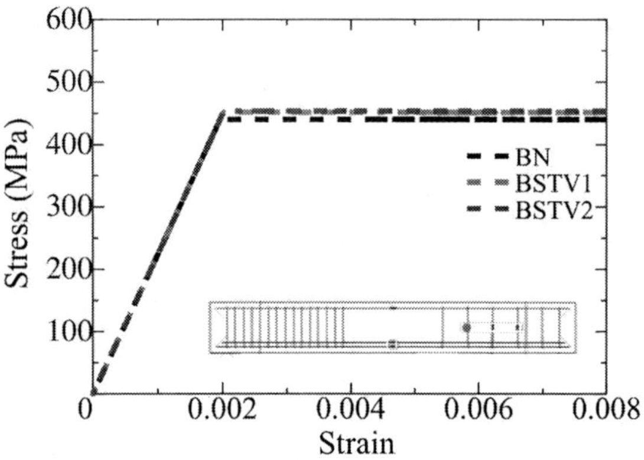

Figure 6. Strain behavior of shear reinforcement.

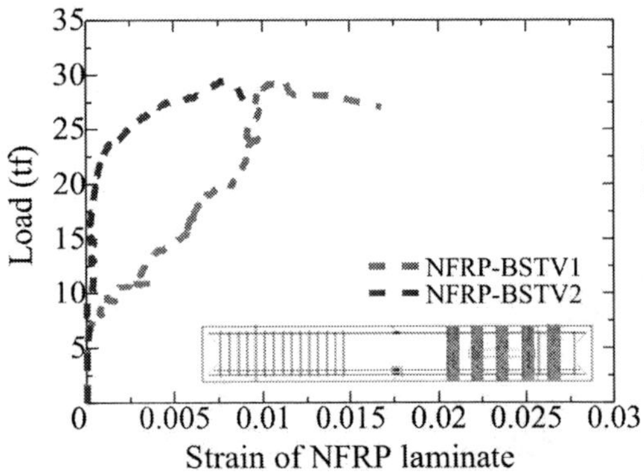

Figure 7. Strain behavior of abaca NFRP.

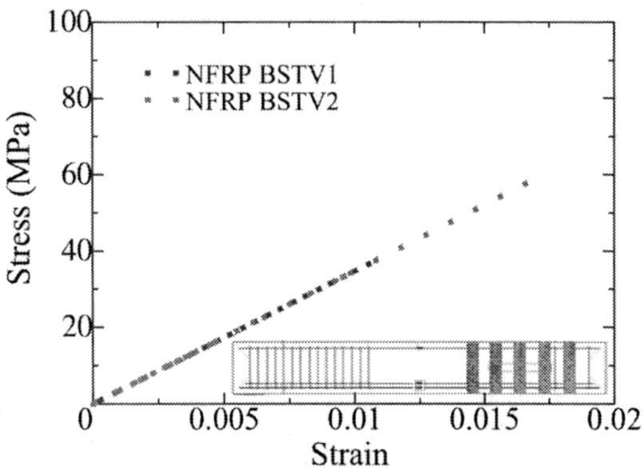

Figure 8. Stress-strain relationship of abaca NFRP.

Figure 7 shows the strain development of abaca NFRP at the same location with shear reinforcement in Figure 6. The strain of abaca NFRP laminates were found to be exceedingly small before the shear crack occurred. The development strain of abaca NFRP laminate accelerated after shear crack initiated as can be seen in BSTV1. This behavior has the same trend as [15]. However, different behavior was found in BSTV2 abaca NFRP laminate. The elongation of abaca NFRP accelerated after the yield of shear reinforcement. Therefore, the number of layers affect the strain development of abaca NFRP. Abaca NFRP laminate showed brittle behavior as can be seen in Figure 8. However, abaca NFRP in both strengthened beams had reached its maximum tensile strength.

 doi:10.1088/1757-899X/1087/1/012006

3.2. Abaca NFRP material contribution of the beam internal force

The load-internal force behavior of the strengthened beam is presented in Figure 9. The calculation of NFRP shear capacity was according to [18]. However, this code is presented for synthetic FRP and may affect the result of NFRP. It was found that the internal shear capacity of shear reinforcement at the strengthened side of the beam was 4.46 ton, while abaca NFRP contributed by 1.69 ton in BSTV1 which was 11% of the total maximum shear load. Meanwhile, from Figure 10, the internal shear capacity of shear reinforcement at the strengthened side of the BSTV2 beam was 4.49 ton while abaca NFRP could carry 2.73 ton which was 18.57% of the total maximum shear load. The contribution of shear reinforcement and abaca NFRP initiate when the first crack occurs. Higher number of abaca NFRP layers gives higher shear capacity due to the increase of the NFRP area.

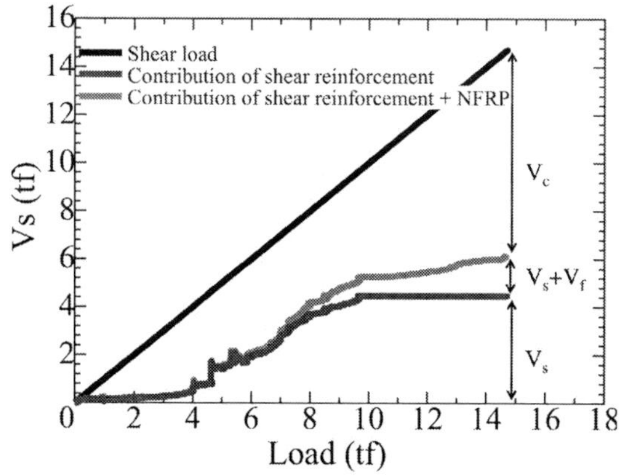

Figure 9. Contribution of abaca NFRP in BSTV1 shear capacity.

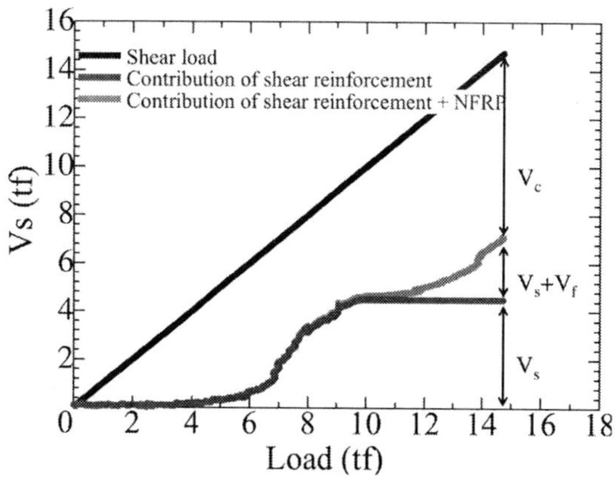

Figure 10. Contribution of abaca NFRP in BSTV2 shear capacity.

3.3. Crack pattern and failure mechanism

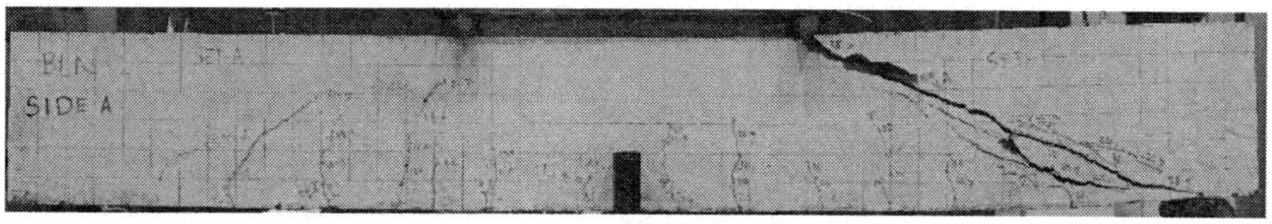

(a) BN

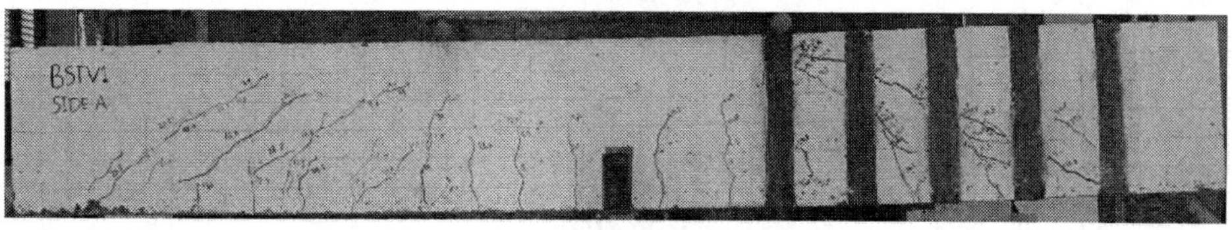

(b) BSTV1

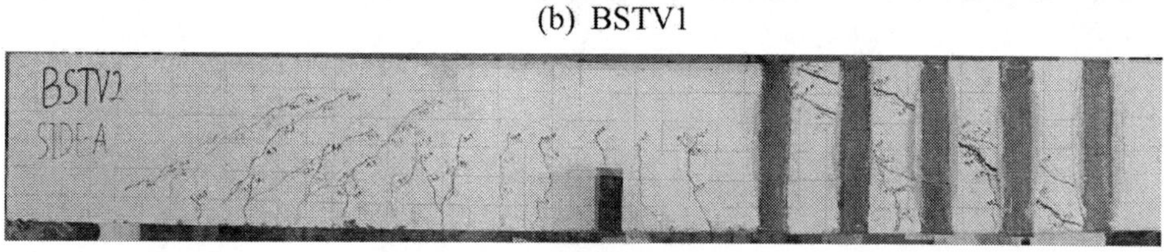

(c) BSTV2

Figure 11. Failure mode and crack pattern of tested beams.

4. Conclusion

This paper evaluates the number of abaca NFRP layer effect for the shear-strengthened RC beam. The maximum tensile strength of abaca NFRP was 54.58 MPa. Abaca NFRP material contributed 11% and 18.57% of the total maximum shear load for one-layer and two-layer laminates respectively. The abaca NFRP slightly increased the shear capacity of the RC beam. However, debonding failure of abaca NFRP was found in BSTV2 specimen, thus might be affected the results. The study showed that a shear-strengthened beam with higher number of abaca NFRP layers can reduce the number of cracks.

5. Acknowledgement

The research work of this study was conducted in Construction and Building Materials Laboratory, Civil Engineering Department of Syiah Kuala University. The authors would like to express their honest gratitude to the staff and students for their support of the data collection and processing. The authors also acknowledge the supports of Research and Community Service of Universitas Syiah Kuala for the research funding (contract number of 254/UN11.2.1/PT.01.03/PNBP/2020).

References

[1] Saidi T, Putra R, Amalia Z, Munawir M 2020 *Aceh International Journal of Science and Technology.* **8(3)** 127–36.
[2] Fukuyama H, Nakai H, Tanigaki M, Uomoto T 1991 *JCI Proceeding of the Third International Symposium,* **1** 602–15.
[3] Nanni A 2003 *Construction and Building Materials,* **7(6-7)** 439-46.
[4] Ghafoori E, Dawood M, Hosseini A 2017 *Fourth Conference on Smart Monitoring, Assessment and Rehabilitation of Civil Structures.* Zurich, Switzerland.
[5] Benjamin M, Tang P E 2003 *Proc. Publ. 2nd Int. Work. Struct. Compos. Infrastruct. Appl.*
[6] Shrivastava R, Gupta U, Choubey U B 2010 *Indian Concrete Journal,* **84(8)** 37–9.
[7] Pangestuti E K 2009 *Dinamika Teknik Sipil,* **9(2)** 180-8.
[8] Baan R A, Grosse Y 2004 *Mutation Research - Fundamental and Molecular Mechanisms of Mutagenesis,* **553(1-2)** 43-58.
[9] Tong F S, Chin S C, Doh S I, Gimbun J 2017 *Indian Journal of Science and Technology,* **10(2)** 1-5.
[10] Athijayamani B P A, Thiruchitrambalam M, Manikandan V 2010 *Polymer-Plastics Technology and Engineering,* **55(6)** 104-16.
[11] Ku H, Wang H, Pattarachaiyakoop N, Trada M 2011 *Composites Part B: Engineering,* **42(4)** 856-73.
[12] Sen T, Jagannatha H N R 2013 *International Journal of Sustainable Built Environment,* **2** 41-55.
[13] Alam M A, Alriyami K, Jumaat M Z, Noori K 2015 *Indian Journal of Science and Technology,* **8(30)** 1-6.
[14] Alam M A, Hassan A, Muda Z C 2015 *Materials and Structures,* **49(3)** 795-811.
[15] Alam M A, Alriyami K 2018 *Construction and Building Materials,* **162** 683-96.
[17] ASTM D3039. 2014. *Standard Test Methods For Tensile Properties of Polymer Matrix Composite Materials.* Annu. B. ASTM Stand.
[17] Imjai T, Guadagnini M, Garcia R, Pilakoutas K 2016 *Engineering Structures,* **126** 353-64.
[18] ACI Committee 440 2008 ACI 440.2R-08 *Guide for the Design and Construction of Externally Bonded FRP Systems.*

IOP Conf. Series: Materials Science and Engineering 1087 (2021) 012007 doi:10.1088/1757-899X/1087/1/012007

Mode choice analysis among motorcycle and Trans Koetaradja urban bus and its contributing factors using revealed preference (RP) data

M Irza[1], S Sugiarto[2,*], S M Saleh[2]
[1]Master Student, Department of Civil Engineering, Syiah Kuala University, Banda Aceh, Indonesia
[2]Department of Civil Engineering, Syiah Kuala University, Banda Aceh, Indonesia.

*E-mail: sugiarto@unsyiah.ac.id

Abstract. The random utility maximization (RUM) is a widely used principle to study customer behaviors derived from econometric theory. The mode choice problem in transportation planning has been intensively studied to explore the choice behavior of travelers. In the demand analysis, market shares among transport modes are substantially used for public transport planning and operation. Therefore, this study is a preliminary analysis of the modeling complex mode choice behaviors among motorcycle and bus system so-called Trans Koetaradja by considering specific attributes of mode, households, and individual sociodemographic attributes. The developed model of mode choice is used to explore the effect of specific attributes of mode, households, and individual sociodemographic attributes on traveler's mode choice behaviors and its contributing factors. Using the Revealed Preference (RP) data collected in Banda Aceh, we formulate Binary Logit (BL) to identify and evaluate the effects specific attributes of mode, households, and individual sociodemographic on mode choice. These results confirm that there is a substantial contribution of the specific mode attributes of the individual in determining selected mode choice. The specific attribute of motorcycle mode preference such as travel time (TT) and distance of destination (DD). Findings further confirm that individual sociodemographic variable including dummy young Age (DYA), dummy has elementary education background (DEB), dummy has low-medium income (DIN), dummy owned motorcycle (DOM) and dummy has driver license (DDL) have a statistically significant alert to the selecting mode choice of motorcycle.

1. Introduction

Public transport planning and operation become the major challenger of transport planners, particularly in developing country perspective. It is widely known that private modes such as cars and motorcycles are the most common mode used in emerging countries, including medium-big cities in Indonesia [1-3]. Thus, understanding the behavioral traveler, specifically related to the mode choice is a crucial issue in transit planning. Realizing the total market shares of the mode choice and its influencing factors is an essential problem for any investigation on the demand analysis.

In the context of the high demand for private mode usage, planning for public transport is crucial needs to be deliberate a policy that encourages public mode usage. One promising of transportation demand management (TDM) strategies aimed at alleviating private mode effects is promoting public transportation usage. In such a case, public transport (PT) such as the urban bus system is viewed as a prospective policy aiming at achieving efficiency of infrastructure use. Several implementations of urban bus policy such as bus reform policy [4-5] and Bus Rapid Transit [6-7]. In Banda Aceh, the Trans Koetardja (TK) is a bus reform policy that has been operated to promote bus ridership. Previous studies by [4-5] explore that the TK is acceptable in a policy aiming at reducing congestion within the city center, but people somehow do not want to use the bus due to highly autos dependency. Regarding the demand of the TK users, forecasting the demand for bus ridership relates the choice behavior of travelers to the characteristics of private and TK mode. The accuracy of the mode share will provide the overall market shares among private mode and the bus TK, which are substantially used for the planning and operation of the bus.

The random utility maximization (RUM) is a widely used principle to study customer behaviors derived from econometric theory. The mode choice problem in transportation planning has been intensively studied to explore the choice behavior of travelers [8]. Further implementation BL model by [9-10] to explore travel behavior related to individual preferences in the voting congestion charging policy in Jakarta. Perhaps a more recent study by [8] conducted BL analysis to assess the effects of numerous factors and to identify the leading factor that exaggerated evacuation mode choice in Tohoku, Japan. In the demand analysis, market shares among transport modes are substantially used for public transport planning and operation. Therefore, this study is a preliminary analysis of the modeling complex mode choice behaviors among private (car and motorcycle) and public mode (Trans Koetaradja) by considering household and individual household and individual sociodemographic attributes. The developed model choice is used to explore the effect of household and individual household and individual sociodemographic attributes on traveler's mode choice behaviors and its contributing factors. Using the Revealed Preference (RP) data collected in Banda Aceh, we formulate Binary Logit (BL) based on the RUM principle to identify and evaluate the effects of household and individual sociodemographic factors on individual mode choice. The rest of this paper is organized as follows.

Therefore, this study is a preliminary analysis of the modeling complex mode choice behaviors among motorcycle and bus system so-called Trans Koetaradja by considering specific attributes of mode, households, and individual sociodemographic attributes. The developed model of mode choice is used to explore the effect of specific attributes of mode, households, and individual sociodemographic attributes on traveler's mode choice behaviors and its contributing factors. Using the Revealed Preference (RP) data collected in Banda Aceh, we formulate Binary Logit (BL) to identify and evaluate the effects specific attributes of mode, households, and individual sociodemographic on mode choice. In the next section gives a brief description of the RP survey, the distribution of the data, and describe the BL formulation. The finding, discussion, and conclusion of this work are presented at the end of this paper.

Table 1. Summary of SP questionnaire.

Description	Detail
Time of survey	Juni 2020
Target responden/location	Citizen/Banda Aceh
Distribution methods	Online based questionnaire using google form
Number of questionnaires/valid samples	497/400 valid samples

2. Study area

This preliminary study is focused on the TK bus lines located in the city of Banda Aceh, a capital province of Aceh. The city centers of Banda Aceh attract many visitors and are served by urban arterial roads that experience serious autos traffic. The target bus lines are within the city center of Banda Aceh, as illustrated in Figure 1. The questionnaires were distributed within a city of Banda Aceh to the individual who has experienced using both motorcycle and the TK bus as their mode. Among 497 samples, 400 samples were used for the analysis of mode choice among motorcycle and TK bus.

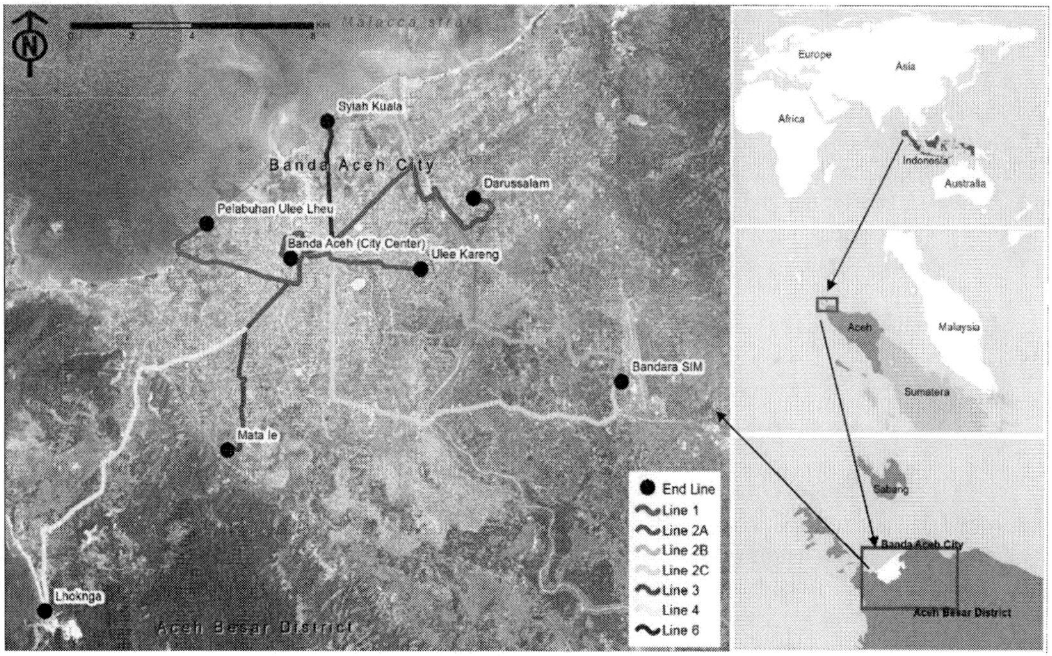

Figure 1. Map of the study area.

Table 2. Description of respondents' socio-demographics.

Item	Category	Share	Item	Category	Share
Gender	Male	42%	Monthly	IDR 1 or less	14%
	Female	58%	Income	IDR 1 – 2.9	24%
Age	17-19 years	15%	(in	IDR 3 – 4.9	25%
	20-29 years	76%	million)	IDR 5 – 6.9	19%
	30-39 years	8%		IDR 7 – 9.9	8%
	40-49 years	1%		More than IDR 10	10%
	50-59 years	0%			
	60 years or more	0%			
Occupation	Working	17%	Driver's	Has a driver's license	84%
	Student	61%	license	Has no driver's license	16%
	Unemployed	22%			

The respondents were asked general features of mode choice preference for their daily activities for both commuting and non-commuting by competing for motorcycle and TK bus as their mode preferences. Furthermore, the question related to a specific attribute of the selected mode, travel behavior attribute, individual socio-demographic characteristics such as gender, age, income, marital status, household membership, the number of a car owned, and the number of motorcycles owned. The questionnaires were distributed using the online system of google form as much as 480 samples were collected and a total of 400 samples valid and used in this study. The RP survey is summarized in Table 1, which gives the time of the surveys, target respondents/locations, distribution methods, number of samples, and valid samples.

Table 2 shows the distribution of respondent's socio-demographics. It can be seen that young people (≤30 years) is dominated sample set, accounting for more than 90% of respondents. The gender distribution is slightly skewed toward females (58%). As regards employment status, our sample indicates 17% of employed respondents as workers, and less surprisingly more than 60% of the sample consists of students. As for the income, it is found that up to 49 % has a medium monthly income. Additionally, the distribution of samples reveals that close to 85% of respondents are having a driving license.

3. Model formulation and variable setting

The RP questionnaire was employed to capture the effect of specific attributes of mode, households, and individual sociodemographic on mode choice on the mode preference among motorcycle and TK bus. The dependent variable is set as the preference of the mode choices used by respondents for their daily traveling. Two choices of the mode are offered to the respondent to select one between two choices (i.e., motorcycle and TK bus).

Table 3. Distribution of the dependence variable (mode choice for daily traveling).

Variable	Observed Mode Choice Preference	
	Motorcycle	Trans Koetardja (Bus)
Mode Choice Set	80%	20%

Table 3 depicts that the dependent variable looks like to be normally distributed. A Kolmogorov-Smirnov test is conducted to ensure the data normally distributed with a 5% significant level of error. The mean of selecting a motorcycle as a preferred mode in the dataset was 0.80 (SD = 0.402). It depicts that people in Banca Aceh have a strong preference to use a motorcycle for their daily travel.

4. Binary logistic model

A modeling approach of a discrete choice model based on the random utility maximization (RUM) is used to explore the effect of specific attributes of mode, households, and individual sociodemographic on mode choice on the mode preference among motorcycle and TK bus. A simple formulation of binary logistic (BL) regression of the odds ratio is used to delve more deeply related respondent's intentions to choose their mode for their daily travel.

The modeling BL choice behavior is formulated for the following choices: respondents to select one between two choices among motorcycle and TK bus for their travel. The probability that an individual i chooses motorcycle π_i is modeled based on the RUM theory as shown in Equation (1) and (2). More advance and detail of the binary choice and maximum likelihood estimator (MLE) can be referred to [8-10]. In this study, unknown parameters are estimated using MLE and implemented in NLogit 5.

$$logit(\pi_i) = \log\left(\frac{\pi_i}{1-\pi_i}\right) = \boldsymbol{\beta x_i} \quad (1)$$

$$\pi_i = Pr(Y_i = 1|\mathbf{x}_i) = \frac{exp(\boldsymbol{\beta x_i})}{1+exp(\boldsymbol{\beta x_i})} \quad (2)$$

The BL regression is a conventional method to determine the best fit model among dependence variables and predictors using the calibration data set as described in Table 3. The exogenous variables were set as binary choice as Yi = {1: used motorcycle, 0: used TK bus}. According to the equation (2), Y represents response variables, and xi is a vector of explanatory variables namely Travel Time (TT), Distance of Destination (DD), Age (dummy young AG), Academic Background (dummy elementary education AB), Income (dummy low-medium income INC), Motorcycle Owned (dummy has motorcycle MO), and Driver's license (dummy has driver license DL). Variable notation, definition, and empirical settings for both response and regressor used in BL calibration of mode choice model are summarized in Table 4.

Table 4. Distribution of the dependence variable (mode choice calibration).

Variable	The setting of variables (parameter coding)
Mode Choice (MC)	Binary choices: 1 used motorcycle; 0 used TK bus
Travel Time (TT)	Dummy variable of Travel Time: 1 if <30 minute; 0 if more >30 minute
Distance of Destination (DD)	1 if < 4 km or less; 2 if 4-6,9 km; 3 if 7-9,9 km o 4 if 10-12,9 km; 5 if > 12 km or more
Dummy Young Age (DYA)	Dummy variable of Age: 1 if has < 30 years; 0 if has more >30 years
Dummy has elementary education background (DEB)	Dummy variable of academic background: 1 if elementary education; 0 Otherwise
Dummy has low-medium income (DIN)	Dummy variable of monthly household' income: 1 if low-income household (income < 5 million IDR/month); 0 Otherwise
Dummy owned motorcycle (DOM)	Dummy of having a motorcycle within the household: 1 if has a motorcycle; 0 otherwise
Dummy has a driver license (DDL)	Dummy has a driver license: 1 if has driver's license; if no driver's license

5. Result and discussion

The stepwise calibration method is used to select the best fit of the model. The maximum likelihood estimator (MLE) is implemented to calibrated BL parameters. To ensure our model is statistically acceptable the 5 % level of error is applied for t-test significantly. However, several variables still are selected within a 10% level of error as it is for behavioral analysis rather than for the predicted model. The result of the calibrated parameters and its significant level can be seen in Table 5 including the Goodness of Fit (GoF) calibrated model.

Table 5. Calibrated parameters and goodness of fit of mode choice model.

Predictors	Coefficient (β)	Odds Ratio (exp β)	T-value	P-Value
Travel Time (TT)	0.911	1.140	3.03	0.002
Distance of Destination (DD)	0.131	2.487	1.26	0.209
Dummy Young Age (DYA)	2.433	11.394	5.36	0.000
Dummy has elementary education background (DEB)	0.795	2.214	4.92	0.000
Dummy has low-medium income (DIN)	0.356	1.428	3.18	0.001
Dummy owned motorcycle (DOM)	0.312	1.366	2.23	0.026
Dummy has driver license (DDL)	0.925	2.520	2.63	0.009
Dependent Variable				MC
Number of Samples				400
LL (0)				-201.53
LL(β)				-165.69
χ2 (5df)/χ2 Significance level				71.82/0.00
McFadden Pseudo R-Squared				0.19
A Receiver Operating Characteristic Curve (ROC)				0.79

Table 4 exhibits the parameter results of the estimation of mode choice model considering explanatory variables such as Travel Time (TT), Distance of Destination (DD), Dummy Young Age (DYA), Dummy has elementary education background (DEB), Dummy has low-medium income (DIN), Dummy owned motorcycle (DOM), and Dummy has driver license (DDL). Calculated indicates of model performance include log-likelihood at convergence (LL (β)) and McFadden Pseudo R-Squared, and a receiver operating characteristic curve (ROC). The values of several goodness of fit indices as shown in the bottom of Table 4 show that the LL0 (-201.53), LLβ (-165.69), $\chi2$ (5df)/$\chi2$ Significance level (71.82 at sig. 0.000), McFadden Pseudo R-Squared (0.19), and a Receiver Operating Characteristic Curve (0.79). These indices depict substantially the best fit for the calibrated model in Table 4.

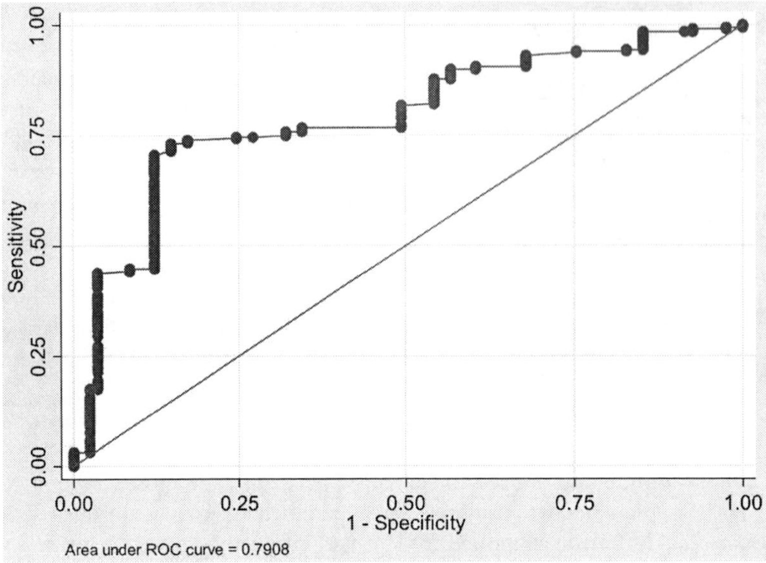

Figure 2. The ROC curve.

Examining the calibrated parameter results presented in Table 5, the parameters of the specific attribute of motorcycle mode preference such as travel time (TT) and distance of destination (DD) have shown a significant positive correlation to the selected motorcycle mode choice. The odds ratio (exp β) indicates that 1.14 and 2.48 for travel time (TT) and distance of destination (DD), respectively, meaning that increasing 0.9 minutes of travel time will lead to an increase in the probability of selecting a motorcycle 1.4 times compared to the probability of selecting bus Trans Koetaradja. Furthermore, the probability of selecting a motorcycle will increase 2.48 times compared to the probability of selecting bus Trans Koetaradja as a distance increase of 0.13 km travel. These results confirm that there is a substantial contribution of the specific mode attributes of the individual in determining selected mode choice.

Pertaining the individual sociodemographic on selecting mode choice, all the variables including dummy young Age (DYA), dummy has elementary education background (DEB), dummy has low-medium income (DIN), dummy owned motorcycle (DOM) and dummy has driver license (DDL) have a statistically significant alert to the selecting mode choice of motorcycle. The variables of dummy young age, dummy has an elementary education background and dummy has low-income have shown the most significant contributing factors. For instance, it seems that young age respondents tend to use motorcycle compared to Trans Koetaradja as the value of odds ratio (exp β) 11.3 times compared to the probability of selecting Trans Koetaradja. This indicates that young people play a crucial role in influencing the used of motorcycle mode in Banda Aceh. Moreover, a positive correlation also linked to other individual's socio-demographic attributes of low-income households and has an education background. It appears that low-income households tend to be used motorcycles to transport as they may feel a motorcycle is a convenient mode and easy to use.

6. Conclusion

This work investigates how individual choice behavior relating to the mode choice used for daily travel by considering revealed preference (RP) data collected in Banda Aceh. A binary choice based on logistic distribution formulated according to random utility maximization theory to explore individual choice behavior of the mode choices among motorcycle and Trans Koetaradja bus system. The effect of specific attributes of mode, households, and individual sociodemographic attributes on traveler's mode choice behaviors and its contributing factors then revealed using developed mode choice model in this study. These results confirm that there is a substantial contribution of the specific mode attributes of the individual in determining selected mode choice. The specific attribute of motorcycle mode preference such as travel time (TT) and distance of destination (DD) have shown a significant positive correlation to the selected motorcycle mode choice. Findings further confirm that individual sociodemographic variable including dummy young Age (DYA), dummy has elementary education background (DEB), dummy has low-medium income (DIN), dummy owned motorcycle (DOM) and dummy has driver license (DDL) have a statistically significant alert to the selecting mode choice of motorcycle. The variables of dummy young age, dummy has an elementary education background and dummy has low-income appears as the most significant contributing factors. These empirical findings may present broad assistance in efforts to the government for the suggested establishment of the Trans Koetaradja bus policy as previous work done by [11-12]. Furthermore, this transport-related policy to mitigate motorcycle dependency and promote Trans Koetardja bus system ridership.

7. Acknowledgement

The Syiah Kuala University financially supported this study under Contract No. 266/UN11/SPK/PNBP/2020 (Penelitian Lektor Kepala).

References

[1] Saleh S M, Sugiarto S, Hilal A, Ariansyah D 2017 *AIP Conference Proceedings*, **1903** 060005.
[2] Aggraini R, Sugiarto S, Pramanda H 2017 *AIP Conference Proceedings*, **1903** 060011.
[3] Sugiarto S, Miwa T, Sato H, Morikawa T 2020 *Case Studies on Transport Policy*, **8(1)** 143-152.
[4] Saleh S M, Sugiarto S, Anggraini R 2019 *The Open Transportation Journal,* **13** 17-24.
[5] Sugiarto S, Saleh S M, Anggraini R, Merfazi M 2019 *IOP Conf. Ser. Mater. Sci. Eng.,* **523** 012036.
[6] John P E 2005 *Transportation Research Record,* **1903(1)** 20-26.
[7] Yusak O S, Wimpy S, Tri B J, Danang P 2007 *IATSS Research,* **31(1)** 59-68.
[8] Sugiarto S, Lulusi, Isya M, Taufik, Haiqal M 2019 *IOP IOP Conf. Ser. Mater. Sci. Eng.,* **523** 012032.
[9] Sugiarto S, Miwa T, Morikawa T 2018 *Transportation Letters: The International Journal of Transportation Research*, **10(1)** 52–63.
[10] Sugiarto S, Miwa T, Sato H, Morikawa T 2017 *Journal of Urban Planning and Development,* **143(2)** 1-12.

The influence of architecture of *Umah Pitu Ruang* on Gayonese modern housing in Takengon

R A Putra*, **A Zahrah, C Dewi, Izziah**
Department of Architecture and Planning, Faculty of Engineering, Universitas Syiah Kuala, Indonesia

*E-mail: rizaauliaputra@unsyiah.ac.id

Abstract. Takengon is the capital of Central Aceh district located in Aceh highland. The geographical location of Takengon in the highlands certainly distinguishes the climate in this city from the cities in the lowlands and the coast in Aceh. This condition certainly affects the architecture in Takengon. The housing owned by the people is certainly constructed in such a way as to be responsive to climatic conditions in this city. Besides, housing is likely influenced by the customs and culture of the local people as well. This study aims to examine to what extent *umah pitu ruang*, which is the traditional architecture of the Gayonese, influences the modern housing architecture that exists today. Based on this study, the influence of *umah pitu ruang* as a traditional Gayonese house on the housing of modern society is almost no longer exist in terms of spatial arrangements and ornaments. This makes the values in the Gayonese traditional architecture could disappear over time.

1. Introduction

Architecture is the result of the culture of the society. The architecture has emerged from the need of the people for a shelter [1]. As a place to live, people will certainly build houses with different elements that can accommodate their activities and provide comfort to those who live in them. The ability of humans to create comfortable housing has been around for centuries [2]. This means that the housing in each region and even in the world will vary depending on the conditions under which the housing is built.

Several aspects may affect humans or the community in the construction of a house or living space, including the geographical, historical, and cultural aspects that develop within that community. These three aspects are generally very influential in housing architecture in every community. Geographical aspects play a role in creating comfort in housing, to give an example, thermal comfort in buildings will be achieved when the architectural design has a good response to local climate conditions. While historical and cultural aspects are generally influenced by people's belief in local cultural values, which are considered sacred and need to be maintained. Traditional architecture is one of the factors that influence society in creating new housing.

This paper focuses on a study of the Gayonese traditional architecture or better known as *umah pitu ruang* and its influence on the modern housing of the people in the city of Takengon today. The author wants to see to what extent the Gayonese maintains the values that exist in the *umah Pitu Ruang* architecture of its housing at present.

Umah pitu ruang is a term for the traditional houses of the Gayo tribe, which were born and formed from people activities in the hereditary agricultural sector [3]. The Gayo tribe is a tribe that inhabits

IOP Conf. Series: Materials Science and Engineering 1087 (2021) 012008 doi:10.1088/1757-899X/1087/1/012008

most of the central part of the province of Aceh, which is a highland area [4]. *Umah pitu ruang* in Indonesian means a seven-room house. However, in reality, there are also houses with only three or five rooms. This means that the traditional house of the Gayo people is a house with a balanced and long layout that protects and supports several families [2].

Umah pitu ruang has an elongated shape extending from east to west. In general, *umah pitu ruang* is divided into 3 (three) large parts which are divided into small spaces or what is known as *umah rinung. Umah rinung*, which separates the two porches on the right and left side of the house, has a higher floor than the porch floor. In the front part of this house, there is a *lepo* or in Indonesian called a terrace. Each of these spaces has several different philosophical functions and values.

Figure 1. Umah Pitu Ruang in Toweren.

One of Gayonese's traditional houses that still exists is *umah pitu ruang* located in Toweren, Lut Tawar sub-district, Central Aceh district, as shown in Figure 1. This building was built in 1860 as a heritage of Reje Baluntara. At first glance, this traditional Gayo house is not very different from the traditional Aceh house or Rumoh Aceh, in which both of them are made of wood and are typical of stilt houses. *Umah pitu ruang* in Toweren consists of five rooms, two *serambi* and one *lepo* at the front side of the building. There are three stairs for access to the house, two at the back which goes directly to the right *serambi* and the left *serambi*, and one stair at the front of the *lepo* (terrace). Figure 2 and Figure 3 portray the layouts.

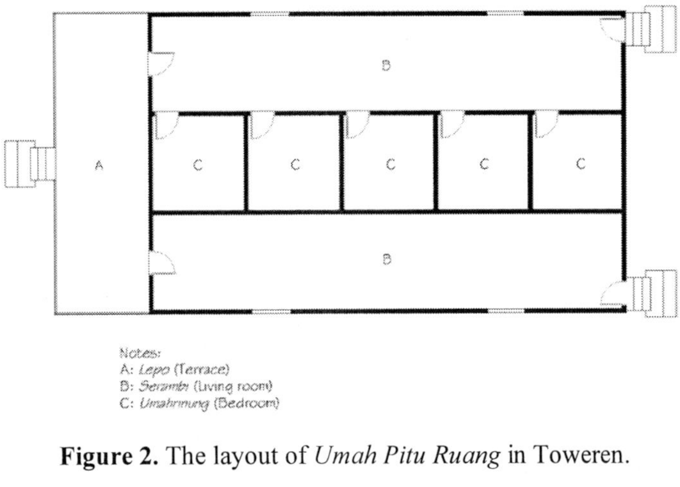

Figure 2. The layout of *Umah Pitu Ruang* in Toweren.

Figure 3. Interior (*Serambi*) of *Umah Pitu Ruang* in Toweren.

2. Methodology

The data collection process in this study was carried out through direct observation, interviews, and literature studies related to *umah pitu ruang*. The observation was carried out by surveying *umah pitu ruang* in Toweren, Takengon, Central Aceh, and observing various elements of *umah pitu ruang*, including layout, facade, and building materials. Observations have also been made on 15 modern housing in Takengon to obtain information regarding the impact of *umah pitu ruang* architecture on the modern housing that currently occupied by the community. For this study, interviews were conducted with gayo traditional and cultural experts and also the homeowners. Observation and results of the interview were then compared by referring to some related literature.

3. Results and discussion

3.1. Building layout

The most basic part of the housing is the layout which is the arrangement of the rooms in the building. The layout is formed from the needs of the mobility of its users. Generally, historical and cultural influences often influence the pattern of home layouts these days. Historical and religious principles form the framework for society in designing their housing arrangement. But, currently, these principles are no longer found in any houses owned by the Gayo people. In the observations made at that time on 15 houses in Takengon, almost no house layout was found to be identical to the traditional house layout of Gayo *Umah Pitu Ruang*. The layout of the house in the modern housing of the Gayonese is generally formed from the needs of the community based on the activities and influences of the existing land lots. In addition, the influence of climate also still dominates the layout arrangement of the modern housing of the Gayonese at present.

Figure 4 and Figure 5 are the 2 (two) out of the 15 Gayonese houses currently observed. Figure 4 is a layout of the house of Mr. Muhammad (65 years old) in the form of a rectangular backward construction with 2 (two) floors. This is because of the form of land plots that extends to the back so that makes the homeowner does not have choices of other layout forms. While Figure 5 is a house plan of Mr. Studi Gunawan which also has a rectangular shape but has a longer width than the house plan in Figure 4. Both houses were built taking into account the homeowner's room needs, consisting of a living room, middle room, kitchen, and bedroom. A fascinating feature in the modern Gayonese house layout is the living or middle room that has different floor heights. Today, nearly all Gayonese houses have a living room where the floor is elevated with wood or concrete material which is then covered with carpet. Floor elevation in the middle room is due to the cold weather in Takengon. The living room that is often used as a space for family gatherings as a communal tradition of the Gayonese who used to live together will become warmer when the floor is elevated and then covered with wood or carpet material, as shown by Figure 6.

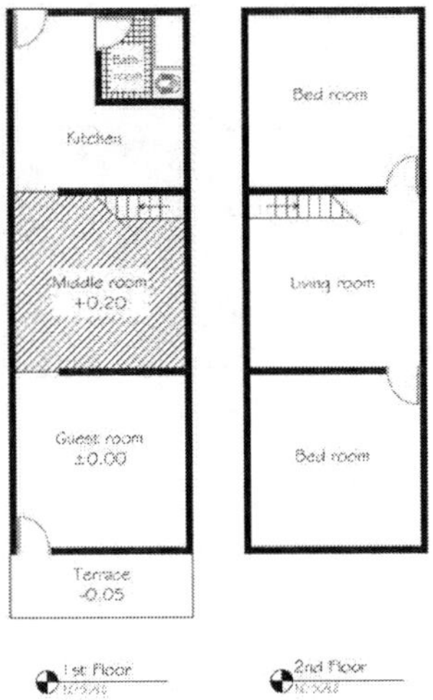

Figure 4. The layout of Mr. Muhammad's House.

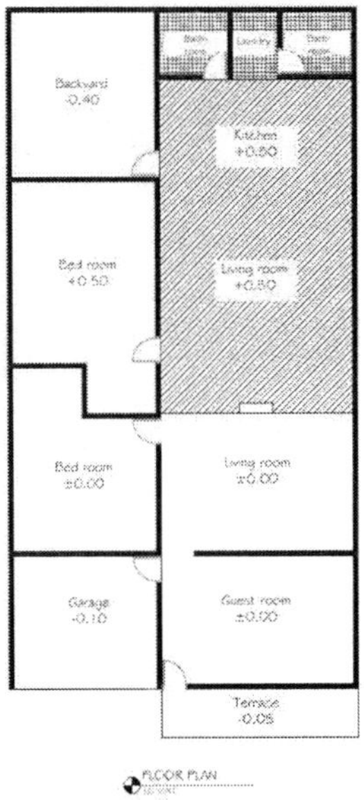

Figure 5. The layout of Mr. Studi Gunawan's House.

Figure 6. Living room with an elevated floor from wood material called *pepanteren.*

3.2. Facade and building material

When discussing architecture, it certainly cannot be separated from the facade of the building. The facade is a display of the building that can be seen directly by anyone who enjoys architectural works. The first impression of the architectural work will be determined by the facade of the building. The facade often becomes the identity of a building [5].

Like the layout of the building, the facade is often influenced by the historical and religious values in which the building was built. In general, the facade reflects the characteristics and culture of the region. Traditional houses are often used as a reference source for the design of facades.

Figure 7. Reje Naggroi Bebesen house.

Based on observations made on several modern houses in Takengon, this study found a variety of different types of facades. In general, the facade is spontaneously formed by following the function of the building. It is rare to find a house building that adopts the facade of a traditional house building. Figure 7 shows a house with a facade influenced by colonial architectural forms without the slightest element of *umah pitu ruang*. Similarly, in Figures 8 and 9, the facade seen in the building is also a relatively common and simple architectural style. There are no carvings like those found in *umah pitu ruang* in both houses. The use of materials in buildings depends relatively on the economic conditions of the homeowner. In general, people of middle to upper economic levels build their houses with concrete materials, while others build their houses with wooden materials.

Figure 8. Facade of Mr. Muhammad's House.

Figure 9. Facade of Mr. Ilyadi's House.

4. Conclussion

This study found that the influence of *umah pitu ruang* as a traditional Gayonese house on the housing of modern society is almost no longer exist in terms of spatial arrangements and ornaments. But in terms of the need for communal space for an extended family gathering, some modern houses have adopted it as an elevated floor called *pepanteren* in the living room or dining room, next to the kitchen. Therefore, the *pepantren* made of wood will make the living room or dining room, which is often used as a gathering place, warmer. Based on several houses observed, almost all of them were built without considering and incorporating the values of the *umah pitu ruang* architecture especially in spatial arrangements and ornaments. Gayonese housing in modern times is formed by the need for space functions for activities. The facade of the building is formed according to the function of the room, without considering the needs of the ornaments. Room functionality is the main reason for the Gayonese to no longer adopt *umah pitu ruang* architecture in modern house design at present.

5. Acknowledgement

This research is funded by Universitas Syiah Kuala through PNBP (PHI) scheme 2020.

References

[1] Putra R A, Ekomadyo A S 2015 *Tesa Arsitektur,* **13(1)** 1-14.
[2] Sari L H, Izziah, Meutia E, Irwansyah M 2016 *The 4th Int. Conf. on Sustainable Built Environment: Sustainable Bulding And Environment for Sophisticated Life,* 68-77.
[3] Ifani S M 2015 *Kajian Arsitektur Tradisional Sebagai Acuan Desain Rumah Tinggal Kontemporer*. Thesis, Universitas Syiah Kuala, Indonesia.
[4] Salwin S, Hendra A 2019 *Jurnal Arsitek,* **1(1)** 14-25.
[5] Lisa N P, Fithra H, Siska D, Dafrina A 2019 *International Journal of Engineering & Technology,* **8(1.9)** 560-564.

IOP Conf. Series: Materials Science and Engineering 1087 (2021) 012009 doi:10.1088/1757-899X/1087/1/012009

Study on natural ventilation performance in flats design at Banda Aceh

M Iqbal[1,*], A Munir[2], L H Sari[2]
[1]Student of Master's Degree Program of Architecture Department, Faculty of Engineering, University of Syiah Kuala, Darussalam, Banda Aceh 2311, Indonesia
[2]Department of Architecture and Planning, Faculty of Faculty of Engineering, University of Syiah Kuala, Darussalam, Banda Aceh 2311, Indonesia

*E-mail: iqbalmuhammad95@yahoo.co.id

Abstract. Flats are one of the solutions to housing problems faced by all countries in the world, especially developing countries like Indonesia. Flats are intended for low-income society. But ventilation problems often occur in some vertical housing due to non-optimal layout and lack of openings in each unit. The cheapest solution is to optimize the natural ventilation to expediting air circulation from inside to outside. The purpose of this study is to review natural ventilation performance in flats design at Banda Aceh. Natural ventilation offers healthy and comfortable air circulation without additional energy. The performance of natural ventilation is affected by the building's layout and orientation, the opening's orientations, locations, dimensions, types and the air-flow controller. The methodology used is the observation method with a descriptive survey that aims to look for symptoms with an observation method. The study begins with a field study and then conducts an analysis to compare existing conditions with theoretical studies. The results obtained are the design of these flats has been designed to be able to maximize the natural ventilation system, but the air movement cannot be felt by the occupants due to the opening's dimensions which is not following the existing standards.

1. Introduction

Flats are one of the solutions to housing problems faced by all countries in the world, especially developing countries like Indonesia [1]. Flats are also one strategy to optimize land use. The flats construction is intended for low-income society, therefore the flat's operational costs must be cheap. Even though the operation is cheap, it must still provide comfort for its residents, because living comfort is a basic requirement that must be felt by everyone.

Problems that often occur in part occur in vertical housing are ventilation system problems, this problem also occurs in flats. The Ministry of Public Works and Public Housing has also formulated 10 Flats problems in Indonesia at the Indonesia Property Expo in 2018, one of the flats problems in Indonesia is the problem of the ventilation system which causes a lack of air circulation due to non-optimal layout space and lack of openings from each unit. So, the residents didn't get sufficient natural lighting and adequate air circulation. Besides, each residential unit has an area of 24 m^2, which is inhabited by more than 2 people and filled with household furniture. The solution to overcome this problem is to optimize the use of natural ventilation to expediting air circulation in residential units. Air circulation has good benefits for buildings as an effort to sanitation, comfort and cooling the space if it can be channeled properly into the building [2-3]. One way to improve the air-flow into the building is

IOP Conf. Series: Materials Science and Engineering 1087 (2021) 012009 doi:10.1088/1757-899X/1087/1/012009

by maximizing the natural ventilation system for air changes to the building. The extent of ventilation relates to the air volume and circulation in the building, if the size of the ventilation is too small, it will cause the air to not circulate properly [4].

The purpose of this study is to conduct a review and research related to the natural ventilation performance in the flat design at Banda Aceh and to find out how the performance of natural ventilation in the flats itself.

2. Natural ventilation

Natural ventilation is an air circulation system that exchanges the air naturally (does not involve mechanical equipment, such as fans, exhaust fans, air conditioning, and so on). Natural ventilation is needed so the air in the room remains healthy and comfortable. Natural ventilation offers healthy and comfortable air circulation without additional energy [5]. Air circulation in natural ventilation system can be occurring because there is a difference in air pressure outside and inside the building.
The performance of natural ventilation is influenced by several factors that cause air to enter the building. These factors are building's layout and orientation, opening's orientation, opening's location, opening's dimensions, opening's type, and air-flow controller.

2.1. Building's layout and orientation

The building's layout in an area will affect the air movement. The location of the building which forms an angle to the direction of the coming wind will reduce wind speed by 50-60% [6]. The best building's orientation is perpendicular to the coming wind direction.
Building's layout and orientation also affect the air movement around the building. Air movement is formed due to differences in pressure in the air. Because of the difference pressure, the air moves and tries to balance the pressure. This pressure difference can be utilized by natural ventilation to channel air from outside into the building.

2.2. Opening's orientation

The opening's orientation will not only affect the air velocity, but also air-flow's pattern in the room. The position of the inlet and outlet can be divided into three, namely face to face, side by side and on the same side, as shown in Figure 1.

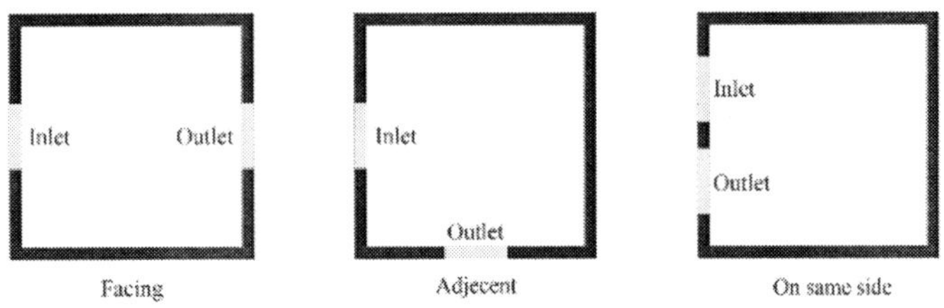

Figure 1. Inlet and outlet orientation [5].

2.3. Opening's location

One of the requirements for a good natural ventilation system is an attempt to create cross ventilation. It can be done with providing openings on both sides of the room so the opportunity for air to flow in and out can be maximized. It is portrayed by Figure 2.

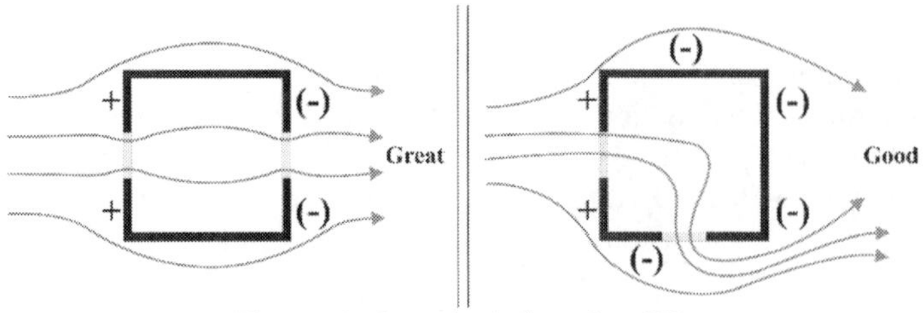

Figure 2. Opening's location [7].

2.4. Opening's dimensions

To make air circulation run properly, it is necessary to have an opening's area with a certain value. The calculation of the minimum opening's area in the room facade is 40% -80% of the wall area and 20% of the floor area and for the ventilation's area, according to the SNI Department of Public Housing the minimum of fixed ventilation's area are 5% of the room's floor area, while the incidental ventilation area (can be opened and closed) minimum 5% of room's floor area. That numbers are the requirement of the opening's area so the air can circulate properly in a humid tropical climate with normal air velocity conditions (0.6 m/sec to 1.5 m/sec).

2.5. Opening's type

The opening's type determines the volume of air distribution at indoor space, as shown by Figure 3. Every opening's type has a different level of effectiveness. Opening's type plays main role in circulating air to the building. Therefore, openings should be able to function as follows:

- As an air director in the space as much as possible.
- Must be optimized to support airflow and air change.
- Must be flexible to open and close as needed.

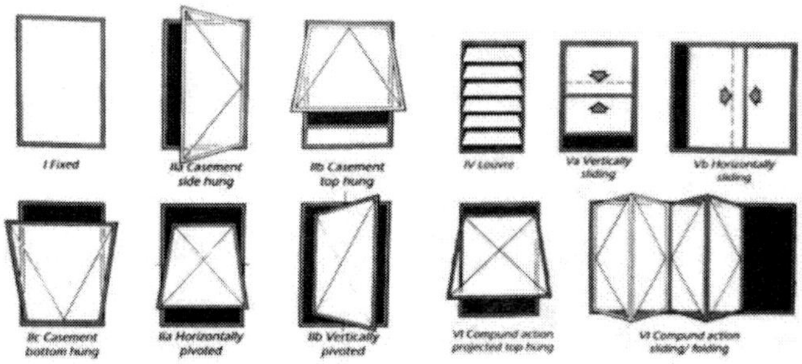

Figure 3. Opening's types [7].

2.6. Air-flow controller

The air-flow controller also plays a role in the use of natural ventilation system. The air-flow controller can determine the air motion, direction, and pattern in the building. Different air-flow controller's locations and types will make different results too. The air-flow controller can be a canopy, louvers even shutters.

3. Research Method

The research method used is the observation method with descriptive surveys. The object of this research is a type 24 apartment building in Banda Aceh. A descriptive survey aims to look for symptoms using observation methods. The observation method is a method carried out by direct observation and recording of the symptoms of the phenomena under study [7]. This research begins with a field study on the flats building, then analyzes the building design and compares the existing conditions with a theoretical study. The analysis is more emphasized on the existing opening's conditions and other design elements that influence natural ventilation performances.

4. Result and discussion

4.1. Flats building
The flats in this study are rental flats managed by the Banda Aceh City Government, located on Rama Setia, Merduati, Kuta Raja, Banda Aceh. This flat consists of 4 residential blocks named Block A, Block B, Block C and Block D. Figure 4, 5, and 6 depict the flat.

Figure 4. Flat site plant and building's perspective.

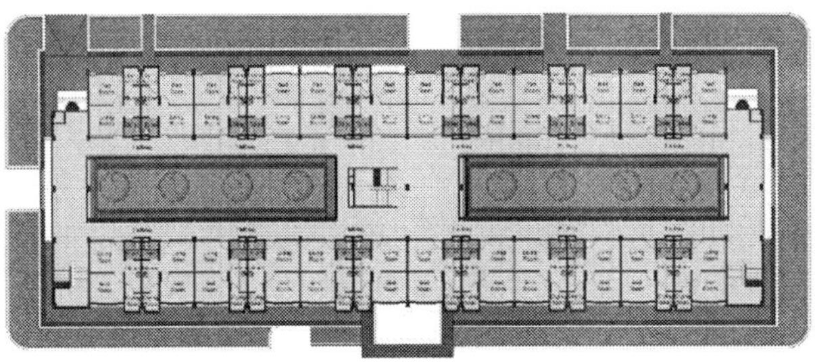

Figure 5. 2nd to 5th floor plan.

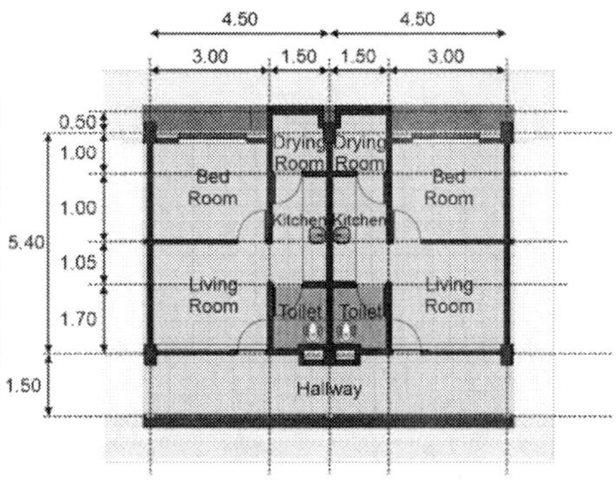

Figure 6. Unit floor plan.

This flats building has only one type of residential's unit, namely, type 24 with 24 m^2 area. Type 24 residential units consist of 1 bedroom, 1 family room, 1 bathroom, and 1 clothes drying room.

4.2. Building's layout and orientation
This building has been arranged parallel to the direction of the site (Figure 7). Buildings that are positioned like this will create some turbulence that contains small air movements and create unusual air-flow patterns. Besides, the space between buildings causes the air-flow to become narrow, which results in the building not being able to receive air movement with sufficient speed on each side (Figure 8). The parallel structure is not effective for the use of wind in humid tropical climates such as in Indonesia [6].

Figure 7. Layout plan.

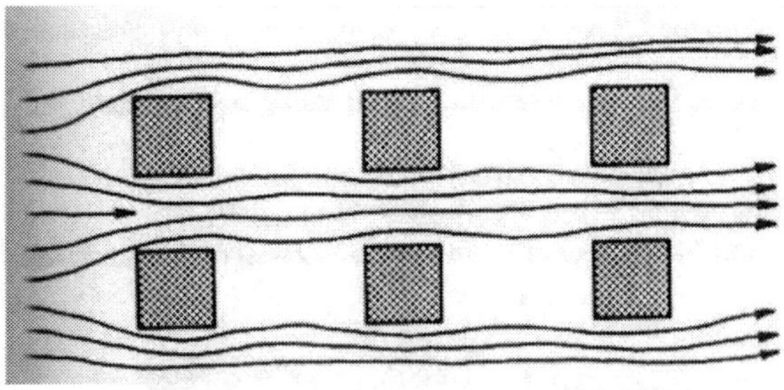

Figure 8. Linear air-flow pattern [6].

However, these flat's residential units are placed in the part that is directly facing the air-flow, so that air can be accessed by residential area (Figure 9).

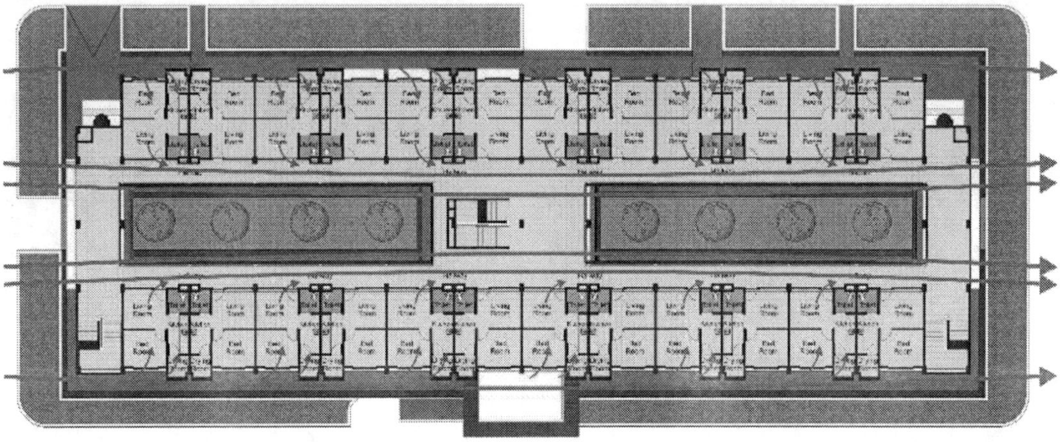

Figure 9. Air-flows at flats.

The building was also designed using courtyard and voids as a separator between residential unit's rows (Figure 10 & 11). Courtyard and voids also function as a path for air circulation horizontally and vertically.

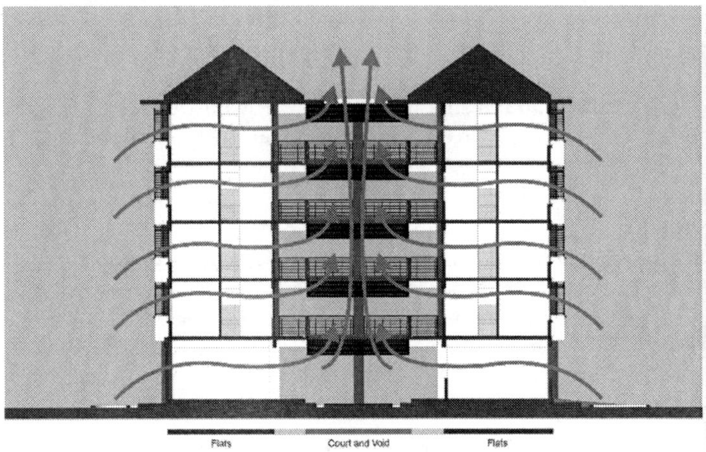

Figure 10. Flats section.

Figure 11. Courtyard and void

4.3. Opening's location

Placing openings facing both sides can create a cross ventilation system, as demonstrated by Figure 12. This ventilation system is a very effective way to circulate air in the room, because cross ventilation can provide high wind speeds and the best air distribution patterns in the room [6].

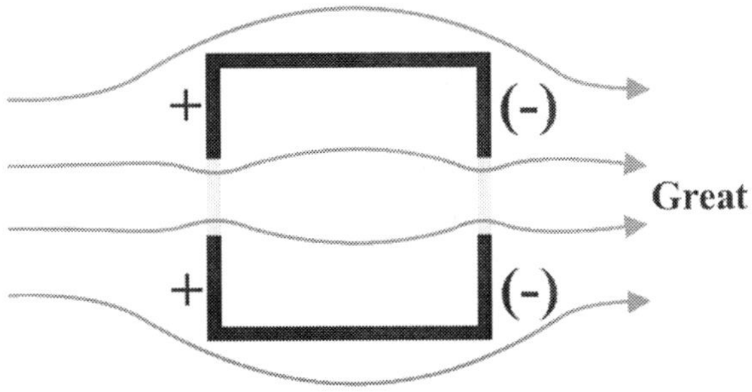

Figure 12. Cross circulation air-flows [6].

Each residential unit has been designed by optimizing cross ventilation system, but the location of the inlet and outlet of the wind does not face frontally, therefore the wind does not move directly towards the outlet and results in non-optimal airflow. If air enters the unit through the hallway, the air that entering the bedroom can only be streamed through ventilation above the door and door if they are open. These can be viewed in Figure 13, 14, and 15.

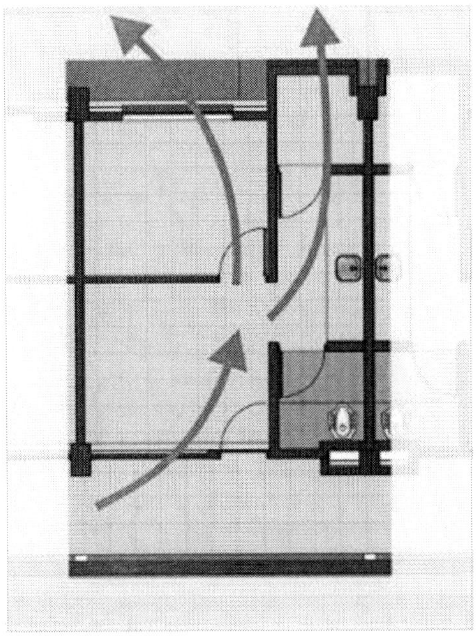

Figure 13. Indoor air circulation pattern. **Figure 14.** Bedroom door and ventilation.

The inlet ventilation position has the same elevation as the outlet, while the inlet coming from the window has a different elevation to allow a cross ventilation system to occur.

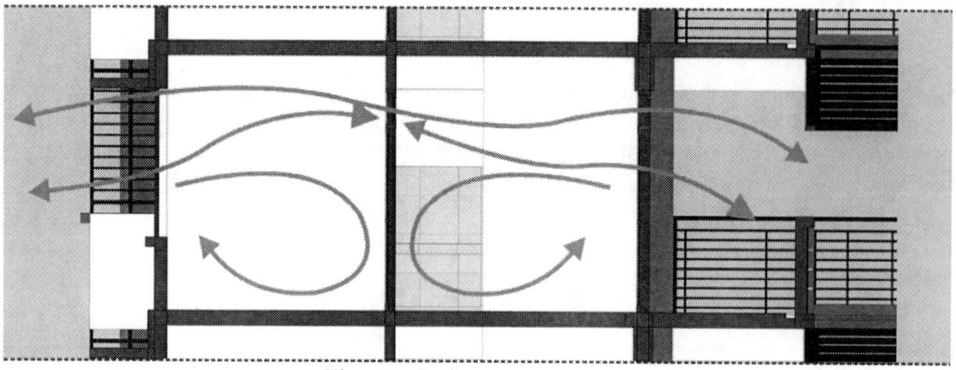

Figure 15. Cross ventilation.

4.4. Opening's dimensions

In this type 24 flats have 4 openings that affect air circulation for the residential unit, as displayed in Figure 16.

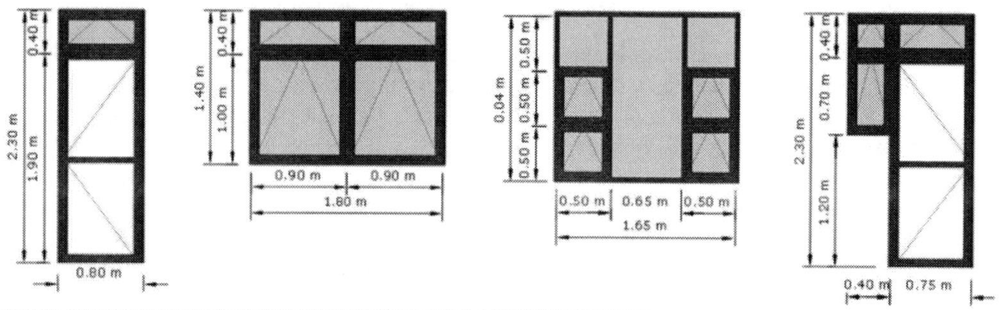

Figure 16. Opening types.

The opening dimensions found in this apartment do not meet the specified requirements. From the calculation of the minimum opening area based on the floor area (20% of the floor area) the opening area does not meet the requirements. The opening area that should be available is 4.80 m^2 while only 4.58 m^2 is available. When reviewed based on the front and rear façade wall area (40-80%) the opening area does not meet the minimum requirement of 40%, the opening area that should be available on the façade side of the residential unit is 5.04 m^2 while only 2.84 m^2 is available, as well as the rear side of the residential unit only available 1.74 m^2. If reviewed based on SNI the Department of Public Housing, the fixed ventilation remains not available at all in the residential units, except for the bathroom. As for incidental ventilation requirements, ventilation in these flats meets a minimum standard of 5%. It is summarized in Table 1.

Table 1. Opening's area analysis.

Building's Part	Area	Standard		Existing condition	note
Floor area	24 m^2	20%	4,80	4,58	Does not meet standards
Façade area	12,6 m^2	40%	5,04	2,84	Does not meet standards
Rear wall area	12,6 m^2	40%	5,04	1,74	Does not meet standards
Fixed Ventilation	24 m^2	5%	1,20	0	Does not meet standards
Incidental Ventilation	24 m^2	5%	1,20	4,58	Meet the standards

4.5. Openings type

Air openings available in this flats building are dominated by awning opening type, awning opening has the ability to steamed air effectively by 75% [9]. This opening type is one type of openings that is effective for entering air into the room, as Figure 17 and 18 reveals. However, the air volume that circulates in the building also affected by wind speed and temperature.

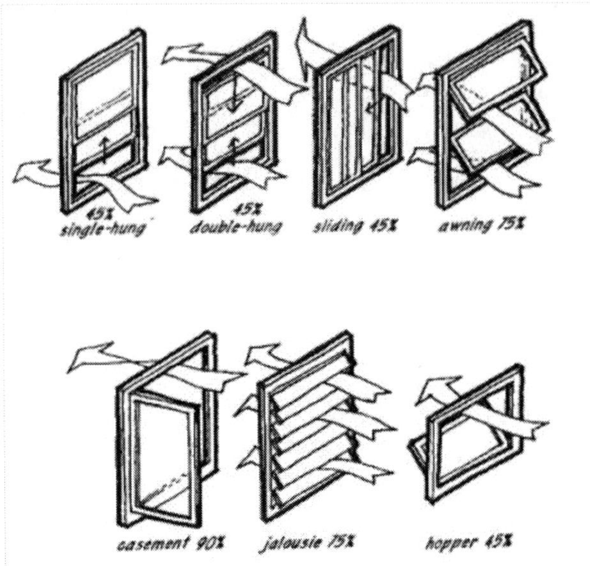

Figure 17. Opening's effectivity [8].

Figure 18. Openings type.

4.6. Air-flow controller

In this flat's design uses a lot of design elements that has a function to control the air-flow to enter the residential unit.

Figure 19. Rear side of flats.

The columns are in the flats prominent than the walls used for rooms. This can direct the flow of air into the bedroom, this also applies to the walls of the laundry room and the canopy which is located above the bedroom opening (Figure 19).

Figure 20. Inside view.

The principle of directing the air-flow is also used on the inside of the flats building. The columns in the hallway area are also designed to be more prominent than the walls from the residential units (Figure 20). In the hallway there are lattices and parapet walls that function as air-flow controllers.

Figure 21. The left side of flats building.

The left and right-side view of the building only uses an iron railing and does not use the walls as a barrier, as seen in Figure 21. This can steam the air optimally because there is no air-flow barrier.

5. Conclusions

Based on the results of field analysis and theoretical studies on the object of this study, it can be seen that the shape of the flats is designed to be able to maximize natural ventilation. This can be seen from:

- Layout and orientation of the buildings arranged as possible to maximize the air movement into the residential unit
- Every residential unit designed using a cross ventilation system that can maximize air circulation
- Every residential unit uses an effective opening type, which is awning type which can effective circulating air by 75%
- There is an air-flow controller concept like column location, wall and canopy that can direct the air-flow to the residential unit

However, the maximum air movement cannot be felt by the occupants of the residential units due to the opening's dimensions which do not meet the standard 20% of the floor area, 40% of the wall area and also do not meet the SNI requirements for Public Housing to provide for fixed ventilation.

References

[1] Ritohardoyo S 2015 *Patrawidya,* **16** 39-406.

[2] Manley G 2009 *Riba Journal,* **156** 317-323.

[3] Kussoy W J 1998 *Studi Kenyamanan Ruang Dengan System Ventilasi Alami Pada Bangunan Type 54 Di Kompleks Perumahan Btn Paniki Bawah Manado.*

[4] Lindawaty 2010 *Particulat (PM10) Udara Rumah Tinggal Yang Mempengaruhi Kejadian Infeksi Saluran Pernapasan Akut (Ispa) Pada Balita (Penelitian Di Kecamatan Mampang Prapatan, Jakarta Selatan.* Skripsi, Universitas Indonesia, Indonesia.

[5] Latifah N L 2015 *Fisika bangunan I* Jakarta: Griya Kreasi.

[6] Boutet T S 1987 *Controling Air Movement, Manual for Architect And Building*. New York: McGraw-Hill.

[7] Arikunto S 1988 *Prosedur Penelitian Suatu Pendekatan Praktek.* Jakarta: Rineka Cipta.

[8] Moore F 1993 *Environmental Control Systems*. New York: McGraw-Hill.

[9] ASHRAE Standard 2003 *Ventilation For Acceptable Indoor Air Quality*. Atlanta: American Society of Heating, Refrigeration amd Air Condotioning Engineer.

Integration of public open space and pedestrian pathway to the Trans Koetaradja shelter in the Masjid Raya Baiturrahman area

C Zulkarnaen[1], S Sugiarto[2,*], I Caisarina[3]
[1]Civil Engineering Postgraduate Program, Department of Civil Engineering, Universitas Syiah Kuala, Banda Aceh, Indonesia
[2]Department of Civil Engineering, Universitas Syiah Kuala, Banda Aceh, Indonesia
[3]Department of Architecture and Planning, Universitas Syiah Kuala, Banda Aceh, Indonesia

*E-mail: sugiarto@unsyiah.ac.id

Abstract. The area of Masjid Raya Baiturrahman is an area that is integrated with public space, pedestrian pathway, and public transportation infrastructure namely Halte Trans Koetaradja. This area is always crowded because of the Masjid Raya Baiturrahman besides as a center of Islamic worship also as a center of Islamic tourism in Banda Aceh. This area is always crowded by motorists and pedestrians who visit and perform activities in the area. Because the still mix of motor vehicles and pedestrians makes the area at certain times very dense so it is not safe and comfortable for pedestrians. This research aims to identify the existing conditions of the pedestrian street in the area of Masjid Raya Baiturrahman Grand and also identify the characteristics of pedestrian movements to determine the level of service from the pedestrian path. This study was conducted using a table based on the list of Ministerial Regulation of Public Works No. 03 the Year 2014 and also the Department of Transport (DOT) to identify the existing conditions of the pedestrian path in the area of Masjid Raya, then conducted a review of the characteristics of pedestrian movements using the formula of pedestrian speed, pedestrian flow and also pedestrian density. The results of this research appointment that the pedestrian pathway in the area of Masjid Raya Baiturrahman when reviewed from the existing service level, it is still a level of service A, but if the review of existing conditions then there is an improvement to the pedestrian pathway. This is because the supporting facilities have not been compactor to the pedestrian pathway by the standard guidelines.

1. Introduction

As the center of the activity, Masjid Raya Baiturrahman (MRB) can not only stand-alone, is of course supported by the existence of the surrounding area that is connected to the pedestrian pathway and public open space around the area of the mosque. The current pedestrian path is still a sidewalk along the mosque's fence, while the public open space in the form of a park called the Aman Kuba park is connected to the public transportation infrastructure. This relates to Linkage theory which means the pseudo-line linking between the elements of one another, the nodes that are one with the other nodes, or the district with one another [1]. Meanwhile, according to Maki, the linkage is a city adhesive that brings together activities and produces the physical form of the city [2]. The pedestrian pathway in this area is used as connectivity to the means of movement or transfer of people or groups of people from the Grand Mosque to the public open space is also the Aman Kuba Park to the surrounding trade and services area.

According to Shirvani Pedestrian is part of the public space and is an important aspect of urban space, both field (square-open space) and road-corridor (street) [3]. Public open spaces as a shared space, where communities perform functional and ritualized activities in a community bond, both daily life and in periodical celebrations that have been set as open, where people do personal activities and groups [4].

But unfortunately, the current pedestrian path used as a place to sell by street vendors and also as a parking lot for motor vehicles and still can be traversed by motor vehicles. This of course inhibits the movement of pedestrians who will use the pedestrian as a transit line that descends from the Trans Koetaradja shelter and towards the stop, therefore it is required that the pedestrian path meets the standards and fulfills the movements of all regions. Also, it needs to apply the concept of walkability and Non-Motorized Transport. The walkability broadly refers to the feature variation of a community or surrounding environment that creates a quite easy place to move around without using a motor vehicle [5]. Non-Motorized Transport (NMT) is a type of transport where the model used does not depend on the engine or motor to move. It includes walking, biking, wheelchairs, and other variations of personal vehicles such as roller-skating, skateboards, and Carts [6].

A safe, comfortable, and humane pedestrian network in urban areas is an important component that must be provided to improve the effectiveness of urban mobility. The pedestrian network is a pedestrian area, either integrated or separate from the road, which is destined for the infrastructure and pedestrian facilities and connects the activities centers and/or the mode change facility [7]. According to the Government of Western Australia, pedestrians can be interpreted as walking persons or persons using the equipment running with wheels such as a person who sits in a wheelchair, a person who encourages a stroller, or runs using a skateboard [8].

In the area of Masjid Raya, the pedestrian path and the public room still has a function transition because it is still used by traders and parking of motor vehicles. This causes pedestrians to become unsafe and not inconvenient when using a pedestrian path. Also, the lack of completeness of supporting street furniture, vegetation, signage, undirected circulation, and others. This relates to the road users who will be heading for the Trans Koetaradja shelter or vice versa. Pedestrian facilities need to be well prepared following existing standards so that the environment for pedestrians surrounding the historical area (MRB) in Banda Aceh can be created and not only seen from the design aspect but also the security aspects [9].

Safety, safety and health criteria are at top priority, the criteria of attraction and comfort are at the second priority, as well as accessibility criteria and beauty are at a third priority [10]. Various land-use planning strategies and road plans to improve road safety have been created and implemented in many countries [11]. With the evaluation of the conditions of public open spaces and pedestrians in Masjid Raya Baiturrahman, it is expected to later be a recommendation for the Government to realize a pedestrian path that is safe, comfortable, effective and pedestrian-friendly.

2. Area of study, data collection, and processing

Figure 1 and Figure 2 show the area of study in which in the area of Masjid Raya Baiturrahman Banda Aceh. It is a pedestrian path as well as a public open space integrated with Trans Koetaradja. The pedestrian path is a major review for viewing the conditions of the pedestrianized pedestrian path as well as the characteristics of the pedestrian movements on this pedestrianized trail.

Regarding the data collection method, data retrieval to see the existing condition is by observing directly using the Checklist table to assess the current condition of the pedestrian path to see the conformity and inconsistency according to the existing guidelines. As for the data retrieval characteristics of pedestrians, movements use the stopwatch to calculate the speed of pedestrians from the west and east sides of each segment and manually calculate the number of pedestrians per segment for three days (Monday, Wednesday, Saturday) from morning to night. The number of pedestrians took every 15 minutes in the morning peak (07:00-10:00), afternoon (13:00-16:00), and evening (18:00-22:00). A sample pedestrian is a male and adult woman using a pedestrian path on every research segment.

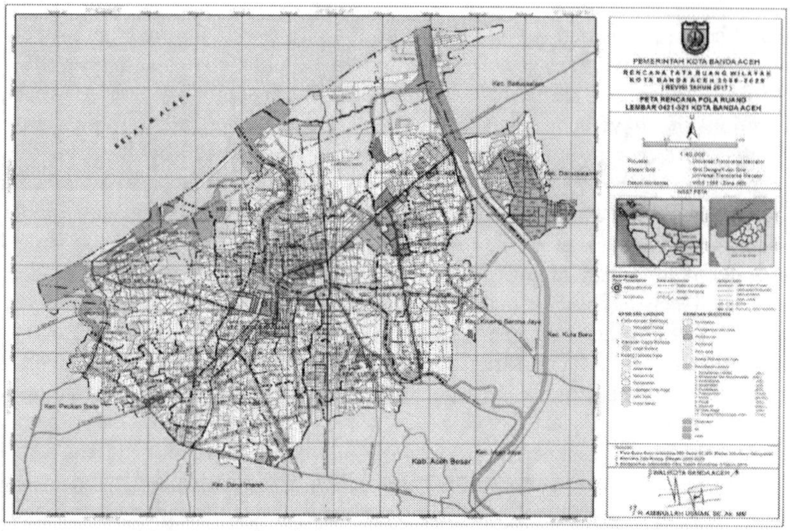

Figure 1. Map of Banda Aceh city.

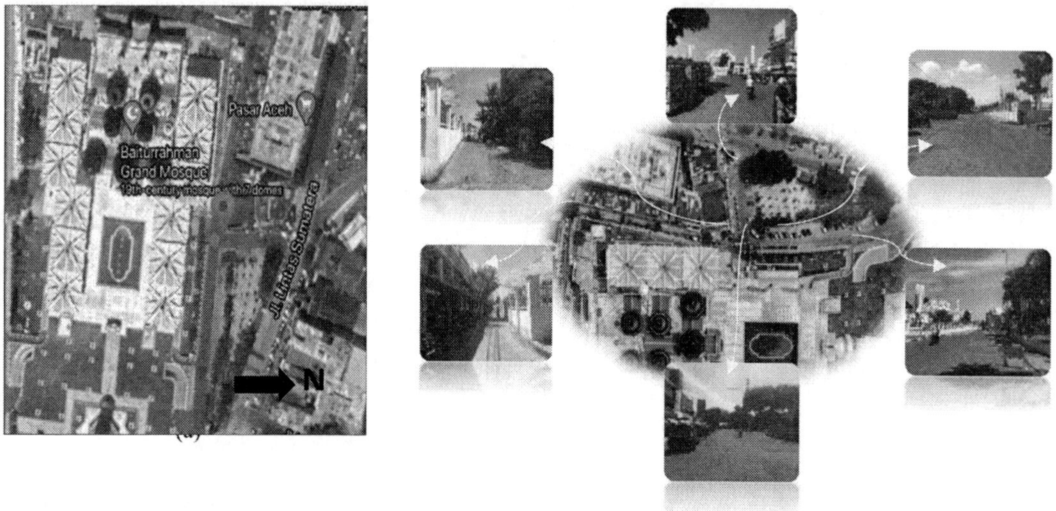

Figure 2. Map of (a) Area of study, (b) Existing condition of the study area.

All data of existing pedestrian and pedestrianized data are collected for processing so that later can be used in the analysis. Pedestrian data is both for pedestrian speed and the number of pedestrians is used in Ms. Excel, where it is calculated using the characteristic formula of the pedestrian movement so that the final result of the pedestrian movement characteristics will determine the level of service of this pedestrian path. As mentioned before, the analysis in this study is focused on the level of services pedestrian and improvement of existing conditions of the pedestrian path (improvement of an existing. The method used for the analysis will be described in the following section).

3. Analysis method

In this study, as mentioned above, the data analysis to get two main objectives of the (1) identification of the condition of the pedestrian path (2) Identification of the characteristics of pedestrian movements i.e. speed, pedestrian flow, and pedestrian density. The table of service identification according to the ministerial regulation of Public Works No. 3 of 2014 and also DOT (planning and designing for pedestrian guidelines) is used when observing existing conditions of the pedestrian path at the research site, as indicated by Table 1. From the results of the list can then be a recommendation for the improvement of the pedestrian path following the predefined standards. As for the number of pedestrians and pedestrians speed data is taken using the stopwatch to make the speed of pedestrians both from the west and the east side of each existing segment. The number of samples taken to see the speed of pedestrians in their segments amounted to 60 people consisting of 30 men and 30 the women. While the number of pedestrian data is calculated manually by viewing and writing the number of pedestrians for 3 days of research (Monday, Wednesday, Saturday) to see the difference in the number of pedestrians through the pedestrian path on weekdays and weekends. The number of pedestrians is subsequently processed using MS. Excel uses a characteristic formula of pedestrian movements.

Table 1. Existing pedestrian conditions

Conformity	Description
Track width	At least 1.5 meters
Surface Line	Stable, strong, and flat
Floor Textures	Smooth and not slippery
Degree of tilt (ramp)	1:12 (8%)
Lighting	Space required 60-75cm
Signs and markings	Space required 60-75 cm
Trash Can	Space required 90cm per 20m
Shade Plants	Planted greenery on furniture lines 1, 5m
Bench/Seat	Requires Room 1, 5m

Source: The Decree No.3 2014 [12] and DOT [8]

Table 2. Pedestrian Level of Service.

Service level	Space m²/pedestrian	Current and expected speed		
		speed m/min	Pedestrian flow/min/m	Volume/Cap
A	≥ 12	≥ 79	≤ 6,5	≤ 0,08
B	≥ 4	≥ 76	≤ 23	≤ 0,28
C	≥ 2	≥ 73	≤ 33	≤ 0,40
D	≥ 1,5	≥ 69	≤ 46	≤ 0,60
E	≥ 0,5	≥ 46	≤ 82	≤ 1
F	≥ 0,5	≥ 46	Vary	Vary

Source: Transportation Research Board [14]

To analyse the characteristics of pedestrians on the pedestrian path of the Masjid Raya Baiturrahman Use the appropriate method of formula as a parameter in analysing. To calculate the flow of pedestrians, pedestrian speed, pedestrian density using the characteristic formula of pedestrian movements [13]. To determine the level of service of the pedestrian path in the mosque of Masjid Raya Baiturrahman using Level of Service (LOS) analysis. Service level is the classification of traffic flow quality at the maximum range of capacity fraction [14] (Transportation Research Board, 2000). The concept of service level relates to the comfort factor. Like the ability to choose a walking speed, precede the slower pedestrians avoiding conflicts with other pedestrians. The level of service can be classified in service level A to F service level, all of which reflect the conditions on the specific needs or currents of service. The details of these service levels can be seen in Table 2.

4. Analysis of results and discussion

After the field observation to see the existing condition of the pedestrian path in the area of Masjid Raya Baiturrahman obtained analysis made in the form of a table containing the suitability of the guidelines adapted to existing conditions in the research location that has been determined in determining analysis with the description of the mark (√) has a meaning fulfilled, while the sign (x) is still not being fulfilled, then the following analysis results in each segment has been determined on the pedestrian route in the area of Masjid Raya Baiturrahman. The result is presented in Table 3.

Table 3. Existing condition Analysis Pedestrian pathway.

Conformity	Description	Segment 1	2	3
Track width	At least 1.5 meters	√	√	√
Surface Line	Stable, strong, and flat	√	√	√
Floor Textures	Smooth and not slippery	√	x	√
Degree of tilt (ramp)	1:12 (8%)	√	x	x
Lighting	Space required 60-75cm	√	√	x
Signs and markings	Space required 60-75 cm	x	√	√
Trash Can	Space required 90cm per 20m	√	x	x
Shade Plants	Planted greenery on furniture lines 1, 5m	√	x	x
Bench/Seat	Requires Room 1, 5m	x	x	x

Based on the existing condition analysis of the pedestrian path in segment 1, the results of the suitability have been much fulfilled (√) only 2 parts that are not met (x). As for segment 2 and Segment, 3 are more incompatibilities (x). Hence it can be concluded that the pedestrian path in this area still needs improvement according to the existing guidelines so that it has a comfortable and safe pedestrian path for pedestrians. The calculation of the pedestrian data in the current can be after obtaining the number of pedestrians passing through the location of the research in the survey for 3 days in each segment of segment 1 to segment 3 in the Masjid Raya Baiturrahman. Calculations in the can of the number of pedestrians and divided by the effective width of the pedestrian path.

Table 4. Pedestrian characteristics.

Segment	Flow Pedestrian/min/m	Speed m/min	Density Pedestrian/m^2	Space m^2/pedestrian
1 West	0,312	416,76	0,343	3291,24
1 East	0,358	361,01	0,000	3416,18
2 West	0,218	633,35	0,000	3830,96
2 East	0,240	1345,80	0,000	6484,92
3 West	0,706	1051,71	0,001	1800,30
3 East	0,755	796,28	0,001	1332,20

The Data used in the pedestrian speed calculation is the travel time of the pedestrians passing through the observation. The implementation of the travel time survey is conducted in each segment, namely segments 1 to segment 3 on the west and east side. In this study, long-term observation is 10 meters. The travel time is counted in seconds. The speed unit used is a meter per minute. Since within 1 minute there are 60 seconds, the T should be divided by 60. Density is derived from the advanced calculations sought in pedestrian flow calculation and the average speed calculation of the pedestrian space. Density is calculated from the results for both variables. The result of the density calculation will be used in fulfilling one sub variable in determining the service level. The pedestrian space Data is a division of 1 divided by the density of each segment. Pedestrian characteristics include pedestrian flows, pedestrian speed, pedestrian density, and pedestrian space. Table 4 is the result obtained from each of the characteristics in segments 1 to 3 segments on the west and east.

Service level analysis is an advanced analysis of the first goal. The analysis is divided into 2 (two), namely the analysis of the service level based on the average value is 1 (one) day. And the level of service based on the highest score in 15 minutes, namely looking at the characteristics of pedestrian movements (current, speed, and space). Table 5 is the result of Level of Service (LOS) that is obtained in each segment of the pedestrian path in the area Of Masjid Raya Baiturrahman.

Table 5. Pedestrian level of service

Segment	Current and expected speed				Level of Service
	Space m2/pedestrian	Speed m/min	Flow pedestrian/min/m	Ratio (Vol/cap)	
Segment 1 West	1542,81	416,76	0,160	0,0004	A
Segment 1 East	1514,97	361,01	0,158	0,0004	A
Segment 2 West	3830,96	633,35	0,218	0,0003	A
Segment 2 East	6484,92	1345,80	0,240	0,0002	A
Segment 3 West	1800,30	1051,71	0,706	0,0007	A
Segment 3 East	1332,20	796,28	0,755	0,0009	A

To determine the level of service based on the characteristics of movement on the pedestrian path in the area of Masjid Raya Baiturrahman by looking at the highest value of an analysis that has been done before in the calculation of current, speed, and space, as follows:

- Based on the largest 15-minute pedestrian flow interval, where the maximum number of pedestrians occurred in segment 1 at 15.15-15.30 WIB, which is 40 pedestrians, and the width of the pedestrian line 2.2 meters. So the current magnitude as follows:
  $$Q15 = \frac{Nm}{15\,WE} = \frac{40}{15x\ 2{,}2}$$
  Q15 = 1,212 pedestrian/min/m
- Based on speed, the length of the observation is 10 meters. The travel time is counted in seconds. The speed unit used is a meter per minute. Since within 1 minute there are 60 seconds, the T should be divided by 60. For more details is stated in the formula:
  With L = 10 meters, Then the formula is changed to:
  $$V = \frac{L}{T/_{60}} = \frac{600}{1{,}20} = 500 \text{ meter/minute}$$
- Based on the largest pedestrian 15-minute interval space, where the value of density occurs in segment 1 at 09.15-09.30 WIB on the west side of 1.503 pedestrian/m^2 as follows:
  $$S = \frac{1}{D} = \frac{1}{1{,}503} = 0{,}665\ \text{M}^2\text{/pedestrian}$$

From the above calculation, based on pedestrian flow, the level of service is A, based on the speed of service level is A, based on space, the service level is E.

5. Conclusion

Based on the analysis of service levels on the pedestrian route reviewed from the pedestrian flow, speed, and density in one day for each segment it meets the A service level A. Whereas if the review of the highest value in 15 minutes is known that based on the pedestrian flow, the level of service is A, based on the speed of service level is A, based on the density of service level is E. This journal is expected to be input for government and related for pedestrian pathways such as shade plants, trash cans, lighting lamps, seating, directions, and other supporting facilities are well organized according to existing standards. Also, it is expected that the government can apply the concept of walkability and Non-motorized transport so that pedestrians can realize the importance of the pedestrian path and can use this pedestrian path with a sense of safety and comfort.

References

[1] Trancik R 1986 *Finding Lost Space: Theories of Urban Design*. New York: Van Nostrand Reinhold Company.
[2] Maki F 1964 *Investigations in Collective Form*. St. Louis: The School of Architecture, Washington University.
[3] Shirvani H 1985 *The Urban Design Process*. New York: Van Nostrand Reinhold Company.
[4] Carr S, Stephen C, Francis M, Rivlin L G, Stone A M 1992 *Public Space*. Cambridge University Press.
[5] Mcnally K 2010 *Design Guidelines for Walkable Communities*. Ohio: Niehoff Studio.
[6] Yazid M, Ismail R, Atiq R 2011 *Procedia Engineering,* **20** 125–134.
[7] The Decree No 13 2014 The Ministry of Public Work, Republic of Indonesia.
[8] Australian Federal Government 2012 *Planning and Designing for Pedestrians: Guidelines. Department of Transport*. Department of Planning, Public Transport Authority, 221.
[9] Caisarina I, Harahap S A, Rani H A 2018 *ICRP 2018 3rd International Conference on Rebuilding Place*. 13-14
[10] Lang J 1994 *Urban Design: the American Experience*. New York: Van Nostrand Reinhold.
[11] World Health Organization (WHO) 2013 *Safety Pedestrian Manual Road Safety*.
[12] The Decree No. 3 2014 The Ministry of Public Work, Republic of Indonesia.
[13] Mannering F L, Kilareski W P 1990 *Principles of Highway Engineering and Traffic Analysis*. John Wiley and Sons.
[14] Transportation Research Board. 2000. *Highway Capacity Manual (HCM)*. National Research Council.

Disaster mitigation in Indonesia: between plans and reality

M Fuady[1,*], R Munadi[2], M A K Fuady[1]
[1]Department of Architecture and Planning, Faculty of Engineering, Universitas Syiah Kuala, Indonesia
[2]Department of Electrical Engineering, Faculty of Engineering, Universitas Syiah Kuala, Indonesia

*E-mail: mirzafuady@unsyiah.ac.id

Abstract. Indonesia is one of the most vulnerable countries in the risk of natural disasters from the standpoint of geography, geology, climatology and demography. The geographical position of Indonesia is located at the confluence of 3 main tectonic plates and includes the Pacific Ring of Fire, and has more than 500 volcanoes, of which about 128 are still active. Some of the disasters that caused severe damage, large casualties and high losses included the earthquake and tsunami in Aceh (2004), the earthquake in Yogyakarta (2006), the earthquake in Padang (2009), the earthquake in Lombok (2018) and earthquake, tsunami and liquefaction disasters in Palu (2018). For this reason, it has become a necessity for the Indonesian people to learn from this disastrous experience of disaster by identifying all aspects related to risk and vulnerability to improve the ability of communities to cope with disasters. This study, which is based on the results of a study of a number of disaster studies, seeks to revisit several disaster events in Indonesia with a special review of the concept of disaster mitigation and its application to minimize disaster risk in urban areas. The results of the study show that the concept of disaster mitigation has been planned in several cities in Indonesia but there are still weaknesses in its application when a disaster occurs. This result is expected to be one of the references in improving the quality of urban disaster mitigation planning in Indonesia.

1. Introduction

From the documentation of the National Disaster Management Agency (BNPB) it is known that 2,564 disasters have occurred throughout 2018. The disasters that occurred in Indonesia throughout 2018 have caused the death of 3,349 people. This amount is still lower than in 2017 which reached 2,862 disasters [1]. However, the impact of the disaster in 2018 was greater and claimed more lives than the previous year.

There are 11 earthquakes with magnitudes above 7 on the Richter Scale (SR) that have rocked Indonesia in the last 15 years. In addition to the death toll, there were also 21,064 people injured, 1,432 people missing and as many as 10.2 million people displaced by the disaster. Three major disasters in 2018 that occurred in several parts of Indonesia with the emergence of large numbers of casualties were (a) successive earth shaking which occurred in West Nusa Tenggara; (b) earthquake, tsunami and liquefaction in Central Sulawesi; and (c) tsunamis in the Sunda Strait due to underwater avalanches and eruptions of Mount Anak Krakatau. Figure 1 is the diagram of the number of disaster events in Indonesia and the large earthquakes that have occurred in the last 15 years [2].

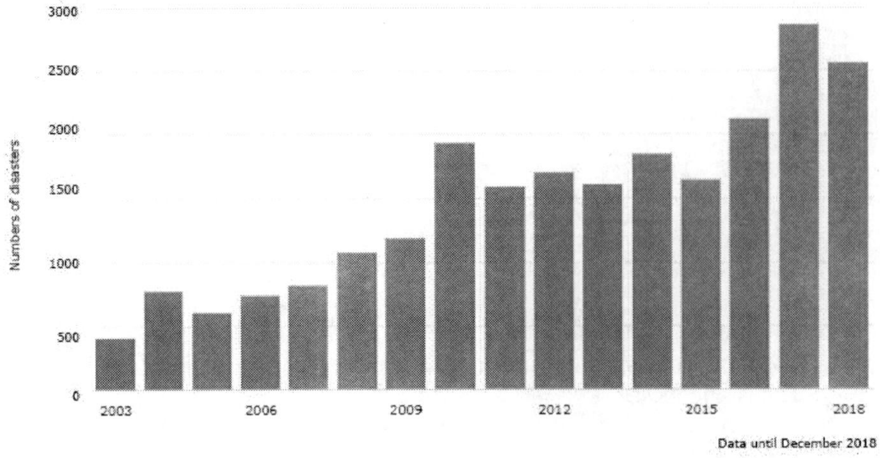

Figure 1. Number of Disasters in Indonesian in 2003-2018.

In 2004, several districts and cities in Aceh were severely damaged by earthquake with accompanied by tsunami. Victims who died from the biggest earthquake in Indonesia in the last few decades claimed more than 160,000 lives. The earthquake also caused tsunami in Thailand and even in Sri Lanka. One year later, the Nias archipelago had a turn to be shaken by earthquake with magnitude 8.7 SR. Most of the earthquakes that hit Indonesia have the potential to cause tsunami, but only a few really did have tsunami. It is summarized in Figure 2.

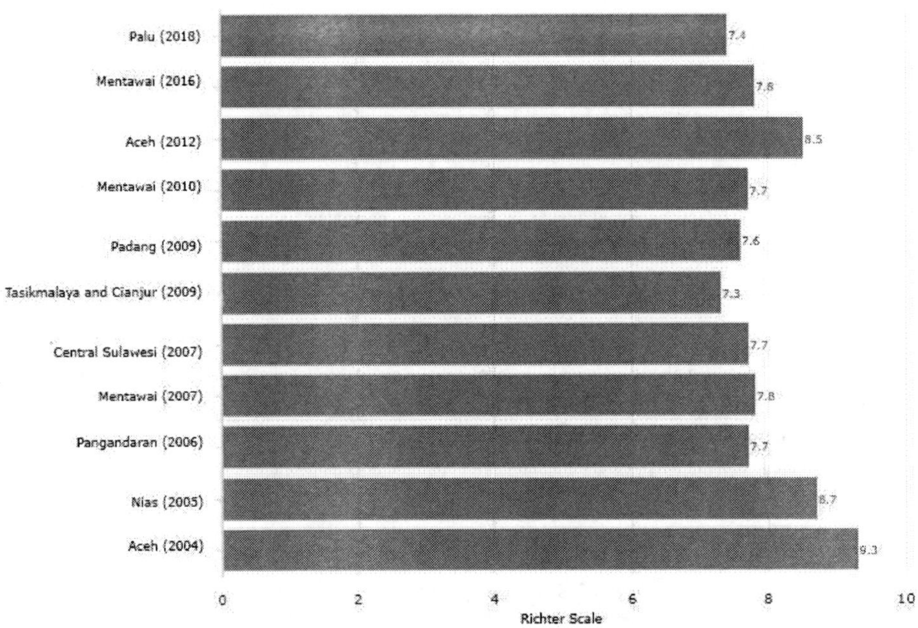

Figure 2. Earthquake Disaster in Indonesia in 2004-2018 with a magnitude above 7 SR.

The 2004 Aceh earthquake was the largest in Indonesia with magnitude of 9.3 on the Richter Scale (RS) and had the worst impact in the last 15 years. After that Indonesia seems to continue to shake and it was not long before Situbondo, East Java, was hit by earthquake with magnitude 6.4 SR. Still fresh in memory, on September 28, 2018, Palu and Donggala were rocked by an earthquake with a magnitude of 7.4 SR that was accompanied by tsunami. More than 2,000 people died as a result of the disaster. Still in the same year, Lombok was also hit by an earthquake with a magnitude of 6.9 SR [2]. From the factual explanation above, it is realized that disaster as a series of events that threaten and disrupt human life, has become a familiar phenomenon that is known by the people of Indonesia. For this reason, the government has also prepared various regulations and policies in disaster management as stated in Law No. 24 of 2007 concerning Disaster Management. Disaster management mandated in the law contains activities such as prevention, mitigation, preparedness, early warning, emergency response, rehabilitation, and reconstruction [3]. All of these activities are carried out in a series of holistic-continuous work with a framework for the success of development.

However, because this disaster management policy is included in the imperative policy model, namely centralized social policy, where all social objectives, types, sources, and number of social services, all have been determined. So often the role of development planning is mostly carried out by the government and feels foreign in its application in the community. Although various efforts to disseminate information and disaster education continue to be improved, there are still some weaknesses in the implementation of disaster mitigation plans that must be improved. For this reason, research is conducted on the implementation of disaster mitigation plans to determine the reality of the implementation of the plan in the community. The study based on the results of a number of studies on disasters in several cities in Indonesia seeks to re-express disaster events in Indonesia with a special review of the concept of disaster mitigation and its application in urban areas. The results of the study show that the concept of disaster mitigation has been planned in several cities in Indonesia but there are still weaknesses in its application when disaster occurs. The benefits and contributions of this research can be one of the references in improving the quality of urban disaster mitigation planning in Indonesia.

2. Literature review

As an archipelagic country, Indonesia is formed by the interaction of the complex earth's main plates, namely the Indo-Australian Plate, the Eurasian Plate and the Pacific Plate. From the south side the Indo-Australian Plate moves north at a speed of 6-7cm / year pounding the stable Eurasian Plate along the Sunda Trench. While the Pacific Plate from the east side, pounding eastern Indonesia along the Trench of New Guinea-Pacific with speeds up to 11cm / year. The interaction between the earth plates results in the formation of faults of the earth's crust or faults either in the sea (Mentawai Fault in Sumatra; Sula-Sorong Fault in Maluku-Papua; Flores Fault north of Flores) or on land (Sumatra Fault; Cimandiri-Baribis Fault and Opak fractures, on Java Island, Palu-Koro Fault, in Sulawesi and Sorong Fault and Tarera-Aiduna, in Papua) as displayed by Figure 3 [4].

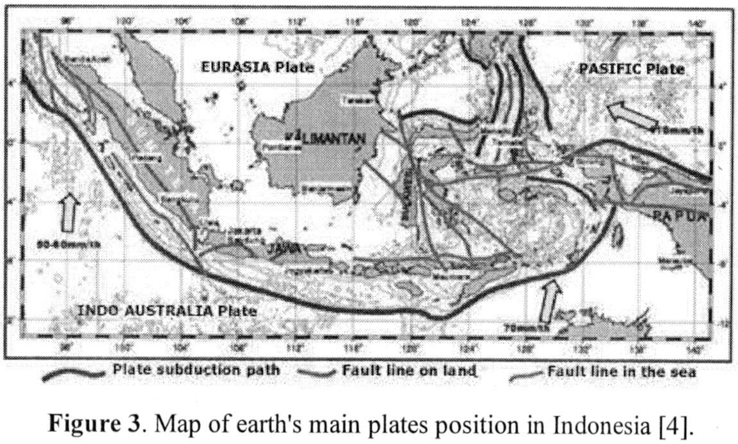

Figure 3. Map of earth's main plates position in Indonesia [4].

Places where the plates interact, along the trough and fractures of the earth's crust, are the source of earthquakes, both on land and at sea. If the earthquake occurs under the sea, it can trigger a tsunami if the epicenter is relatively shallow (<65Km) with an earthquake strength of more than 6.5 SR. As a result of the underwater earthquake that occurs due to the shift of the plate, especially along the subduction zone of the plate or fault, it can generate tsunami. Therefore, several Indonesian waters have the potential to be hit by tsunamis such as the west coast of Sumatra, southern Java, the west coast of Sulawesi, Gorontalo bay, northern part of North Sulawesi, the west coast of Sangihe-Talaud archipelago, west coast, north and east Halmahera and in the northern Papua. Figure 4 shows the coastal areas that have the potential to be hit by the tsunami in Indonesia.

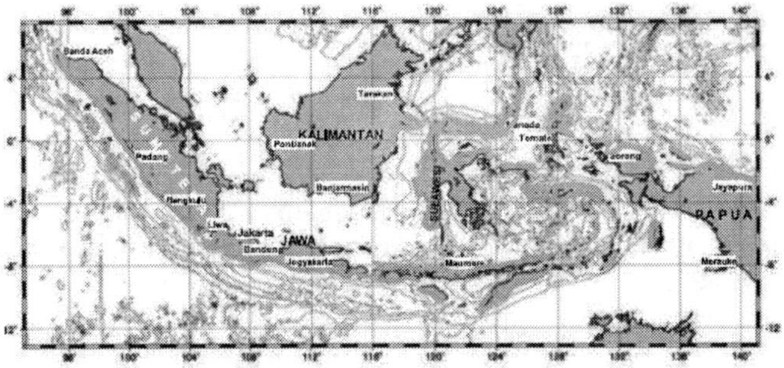

Figure 4. The coastal areas that have the potential to be hit by the tsunami in Indonesia [4].

Earthquakes are geological phenomena in the form of vibrations on the surface of the earth due to collisions of tectonic plates, or in the form of volcanic eruptions that cause material eruptions or melt magma. These earthquakes are known as tectonic earthquakes and volcanic earthquakes. Both of these earthquakes often cause harm to humans, both in the form of property victims or fatalities, so it can be a disaster for humans. In Indonesia there are 7000 earthquakes per year, including 50-60 times the earthquake that can be directly felt by humans. The biggest problem with the earthquake is that humans with their knowledge and technology to date have not been able to predict when an earthquake will occur [5].

A number of earthquake parameters can be measured and examined, then from recorded earthquake data will be analyzed for the return period of the earthquake. This return period is usually in multiples of 50 years. However, this calculation is especially for large earthquakes, such as those that occurred in Aceh in 2004. For small earthquakes almost every day occurs. The facts mentioned above provide an overview of a potential disaster. Disaster victims usually occur as a result of secondary impacts, namely from the collapse of buildings and landslides of slopes that afflict buildings and vehicles. So the most important thing about earthquake evaluation is disaster mitigation, and not the period of the earthquake that will occur.

Disaster mitigation can be interpreted as an effort to review the condition of disaster-prone areas, so that the potential for disasters and vulnerability of the region can be evaluated. Re-examination of an area may start from the distribution of earthquake areas such as dividing Indonesia into 6 earthquake regions. This division is based on the distribution of recorded tectonic earthquakes, starting from region 1, the lowest seismic region, to region 6, the highest seismic region. The division of the earthquake area above is mainly based on the distance of the region to the epicenter.

Large earthquakes can cause tsunamis, which are sudden movements of water caused by changes in sea level. These changes are caused by three main sources, namely tectonic earthquakes, volcanic eruptions, and avalanches that occur on the seabed. Tsunamis are categorized as long waves because the ratio between water depths to wavelengths is very small, so the vertical acceleration of water particles is ignored because the value is smaller than the acceleration of gravity. The vertical movement of water particles has no effect on the pressure distribution. Tsunami wavelengths can reach 240 km in open sea such as the Pacific Ocean with an average wavelength of 4600 m with wave velocities reaching 760 km / hr. When the tsunami waves are on the high seas, the tsunami height is less than 0.4 meters and the wavelength on the high seas ranges from 10 - 500 km causing the ship sailing not to feel the tsunami [6].

3. Methodology

This research is a type of qualitative research conducted by evaluating disaster events in several regions in Indonesia and conducting a review of the implementation of disaster mitigation plans to find out the implementation of mitigation plans when a disaster occurs. This study examines two case locations, namely (a) disaster that occurred in Banda Aceh in 2004 and the application of a post-disaster mitigation plan; and (b) the disaster that occurred in Palu in 2018. The data used is disaster data from official government agencies and notes of a number of studies which are listed in the reference list.

4. Discussion

4.1 Earthquake and tsunami disasters in Aceh

After the massive earthquake that followed by tsunami on December 26, 2004, many small earthquakes still occur in Aceh for up to 15 years. Even noted, in three days there could be earthquake up to 140 times in the province of Aceh. This tectonic earthquake occurs because of the movement of the earth's plate, where Aceh is on the path of the meeting of two earth plates, namely the Indo-Australian and Eurasian plates. Aftershocks continue to occur now to stabilize the earth's plate after a strong earthquake at the end of 2004.

The major disaster that struck Indonesia on December 26, 2004, began with an earthquake measuring 9.3 Mw (magnitude) centered on the Indian Ocean, west of the island of Sumatra, Indonesia. This earthquake caused a tsunami with a height of 30 meters and resulted in 230,000 deaths in 14 countries, of which 130,000 occurred in Banda Aceh, Indonesia. As the capital city of Aceh Province, Banda Aceh City is the city with the largest population where before the tsunami there were around 239,000 people spread across 9 sub-districts [7]. The four sub-districts of Banda Aceh City are directly adjacent to the Andaman Sea and the Indian Ocean, and two of these sub-districts, namely Meuraxa and Kutaradja Sub-district, suffered severe damage due to the 2004 tsunami disaster. These two regions are also sub-districts that have a high population in the coastal area [8]. Figure 5 illustrated the before-after comparison.

The combination of the earthquake and tsunami at the end of 2004 caused the biggest human casualties and losses in the history of human life on earth to date, with the most fatalities occurring in Aceh. On that day, like a nightmare, this region was devastated by the devastating earthquake and tsunami. Hundreds of thousands of human lives floated, houses were destroyed. Road and bridge infrastructure are in ruins. It has been almost 15 years. Now, Aceh is already improving. Damaged buildings were rebuilt. However, how is the readiness of the people of Aceh to face the earthquake and tsunami after life has returned to normal [9].

Figure 5. Satellite image shows part of Aceh Province before and after Tsunami in 2004 [7].

At the beginning of Aceh's rehabilitation and reconstruction, the government made a development blueprint for Aceh after the tsunami. One recommendation, the coastal area of Aceh that was hit by the tsunami became a green belt and was forbidden to occupy. But when the government offered the community to move to a safe area from the tsunami, residents refused. Coastal residents do not want to be relocated to places far from the sea. As fishermen who go out to sea as livelihoods, they refuse to become farmers.

Finally, the government decided they could stay there. In 2007, assistance for fishermen began to arrive. Some help boats and fishing gear. Some other residents began to return to the fields, planting rice and watermelons. The fields were rehabilitated with the help of NGOs. Similarly, the houses were rebuilt with assistance from abroad. When remembering the earthquake, the minds of most of the people of Aceh returned to the 2004 tsunami. Again, the horrors of the waves would roll them up so that they could cause even panic, pale and back-minded thoughts of the 2004 disaster. But in panic, they could not think of saving themselves. For this reason, the Banda Aceh city government continues to actively prepare the psychology of the community to face disasters and not panic when saving themselves. Community psychology readiness is very important to reduce disaster risk.

This is different from the people in Simeulue District, Aceh. Residents in Simeulue are better prepared and understand the earthquake and tsunami. If a strong earthquake, women, parents and children immediately fled to high ground, while a number of young people will move to the seafront to monitor sea water. Awareness of life in areas prone to earthquakes and tsunamis has been given since babies passed songs on swings. After the 2004 tsunami, the government built around 50 units of tsunami warning sirens in all vulnerable areas in Indonesia. Of these, six are in Banda Aceh and Aceh Besar. However, most Acehnese cannot hold on to this warning system. Tsunami sirens have not been used as a benchmark, because tsunami warning sirens have had several problems. Once a tsunami siren sounded loud in June 2007 but there was no earthquake. Hearing the tsunami sirens, the people in Baitussalam and Darussalam, Aceh Besar and the people in Syiah Kuala, Banda Aceh panicked and ran away from the sea. After successfully escaping a few kilometers, it was discovered that the siren sounded itself because of a system error.

The error occurred when an earthquake with a magnitude of 8.5 SR on April 11, 2012. Although the Meteorology, Climatology and Geophysics Agency (BMKG) said the earthquake had a potential tsunami, but none of the sirens sounded, even though in each simulation, nine minutes after the earthquake occurred if potentially tsunami, the siren will ring. Of the six sirens in Aceh, at that time, it was only heard at the Aceh Governor's office. Even then it just rings after the officer turns on manually. These various errors make the people not trust the sirens. If a strong earthquake occurs, even though the siren does not ring, residents will immediately save themselves to a high place [10].

4.2 Disasters in the city of Palu

Palu is the capital of Central Sulawesi Province as well as an economic and activity center in Central Sulawesi Province. The position of Palu City around the Palu Koro Sesar makes this city very vulnerable to the risk of earthquakes and tsunamis. As one of the cities in Indonesia that has a high level of vulnerability to tsunami hazards, several tsunamis hit around Palu City, namely: (a) December 1, 1927 tsunami triggered by an earthquake at the hammer bay with a wave height reaching 15 meters which resulted in 14 deaths, 50 injuries and severe damage in Palu City, Biromaru and its surroundings; (b) Parigi tsunami earthquake May 20, 1938; (c) the Tambu earthquake and tsunami on August 14, 1968; and (d) the 1996 Tolitoli earthquake and tsunami [11].

Furthermore, on September 28, 2018, a magnitude 7.4 earthquake occurred again followed by a tsunami that hit the west coast of the northern part of Sulawesi Island. The epicenter was 26 km north of Donggala and 80 km northwest of Palu with a depth of 10 km. Earthquake shocks were felt in Donggala Regency, Palu City, Parigi Moutong Regency, Sigi Regency, Poso Regency, Tolitoli Regency, Mamuju Regency even to Samarinda City, Balikpapan City, and Makassar City. The earthquake triggered a tsunami up to 5 meters in the city of Palu.

In general, the earthquake feels strong intensity for 2-10 seconds. Taking into account the location of the epicenter and the depth of the hypocenter of the earthquake, it appears that this shallow earthquake occurred due to activity in the Palu Koro fault zone. This fault is the most active fault in Sulawesi and can also be called the most active in Indonesia with a movement of 7 cm per year. The earthquake that occurred was a type of shallow earthquake due to the Palu Koro fault activity, which was generated by deformation with the movement mechanism of the strike-slip synistral structure.

As a result of the earthquake shock, some time after the peak of the earthquake, symptoms of liquefaction occurred (liquefaction of the soil) which took a lot of casualties and material. The two most obvious places to experience this disaster are the Petobo Village and Balaroa Housing in Palu City. Balaroa is located in the middle of the Palu-Koro fault.

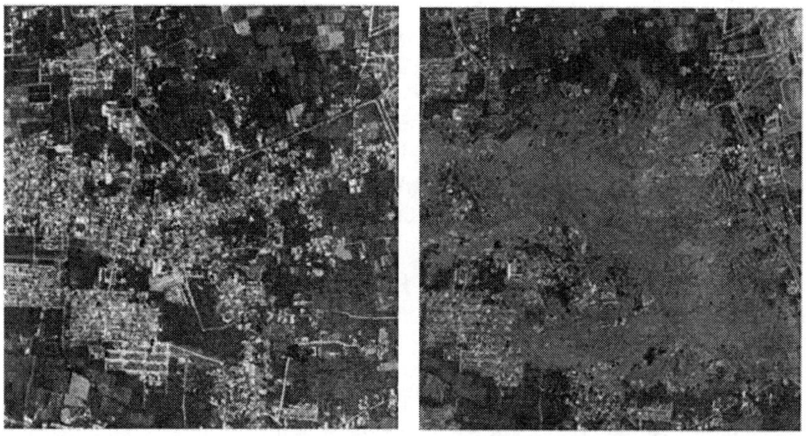

Figure 6. Petobo Village Area in Palu City before and after the disaster of liquefaction in 2018 [12].

When liquefaction occurs, there is an increase and decrease in land surface. Some parts collapsed 5 meters, and some parts rose to 2 meters. In Petobo, hundreds of houses are buried in black mud with a height of 3-5 meters. After the earthquake, the land in the area quickly turned to mud which immediately dragged the buildings above it, as captured in Figure 6. In Balaroa, the house collapsed, like being sucked into the ground. Reports and likuefaction recordings also emerged from the border of Sigi Regency with the City of Palu. Mud emerges from beneath the surface of the ground and shifts the ground to tens of meters and finally sinks the building and victims alive, as can be seen in Figure 7. According to the data, liquefaction that occurred at Balaroa Housing sank around 1,747 housing units, while in Petobo Urban Village around 744 houses were drowned [12].

Figure 7. Balaroa Housing Area in Palu City before and after the disaster of liquefaction in 2018 [12].

The earthquake that occurred in Central Sulawesi was then declared potentially tsunami by the Meteorology, Climatology and Geophysics Agency (BMKG) so that a tsunami early warning was issued for the coastal areas of Donggala Regency, Palu City and part of the northern coast of Mamuju Regency. The tsunami is predicted to have a height of 0.5 - 3 meters with the arrival time in Palu City at 18.22 WITA. At 6:27 p.m. there was a 6 cm sea level rise on the coast of Mamuju Regency.

According to BNPB, this tsunami is caused by the presence of sediment in the sea which reaches 200-300 meters. The sendiment has not been consolidated so strongly that when it was rocked by the earthquake there was a landslide. In other places besides Donggala, there was a local earthquake that made the tsunami not as big as in Donggala. In Palu Bay, which is closer to the epicenter, it is estimated that it first experienced a 1.5 m high tsunami. At 18.37 WITA, BMKG ended the tsunami early warning due to this earthquake.

The latest fact states that the highest tsunami point was recorded at 11.3 meters, occurring in Desa Tondo, Palu Timur, Kota Palu. While the tsunami lowest point was 2.2 meters, occurred in Mapaga Village, Donggala Regency. Both at the highest and lowest points, tsunamis hit the coast, hit settlements, to buildings and public facilities [13]. Long before this disaster occurred, the city of Palu actually had an earthquake and tsunami contingency plan document, which was planned in 2012 [13]. In this plan a scenario of a 7-magnitude earthquake with a depth of 10 km in Donggala was made, which would cause people to panic, and run up the hill. It was also stated that the estimated number of victims was 2,000 people, the bridge collapsed, roads were damaged, electricity was damaged, clean water was blocked, telecommunications was cut off, logistics was limited. It turns out that the scenario occurred on September 28, 2018.

Based on contingency plans, every year there should be a disaster simulation based on that scenario. This simulation will then be evaluated. However, the simulation was not carried out so that there was no evaluation of the contingency plan, even though contingency planning had been made at no small cost. Disaster-related policies sometimes become unpopular and avoided due to consideration of uncertainty about the arrival of disasters. Although the determination of an area in a risk map or disaster map is based on regional scientific research, there are still policy makers who do not heed it. Sometimes investment considerations are more prioritized than safety, so areas at risk can turn out to be feasible to build.

In the Palu City Spatial Plan, the area along Palu Bay is included in an area that has the potential for natural disasters. However, in the 2030 Space Pattern Plan the Palu Bay Area is dominated by Settlement Areas. The concentration of economic growth in the potential disaster area is agreed not to be used again and proposed as a protected area. Suggestions submitted are for areas included in disaster-prone areas so that the Spatial Pattern map is converted into a Protected Area section with specifications for Earthquake and Tsunami Disaster Prone Areas.

A review and revision of the Palu City Spatial Plan was carried out to anticipate the potential impact of the disaster. A review of more than 1 (one) time in 5 (five) years in major force conditions, namely, large-scale natural disasters, massive economic development in disaster-prone areas, changes in the country's territorial boundaries stipulated by legislation. Regulations need to be formulated with strict (criminal) sanctions stating that Protected Areas with Earthquake and Tsunami Hazard Prone specifications are no longer permitted as built-up areas, especially Strategic Areas of Economic Growth.

4.3 Disaster mitigation in Japan

Disaster education to the community is an important thing in minimizing disaster risk. An example is the experience of the Chilean state in educating and raising awareness of citizens of disasters. This can be seen in the declining number of deaths of 13 people in the 8.4 SR earthquake and 4.5 meter tsunami that occurred in 2015. The number of victims was very small compared to 500 deaths in the 8.8 SR earthquake five years earlier. The small number of victims is supported by the fact that the Chilean people already know how to deal with disasters thanks to the disaster knowledge that is included in the education curriculum.

In line with the Chilean state, the Japanese government also realizes the importance of disaster education for its citizens. However, when the earthquake that occurred in Honshu, Japan on March 11, 2011, it recorded a lot of damage and casualties. The earthquake was the biggest earthquake that ever happened in Japan. The Japanese disaster response system in this case looks inefficient, compared to China's experience in dealing with the 2008 Wenchuan earthquake, namely the importance of centralized strength in dealing with large-scale disasters.

In this earthquake-tsunami disaster, the Japanese government did not seem to solve many problems, especially with regard to the nuclear crisis. Coordination between the Government (emergency headquarters response), Tokyo Electric Power Company, and nuclear and industry security agencies are not organized enough. Information is not simultaneously shared after a disaster, delayed decision making. Finally, Japan, and perhaps all nuclear countries in the world, are not really ready for the nuclear crisis [14].

Experience in other countries shows that large scale disasters cannot be overcome only by local capacity but assistance from outside the affected area is indispensable. During this earthquake, the area that was damaged / affected was so extensive that local government groups for cities and prefectures became paralyzed. Not only the public sector, but also many private sectors are unable to provide services during disasters due to damaged infrastructure. These services include providing energy, food and water, and medical care.

In the future, there is a need for thorough preparation in disaster management, including: (a) Information systems. All parties need to understand hazard information all the time. This guarantee of continuous and valid information distribution is the duty of official functional institutions such as the Meteorology and Geophysics Agency. This information is also expected to be a form of early warning for the parties concerned, so that each party will be able to determine the follow-up of the information correctly; (b) Warning system. Efforts to create disaster preparedness in the public should also be equipped with an effective warning system for the community, in the sense that it is guaranteed to be obeyed by the community. Warning methods directed at safeguarding public memory should also be created as a way to stay in control of various social behaviors and create self-awareness; (c) Response mechanism. Planning disaster response mechanisms must be carefully and comprehensively established. Evacuation and SAR procedures, the condition of the affected area, the assessment team, the mechanism of activation of the installation and special infrastructure, preparation of distribution of aid, shelter and planning for the activation of emergency programs for airports, ports and land transportation are minimal matters that must be covered in response mechanisms disaster [15].

5. Conclusion

As a disaster-prone country, Indonesia is very late in anticipating and handling the impact of disasters. This can be proven by the government's delay in issuing regulations in the form of laws that regulate disaster management in a sustainable or sustainable manner. Previously, regulations were only ad hoc so that disaster management was carried out partially and was not well coordinated according to the disaster management cycle.

However, with the presence of the disaster law, it can treat the needs of civil society who add integrated disaster management. Because, there have been several paradigm changes in disaster management, including a linear to cycle paradigm, from responsiveness to management, from caricative to empowerment, and from managing impacts to reducing risk.

With the presence of this policy, it is hoped that disaster management will be better in the future, the capacity of the community will increase, the lives of victims will recover quickly and be better, the community empowered and responsive to disasters can be achieved rather than seeing the disaster as destiny. The monitoring and evaluation efforts of all interested parties, especially the community, are very important elements related to the budget and implementation of policies for the welfare of the Indonesian people.

References

[1] Sanjaya J 2018 *Journal of the Civil Engineering Forum,* **4(2)** 69-180.

[2] katadata.co.id 2018 11 Gempa Dengan Kekuatan di Atas 7 SR Yang Mengguncang Indonesia. Available at https://databoks.katadata.co.id/datapublish/2018/10/11/11-gempa-dengan-magnitudo-di-atas-7-sr-yang-mengguncang-indonesia.

[3] Faturahman B M 2018 *Publisia: Jurnal Ilmu Administrasi Publik,* **3(2)** 121-134.

[4] Triyono P 2010 *Bertahan dari Gempa Bumi dan Tsunami.* UNESCO Office Jakarta dengan Lembaga Ilmu Pengetahuan Indonesia.

[5] Simamora B, Ariwibowo S, Monica T 2018 *Sekolah Tinggi Meteorologi Klimatologi dan Geofisika.*

[6] Irwan 2012 *Pertemuan Ilmiah Tahunan Masyarakat Ahli Penginderaan Jauh Indonesia* (PIT-MAPIN XIX), Universitas Hasanuddin, Makassar 7 Juni 2012.

[7] BRR NAD-Nias 2009 *Membentang Atap Berpilar Asa.* Banda Aceh: BRR.

[8] Fuady 2015 *Journal of Architecture and Built Environment Dimensi.* **4(2)** 51-58.

[9] Fuady 2016 *Proceedings of International Conference on Engineering and Science for Research and Development.* **117**.

[10] TDMRC. 2010. *Perencanaan Pembangunan Berbasis Mitigasi Bencana (Aceh Disaster Risk Map).* Banda Aceh: TDMRC-Unsyiah.

[11] Fauzi 2018 *Resume Gempa Palu, Sulawesi Tengah, Mata kuliah Sistem Mitigasi dan Diseminasi Informasi Kebencanaan (GD6104).* Bandung: Program Studi Teknik Geodesi & Geomatika, Fakultas Ilmu Dan Teknologi Kebumian, Institut Teknologi Bandung.

[12] Utomo K P 2018 Pemotretan satelit kondisi Palu sebelum vs sesudah tsunami, ngeri. Available at https://m.brilio.net/serius/6-pemotretan-satelit-kondisi-palu-sebelum-vs-sesudah-tsunami-ngeri-1810029.html.

[13] Abdulsalam H 2018 Ramai-Ramai Menista Sains: Potret Bencana Alam Indonesia 2018. Available at https://tirto.id/ramai-ramai-menista-sains-potret-bencana-alam-indonesia-2018-dcGM.

[14] Supardiono 2014 *Jurnal Fisika* **3(3)** 11-15.

[15] Larasati, Aisharya, Medha 2016 Mata Kuliah Perencanaan Wilayah. Surabaya: Program Studi Perencanaan Wilayah dan Kota, Institut Teknologi Sepuluh Nopember.

Testing circular weaving (roundabout) to improve traffic performance and level of service of the unsignalized intersection

F Amir[1], S M Saleh[2], S Sugiarto[2,*]
[1]Civil Engineering postgraduate program, Department of Civil Engineering, Universitas Syiah Kuala, Banda Aceh, Indonesia
[2]Department of Civil Engineering, Universitas Syiah Kuala, Banda Aceh, Indonesia

*E-mail: sugiarto@unsyiah.ac.id

Abstract. Unsignalized intersection is the most operated traffic facility in Indonesia including several small-medium cities like Bireuen. The unsignalized intersection is mainly characterized by the complex maneuver mechanism by road users. The RSU Fauziah intersection is one of the most crowded intersections within the city center. This unsignalized intersection seems does not fit with current traffic demand. Indiscipline driver's movement and the rejection gap by the driver leads to deteriorating the capacity and level of service (LOS) of the intersection. Therefore, this work proposes to investigate the deterioration of capacity and level of service of targeted unsignalized intersection considering current traffic facility (unsignalized) and tested roundabout type traffic operation. The study was conducted at RSU Fauziah unsignalized intersection which located at the city center of Bireuen, Aceh Province, Indonesia. Using peak hour traffic data recorded using a video camera, the capacity was evaluated using the Indonesian Highway Capacity Manual (IHCM, 1997) for the existing traffic performance and simulating using VISSIM for trial implementation of roundabout. While the level of services (LOS) was examined using the Ministry of Transportation Decree (2006). The result of analysis demonstrates that the capacity of unsignalized slightly increases by 24.5% while the delay and level of service are significantly reduced to 39.2%, and improved from E to C, for delay and LOS, respectively. It can be concluded that the improvement of the existing condition by implemented a roundabout could improve the capacity and LOS of an existing unsignalized intersection.

1. Introduction

Capacity and level of service are the most important metrics to explain the traffic performance measures for traffic facilities including unsignalized intersection. The intersection is mainly operated in three ways namely by implementing circular weaving (roundabout), unsignalized and signalized intersection. The accuracy in predicting capacity is considered as a vital issue for traffic engineers. It because inaccurate capacity estimation could lead misleading in defining the quality of service of such facilities. For unsignalized intersection, for most cases in emerging cities within developing countries, road user's behaviors are substantially more complex in terms of lane discipline, lane changing, and the rejection of the gab from other drivers. Furthermore, the heterogeneous traffic composition has also affected the traffic performance significantly particularly within the city center and mostly repeatably during peak hour traffic.

Previous works have mentioned that the traffic performance of urban areas exceedingly affected by the mechanism of traffic flow and its disturbance due to drive's performance and caused traffic congestion within the corridor [1]. Studies have shown that traffic breakdown mechanism and its effect on the capacity and speed dropped within the urban arterial road [2-4]. Traffic performance within roundabout including effect driving maneuver and vehicle composition on capacity determination has investigated in Banda Aceh [5]. These findings decided that the U-turn and on-street parking behaviors are drastically influenced by the traffic mechanism on its segments.

In heterogeneous traffic, traffic rules such as lane discipline and accepted the gap are completely neglected by the driver in most cases. Vehicles could occupy any available lateral space on the road space. If the total conflicting volume is not considerably high the unsignalized intersection could perform very efficient operated facilities [6]. Uncontrolled intersections are commonly operated as first-in-first-out (FIFO) to deal with none of the traffic streams that have driving priority to enter the intersection. Furthermore, the capacity at two-way stop-controlled (TWSC) intersections is generally treated as gap acceptance theory. The Gap acceptance method is the extensive approach adopted to develop a highway capacity manual in most countries including Indonesia.

The Indonesian Capacity Highway Capacity Manual (IHCM) 1997 [7] is a guide for engineers and planners for planning, designing and operating traffic facilities in Indonesia, which is adopted from the US highway safety manual 1985 [8]. As regard to unsignalized intersection, the methodological for data analysis and its examples are described in chapter 2. It should be noted that the IHCM is only can be used for evaluating traffic performance for certain traffic facilities. The level of service (LOS) can be determined using the Ministry of Transportation Decree 2006 [9]. As mentioned above, since the drivers mostly always rejected the and role of priority movement at unsignalized intersection they ignored, and this situation is repeatable and occurs most of the time particularly during high traffic demand. Therefore, this work proposes to investigate the deterioration of capacity and level of service of targeted unsignalized intersection considering current traffic facility (unsignalized) and testing weaving section (roundabout) type traffic operation into existing unsignalized intersection. The study was conducted at RSU Fauziah unsignalized intersection which located at the city center of Bireuen, Aceh Province, Indonesia. Using peak hour traffic data recorded using a video camera, the capacity was evaluated using the IHCM and then LOS is determined using the Ministry of Transportation Decree 2006.

The remainder of this paper is organized as follows. The following section describes the area study, data collection, and processing. The next section explains the analysis method using IHCM and the setting of simulation using VISSIM. Lastly, discussions and conclusions are presented at the end of the paper.

2. Area of Study, Data Collection and Processing

The RSU Fauziah unsignalized intersection is targeted as a case study in this paper. The location of unsignalized intersection was located within the city center of Bireuen. This intersection has four-legged approaches and has considerably complex traffic movements with mixed traffic conditions. The existing facility has not operated priority role for traffic movements. The study location is displayed in Figure 1 and Figure 2.

This work was utilized primarily data set such as geometric features of the targeted intersection including approach width, number of lanes, lane width and conflict area dimension. The geometric condition data is collected directly by surveyor within off peak hour at the targeted intersection. The second primarily data is traffic data including traffic flow, traffic compositions, and side frictions. All the traffic data was collected on site by the surveyors. Moreover, it is observed that the side friction was considerably had medium to high side friction. This partially due to the location of intersection is in the dense urban road which pedestrian, entry-exit vehicle at parking lots, the slow-moving vehicle is quite significant even if non-motorized transport is less observed.

Traffic stream data such as trajectory movements, signs, and typical potential traffic conflict were observed directly at the location. Traffic data such as volume, vehicle composition, and type of vehicle movements were recorded using a video camera for each approach. The cameras were mounted on the high-rise building at the location of the survey. Traffic stream characteristic was recorded during peak hour traffic within the morning peak (07:15-08:50), afternoon (12:00-14:00) and evening (16:30-18:00).

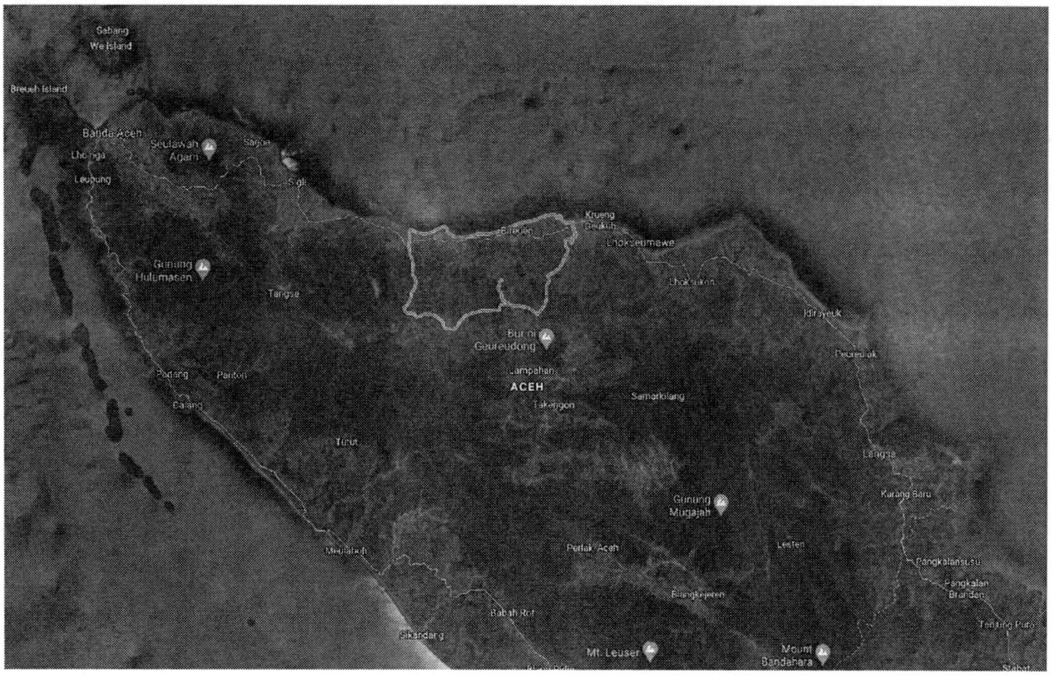

Figure 1. Bireuen city.

Figure 2. Map of the study area.

Recorded video data was extracted in the traffic system laboratory to obtain hourly traffic volume and vehicle composition of each movements. Firstly, recorded video data was extracted to get hourly traffic volume and composition for each approach of the intersection. Secondly, all processed data related to traffic stream characteristics within the intersection were tabulated in Ms. Excel to get an easy in data analysis. The analysis in this study is focused on capacity and level of services considering existing condition which is unsignalized intersection facility. The method used for the analysis will be described in the following section.

3. Analysis Method

In this paper, the analysis method used for determining traffic performance is based on the IHCM [7]. The level of service is then determined using the Ministry of Transportation Decree [9], as shown in Table 1. The level of service (LOS) for unsignalized intersection is based on the average delay (second/PCU) and classified them as LOS A to F as illustrated in Table 1. Furthermore, according to the Ministry of Public Work Decree [10] the upper limit of degree of saturation (DS) for arterial/collector, and local road is 0.85, and 0.9, respectively.

Table 1. The Level of Service (LOS) Criteria [10].

Criteria	Average Delay (Second/PCU)
A	< 5
B	5-10
C	11-20
D	21-30
E	31-45
F	> 45

According to IHCM [7] The capacity of unsignalized intersection consists of a multiplicative model among base capacity (C_0)with adjustment factors such as adjustments factor for approach width (F_W), type of median (F_M), city size (F_{CS}), the ratio of unmotorized (F_{RSU}), the ratio left turn vehicle (F_{LT}), the ration right turn vehicle (F_{RT}) and the ratio of flow from the minor road (F_{MI}), as it is written in Equation (1).

$$C=C_o \times F_W \times F_M \times F_{CS} \times F_{RSU} \times F_{LT} \times F_{RT} \times F_{MI} \tag{1}$$

$$DG = (1\text{-}DS) \times (PT \times 6 + (1\text{-}PT) \times 3) + DS \times 4 \text{ (second/PCU)} \tag{2}$$

$$DG = 4 \tag{3}$$

Where C represents the capacity (PCU/h), and C_0 is base capacity defined using default value in IHCM 1997 chapter 2 (unsignalized intersection). As for the delay, the IHCM has consisted of two delay namely delay due to traffic stream (DT) and delay due to geometric condition (DG). The DT must be determined using the delay curve versus the degree of saturation Figure C-2:1 and Figure C-2:2 (IHCM, Chapter 2, page 3:40-41). The DG can be calculated using equation (2) for DS < 1, and equation (3) for DS ≥ 1.

As for the assessment of the future demand scenario using the traffic microsimulation so called VISSIM. Microscopic models predict the state of individual vehicles and primarily focus on individual vehicle speeds and locations. Macroscopic models aggregate the description of traffic flow, and the measures of effectiveness, which are speed, flow, and density [11].

The VISSIM has been widely used to simulate traffic performance based on the improvement scenario, for example testing roundabout improvement [12-15]. The VISSIM is a microscopic, time step and a behavior-based simulation model developed to model urban traffic and public transit operations. The VISSIM stands for "Verkehr Stadten–SIMulationsmodell" which is developed by Planung Transport Verkehr AG (PTV-AG) in Karlsruhe Germany. The program can analyze traffic and transit operations under constraints such as lane configuration, traffic composition, traffic signals, transit stops and it is making uses tool for the evaluation of various alternatives based on the transportation engineering and planning measures of effectiveness.

4. Analysis of Results and Discussion

After conducting field observation, the aggregation of traffic flow data has shown maximum flow obtained during evening peak hours. Therefore, the analysis explained in this section is based on the evening peak hours data set observation. The composition of the vehicle is classified into three types of common vehicles namely motorcycle (MC), light vehicle (LV) and heavy vehicle (HV). To convert the mixed traffic condition into the homogenous one, the term of passenger equivalent unit (PCU) is used. The PCU is a unit to measure various vehicles into a unit of LV by multiplying them with passenger car equivalent (PCE). The PCEs used in this study are based on the value in IHCM [7].

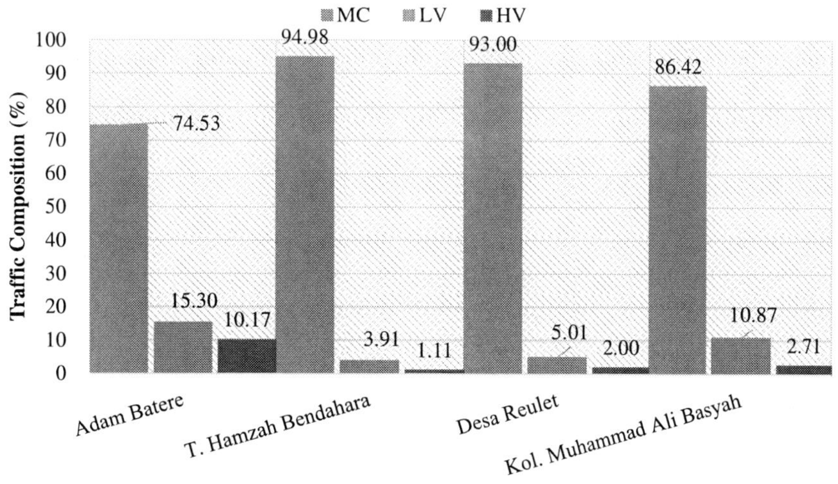

Figure 3. Vehicle composition distribution.

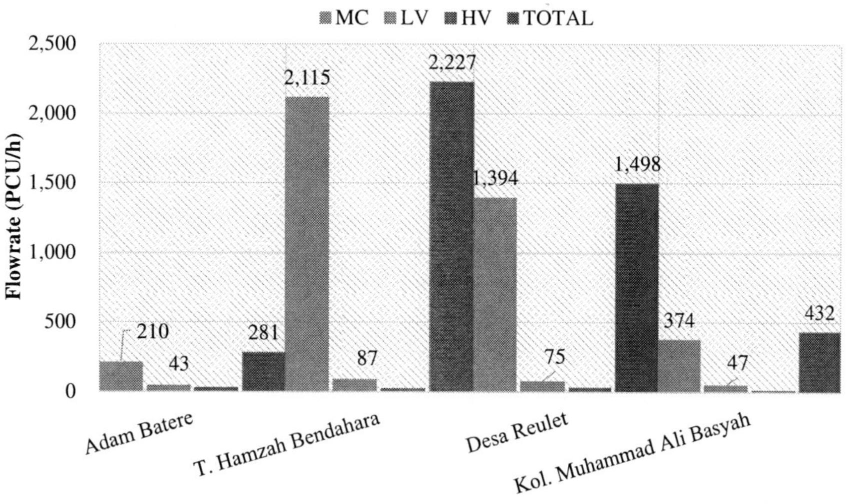

Figure 4. Traffic volume distribution.

Figure 3 illustrates the distribution of vehicle composition namely MC, LV, and HC. The composition of the MC has dominated the traffic about 75, 95, 93, and 86% for Adam Batere, T. Hamzah Bendahara, Desa Reulet and Kol. Muhammad Ali Basyah approaches. Following by LV approximately accounted for an average 10-15 % for major approaches of Adam Batere and Kol. Muhammad Ali Basyah. Furthermore, Figure 4 shows the traffic volume at the study area. It has shown about 2,227 PCU/h and 1,498 PCU/h for major road of T. Hamzah Bendahara and Desa Reulet, respectively.

Table 2. The Capacity of Unsignalized Intersection (Existing).

*Performance Measures Descriptions	Average Approach Width (m)	Base capacity C_0 (PCU/h)	Calculated Capacity	Hourly Volume (PCU/h)	Calculated Total Delay (sec)	Degree Of Saturation (DS)	Level of Services (LOS)
Value of Performance Measures	3.65	2,900	3,098	4,438	33.06	>1	E

**traffic performance calculated using [7] and LOS using [8]*

As mentioned in section 3, the capacity was calculated utilizing the IHCM formula [7] as it is shown in equation (1) and delay calculation using equation (2-3). The calculated delay is utilized to determine the level of service using the Ministry of Transportation Decree [8]. Table 2 summarizes the calculation of capacity, delay, and level of service in existing conditions. Considering an existing condition with an unsignalized intersection the capacity can be obtained as much as 3,098 PCU/h and the total delay has shown 33.06 sec. The DS value can be computed by dividing the observed hourly volume to capacity, the DS value is larger than 1. Using the Ministry of Transportation Decree [8], then the LOS criteria of E are determined. The LOS of E has meant that the traffic performance using unsignalized intersection has been considerably oversaturated and certain improvement must be proposed.

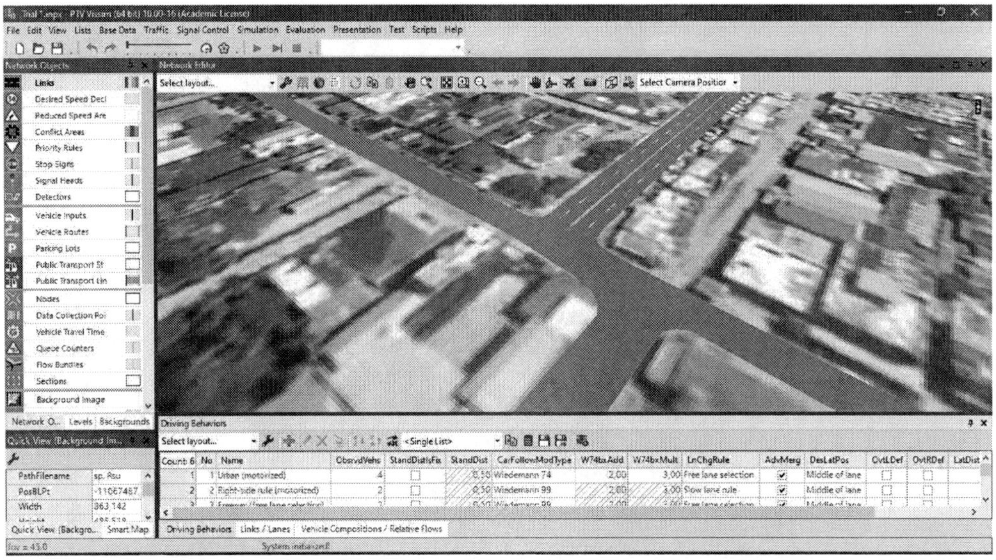

Figure 5. Simulation using VISSIM (unsignalized scenario).

According to the Ministry of Public Work Decree [9] the upper limit of degree of saturation (DS) for arterial/collector, and local road is 0.85, and 0.9, respectively. The obtained DS (>1) is exceeded the upper limit permitted DS. Therefore, improvement of traffic facility on the existing condition is needed. In this study, the mini roundabout is tested and examined using VISSIM simulator. Figure 5 and 6 depict the simulation set up and run using VISSIM. The VISSIM using in this study is version 10.00-16 academic licensed by Universitas Syiah Kuala.

Prior to the simulation, several procedures involved such as (1) proposed a new type of traffic operation, (2) coding network/calibrated geometric condition, (3) establish the measurement of effectiveness (MOEs), (4) generating several numbers of seeds for multiple runs, usually 5 to 10 times, (5) established a value of simulation time, the default value of 3600 seconds for VISSIM, (6) running simulations to obtain the MOEs. These steps practically can be done by several instructions namely (1) input background, (2) network coding, (3) building link-connector, (4) vehicle/traffic input, (5) inputting desire speed, (6) inputting vehicle compositions, (7) assigning vehicle route, (8) calibrating driving behaviors, (9) setting simulation runs and random seeds, (10) simulation runs, (11) establishing MOEs, (12) evaluating assessment of new proposed traffic operation.

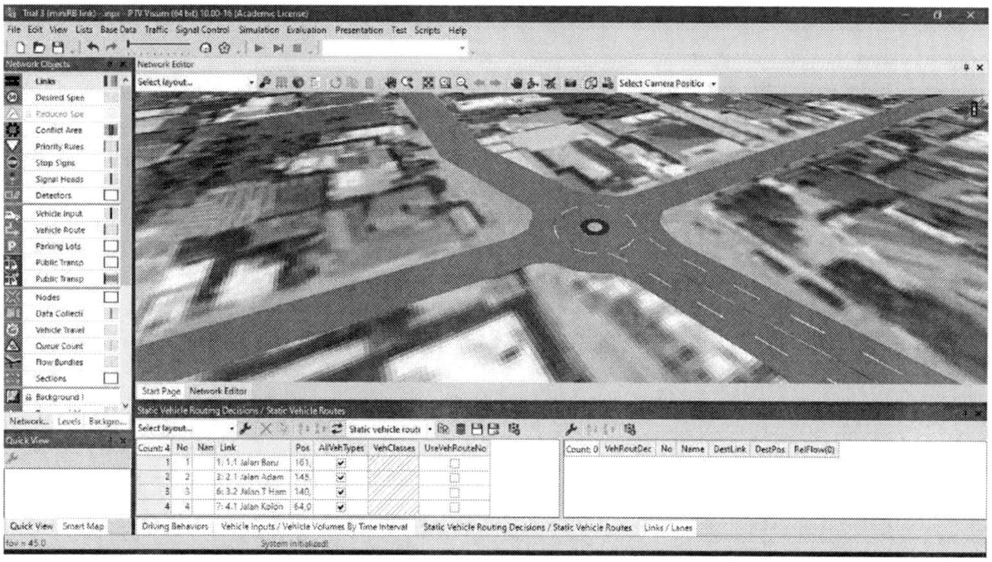

Figure 6. Simulation using VISSIM (improvement scenario using mini roundabout).

Table 3. Simulation result using VISSIM based on roundabout facility improvement.

Simulation Run/Seed Number	Simulation Interval (sec)	Queue Length (QlenMax)	Dynamic capacity (PCU/h)	Total Delay (sec)	Level of Services (LOS) Value	Level of Services (LOS)
1	900-4500	17.27	3,825	20.82	3	C
2	900-4500	20.44	3,910	20.29	3	C
3	900-4500	21.44	3,889	19.16	3	C
4	900-4500	21.72	3,789	19.92	3	C
5	900-4500	15.34	3,886	20.39	3	C
Average		19.68	3,859	20.08	3	C
Standard Deviation		2.40	1,561	0.61	0	C
Maximum		21.72	3,910	20.82	3	C

The weaving section of the roundabout then virtually implemented in the study area. Due to limited land for ideal roundabout and the highly cost of land acquisition, a roundabout with a diameter of 5 m was tested in this study. According to Sugiarto *et al.* [2-3] the simulation must be considered the randomness of stochastic nature. Therefore, the simulation was conducted 5 runs with 5 random seeds is randomly selected to run simulation. The average value of traffic performance and LOS is then selected as the measurement of effectiveness (MOE) as it is described in Table 3. The MOEs consists of a maximum of queue length, the capacity, and obtained LOS.

Comparing the performance measures among existing (unsignalized intersection) with the improvement (mini roundabout) has considerably improved the performance in terms of capacity and LOS. The capacity seems to increase from 3,098 PCU/h to 3,859 PCU/h or improve about 24.6% compared to the existing condition. While the delay has dropped for about 39.2% (33.06 sec to 20.08 sec). This enhances the value of delay leads to an increase in the LOS from E to C. The LOS of C is considerably excellent services for the urban street facility. Therefore, it can be concluded that the improvement of the existing condition by implemented a dedicated mini roundabout could improve the capacity and LOS in the study area.

5. Conclusion

This paper investigates the impact of dedicated mini-roundabout on the capacity and level of services at a four-legged unsignalized intersection. The Indonesian highway capacity manual (IHCM, 1997) is used to evaluate the capacity and level of service in existing conditions while the simulation approach using VISSIM is then performed to assess the implementation dedicated mini roundabout. The result of analysis demonstrates that the capacity of unsignalized slightly increases by 24.5% while the delay and level of service are significantly reduced to 39.2%, and improved from E to C, for delay and LOS, respectively. It can be concluded that the improvement of the existing condition by implemented a weaving section (roundabout) could improve the capacity and LOS of the existing unsignalized intersection.

References

[1] Saleh S M, Sugiarto S, Hilal A, Ariansyah D 2017 *AIP Conf. Proc,* **1903(1)** 06005.
[2] Sugiarto S, Limanond T, Nakatsuji, T 2012 *Aceh International Journal of Science and Technology,* **3** 86-92.
[3] Sugiarto S, Limanond T 2013 *Aceh International Journal of Science and Technology,* **2(1)** 63-69.
[4] Yusmadi Y, Sugiarto S, Renni A 2019 *Jurnal Arsip Rekayasa Sipil dan Perencanaan,* **2(3)** 180-189.
[5] Sugiarto S, Apriandy F, Faisal R, Saleh S M 2018 *Aceh International Journal of Science and Technology,* **7(2)** 77-84.
[6] Gowri A, Chepuru A 2016 *Journal of Modern Transport,* **24(4)** 304–313.
[7] MKJI 1997 *Indonesian Highway Capacity Manual 1997*. Jakarta: Directorate General Highways, Ministry of Public Works.
[8] HCM 1985 *Highway Capacity Manual.* Washington D.C.: Transportation Research Board, National Research Council.
[9] The Ministry of Transportation 2006 The Decree No. 14. 2006. Jakarta: The Ministry of Transportation.
[10] The Ministry of Transportation 2011 The Decree No. 19. 2011. Jakarta: The Ministry of Public Work.
[11] Boxil S A, Yu L 2000 *An Evaluation of Traffic Simulation Model for Supporting ITS Development. Center of Transportation Training and Research.* Texas Southern University, USA.
[12] Faisal R, Sugiarto S, Syara A 2017 *Jurnal Teknik Sipil,* **6(2)** 183-194.
[13] Rosadi D T, Sugiarto S, Anggraini R 2019 *Teras Jurnal,* **9(2)** 125-132.
[14] Sugiarto S, Ruhdi F, Muhammed R 2019 *Teras Jurnal,* **8(2)** 416-425.
[15] Defry B, Sugiarto S, Renni A 2017 *Jurnal Teknik Sipil,* **1(1)** 17-28.

An evaluation of the tropical architectural concept on the building design for achieving thermal comfort (Case study: engineering faculty of Syiah Kuala university)

M L Ghassan[1,*], L H Sari[2], A Munir[2]
[1]Student of Master's Degree Program of Department of Architecture and Planning, Faculty of Engineering, University of Syiah Kuala, Darussalam, Banda Aceh 2311, Indonesia
[2]Department of Architecture and Planning, Faculty of Engineering, University of Syiah Kuala, Darussalam, Banda Aceh 2311, Indonesia

*E-mail: luthfighassan10@gmail.com

Abstract. Thermal comfort is one of the main goals in architectural design. For the tropics, thermal comfort can be achieved with passive cooling and active cooling. Buildings in the tropics should be designed to maximize the application of passive cooling to achieve thermal comfort by adapting to the climate using the concept of tropical architecture. The study was performed at the building of the Faculty of Engineering of Syiah Kuala University seen to be applying tropical architecture. The research aims to find out how the application of tropical architectural concepts in building designs to achieve thermal comfort. The methodology used was a qualitative method by observing the mass and orientation of the building, roofs, ceilings, walls, openings and vents, shading, and landscaping. The results indicated that the rectangular building with the shortest side faces east-west aiming to reduce the hottest exposure to solar radiation in the east and west. There are many openings on the north-south side to maximize air flow. The gable shape with a slope of 30° is a form of adaptation to the tropical climate to shade the building from the sun and rain. The use of vertical, horizontal, and hallway shading devices as shade can protect direct light from the sun. The building maximizes cross air flow with many openings. It is recommended to conduct further research in this study which is accompanied by field measurements and thermal comfort simulation tests.

1. Introduction

Every building ideally is designed to be functional, safe, healthy, beautiful, having aesthetic value, environmentally friendly, and comfortable [1]. Comfort, especially thermal comfort, is one of the main goals in architectural design. Thermal comfort can be achieved with passive cooling and active cooling strategies [2]. Szokolay mentioned that thermal comfort depends on climate variables (sun/radiation, air temperature, humidity, and wind speed) and several individual factors such as clothes, age and gender, obesity level, type of food and drink consumed, level of health, and skin color [3]. However, active cooling which is by mechanical airing uses electricity which mostly uses fossil energy that can have a negative impact on the environment and sustainability. The design of the building should apply a passive cooling design by adjusting the building to the climate and environment which achieves thermal comfort in order to not have a negative impact on the environment. Building construction and operation accounts for around 50% of all energy resources consumed across the planet. It makes the construction industry the 'most unsustainable industry on earth' [4].

Climate has a significant impact on building performance and energy consumption. In building design it is important to identify, understand and regulate climate because it has the advantage of reducing the burden of energy costs, utilizes natural energy, as well as provides a healthy and comfortable environment [5]. Indonesia is a tropical country with humid characteristics. The humid tropical climate has temperatures close to the comfortable temperature range for humans in tropical climate regions. Hence, it provides opportunities to design buildings with a local tropical climate approach. In tropical climates the building should be designed to adapt to the environment and climate by using the concept of tropical architecture to achieve thermal comfort. Tropical architecture is a specific design of architectural work that leads to problem solving caused by tropical climate [1].

Each region has a characteristic building design that presents the characteristics of buildings that have been adapted to the local climate. The case study taken is the Engineering Faculty of Syiah Kuala University (FT Unsyiah), because it is considered to apply a tropical architectural approach based on visible characteristics to find out how the concept of Tropical Architecture is applied to the building. In general, the building characterizes the vernacular architecture of Aceh culture with a broad saddle roof and elongated mass forms. The building has three floors and operates from morning to afternoon. The Faculty of Engineering of Unsyiah is on St. Teuku Syiah Abdul Rauf, Kopelma Darussalam, Banda Aceh. The research aims to find out how the application of tropical architectural concepts in building designs to achieve thermal comfort. To find out how the concept of Tropical Architecture is implemented to the building design to achieve thermal comfort, it takes a review and research that can be used as knowledge and can also input for tropical buildings.

2. Tropical architectural concepts and thermal comfort

The tropical area is defined as areas that lie between the isotherm line of 20° to the north and south of the earth [2]. The humid tropical climate has the characteristics of having rainforest areas in the lowland, having soil which is usually red or brown with a green soil surface from vegetation, having a variety of types of vegetation throughout the year, having a slight seasonal difference from the hottest months which is from moist heat to wet and the coldest months which is medium heat as well as humid, wet, cloudy as well as foggy throughout the year, having annual rainfall that is 500-1250 mm, having the dry season which has little rain, having relative humidity which ranges from 20-85% depending on the season, having a strong and constant wind where in the jungle it gets slower and it gets faster when it rains. Usually there are one or two main wind directions [2].

The tropical architecture is a specific design of architectural work that leads to problem solving caused by tropical climate [1]. The tropical architecture is a concept of architecture that provides answers/adaptation of building forms to the influence of tropical climate, where tropical climate has certain characteristics caused by solar heat, humidity, rainfall, wind movement, and others. Moist tropical architecture is an architectural concept that provides answers to the adaptation of buildings to the influence of the humid tropical climate with the conditions that influence it. Tropical architectural design for the aim of adapting to climate and environment is done by maximizing the application of passive cooling in the building. Several conditions affect buildings in humid tropical climates. They are thermal comfort, air flow through buildings, natural glow and lighting, air temperature, rainfall, and air humidity [6]. There are several characteristics of humid tropical architectural design. They are the orientation of the main function room placed in the south and north, the longest side of the building facing north-south, the sloping roof design which has a broad roof space, having a wide roof terraces, maximizing openings on the south-north side and minimizing on the east-west side, utilizing the flow of the wind with the concept of cross ventilation, the shade of buildings, optimizing lighting and natural ventilation with lean buildings, choosing the type of material affecting the heat distribution of the room, and structuring outdoor space [1].

Comfort in a building can be defined as a condition that can provide a comfortable and pleasant feeling for its inhabitants [1]. Thermal comfort is a condition related to nature that can affect humans and can be controlled by architecture [7]. According to Idham the main principle of design consideration with passive cooling to tropical climates to achieve thermal comfort is to eliminate or reduce external heat sources during the day by building envelope designs. There are several main elements in the design of passive cooling. They are the orientation of buildings to the wind and sun, zoning plans by maximizing day and night activities, proper window and glass openings, the use of effective shading, the use of appropriate insulation, the use of high thermal mass construction in areas with significant diurnal temperature ranges, and the use of low thermal mass construction in areas with low diurnal temperature ranges [5].

3. Research method

This study used a descriptive qualitative method carried out by observing the building of the Engineering Faculty of Syiah Kuala University, by recording data and physical conditions in the field. Observations focused on observing the mass and orientation buildings, roofs, ceilings, walls, openings and ventilation, shading, and landscaping (Table 1). This parameter is used in Lisa's research [8]. In addition, the writer also observed the conditions that could interfere with comfort. Observations were made by analysing, documenting, and measuring elements of research objects. The Faculty of Engineering consists of the FT-1 and FT-2 buildings. The object of the research is carried out only at the FT-1 Building, because the FT-1 and FT-2 buildings have almost the same shape.

Table 1. Research scope.

	Element	Indicator
1	Building mass and orientation	Sun path Form Dimension Position
2	Roof	Roof shape Roof slope Dimension Material The volume of roof space
3	Ceiling	Ceiling elevation Ceiling material
4	Wall	Material type and dimensions Material properties
5	Openings and Ventilation	Orientation and location Opening's type Material Dimensions and quantities Air movement
6	Shading	Sun path Type Material
7	Landscape	Potition Ground cover material

The method for data collection used was descriptive analysis method. Data collection begins with direct observation of the field by data recording, measurement, and analyzing the observed object elements. Then the data obtained were juxtaposed and compared with theories and literature to find out the relation and description of the tropical architecture concept's application in buildings. After knowing the relation, then describe each corresponding element.

4. Results and discussion

Generally, the building of the Engineering Faculty of Syiah Kuala University applies the concept of tropical architecture which aims to adapt the building design to the local climate and environment, namely a tropical climate with a humid character. The shape and mass of the building in this building characterizes the application of the concept of vernacular architecture with the characteristic of Aceh's culture (Figure 1). The vernacular architecture itself is an architecture that is formed from a process that is long and repetitive following the behavior, habits, and culture in its place of origin which also aims to anticipate and respond to local climate problems [9]. The observations can be seen in Table 2.

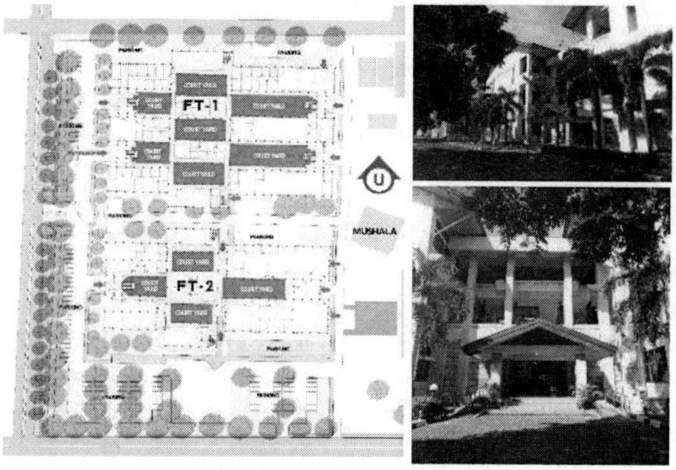

Figure 1. Engineering Faculty of Unsyiah and photos of the buildings.

4.1. Mass and orientation of the building

The building of the Faculty of Engineering of Syiah Kuala University consists of several rectangular building masses with different floors story. In the three west building masses, it has three building stories, and the rest of the building has two stories. The total mass of the building is connected by corridors and stair circulation paths (Figure 2). The three-story mass of the building provides shadowing of the lower building mass and courtyard when exposed to sunlight from the west. The masses with a rectangular shape are positioned with the longest side of the building facing north-south, and the shortest side facing west-east. The shape and orientation are appropriate and optimal in the humid tropical area [10]. The west-east side of the tropics gets more heat than the north-south side, so on the west side it is optimized to reduce the width of the building surfaces.

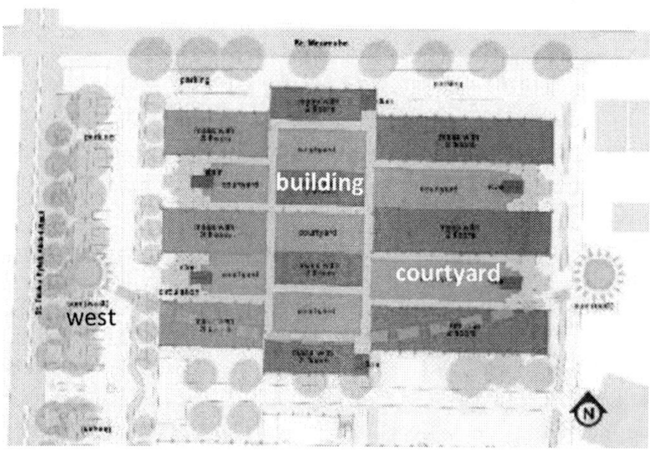

Figure 2. The position and orientation of the building mass.

The building of the Engineering Faculty of Syiah Kuala University has 13 meters width for 3 masses in the west and 3 masses in the east, and 9.8 meters width in 4 masses in the middle. This allows air and light to pass through the room more optimally. The sunlight and wind can reach the farthest distance of about 6 meters from the window to a room height of about 3 meters accompanied by open the openings. Therefore, the optimal width of the shortest side of the building is no more than 12 meters [1]. The masses of the buildings which are intermittently connected through the hall provide space for an open courtyard in the middle of the building. The intermittent building mass also allows air to flow heavily and can be captured by each building mass [11].

4.2. Roof

According to Lipssmeier, the roof is the most important element in tropical design [3]. The roof determines the temperature of the air space under the roof, because the solar thermal radiation exposed to the roof is partially reflected, partly absorbed by the roof material and channeled into the space under the roof [1]. The building of the Engineering Faculty of Syiah Kuala University uses a saddle roof with ceramic tile material with a slope of approximately 30 degrees. The slope of the roof serves to protect the building from direct sunlight and rain that often occurs in tropical areas. Compared to flat roofs and shielding roofs commonly used in tropical buildings, the use of gable roofs results in a greater volume of space under the roof (between the roof and ceiling layers) supported by the slope. However, on the gable, the roof cannot protect the entire wall of the building. Roofs with ceramic tile material allow heat reflection and slow heat absorption. Under the ceramic tile roof there is a layer of aluminum foil that can prevent heat from entering the roof space, to reduce heat.

4.3. The ceiling

The storey height of the building of the Engineering Faculty of Syiah Kuala University varies between floors. The storey height of the 1st floor to the 2nd floor is 4.8 meters, the 2nd floor to 3rd floor is 5 meters, and the second-floor height to the 3rd floor is 4.55 meters. This condition is used to use a high ceiling from the floor of each floor. The average ceiling height is almost 3.7 meters. The high ceiling allows air to flow more freely so that it can reduce humidity and heat (Figure 3a). Indoor ceilings in tropical climates are very important so that heat radiation from the roof layer above it can be retained in the roof space and not directly enter the room below it [1].

4.4. Wall

The main construction building's materials using concrete and brick. The wall material is brick that has a thickness of 15 cm. Concrete and brick material has the ability to keep heat and has a heat release delay that is longer. With the delay in the release of heat in the material, it removes the peak heat flow conditions at night so the room during the day stays cool when equipped with good air movement and shade. It is very suitable to be applied in the study room which is only used from morning to evening in the tropical climate. Cooling and heating by utilizing the appropriate time shift can reduce the cost of electrical usage [2]. Lippsmeier explained that concrete with a thickness of 15 cm has a time lag of 3.8 hours, a thickness of 20 cm has a time lag of 5.1 hours, while bricks with a thickness of 10 cm have a time lag of 2.3 hours, and thickness 20 cm has a time lag of 5.5 hours [2]. The walls of the building are painted in white and the columns beige. White is the color with the best reflective properties [12]. White walls allow a large reflection of sunlight and reduce absorption [12].

4.5. Openings and ventilation

In the laying of openings, it is necessary to pay attention to the direction of orientation so that direct sunlight does not enter the room, which is utilized indirect light from the sun (Figure 3). In the tropics, the west and east sides are the most receptive to solar radiation, while the north-south is the least. In the tropics, the light utilized is indirect light from the sun. In the building of the Faculty of Engineering of Syiah Kuala University, the north-south side of the building maximizes many openings (Figure 5a). Conversely on the west-east side minimize openings (Figure 5b). The east-west side with few openings aims to block direct sunlight from entering the room. Side opening of the north-south side allows the flow of air in and out crossing the room without direct sunlight from the west and east.

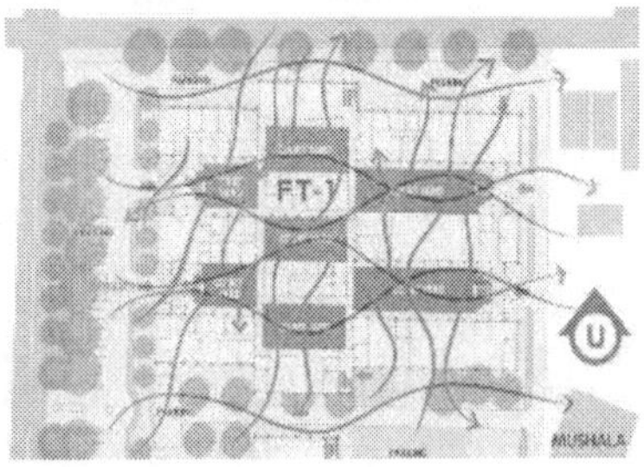

Figure 3. Layout, circulation of air flow movement.

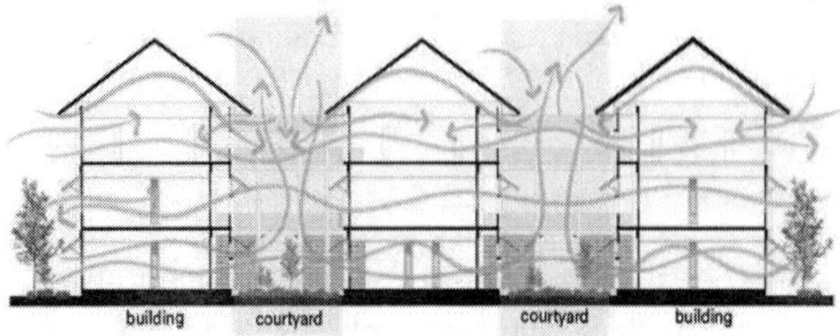

Figure 4. Building section, cross air movement, courtyard help air distribution.

The concept of cross ventilation is also supported by the open space in the middle of the building as one of the many spaces flowed by air (Figure 4). Some rooms only have one side opening for air in and out with no air exchange crossing. In addition, the ventilation has been closed for AC use (Figure 5c). The flow of air can no longer flow into the room. The position of the window in the corridor has a distance of 1.85 meters from the floor. This is less beneficial because the air flows above the human body when sitting indoors (Figure 5d). The body is only slightly passed by the wind. The type of opening used is the window with the type of awning hung-up, and the type of ventilation above the window and door used are the horizontal fin (Figure 5d). Window with awnings hung-up type has effectiveness 75%, whereas according to Bell and Paskins, horizontal fin ventilation has an effectiveness of 85-90% [13] [14]. The windows use dark tinted glass so that it reduces light entering the room and it is able to absorb some of the heat.

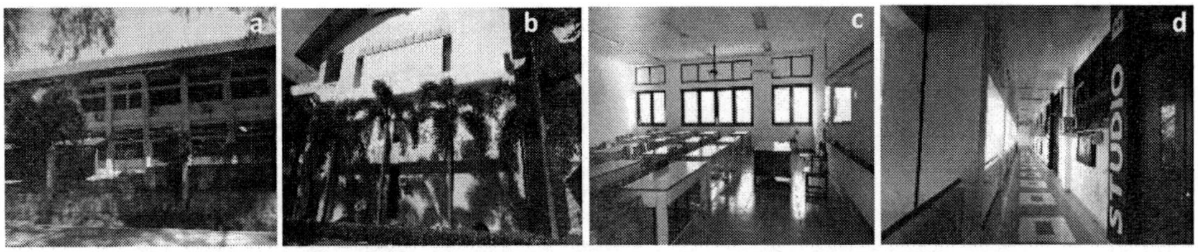

Figure 5. (a) west view, (b) north view, (c) classroom, (d) corridor

On the stairs and lobby there is a hole made of a concrete roster that allows air to get in and out and also can dispel the light and rain splash (Figure 6a). Holes of the concrete roster are also found in railings in corridors and walls above corridor railings (Figure 6b).

4.6. Shading

In tropical area, sunlight occurs from morning to evening throughout the year. Therefore, shading is needed to provide shade to buildings for reduce the heat. The main roof of the building has a roof over-hang with a width of up to 2.20 meters which can provide a lot of shading to the building walls, especially the wall on the story where the roof is located (Figure 5a). Sun shading with type of over-hang in the form of a ceramic tile-coated roof located above the window and above the corridor also provides shade and dispels direct sunlight (Figure 5a). Sun shading with over-hang type is effective on

the building with openings facing to the north and south, the effectiveness depends on changes in the movement of the angle of arrival of sunlight [15]. Prominent beams also give shading to the wall as vertical shading. Vertical sun shading is effectively placed on the west-east side to block sunlight from the south-north side [16]. Several types of shading are applied according to the tropical climate.
Corridors or hallways provide shelter to the sun's rays so that the light entering the room is indirect light (Figures 6a, 6c). The previously mentioned roster grille allows light to enter with shade (Figure 6b). Poles that are advanced out of the building can also function as shading vertical blocking sunlight from east-west towards the north-south side (Figure 5a).

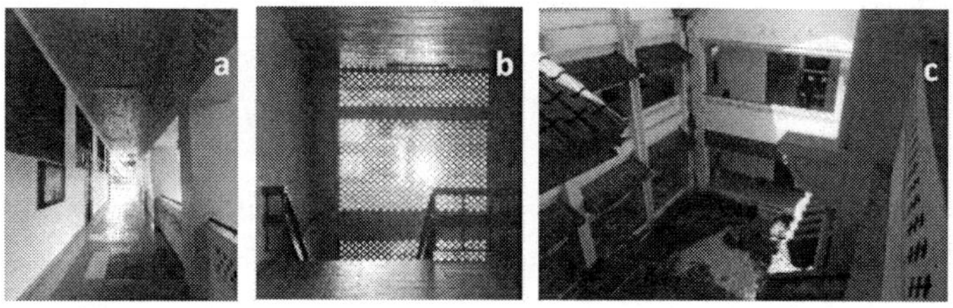

Figure 6. (a) Corridor, (b) Stair, (c) Corridor and Courtyard

4.7. Landscape

The efficient use of the courtyard is applied as a bioclimatic thermal controller in the built environment with natural air comfort control [17]. The courtyard will let the air move throughout the house plans that surround the courtyard especially those directly facing the open courtyard. High indoor air temperatures will be solved through shade from plants planted in the courtyard and shade from subsequent rooms [18]. The Courtyard is able to suck air from outside and also circulate air [5]. Based on Kubota's research, the small yard that is usually built in Chinese shop houses is effective for cooling the high thermal structures of building masses through night ventilation and radiation coolers [19]. Trees in the central courtyard of the building can also provide fresh air and help lowering the temperatures (Figure 7). Plants on the west and east sides can provide shade to buildings (Figure 8). There are many trees between buildings that provide shade and can help lowering the temperatures.

Figure 7. Courtyard in the middle of the building.

Figure 8. Vegetation on the west side of the building.

4.8. Conditions that interfere with thermal comfort

Some rooms in the building of the Engineering Faculty of Syiah Kuala University use artificial air conditioning system named air conditioner (AC). The use of air conditioners in rooms that were previously designed to use natural ventilation is less beneficial. High ceilings in the room can aggravate the AC workload because it has a large volume of space and is not optimal for cooling the room. For air conditioning usage, most of the ventilation is closed so that air cannot flow into and out of the room (Figure 9). Some placements of AC machines are in the corridor and close to the circulation path of people, and some are at the top of the corridor. Placement of the AC compressor engine generates heat and causes an increase in temperature around its location (Figure 10). This is not beneficial in thermal comfort.

Table 2. Observation Results.

	Element	Indicator	Theory	Function	Observation Results
1	Building mass and	Sun path	The shortest side of the building faces west-east, the longest side of the building faces north-south	reducing the sun's hottest exposure on the east-west side	The shortest side of the building faces west-east, the longest side of the building faces north-south
	Building orientation	Form Dimension	Rectangle (optimum in a ratio of 1: 1.7)	reducing the sun's hottest exposure on the east-west side	Rectangle
		Position	Not parallel to each other	air flow moves a lot in each mass	Not parallel to each other
2	Roof	Roof shape	Not a flat roof (over 20o), wide or narrow terraces, but consider the wind	Protect from rain, sunlight. Roof space reduces heat	Gable roof
		Roof slope	The roof slope is more than 30°	Drains rain well, reducing wind suction	The roof slope is more than 30°
		Dimension	Wide or narrow terraces, but consider the wind	Give shading to the building	The width of the eaves is 2.20 m, the roof eaves are damaged and replaced with PVC, the roof gives a good shading effect on the walls
		Material	A strong material, not lightweight, waterproof coating, reduce heat	Consideration of strong winds, can reduce heat not flowing directly into the cellar, protecting against rain	Using a ceramic tile roof layer, with radiation and heat absorption 62-66%
		Volume of roof space	spacious roof space (space between the roof and ceiling)	Give room so that air can freely flow	spacious rooftop space
3	Ceiling	Ceiling elevation	High ceilings	Gives room so that air can freely cross (cross ventilation)	The average ceiling height of 3.8 meters
		Ceiling material	material with the ability to reduce heat	reduce heat	Acoustic gypsum tile ceiling material
4	walls	Material type and dimensions	Bricks and concrete (massive wall), waterproof, moisture resistant, noisy	using material and color properties	Bricks and concrete
		Material properties	Time Leg, slow to absorb heat	Release the heat at the desired time	release heat at night
5	Openings and Ventilation	Orientation and location	Many openings in north-south, few openings in east-west	Blocking east-west sunlight	Many openings in north-south, few openings in east-west
		Type	Maximizing air flow	Maximizing air flow	Using a hung-up window, horizontal fin ventilation, a partial ventilation was closed for the use of Air Conditioner.
		Material	A strong material, waterproof, moisture resistant	Durable, avoid mold, water and termites	Wood with dark glass
		Dimensions and quantities	Maximize opening area	Maximizing natural air flow	Has a wide and many openings
		Air movement	Using the concept of cross ventilation	Maximizing natural air flow	Most use the concept of cross ventilation,
6	Shading	Sun path	shadowing the wall on the east-west side	Block out direct sunlight	A little shading on the east-west side, a lot of shading on the north-west side
		Type	Vertical shading on the west-east side, horizontal shading on the north-south	Block out direct sunlight	Horizontal shading (overhangs) of ceramic tile roofs on the south-north side
		Material	waterproof	Not leak, not mold, not damp, and durable	Gable roof
7	Landscape	Position	Shade vegetation especially in the east-west, not too close to the building, giving an area for air movement towards ventilation	Shadowing and producing fresh air	Vegetation on the west side gives a shadow in the afternoon, there is vegetation around the building

Most of the ceiling on the 3rd floor has been damaged causing a hole between the roof space and the space under the ceiling (Figure 11,12). This leads the loss of divider between the roof spaces and the space below the roof so that it can increase the heat of the room because of the heat of solar radiation from the roof is channeled directly to the room below.

Figure 9. Ventilation is closed with white plywood.

Figure 10. The location of the Air Conditioner machine is not quite right in the corridor.

Figure 11. Broken and perforated ceiling.

Figure 12. the broken ceiling on the 3rd floor

5. Conclusions

Overall, the building of the Engineering Faculty of Syiah Kuala University has applied the building design rules to adapt to the humid tropical climate based on a tropical architectural approach to achieving thermal comfort. The results indicated that the rectangular building with the shortest side faces east-west aiming to reduce the hottest exposure to solar radiation in the east and west. There are many openings on the north-south side to maximize air flow. The shape of the gable with a slope of 30° is a form of adaptation to the tropical climate to shade the building from the sun and rain. The brick wall is suitable for lecture buildings in the tropics moving from morning to evening because it releases heat at night. The use of vertical, horizontal shading and hallway as shade can protect direct light from the sun. The building maximizes cross air flow with many openings, high ceilings, and the presence of a courtyard. Parks with trees around the building provides a shading device to the building and can reduce heat. However, the use of air conditioners, closed ventilation, and broken ceilings under the main roof can interfere the thermal comfort.

This study only analyzes the application of tropical architecture concept in buildings to achieving thermal comfort. Consequently, a further research is needed which is accompanied by field measurements and thermal comfort simulation tests to determine the effect of applying tropical architecture to the level of thermal comfort in buildings.

References

[1] Karyono T H 2016 *Arsitektur Tropis Bentuk, Teknologi, Kenyamanan, & Penggunaan Energi* Penerbit Erlangga.

[2] Lippssmeier G 1980 *Bangunan Tropis*. Penerbit Erlangga.

[3] Szokolay S V 1973 *Manual of Troopical Housing and Building*. Bombay: Orient Langman.

[4] Abbey I, Heartfield J 2001 *Sustaining Architecture in the Anti-machine Age*. Britain: Willey Academy.

[5] Idham N C 2016 *Arsitektur dan Kenyamanan Termal*. Yogyakarta: Penerbit Andi.

[6] Prianto E 2018 *Jurnal PPKM*, **I** 24-41.

[7] Snyder, James C, Catanese, Aanthony J 1989 *Introduction to Architecture*. Jakarta: Erlangga.

[8] Lisav N P. 2017. *Seminar Ikatan Peneliti Lingkungan Binaan Indonesia (IPLBI)*, **1** 111-118.

[9] Wikipedia. Arsitektur vernakular. Available at: www.wikipedia.org/wiki/Arsitektur_vernakular.

[10] Olygay V 1963 *Design with Climate; Bioclimatic Approach to Arvhitectureal Regionalisme.* New Jersey: Princeton University Press.

[11] Errel E, Pearlmutter D 2011 *Urban Microclimatie-Designing the Space Between buildings.* London: Earthscan.

[12] Bansal N K 1992 *Building Environment,* **27(1)** 31-37.

[13] Bell S, Paskins J 2013 *Future-proofing London: Mitigation strategies for a changing global climate in Imagining the Future City: London 2062.* London: Ubiquity Press.

[14] Moore F 1993 *Environmental Control Systems, Heating, Cooling Lighting, Architecture and Urban Planning.* New York: McGraw Hill.

[15] Pelsmekers S 2012 *The Environmental Design Pocketbook.* London: RIBA Publishing.

[16] Olgyay A, Olgyay V 1957 *Solar Control and Shading* Devices. Princeton: Princeton University Press.

[17] Sharif, Salem M, Zain M, Surat M 2010 *Australian Journal of Basic and Applied Science,* **4(8)** 4029-4037.

[18] Sari L H 2017 *The 7th AIC-ICMR on Science and Engineering.*

[19] Kubota T, Toe D H C, Ossen D R 2014 *Malacca Journal of Asian Architecture and Building Engineering,* **13** 247-254.

The evaluation of green open space of disaster prone coastal areas

A Abubakar*, M Fuady, A Achmad
Architecture and Planning Department, Engineering Faculty, Universitas Syiah Kuala, Banda Aceh 23111, Indonesia

*E-mail: andrian73@mhs.unsyiah.ac.id

Abstract. The utilization of Green Open Space (GOS) is one of the aspects of urban environment that has an important value to the human life in urban areas considering its function in providing benefits ecologically, socially, culturally, aesthetically, and economically. An aappropriate utilization aims to improve the quality and quantity of green space, especially in the coastal areas, the use of green space is interesting to study because most of the land planned as green space is considered as a potential land and tends to be often misused on its use. This study aims to identify the existing land use andevaluate the use of GOS to increase the potential development of GOS utilization in coastal areas in Banda Aceh as stipulated in the Regional Spatial Plan (RSP) of Banda Aceh. The evaluation starts from analyzing the existing land use of GOS that has been determined in the SRP of Banda Aceh City through the processing of satellite images in 2016 by using the geospatial information systems application as an analytical aid. The image illustrates the existing land use conditions. The outcome obtained indicates that there is a change in the function of GOS that has been designated as a residential area and trade and services as well as the utilization of GOS that is not optimal yet. The results are expected to be a reference in optimizing the use of green open space in disaster-prone coastal areas in the city of Banda Aceh.

1. Introduction

Post-tsunami spatial use in Banda Aceh has been running rapidly due to high rehabilitation/reconstruction activities, emphasizing the reconstruction of damaged buildings and infrastructure, capacity building and community empowerment, economic, social, and cultural activities [1]. At the beginning of rehabilitation/reconstruction, most of the northern or coastal areas did not allow for the physical development of the City, so development activities tended to develop towards the East and South, which were relatively safe from the tsunami and had the potential to become new development areas. In line with the trend of urban development, the Banda Aceh City Government has stipulated the Regional Spatial Plan (RSP) Regulation Number 4 of 2009 on the Banda Aceh City Regional Spatial Plan 2009-2029 as the basis for implementing development in Banda Aceh City. In the RSP, most of the northern part of the City or coastal area is designated as GOS.

After enacting the RSP regulation, rehabilitation/reconstruction activities continue, housing construction, office buildings, trade, infrastructure, and other urban services began to increase [1]. Increased development activities also occurred in the northern part of the City, a coastal area devastated by the tsunami waves. The utilization of land for these activities changes the natural form of urban land and uses green open space, which has raised concerns that it can make cities less livable [2]. This condition has resulted in changes in spatial use as stipulated in the spatial plan so that the Banda Aceh

Government has revisited the RSP by considering various aspects of urban physical development while continuing to prioritize the sustainability of elements of a sustainable urban environment. Determination of changes in spatial patterns in Banda Aceh is based on considerations to maintain the sustainability of urban protected areas by applying the concept of development based on the arrangement of green cities, sustainable environmental sustainability, optimization and efficiency of spatial use, and disaster mitigation.

Determination of GOS in the RSP is mostly located in the coastal area of the City, which aims as a buffer zone through the development of mangrove and other coastal plants by prioritizing ecological, economic, social, and cultural aspects according to community needs and can be implemented sustainably by not ignoring architectural elements [3]. This is in line with Fuady's opinion that green space has a significant role in tsunami disaster mitigation in the City's coastal areas as a rescue and protector [4].

The coastal area is an attractive area due to natural characteristics, microclimate, and economic factors, which includes the place where the influence of water and land meets, to the land it covers the administrative area of the sub-district while towards the sea as far as 12 (twelve) nautical miles is measured from the coastline towards open ocean [5] so that it can be utilized as a variety of recreational activities [6]. Spatial use in coastal areas as green open space is a step to use as a buffer zone for coastal regions from the tsunami disaster. At the same time, for the tide do the construction of coastal protection in the form of the revetment.

Related to development in coastal areas, coastal cities have city characteristics as a concentration of development activities because of their strategic position; this will affect urban green space [7]. The land ownership that has been determined as a GOS in the RSP is mostly community-owned land. This impacts the limited use of land so that the land seems neglected, and there is no effort to make it have more value. The limited budget of the City Government to conduct the acquisition of community land designated as green space can impact the occurrence of land conversion.

As land prices continue to rise, rental prices, distance from work locations, and family connections have caused landowners and developers to glance at coastal areas dominated by GOS. The history of the tsunami event is no longer a consideration of the community in choosing the location of residence. This can be a trigger for land conversion from the GOS to the built area.

Based on the above background, it is deemed necessary to evaluate the use of green open space in disaster-prone coastal areas in Banda Aceh City to find out the progress of space utilization activities and potential development of green open space utilization to achieve a balance between economic growth and social welfare with environmental protection and disaster mitigation [8]. This study contributes to information on the utilization and development of the potential of GOS, as a guideline for determining activities in the GOS area and form the basis for policy recommendations for the City Government in determining spatial planning policies, especially for the arrangement and utilization of GOS land use in disaster-prone coastal areas.

2. Materials and methods

2.1. Study area

This research was conducted in Banda Aceh the capital of Aceh Province, a city on the northern tip of Sumatra Island, consisting of 9 districts, 90 Villages with an area of 6,136 ha. The geographical position is flanked by the Malacca Strait on the north side while on the west, east and south sides are limited by the territory of Aceh Besar Regency. Astronomically, Banda Aceh is located between 05°16 '15"- 05°36 '16" Northern Latitude and 95°16 '15" - 95°22 '35" East Longitude [3]. The altitude of the urban areas ranges from -0.45 m to +1.00 m above sea level, with an average height of 0.80 m above sea level.

Based on the provisions of the scope of regulation of coastal areas, where in land direction covers the administrative area of the sub-district [5]. The coastal area of Banda Aceh City is in 4 sub-districts, Meuraxa, Kutaraja, Kuta Alam and Syiah Kuala Sub-District, with an area of 3,592.53 Ha (Figure 1).

The research location is focused on the area that has been designated as the designation of green space in the RSP of Banda Aceh City.

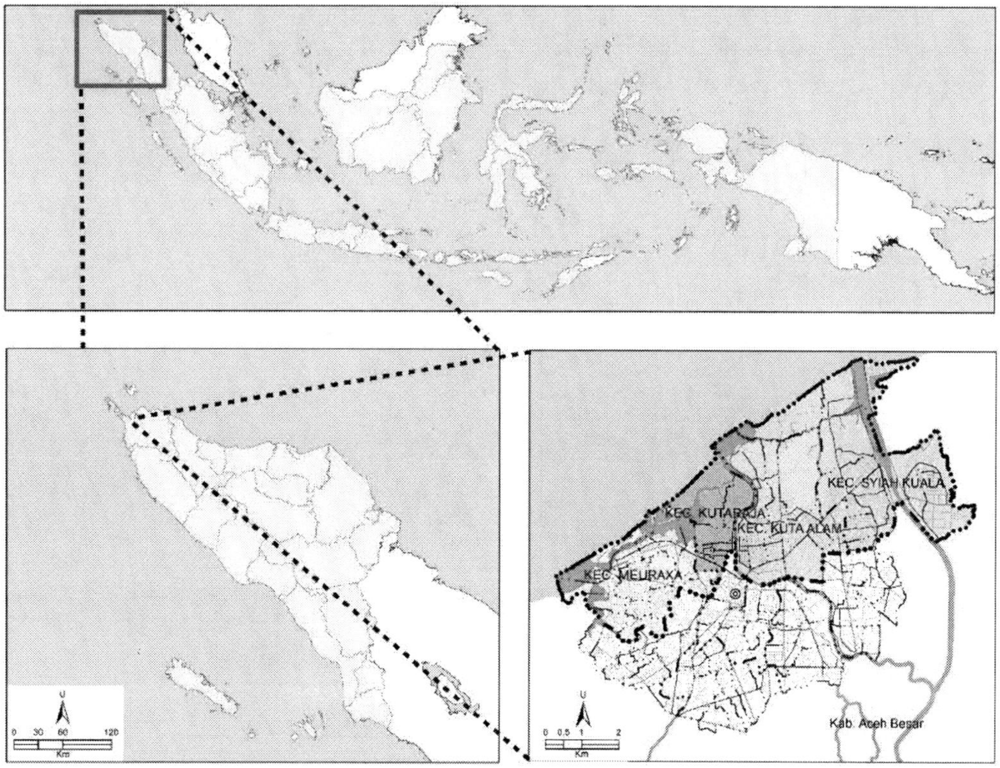

Figure 1. Location of study area.

2.2. Methods

In this study the research approach used is a combination of quantitative and qualitative with descriptive approaches that describe the relationship between the phenomena under study. The quantitative approach is used in describing existing land-use while the qualitative approach is used in describing the potential development of green space utilization.

2.2.1. Identification of green open space plans in coastal areas

urban GOS is an open space or area within a city planted with trees, shrubs, bushes, and grass [8]. Land planning as GOS aims to present beauty, comfort and safety element for the owner or user which has aesthetic value and functions as an open space with diverse functions [9].

The coastal area of Banda Aceh City consists of low-lying areas and part of it is an area of mangrove forests, water body and ponds. In the RSP, the establishment of GOS in coastal areas is planned to be 992.58 Ha. (Table 1) or 84.07% of the total green space planning in Banda Aceh City (Figure 2).

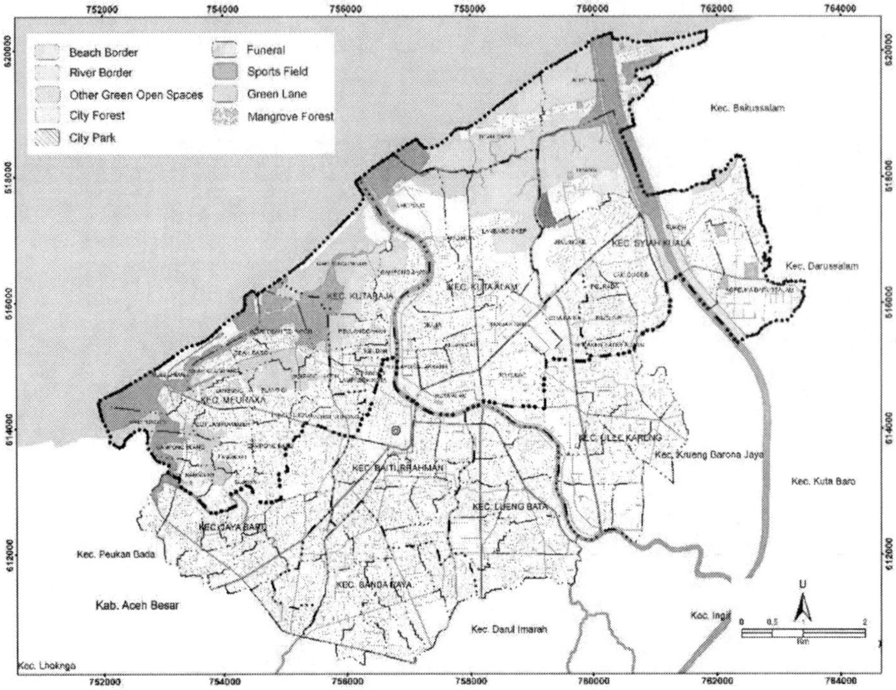

Figure 2. Map of GOS planning in the coastal area of Banda Aceh City.

Table 1. GOS planning in the coastal area of Banda Aceh City.

GOS planning	Area					
	Kuta Alam	Kutaraja	Meuraxa	Syiah Kuala	Ha	%
Beach border	0.00	4.52	2.15	33.61	40.28	4.06
River border	8.67	2.48	2.08	50.87	64.10	6.46
City Forest	0.00	4.57	0.00	25.07	29.65	2.99
City Park	2.24	0.00	6.52	5.63	14.38	1.45
Mangrove forest	32.04	87.49	162.95	29.57	312.06	31.44
Cemetery	2.01	1.63	3.56	0.88	8.08	0.81
Sports field	2.23	0.16	1.02	17.29	20.70	2.09
Green lane	13.44	2.48	9.80	14.36	40.07	4.04
Other open spaces	120.52	58.58	9.75	274.40	463.25	46.67
Total	181.16	161.93	197.82	451.68	992.58	100.00

2.2.2. Analysis of existing land use

To find out the existing land use, a land use analysis technique is used through satellite image processing in 2016 using the ArcGIS application, the results of field observations are the primary data source and are supplemented by some secondary data as a complement. So that the description of the existing land use of green space has been determined in RSP of Banda Aceh City. The land use classification refers to the Indonesian National Standard (SNI) 7645-1: 2014. The analysis was carried out descriptively by describing the distribution of land uses.

3. Results and Discussion

3.1. Area and GOS plan in the coastal area of Banda Aceh City

Banda Aceh is a very vulnerable city to several types of coastal disasters, such as tsunamis and tidal waves. It is also vulnerable to earthquakes and floods. This is due to Banda Aceh being traversed by two faults which are a branch of the Sumatra Fault, the Aceh Fault and the Seulimum Fault. Both of these faults have the potential to produce an earthquake with the potential for a Tsunami. Determination of GOS plan in tsunami-prone coastal areas is the right step towards disaster mitigation as a rescue and protective area (Figure 3).

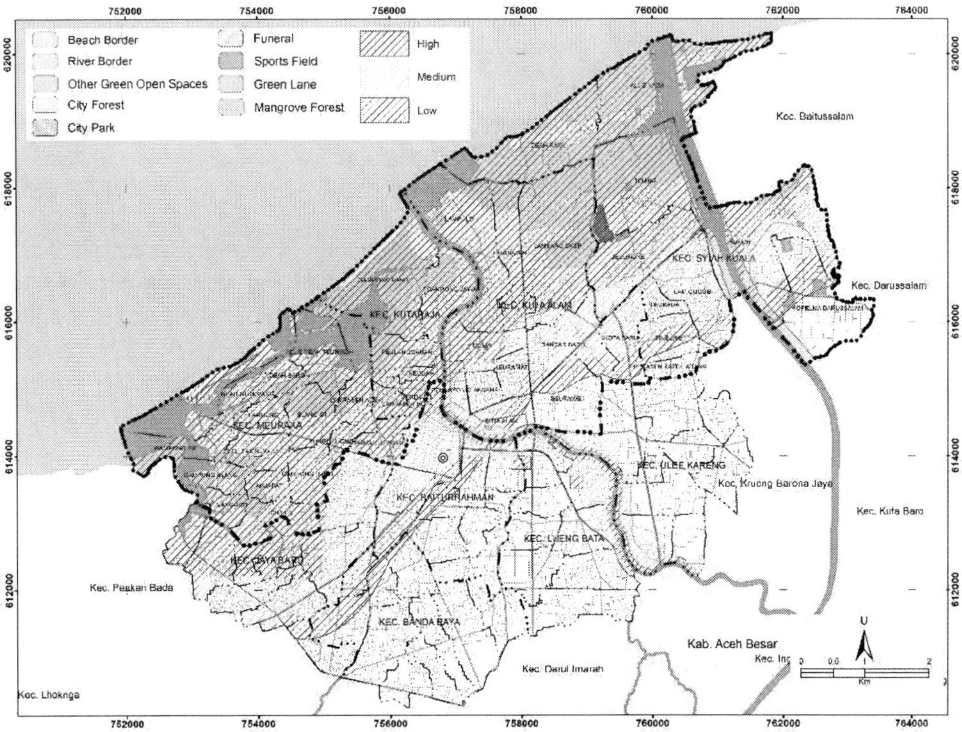

Figure 3. Map of GOS plans in disaster prone coastal areas.

In accordance with the provisions of spatial patterning, in the coastal area a local protected area is planned in the form of a coastal border and river border area. The border area of the coast serves to protect the coastal area from activities that interfere with the preservation of the coast which is located along the banks which width is proportional to the shape and physical condition of the beach, which is 50 m - 100 m from the highest tide point towards the land except for the beach area used for public purposes, such as existing ports/docks, open spaces, public spaces, tourist attractions and fishing settlements [3, 4]. The planned beach boundary covering an area of 40.28 Ha, established along the coast in some village, such as Pande Village, Lampaseh Village, Alue Deah Teungoh Village, Deah Raya Village and Alue Naga Village. The beach borders on the coast of Pande Village which are integrated with the tourism area plan look untreated and the utilization of the coastline has not been maximized, only overgrown with mangroves, shrubs and grass, also the condition of the border beach at other locations. While the river border is planned to be 8 m - 30 m wide to protect the river from activities that can disrupt or damage the river water flow function. The river border in the coastal area of the city is planned to cover 64.10 ha, which are located along the river Krueng Aceh, Krueng Doy, Krueng Neng, Krueng Titi Panjang, Krueng Lueng Paga, Krueng Daroy, and Krueng Cut.

In the coastal areas, mangrove forest green space is also set as large as 312.06 ha, mangrove forest area is directed at the northern coastal area of Banda Aceh City which is spread in Meuraxa Sub-district as large as 162.95 ha, Kuta Alam Sub-district covers an area of 32.04 ha, Kutaraja 87.49 ha and Syiah Kuala 29.57 ha. This determination is in accordance with the characteristics of the area and city development based on disaster mitigation [4]. The condition of mangrove forests has not been maximally utilized. The utilization of mangrove forest green space is only in the form of mangrove plant growth and utilization as a pond area, so that it only has a limited function to fulfill ecological functions, there is no economic, socio-cultural or architectural function. Furthermore, GOS with other green space designation is set at 463.25 ha, with existing land use dominated by mixed plantations, ponds and vacant land areas that have not been planned, well-maintained, thus creating unattractive visual impressions. For this other green space designation.

3.2. Analysis of existing land use of GOS in the coastal areas

Land use is the result of changes in spatial patterns and the magnitude of human activity which is a dynamic process. Land use is all human activities on the land to use the land to meet all their needs [9]. Analysis of existing land use aims to get a picture of land use and as a basis for evaluating the suitability of spatial use [10].

Based on the results of the analysis, land use in the designation of GOS in the coastal area of Banda Aceh City in 2020 consists of built areas namely residential and trade areas and services, mangrove forests, urban forests, city parks, green lanes, cemeteries, sports fields, green open spaces others, grass / vacant land, ponds, swamps, riverbanks, coastal banks and water bodies (Figure 4) with the area of each sub-district use shown in Table 2.

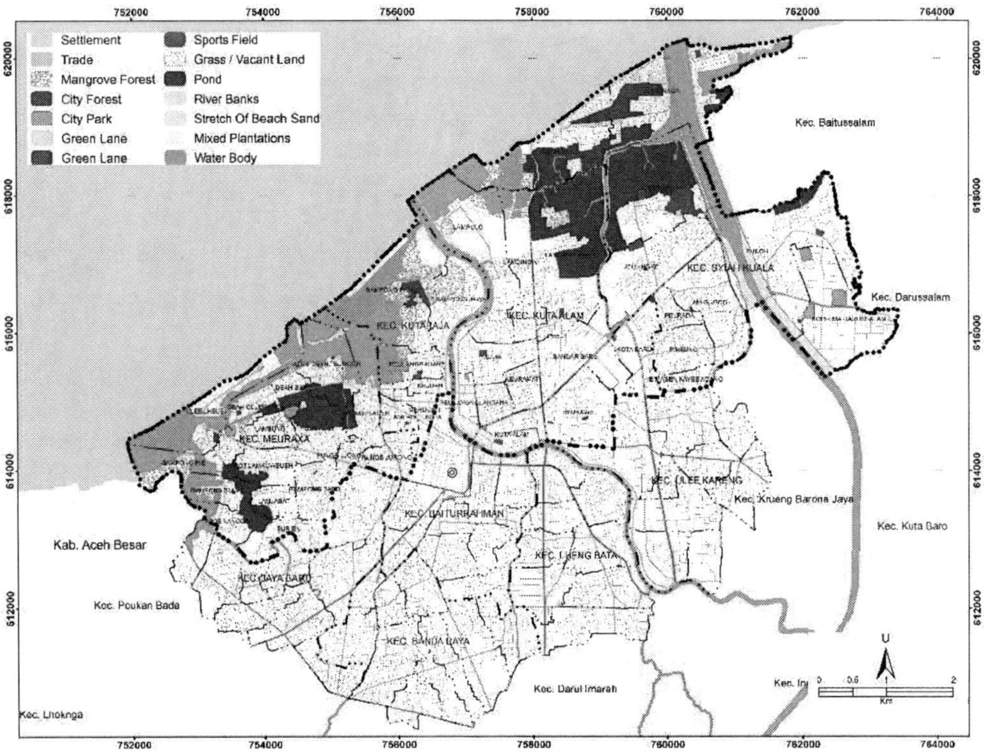

Figure 4. Map of existing land use in coastal areas.

Table 2. Land use exists in the coastal areas of Banda Aceh City.

Existing land use	Area					
	Kuta Alam	Kutaraja	Meuraxa	Syiah Kuala	ha	%
Settlement	0.28	1.28	1.76	9.13	12.44	1.25
Trade	-	-	-	0.47	0.47	0.05
Mangrove forest	22.12	71.05	34.75	10.80	138.71	13.98
City forest	-	-	-	13.63	13.63	1.37
City park	1.28	5.46	5.91	13.78	26.43	2.66
Green lane	10.85	0.64	9.79	9.41	30.68	3.09
Cemetery	2.06	1.63	3.56	0.87	8.12	0.82
Sports field	1.67	-	-	5.98	7.65	0.77
Mixed plantations	0.29		2.16	2.62	5.07	1.52
Grass / vacant land	9.76	3.54	3.86	63.01	80.18	8.08
Pond	111.36	9.32	95.34	205.17	420.97	42.41
River border	2.69	0.42	0.21	46.37	49.68	5.01
stretch of beach sand	-	3.19	6.50	22.16	31.85	3.21
water body	18.78	65.23	33.63	38.82	156.47	15.76
total	162.37	96.51	163.82	413.40	992.58	100.00

The results of the analysis between the existing land use and the green space plan shows that 12.91 ha of GOS plan has changed its function as a built area such as a residential area and a trade and service area. The existing land use that is appropriate for the allocation of GOS covering an area of 225.22 ha consists of mangrove forests, urban forests, city parks, green lanes, cemeteries and sports fields. While other land area of 754.45 ha which consists of mixed plantations, grass/vacant land, ponds, riverbanks, coastal banks and water bodies has not been used optimally according to its purpose.

3.3. Potential utilization of GOS in disaster prone coastal areas

GOS is an important component of urban landscape and provides many benefits as a city lung, improving physical, psychological and increasing social harmony through social interaction and recreation. Meanwhile, unplanned expansion and development have directed the city to risky hotspots from various environmental, social and disaster perspectives. In this situation, urban green space is considered an important natural contributor that supplies a variety of ecosystem services to the humanity who live in cities and can be an important part of achieving sustainable development goals [11].

The existence of GOS is not only has an ecological function, but also have an additional function, namely as a social and cultural, economic, aesthetic and the function of tsunami disaster mitigation in the coastal areas of the city [4, 12]. This is a consideration in determining the potential types of utilization of GOS. Utilization of GOS should be in accordance with the conditions and characteristics of a city and provide an effective impact according to the purpose and function of allocating space for GOS so that it provides benefits of comfort, security, beauty and prosperity [13].

GOS of beach border which has the main function as a barrier to the growth of settlements or other activities so as not to interfere with coastal sustainability, can be utilized for non-green activities by 10% of the planned area This refers to the stipulation that the area planted with plants (green space) on the coastal boundary is at least 90% of the planned beach border areas. Provision of activities and facilities in the coastal boundary area should not cause disturbance to the sustainability of the coastal ecosystem, including disturbance to visual quality, planting vegetation prioritizes local vegetation with a cropping pattern aimed at preventing abrasion, erosion, protecting against the threat of tidal waves, and reducing strong winds. The potential for developing the utilization of coastal borders can be in the form of planting mangrove vegetation, as well as the development of ecotourism integrated with the planned tourism area in Pande Village and GOS mangrove forests in Syiah Kuala Sub-District. Ecotourism is a form of natural-based sustainable tourism with a primary focus on nature conservation, learning and education as well as locally oriented [14].

River border areas can be used for outdoor activities that are recreational in nature which can increase the intensity of community socio-cultural interactions and limited use for activities that function as utility buildings and infrastructure related to river management [3]. The GOS for mangrove forests has the potential to be used as a mangrove conservation area that can be combined with other plants as a buffer zone, tourist area [8] and the development of a Silvofishery pond system. Silvofishery or also known as Wanamina is a system of aquaculture which is a combination of aquaculture with the activities of planting, maintaining, managing and conserving mangrove forests that can reduce negative impacts on the environment [15]. The pattern of this technical approach seeks to overcome the problem of mangrove forest sustainability and community welfare. Fishery activities can continue to be carried out in the mangrove forest without disturbing its function as a natural coastal protector from waves. This pond system has a simple technology that can be operated without damaging existing mangrove plants and can be done as an activity to reforest coastal areas. The development of the concept of ecotourism and Silvofishery pond system in the green open space plan of the mangrove is expected to have a positive impact on the community, especially in the economic, socio-cultural, educational fields and the development of the concept of ecotourism will present the aesthetic impression of the area by not reducing the main function of green space as a provider of ecological functions.

For another GOS dominated by existing land use as ponds and empty land area, this GOS can be developed to the maximum with a variety of potential uses that can provide more value to the environment and society. Several potential uses that can be developed in other GOS, those are: as an ecotourism area, especially in some villages, such as: Deah Raya, Tibang and Alue Naga in Syiah Kuala sub-district. in Pande Village and Jawa Village Kutaraja sub-district which is directly adjacent to the mangrove forest GOS. The utilization as a breeding ground for fauna, especially birds, plantations with the concept of tourism in Blang Oi Village Meuraxa sub-district, Lambaro Skep Village Syiah Kuala sub-district and in Peulanggahan Village, Kutaraja sub-district. Besides that, it can also be used as a pond with Silvofishery system, especially in areas bordering the mangrove forest GOS in Lambaro Skep Village Syiah Kuala Sub-district and the development of thematic GOS such as City Forest, City Park, Sports Field, under the potential of the area. allocating space for GOS so that it provides benefits of comfort, security, beauty and prosperity [10].

4. Conclusion

GOS in the coastal area of Banda Aceh which has been set in the Banda Aceh City Spatial Plan is dominated by green space borders of the coast, river borders, mangrove forests and other open green areas as a whole are still not optimally utilized yet, not treated properly and presents a less attractive visual impression, this is illustrated from the existing land use and visual impressions that are present at the site.

The potential use of GOS in coastal areas that can be applied include development of mangrove forest conservation, development of ecotourism areas, development of Silvofishery system ponds, fauna breeding sites and tourism-based plantations and thematic green space development including city forests, city parks, sports fields following the characteristics and potential of the green space.

GOS will be valuable in every part of the urban area, if it can be planned for the welfare of urban residents, both from ecological, economic, socio-cultural functions and aesthetically designed through architectural touches that are responsible and responsive to disasters to create a balance of the environment and environment the ongoing development.

5. Acknowledgement
Thank you to Syiah Kuala University and all parties who have contributed to this writing, so that it can be well done.

References
[1] Achmad A, Hasyim S, Dahlan B, Aulia D N 2015 *Applied Geography,* **62** 237-246.
[2] Achmad A 2015 *International Journal of Ecology and Development,* **32(3)** 67-74.
[3] Pemerintah Kota Banda Aceh. 2018. *Qanun Kota Banda Aceh Nomor 2 Tahun 2018 Tentang Perubahan Atas Qanun Kota Banda Aceh Nomor 4 Tahun 2009 Tentang Rencana Tata Ruang Wilayah Kota Banda Aceh Tahun 2009- 2029.* Banda Aceh.
[4] Fuady M 2015 *Journal of Architecture and Built Environment,* **42** 51-58.
[5] Pemerinah Republik Indonesia. 2007. *Undang-Undang Republik Indonesia Nomor 27 Tahun 2007 tentang pengelolaan wilayah pesisir dan pulau-pulau kecil.* Jakarta.
[6] Malkoc E, Kilicaslan C, Ozkan M B 2010 *Indoor and Built Environment,* **19(5)** 520-537.
[7] Sari R, Kustiwan I 2004 *Jurnal Perencanaan Wilayah dan Kota B SAPPK V2N.* 1.
[8] Achmad A, Irwansyah M, Ramli I 2018 *IOP Conference Series: Earth and Environmental Science,* **126** 012166.
[9] Arsyad S 2010 *Konservasi Tanah dan Air.* Edisi Kedua Cetakan Kedua. Bogor: Serial Pustaka IPB Press.
[10] Fuady M 2016 *International Conference on Engineering and Science for Research and Development (ICESReD)*, 109-116.
[11] Pokhrel S 2019 *Environmental Research Communications,* **1(10)** 105003.
[12] Departemen Pekerjaan Umum. 2008. *Peraturan Menteri Pekerjaan Umum Nomor: 05/PRT/M/2008 Tentang Pedoman Penyediaan dan Pemanfaatan Ruang Terbuka Hijau di Kawasan Perkotaan.* Jakarta.
[13] Hakim, Rustam 2007 *Ruang Terbuka Hijau.* Available at: http://rustam2000.wordpress.com/ruang terbukahijau.htm.
[14] Tanaya D R, Rudianto I 2014 *Jurnal Teknik PWK,* **3(1)** 71-81.
[15] Paruntu C P, Windarto A B, Mamaseh M 2016 *Jurnal LPPM Bidang Sains dan Teknologi,* **3(2)** 1-25.

Walkability level study in Kota Tua area of Meulaboh district

L Juhardi[1,*], Izziah[2], R Anggraini[3]
[1]Civil Engineering Postgraduate Program, Department of Civil Engineering, Universitas Syiah Kuala, Banda Aceh, Indonesia
[2]Department of Architecture and Planning, Universitas Syiah Kuala, Banda Aceh, Indonesia
[3]Department of Civil Engineering, Universitas Syiah Kuala, Banda Aceh, Indonesia

*E-mail: lizwarjuhardi6@gmail.com

Abstract. The pedestrian pathway is a facility for pedestrians and also as a means of connecting one region to another on the side of the main road. The case study in this research is in the Kota Tua area of Meulaboh National Road and Teuku Umar Street. As an old city this location has historical value so it attracts people to come. At this location there is also a hawker and culinary center so this location will be crowded to visit. This study was conducted to determine the walkability index of the pedestrian pathway to be studied. Based on the results of observations made, the pedestrian lane in this Kota Tua has a lack of facilities and infrastructure that supports comfort for users. This study uses a mixed-method method where field observations will be conducted to assess the conditions of the pedestrian path. Based on observations made, the results of the walkability index on the National Road, Jalan Iskandar Muda, and Jalan Teuku Umar were declared uncomfortable. Whereas the Area Walkability Index value of 45.75 means it is uncomfortable. Based on the above calculation, the pedestrian path is declared uncomfortable to pass. In order for pedestrian lanes to be comfortable when traversed it is necessary to arrange pedestrian lanes so that pedestrian lanes are comfortable to pass. The structuring carried out refers to Minister of Public Works Regulation No. 3 of 2014.

1. Introduction

Walking is a major and fundamental mode of transportation for almost all humans. Pedestrian lane is access for pedestrians and a means of connecting between one region and another separated from the main road as a motorized vehicle movement lane for motorbikes or cars. [1], Rapid development of the city needs to be balanced with the provision of pedestrian facilities in support of the realization of a comfortable city (Livable City) and pedestrian-friendly (Walkable) [2], Pedestrian path and environmental conditions are the main components to support the mode of walking transportation. A functional pedestrian path has supporting factors that shape it, including physical dimensions or factors (which include the length, width, and height of the pedestrian area itself), pedestrian accessibility, actors or users, the frequency of activities that occur, the relationship with the surrounding environment (area settlements, offices, trade, trade, and city magnets that support social interaction [3], is a way to assess the environmental quality of pedestrian walking activities. Based on preliminary observations made, the pedestrian path in the Kota Tua area, Meulaboh has a level of comfort that is not good. [4]. In urban areas, pedestrian paths are an important element of urban planning to be considered. [5]. The comfort factor greatly influences the quality of the pedestrian path. Figure 1 are displayed pictures of pedestrian path location conditions in Kota Tua, Meulaboh.

Figure 1. Photo of pedestrian path conditions in Kota Tua, Meulaboh.

The research location is in the area of the old town of Meulaboh, which is along the area of the National Road and Jalan Teuku Umar in the city of Meulaboh, Johan Pahlawan, West Aceh Regency. The selection of this route is done because this path is a shopping center and culinary business area and has a historical side as the Kota Tua that must have a pedestrian path that is comfortable for its users. From the problems above, it is necessary to know how the exciting conditions of the pedestrian facilities and the feasibility of the pedestrian path in the Kota Tua area, how the pedestrian path walkability index in the Kota Tua area, and how to plan a good pedestrian location and in accordance with the Planning Principles. So that the pedestrian lane users are comfortable when passing through this pedestrian lane.

2. Area of study, data collection, and processing

Research conducted is on the national road and Teuku Umar Street, as shown in Figure 2. The length of the road under study for the National road is 200 Meters divided into 2 lanes namely left and right lanes and divided into 2 STA for one lane, the length of 1 STA examined is 100 meters, meaning that on the National road there are 4 STA examined. While the Teuku Umar roadway survey that was investigated was 700 meters divided into 2 lanes namely left and right lanes divided into 14 STA with each STA having a distance of 100 Meters. In this research, data processing and analysis were carried out using the mix method method with a descriptive approach. For the assessment of the lane in the pedestrian lane segment studied, a scoring system with a score of 1 to 4 is used where 1 is not very good, 2 is not good, 3 is good, and 4 is very good. For the type of facilities and infrastructure to be observed there are 6 items, as for the items as follows: 1. Availability of lanes, 2. Traffic signs, 3. Seats, 4. Availability of ram, 5. Surface conditions of pedestrian lines, and 6. Safety side.

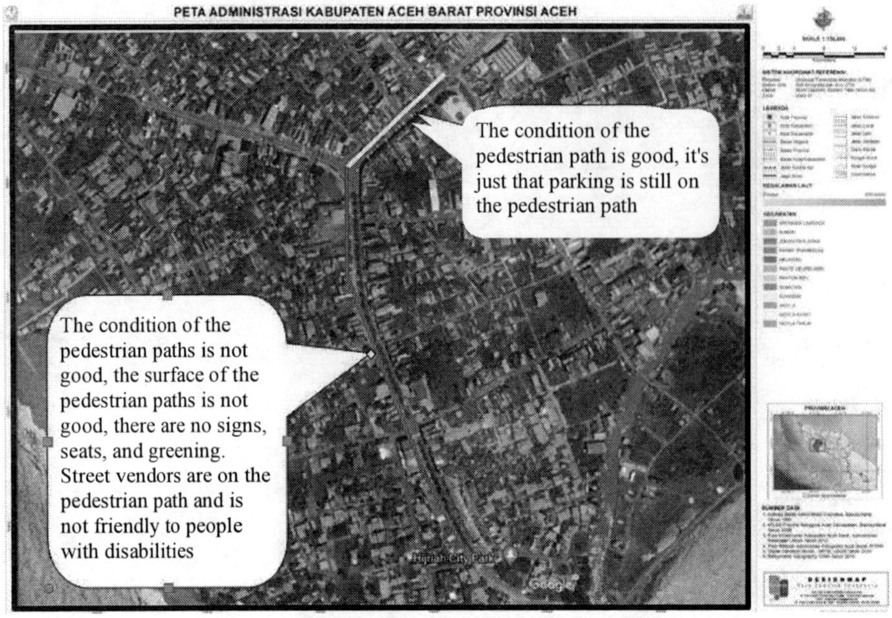

Figure 2. Study area.

3. Analysis Method

The study was conducted by observing and evaluating the pedestrian path in the Kota Tua area along the National Road, and Teuku Umar Street, Meulaboh City, Johan Pahlawan District, West Aceh Regency. The stages in this study are as follows:

1. Preparation Stages
   - Formulate background
   - Gather literature in accordance with the research to be conducted
   - Preparation of research tools such as cameras, stationery, meters as a field observation tool as well as checklist forms and questionnaires.
2. Implementation Stages
   - Field observations to see pedestrian path conditions that are located in the shopping area of National Road and Jalan Teuku Umar, Meulaboh City, Johan Pahlawan District, West Aceh Regency.
   - After further observations, walkability index calculations of the pedestrian track are examined.

After observing the scoring for each answer and then it will be averaged and the percentage is made with the formula Walkability Index [6] as equation 1 states.

$$Segment\ Score = (value * weight) \tag{1}$$

Then the value of the segment score is divided by the number of walkway segments examined. So that the index of each region that can be calculated using equation 2.

$$Utility\ Area = \frac{\text{Skor Segmen}}{\sum i} \tag{2}$$

Information:

n = total parameter
i = segment
I = parameter

To get the weight value of each value in the survey form in this study, the assumptions are made as in the equation 3.

$$Weight = \frac{\text{Area Percentage value of each question}}{\text{Number of questions on the survey form}} \quad (3)$$

So we get the weight value for each score as Table 1 shows.

Table 1. Scoring weight value.

No	value	weight
1.	1	4,17
2.	2	8,33
3.	3	12,5
4.	4	16,67

4. Analysis of Results and Discussion

After field observations, an assessment of the conditions of the pedestrian path on the National Road and Teuku Umar Road was obtained. After evaluating the pedestrian path, then the walkability index will be calculated based on the assessment of the pedestrian path condition. An assessment of the conditions of this pedestrian path is shown in Table 2:

Table 2. Walkability assessment.

Street Name	Total Answers				Weight value of each question				*Road segment score*
	1	2	3	4	1	2	3	4	
National Street	13	3	8	0	4,17	8,33	12,5	16,67	179,16
Teuku Umar Street	40	15	29	0	4,17	8,33	12,5	16,67	654,16
Total walkability index value *(wi)*									833,32

After obtaining the segment score of each road that is observed then the walkability index value is calculated for each of the studied roads.

- National Street

$$score\ segment = \frac{street\ segment\ score}{\text{Total STA}}$$

$$score\ segment = \frac{179{,}16}{4}$$

$$score\ segment = 44{,}79$$

- Teuku Umar Street

$$segment\ score = \frac{street\ segmen\ score}{\text{Total STA}}$$

$$segment\ score = \frac{654{,}16}{14}$$

$$segment\ score = 46{,}72$$

The Walkability Index values for the old town of Meulaboh is:

$$WI\ Region = \frac{WI\ National\ street + WI\ Teuku\ Umar\ street}{2}$$

$$WI\ Region = \frac{44{,}79 + 46{,}71}{2}$$

$$WI\ Region = 45{,}75$$

Based on field observations and walkability calculation of pedestrian lane facilities and infrastructure, it is stated that the pedestrian pathways studied did not have comfort, whereas the area walkability index score was 45.75, which according to the walkability index score table issued by ADB stated that it was uncomfortable. Table 3 summarizes walkability levels based on research that has been conducted [6].

Table 3. The walkability index value is based on research conducted.

Score	Information	Rating
>70	In carrying out daily activities it does not require a motorized vehicle or most activities are carried out on foot	*Highly Walkable*
50-70	Some facilities can be reached on foot	*Waiting to Walk*
<50	Few facilities can be reached by foot or almost all activities requiring motorized vehicles	*Not Walkable*

Based on the calculation that has been done, it is necessary to improve the pedestrian path in the Old Town of Meulaboh. The repairs made must follow the standards issued by the government through Ministry of Public Works Regulation No. 3 of 2014 related to pedestrian paths. The pedestrian path structuring concept can be seen in Figure 3.

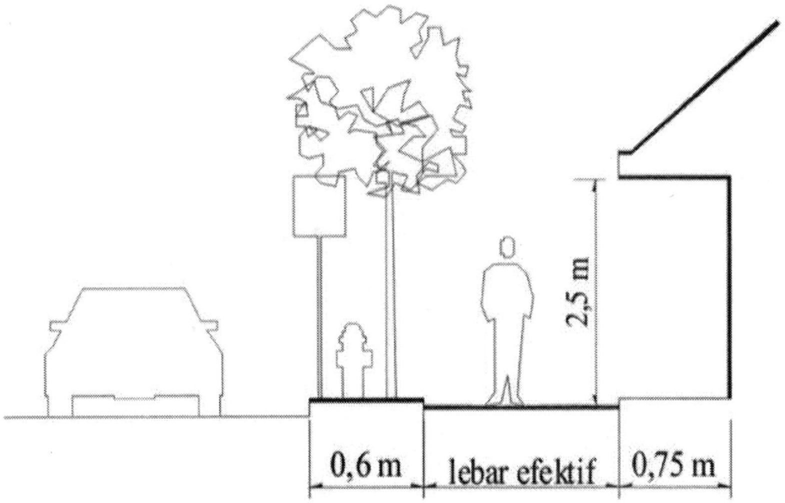

Figure 3. The pedestrian path concept is based on Minister of Public Works Regulation No. 3 of 2014.

5. Conclusion

Based on the results of the calculation of the Walkability Index pedestrian path on the National road has a Walkability Index Value of 44.79 or uncomfortable, and Jalan Teuku Umar of 46.72 = uncomfortable. While the Walkability Index value for the Old Town Area, Meulaboh is 45.75 = Uncomfortable. Based on the results of observations made The conditions of the pedestrian track that were examined were not good because the pedestrian path was used as a place for placing merchandise from shop traders, did not have signs, did not have a good Ram for people with disabilities, the condition of the path surface was not good and not good have side security that can cause pedestrian lane users to feel threatened. Based on the results of calculations that have been carried out, it is necessary to arrange pedestrian lines in accordance with Minister of Public Works Regulation no. 3 of 2014. For readers, the results of this study are expected to add insight to knowledge related to the Walkability Index in the pedestrian path of the old town, Meulaboh. Especially those who are interested to know more about the Walkability Index (conducting research), it is necessary to modify the independent variables to either add variables or add time series data. Hence it will be more objective and varied in conducting research. For the Government of West Aceh Regency, as a reference in future policy making in the efforts to Structure the Pedestrian Path in the Old Town, Meulaboh or the whole Pedestrian pathway in the City of Meulaboh.

References

[1] Setianto S, Joewono T B 2018 *The 19th International Symposium of Indonesian Inter University Transportation Studies Forum At Islamic University of Indonesia,* **19** 396-405.
[2] Wowor V D, Kumurur V A, Lefrandt L I R 2019 *SPASIAL,* **6(1)** 178-186.
[3] Finanda A N, Suharyadi R 2015 *Pathway Assessment Assessment Pedestrian Using The Walkability Index (Pedestrian Comfort) In The Yogyakarta Education Area,* 347–355.
[4] Agradiana P 2020 *MARKA (Media Arsitektur dan Kota): Jurnal Ilmiah Penelitian,* **3(2)** 68–82.
[5] Septika D E 2016 *Prosiding Temu Ilmiah IPLBI,* 111–116.
[6] Tanan N, Wibowo S S, Tinumbia N 2017 *Jurnal Jalan-Jembatan,* **34(2)** 115–127.

Study of coffee husk ash addition for clay soil stabilization

R P Munirwan[1,*], D Sundary[1], Munirwansyah[1], Bunyamin[2]
[1]Department of Civil Engineering, Faculty of Engineering, Universitas Syiah Kuala, Banda Aceh, Indonesia, 23111.
[2]Department of Civil Engineering, Faculty of Engineering, Universitas Iskandar Muda, Banda Aceh, Indonesia, 23234

*E-mail: r.munirwan@unsyiah.ac.id

Abstract. Industrial solid wastes have become a difficult problem in some developing countries, especially in Indonesia. The environmentally friendly issue regarding how to decrease industrial wastes by reuse or recycling process has become famous in recent years. Coffee husk ashes in this research were attained from burned waste coffee husk which can be considered as solid waste material from the coffee processing industry. In this study, clay soil obtained from Cot Bagie Village, Blang Bintang, Aceh was mixed with coffee husk ashes. Several soil physical properties test was conducted such as specific gravity, Atterberg limit, and grain size distribution by following ASTM standard. The physical parameter of soil was evaluated with the addition of 3%, 6%, 9%, and 12% coffee husk ashes by soil dry weight. The result shows that the physical properties of soil were improved if compared to untreated soil. A further mechanical laboratory test in soil mechanics was suggested for more advanced analysis of effect in soil-coffee husk ashes mixing.

1. Introduction

Indonesia as an archipelago is geographically vulnerable to tsunami natural disasters. Most earthquakes epicenter on the bottom of the Indian Ocean and some of them caused large ocean waves (tsunami). Environmentally friendly alertness has been increasing nowadays around the world. As a result, efforts in developing and recycling environmentally friendly materials to be reused again is also inclined. Moreover, many studies regarding industrial waste materials for construction have been developed [1] and [2]. In developing countries, industrial solid waste to become problematic recently, which results in a hazard of nature. The geographical position of Indonesia which are prone to earthquake disaster requires the use of good surround material for construction [3].

Stabilizing soil with industrial waste materials to obtain better soil parameter are also conducted by several types of research. Coir fiber as industrial waste material was added to clay soil for pavement construction [4]. The result shows an increment of CBR soil value with the adding percentage of coir fiber. Palm oil clinker as waste material to be used as light material in constructing a floating foundation in soft soil was also experimented [5]. Furthermore, the use of waste marble dust production for cement and concrete production was also introduced by [6]. Several other industrial waste material for soil stabilization was conducted by [7]. They use a recycled tire and fiberglass as substituted material for soil improvement. The use of nanoparticle as soil additive was also conducted by [8]. The research attempts to evaluate the behavior of silty clay geotechnical parameters by adding lime and nano lime. The nano lime presence can give a decent effect on the soil properties. Another lime-clay stabilization was also performed by [9]. The swelling potential of stabilized soil was reduced by the addition of lime.

Coffee husks are industrial waste material from the coffee industry. Figure 1 shows the fruit coffee structure. The size of the coffee fruit is approximately 10 mm with an oval shape. Nowadays coffee has become a basic need for people and has already become top commodities in the world. According to [10] coffee is the number two most treaded goods worldwide after petroleum. Brazil is the number one country in producing coffee which oversees more than 35% of worldwide production in the last 5 years.

Every 1 kg of coffee bean production will result in around 1 kg of coffee husk waste [10]. In Indonesia, twelve million 60 kg bags production yearly as the number four largest coffee-producing countries in the world [11]. Solid waste produces from coffee industries is a coffee husk, coffee silver skin, and spent coffee ground. These wastes have been used for limited applications like livestock feed, compost and fertilizer, and some other which remain a large amount of idle coffee waste. This unused waste will lead to a bad environmental effect on the world.

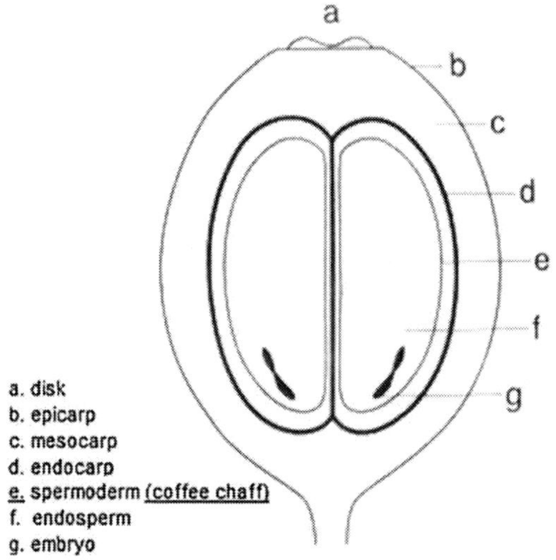

Figure 1. The structures of coffee fruit [12].

Several advances used for coffee waste material namely as bioprocessing products, biodiesel, fertilizer, animal feed, thermochemical conversion products, antioxidants, bio sorbents, biodegradable films in food packages, material for construction, and energy storage devices [13].

The use of coffee products in construction was developed by some researches. [12] introduces coffee chaff characterization for material in building applications in terms of environmental purposes. They investigate sound insulation material by coffee chaff in construction. The result shows that coffee chaff is may be suitable for insulation panel material for industry. [14] conducted a study in untreated coffee husk ashes for flux ceramic tiles. The result shows that coffee husk ashes can be used to replaced feldspars for fluxing material in the ceramic industry. Moreover, the use of coffee husk ashes can also reduce waste disposal and consumption of ceramic raw material.

In geotechnical engineering, [15] already studied about expansive soil behavior of compressibility and strength after stabilized with coffee husk ash. The results specify the good strength parameter of soil after treated by coffee husk ashes. Furthermore, the indication of hydrates particle formation and cementitious compound show after stabilization leads to the possible use of coffee husk ashes as stabilization material and subsequently. The chemical properties of coffee husk ash are shown in Table 1. Potassium oxide (K_2O) was the dominant chemical properties with 46.46 %.

Table 1. Chemical properties of coffee husk ashes (CHA) [15].

Chemical Properties	Composition (%)
Silica (SiO_2)	1.24
Alumina (Al_2O_3)	0.58
Iron oxide (Fe_2O_3)	0.56
Calcium oxide (CaO)	17.7
Magnesium oxide (MgO)	4.51
Sodium oxide (Na_2O)	0.14
Potassium oxide (K_2O)	46.46
Manganese oxide (MnO)	0.06
Titanium dioxide (TiO_2)	0.08
Phosphorpentoxid (P_2O_5)	3.85
Sulphur oxide (SO_3)	3.75
Loss on ignition (LOI)	21.07

2. Experimental method

The soil used for the experiment was obtained from the local site of Cot Bagie Village in Blang Bintang, Aceh Province. This site has been used as a quarry for construction projects for around Banda Aceh area recently. Disturbed soil was collected from around 0.5 m below ground surface and taken to the Soil Mechanics Laboratory of Universitas Syiah Kuala. The soil was cleared from all roots and stones, then dried in an oven for 24 hours in 105°C temperature and then sieved. The Cot Bagie soil is classified as OH and A-7-5 for USCS and AASHTO methods respectively. The gradation curve of the soil is given in Figure 2 where 67.42 % of original soil particles are pass sieve #200.

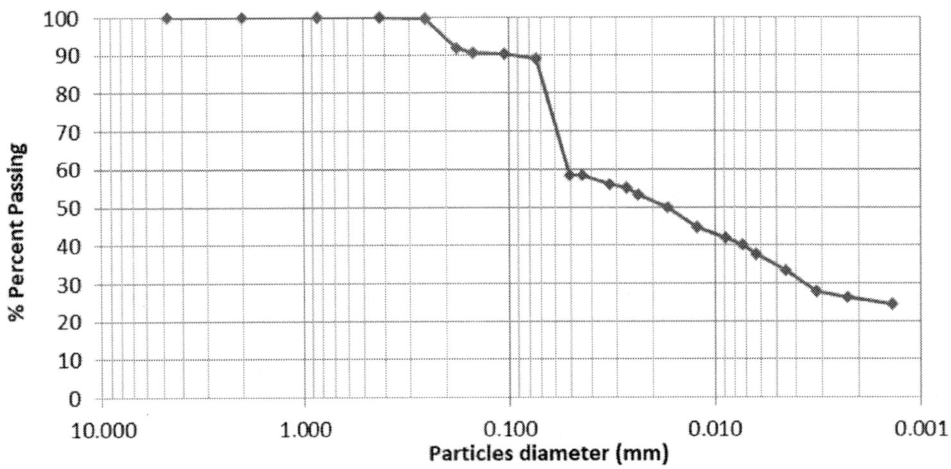

Figure 2. Grain sizes distribution for untreated soil.

The soil was mixed with 3%, 6% 9%, and 12% of coffee husk ashes by the dry weight of soil. The coffee husk was collected from the local coffee farm area in Bathin Weh Pongas Village, Bener Meriah, Aceh, then burned into ashes grey in color. Figure 3 shows the coffee husk ashes use in this study.

Figure 3. Coffee husk ashes (CHA) used for soil mixing.

The mixing percentage was chosen according to some previous research. Untreated samples (0%) were also prepared for comparison purposes. Some physical tests i.e., specific gravity, Atterberg limit, and grain size distribution were carried out in this study by using the American Society for Testing and Material (ASTM) standard.

3. Result and discussion

Table 2 shows the recapitulation of the physical properties of soil mix with coffee husk ashes for different percentage of addition. The test parameter analyze are specific gravity, liquid limit, plastic limit, plasticity index, and sieve passing #200. Generally, the plastic limit of treated soil showed good decrement with the increasing percentage of coffee husk ashes. The specific gravity value of untreated soil is 2.32, then there is a slight decrease with and addition of coffee husk ashes from 3% to 12% until 2.11.

Table 2. Physical properties of soil mix with coffee husk ashes.

No	Test Parameter	Mixing Percentage				
		0%	3%	6%	9%	12%
1	Specific Gravity (SG)	2.32	2.29	2.20	2.12	2.11
2	Liquid Limit (LL) (%)	63.00	59.00	58.00	55.40	52.2
3	Plastic Limit (PL) (%)	41.76	41.80	42.41	41.37	44.11
4	Plasticity Index (PI) (%)	21.24	17.20	15.59	14.03	8.09
5	Sieve Passing #200 (%) F	67.42	71.39	69.53	61.75	55.22

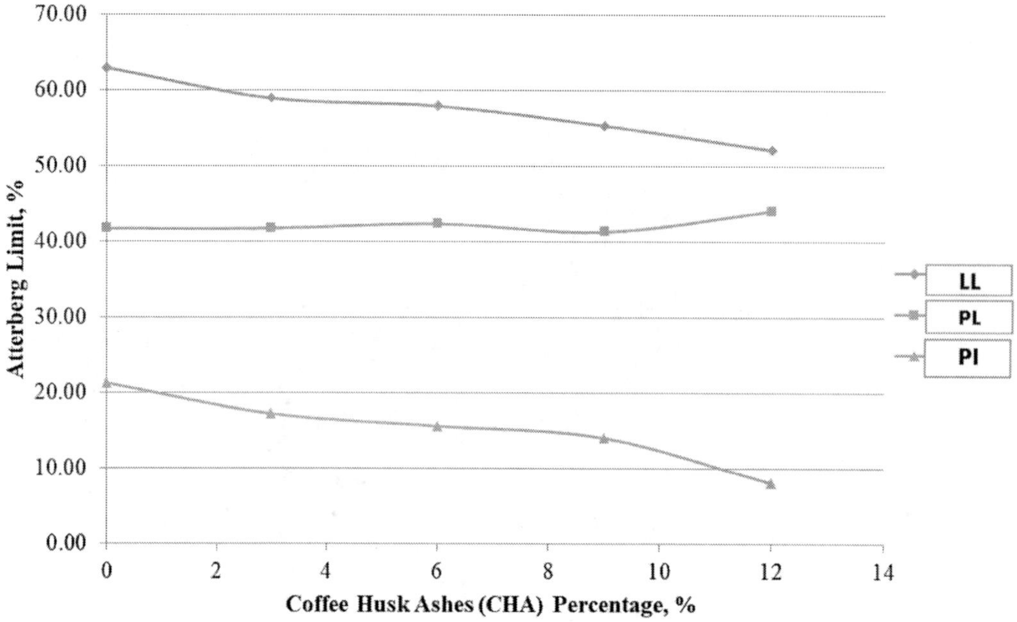

Figure 4. Effect of coffee husk ashes on Atterberg limit.

Figure 4 shows the effect of coffee husk ashes in addition to the soil in terms of the Atterberg limit parameter. The liquid limit of untreated soil is 63.0%, with the addition of coffee husk ashes, the liquid limit value then decreases up to 52.2 %. On the other hand, the plastic limit value increases slightly from 41.76 % for original soil to 44.11 % in 12% addition, even though there is a slight decrease from plastic limit 6% is 42.41 % to 8 % addition at 41.37 %. The index plasticity parameter decreases constantly from 21.24 % of untreated soil to 8.09 % for 12 % of coffee husk ashes. Commonly the Atterberg limit trend of treated soil showed good results with the cumulative percentage of coffee husk ashes. These results are like the literature studies of general solid waste materials which also found that the decrease of plasticity index with an increase of additive materials.

4. Conclusion

This article present laboratory results from work done at Universitas Syiah Kuala (UNSYIAH) for coffee husk ash application for soil improvements. Coffee based waste is readily available in huge amount, therefore it is also natural, economical, and environmentally friendly materials which can be used as an alternative material for construction purposes. From obtained results, the combinations of coffee husk ashes and soil enhance the physical properties of soil and are potential materials to be used for soil stabilization. For example, the index plasticity of treated soil exhibits a good reduction at 8.09 for 12 % of coffee husk ashes compared with 0% soil at 21.24%. It can be expected that with further laboratory tests in soil mechanics, more soil parameters can be improved with the soil-coffee husk ashes mixing.

Acknowledgement

The authors appreciatively acknowledged the "Indonesia Ministry of Research and Technology and Lembaga Penelitian dan Pengabdian Masyarakata (LPPM) Universitas Syiah Kuala Banda Aceh Indonesia" under Lektor Research Scheme 2020 No. 270/UN11/SPK/PNBP/2020 for partially supporter for this research.

References

[1] Aamir M, Mahmood Z, Nisar A, Farid A, Khan T A, Abbas M, Ismaeel M, Adnan S, Shah R, Waseem M 2019 *Processes,* 71–16.

[2] Thirumalai R, Babu S S, Naveennayak V, Nirmal R, Lokesh G 2017 *Engineering,* **9** 1008–1017.

[3] Munirwansyah M, Munirwan R P, Yunita H 2018 *International Journal on Advanced Science Engineering Information Technology*, **8(3)** 870-875.

[4] Munirwan R P, Munirwansyah, Marwan, Ramadhansyah P J, Kamchoom V 2020 *IOP Conf. Ser. Mater. Sci. Eng.,* **712(1)** 012009.

[5] Sulaeman A, Fulazzaky M A, Haroen M, Bakar I 2018 *KSCE Journal of Civil Engineering,* **22** 2232–2240.

[6] Aliabdo A A, Elmoaty A, Elmoaty M A, Auda E M 2014 *Construction and Building Materials,* **50** 28–41.

[7] Ahmad F, Yahaya A S, Safari A 2016 *IOP Conf. Ser. Mater. Sci. Eng.,* **136(1)** 012003.

[8] Taha M R, Govindasamy P, Alsharef J 2019 *E3S Web Conference,* **11005** 1–6.

[9] Munirwansyah, Munirwan R P 2016 *International Conference of Engineering and Science for Research and Development,* **1(1)** 63–68.

[10] Carvalho F De, Srinivas K, Helms G L, Isern N G, Cort J R, Roberto A, Kiær B 2018 *Bioresource Technology,* **257** 172–180.

[11] Banu J R, Kavitha S, Kannah R Y, Kumar M D, Atabani A E 2020 *Bioresource Technology,* **302** 122821.

[12] Ricciardi P, Torchia F, Belloni E, Lascaro E, Buratti C 2017 *Construction and Building Materials,* **147** 185–193.

[13] Massaya J, Prates A, Mills-lamptey B, Benjamin J, Chuck C J 2019 *Food Bioproducts Processing,* **118** 149–166.

[14] Acchar W, Dultra E J V, Segadães A M 2013 *Applied Clay Science,* **75** 141–147.

[15] Atahu M K, Saathoff F, Gebissa A 2019 *Journal of Rock Mechanics And Geotechnical Engineering,* **11** 337–348.

Sun-path model as a simple helping tool for architecture students in understanding saving energy building design

L H Sari*, E N Rauzi, Muslimsyah, M Mahmud
Architecture and Planning Department, Universitas Syiah Kuala, Darussalam, Banda Aceh, Indonesia

*E-mail: laina_hilma@unsyiah.ac.id

Abstract. Understanding the sun-path in designing an energy-saving building is essential. Saving energy is approached by daylight provision, which needs access for lighting the room during the day. In some latitudes, the building also needs the thermal mass to warm up the building during the cold time, requiring the sun's heat radiation. The sun path on the regions creates various ways to design buildings. There is multiple architecture software embedding the sun-path feature. However, the software is not easily installed for free, and some other constraints become the challenges for students to learn it. Therefore, this article figures out the project-based learning in utilizing sun path in 3D models for architecture students. The methods comprise the small project designing sun path in 3D models in various latitudes supported by Andrew sun-path software. This study aims to understand the position of the sun throughout the year, which in building design is useful for saving energy in ventilation and lighting applies. The result is the students' knowledge achievement for understanding the sun position and the site and building analysis toward the sun for obtaining the saving energy building design.

1. Introduction

Sun-path is the apparent significant seasonal-and-hourly positional changes of the sun (and length of daylight) as the Earth rotates and orbits around the sun [1]. Understanding the sun path is essential in protecting the environment and saving energy through passive building design. It is closely linked to the actions identified by the World Business Council for Sustainable Development, i.e., to increase and train workforce capacity and to evolve energy-efficient designs and technologies that use passive and active approaches [2]. Sun path is adopted in solar design that architects use at an Early Design Phase (EDP). According to Pfitzner et al. [3], EDP included a cost-benefit analysis of solar solutions that include future building life-cycle considerations. It is about 80% of design decisions affecting a building's energy performance at the early design stage. Therefore, designing buildings with an initial approach on sun path is essential. It is about 40% of the energy produced worldwide consumed by buildings, equal to 2500 Mtoe "million-ton oil equivalents" per one year [4]. In architecture, the solar design has been adopted in some software to predict energy use in the building. Nowadays, the 3D solar machine has also been developed, which can give a good insight for predicting the daylight entering the building at any particular time. However, technology runs sometimes much faster than what we have prepared. Sometimes the lack of ability to run the technology or the availability to have the technology become the problem.

In this study, we conduct project-based learning (PjBL) on understanding the sun path related to saving energy building design. The project developed a simple 3D model of sun path in maquette form, designed in some latitudes. The 3D design is proposed to make the students easier to understand the sun's way of illuminating the room. This study also creates knowledge for developing an understanding of the sun position and the site and building analysis toward the sun for obtaining the saving energy building design. The sun-path also develops the students' insight and creativity in designing apertures for daylight access or, conversely, avoiding excessive sun heat during summer or hot climate by providing shadings or building masks. Project-Based Learning (PjBL) is one of the learning models appropriate for learning with the specific product as an output, such as creating learning media. This model is suitable for students from the elementary school level to the university level [5, 6]. PjBL accommodates students to understand the concept and prior knowledge to obtain the skils by creating the project [7].

1.1 Sun path and solar design

Sun path related to buildings has been previously well-known studied by Szokolay [8], who worked with the northern hemisphere and Nobert Lechner [9] on the southern hemisphere. The use of a sun-path diagram is that the solar azimuth and the altitude can be read, and therefore the position can be precisely determined. By identifying the solar-windows of a particular location, one can design the building such that there is the maximum utility of the solar energy by placing thermal mass required for indirect heat gain in the right orientation. The building meets the visual and thermal comfort level by tracing down the surface area illuminated to apertures' location. Shading devices also can be designed similarly [10].

In understanding the sun-path, some diagrams would identify the sun's positions. The most popular ones are polar and cartesian sun path diagrams. Polar sun path diagrams are designed following the scale of altitude circles. Polar sun path has azimuthal, altitude, date and month, and hour lines [10]. Azimuthal lines are angles running around the edge of the diagram in 15° increments. A point's azimuth from the reference position is measured in a clockwise direction from True North on the horizontal plane. True North on the stereographic diagram is the positive Y-axis (straight up) and is marked with an N. While altitude lines are represented as concentric circular dotted lines that run from the center of the diagram out, in 10° increments from 90° to 0°. A point's altitude from the reference position is measured from the horizontal plane up. Date and month lines represent the sun's path through the sky on one particular day of the year. They start on the eastern side of the graph and run to the western side. There are twelve of these lines shown for the 1st day of each month. The first six months are shown as solid lines (Jan-Jun), while the last six months are shown as dotted (Jul-Dec) to allow a clear distinction even though the sun's path is cyclical. Hour Lines represent the sun's position at a specific hour of the day, throughout the year [10] (see Figure 1).

1.2 Sun path projections

The sun path projections are classified into three types: spherical, equidistant, and stereographic [11], as shown in Figure.

- Spherical (Orthographic) :
  In this method, the radial distance from the center is the cosine of the altitude angle. This sun-path diagram is utilized in making the sun-path 3D model.
- Equidistant
  In this graphic, the radial distance is simply a linear factor of the altitude angle. Thus the relative change in radius between all angles is the same.
- Stereographic.
  This graphic is a more complicated projection in which azimuth lines are first projected back to a reference point located a distance of 1 radius beneath the circle center. The point where each of these lines intersects the zero axis gives the radial distance.

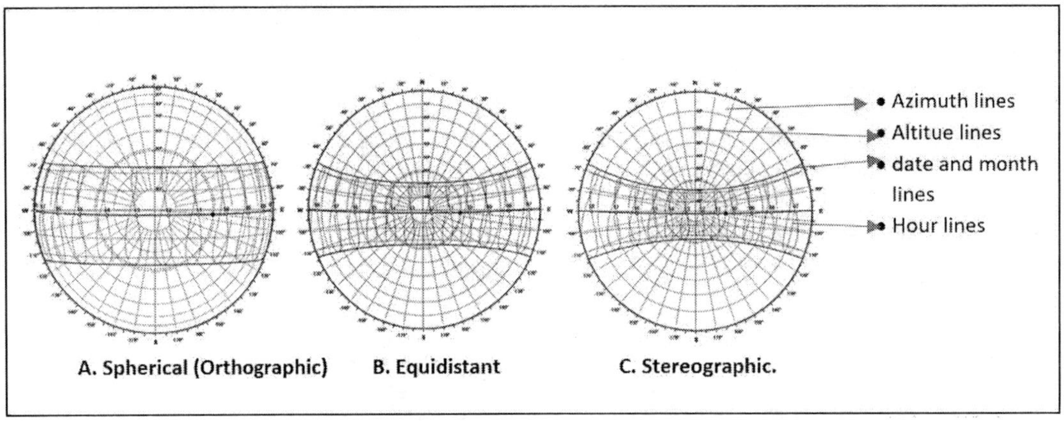

Figure 1. Polar sun path [11].

2. Method

This study is a descriptive study based on the observation of the sun-path workshop. The study aims at expressing the implementation of project-based learning (PjBL) based on the sun-path models designed. The study involved 24 architecture students of Universitas Syiah Kuala from Architectural Science class. Those students are divided into four groups, which refer to 4 zones of latitude range based on sun climate, as shown by Figure 2. Those ranges are tropical zone (23,5°N-23,5°S); Subtropic zone (23,5°N-40°N and 23,5°S-35°S); Tempered zone (40°N- 60,5°N and 35°S-66,5°S); and cold zone (60,5°N- 90°N and 66,5°S-90°S). Each group performs four sun-path models based on the latitude zones. This grouping zones will give an understanding of the participants where the sun path runs on each latitude zones.

In this study, the PjBL procedure comprises:

a. Introductory about the sun path to the students. In this case
b. The preparation for designing sun path 3D models
c. The making process of sun path models
d. Examining the result
e. Evaluation

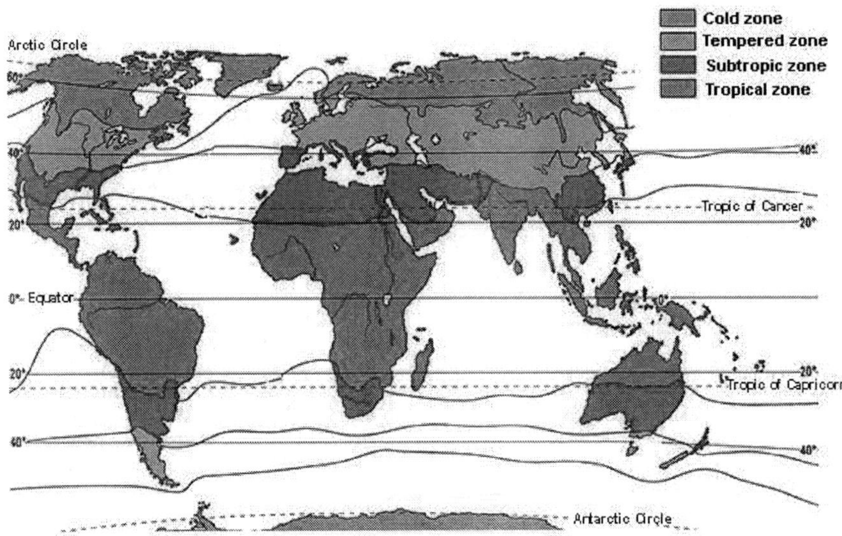

Figure 2. Climate zones [12].

3. Results and discussion

This study applied the project in March 2020 for about three times of meetings. In the first meeting as the 1st procedure, we introduced the students about the sun path and its eligibility to save energy building design. In this time, as previously mentioned, we divided the students into four groups of latitudes. In week two, we performed the 2nd procedure that the students brought all stuff for designing the sun path models. The steps of developing sun path models adopt the directions from a book Heating, Cooling, and Lighting for constructing a sun-path model [9]. We asked the students to prepare the board as a base (65mm thick), and three pipe cleaners, or two chenille pieces soft wire, such as copper or aluminum. Procedure 3, namely the making process, was conducted on three steps. The first step was to copy or draw the orthographic projection closest to the latitude of interest and glue it on a piece of foam board of the same size (see Figure 3).

The size of the model is 30cm in diameter. In modeling the 3D sun-path model, we use orthographic to identify the sun path and locate the wire on the model. The orthographic sun path is similar to the spherical sun path, which shows how points of the hemisphere (shown at 150 altitude increments) would be projected onto the horizon plane, giving the positions of the corresponding altitude circles on the horizon plane [8] (see Figure 4).

Step 2 requires students to cut a deep slit from A to B on the projections' north-south line. Then they support quadrant on a piece of reasonably thick transparent plastic film Trace the support quadrant on the part of fairly thick transparent plastic film (see Figure 5).

Figure 3. The students drawing the Orthograpic sun path.

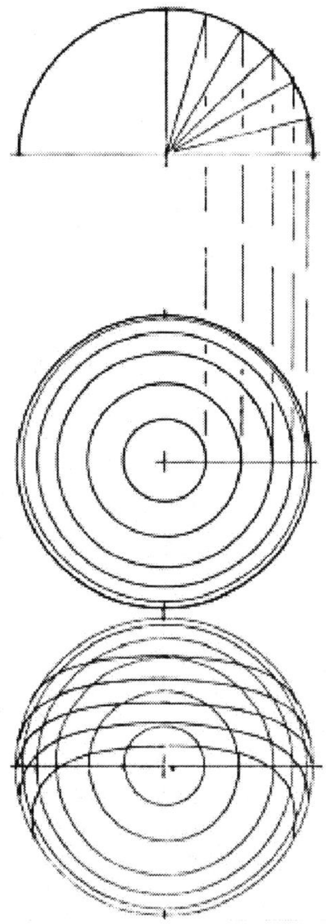

Figure 4. Orthographic projection Source [8]

Figure 5. The Installation of the support quadrant for defining the annual sun path.

The support quadrant loads the sun path annual shiftings i.e. mid-summer, equinox, and mid-winter (see Figure 6). On equinox days the sun appears to rise at due east and set at due west, (at exactly 6:00 and 18:00 h respectively) and at noon it reaches an altitude of ALT = 90 - |LAT|, i.e. a position when the zenith angle is the same as the latitude (ZEN = |LAT|). Here LAT is taken as its absolute value. At mid-summer noon the sun would be 23.45° higher (ZEN = LAT - 23.45° or ALT = 90° - LAT + 23.45°). At mid-winter 23.45° lower: ZEN = LAT + 23.45° or ALT = 90° - LAT - 23.45°.

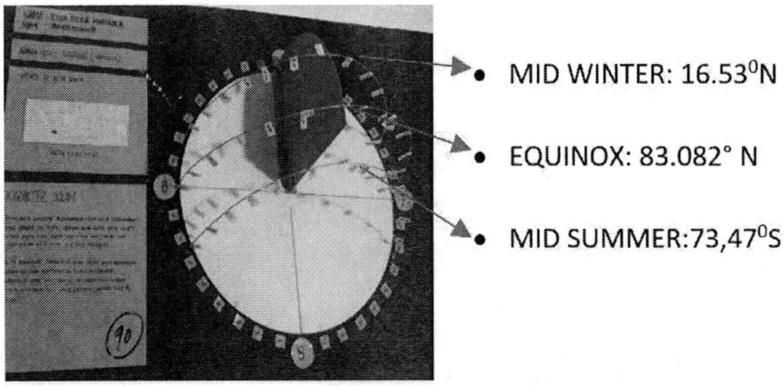

Figure 6. The sun path annual shiftings of Bandung.

As an example, Bandung city was identified through its latitude i.e., 6,9175°S (see figure 7). The students defined the mid-winter, equinox, and mid-summer lines (annual shiftings) based on the latitude.

Mid-winter : 90°- 6,9175°S- 23.45°N (to the north)
: 16,53°N
Euinox : 90° - 6,9175°S (to the north)
: 83,082⁰N
Mid-summer: 90° - 6,9175°S + 23.45°S (to the south)
: 73,47°S

Step 3, the students insert three wires on the three annual shiftings. They bend the first wire on the equinox across the support quadrant and add the others end in the sunset hole. Then they repeat this procedure for the other two sun paths.

Figure 7. The installation of the wire on three sun paths.

Latitude	orthographic	Sun path model (macquette)	3D sun path
Bandung 6.9175° S, 107.6191° E	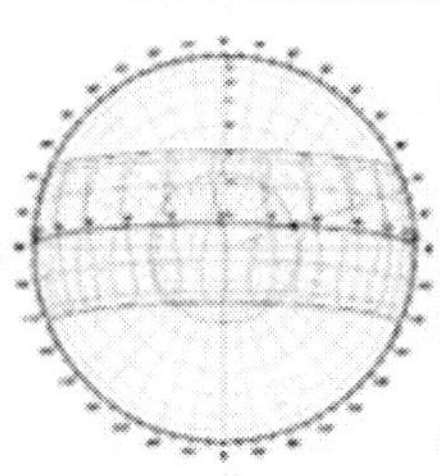		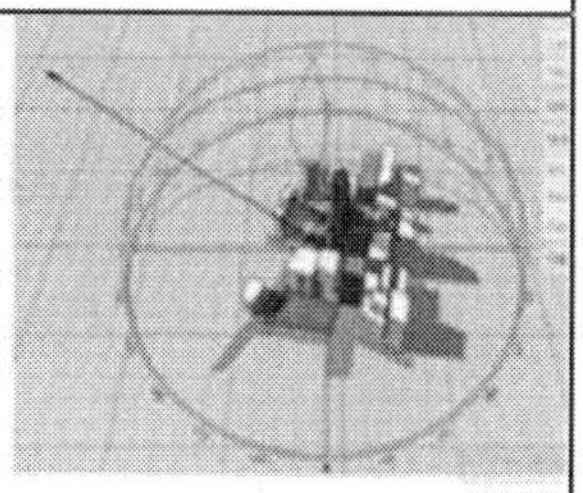
Jeddah 21°32'32.5 7"N, 39°11'52.6 9"E	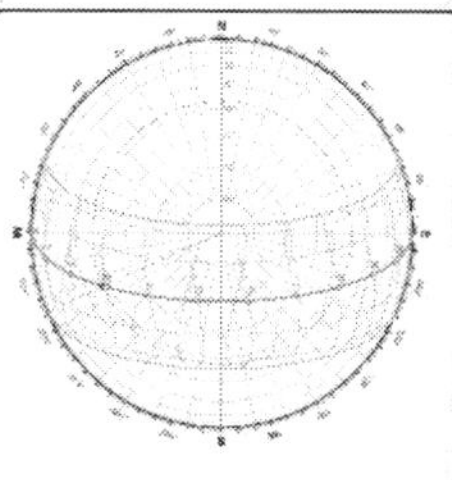	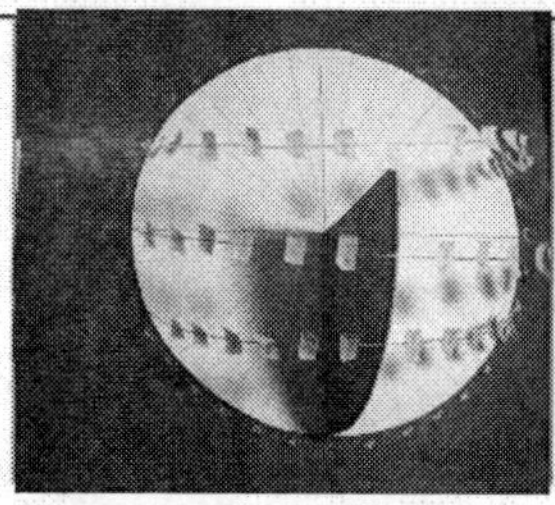	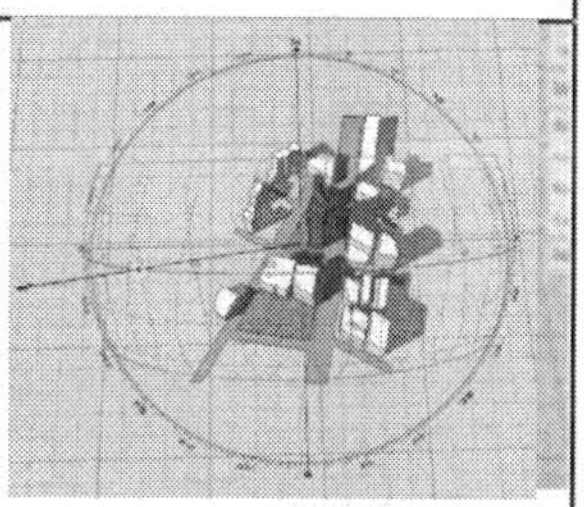
Wellington, 41.2865° S, 174.7762° E	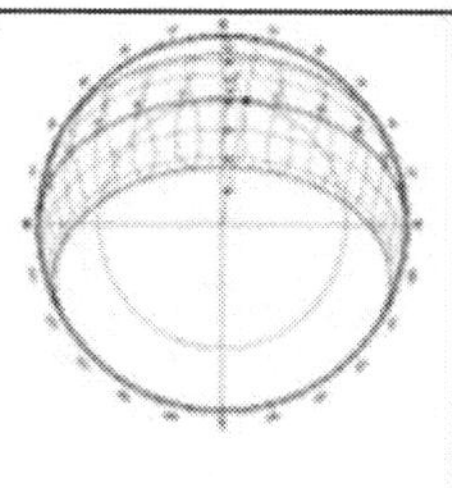	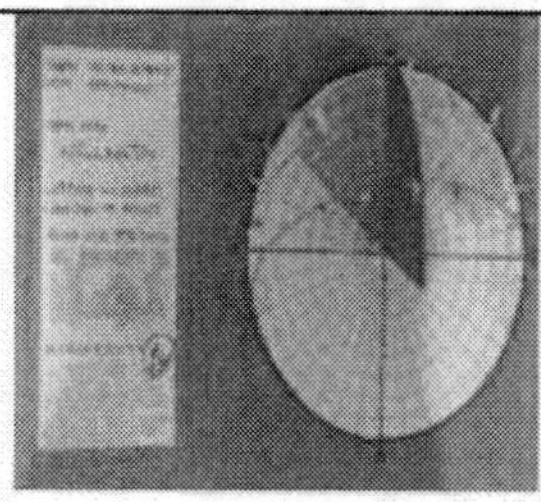	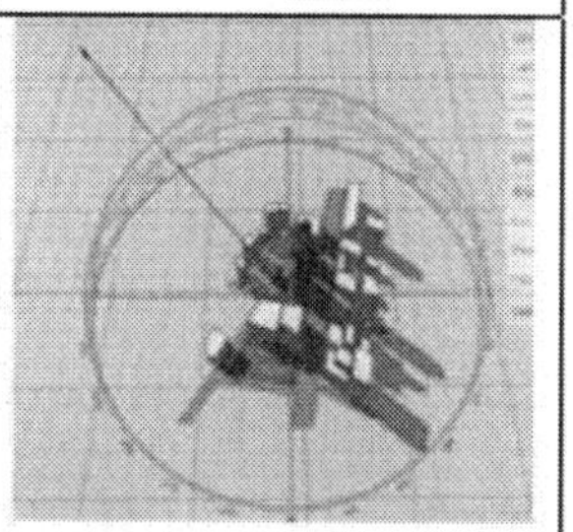
Nuuk, 64.1814° N, 51.6941° W	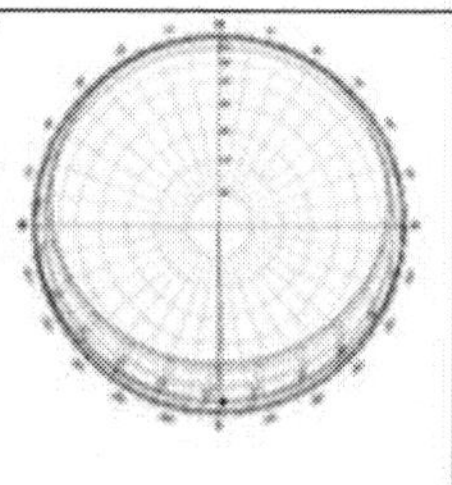	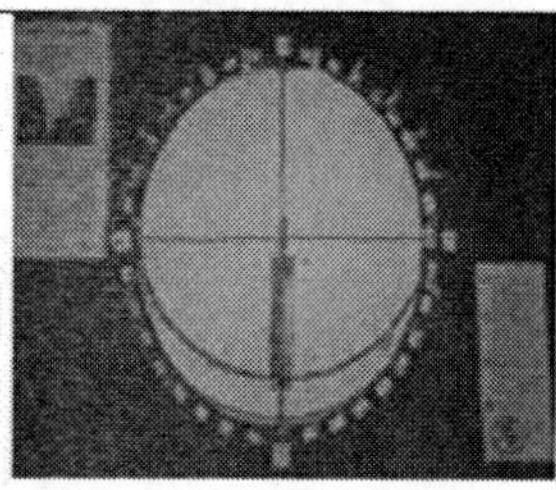	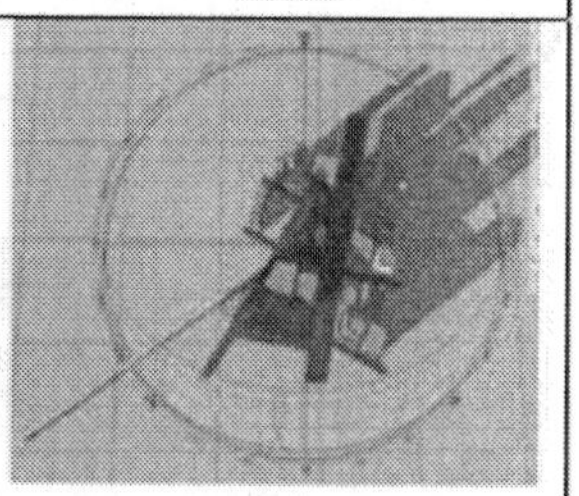

Figure 8. Sun path diagram 3D models on various latitudes designed by the students.

The sun-path influences the architecture through the daylight access and the shadings. Since the sun path diagram was made on a small scale, which is about 30 cm in diameter, therefore the building model should be made in a very tiny size for modeling daylight access and shading. Instead of modeling the small scale of building maquette, the students are directed to use the free application of sun path, i.e., Andrewmarsh sunpath3, for keeping students understand about the further use of sun path in building design.

As procedure four, which aims at examining the result, the application of Andrewmarsh sunpath3 [13] guided the students to understand how the sun path creates the shadings as shown by Figure 8. As we divided the groups into four latitude ranges referring to the specified climate zones, we took four samples representing those latitudes. Figure 8 shows the sun path diagram 3D models on various latitudes designed by the students combined with the orthographic sun path and the 3D model sun path software. The 3D model of shadings is setting simultaneously as a solar time on 21 March 2020 at 15.02.

a. *Bandung (tropical zone).*
   Bandung is located at 6,9175°N latitude. It has a warm-humid climate, which is high in Relative Humidity. The latitude shows the location which is close to the equator. Being close to the equator will give the not significantly different sun running hours on its path throughout the year. Figure 8 shows that the afternoon shading in Bandung at 15.02 is precisely near the east. Conversely, in the morning, the shading will go to the west. The equator's position gives the high radiation, which is straight from the east and the west. Therefore, we recognize that the traditional tropical house has more apertures, such as windows on the southern and northern side, aiming to achieve sunlight without excessive heat.
b. *Jeddah (subtropic zone).*
   Jeddah is located at 21,32°N latitude. The city, located in Saudi Arabia, has an arid climate assigned to extremely hot weather during summer. Due to the arid and desert environment, the shading against the sun is essential for creating a thermally comfortable microclimate. The sun path performs the shortest day that occurs in December and January. During these months, the shading will depend more on the northern side. The longest day is in September, which has the shading just close to the east and the west.
c. *Wellington (tempered zone)*
   Wellington is located at 41, 2865°S. The position on the southern atmosphere gives access to the sunlight on the northern side. The region has the summer and most extended daytime in January and December, which contrasts with the north atmosphere. Therefore, the shading works more on the southern side (south-west, south, south-east) (see Figure 8). This recommends creating more apertures on the north side of the buildings for giving optimal daylight.
d. *Nuuk (cold zone)*
   Nuuk is located at 64,18°N. This region has a more cold time throughout the year. Due to the position on the northern side, Nuuk sees the sun path on the southern side. Therefore, the buildings are built with more openings on the south of sides to get the daylight. The sun only appears best in June and July, while in January and December, no daylight is performed.

Here the students can understand that the place in the northern atmosphere will see the sun at the southern atmosphere and vice versa. The shading positions will figure out where the buildings' openings or apertures should be located for obtaining daylight. The students also could understand why the buildings in the northern part, which have latitude more than 40°N tending to have more cold times, should orient the large aperture to the south. The orientation will give plenty of daylight and create a passive heating design. Directing building opening to the north will cause a lack of daylight and invite freezing wind from the north pole. The latitude toward the south pole vice versa should orient the building apertures toward the north. The objective is to avoid the cold breeze from the south and gain precious daylight from the north.

In comparison, the buildings located in the sub-tropic zone (LAT: 23,5-40°S/N) are built typically with small openings, compact, and dense with other buildings. Small apertures aim to avoid the excessive solar radiation and gain more shades by the dense buildings nearby. Tropic zone ranges in latitude 0-23,45° S/ N suffers direct solar radiation from west to east on the equator. Tropics also have an ample amount of rains annually and high relative humidity. Therefore the buildings are recommended to have large openings toward north and south for gaining undirected sun radiation and air circulation for reducing the relative humidity.

4. Conclusion

The article presents the study on the sun-path diagram related to building design. It is developed on Project-Based Learning, i.e., a workshop involving Architecture students of Universitas Syiah Kuala. The workshop taught the students that the sun path on various latitudes would create different daylight access and shadings. The workshop guided the students to create a 3D model of sun path diagram. The model also gives insight the closer the latitude toward the north or the south pole, the further the equator's zones. The northern latitudes should orient the opening to the south and vice versa, aiming to deliver more access to daylight. In building design, it is useful for saving energy in receiving the light throughout the day. The sun path's 3D model will create a more understanding of the sun path and its relationship with building design. This learning method is straightforward to be made and learn compared with the paid software embedding the sun-path application. The software usually takes time to learn and run. The cost of registering and installing the program is also high. Therefore, the 3D model of sun-path would be a helpful tool for studying and understanding the sun-path in architecture.

5. Acknowledgement

The authors would like to express sincere thanks to Universitas Syiah Kuala for the funding support through ‘Hibah Lektor.’ Our gratitude also goes to the Architecture students involved in the workshop.

References

[1] Bloom J W 2011 *The really useful elementary science book.* New York: Routledge.
[2] World Business Council for Sustainable Development (WBCSD) 2009 *Energy efficiency in buildings: transforming the market.* WBCSD. Available online at www.wbcsd.org.
[3] Pfitzner M, Bögl M, Neuberg F, Tulke J, Hochtief, Nummelin O, Benning P, Bouygues 2007 *Analysis of existing software tools for early design.* Inpro report. European Commission, sixth framework programme.
[4] Attmann O 2010 *Green architecture: advanced technologies and materials.* New York: McGraw-Hill.
[5] Jacques L A 2017 *Am. J. Educ. Res.,* **5 (4)** 428-433.
[6] Burlbaw L 2013 *STEM project based learning: an integrated science, technology, engineering, and mathematics (STEM) approach.* Rotterdam: Sense Publishers.
[7] Capraro R, Slough S 2013 *Why PBL? Why STEM? Why now?*. Rotterdam: Sense Publisher.
[8] Szokolay S V 2007 *Solar geometry.* Passive and Low Energy Architecture (PLEA).
[9] Lechner N 2008 *Heating, cooling, lighting: sustainable design methods for architects.* 3rd. Ed. New York: John Wiley & Sons.
[10] Sriram K K 2015 Different forms of sunpath diagram and their uses in functional design of buildings. Available at: https://www.slideshare.net/3064026/sunpath-diagrams-different-forms-and-their-uses-in-functional-design.
[11] Andrew J M 2014 Sunpath2d. Available at: http://andrewmarsh.com/apps/releases/sunpath2d.html.
[12] Climate zones 2020 Available at https://content.meteoblue.com/nl/meteoscool/general-climate-zones.
[13] Andrew J M 2014 Sunpath3d. Available at: http://andrewmarsh.com/apps/staging/sunpath3d.html.

Analysis of the effect of using wire mesh layers on hot asphalt mixtures with modulus of elasticity

R Ismy[1,2], Husaini[3,*], S M Saleh[4], M Isya[4]
[1]Doctoral School of Engineering Science, Universitas Syiah Kuala, Banda Aceh, Indonesia
[2]Department of Civil Engineering, Universitas Almuslim, Matang Glumpang Dua, Indonesia
[3]Department of Mechanical Engineering, Universitas Syiah Kuala, Banda Aceh, Indonesia
[4]Department of Civil Engineering, Universitas Syiah Kuala, Banda Aceh, Indonesia

*Email: husainiftm@unsyiah.ac.id

Abstract. The value of the modulus of elasticity is very important to know for analyzing the strength of a hot asphalt mixture because the increase in the modulus of elasticity indicates an increase in the ability of a mixture to withstand loads by itself. The use of the wire mesh layer as reinforcement in hot asphalt mixture is considered capable of increasing the modulus of elasticity. The aim of this study was to analyze the relationship between the use of wire mesh layers in the hot asphalt mixture with the modulus of elasticity. The research was conducted using the three-point flexural test apparatus. Based on experiment methode, it was found that hot asphalt mixture with a layer of wire mesh laying 30 mm from the surface of the specimen was the best type of specimen modeling with a bending resistance value of 291.85 KN. The higher the flexural resistance value that can be withheld, the higher the modulus of elasticity, namely 3,800.13 MPa. The best asphalt mixtures are characterized by a high modulus of elasticity, high stress, and low strain.

1. Introduction

In general, road constructions in Indonesia use hot mix asphalt as the surface course for the pavement. The application of asphalt as the main material in pavement construction is classified as a flexible pavement [1-5]. Flexible pavement has several courses, which are-; surface, base course, and subbase course located on subgrade [6,7]. Strength dependency of flexible pavement in base course and subgrade causes this pavement type becomes strongly not recommended to apply on unstable soil conditions [5-6, 8-11]. The flexural strength is a common treatment for rigid pavement. The flexural strength of concrete is the capability of a concrete beam located on two laying to hold force with perpendicular axis of test specimen which is given until the test specimen is broken and stated in Mega pascal (Mpa) force per unit area [12-15].

In application of rigid pavement, wire mesh is used as cracked and pumping controller due to the controlling of stress and deflection values that increase service index [2, 7, 16-18]. Stiffness owned by wire mesh is also considerably capable to increase modulus of elasticity value [16-18]. Wire mesh is kind of reinforcements of supporting that commonly used in rigid pavement [13, 16]. Using reinforcement considerably can make the flexible pavement less dependent to the foundation course and base soil so it can be applied on the unstable soil [16-19]. High modulus of elasticity in a pavement will cause structure strength placed in its pavement [1, 16–18]. Several previous studies regarding to the method used to control the crack and hold pull force implemented checkered pattern which are available in a pavement [20].

Based on the maximum load gained by using flexural test tool with three points, this study develops the analysis the distributions of stress, strain and moment to the maximum load transpiring in hot mix asphalt with variation of Wire mesh; 20 mm from the surface of test specimen, 20 mm laying from the base of test specimen, from the centre or 30 mm from the surface of test specimen with high of test specimen is 60 mm [21]. The aims of this study was to analyze the effect of using wire mesh layers on the modulus of elasticity.

The test specimen used is Marshall Test to gain optimum asphalt content, while to analyse the flexural strength, flexural test tool with three points is used for this study [22-23]. To facilitate data analysis in order to get the required result, several parameters such as Marshall Characteristic and flexure performance in the form of maximum load values and deflection or deformation are used [2, 7, 16, 24, 25].

This study is conducted by applying experimental and empirical methods using wire mesh as reinforcement material. [26]. The test specimen used are hot mix with mixed aggregate and asphalt as well as Wire mesh. The specification of used wire mesh is Wire mesh M4 with 4 mm diameter and its hole size is 150 mm. Total amount of test specimens are 32; 24 of test specimens for obtaining Optimum Asphalt Content with Marshall Characteristic analysis and 8 test specimens for analysing the flexural strength value of hot mix asphalt in wire mesh course as reinforcement.

To analyse the flexural strength values on the test specimen, flexure test tool with three points is implemented with four types of wire mesh course placement, they are:

a. Hot mix asphalt without wire mesh as comparison
b. Hot mix asphalt with wire mesh placed 20 mm from the surface of test specimen
c. Hot mix asphalt with wire mesh placed 30 mm from the surface of test specimen
d. Hot mix asphalt with wire mesh placed 20 mm from the base of test specimen

2. Literature review and theory

2.1 Flexible pavement

Flexible pavement is a pavement course consists of subgrade, subbase course, base course dan surface [7, 27], as Figure 1 illustrates. Hot mix asphalt is a course in a road construction composed from the mixing of asphalt cement and well graded aggregate, it is mixed, overlaid and compacted while it is hot in certain temperature [7, 9, 27]. The material of the aggregate consists of coarse aggregate, fine aggregate and well graded filler mixed with asphalt [3-5, 25, 27].

In asphalt mixing, asphalt works as binder or adhesive between aggregate particles and the aggregate works as supporting. The mechanism characteristics of asphalt in asphalt mixing are obtained from friction and cohesion from its forming materials [3-5, 7].

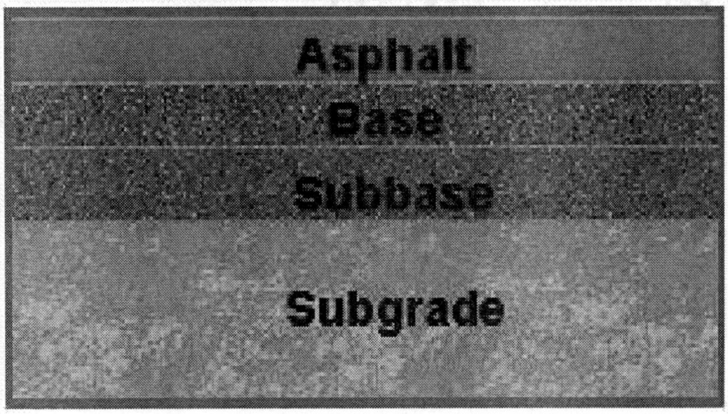

Figure 1. Composition of flexible pavement components.

2.2 Wire mesh

Wire mesh is a type of reinforcement from supporting pattern which commonly used in Rigid Pavement [13, 16]. In rigid pavement, the use of wire mesh is also used as crack control and pumping control due to the control of stress and deflection values so as to increase the service index with increasing the value of the modulus of elasticity [16, 28]. Wire mesh or usually called as woven wire is the iron chains resemble iron net that usually used to reinforce roof concrete. Wire mesh look like iron boxes that in each connection is glued together [13, 16], as shown in Figure 2.

Figure 2. Wire mesh.

2.3 Marshall test

Marshall Test, displayed in Figure 3, is the important stage in determining the characteristic of asphalt mixing. Marshall parameters used for this stage are stability, durability, flow, density, Marshall Quotient, void in mix, void in mineral aggregate and void filled with asphalt (VFA). Using Marshall test as well as analysing Marshall characteristic will obtain the optimum asphalt content [6, 9, 10, 25, 29, 30].

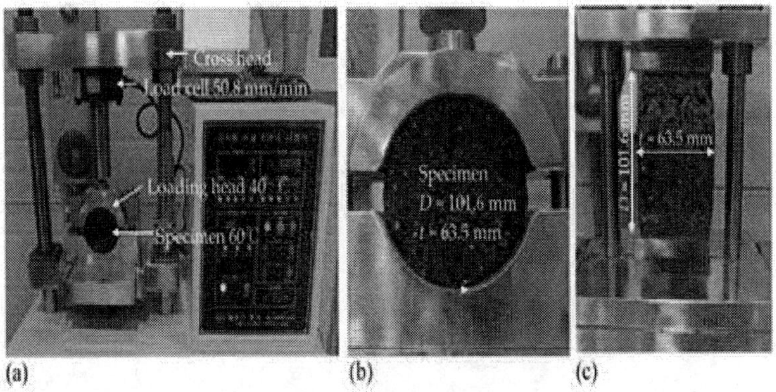

Figure 3. Marshall test.

2.4 Three points flexure test

Flexural test with three points is accordingly conducted to standardization of measuring flexural performance of concrete asphalt in temperature 10° C, when the asphalt mixing is susceptible to crack [31]. Loading rate is controlled in 50 mm/minute. Spesimen and test setting are illustrated in Figure 4.

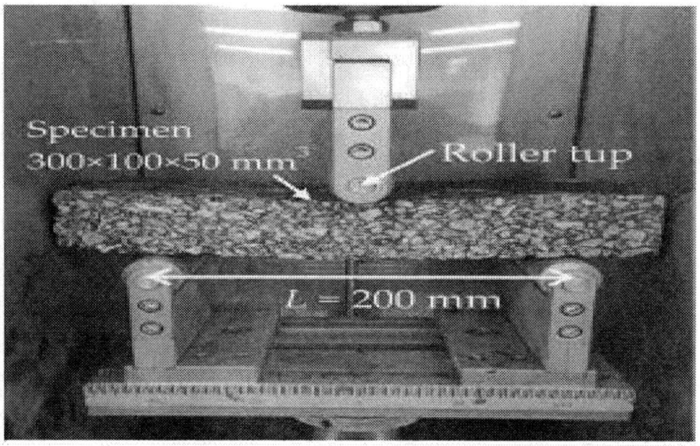

Figure 4. Three points flexure test.

3. Result and discussion

3.1 *The optimum asphalt content*

The first step of the study in hot mix is determining the Optimum Asphalt Content (OAC). After physical control of the aggregate and asphalt, then it is continued by preparing specimen test to analyse Marshall Characteristic. Based on the evaluation of Marshall characteristic in test specimen using Marshall test, it is found that OAC value is 5,75%.

3.2 *Correlation between loading and placement model of wire mesh*

The analysis of correlation between loading and placement model of wire mesh gained by hot mix asphalt is conducted by using flexural test tool with three points. The specimen test used is hot mix asphalt in Optimum Asphalt Content (OAC) that is obtained from Marshall characteristic test which is 5,75%. The test specimen is made with four types of wire mesh M4 placement in hot mix asphalt, as shown in Table 1.

Table 1. Asphalt mixed flexural testing results.

Number Of Test Object	Average KN	Loas (P) N	Deflection mm	Stress Mpa	Strain	Modulus Of Elasticity Mpa	Poison Ratio
1	188,5	188.500	9,5	78,54	0,038	2.066,89	0,15-0,45
2	197,7	197.700	9	82,38	0,036	2.288,19	0,15-0,45
3	291,85	291.850	8	121,60	0,032	3.800,13	0,15-0,45
4	251,3	251.300	8	104,71	0,032	3.272,14	0,15-0,45

Note:

Type of test object:

1. *Hot asphalt mixture without wire mesh layer (as comparison standard);*
2. *Hot asphalt mixture by laying 20 mm wire mesh layer from the surface of the test object;*
3. *Hot asphalt mixture by laying 30 mm wire mesh layer from the surface of the test object;*
4. *Hot asphalt mixture by laying 20 mm wire mesh layer from the base of the test object;*

Based on the result of the study, it is found that the highest loading which can be hold by hot mix asphalt is in test specimen type no. 3 where the hot mix with wire mesh course placed 30 mm from the surface of test specimen is 291,85 KN. The value of deflection occurs is 8 mm.

3.3 Stress, strain and modulus of elasticity

Looking at the relationship between loading and stress, strain and modulus of elasticity will be presented in a graphic form which is the result of the exposure from Table 1.

3.3.1 Stress

Stress is a reaction that occurs in all parts of the specimen in order to withstand a given load [18, 32, 33]. The description of the relationship between loading and stress is illustrated in Figure 5.

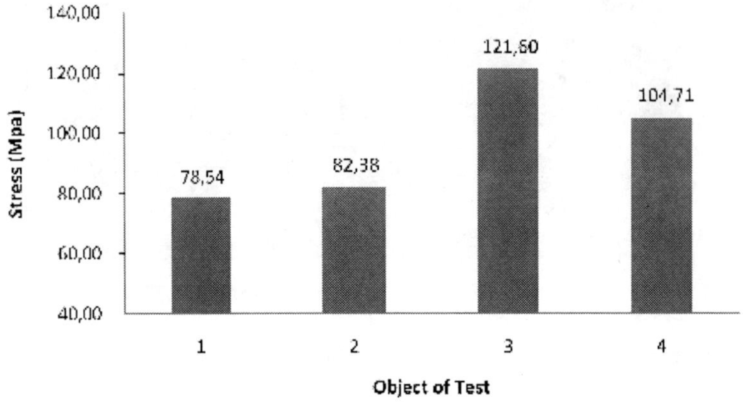

Figure 5. Graph of the relationship between test object loading and stress.

Based on Figure 5 the results obtained are:

a. The highest/maximum stress is 121.60 Mpa on the type 3 of the specimen, which is a hot asphalt mixture with a layer of wire mesh placed 30 mm from the surface of the specimen.
b. The lowest/minimum stress is 78.54 Mpa on the type 1 of the specimen, namely hot asphalt mixture without wire mesh layer.
c. The maximum stress that occurs between the type 1 and type 2 of the specimens has a value that is not much different, namely 78.54 Mpa and 82.38 Mpa.

3.3.2 Strain

Strain is a relative change in the size or shape of an object that is subjected to stress [18, 28, 33]. The magnitude of the strain value according to Table 1 that occurred is described in Figure 6.

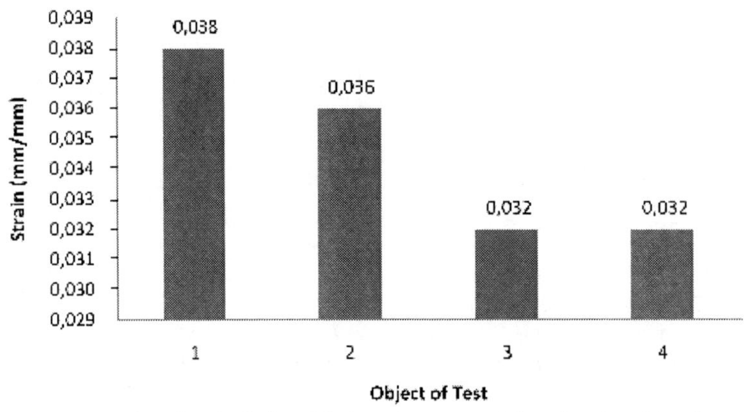

Figure 6. Graph of relationship between test object load and strain.

Figure 6 explains that the results obtained are:

a. The lowest/minimum strain is 0.032 mm/mm on the type 3 of the specimen, which is a hot asphalt mixture with a layer of wire mesh placed 30 mm from the surface of the specimen
b. The highest/maximum strain is 0.038 mm/mm in the type 1 of the specimen, namely hot asphalt mixture without wire mesh coating
c. The value of strain is very dependent on the amount of deflection that occurs
d. The maximum strain value that occurs between type 3 and type 4 of the specimens has the same value because they both have the same deflection amount of 8 mm

3.3.3 Modulus of elasticity

Modulus of elasticity is the ratio between the stress and strain of an object [18, 30]. The modulus of elasticity is also known as Young's modulus and is denoted by E.

$$E = \frac{\sigma}{\varepsilon} \tag{1}$$

E = Modulus of elasticity (MPa atau Psi),
τ = Stress (MPa), and
ε = Strain.

The modulus of elasticity is very important in analyzing the mixed strength of a construction. The high modulus of elasticity indicates that the ability of a mixture is good enough to withstand loads by itself.

Based on Table 1, it will be drawn in Figure 7.

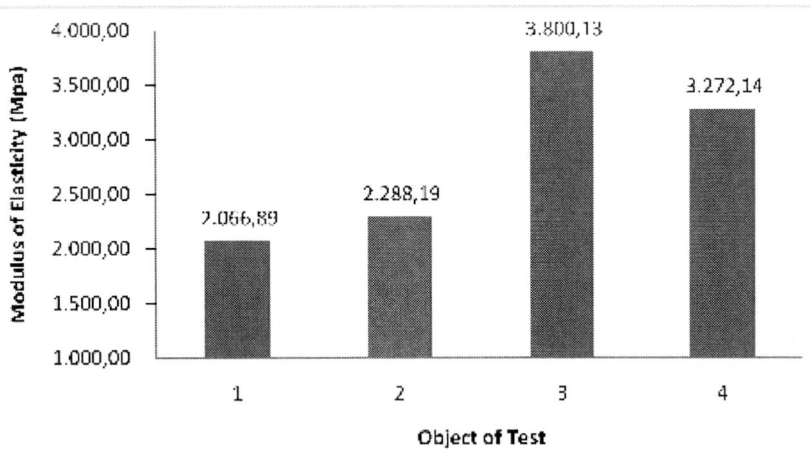

Figure 7. The relationship between test object loading and modulus of elasticity.

Figure 7 explains that the results obtained are:

a. The highest/maximum modulus of elasticity is 3,800.13 MPa in type 3 of the specimen, namely hot asphalt mixture with wire mesh layers placed 30 mm from the surface of the specimen
b. The lowest/minimum modulus of elasticity is 2,066.89 MPa in the type 1 of the specimen, namely hot asphalt mixture without wire mesh coating
c. The value of the modulus of elasticity is very dependent on the value of Stress and Strain
d. The value of the modulus of elasticity also depends on the placement of the wire mesh layers in the hot asphalt mixtures.

4. Conclusion

Based on the Marshall test and three-point bending test, several things can be concluded. The use of wire mesh layers in hot asphalt mixtures can increase the flexural resistance of the asphalt mixture.

The highest load that can be withheld as a flexural resistance value is obtained in the hot asphalt mixture with a layer of wire mesh placed 30 mm from the surface of the test object, namely 291.85 KN. The maximum stress is 121.6 Mpa, the minimum strain is 0.032 mm/mm, the maximum elastic modulus is 3,800.16 Mpa. The amount of deflection that occurs is between 8 mm

The value of the modulus of elasticity is very dependent on the value of Stress and Strain. The increase in voltage value depends on the increase in load held. The value of the strain depends on the amount of deflection that occurs

The modulus of elasticity is directly proportional to stress and inversely proportional to strain. Increasing the value of the modulus of elasticity indicates the high ability of a mixture to withstand loads by itself.

References

[1] Behiry A E A E M 2012 *Ain Shams Eng. J.,* **3 (4)** 367-374.
[2] Suprapto T M 2006 *Bahan dan struktur jalan raya.* 3rd Edition. Yogyakarta: KMTS FT UGM. *(in Bahasa Indonesia).*
[3] Isnanda I, Saleh S M, Isya M 2018 *Jurnal Teknik Sipil,* **1(3)** 637-646. *(in Bahasa Indonesia).*
[4] Razuardi R, Saleh, S M, Isya M 2018 *Jurnal Teknik Sipil,* **1 (3)** 715-724. *(in Bahasa Indonesia).*
[5] Saleh S M, Anggraini R, Aquina H 2014 *Jurnal Teknik Sipil,* **21 (3)** 241-250. *(in Bahasa Indonesia).*
[6] Bukhari R A 2007 *Rekayasa bahan dan tebal perkerasan jalan raya.* Banda Aceh: Engineering Faculty of Universitas Syiah Kuala. *(in Bahasa Indonesia).*
[7] Sukirman S 2003 *Beton aspal campuran panas.* Jakarta: Yayasan Obor Indonesia. *(in Bahasa Indonesia).*
[8] Hassani A. Taghipoor M, Karimi M M 2020 *Constr. Build. Mater.,* **253** 119196.
[9] AASTHO 1990 *Standard specification for transportation materials and methods of sampling and testing.* 15th ed. Washington, DC: AASHTO.
[10] Ismy R, Saleh S M, Isya M, Husaini. 2020. *Int. J. Psycho. Rehab.,* **24 (2)** 502-509.
[11] Afonso M L, Dinis-Almeida M, Pereira-de-Oliveira L A, Castro-Gomes J, Zoorob S E 2016 *Constr. Build. Mater.,* **102** 155-161.
[12] Wikipedia 2019 Material Description list. Available at https://id.wikipedia.org/wiki/Kekuatan_tarik. *(in Bahasa Indonesia).*
[13] Huang Y H 1993 *Pavement analisys and design.* London: University of Ketucky.
[14] Huang Y H 2004 *Pavement analisys and design.* Vol. 2, 2nd Edition. New Jersey: Pearson Prentice Hall.
[15] Saputra E, Putra E, Rachman M 2019 *IOP Conf. Ser.: Mater. Sci. Eng.,* **523 (1)** 012073.
[16] Suryawan A 2005 *Perkerasan beton semen portland (rigid pavement).* Jakarta: Beta Offset. *(in Bahasa Indonesia).*
[17] Saleleubaja S, Warman M, Farni I 2015 *Perencanaan perkerasan kaku (rigid pavement) pada ruas jalan "batas Kota Padang–Simpang Haru.* Thesis, Universitas Bung Hatta, Indonesia. *(in Bahasa Indonesia).*
[18] Lubis A S, Muis Z A, Iskandar T D 2018 *IOP Conf. Ser.: Mater. Sci. Eng.,* **309(1)** 012113.
[19] Elleboudy A M, Saleh N M, Salama A G 2017 *Alexandria Eng. J.,* **56 (3)** 319-326.
[20] Walubita L F, Nyamuhokya T P, Komb, J J, Tanvir H A, Souliman M I, Naik B 2018 *Constr. Build. Mater.,* **191** 726-735.
[21] Implementation Team 2019 *Laboratory Center for National Road Works II.* Medan.
[22] Kim M J, Kim S, Yoo D Y, Shin H O 2018 *Constr. Build. Mater.,* **178** 233-243.
[23] Sesaria D A 2018 P*engaruh gradasi crumb rubber pada karakteristik campuran aspal.* Universitas Sumatera (USU) Repocitory. *(in Bahasa Indonesia).*

[24] Widyantara I G N, Suparma L B, Muthohar I 2018 *Informasi dan Ekspose hasil Riset Teknik Sipil dan Arsitektur,* **14 (1)** 48-61. *(in Bahasa Indonesia).*
[25] Ismy R 2015 *Jurnal Variasi,* **3 (8)**. *(in Bahasa Indonesia).*
[26] The latest list of prices for goods in May 2019 2019 Available at: http://cekhargabahan.com/daftar-harga-bahan-bangunan-terbaru/. *(in Bahasa Indonesia).*
[27] National Standardization Body (BSN) 1989 *SNI 03-1732-1989 Tata cara perencanaan tebal perkerasan lentur jalan raya.* Jakarta: National Standardization Body (BSN). *(in Bahasa Indonesia).*
[28] Correia N S, Zornberg J G 2018 *Geotextiles and Geomembranes,* **46 (1)** 111-120.
[29] Direktorat Jendral Bina Marga 2014 *Spesifikasi Umum Direktorat Jenderal Bina Marga 2010.* Edition Revision 3 Division 6. Jakarta: Ministry of Public Works of Indonesia. *(in Bahasa Indonesia).*
[30] Widodo S, Setyaningsih I 2011 *Simposium Nasional RAPI X FT UMS.* S13-S18. *(in Bahasa Indonesia).*
[31] Wu S, Ye Q, Li N 2008 *Constr. Buildi. Mater.,* **22 (10)** 2111-2115.
[32] Husaini H, Yudha S B Z, Akhyar H 2018 *AIP Conference Proceedings,* **1983 (1)** 030001.
[33] Hasanuddin I, Husaini M, Akhyar H 2018 *IOP Conf. Ser.: Mater. Sci. Eng.,* **352** 1-7.

IOP Conf. Series: Materials Science and Engineering 1087 (2021) 012019 doi:10.1088/1757-899X/1087/1/012019

The unit price of building changes to different design of seismic importance factors

C A W S Mastura[1,*], M Mubarak[2], A Abdullah[2]
[1]Master of Civil Engineering Study Program, Universitas Syiah Kuala, Banda Aceh, Indonesia
[2]Department of Civil Engineering, Universitas Syiah Kuala, Banda Aceh, Indonesia

*Email: widya94@mhs.unsyiah.ac.id

Abstract. The design of a building needs to accommodate the potential for earthquake risk. The function and utilization of buildings are closely related to the risks caused by earthquakes, especially for people using the building. In the design process, load changes due to utilization are accommodated through an index value called seismic importance factors (*SIF*). The application of the index will directly affect the structural dimensions of a building. Thus, the amount of building unit price (*BUP*) needed will also vary according to the function of the building. This research was conducted to analyze changes in *BUP* on some seismic importance factors in a building structure design. Changes in unit prices were explained by applying the price index approach in the form of the building unit price index (*BUPI*). This study was applied to a six-story building using reinforced concrete structures and following earthquake load conditions in four seismic zones in Aceh Province. The *BUPI* was analyzed based on comparing a building unit price (*BUP*) to the reference *BUP* that was set from the building designed with the lowest *SIF* index in the initial seismic zone. The analysis results show that the increase in the *SIF* indices will potentially increase the *BUP* to 6 % (IDR. 105,061.25) for *SIF* 1.00 to 1.25, while the increase in costs for *SIF* 1.25 to 1.50 was 4 % (IDR. 84,507.32). The *BUPI* values have clearly explained how the building unit price increases following the *SIF* indexes increase.

1. Introduction

Earthquake is one of the common natural disasters in Aceh, and the intensity is increasing. High intensity of earthquakes requires building design that can withstand earthquake loads. From the recent earthquake, many types of buildings were damaged or even collapsed. Hence, building infrastructure needs to be able to survive when an earthquake occurs. Structures are designed not only to meet the building requirements but also to consider economic aspects. Costs required for structural components are needed to consider the potential for earthquakes [1, 2-3]. The building is designed in a physical form as the results of construction works, integrate with the place of domicile, and serves as a place for people to carry out various activities, both for residence, religious (worship), business and social activities, as well as culture, and special activities, as stated in Indonesian Regulation No. 28 of the Year 2002 [4]. Therefore, a building must have the resilience and strength to ensure the safety of people who utilized it. The loss of building functions can be caused by the human activity itself or nature beyond previous estimates. Natural building failure due to natural disasters can occur due to earthquakes, volcanoes, floods, fires, and various other types of natural disasters.

The magnitude of the earthquake load will cause changes in building costs. The exact cost will be different based on earthquake zoning and dimensions of the building structure. The structural component is a component that has the most substantial portion of building costs. These components include the construction of foundations, columns, beams, slab, and roof structures [1].

Indonesia is vulnerable to earthquakes, so it is vital to understand the seismic potential in a building. Changes in building loads modify the dimensions of the structural components. The higher risk of seismicity in an area results in the more critical requirement for strengthening structural elements in a building to withstand the earthquake loads. All design of structures must consider the seismic importance factors (*SIF*) based on the risk category of the building (or other structure). *SIF*, applied in the seismic response coefficient (*SRC*) equation, aims to increase yield levels for essential structures (such as hospitals, fire stations, central emergency operations, dangerous facilities, and other essential structures). The use of a *SIF* higher than 1.00 is intended to provide lower inelastic demand for a structure that will result in a lower level of structural and nonstructural damages, which lower the risk to the building and the people using the building. There are three risk criteria for seismic importance factors based on SNI 2847: 2013, i.e., risk categories I and II with a *SIF* value of 1.00, risk category III (high occupancy) with a *SIF* value of 1.25, and risk category IV (important facility) with a *SIF* value of 1.50 [5, 6]. The seismic load is an essential factor that must be considered in cost estimation. The cost estimation process must be carried out by considering aspects of variations in seismic load in a building if planned to be built in a different location [7].

Cost estimation is a predictive process used to determine the size, cost, and price of resources needed by the scope of investment, work, or project choices [8]. Cost estimation results can be used as a reference to cost control so that the possibility of deviations can be reduced as much as possible. The accuracy of the estimation is mainly determined by the availability of information in the form of a relationship between cost estimation patterns or cost models [9]. Cost modeling in previous research was applied as a trend and recognizes the need for long-term construction costs so that it can benefit project stakeholders [10, 11]. Cost modeling is also more accurate to infer location adjustment factors and has the potential to improve industrial construction performance with more accurate and efficient costs [12].

The price index becomes an indicator by referring to the statistical index of the general level of price changes. In previous studies, the price index is used as measurement methods in compiling the price to reflect changes in quality and productivity and the introduction of new products [13]. Then the price index is also used to adjust housing sales [14]. Changes to prices can also be caused by other factors other than the economic factors, as explained in the previous section. Changes in the price of a building are closely related to the potential for earthquake vulnerability in each region. This finding has been shown by the results of the study of the unit price of work [7] and the total price of the building [15]. These changes arise due to structural dimensions and material quantities changes to adjust all loads received by the building [16]. The building unit price (*BUP*) is defined as the cost per square meter for the construction of buildings based on the classification, location, and time of construction [5].

By considering the above background, this research aims to analyze the changing patterns of the needs of building unit prices (*BUP*) using a price index approach referring to the seismic importance factor (*SIF*). The price index represents the unit price of a building and is defined by the term "Building Unit Price Index (*BUPI*)."

2. Object of the research

The location of the design of building structures studied was determined in four cities in Aceh Province. They represent four earthquake zones referring to the Indonesian earthquake zoning map. The object used in the study was a six-story building structure and was classified as a non-simple building as defined by the Minister of PUPR Regulation No. 22/PRT/M/2018 [17]. The building has a dimension of 40 m in length and 11 m in width on each floor and has a total floor area of 440 m^2. The structure of the building was designed using reinforced concrete material. The shape of the building structure is shown in Figure 1.

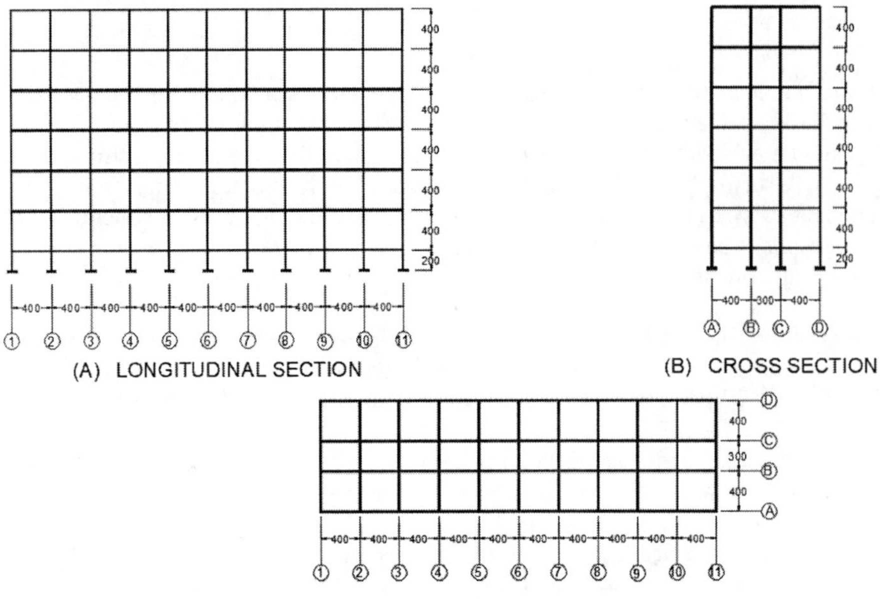

Figure 1. Design of building structures.

3. Data

The data was secondary data obtained from the building's technical manager at the relevant agency, as presented as follows.

1) The spectral response from the observed seismic zones, as can be seen in Table 1.

Table 1. Spectral response (*SR*) values

Zone	*City*	*SR (*in gravity, *g)*
10	Langsa	0.691
12	Sigli	0.873
14	Sabang	1.192
16	Kutacane	1.567

2) The data of the basic price of material and labor, and the unit price of works as referring to the standard prices established by the province government [15].
3) The data of dimension for all reinforced concrete structural components consist of foundation, columns, beams, and slab. This data was obtained from SAP2000 v.20.0.0 International Ultimate Academic software.

4. Analysis

The analysis process was carried out as follows.

1) Building volume analysis in this study was determined from the result of dimensions design for all structural components of the building.
2) The unit price of work is analyzed for three jobs related to reinforced concrete construction: concrete work, reinforcement work, and formwork.
3) The analysis of the cost of building components was based on the *SIF* by adding up all building cost components, both based on the type of reinforced concrete works and the type of structural components.
4) The building unit price (*BUP*) was analyzed based on the total building price of the total building area. The building unit price (*BUP*) can be calculated using Equation (1).

$$BUP = \frac{TBP}{TBA} \tag{1}$$

Where, BUP = The Building Unit Price (in Indonesian Rupiah, IDR);
TBP = Total Building Price (in Indonesian Rupiah, IDR); and
TBA = Total Building Floor Area (in m^2).

5) The analysis of building unit price index based on FUG was done by comparing a unit price value of buildings with the reference price value on buildings in FUG 1.00. The building unit price index (*BUPI*) is one form of price index application to explain the pattern of HSB changes to a variable [18]. This study applies FUG changes to building unit prices. Changes in the building price index refer to one reference unit price (TBP_R) set for buildings designed with FUG 1.00. Mathematically, the building unit price index can be calculated using equation (2).

$$BUP_z = \frac{TBP_z}{TBP_R} \quad (2)$$

Where, $BUPI_Z$ = Building Unit Price index of seismic importance factor indices.
TBP_Z = Unit price of seismic importance factor indices; (in Indonesian Rupiah, IDR).
TBR_R = Reference unit price (rupiah).

5. Results and discussion

5.1 Total costs of structural components

The total cost of the structure is defined as the total cost of all structural components of a building consisting of foundations, beams, columns, and slab. The price for each component changes following the seismic importance factor (*SIF*), as in Table 2. These changes are distributed according to the price of work consisting of concrete work, reinforcement, and formwork.

Table 2. Changes in the cost of structural component.

Structural components	Seismic zone	Component cost based on *SIF* (in IDR Million)		
		1.00	1.25	1.50
Foundation	10	306	306	373
	12	306	373	392
	14	347	414	426
	16	347	426	426
Average		326	380	404
Std. Dev.		24	54	26
Beams	10	1,337	1,386	1,474
	12	1,349	1,410	1,484
	14	1,367	1,493	1,65
	16	1,564	1,689	1,856
Average		1,404	1,495	1,618
Std. Dev.		107	138	218
Columns	10	1,134	1,134	1,173
	12	1,134	1,134	1,195
	14	1,231	1,231	1,430
	16	1,231	1,430	1,430
Average		1,183	1,233	1,307
Std. Dev.		56	140	142
Slab	10	1,853	2,006	2,006
	12	1,853	2,006	2,006
	14	1,976	2,006	2,006
	16	2,006	2,006	2,006
Average		1,922	2,006	2,006
Std. Dev.		81	0	0

Table 2 shows that the average change in the cost of structural components on the foundation with the highest average cost is *SIF* 1.50 (IDR. 404,972,353), while the lowest value is *SIF* 1.00 (IDR. 326,736,425). As with the beam component, the highest average cost change is at *SIF* 1.50 (IDR. 1,618,713,632) and the lowest at *SIF* 1.00 (IDR. 1,404,752,317). The highest value column is also at *SIF* 1.50 (IDR. 1,307,311,091) while the lowest at *SIF* 1.00 (IDR. 1,183,335,363). while for the Slab, the highest is at *SIF* 1.25 and *SIF* 1.50 (IDR. 2,006,439,615, no change between *SIF* 1.25 and *SIF* 1.50), while the lowest cost at *SIF* 1.00 (IDR. 1,922,151,545).

5.2 Total cost of works

The total price of the building is obtained from the sum of costs from all structural components of the building, or in another way, the total price of the building could also be generated from adding up the cost of works. The total cost of works for the three *SIFs* is provided in Table 3. The table indicates that the total cost of concrete work, reinforcement, and formwork increases from each *SIF*, the average cost increase is IDR. 99,602,470.93 (12%), IDR. 305,177,561.44 (14%) and the IDR. 95,681,008.37 (5%) in concrete work, reinforcement works, and formwork work, respectively.

Table 3. The total cost of works.

Seismic Zone	Cost of Works based on *SIF* (in IDR million)								
	Concrete			Reinforcement			Formwork		
	1.00	1.25	1.50	1.00	1.25	1.50	1.00	1.25	1.50
10	820	820	868	1,976	2,179	2,295	1,835	1,835	1,865
12	820	856	873	1,988	2,217	2,330	1,835	1,853	1,876
14	827	873	970	2,261	2,396	2,580	1,835	1,876	1,974
16	834	960	989	2,480	2,638	2,720	1,835	1,954	2,009
Average	825	877	925	2,176	2,357	2,481	1,835	1,879	1,931
Std. Dev.	6,64	59,38	63,39	241,63	209,72	203,51	0,00	52,25	71,52

5.3 Building unit price index

The total building price (*TBP*) was obtained from the sum of costs from all structural components of the building, or from adding up the cost of works. The building unit price (*BUP*) explains the cost requirements of the building structure for each square meter area. The *TBP* and the *BUP* are shown in Table 4.

Table 4. The total building price and the building unit price.

Seismic Zone	Total Building Price (in IDR million)			Building Unit Price (in IDR million)		
	1.00	1.25	1.50	1.00	1.25	1.50
10	4,631	4,834	5,027	1.754	1.831	1.904
12	4,643	4,926	5,079	1.758	1.865	1.924
14	4,923	5,145	5,524	1.864	1.949	2.092
16	5,149	5,552	5,718	1.950	2.102	2.165
Average	4,836	5,114	5,337	1.857	1.937	2.021
Std. Dev.	248	319	337	0.096	0.121	0.128

Table 4 shows that the total price of the building structure changes with every 0.25 increase in the earthquake priority factor. The average price increase for the building structure from *SIF* 1.00 to 1.25 is IDR. 277,361,703 (6%), while the average increase in costs from 1.25 to 1.50 FUG was IDR. 223,099,337 (4%). The building unit price also increases in each 0.25 of the earthquake priority factors. The average increase in *BUP* from FUG 1.00 to 1.25 is IDR. 105,061.25 (6%) while the average unit price increase from FUG 1.25 to 1.50 is IDR. 84,507.32 (4%). The highest standard deviation for the total unit price is at *SIF* 1.50 (IDR. 337,645,553), while for the *BUP*, the highest standard deviation is at FUG 1.50 (IDR 127,896,043).

The change in the value of *BUP* for each type of building based on the parameters of the seismic priority factor is indicated by a price index value. The index is analyzed by comparing the value of a building from each parameter with the reference *BUP* value. The reference *BUP* determined in this study is the *BUP* value at *SIF* 1.00. *BUP* changes for the types of structural components and types of work can be seen in Figure 2 and Figure 3.

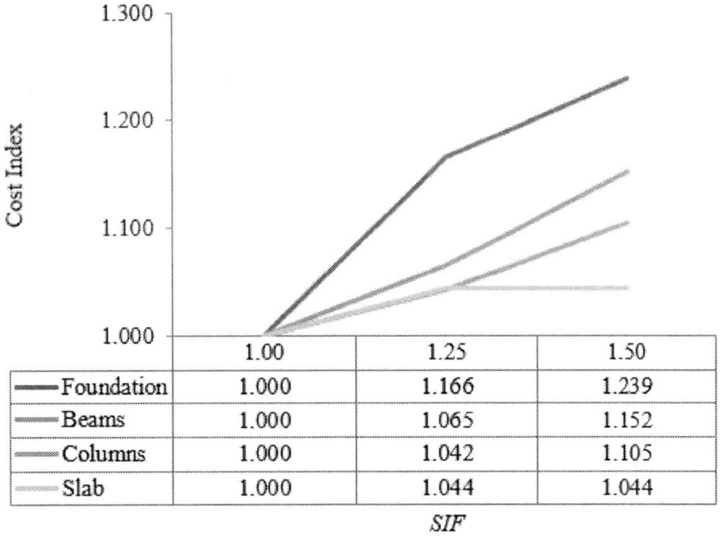

	1.00	1.25	1.50
Foundation	1.000	1.166	1.239
Beams	1.000	1.065	1.152
Columns	1.000	1.042	1.105
Slab	1.000	1.044	1.044

Figure 2. Cost index for structural components.

Figure 2 reveals that the type of structural component increase in the index value for each *SIF*. The lowest *SIF* of 1.00 was set as the reference cost index of 1.000, while for the highest *SIF* of 1.50, the cost index raises to 1.239; 1.152; 1.105; and 1.044 for foundation, beams, columns, and the slab, respectively.

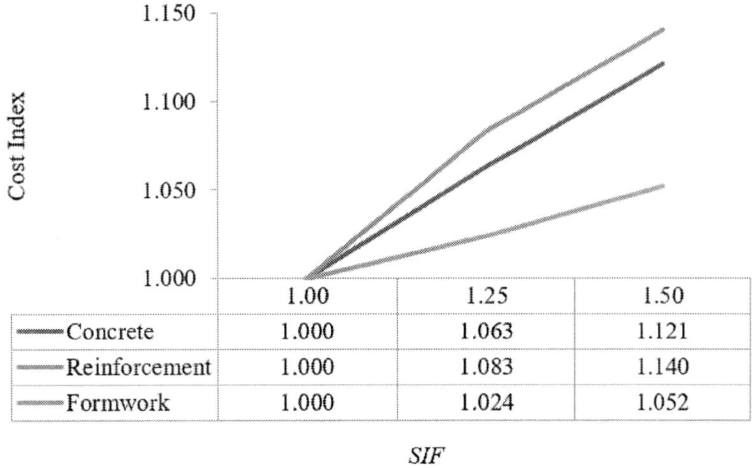

	1.00	1.25	1.50
Concrete	1.000	1.063	1.121
Reinforcement	1.000	1.083	1.140
Formwork	1.000	1.024	1.052

Figure 3. Cost index for the type of works.

According to Figure 3, the changes in the cost of works varies for the three observed *SIF*. The cost index of all works for SIF 1.00 to SIF 1.50 increases from an initial index (1.000) to 1.121 (concrete work), 1.140 (reinforcement work), and 1.052 (reinforcement work). This explains that the higher the seismic priority factor, the higher the cost of building structure.

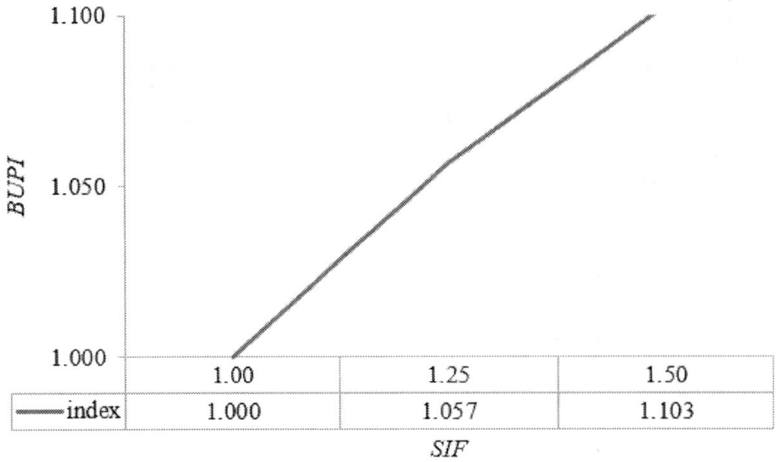

Figure 4. The *BUPI* patterns for changes of *SIF*.

Figure 4 shows that for *SIF* 1.00, the lowest average index value of 1.000, while the index value for *SIF* 1.25 is 1.057, and for the *SIF* 1.50 is 1.103. This shows that the higher the seismic priority factor in a study area, the higher the building unit index value.

6. Conclusions

Based on the results of the analysis and the research objectives, it can be concluded that:

1) The building unit price indices (*BUPI*) indicate the increases in cost in responding to the change of building utilization as indicated by the seismic important factors (*SIF*) index. The increase in the building unit price (*BUP*) for every increase of the SIF index is about 6% for the change of *SIF* 1.00 to 1.25, and 4% for the change of *SIF* 1.25 to 1.50. On average, the cost of the *BUP* of structure potentially rises up to IDR. 81,958 for each 0.25 increase of the *SIF* index.
2) Changes in building functions that are marked the SIF index value directly affect the increase in the cost of the structural component and the cost of works. The largest changes in the cost of structural components are found in the foundation components, followed by beams, columns, and slabs. From the aspect of work costs, the largest change in costs occurs in reinforcing works, followed by concrete and formwork works.

References

[1] Tripoli, Mubarak, Nurisra, Mahmuddin 2018 *IOP Conf. Ser.: Mater. Sci. Eng.,* **352** 012027.
[2] Lozano-Ram´ırez N E, Muñoz-D´ıaz E E 2015 *J. Arch. Eng.,* **21 (2)** 5015001.
[3] AlHamaydeh M, Galal K, Yehia, S 2013 *Earthq. Eng. Eng. Vib.,* **12 (3)** 385-397.
[4] Government of Republic of Indonesia 2002 *Indonesian Regulation No. 28 the Year 2002 concerning the buildings.* Jakarta: Government of Republic of Indonesia. *(in Bahasa Indonesia).*
[5] *ASCE 7 on Minimum Design Loads for Buildings and Other Structures* 2010.
[6] National Standardization Board 2013 *SNI 2847:2013 on The Concrete Structural Requirements for Buildings. (in Bahasa Indonesia).*
[7] Mubarak M, Abdullah A, Azmeri A, Hayati Y 2019 *Advances in Civil Eng.,* 7357913
[8] *AACE on AACE International Recommended Practice No. 10S-90: Cost Engineering Terminology* 2017.
[9] Kerzner H 2017 *Project management: a systems approach to planning, scheduling, and controlling*. New Jersey: John Wiley & Sons.
[10] Swei O 2018 *J. Constr. Eng. Manag.,* **144 (7)** 04018058.
[11] Faghih S A M, Kashani H 2018 *J. Constr. Eng. Manag.,* **144 (8)** 04018075.

[12] Zhang S, Bogus S M, Lippitt C D, Migliaccio G C 2017 *J. Constr. Eng. Manag.,* **143 (1)** 04016087.
[13] Jany-Catrice F 2018 *Cambridge J. Econ.,* **42(4)** 963-986.
[14] Guo X, Zheng S, Geltner D, Liu H 2014 *J. Hous. Econ.,* **25** 20-38.
[15] Mubarak, Abdullah, Riza M, Hayati Y. 2019. *IOP Conf. Ser.: Mater. Sci. Eng.,* **523** 012035.
[16] Mubarak, Abdullah, Riza M, Hayati Y 2019 *Rekayasa Sipil,* **13 (1)** 1-8. *(in Bahasa Indonesia).*
[17] Ministry of Public Work 2018 *Ministry of Public Work Regulation No. 22/PRT/M/2018 on the Construction of State Building.* Jakarta: Ministry of Public Work. *(in Bahasa Indonesia).*
[18] Ostwald P F 2001 *Construction cost analysis and estimating*. New York: Pearson College Division.

An experimental study on bond strength of abaca fiber as natural FRP material

T Saidi[1,*], Z Amalia[1], M Hasan[1], I Hasanuddin[2], W Salvana[3], Akram[4]
[1]Department of Civil Engineering, Universitas Syiah Kuala, Banda Aceh, Indonesia
[2]Department of Mechanical Engineering, Universitas Syiah Kuala, Banda Aceh, Indonesia
[3]Civil Engineering Postgraduate Program, Department of Civil Engineering, Universitas Syiah Kuala, Banda Aceh, Indonesia
[4]Engineering Doctoral Program, Faculty of Engineering, Universitas Syiah Kuala, Banda Aceh, Indonesia

*E-mail: taufiq_saidi@unsyiah.ac.id

Abstract. The use of fiber reinforced polymer (FRP) has been implemented as an alternative for strengthening and repairing methods. Nowadays, the use of natural material for FRP has been developed in order to minimize the disadvantage effects to nature due to synthetic FRP material and economic reason. In this paper, an experimental study was carried out to evaluate the bond strength of abaca fiber as natural reinforced polymer (NFRP) material in reinforced concrete (RC) beams. The test specimen was a beam that had cross-section area of 100 x 100mm^2 and 300mm length. Single rebar was used in this study with 10mm diameter of rebar. Artificial crack was applied in order to consider the initial crack by using cardboard between the concrete. Two externally bonded strengthened beams with a different type of abaca fiber arrangement, bond length, and thickness were applied on the concrete surface. The test was conducted by applying a tension load on the beam until the specimen reach its failure. The results showed that the bond strength decrease as the bond length becomes longer because the maximum load was almost constant for different bond length. The maximum load was approximately around 4 tf for short and long bond length. The compatibility of abaca fiber and rebar was also monitored. Both abaca fiber and rebar able to stand the load compactly. Abaca fiber composite laminate had a similar trend with rebar at the same location where an artificial crack was made. Furthermore, the arrangement and thickness of the abaca fiber composite laminate affected the results.

1. Introduction

Degradation of RC structures can cause by environmental factors, ages, natural hazards, improper initial design, and maintenance. Strengthening and repairing existing RC structures is a common solution to improve and extend its service life. The use of FRP has been implemented as an alternative for strengthening and repairing methods. However, synthetic FRP material has a relatively high cost and may harm the environment due to its difficulties to recycle. In addition, during the production and application process may affect human health [1]. Therefore, the use of natural fiber has been developing nowadays. The advantages of natural fiber such as high tensile strength, environmentally friendly, light, and non-toxicity make it a possible alternative for FRP material [2]. Furthermore, natural fiber has a relatively low cost at production and easy to find in nature. In [2], various natural fibers were explored in order to know its' possibility to produce a composite material. The tensile strength of various kinds of fibers such as sisal, kenaf, jute, and bamboo was found to be approximately 227-700 MPa, 930 MPa, 400-800 MPa, and 140-800 MPa respectively. Meanwhile, abaca fiber has tensile strength of approximately 970 MPa [3]. The strength of natural fiber has a wide range because it depends on the composition of its constituent.

There have been many studies of FRP application in order to update the knowledge [4-6]. Based on their studies, the failure mode of RC structures coupled with debonding failure of FRP laminates. , The bond between FRP and concrete, is one of the important parts to achieve high load capacity. Some factors contribute to bond strength. Primary considerations include mechanical and physical properties on concrete, epoxy quality, surface preparation, application method, FRP material quality, and bond length [7-11]. Indeed, a number of researchers investigate the bond strength of FRP [12-16]. However, the synthetic material of FRP was used in their studies. Therefore, this study aims to evaluate the performance of abaca fiber composite laminate as a natural FRP material bonded to concrete. Influenced parameters such as fiber arrangement, bond length, and thickness of the laminate were discussed.

2. Experimental programme

2.1 Specimen design and material properties of concrete

The specimen used in this study was a beam with cross-section area 100 x 100mm^2 and 300mm length. Single rebar was put in the center of the specimen and had a diameter of 10mm. Artificial crack was made by applying cardboard between the concrete, as can be seen in Figure 1. Strain gauges were applied on the abaca fiber composite laminate every 50mm from the artificial crack and on the rebar. A total of six specimens was made in order to evaluate the effect of bond length, the thickness of abaca fiber composite, and abaca fiber arrangement for bond strength of abaca fiber. The detailed study parameter can be seen in Table 1 and Figure 2.

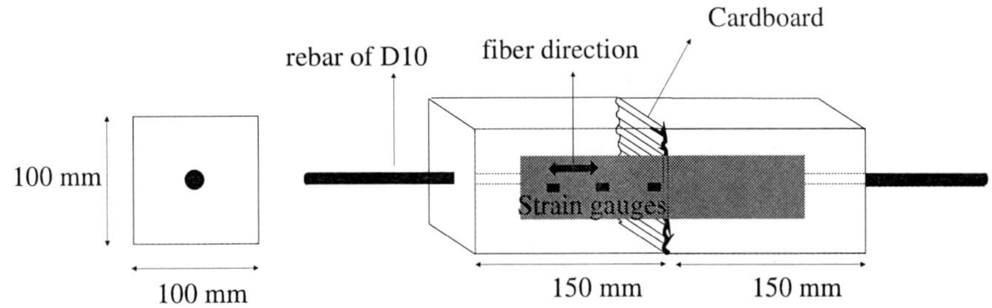

Figure 1. Specimen geometry.

Table 1. Parameter of the study.

No	Specimen	Abaca fiber arrangement	Thickness of abaca fiber composite (mm)	Bond length (mm)
1	SAA-LP1-L	Randomly (Fig 2a)	0.78	120
2	SAA-LP1-S	Randomly (Fig 2a)	0.78	75
3	SAS-LP1-L	In the same direction (Fig 2b)	0.78	120
4	SAS-LP2-L	In the same direction (Fig 2b)	1.875	120
5	SAS-LP2-S	In the same direction (Fig 2b)	1.875	75
6	SAS-LP3-L	In the same direction (Fig 2b)	2.150	120

(a) SAA specimen (b) SAS specimen

Figure 2. Type of abaca fiber arrangement.

The concrete was made using ordinary Portland cement. The mixed proportion is shown in Table 2. The maximum used aggregate had a diameter of 19.1mm. The specimens were cured by sealing all its surface using wet fabric, and it was watered periodically to keep the humid condition for 28 days. A compressive test of the concrete cylinder was conducted before the loading test of the beam to controlled the concrete quality. The compressive strength of the concrete was 58.33MPa.

Table 2. Concrete mixture proportion.

W/C (%)	Unit (kg/m^3)			
	Water	Cement	Coarse aggregate	Fine aggregate
43.0	228	651.43	420.17	560.23

2.2 Fabrication and tensile strength of abaca fiber composite laminates

The hand lay-up method using glass mold was used in this study for fabricating the abaca fiber composite laminate as a natural FRP material. Yukalac 157 was the type of polyester used in this study as an adhesive that had resin and hardener. The adhesive ratio between resin and hardener was 1:1. The fiber was pressed gently during the fabrication process to ensure there is no air void inside the laminate. The drying process was 24 hours. Figure 3 shows the fabricated of abaca fiber composite laminate.

Figure 3. Fabricated of abaca fiber composite laminate

Before fixing the laminates on the beam, tensile strength was conducted. Tensile test specimens were prepared based on [17]. The results show that the average of abaca fiber tensile strength was 17.023MPa, 14.054MPa, and 15.906MPa for SAA with 0.78mm thickness specimen, SAS with 1.875mm thickness specimen and SAS with 2.150mm thickness specimen, respectively.

Abaca fiber composite laminate was fixed on both sides of the beam, as shown in Figure 1. The laminate had 50mm width and length depends on the designed bond length. Before fixing the abaca fiber composite laminate, loose particles at the bonding surface of the concrete beam was removed by using a grinder. Abaca fiber composite laminate was fixed on the concrete surface by using adhesive Sikadur 330. The laminates were set for 24 hours. The test was conducted by applying tension loads on the beam, as can be seen in Figure 4.

Figure 4. Bond strength setting test.

3. Experimental Results and Discussion

3.1 Compatibility of abaca fiber composite laminate and rebar on bond strength

The test results on maximum load and bond strength of abaca fiber composite laminate present in Table 3. All specimens almost had the same maximum load, which resulted in a similar bond strength. The adhesive and resin that was used in the composite laminate affect the strength of natural FRP. In this study, the use of polyester in the matrix composite resulted in a brittle composite material, thus easier to break. It was shown in Table 4 that most of abaca fiber composite laminates had strain yield earlier than rebar. However, the strain yield of SAS-LP1-L and SAS-LP3-L was higher than rebar. The strain yield of SAS-LP1-L and SAS-LP3-L were 61.82$\mu\varepsilon$ and 62.88$\mu\varepsilon$, respectively. The homogeneity of resin and fiber in a laminate greatly affects the results.

Table 3. Test results.

No	Specimen	Maximum load (tf)	Bond strength (MPa)	Failure mode
1	SAA-LP1-L	4.00	6.538	Abaca fiber composite laminate breakage
2	SAA-LP1-S	3.99	10.434	Abaca fiber composite laminate breakage
3	SAS-LP1-L	4.05	6.619	Abaca fiber composite laminate breakage
4	SAS-LP2-L	4.10	6.669	Abaca fiber composite laminate breakage
5	SAS-LP2-S	4.42	11.559	Abaca fiber composite delamination
6	SAS-LP3-L	4.20	6.865	Abaca fiber composite laminate breakage

Figure 5 shows the load-strain behavior of SAS-LP3-L specimen. Abaca fiber composite laminate had a similar trend with rebar at the same location where an artificial crack was made (SR2). It concluded that abaca fiber composite laminate and rebar were compatible. Strain development of SR1 and SR3 was similar because it had the same location 50mm from the artificial crack while SR4 was located 100mm from the artificial crack. SR2 had higher strain due to its location near the crack.

Table 4. Strain yield of rebar and abaca fiber composite laminate.

No	Specimen	Strain yield (με)	
		Rebar	Abaca fiber composite laminate
1	SAA-LP1-L	45.05	38.12
2	SAA-LP1-S	18.44	15.74
3	SAS-LP1-L	38.02	61.82
4	SAS-LP2-L	40.23	34.05
5	SAS-LP2-S	43.11	25.68
6	SAS-LP3-L	34.71	62.88

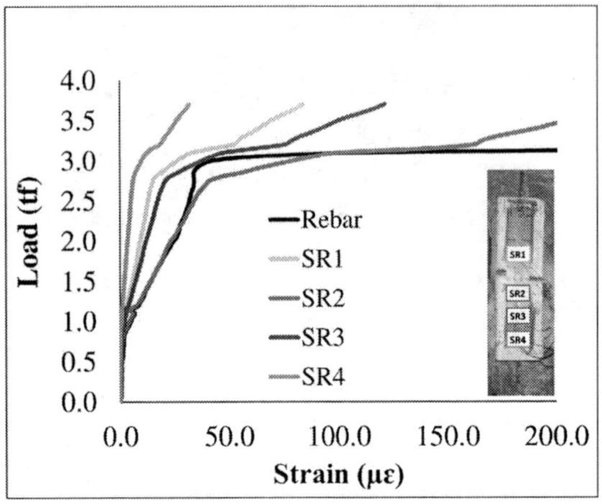

Figure 5. Load-strain behavior of SAS-LP3-L specimen.

3.2 Abaca fiber arrangement of composite laminate effect

Figure 6 shows the load-strain behavior of abaca fiber composite laminate and rebar for SAA-LP1-L and SAS-LP1-L specimens. SAA-LP1-L had higher stiffness compared to SAS-LP1-L. Random distribution of fiber affected the load-strain behavior because most of the load was held by the resin. The consistency of abaca fiber was less in SAA-LP1-L laminate compared to the abaca fiber in SAS-LP1-L laminate, thus made the laminate more brittle than abaca fiber composite laminate in SAS-LP1-L specimens.

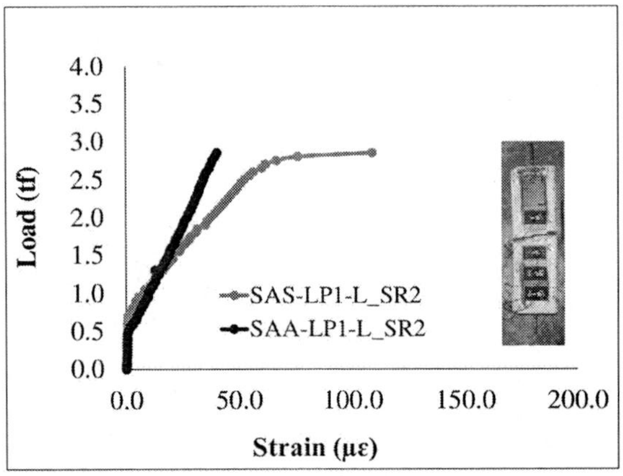

Figure 6. Load-strain behavior of a different type of abaca fiber arrangement.

3.3 Bond length effect

Table 3 informed that the bond strength decreases as the bond length becomes longer because the maximum load is almost constant for different bond lengths. Figure 7 presents the load-strain behavior of SAA-LP1 specimen with different bond length. It was found that shorter bond length showed higher stiffness than longer bond length specimen, although the maximum load was almost the same.

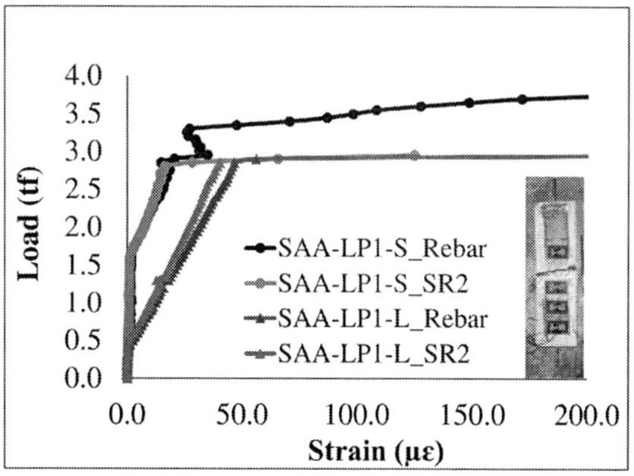

Figure 7. Load-strain behavior of SAA-LP1 specimens.

Strain distribution of SAA-LP1 was shown in Figure 8. In the specimen SAA-LP1 with a bond length of 75mm, the strain distribution shows almost uniform before the abaca fiber composite laminate broke. When applied load is 3tf, the strain increased at 0mm from the crack and linearly decreased at 50mm from the crack. It can be considered that the breakage of abaca fiber composite laminate occurred. In the specimen SAA-LP1 with a bond length of 120mm, strain in the area between 0mm and 50mm had uniform value, and linear distribution is observed in the area between 50mm and 100mm at the early stage of loading. The strain is near to zero as the location far from the crack (100mm). The same behavior was also found in all specimens. Figure 9 shows the strain distribution of SAS-LP3. In the specimen SAS-LP3 with a bond length of 75mm, the strain distribution shows almost uniform at the beginning of loading between 0mm and 50mm. As the load increase, the strain is observed to be increased higher than in the specimen SAA-LP1 at 0mm from the crack and linearly decrease at 50mm from the crack. In the specimen SAS-LP3 with a bond length 120mm, the trend is similar to SAS-LP3-S. It shows that the effective bond length affects the bond strength. The concept of effective bond length is defined through the strain distribution for which the effective bond length is the distance required for the strain to be vanished [12].

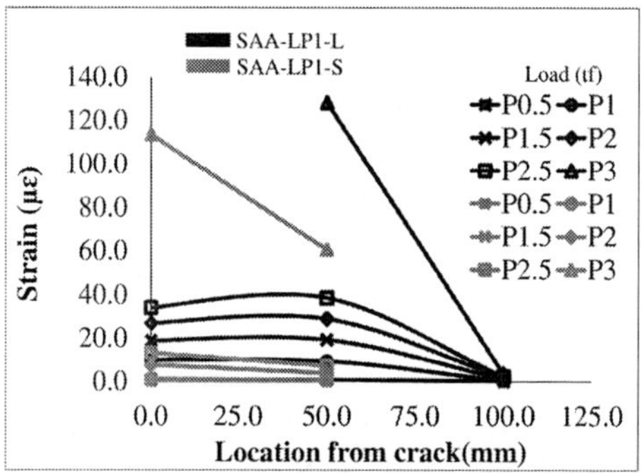

Figure 8. Strain distribution of SAA-LP1 specimens

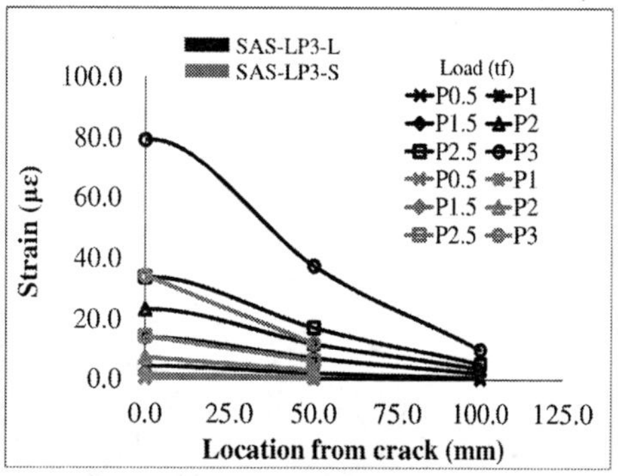

Figure 9. Strain distribution of SAS-LP3 specimens

3.4 Thickness of abaca fiber composite laminate effect

Figure 10 shows the load-strain behavior of specimens with different thicknesses of abaca fiber composite laminate. SAS-LP2-L and SAS-LP3-L had higher stiffness compared to SAS-LP1-L. The thickness of the abaca fiber composite laminate affects the load-strain behavior. However, at the early stage of loading, all three specimens were found to be similar behavior.

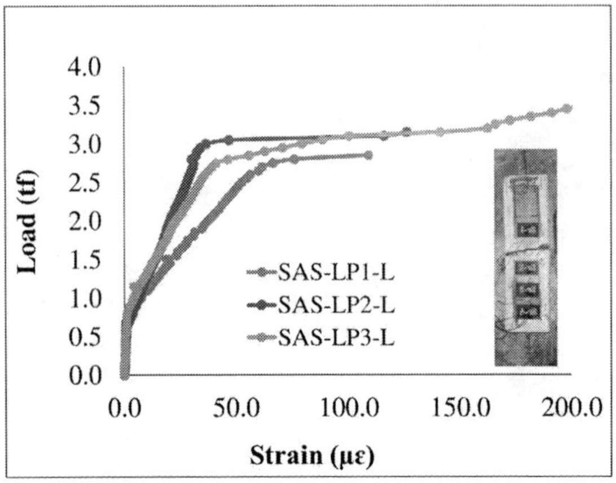

Figure 10. Load-strain behavior of different abaca fiber composite laminate thickness.

4. Conclusion

This paper evaluates the bond strength of abaca fiber composite laminate as a natural FRP material. Both abaca fiber and rebar could work compatibly when load was applied. It was found that the bond strength decrease as the bond length becomes longer because the maximum load was almost constant for different bond length. Maximum load for different bond length was approximately around 4 tf. Effective bond length is required in order to calculate bond strength. Strain distribution for each specimen had the same behavior. The strain value decreased as the location far from the crack. Furthermore, the load and strain behavior of different arrangement and thickness of abaca fiber composite laminate had the same trend at the early stage of loading. However, the stiffness was changed at around 0.5 tf loading. Thus, arrangement and thickness of abaca fiber composite laminate affect the stiffness of the abaca fiber composite laminate.

5. Acknowledgement

The research work of this study was conducted in the Construction and Building Materials Laboratory, Civil Engineering Department of Universitas Syiah Kuala. The authors would like to express their honest gratitude to the staff and Sapriadin for their support of the data collection and processing.

References

[1] Baan R A, Grosse Y 2004 *Mutat. Res. - Fund. Mol. Mech. Mutagen.*, **553 (1-2)** 43-58.
[2] Tong F S, Chin S C, Doh S I, Gimbun J 2017 *Indian J. Sci. Technol.*, **10(2)** 1-5.
[3] Anggoro A T 2012 *Pengaruh Variasi Serat Abaka Terhadap Susut Plastis Fly Ash-Based Geopolymer Mortar*. Thesis, Universitas Gadjah Mada, Indonesia. *(in Bahasa Indonesia).*
[4] Alam M A, Hassan A, Muda Z C 2015 *Mater. Struct.*, **49(3)** 795-811.
[5] Sebastian W M 2001 *J. Struct. Eng.*, **127(7)** 792-798.
[6] Thomsen H, Spacone E, Limkatanyu S, Camata G 2004 *J. Compos. Construc.*, **8(2)** 123-131.
[7] Chen J F, Teng J G 2001 *J. Struct. Eng.* **127(7)** 784-791.
[8] Neubauer U, Rostasy F S 1997 *Proceedings of the 7th Int. Conf. on Struct. Faults Repairs.* **2**.
[9] Wu Z, Islam S M, Said H 2009 *J. Reinforced Plastics and Compos.*, **28 (9)** 2309-2323.
[10] Harajli M H, Rteil A A 2004 *ACI Struct. J.*, **101(1)** 47-56.

[11] Sayed-Ahmed E Y, Bakay R, Shrive N G 2009 *Elect. J. Struct. Eng.,* **9(1)** 45–61.
[12] Diab H M, Farghal O A 2014 *Compos. Part B: Eng.,* **58** 618-624.
[13] Diab H M 2013 *Turkish Online J. Sci. Technol.*, **3(2)** 116–26.
[14] Wu Y, Zhou Z, Yang Q, Chen W 2010 *Eng. Struct.,* **32(3)** 897-905.
[15] Wu Y F, Xu X S, Sun J B, Jiang C 2012 *Compos. Struct.,* **94(11)** 3233-3239.
[16] Caggiano A, Martinelli E, Faella C 2012 *Int. J. Solids Struct.,* **49(17)** 2291-2300.
[17] ASTM D3039 2014 *Standard test method for tensile properties of polymer matrix composite materials*. West Conshohocken, PA: ASTM International.

Trip generation and attraction model and forecasting using machine learning methods

S M Saleh*, Lulusi, F Apriandy, J Fisiani, A Salmannur, R Faisal
Department of Civil Engineering, Universitas Syiah Kuala, Banda Aceh, Indonesia

*E-mail: sofyan.saleh@unsyiah.ac.id

Abstract. The necessity of people and goods to be moved from one place to another place has increased dramatically in recent years. It requires numbers of connectivity among the regions. Besides, policy changes in sea transportation sector including development of sea transport infrastructures as well as shipping/ferry lines to support fulfilling these needs are introduced. The ferry line from Kuala Langsa Port, Indonesia to Penang Port, Malaysia was introduced to encounter the need of mobility between those regions. In consequence, it is important to estimate future transport demand. This study is conducted to familiarize the use of machine learning methods in modelling and forecasting trip generation and trip attraction. Time-series trip generation and attraction data from Kuala Langsa to Penang and vice versa and socio-economic data were employed to develop the model. The result shows that gross domestic regional product (GDRP) and population variables has significant influence to generate trips between these ports.

1. Introduction

The expansion of low-cost air carrier has unlocked opportunities for ever-increasing people and goods movements across regions. With ASEAN-China Free Trade Agreement (ACFTA) in place, the South East Asia (SEA) region has seen border restriction loosen at a previously unseen level. However, since South East Asia Region has abundance of maritime wealth, there is a huge potential to develop seaborne transportation in the region.

United Nations Conference on Trade and Development (UNCTAD) states that approximately 90% of the global trade is attributed to maritime transport. Global ship density analysis by [1] shows that the sea-borne traffic in 2014 has quadrupled that of early 1990s, with the Indian Ocean and Western Pacific Seas experience the largest growth. The rise is related to maritime transport cost reduction, the decrease in the delivery times, and security of shipping [2]. Furthermore, as [3] states, in terms of CO2 emissions per cargo ton, maritime transport is environmentally more friendly. Larger cargo volume, both size and weight-wise, also plays a role.

Transportation in general is susceptible to policy change and technological developments. as a result of these, there is a need to estimate future transport demand. The forecast could be utilized to determine and analyse whether some adjustment, either engineering or management-wise, is needed. A number of factors widely used in transport demand estimation comprises Gross Domestic Product (GDP), population, household expenditure, fuel price, ticket fares, among others. These variables are deemed predominant and play significant roles in reflecting future transport demand. Furthermore, there is an intense correlation between GDP and transport demand, as it is considered to be the most suitable variable to demonstrate transport demand [4].

In this study, the features employed to estimate future trip generation and attraction are population and Gross Domestic Regional Product (GDRP). These variables are deemed exogenous, because they are external to the problem under study. GDRP varies from time to time due to goods and services prices fluctuation and volume changes [5]. The dataset is compiled from Langsa's Centre for Statistics Board (BPS Langsa). Data prior to 2010 are not taken into consideration because, as in line with UN recommendation, The GDP base year calculation has been changed from 2000 to 2010 taking into account such exceptional circumstances as ACFTA, international trade recording system, and capital markets services expansion. Based on the data used, it is considered an economics and time-series based method.

Transport demand modelling predominantly employs statistical approaches most notably regression, gravity model, and time-series method among others [6]. These approaches, which need to be mathematically proven, require the data to satisfy pre-existing assumptions beforehand. Unlike traditional method, machine learning technique exempt the data to follow any presumptive foundations and let the computers to learn from the observations. This offers flexibility for engineers to develop model [7].

Machine learning-based methods can be utilized for descriptive purpose by describing the system current status, predictive objective by predicting the system ensuing condition, or prescriptive aim by proposing system improvement [8]. Machine learning algorithms teaches computers to replicate the simplest way from which human learn: learning from experiences. Researches, such as [8-10], have shown that machine learning approaches often outperforms traditional modelling techniques. A number of most commonly used machine learning method in researches include regression, k-means clustering, Artificial Neural Network (ANN). The growing interest in the use of machine learning methods, in particular ANN, is attributed to its capability to deal with nonlinearities, discontinuities and polynomial aspects [9]. The output of ANN could be a nonlinear function of the variables which is a result of a linear integration of variables in the so-called hidden layers at the first place. It allows the model to learn and adapt by itself [11]. ANN is superior in its ability to train complex models, however a huge amount of data and iteration for training is often a prerequisite [12]. ANN also suits to model greatly varied data, missing data, or data that ignore cause and effect relationship [6].

2. Study area

The study was focused on the international Langsa-Penang sea route, as portrayed in Figure 1. The journey starts from Kuala Langsa Port, Langsa, Aceh, Indonesia, then docks in Penang Port, Penang, Malaysia, and vice versa. It was first opened in 2013, aiming to improve and strengthen bilateral cooperation in politics, economy, social, culture, investment, and tourism. It is the only direct transport mode connecting Langsa and Penang, and the only scheduled passenger ship linking Northern Sumatera Island and Penang Island. The 192-passenger seats Kenangan-3 ferry, owned by Fast Ferry Ventures Sdn Bhd, is the sole ship serving the route. The ship operates twice a week, departing from Penang at 8.30 am on Mondays and Wednesdays, and leaving Langsa every Tuesday and Thursday at 9 am. With a distance of around 300 km, a one-way voyage normally takes 6 to 7 hours, depending on sea and weather condition. A return trip costs Rp. 500,000.

3. Methodology

This study performs linear regression and ANN method to model trip generation and trip attraction given by GDRP and population as independent variables. Both methods are built using Python code. The algorithms are executed in Scientific Python Development Environment (Spyder).

Dmatrices from patsy and statsmodels libraries are employed in linear regression method. Dmatrices is useful in separating independent and dependent variables and adding a constant value in regression analysis. Statsmodels has Ordinary Least Square (OLS) function that can model linear relationship between independent and dependent variables. It returns intercept, coefficient of each independent variable, and important statistical values such as R-squared and p-value. The 5% confidence level test is performed in this study.

ANN uses sklearn neural network's MLP Regressor, sklearn model selection's train test split. MLP Regressor train the data to fit a neural network model using Rectified Linear Unit (ReLU) as activation function for the hidden layer and the family of quasi-Newton methods-based optimizer called 'lbfgs' as the solver for weight optimizer. ReLU activation function returns the output value as $f(x) = \max(0, x)$. ReLU is commonly used to train numerical value-based neural network. Lbfgs explores the most suitable weights for each node in ANN by using iterative method: it trains the model by inserting random weights to initiate training, and then it updates the weights for the next iteration considering previous iteration results until the model converge. Lbfgs works well and converges fast for small dataset. Train test split module split the dataset into two subsets: training and test sets which is necessary in ANN to validate the neural network model. The number of maximum iterations is set manually, but the system will return non-convergence warning should the number is deemed insufficient. Once the model converged, it will return the value of R-squared.

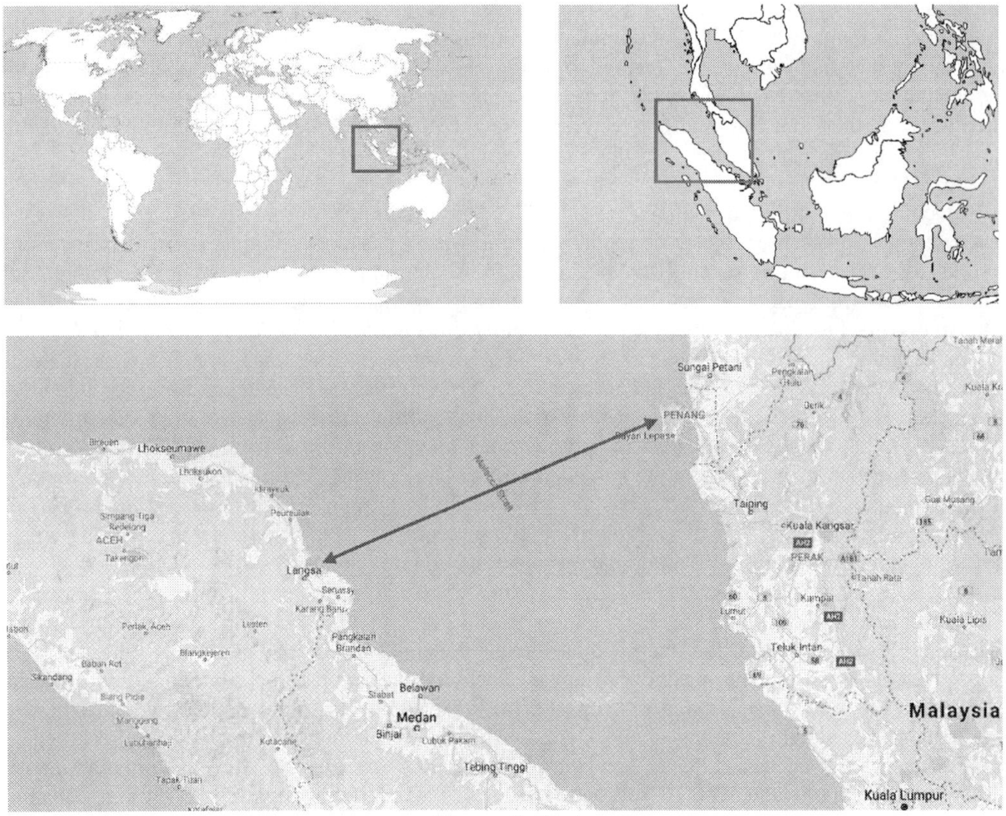

Figure 1. Study area

4. Results

The summary of the linear regression model is shown in Table 1 and Table 2. At first glance, the small number of each independent variable's coefficient appears to indicate its insignificance. However, its p-value is small, suggesting that the independent variables, in this case GDRP and population, do have a strong connection to trip generation and trip attraction. Small coefficient does not necessarily mean meaningless. It might result from an imbalance dataset in the sense that population is a six-digit number, GDRP is a seven-digit number, whereas trip itself is a five-digit number. For every 10000 increase in population, 121 more trips will be generated, and an extra 33 trips will be attracted. On the other hand, one million rupiahs of GDRP growth is expected to result in 2700 and 700 added trip generation and trip attraction respectively. Both trip generation and trip attraction models produce an R-squared of almost 1.

After a few trials in ANN model, setting the maximum number of iterations to 20 does return the model converged. It can be seen from Figure 2 and 3 that the algorithm reaches its convergence at around 17 iterations in trip generation model and 14 iterations in trip attraction model. Fitting the models into the test subset, both models yield an R-squared values of 0.999, similar to those of the linear regression. ANN has no specific coefficient for each variable, because it assigns the weight in hidden layers. The only way to determine whether this ANN model perform well is through the value of R-squared. Hence for simple model, linear regression could be considered perform better because it specifies coefficient for each variable and its attributed statistical values such as p-values.

Table 1. Trip generation models summary

Model	Independent variables	Coefficient	P-value	R-squared
Linear regression	Intercept	-1304.88	0.038	0.999
	Population	0.0121	0.043	
	GDRP	0.0027	0.000	
Artificial Neural Network (ANN)	-	-	-	0.999

Table 2. Trip attraction models summary

Model	Independent variables	Coefficient	P-value	R-squared
Linear regression	Intercept	-358.385	0.038	0.999
	Population	0.0033	0.043	
	GDRP	0.0007	0.000	
Artificial Neural Network (ANN)	-	-	-	0.999

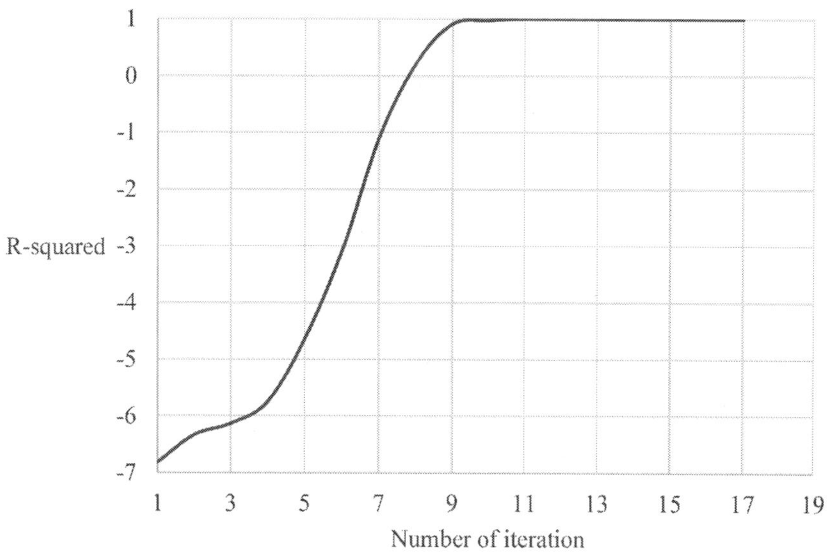

Figure 2. ANN R-squared values for each iteration in trip generation forecasting

Comparing both methods' R-squared values, both linear regression and ANN models fit the data almost perfectly. From statistical perspective, it is considered the ideal model. However, it also hints a possible overfitting. It often happens for a small number of datasets.

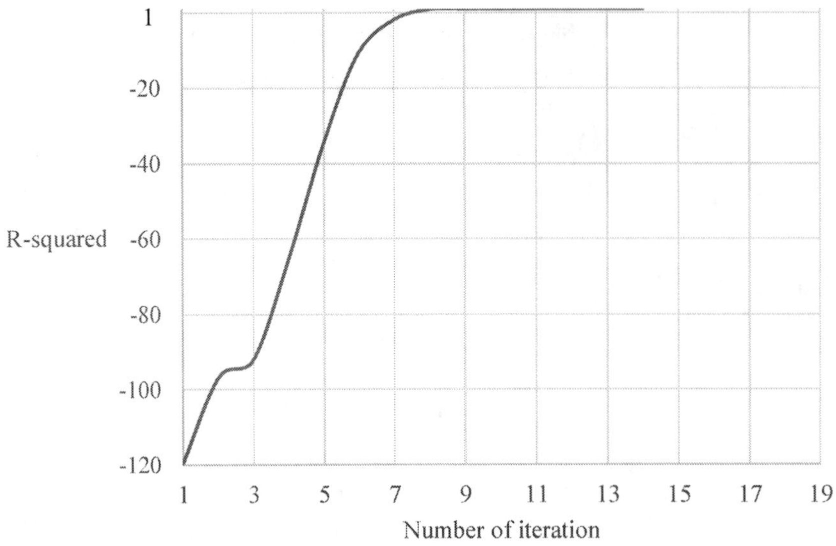

Figure 3. ANN R-squared values for each iteration in trip attraction forecasting

5. Conclusion

In conclusion, the application of machine learning methods in modelling and forecasting trip generation and attraction is confirmed as a beneficial instrument to be used. It helps to analyse the accuracy of developed model by using artificial neutral network (ANN). Socioeconomic data (GDPR and population data) has high influence to determine number of trip generation and attraction between the ports. Given the R-squared resulted, both regression and ANN models perform well in forecasting trip generation and attraction. However, due to the limited number of datasets used in the modelling, there is an indication of a possible overfitting.

References

[1] Tournadre J 2014 *Geophys. Res. Lett.*, **41 (22)** 7924-7932.
[2] Cordón-Lagares E and García-Ordaz F 2020 *Res. Transp. Bus. Manag.*, **10** 100520.
[3] Grant D B, Elliott M 2018 *Ocean Coast. Manag.*, **163** 162-172.
[4] Tsui W H, Fung M K 2016 *J. Air Transp. Manag.*, **50** 1-11.
[5] Langsa's Centre for Statistics Board (BPS Langsa) 2020 *Langsa Municipality in Figures 2020.* Langsa: BPS Langsa.
[6] Profillidis V A, Botzoris G N 2019 *Modeling of Transp. Demand*. Amsterdam: Elsevier.
[7] Zhao X, Yan X, Yu A, van Hentenryck P 2020 *Travel Behaviour and Soc.*, **20** 22-35.
[8] Barua L, Zou B, Zhou Y 2020 *Res. Transp. Bus. Manag.*, **34** 100453.
[9] Vijai P, Sivakumar P B 2018 *Procedia Computer Sci.*, **143** 258-266.
[10] Alekseev K P and Seixas J M 2009 *J. Air Transp. Manag.*, **15 (5)** 212-216.
[11] Liu Y, Zou B, Ni A, Gao L, Zhang C 2020 *Transp. Lett.*, 1-13.
[12] Jiang J, Trundle P, Ren J 2010 *Computerized Medical Imaging and Graphics.*, **34 (8)** 617-631.

Performance assessment of building structures due to Pidie Jaya earthquake using pushover analysis

M Riswandi, M Hasan*, M Afifuddin, R Putra, M Maulana, G R Bilqaini
Department of Civil Engineering, Universitas Syiah Kuala, Banda Aceh, Indonesia

*E-mail: muttaqin@unsyiah.ac.id

Abstract. Pidie Jaya earthquake occurred on December 7, 2016 caused the damage of many building structures. Therefore, it is important to assess the performance of building structures in resisting the earthquake load. In this paper, the performance assessment of two building structures, which were the Multipurpose Hall Building that has been deformed permanently and Prosecutor's Office Buildings that have been experienced some flexural and shear cracks was conducted. The study was started by performing site, design document, and as built drawing assessment. The building structures were modelled as space frames using ETABS software and the data obtained from the site was inputted. The design response spectrum recommended by SNI 1726:2019 was used as spectrum demand. The performance of building structures was assessed by conducting a non-linear push over analysis in two X and Y directions until the structures near to collapse and capacity curves were obtained. By comparing the capacity and demand spectrums in Acceleration-Displacement Response Spectrum (ADRS) format, then the performance points of the structures were obtained based on ATC-40. Based on the inter-story drift at the performance point, then the performance levels of the structures were determined. The assessment results showed that the performance level of the Multipurpose Hall Building structure is Life Safety, while the Prosecutor's Office Building structure is Damage Control.

1. Introduction

According to the National Disaster Management Agency (BNPB), the 6.5 Mw magnitude earthquake experienced in Pidie Jaya Regency with the epicentre at coordinates of 5.25 latitude north, 96.24 east longitude, and a depth of 15 km at 05:03:36 am in West Indonesia Time on Wednesday, 7th December 2016 resulted in 112 people killed by building collapse, more than 600 people wounded and 85,161 people displaced [1-3]. The earthquake also caused the structural failure of 16,238 buildings in Pidie Jaya, Bireuen, and Pidie Regencies [1,2]. It is reported that the most common construction damage due to Pidie Jaya earthquake was failures of reinforced concrete, confined masonry, and timber structures [3]. Therefore, it is necessary to assess the performance of building structures that have been experienced the earthquake.

Performance assessment of building structures can be performed by conducting dynamic time history analysis so that the deformation of the building structures at any time during an earthquake can be obtained [4-7]. However, to perform the time history analysis, it is necessary to have a ground acceleration data during the earthquake. Another simpler method to determine the performance of the structures is to perform a non-linear static pushover analysis [8-14]. This analysis is carried out by

gradually applying lateral loads to the structures until the structures is on the verge of collapse [15, 16].
In this study, pushover analysis has been used for performance assessment of building structures that have experienced the Pidie Jaya earthquake. Two building structures were used as cases in this study, namely the Multipurpose Hall Building and the Prosecutor's Office Building. Those building structures was chosen as the cases in this study because the Multipurpose Hall Building structure has experienced a very large lateral displacement, while the Prosecutor's Building structure has experienced shear and flexural cracks in its floor beams. And what's interesting is that the two buildings have not collapsed yet. Therefore, the purpose of this study is to determine the performance of the structures of the two buildings so that further structural strengthening efforts can be performed.

Table 1. Structural performance level.

Performance Level	Description
Immediate Occupancy	The post-earthquake damage state in which only very limited structural damage has occurred. The basic vertical and lateral force-resisting systems of the building retain nearly all of their pre-earthquake characteristics and capacities. The risk of life-threatening injury from structural failure is negligible.
Damage Control	This term is actually not a specific level, but a range of post-earthquake damage states that could vary from Immediate Occupancy and Life Safety.
Lafe Safety	The post-earthquake damage state in which significant damage to the structure may have occurred but in which some margin against either total and partial structural collapse remain. Major structural components have not become dislodged and fallen, threatening life safety either within or outside the building. While injuries during the earthquake may occur, the risk of life-threatening injury from structural failure is very low.
Structural Stability	This level is the limited post-earthquake structural damage state in which the building's structural system is on the verge of experiencing partial or total collapse. Substantial damage to the structure has occurred, potentially including significant degradation in the stiffness and strength of the lateral force-resisting system. However, all significant components of the gravity load resisting system continue to carry their gravity demands. Although the building retains its overall stability, a significant risk of injury may exist and significant aftershocks may lead to collapse.

Table 2. Deformation limit for the structural performance level.

Interstory drift limit	Performance Level			
	Immediate Occupancy	Damage Control	Lafe Safety	Structural Stability
Maximum total drift	0.01	0.01 - 0.02	0.02	0.33 V_i/P_i
Maximum inelastic drift	0.005	0.005 - 0.015	no limit	no limit

2. Performance of structures

According to ATC-40 [17] the structural performance level can be classified as shown in Table 1. The deformation limit for each performance level is given in Table 2. Maximum total drift is defined as the interstory drift at the performance point displacement, while maximum inelastic drift is defined as the portion of the maximum total drift beyond the effective yield point.

3. Assessment method

The research started with the collection and studying the as-built drawings with a focus on the building size, structural members, and details. Furthermore, field assessments were also conducted, and the structural member dimension was measured by using a meter instrument, permanent structural deformation with a water pass, arrangement and diameter of the reinforcing steel bars using a profometer, while the concrete compressive strength was tested using the non-destructive method with a Schmidt hammer type N/N. Moreover, the data on the tensile strength of the reinforcing and shape steel was obtained from job specifications at the time of building construction.

The building structures were modeled as a space frame using ETABS software. The material properties were then inputted to the software, and a non-linear static pushover analysis was performed. Besides the lateral load, the dead load of the structures and the other components were also considered. The lateral load was applied step by step with the control of deformation at a top level of the structures in two X and Y directions until the structures were on the verge of collapse.

The design response spectrum, as recommended in SNI 1726:2019 [17] with soft soil, was used as the demand spectrum. The soft soil was assumed based on the results of previous studies conducted in the Pidie Jaya region using microtremor analysis, which showed the soil to have a dominant period greater than 0.6 seconds [1]. Moreover, it was also supported by the value of V_s30 found to be between 42.51-178.73 m/s [19] as well as the classification of the SNI 1726:2019 [17]. The capacity curves resulting from the push over analysis and the demand spectrum curve were plotted together in the Acceleration-Displacement Response Spectrum (ADRS) format to obtain performance points.

4. Description of the assessed building structures

4.1 Case 1: prosecutor's office building structure

The type of structure of the building is a reinforced concrete frame. The building was built in 2013 has 3 floors with a building length of 38.00 m, width at 33.50 m, and a total height of 14.00 m. The height of the first, second, and third-floor columns is 4 m each while the size of the columns section is 500 mm x 500 mm with a longitudinal reinforcement of 16D19 mm for each of the floors. The size of the floor beams is 350 mm x 700 mm with the reinforcement in the positive and negative moment areas A_s = 10D19 mm and A'_s = 5D19 mm. The size of the ring beams is 200 mm x 300 mm with the reinforcement in the positive and negative moment areas A_s = 3D16 mm and A'_s = 2D16 mm. The thickness of the floor slab is 12 mm.

Based on the site tests, the concrete compressive strength for the columns is 21.24 MPa, the floor beam is 12.40 MPa, the ring beam is 20.21 MPa, and the floor slab is 20.38 MPa. The yield strengths of reinforcing bars for longitudinal and transversal reinforcements are 360 MPa and 240 MPa, respectively. The structural plan of the building is shown in Figure 1. Due to the Pidie Jaya earthquake, almost all the floor beams have shear cracks, as shown in Figure 2a, as well as flexural cracks, as indicated in Figure 2b.

4.2 Case 2: multipurpose hall building

The type of structure of the building is a steel-concrete composite frame. The building was built in 2014 has 3 floors with a building length of 36.00 m, width at 23.40 m, and a total height of 10.00 m. The height of the first, second, and third-floor columns is 4 m, 4 m, and 2m, respectively. The structural plan of the building is shown in Figure 3. The columns K1' have the size of 500 mm x 500 mm, which are steel WF 400.400.13.20 encased by concrete. The columns K2' have a size of 400 mm x 400 mm, which are steel WF 300.300.10.15 encased by concrete. The beams BL2' have a size of 300 mm x 500 mm, which are steel WF 200.200.8.12 encased by concrete. Based on the site tests, the compressive strength of concrete was 25 MPa. Yield strength and tensile strength of steel are 240 MPa and 370 MPa, respectively, while its modulus of elasticity is 200 GPa. Due to the Pidie Jaya earthquake, the columns in the front row have undergone considerable permanent deformation, reaching 420 mm at the top of the column on the 3rd floor, as shown in Figure 4.

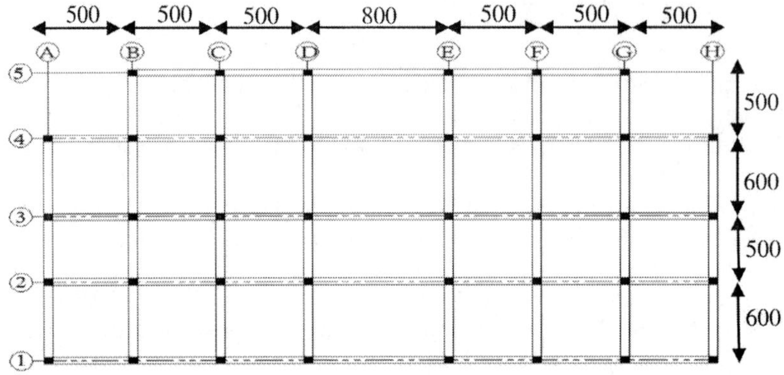

Figure 1. Prosecutor's Office Building structural plan.

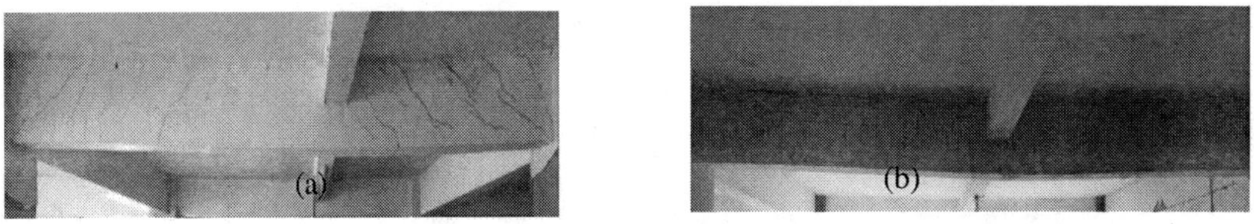

Figure 2. Cracks of the floor beams: (a) shear cracks; (b) flexural cracks.

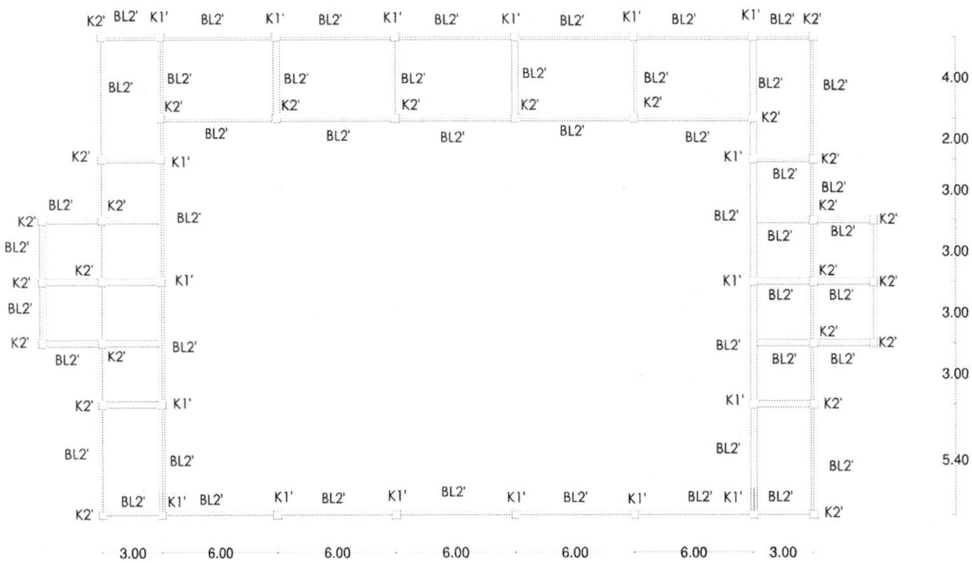

Figure 3. Multipurpose Hall Building structural plan.

Figure 4. Permanent deformation of the building structure.

5. Assessment results

5.1 Case 1: prosecutor's office building structure

Pushover capacity curves in X and Y directions are shown in Figure 5. At the same base shear, the displacement in X direction is greater than in Y direction. This means the stiffness of the structure in X direction is smaller than in Y direction. The maximum base shear was 8,500 kN. The structure gets first to yield at the base shear of 7,100 kN with the displacements of 170 mm and 37 mm in X and Y directions, respectively.

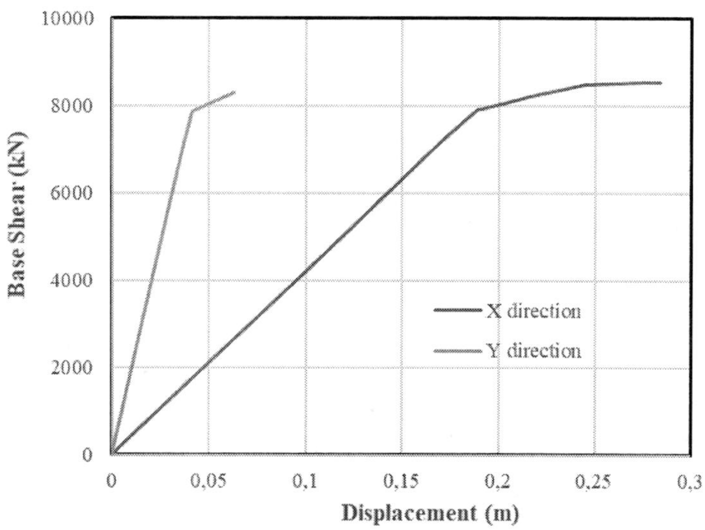

Figure 5. The pushover capacity curves of Prosecutor's Office Building structures.

Since the displacement in X direction is greater than in Y direction, then the pushover capacity curve in X direction was used as the basis to make the ADRS capacity curve. By comparing ADRS capacity curve and ADRS demand curve recommended in SNI 1727:2019 for Pidie Jaya as shown in Figure 6, the performance point was obtained at the spectral acceleration of 0.205 g and spectral displacement of 680 mm. The inter-story drift at the performance point displacement is given in Table 3, and the performance level of the structure can be categorized as Damage Control.

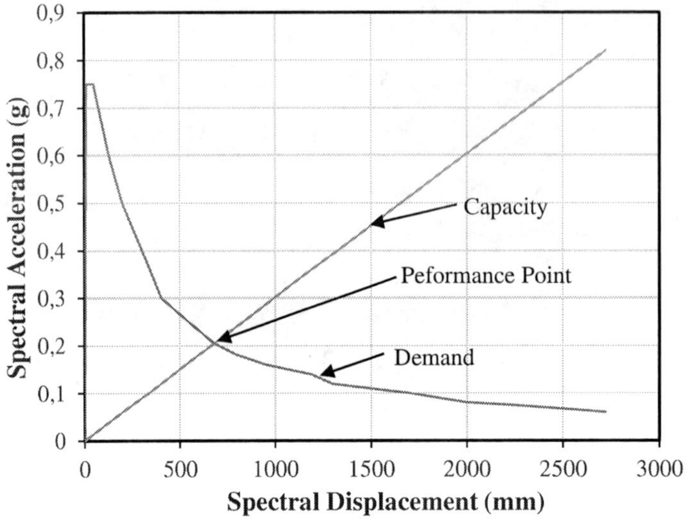

Figure 6. The capacity and demand ADRS curves of Prosecutor's Office Building structure.

Table 3. Interstory drift of Prosecutor's Office Building structure.

Story	Displacement (mm)	Story height (mm)	Interstory drift
1	26	4000	0,007
2	65	4000	0,010
3	105	4000	0,010

5.2 Case 2: multipurpose hall building structure

Pushover capacity curves in X and Y directions are shown in Figure 7. At the same base shear, the displacement in X direction is smaller than in Y direction. This means the stiffness of the structure in X direction is greater than in Y direction. The maximum base shear was 27,500 kN. The structure gets the first yield at the base shear of 9,500 kN with the displacements of 34 mm and 50 mm in X and Y directions, respectively.

Since the displacement in Y direction is greater than in X direction, then the pushover capacity curve in Y direction was used as the basis to make the ADRS capacity curve. By comparing ADRS capacity curve and ADRS demand curve recommended in SNI 1727:2019 for Pidie Jaya as shown in Figure 8, the performance point was obtained at the spectral acceleration of 0.742 g and spectral displacement of 54 mm. The inter-story drift at the performance point displacement is given in Table 4 and the performance level of the structure can be categorized as Life Safety.

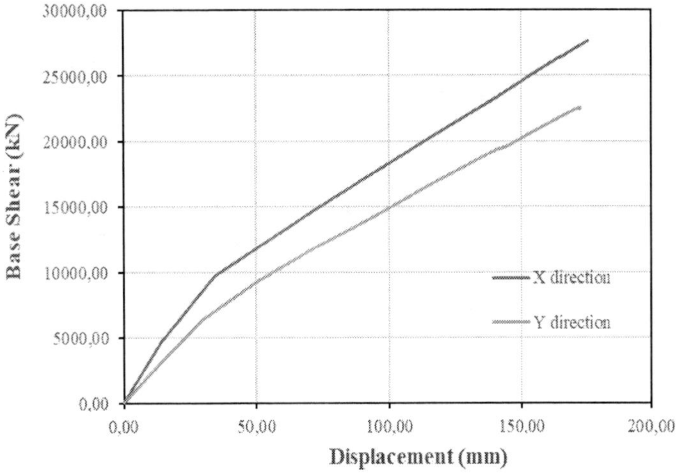

Figure 7. The pushover capacity curves of Multipurpose Hall Building structure.

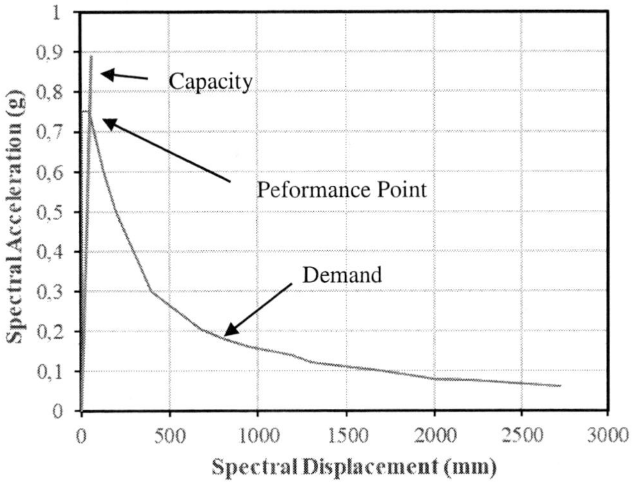

Figure 8. The capacity and demand ADRS curves of Multipupose Hall Building structure.

Table 4. Interstory drift of Multipurpose Hall Building structure.

Story	Displacement (mm)	Story height (mm)	Interstory drift
1	113	4000	0.028
2	244	4000	0.033
3	310	2000	0.033

6. Conclusions

The performance of two building structures subjected to earthquake load was assessed. The assessment was conducted by performing non-linear pushover analysis and the ATC-40 was adopted. The following conclusions can be drawn:

1. The Prosecutor's Office Building structure has no permanent displacement. However, their floor beams have many shear and flexural cracks due to the earthquake. The maximum inter-story drift of this building frame structure is 0.010. The performance level of this building structure is Damage Control.
2. The Multipurpose Hall Building structure has a permanent displacement of 420 mm on the third floor. However, there is no crack on the structure. The maximum inter-story drift of this building frame structure is 0.033. The performance level of this building structure is Life Safety.

7. Acknowledgement

This research was supported by a grant provided by Directorate of Research and Community Service, Ministry of Research and Technology/ National Research and Innovation Agency, Republic of Indonesia (contract number: 070/SP2H/LT/DPRM/2020) and Research and Community Service Center of Universitas Syiah Kuala (contract number: 56/UN11.2.1/PT.01.03/DPRM/2020).

References

[1] Supartoyo, Subandriyo, Amalfi O 2017 *Tanggap darurat gempa bumi Pidie Jaya (frekuensi 1) 6.5 Sr - 7 Desember 2016*. Bandung: Badan Geologi, Kementerian Energi dan Sumber Daya Mineral. *(in Bahasa Indonesia).*

[2] Hasan M, Afifuddin M, Putra P 2020 *IOP Conf. Ser.: Mater. Sci. Eng. (in press).*

[3] Idris Y, Cummins P, Rusydy I, Muksin U, Syamsidik, Habibie M Y, Meilianda E 2019 *J. Earthq. Eng.,* 1-18

[4] Patil A S, Kumbhar P D 2013 *Int. J. Struct. Civil Eng. Res.,* **2(3)** 194-201.

[5] Lestuzzi P, Charif H, Rossier S, Ferriere M, Person J P 2018 *Advances in Civil Eng.,* **2018** 9879101.

[6] Avila J A, Avila-Haro J A 2017 *Key Eng. Mater.,* **754** 349-352.

[7] Lombardi L, De Cula F, Macdonald J 2019 *Eng. Struct.,* **192** 279-295.

[8] Repapis C C. 2016. *The Open Constr. Build. Technol. J.,* **10** 158-179.

[9] Suriyansyah I, Afifuddin M, Hasan M 2020 *Int. J. Civil, Struct. Env. Infrastr. Eng. Res.,* **10 (2)** 1-8.

[10] Bento R, Falcao S, Rodrigues F 2004 *13th World Conference on Earthquake Engineering.* Vancouver, B.C., Canada.

[11] Cardone D 2007 *J. Earthq. Eng.,* **11(6)** 847-875.

[12] Irtem E, Hasgul U 2009 *J. Performance of Constructed Facilities,* **23(6)** 456-466.

[13] El-Ghoulbzouri A, Zakaria E A, Hannoudi S E 2014 *J. Mater. Eng. Struct.,* **1(3)** 146-154.

[14] Saidi T, Aulia T B, Setiawan B, Abdullah N and Hasan M 2018 *MATEC Web Conferences,* **197** 10001

[15] Pan P, Wang T, Nakashima M 2016 *Development of Online Hybrid Testing, Theory and Applications to Structural Engineering*. London: Butterworth-Heineeman.

[16] Khan M A 2013 *Earthquake-Resistant Structures*. London: Butterworth-Heineeman

[17] ATC-40 1996 *Seismic Evaluation and Retrofit of Concrete Building* Volume 1. Applied Technology Council, California Seismic Safety Commission

[18] SNI 1726 2019 *Tata Cara Perencanaan Ketahanan Gempa untuk Struktur Bangunan Gedung dan Non Gedung. (in Bahasa Indonesia).*

[19] Prasetyo D 2017 *Pemetaan Persebaran Nilai Alamiah dan Amplifikasi Menggunakan Metode HVSR dan InVersi HVSR Daerah Pidie*. Lampung: Universitas Lampung. *(in Bahasa Indonesia).*

Shear failure of building floor beams due to December 7, 2016 Pidie Jaya earthquake

A Muyasir[1], M Hasan[2,*] M Afifuddin[2]
[1]Master's Program, Department of Civil Engineering, Universitas Syiah Kuala, Banda Aceh, Indonesia
[2]Department of Civil Engineering, Universitas Syiah Kuala, Banda Aceh, Indonesia

*E-mail: muttaqin@unsyiah.ac.id

Abstract. The 6.5 Mw magnitude earthquake experienced in Pidie Jaya Regency on December 7, 2016 caused the failure of several structures including the District Attorney's Building with almost all the beams on the first floor observed to have shear cracks. Besides that, some flexural cracks were also found, although only in a few places. This study aimed to determine the cause of the shear failure in the building and the amount of deformation observed in the structures due to the earthquake. Therefore, an assessment was conducted in the field to determine the quality of concrete and the adequacy of shear reinforcement installed. A time history analysis was also performed using the Pidie Jaya earthquake history recorded by the Meteorology, Climatology and Geophysics Agency (BMKG). The results showed the required shear strength of the floor beams due to earthquake load was lower than its shear capacity due to a low compressive strength of the concrete in the beam and installation of insufficient shear reinforcement. Furthermore, the flexural strength required in the beams in some places was also greater than the available flexural capacity. The lateral deformations due to the earthquake were also observed to be very large thereby causing the inter-story drift of the building structures to exceed the value allowed by earthquake-resistant building standards (SNI 1726:2012). This subsequently led to the failure of the building floor beams, however, since the strength of the column was greater than the required standard, no crack was observed in the columns.

1. Introduction

The main causes of structural failure during earthquakes include building mass irregularities, low quality of construction materials, soft stories, inconsistent earthquake response, the collision of two buildings constructed side by side due to large lateral deformation, floating columns, unfavourable soil influence, and insufficient foundation [1]. These failures are usually in the form of beam shear failure, building floor collapse due to insufficient strength between the floors, weak column - strong beam, beam-column joint failure, and a very large inter-story drift [2-5]. Moreover, the level of deformation due to insufficient structural stiffness can cause cracking and, with subsequent loads, structural damage [6]. The errors in the choice of an adequate structural system at the design phase also can cause structural failure [4].

The Province of Aceh in Indonesia is located in a very active tectonic zone and apart from the three large plates known globally, there are other nine small ones combined to form the path of meeting complex [7-10]. The 6.5 SR magnitude earthquake experienced in Pidie Jaya Regency with the epicentre at coordinates 5.25 latitude north, 96.24 East longitude, and a depth of 15 Km at 05:03:36 am in West Indonesia Time on Wednesday, 7th December 2016 caused the structural failure of several buildings in Pidie Jaya, Bireuen, and Pidie Regencies. One of these was the District Attorney Building located in the Pidie Jaya Regency Government Office Complex, Manyang Lancok Village, Meureudu District. Almost all the floor beams have shear cracks as shown in Figure 1a as well as flexural cracks as indicated in Figure 1b. The building was built in 2013 with reinforced concrete structures and has 3 floors with a building length of 38.00 m, width at 33.50 m, and a total height of 14.20 m. This means it was only 3 years old at the time of the earthquake.

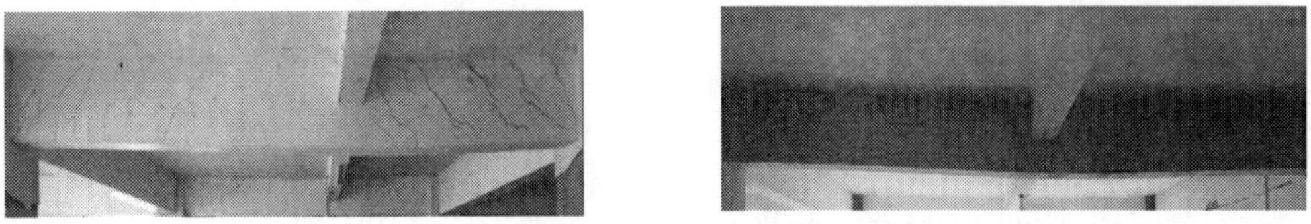

(a) (b)

Figure 1. Cracks of the floor beams: (a) shear cracks; (b) flexural cracks

The purpose of this study was to analyze the causes of the beam's failure by comparing the required structural strength due to earthquake forces and the capacity of the structures. Moreover, the history of deformation magnitude in buildings during earthquakes was analyzed through the provision of the acceleration of ground motion history at the time of the earthquake. Based on the maximum deformation observed, the interstory drift was also evaluated and compared with the value allowed by SNI (Indonesian National Standard) 1726:2012.

2. Research method

2.1 Site assessment

The research started with the collection and studying of the as built drawings with a focus on the building size, structural members, and details. Furthermore, field assessments were also conducted and the structural member dimension was measured by using a meter instrument, permanent structural deformation with a water pass as shown in Figure 2(a), arrangement and diameter of the reinforcing steel using a profometer shown in Figure 2(b), while the concrete compressive strength was tested using the non-destructive method with a Schmidt hammer type N/N as shown in Figure 2(c) [11]. Moreover, the data on the tensile strength of the reinforcing steel was obtained from job specifications at the time of building construction and was found to be 320 MPa for longitudinal reinforcement and 240 MPa for stirrup.

(a) (b) (c)

Figure 2. Structural assessment: (a) measurement of structural deformation; (b) Measurement of reinforcing bar arrangement and diameter; (c) hammer test for concrete compressive

2.2 Structural modeling
The building is modeled as a space frame using SAP 2000 software with the structural plan shown in Figure 3. The height of the first, second, and third-floor columns is 4 m each while the size of the columns section is 500 mm x 500 mm with a longitudinal reinforcement of 16D19 mm for each of the floors. The size of the floor beams is 350 mm x 700 mm with the reinforcement in the positive and negative moment areas A_s = 10D19 mm and A'_s = 5D19 mm. The size of the ring beams is 200 mm x 300 mm with the reinforcement in the positive and negative moment areas A_s = 3D16 mm and A'_s = 2D16 mm. The thickness of the floor slab is 12 mm and modeled as a rigid diaphragm.

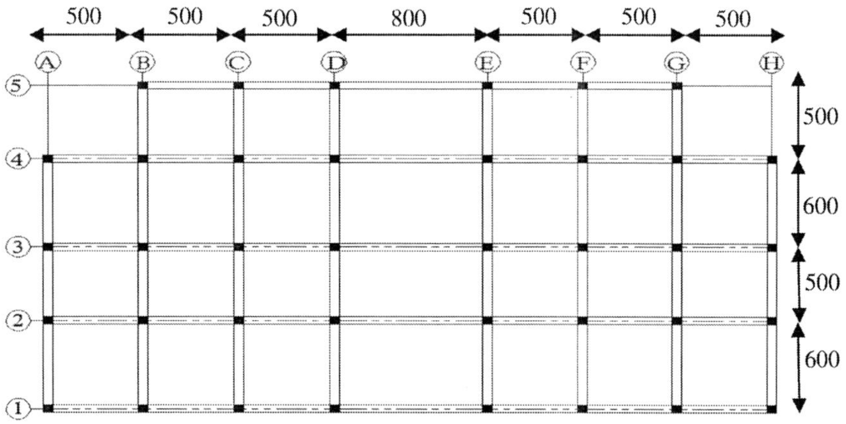

Figure 3. Building structural plan

2.3 Structural analysis
Time history analysis was used to calculate the effect of earthquake loads using the SAP 2000 software and this involved the application of the data on ground movement acceleration recorded during the Pidie Jaya earthquake by the Meteorological, Climatological, and Geophysical Agency (BMKG) as well as the dead and live loads on the structure. Moreover, the concrete and steel strength data were also inputted before analysis and the axial force, shear force and bending moment were obtained afterward. These forces were combined to determine the required strength using different combinations of loads as stated in SNI 1727:2013 standards [12]. Meanwhile, structural cross-section capacity was also calculated based on the SNI 2847:2013 standards [13].

In the time history analysis, the soil on which the structure is built was assumed to be soft based on the results of previous studies conducted in the Pidie Jaya region using microtremor analysis which showed the soil to have a dominant period greater than 0.6 s [14]. Moreover, it was also supported by the value of V_s30 found to be between 42.51-178.73 m/s [15] as well as the classification of the SNI 1726:2012 [16].

3. Results and discussion

3.1 Assessment results
The tests conducted on the field showed the concrete compressive strength for the column was 21.24 MPa, the floor beam was 12.40 MPa, the ring beam was 20.21 MPa, and the floor slab was 20.38 MPa. Meanwhile, the minimum compressive strength for structural concrete according to SNI 2847: 2013 is 17.5 MPa [13]. Therefore, the requirements of SNI 2847:2013 standard were met except for the compressive strength of the floor beams. This means one of the causes of the cracks in the floor beam is the lack of concrete quality during construction.

The steel reinforcement analysis showed stirrups were installed with the same distance along the beam span, at 200 mm. Meanwhile, in the area near the column, the beam receives greater shear force and this requires more stirring. Therefore, the shear failure observed in almost all the beams was also caused by the lack of stirrups in the shear span area. The diameter of the stirrup used was 10 mm which is in accordance to SNI 2847:2013 for building structures to withstand earthquake loads [13]. The reinforcement diameter used on the floor slab was 10 mm and installed every 100 mm distance in accordance with the provisions of SNI 2847:2013 standard, therefore, no crack was found on the floor slab.

3.2 Comparison of required strength and capacity

The required shear strength (V_u) of all loading combinations in SNI 1727:2013 (Comb 1 - Comb 6) was compared with the beam shear capacity (ϕV_n) as shown in Figure 4(a). It was discovered that the shear beam capacity is smaller than the required shear strength in the shear span area, therefore, the failure was due to the inability of the beam to withstand the shear forces caused by the earthquake. However, the small value of shear capacity was observed to be due to the low quality of the concrete and insufficient shear reinforcement (stirrups). Meanwhile, the value was greater than the required shear strength for all load combinations in the middle of the beam span, therefore, no shear crack was recorded in the area.

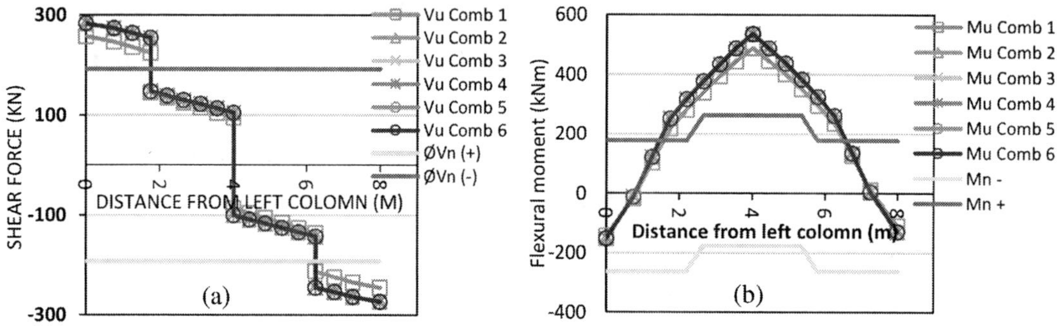

Figure 4. Comparison of: (a) V_u and ϕV_n; (b) M_u and ϕM_n of beam

The required flexural strength (M_u) of all load combinations required in SNI 1727:2013 (Comb 1 - Comb 6) was also compared with beam flexural capacity (ϕM_n) as shown in Figure 4(b). The results showed the beam flexural capacity is smaller than the required flexural strength in the positive moment area and this caused the flexural cracks observed in the middle of the beam span due to its inability to withstand the flexural moments experienced during the earthquake. Moreover, the small flexural capacity of the beam was observed to be due to the low quality of the concrete and inadequate longitudinal tensile reinforcement in the middle of the beam span. However, a contrasting result was obtained at the negative moment part with the beam flexural capacity found to be greater than the required flexural strength for all load combinations. Therefore, no flexural crack was recorded in this area.

The capacity of the column cross-section to withstand a combination of axial compressive load (ϕP_n) and bending moment (ϕM_n) is illustrated using an interaction diagram in Figure 5. The combination of the required axial load (P_u) and the required bending moment (M_u) due to the earthquake for all loading combinations was also plotted on the diagram. The results showed the column cross-section capacity was much greater than the required strength to withstand earthquake loads. Therefore, during the Pidie Jaya earthquake, there was no failure in the column structures.

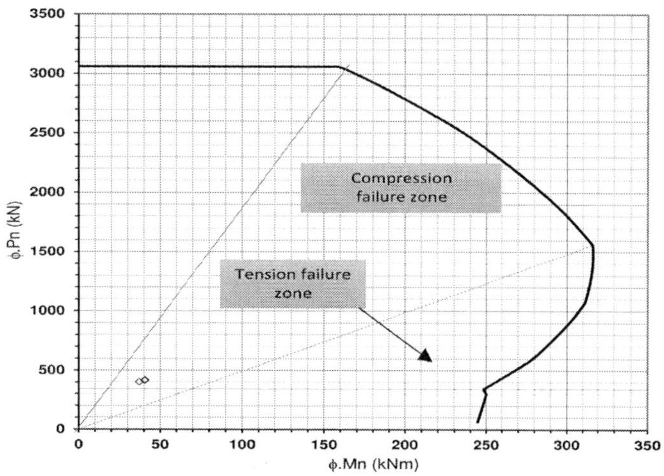

Figure 5. $\phi P_n - \phi M_n$ interaction diagram of the column

3.3 Lateral displacements

The lateral displacements in the X and Y directions at the top of the building structure during the earthquake are shown in Figure 6. The results showed the lateral displacement in the Y direction was much greater than the X direction. The maximum value at the X direction was observed in the 21st second to be 9.88 mm while it was 25th second in the Y direction at 147.50 mm. This shows the stiffness of the building structure in the X direction is much greater than in the Y direction.

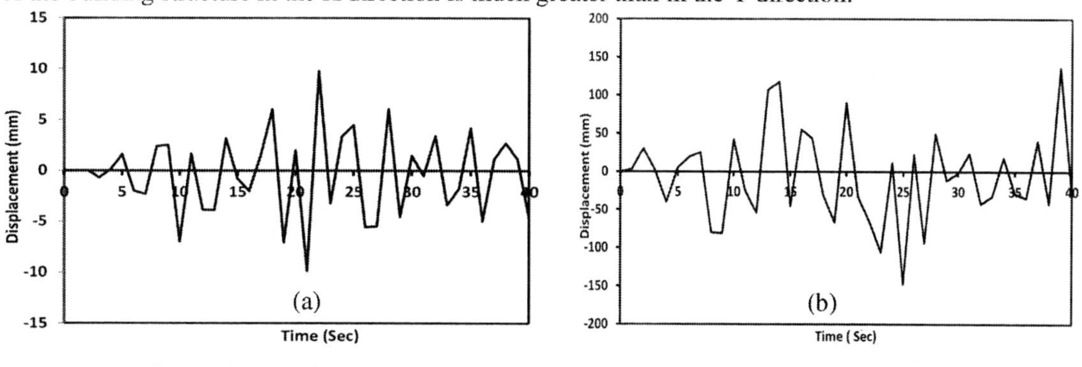

Figure 6. Lateral displacement: (a) in the X direction; (b) in the Y direction

3.4 Inter-story drift

The inter-story drift at 1st, 2nd and 3rd stories in the X and Y directions for all loading combinations, 1 – 6 together with limit value permitted by SNI 1726: 2012, are shown in Table 1. This shows inter-story drift of this structure satisfied the requirement of SNI 1726: 2012 except at the first story in Y direction.

Table 1. Inter-story drift

Story	Inter-story drift (%)		
	X direction	Y direction	SNI 1726:2012 limit
1	0.074	2.178	2.000
2	0.028	2.000	2.000
3	0.095	0.099	2.000

4. Conclusions

A time history analysis was conducted to study the shear failure of beams and deformation of building structures due to December 7, 2016 Pidie Jaya earthquake and the results produced the following conclusions:

1. The shear failure of the floor beams due to the earthquake from the Pidie Jaya District Attorney's Building was due to the very small shear capacity of the beam compared to the shear forces due to the earthquake. However, the small value was due to the low quality of the concrete and lack of shear reinforcement in the shear span of the beam.
2. Flexural cracks were also discovered in the beam due to the smaller value of the beam's flexural capacity compared to the flexural strength required during the earthquake in the positive moment area. The small value was observed due to the low quality of the concrete and lack of longitudinal tensile reinforcement in the middle of the beam span.
3. The capacity of the column cross-section due to a combination of axial compressive force and bending moment is much greater than the load on it due to the earthquake, therefore, there was no column failure.
4. Interstory drift of this building structure exceeded the limit permitted by SNI 1726: 2012, therefore, it is not safe for use and requires structural strengthening.

References

[1] Agarwal P, Thakkar S K, Dubey R N 2002 *ISET J. Earthq. Technol.,* **34** **(3)** 195-217.
[2] Celep Z, Erken A, Taskin B, Ilki A 2011 *Eng. Failure Anal.,* **18** 868-889.
[3] Cosgun C, Dindar A A, Seckin E, Onen Y H 2013 *Gradevinar.,* **65** 743-752.
[4] Damci E, Temur R, Bekdas G, Sayin B 2015 *Case Studies in Construc. Mater.,* **3** 112- 131.
[5] Hasan M, Afifuddin M, Putra R 2019 *Proceed. of the 2nd Aceh Int. Symp. in Civil Eng.,* Banda Aceh.
[6] Hasan M, Husaini, Abdullah N M 2019 *IOP Conf. Ser.: Mater. Sci. Eng.,* **523** 012037.
[7] Bird P 2003 *Geochemistry, Geophysics, Geosystem,.* **4** **(3)** 1027.
[8] Milson J, Masson D, Nicholas D, Sikumbang N, Dwiyanto B, Parson L, Kallagher H 1992 *Tectonics,* **11** 145-153.
[9] Saidi T, Hasan M, Fakhrurrazi 2019 *IOP Conf. Ser.: Mater. Sci. Eng.,* **523** 012039.
[10] Saidi T, Aulia T B, Setiawan B, Abdullah N, Hasan M 2018 *Matec Web of Conferences,* **197** 10001.
[11] ASTM C808/805M-18 2018 *Standard test method for rebound number of hardened concrete.*
[12] SNI 1727:2013 2013 *Beban minimum untuk perancangan bangunan gedung dan struktur lain. (in Bahasa Indonesia).*
[13] SNI 2847:2013 2013 *Persyaratan beton struktural untuk bangunan gedung. (in Bahasa Indonesia).*
[14] Supartoyo, Subandriyo, Amalfi O 2017 *Tanggap darurat gempa bumi Pidie Jaya (frekuensi 1) 6.5 SR - 7 Desember 2016.* Jakarta: Badan Geologi, Kementerian Energi dan Sumber Daya Mineral. *(in Bahasa Indonesia).*
[15] Prasetyo D 2017 *Pemetaan persebaran nilai alamiah dan amplifikasi menggunakan metode HVSR dan InVersi HVSR daerah Pidie.* Lampung: Universitas Lampung. *(in Bahasa Indonesia).*
[16] SNI 1726:2012 2012 Tata cara perencanaan ketahanan gempa untuk struktur bangunan gedung dan non gedung. *(in Bahasa Indonesia).*

Mapping of the depth of hard/dense layer at Banda Aceh-Indonesia and the surrounding areas

H Yunita[1,*], N Al-Huda[1], T Saidi[1], D Sundary[1], H Gunawan[1], A Yulianur[1], B Setiawan[2], F Adrian[2], D Sartika[2], M S Oucherif[3]
[1]Department of Civil Engineering, Universitas Syiah Kuala, Banda Aceh, Indonesia
[2]Program Study of Geological Engineering, Universitas Syiah Kuala, Banda Aceh, Indonesia
[3]Geoexplo, Cheraga, Algiers, Algeria

*E-mail: halida_yunita@unsyiah.ac.id

Abstract. The 26 December 2004 disaster caused enormous casualties and damages in affected regions. Indonesia suffered a loss of 42.7 trillion Rupiah. On the other hand, this disaster has raised awareness among some people to carry out construction following engineering principles, so that the newly constructed buildings have an adequate level of safety. A hard/dense soil layer has a close relationship with the safety level of a building. This paper proposes a map of the depth of hard/dense (stable) layer at one of the worst affected cities, i.e., Banda Aceh (Indonesia) of the boxing day 2004 disaster. A static cone penetration test (CPT) was employed to determine the depth of stable soil. The data from the 35 cone penetration tests were used to develop the zoning map of the depth of the stable soil in the study location. The dominant hard/dense soil depth in the study location is > 10 meters below the existing ground level. In certain places, such as the areas around Lambung, Neusu Aceh, Lambaro Skep, Jeulingke, Doi, Kopelma Darussalam, and Lamjame, the depth of stable soil is from 5 to 10 meters.

1. Introduction

The earthquake and tsunami disaster of 26 December 2004 resulted in a large number of casualties and structural damages [1, 2]. It was reported 127,054 missing and 78,395 deaths in the Provinces of Aceh and Nias Island, North Sumatra. Economically, Indonesia suffered a loss of 42.7 Trillion Rupiah in which 78% destroyed non-public assets. As a result of this disaster, it was estimated that the loss of 2.7% of the Indonesian National Gross Regional Domestic Product (GRDP) or more than 97% of the GRDP of Aceh Province. There was another psychological impact that cannot be measured in real terms [3].

The positive side impact that arises from this disaster is the emergence of public awareness in some parts of the community to carry out construction in accordance with engineering principles [1, 2]. Thus, the newly constructed house/building that they live in has to be built at an adequate level of safety in the event of a disaster. This awareness arises because, during the 2004 tragedy, there were a lot of structural and non-structural failures in their house/building.

An adequate soil bearing capacity is closely related to the subsurface conditions of an area. In a simple way, sufficient bearing capacity will be found on hard/dense soil. The foundation on which a building is based will be placed on a layer of soil that is quite hard or stable [4]. Therefore, mapping the depth of hard/dense soil is very important to do [5].

2. Testing programme

2.1 Quasi-static CPT test

A soil investigation was carried out to map the depth of the hard/dense soil layer. The cone penetration test (CPT), called a quasi-static cone penetrometer, was used in this study. The CPT that was widely used in Indonesia was only the penetration of a static cone (sondir) (Dutch Penetrometer/Dutch Deep Sounding Apparatus/Quasi-Static Penetration Test). This sondir is a static device originating from the Netherlands, where the tip is pressed directly into the ground so that no drill holes are needed [4]. The tip is conical (cone) connected to a series of the internal handlebar, and the outer sleeve (static cone penetration pipe) is pressed into the ground with the help of a frame and a jack anchored to the ground (Figure 1a). There are two types of penetrometer tips that are commonly used, namely the standard type (single cone) and the cone with additional friction sleeve or adhesion jacket type (double-cones). In this study, we used the static cone penetration tool of double-cones type, as shown in Figure 1b [6].

The cone tip of this quasi-static CPT device is a 60^0 cone with a cross-sectional area of 10 cm^2 [6], [7, 8]. In the measurement, the cone tip is pressed downwards with a series of the inner handlebar and outer blanket. The forces required to push the cone without- and with- the sleeve down for 20 cm are measured by a gauge placed on the jack frame at ground level. After the measurement has been taken, the cone, handlebar, and outer blanket are advanced to the point (depth) where the next measurement is made by pressing the outer pipe-casing only. This pressing will automatically return the whole cone to a position ready for the subsequent measurement [6].

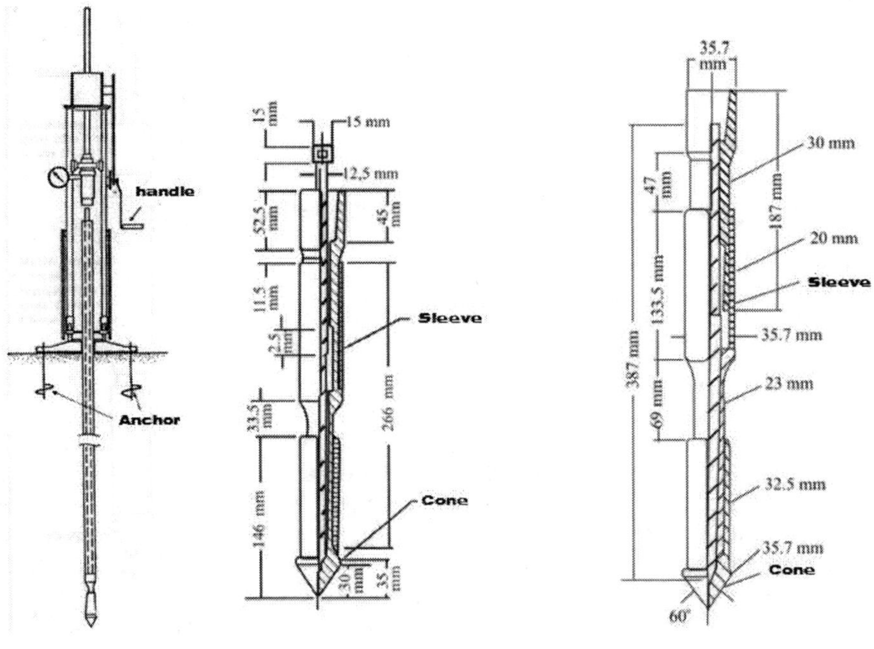

(a) (b)

Figure 1. (a) Installation of the quasi-static CPT for the measurement, and (b) the cone with additional friction sleeve or adhesion jacket type (double-cones) [6-8].

2.2 Spatial testing distribution

The primary target location of this study is in the city of Banda Aceh (Indonesia), which was severely affected by the boxing day 2004 disaster [1]. This city has a total area of 61.36 km^2. The object of the present study is the subsurface soil of Banda Aceh and the surrounding areas. The spatial testing distribution overlaid in Banda Aceh city map can be seen in Figure 2. A total of 35 points of quasi-static CPT was used in this study. Field testing activities are demonstrated in Figure 3.

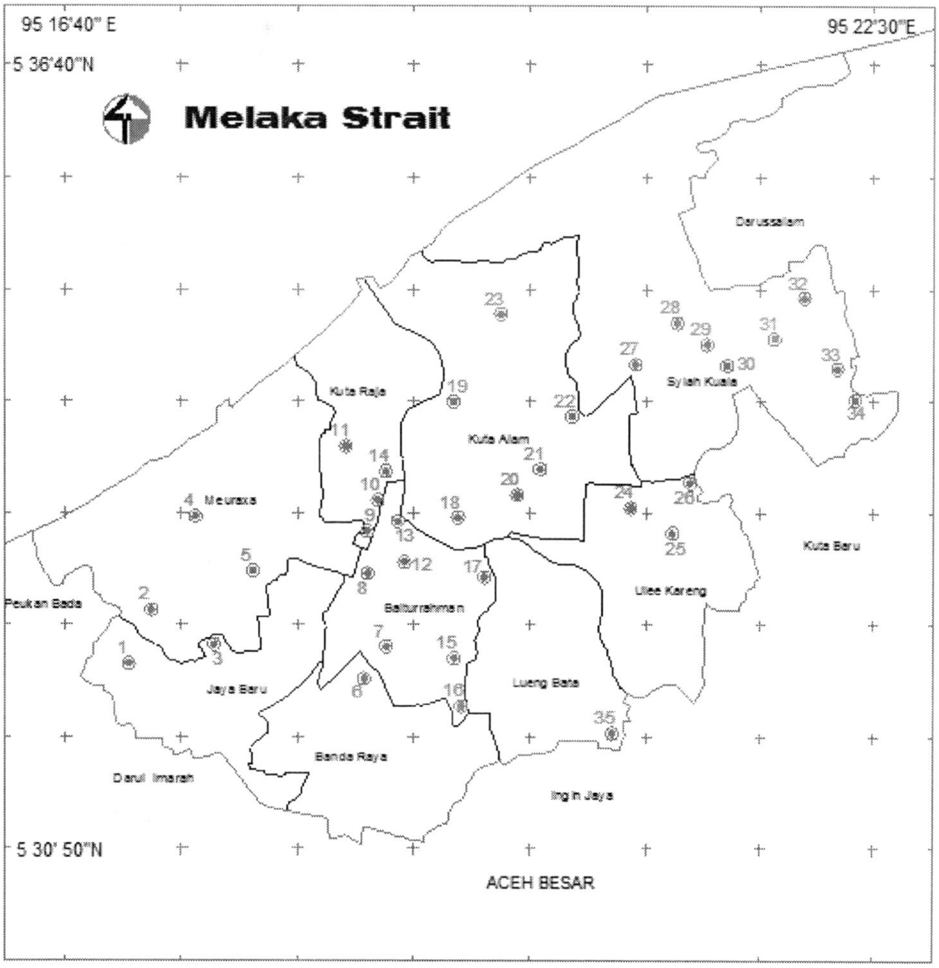

Figure 2. Static cone penetration test points at the research location

Figure 3. CPT field testing activities

2.3 Hard/dense layer criteria

Based on the results of the cone penetration test, i.e., cone resistance data, the relative density/consistency of the soil layer can be interpreted. In the case of granular soil, the interpretation of the relative density can be from very loose to very dense, as shown in Table 1 [7]. For cohesive soil, the classification by [9] can be applied. The justification of cohesive soil consistency is shown in Figure 4.

In the Indonesian common practice [4, 6], the quasi-static CPT is terminated if three successive cone resistance readings reached 150 kg/cm^2. This 150 kg/cm^2 cone resistance can be interpreted as a hard/dense layer [10]. Therefore, this 150 kg/cm^2 is adopted in this study to justify the depth of the hard/dense layer.

Table 1. Interpretation of soil relative density based on cone resistance

No	Granular soil [7]		Low plastic cohesive soil [11]	
	Cone resistance/ CR (kg/cm^2)	Relative density class	Net cone resistance (KPa)	Consistency
1	0 – 16	Very loose	<275	Very soft
2	16 – 40	Loose	275 – 500	Soft
3	40 – 120	Medium	500 – 930	Medium stiff
4	120 – 200	Dense	930 – 1800	Stiff
5	>200	Very dense	>1800	Very stiff/hard

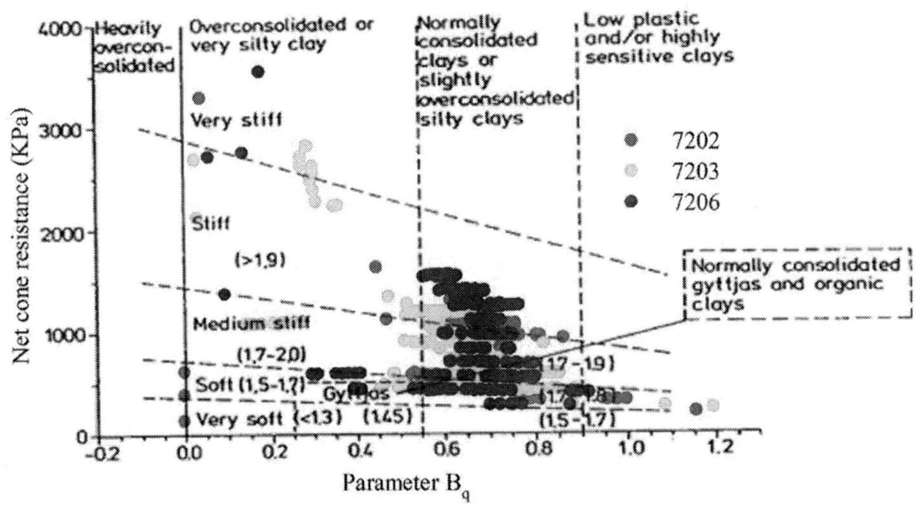

Figure 4. Classification of cohesive soil consistency [9,11]

3. Testing results and discussion

3.1 Results

Data from the reading of the manometer on the CPT tool are cone resistance (CR) and total resistance (TR), which are expressed in units of kg/cm^2. From the two reading values, the sleeve friction (SF) is calculated. Using the hard/dense layer criteria, as mentioned above, the depth of the hard/dense level in each tested location is justified. The depth of the hard/dense soil layer at the study site is tabulated in Table 2.

Table 2. Depth of hard/dense layer of the present study

No	Location name	Depth of hard/dense layer (m)
1	SDN 1 Lamjame	3.2
2	SDN Asoe Nanggroe	>10
3	Jembatan Bitai	>10
4	SMP 5 Banda Aceh	5.2
5	SDN 95	>10
6	SMP Percontohan	>10
7	SDN 96 Banda Aceh	3.6
8	Kejaksaan Tinggi Banda Aceh	6.6
9	SD Muhamadiyah Banda Aceh	>10
10	Kantor Pos at Merduati	>10
11	SDN 17 Banda Aceh	>10
12	Kantor Pemkot Banda Aceh	>10
13	Ruko di Moh. Jam	>10
14	SDN 6 Banda Aceh	>10
15	SDN 64 Banda Aceh	>10
16	Jembatan Ateuk Jawo	>10
17	SDN 30 Banda Aceh	7.0
18	Bappeda NAD	5.0
19	Museum Tsunami	>10
20	Perluasan Gedung Kantor Imigrasi	>10
21	Jamsostek	>10
22	SMP 2 Banda Aceh	>10
23	SDN 45	7.0
24	Asrama Mahasiswa	5.0
25	SDN 76	5.2
26	Perumahan DPRA	>10
27	Kanwil HAM	7.8
28	Rumah Dinas POLDA	6.4
29	Klinik Putro Phang	>10
30	Ruko at Rukoh	5.6
31	Asrama Mahasiswa	>10
32	SDN 83 Banda Aceh	>10
33	Hyugo Perfecture Building	>10
34	Gedung Kuliah Pasca Sarjana	6.0
35	Ruko at Penyeurat	>10

The hard/dense soil depth data can also be plotted in a base map, which is useful for knowing the position of each test point and the depth of stable soil at each location. Then an interpolation is made to determine the zoning of hard/dense depth, as shown in Figure 5. The zoning map shows that most of the tested locations have a stable soil depth of 10 meters below the existing ground surface. In certain places such as the area around Lambung, around Lamjame, around Neusu Aceh, around Lambaro Skep, around Jeulingke, around Doi, and around Kopelma Darussalam exhibited stable depth from 5 to 10 meters. The shallowest hard/dense soil depth is only found around Lamjame of Jaya Baru District.

3.2 Discussion

The depth of stable soil at the study site varies from 3.2 meters to more than 10 meters below the existing ground level. Simple statistical analysis shows that the depth of stable soil is very much dominated by a depth of > 10 meters below the existing ground surface (62.86%). At the study site, no stable soil layer is found at depths from the ground to 2.5 meters. The depth of stable soil ranging from 2.5 to 5 meters has a percentage of 11.43% of the total 35 test points, while for depths of 5 to 7.5 meters, it has a percentage of 22.86%. Stable soil at a depth between 7.5 and 10 meters only has a percentage of 2.86%. A more detailed description can be seen in Table 3 below. This result can be used for the site response analysis to re-estimate the site amplification of the study site [2, 12].

Table 3. Depth of hard/dense layer of the present study

No	Depth of hard/dense layer of qc>150kg/cm^2 (m)	Frequency	Percentage
1	0.00 – 2.50	0	0.00
2	2.51 – 5.00	4	11.43
3	5.01 – 7.50	8	22.86
4	7.51 – 10.00	1	2.86
5	>10	22	62.86
	TOTAL	25	100

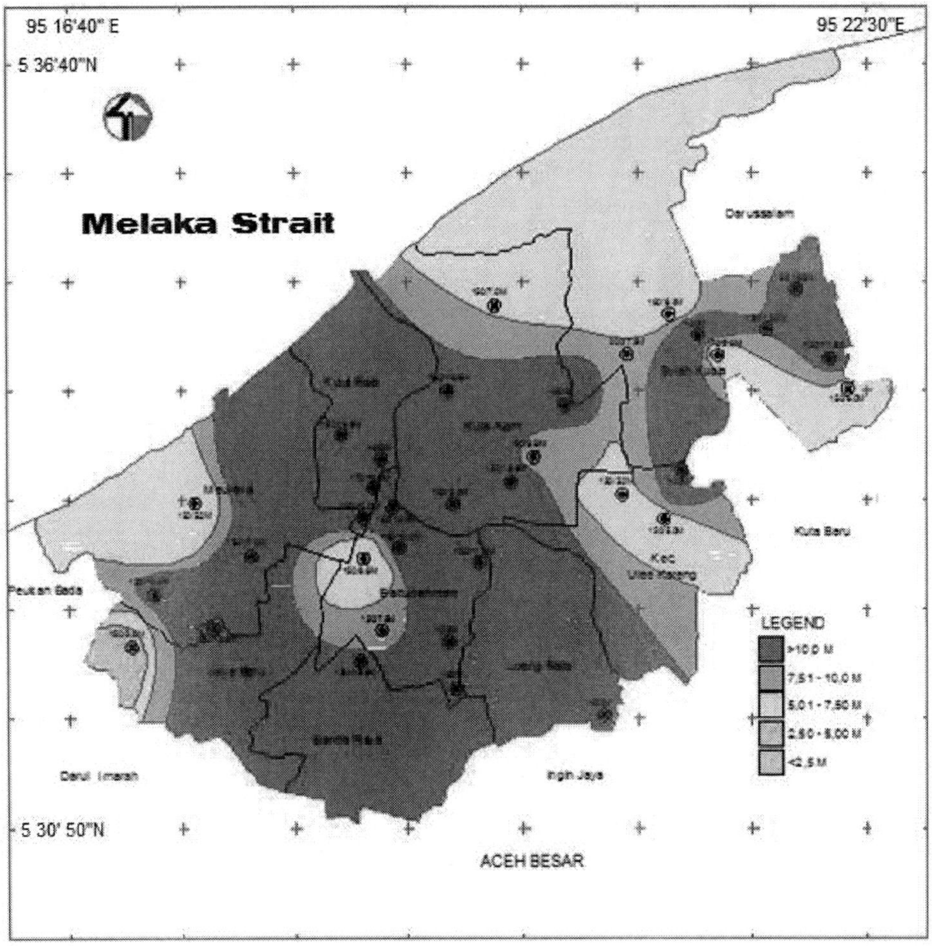

Figure 5. Zoning of the depth of the hard/dense layer at the study area

4. Conclusion

Based on the 35 CPT data, mapping of the depth of the hard/dense layer at Banda Aceh (Indonesia) and the surrounding areas has been carried out. Zoning for stable soil depth at the study site is dominated by more than 10 meters below the existing ground surface. The shallowest hard/dense soil depth is only found around Lamjame-Jaya Baru. This study is still far from perfect considering the very complex conditions of the underground surface at the study site [13-15] so that a couple of suggestion is given, as follows: a) Adding more CPT testing points are suggested. This additional CPT tests will increase the resolution of the map.b). For more detailed results in future research, it is necessary to obtain soil samples through hand/machine drilling or other tools or other methods, i.e., passive seismic data analysis [16, 17].

5. Acknowledgement

This work was funded by Universitas Syiah Kuala (SK Rektor No. 322/UN11/KPT/2020) on the date of 5 February 2020. Also, the authors are grateful to the Faculty of Engineering of Universitas Syiah Kuala for their support.

References

[1] Saidi T, Aulia T B, Setiawan B, Abdullah N, Hasan M 2018 *MATEC Web of Conferences,* **197** 10001.
[2] Yunita H, Setiawan B, Saidi T, Abdullah N 2018 *MATEC Web of Conferences,* **197** 10002.
[3] World Bank 2006 *Hazards of Nature, Risk to Development an IEG Evaluation of World Bank Assistance for Natural Disasters.* Washington DC: The World Bank.
[4] Wesley L D 1977 *Mekanika Tanah.* Jakarta: Badan Penerbit Pekerjaan Umum. *(in Bahasa Indonesia).*
[5] Setiawan B, Jaksa M, Griffith M, Love D 2018 *Eng. Geol.,* **243** 145-159.
[6] Standar Nasional Indonesia 2008 (SNI 2827:2008) 2008 *Cara Uji penetrasi lapangan dengan alat sondir.* Jakarta: Badan Standarisasi Nasional (BSN), ICS 93.020, 23. *(in Bahasa Indonesia).*
[7] Schmertmann J H 1977 *Guidelines for Cone Penetration Test Performance and Design.* Washington DC: U.S. Department of Transportation.
[8] Lunne T, Robertson P K, Powell J J M 1997 *Cone Penetration Testing in Geotechnical Practice.* United Kingdom: Blackie Academic & Professional.
[9] Olsen R S, Mitchell J K 1995 *Proceedings of the International Symposium on Cone Penetration Testing (CPT'95),* Linkoping, Sweden, 251–256.
[10] Robertson P K, Campanella R G, Wightman A 1983 *J. Geotech. Eng.,* **109** (11) 1449-59.
[11] Shahri A A, Malehmir A, Juhlin C 2015 *Eng. Geol.,* **189** 32-47.
[12] Setiawan B 2017 *Indonesian J. Geosci.,* **4**(3) 159-167.
[13] Siemon B, Ploethner D, Pielawa J 2006 *Report C 1,* BGR (Federal Institute for Geosciences and Natural Resources).
[14] Polom U, Arsyad I, Kumpel H 2008 *Advanced Geosci.,* **14** 135–140.
[15] Rusydy I, Ikhlas I, Setiawan B, Idris S, Basyar K, Putra Y A 2020 *Groundw. Sustain. Dev.,* **10** 100372.
[16] Setiawan B, Saidi T, Yulianur A, Polom U, Ramadhansyah P J, Ali M I 2020 *IOP Conf. Ser.: Mater. Sci. Eng.,* **712** 012012.
[17] Setiawan B, Jaksa M, Griffith M, Love D 2019 *Soil Dyn. Earthq. Eng.,* **122** 16-38.

Estimating bearing capacity using static cone penetration test at Banda Aceh area (northern tip of Sumatra)

B Setiawan[1,*], T Saidi[2], M Sungkar[2], H Gunawan[2], A Yulianur[2], Z Zulkifli[3], I Malik[4]

[1]Program Study of Geological Engineering, Universitas Syiah Kuala, Banda Aceh, Indonesia
[2]Department of Civil Engineering, Universitas Syiah Kuala, Banda Aceh, Indonesia
[3]Department of Economic Development, Universitas Syiah Kuala, Banda Aceh, Indonesia
[4]WSP Australia Pty Ltd., Canberra, Australian Capital Territory (ACT), Australia

*E-mail: bambang.setiawan@unsyiah.ac.id

Abstract. The static cone penetration test (s-CPT), or better known in Indonesia as the sondir, has been used as a soil investigation procedure since nearly half a century ago in Europe, especially in the Netherlands and Belgium. This s-CPT is one of the standard practices in soil investigation campaigns and can provide many soil engineering parameters, including soil bearing capacity. This study focuses on the soil bearing capacity analysis in the city of Banda Aceh on the data taken from the 35 points of s-CPT. The analysis of the soil bearing capacity at a depth of 1.4 meters and 2.4 meters below the existing ground level (mbgl) was carried out. Dominantly, at a depth of 1.4 mbgl at the research location, it has a low bearing capacity (<2.5 tons/m^2) with a percentage of 91.43%. There are 2.86% for bearing capacity from 2.5 to 5 tons/m^2, from 5 to 7.5 tons/m^2, and from 7.5 to 10 tons /m^2. No soil layer had a bearing capacity above 10 tons/m^2. At a depth of 2.4 mbgl, the research location has a low bearing capacity (<2.5 tons/m^2) with a percentage of 48.57%, for bearing capacity ranging from 2.5 to 5 tons/m^2 is 17.14%, for bearing capacity of 5 to 7.5 tons/m^2 is 2.86%, for bearing capacity of 7.5 to 10 tons/m^2 of 5.71% and bearing capacity above 10 tons/m^2 is 25.71%.

1. Introduction

The soil bearing capacity and building foundation are closely related to the safety of a building [1]. In the structure, the lower part of a building consists of two main elements, namely the foundation and the soil supporting the foundation. The foundation serves to support all building loads and forward the load of the building into the ground below. A foundation system must be able to guarantee to support the load of the building above it, including external forces such as wind, earthquake, and others (i.e. amplification) [2-4]. For this reason, the foundation must be strong, stable, safe, so that it does not decline, does not break, and does not fail. The occurrence of a break or failure of the foundation system will result in the following things: damage to the walls, cracks, sloping, broken floors, and subsidence of roofs and other building parts [5].

A critical thing closely related to the foundation is soil investigation. The foundation must be placed in a layer of soil with an adequate bearing capacity that is quite hard and dense. It is necessary to conduct a soil investigation to estimate the soil bearing capacity. One of the tools used for soil investigation is the static cone penetration test (s-CPT) or better known in Indonesia as the sondir [6]. This sondir has been used as a soil investigation procedure since nearly half a century ago in Europe, especially in the Netherlands and Belgium [7-9]. The tip is a cone that is pressed into the ground. The results of the penetration of the soil stress can be automatically read in a connected manometer gauge in a unit of kg/cm2. Chandra [10] states that s-CPT is one of the popular geotechnical in-situ tests in Indonesia due to several advantages, including a) Simple use, b) Can provide an overview of the investigated site quickly, and c) Provide a continuous soil strength profile.

2. Testing programme

2.1 Field testing using s-CPT

In this study, we used the double cones apparatus of s-CPT. Wesley [7] explains the use of this double cones s-CPT. In the use of double cones (cone resistance & sleeve friction), the value of both cone resistance and sleeve friction is recorded. This is done by pressing the handlebars. In the beginning, only the cone's tip is pressed down, and thus only the cone resistance (CR) value is measured. When the cone's tip has been moved as far as 4 centimeters, then by itself, it will catch the 'sleeve friction part,' and the cone along with the friction sleeve is then pressed down. So, the value of the cone's tip and the sleeve friction (TR) are measured together. The value of the sleeve friction is obtained later by subtracting the CR value from the total (TR) value. Then by pressing only the outer casing, the cone's tip gear, sleeve friction apparatus, and the handlebars as a whole will be pressed down to a depth where the next reading will be done. Tasks are usually done every 20 centimeters [6]. The s-CPT test was carried out at 35 locations in Banda Aceh and its surroundings. The tabulation of test locations with its village name is presented in Table 1.

Table 1. Testing location of the present study

No	Location name	Village name
1	SDN 1 Lamjame	Lam Jame, Kecamatan Jaya Baru
2	SDN Asoe Nanggroe	Asoe Nanggroe, Kecamatan Meuraxa
3	Jembatan Bitai	Bitai, Kecamatan Jaya Baru
4	SMP 5 Banda Aceh	Lambung, Kecamatan Meuraxa
5	SDN 95	Gampong Baro, Meuraxa
6	SMP Percontohan	Lamlagang, Kecamatan Banda Raya
7	SDN 96 Banda Aceh	Neusu Aceh, Kecamatan Baiturrahman
8	Kejaksaan Tinggi Banda Aceh	Kampong Baro-1, Kecamatan Baiturrahman
9	SD Muhamadiyah Banda Aceh	Merduati-1, Kecamatan Kutaraja
10	Kantor Pos at Merduati	Merduati-2, Kecamatan Kutaraja
11	SDN 17 Banda Aceh	Peulanggahan, Kecamatan Kutaraja
12	Kantor Pemkot Banda Aceh	Kampong Baro-2, Kecamatan Baiturrahman
13	Ruko di Moh. Jam	Kampong Baro-3, Kecamatan Baiturrahman
14	SDN 6 Banda Aceh	Keudah, Kecamatan Kutaraja
15	SDN 64 Banda Aceh	Ateuk Jawo-1, Kecamatan Baiturrahman
16	Jembatan Ateuk Jawo	Ateuk Jawo-2, Kecamatan Baiturrahman
17	SDN 30 Banda Aceh	Ateuk Pahlawan, Kecamatan Baiturrahman
18	Bappeda NAD	Kuta Alam, Kecamaan Kuta Alam
19	Museum Tsunami	Lampulo, Kecamatan Kuta Alam
20	Perluasan Gedung Kantor Imigrasi	Beurawe-1, Kecamatan Kuta Alam
21	Jamsostek	Beurawe-2, Kecamatan Kuta Alam
22	SMP 2 Banda Aceh	Bandar Baru, Kecamatan Kuta Alam
23	SDN 45	Lambaro Skep, Kecamatan Kuta Alam
24	Asrama Mahasiswa	Doi-1, Kecamatan Ulee Kareng
25	SDN 76	Doi-2, Kecamatan Ulee Kareng

26	Perumahan DPRA	Doi-3, Kecamatan Ulee Kareng
27	Kanwil HAM	Jeulingke-1, Kecamatan Syiah Kuala
28	Rumah Dinas POLDA	Jeulingke-2, Kecamatan Syiah Kuala
29	Klinik Putro Phang	Jeulingke-3, Kecamatan Syiah Kuala
30	Ruko at Rukoh	Rukoh-1, Kecamatan Syiah Kuala
31	Asrama Mahasiswa	Rukoh-2, Kecamatan Syiah Kuala
32	SDN 83 Banda Aceh	Rukoh-3, Kecamatan Syiah Kuala
33	Hyugo Perfecture Building	Kopelma Darussalam-1, Kecamatan Syiah Kuala
34	Gedung Kuliah Pasca Sarjana	Kopelma Darussalam-2, Kecamatan Syiah Kuala
35	Ruko at Penyeurat	Penyeurat, Kecamatan Lueng Bata

2.2 Bearing capacity

Chandrawanshi & Kumar [11] stated that soil bearing capacity is the maximum strength of the soil to withstand pressure (due to building loads) well without causing failure. Failure in the soil is an excessive settlement or shear failure (inability) of the soil to resist shear forces. Omar & Sadrkarimi [12] states that if the soil is loaded by any load, i.e., buildings, roads, dams, it will result in shear stress. If the load increases, the shear stress reaches a limit where the soil mass will experience deformation and tend to collapse (failure). The load that results in failure is called the failure load, and the pressure that occurs is called the ultimate soil bearing capacity.

In the bearing capacity analysis using static cone penetration test data, Wesley [7] warns that the cone value obtained by the cone penetration device cannot be equated directly with the bearing capacity of the soil concerned. Thus, an empirical approach is suggested. Schmertmann [8] proposed the bearing capacity of shallow foundation soils using s-CPT data, namely Meyerhof [13] formula, as presented in Equation 1.

$$qa/qc = B/40 \times (1+ D/B) \quad (1)$$

where, qa = allowable bearing capacity (kg/cm^2), B = footing width (m); D = footing depth (m), and qc = cone resistanace (kg/cm^2).

The calculation of bearing capacity analysis used in this study is a shallow square footing with a dimension of 1 x 1 meter. Two foundation depth of 1.40 mbgl and 2.40 mbgl are considered. The bearing capacity of footing at all locations in this study is calculated using the formula mentioned above.

3. Testing results and discussion

3.1 Results

As aforementioned, soil bearing capacity analysis is based on the formula from Meyerhof [13], as in Equation 1. In the following section, more emphasis is placed on the results of soil bearing capacity analysis of a 1 square meter foundation with a foundation depth of 1.40 mbgl and 2.40 mbgl, as shown in Table 2.

The study area has a very variable bearing capacity at a depth of 1.4 mbgl. The results of bearing capacity analysis at 1.4 mbgl, if plotted on the map and then interpolated, is presented in Figure 1. The zoning map of the bearing capacity of the soil at a depth of 1.4 mbgl shows that the study area is very much dominated with bearing capacity <2.5 ton/m^2 (red color). The area which has a bearing capacity of between 2.5 to 5 tons/m^2 only occupies a small part in the Neusu Aceh area, Baiturrahman District and around Lamjame, Jaya Baru District. The bearing capacity of the soil that ranges from 5 to 7.5 tons/m^2 at a depth of 1.4 mbgl is found in the Neusu Aceh area, Baiturrahman District at 6 tons/m^2 and in the Lamjame area, Jaya District New with a value of 9 tons/m^2.

The zoning map of the bearing capacity of the soil at a depth of 2.4 mbgl, as in Figure 2 shows that most of the areas in the study location, also, have a bearing capacity of <2.5 tons/m^2 (red color). The bearing capacity between 2.5 to 5 tons/m^2 (orange color) occupies the area around Neusu Aceh, Baiturrahman District, around Gampong Baro, Meuraxa District, around Keudah, Kuta Raja District, around Doi, Ulee Kareng District, around Lam Gugop, Syiah Kuala District and around Lamjame, Jaya Baru District. The soil bearing capacity from 5 to 7.5 tons/m^2 (yellow color) is found around Neusu Aceh, Baiturrahman District, around Gampong Baro, Meuraxa District, around Keudah, Kuta Raja District, around Rukoh, Syiah Kuala District, around Lam Gugop-Simpang Mesra, Syiah Kuala District and around Lamjame, Jaya Baru District. The bearing capacity of 7.5 to 10 tons/m^2 (blue color) is obtained around Neusu Aceh, Baiturrahman District, around Gampong Baro, Meuraxa District, around Rukoh, Syiah Kuala District, around Lam Gugop-Simpang Mesra, Syiah Kuala District and around Lamjame, Jaya Baru District. The bearing capacity above 10 tons/m^2 (green color) at a depth of 2.4 mbgl is found around Gampong Baro, Meuraxa District (14.17 tons/m^2), around Neusu Aceh, Baiturrahman District (24.08 tons/m^2), around Keudah, Kutaraja District (11.33 tons/m^2), around Lambaro Skep, Kuta Alam District (28.33 tons/m^2), around Simpang Mesra, Syiah Kuala District (19.83 tons/m^2), around Kopelma Darussalam, Syiah Kuala District (14.17 tons/m^2) and around Lamjame (32.58 tons/m^2).

Table 2. Bearing capacity of the present study

No	Location name	Bearing capacity (Ton/m^2)	
		At 1.4 m depth	At 2.4 m depth
1	SDN 1 Lamjame	9,00	32,58
2	SDN Asoe Nanggroe	0,20	0,57
3	Jembatan Bitai	0,20	0,28
4	SMP 5 Banda Aceh	0,20	1,42
5	SDN 95	2,00	14,17
6	SMP Percontohan	1,60	2,83
7	SDN 96 Banda Aceh	6,00	24,08
8	Kejaksaan Tinggi Banda Aceh	0,20	0,57
9	SD Muhamadiyah Banda Aceh	2,00	4,25
10	Kantor Pos at Merduati	0,20	2,83
11	SDN 17 Banda Aceh	1,00	3,40
12	Kantor Pemkot Banda Aceh	0,20	0,57
13	Ruko di Moh. Jam	1,60	0,28
14	SDN 6 Banda Aceh	2,00	11,33
15	SDN 64 Banda Aceh	1,60	9,92
16	Jembatan Ateuk Jawo	0,20	0,28
17	SDN 30 Banda Aceh	2,00	2,83
18	Bappeda NAD	0,20	0,28
19	Museum Tsunami	0,20	0,57
20	Perluasan Gedung Kantor Imigrasi	0,20	0,28
21	Jamsostek	0,20	0,28
22	SMP 2 Banda Aceh	0,20	0,28
23	SDN 45	0,20	28,33
24	Asrama Mahasiswa	0,20	1,42
25	SDN 76	0,40	5,67
26	Perumahan DPRA	0,20	1,98
27	Kanwil HAM	0,40	1,70
28	Rumah Dinas POLDA	1,40	1,13
29	Klinik Putro Phang	0,20	0,28
30	Ruko at Rukoh	0,40	19,83
31	Asrama Mahasiswa	0,40	14,17
32	SDN 83 Banda Aceh	2,00	14,17
33	Hyugo Perfecture Building	2,00	8,50
34	Gedung Kuliah Pasca Sarjana	3,00	14,17
35	Ruko at Penyeurat	0,20	2,83

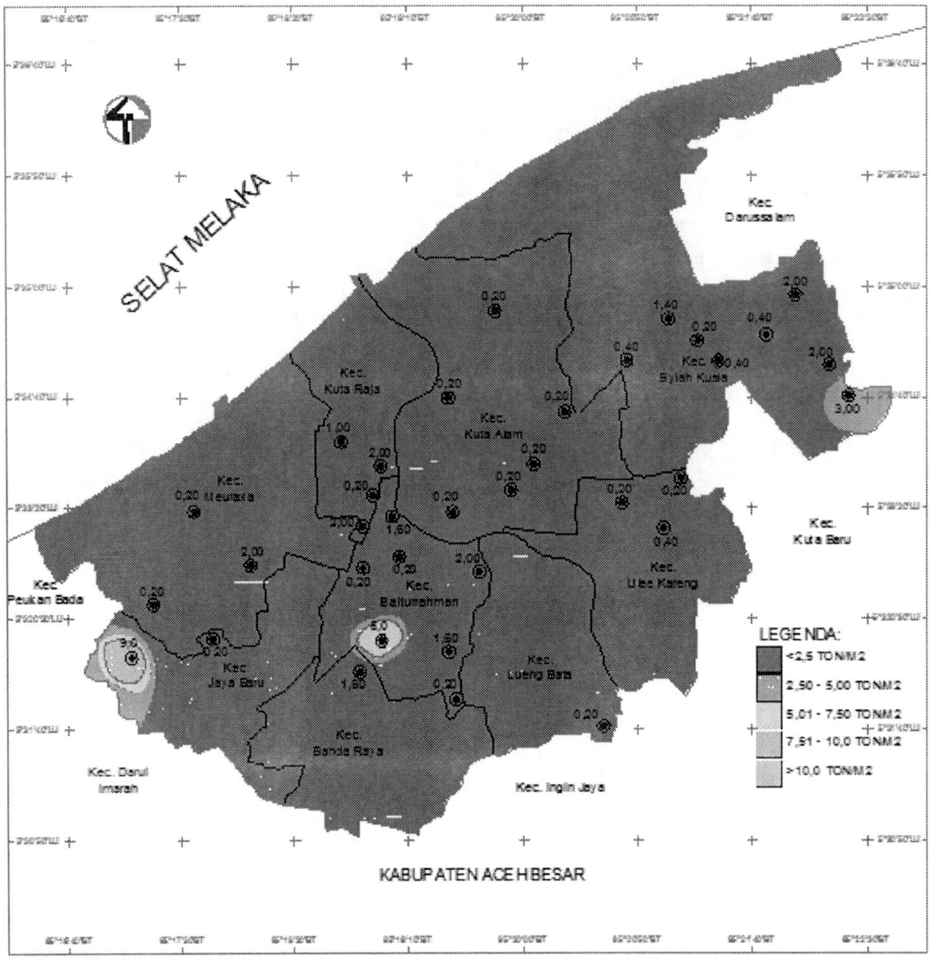

Figure 1. Zoning of soil bearing capacity at a depth of 1.4 mbgl.

3.2 Discussion

Overall, if divided into 5 groups, the bearing capacity obtained in the study area can be clustered, as shown in Table 3. In this table, it can be seen that the very dominant in the study area has a low bearing capacity (<2.5 tons).

Dominantly, at a depth of 1.4 mbgl at the study area, it has a low bearing capacity (<2.5 tons) with a percentage of 91.43%, bearing capacity from 2.5 to 5 tons/m^2 2.86%, bearing capacity from 5 to 7.5 tones/m^2 by 2.86%, bearing capacity from 7.5 to 10 tons/m^2 by 2.86%. No soil layer had a bearing capacity above 10.00 tons/m^2.

From Table 3, it can be seen that at a depth of 2.4 mbgl, dominantly, the study area has a low bearing capacity (<2.5 tons/m^2) with a percentage of 48.57%. The percentages for bearing capacity ranging from 2.5 to 5 tons/m^2 and 5 to 7.5 tons/m^2 are 17.14% and 2.86%, respectively. The bearing capacity of 7.5 to 10 tons/m^2 is 5.71%. At this 2.4 mbgl depth, a percentage location of a layer of soil with a bearing capacity above 10 ton/m^2 is 25.71%.

A complex sub-surface conditions of the study area has been recodnised by [14, 15]. A further sub-surface investigation using passive seismic approach as suggested by [16], or site response analysis by [17], is interesting topic to explore.

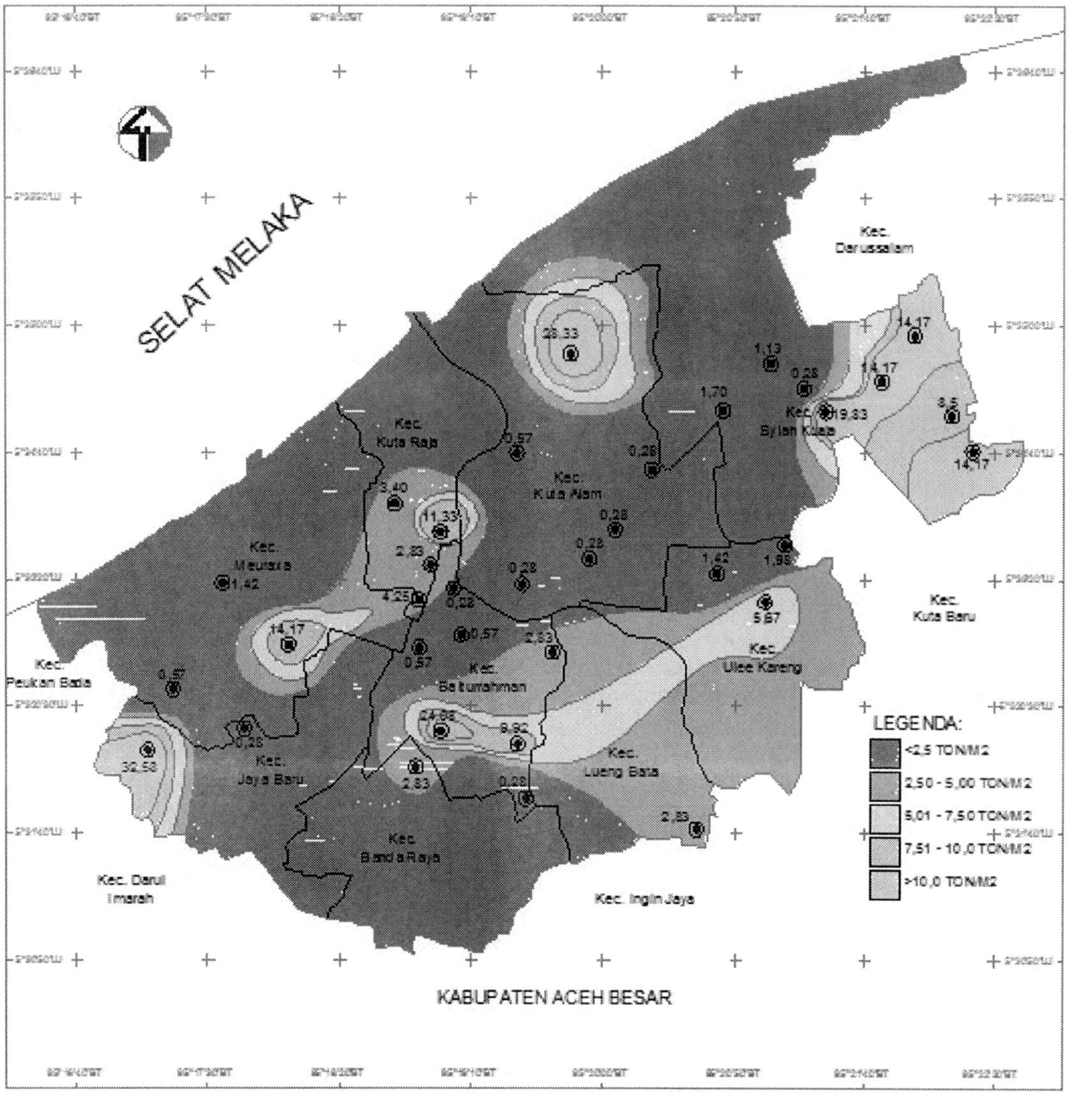

Figure 2. Zoning of soil bearing capacity at a depth of 2.4 mbgl

4. Conclusion

A bearing capacity analysis of the soil in the city of Banda Aceh based on the data from the s-CPT shows that the zoning of the soil bearing capacity and the depth of the soil bearing capacity varies. Zoning bearing capacity at a depth of both 1.4 mbgl and 2.4 mbgl shows that the study area is dominated with a bearing capacity of <2.50 tons/m^2. For the depth of 1.4 mbgl, areas that have a bearing capacity of > 2.50 only occupy the Neusu Aceh and Lamjame areas. For the depth of 2.4 mbgl, areas that have a bearing capacity of between 2.5 to 5 tons/m^2 occupy the Neusu Aceh, Gampong Baro, Keudah, Doi, Lam Gugop, and Lamjame areas. The soil bearing capacity ranging from >5 tons/m^2 at a depth of 2.4 mbgl, was found around Neusu Aceh, Gampong Baro, Keudah, Lambaro Skep, Simpang Mesra, Rukoh, Lam Gugop Kopelma Darussalam and Lamjame.

Table 3. Tabulation of bearing capacity of the present study

No	Bearing capacity (Ton/m2)	At 1.4 mbgl depth		At 2.4 mbgl depth	
		Frequency	Percentage	Frequency	Percentage
1	0,00 - 2,50	32	91.43	17	48.57
2	2,51 - 5,00	1	2.86	6	17.14
3	5,01 - 7,50	1	2.86	1	2.86
4	7,51 - 10,00	1	2.86	2	5.71
5	>10	0	0.00	9	25.71
	TOTAL	35	100	35	100

5. Acknowledgement

The authors wish to acknowledge Universitas Syiah Kuala for providing research funding (Contract No. 267/UN11/SPK/PNPB/2020) on the date of March 17 2020. Also, the authors are grateful to the Faculty of Engineering of Universitas Syiah Kuala for their support.

References

[1] Dey A, Acharyya R, Alammyan A 2019 *Int. J. Geotech. Eng.,* 1-14
[2] Saidi T, Aulia T B, Setiawan B, Abdullah N, Hasan M 2018 *MATEC Web of Conferences,* **197** 10001.
[3] Setiawan B 2017 *Indonesian J. Geosci.,* **4 (3)** 159-167.
[4] Yunita H, Setiawan B, Saidi T, Abdullah N 2018 *MATEC Web of Conferences,* **197,** 10002.
[5] Parvin J 2005 *J. Build. Apprisal.,* **1(2)** 153-163.
[6] Standar Nasional Indonesia 2008 *SNI 2827:2008 Cara Uji penetrasi lapangan dengan alat sondir*. Jakarta: Badan Standarisasi Nasional (BSN). ICS 93.020. *(in Bahasa Indonesia).*
[7] Wesley L D 1977 *Mekanika Tanah.* Jakarta: Badan Penerbit Pekerjaan Umum. *(in Bahasa Indonesia).*
[8] Schmertmann J H 1977 *Guidelines for Cone Penetration Test Performance and Design.* Washington: U.S. Department of Transportation.
[9] Lunne T, Robertson P K, Powell J J M 1997 *Cone Penetration Testing in Geotechnical Practice.* United Kingdom: Blackie Academic & Professional.
[10] Chandra Y P 2000 *Pengujian Lapangan: Diktat Workshop Sertifikasi Himpunan Ahli Teknik Tanah Indonesia.* Jakarta: Himpunan Ahli Teknik Tanah Indonesia (HATTI). *(in Bahasa Indonesia).*
[11] Chandrawnshi S, Kumar R 2015 *Advances in Soft Ground Eng.*
[12] Omar T, Sadrekarimi A 2015 *Int. J. Geo-Eng.*, **6 (5)**.
[13] Meyerhof G 1965 *J. Soil Mech. Found. Div.*, ASCE, **92** (SM-2): 21-31.
[14] Rusydy I, Ikhlas I, Setiawan B, Idris S, Basyar K, Putra Y A 2020 *Groundw. Sustain. Dev.,* **10** 100372.
[15] Setiawan B, Saidi T, Yulianur A, Polom U, Ramadhansyah P J, Ali M I 2020 *IOP Conf. Ser.: Mater. Sci. Eng.*, **712** 012012.
[16] Setiawan B, Jaksa M, Griffith M, Love D 2019 *Soil Dyn. Earthq. Eng.*, **122** 16-38.
[17] Setiawan B, Jaksa M, Griffith M, Love D 2018 *Boll. Geofis. Appl.,* **59 (1)** 27 – 46.

Assessment of regional spatial plan (RSP) of areas along the active fault of Seulimeum segment in Aceh Province

A Darmi[1], N Ismail[1,2,*], T B Aulia[3], M Nanda[4], M Yanis[2]

[1]Department of Disaster management, Universitas Syiah Kuala, Darussalam, Banda Aceh 23111, Indonesia.

[2]Department of Geophysical Engineering, Faculty of Engineering, Universitas Syiah Kuala, Darussalam, Banda Aceh 23111, Indonesia.

[3]Department of Civil Engineering, Faculty of Engineering, Universitas Syiah Kuala, Darussalam, Banda Aceh 23111, Indonesia.

[4]Graduate School of Mathematic and Applied Science, Universitas Syiah Kuala, Darussalam, Banda Aceh 23111, Indonesia.

*E-mail: nazli.ismail@unsyiah.ac.id

Abstract. The Seulimeum Segment is the northernmost branch of the Great Sumatran Fault. This active fault crosses settlements in Aceh Besar Regency and Pidie Regency in Aceh Province. Tectonically, this situation can be a threat along the active fault line. In this study, active fault lines of the Seulimeum Segment were mapped. Based on the active fault lines, residential areas, infrastructure distribution, and Regional Spatial Plan (RSP) were reviewed for eligibility. Some areas, e.g., Tangse, Lamtamot, and Krueng Raya, were the focus of this study. Avoidance zones were categorized based on their distances from the active fault line, which were 20 m from the fault, 100 m from the fault, and 300 m from the fault. According to an active fault line analysis on the RSP map, infrastructure map, and demographics map, the Krueng Raya area would be the most affected by losses due to disasters in the future. Some buildings, hospitals, mosques, government offices, settlements, and industrial areas in Krueng Raya are situated along an active fault line. It is because the RSP of this area has not yet considered the potential for geological disasters in the infrastructure development process. Some recommendations have been proposed to reduce the impact of losses in the future.

1. Introduction

So far, the establishing of the Regional Spatial Plan (RSP) in Aceh Province has not fully considered the geological threat aspects. Previous geological disaster activities have caused a lot of material loss and threatened human lives. This should be the basis for realizing the importance of disaster aspects in infrastructure development in the future. Meanwhile, the government has issued a map of active fault distribution and potential earthquake hazards in Indonesia, including the Aceh region. Along these fault lines, especially in the southern part of Sumatra, there have been quite a lot of seismic activities, while the northern part has not shown significant seismic activity since 170 years ago [1,2]. The north side of the Great Sumatran Fault is divided into two branches that extend to the Andaman Islands, namely, the Aceh Segment that heads to Aceh Island and the Seulimeum Segment that heads to Weh Island, Sabang [2]. Figure 1 shows the seismic activities with Mw ≥ 3 recorded by the United States Geological Survey (USGS) from 1980 to 2020 along the Aceh Segment and Seulimeum Segment [3]. If the energy stored in the fault zone is not released, referring to [4], then the Aceh region has the potential to experience an earthquake with Mw $\geq$ 7. This is a serious threat to the Aceh region in the future. Besides, theoretically, each segment of the GSF ruptures every 100-200 years [1,2]. This seismic gap has the potential to generate earthquakes of around ~ 7.4 Mw due to the absence of energy released so far [4].

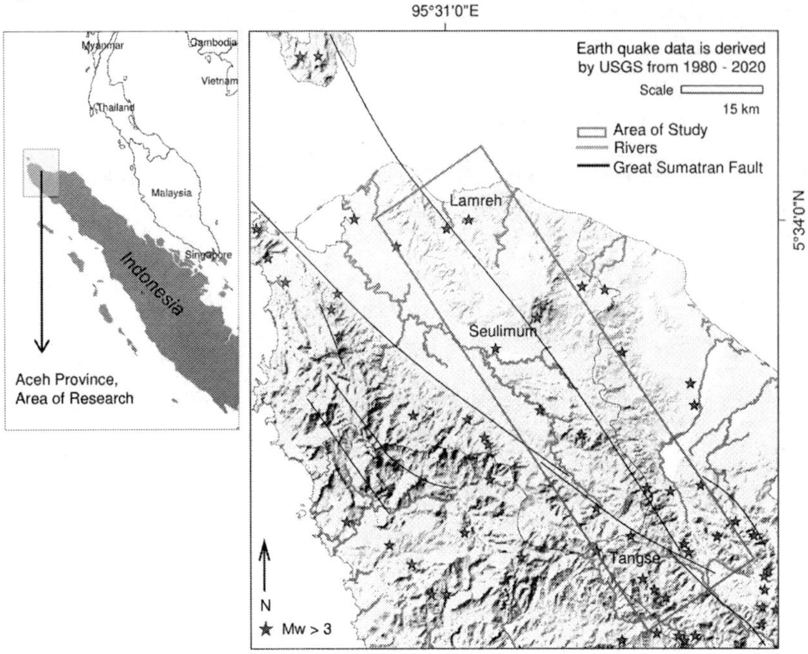

Figure 1. Seulimeum Segment is marked as a research location (red box).

Referring to Figure 1, along the Seulimeum fault, there are several public infrastructures, such as schools, places of worship, hospitals, and residential areas [2,5]. Besides, there are also several historical sites located right on the Seulimeum Segment, such as Lamtamot [6] and Lamreh [7]. Therefore, several studies have been conducted to map faults along the Seulimeum Segment, for example, seismology [8], geophysics [2,9–11], remote sensing [12], and geomorphological observations using DEMNas (BIG, 2018). Based on the information on active faults of the Seulimeum Segment, as a disaster mitigation initiative in Aceh Province, the appropriateness of the Regional Spatial Plan along that segment is an important matter to be reviewed. This research outlines studies and applications to prevent or reduce the risk of disasters and damage from future earthquakes by considering safety, appropriate planning, and finishing processes.

2. Data and methodology

In this research, some information regarding the Seulimeum Segment fault based on geophysical, geomorphological, and remote sensing research was used. The data on the fault becomes the base map for adjusting the map of the Regional Spatial Planning of the Aceh regional government. The implementation of the Aceh region as an independent region, taking mandates, and making decisions to regulate and manage the Aceh region has been regulated since 2006 [13]. Then, based on the provisions of Spatial Planning (Law Number 26 of 2007) and based on the provisions of the National Spatial Plan (Government Regulation Number 26 of 2008), each regency or city administrative region, via the regional head, prepares a Regional Spatial Plan. At this stage, the map of the Regional Spatial Plan was obtained from regencies along active fault lines, which are Aceh Besar Regency (Qanun (Local Law) of Aceh Besar Regency Number 4 of 2013) and Pidie Regency (Qanun of Pidie Regency Number 5 of 2014).

However, the spatial planning map was only analyzed partially. The analysis was only conducted on zones through which the fault line passes. In addition, sampling was also conducted to analyze areas representing public activity along active faults. Sampling was expected to be able to describe the overall state of activity along the fault line. This study employed a quota sampling method. Sampling was conducted on three samples with an area of 800 x 4,000 m per sample. Two samples were in Aceh Besar Regency and one sample was in Pidie Regency. These locations were chosen caused to cover all of the Seulimum segments, i.e. Tangse area which characterized as the segment in the tip of the South-West side, while Lamreh is corresponds to the tip of the Northeast side and the Lamtamot as the middle of the Seulimum segment.

3. Result and conclusion

Information about active faults that were obtained from various sources was used as a base map, and then the distance between an object and the fault was classified. Figure 2.a. shows the avoidance zone of Seulimeum Segment, which was divided into 3 zones, namely: 20 m from the fault, 100 m from the fault, and 300 m from the fault. In general, there are several villages in Aceh Besar Regency and Pidie Regency that are in parallel areas with the avoidance zone of Seulimeum Segment. The data showed that there were 77 villages from 2 regencies, which were distributed into 34 villages in 5 sub-districts in Aceh Besar Regency and 43 villages in 4 sub-districts in Pidie Regency. According to statistical data from Statistics Indonesia in 2010 (Figure 2.b), 21,183 people were in locations that intersected with active faults. A total of 13,713 people were in Aceh Besar and 7,470 people were in Pidie. The 2010 data were chosen because the population survey in Indonesia is conducted every 10 years (Law Number 6 and Law Number 7 of 1960). Of course, the population in those areas has probably increased by now. Based on the projection results, population growth in Aceh in 2020 will reach 0.19%.

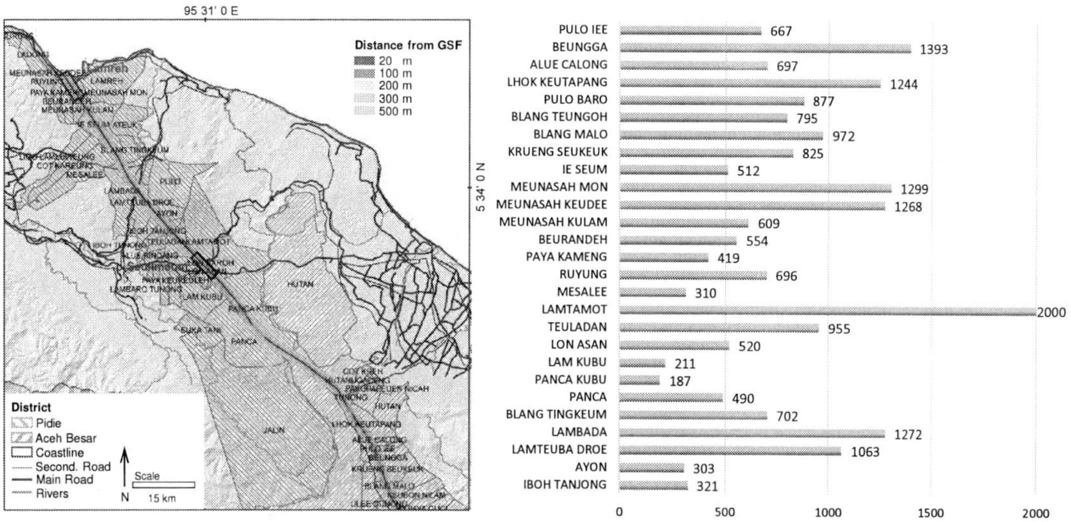

[A]. Fault Distance from Public Infrastructure [B]. Populations

Figure 2. (a) Distribution of areas that are adjacent to the active fault line of the Seulimeum Segment, (b) population in villages that intersect with the main fault of the Seulimeum Segment (modification based on Statistics Indonesia (BPS), 2010).

Besides, statistical data showed that Lamtamot was the village that was close to the fault and had the largest population, which was ~ 2,022 people, while as many as ~1,022 people lived in Lamteuba Droe Village, Lambada Village, Meunasah Keudee Village, Meunasah Mon Village, Lhok Keutapang Village, Alue Calong Village, and Beungga Village. In the locations that have been mentioned, a good mitigation initiative must be implemented because greater losses will be potential for densely populated areas. Based on experience, even relatively small-magnitude earthquakes caused very large losses, for example, the Central Aceh earthquake in 2013 with Mw 6.1 [5], the Pidie Jaya earthquake with Mw 6.5 in 2016 [4,8], and two earthquakes near the Seulimeum Segment in 1936 with Mw 7.1 - 7.3 [14] and in 1964 with Mw 6.5 [2]. This was because the epicenters of the earthquakes were located close to the residential area.

To study the potential area of the surface structure, focusing areas in several places that were in direct contact with the Seulimeum Segment, such as Lamreh in Krueng Raya, and Lamtamot in Seulimeum, Aceh Besar Regency, and Tangse in Pidie Regency, were analyzed. Figure 3a shows development activities, infrastructure, and settlements in the area of Lamreh, Aceh Besar. Public activities, in the form of settlements, were in the middle on the east side, heading towards the middle to the end of the northwestern part, while the rest were landscapes, such as shrubs, agricultural land, and rivers.

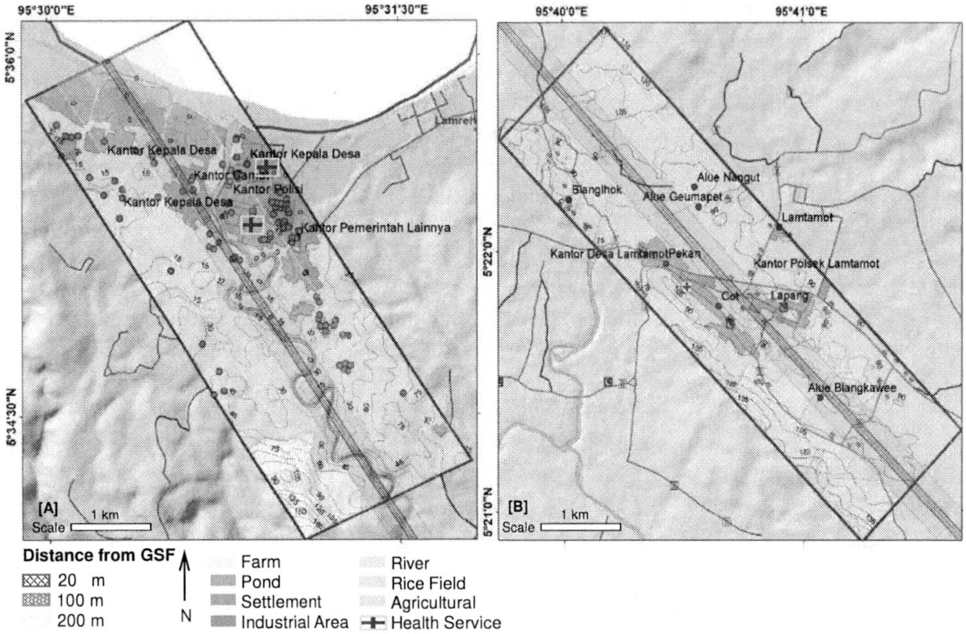

Figure 3. is the distance of the Seulimeum Segment fault area to public activities and public infrastructure in Aceh Besar, which are specifically located at (a) Lamreh, Krueng Raya, and (b) Lamtamot, Seulimeum, Aceh Besar.

The map shows that several places intersect with the fault, namely rice fields, ponds, and agricultural land. Important areas that are in contact with active faults are settlements, buildings, and places of worship. Several other places, from health care facilities, hospitals, buildings, houses of worship, government offices, to residential and industrial areas, are still in the avoidance zone. The area of Lamtamot, Seulimeum (Figure 3.b), is still in the avoidance zone, although this area is generally located in forest and shrubland and is used as rice fields and agricultural land in general. Based on the maps, infrastructure development is right on the fault trail of the Seulimeum Segment. The infrastructure consists of residential houses, places of worship, health care facilities, administrative offices, and markets, which are located in the avoidance zone that is relatively close to the Seulimeum Segment.

As a Lamtamot area, the Tangse in Pidie Regency is also in geomorphology of the forest, agricultural land, and several rivers. Figure 4 shows that the centralization of development is in the northwest direction and concentrated in the southeast. Several important places, namely, houses of worship, health care facilities, government offices, and residential areas, intersect with active faults. Figure 4.b shows that some public infrastructures, such as hospitals and settlements, are on the Seulimeum Segment line (marked by the white line in the figure). Referring to [15], the Acehnese people tend to evacuate themselves to places of worship when a disaster occurs, especially a tsunami or an earthquake disaster. However, the analysis of the distance of Seulimeum Segment with several infrastructures indicated that most of the places of worship, health care facilities, and *puskesmas* (community health centers) were right on the fault line or were close to the avoidance zone, even though these places are the first places to await for evacuation when a disaster occurs. As shown in Figure 4.c, the position of the mosque in Tangse Village is close to the Seulimum fault.

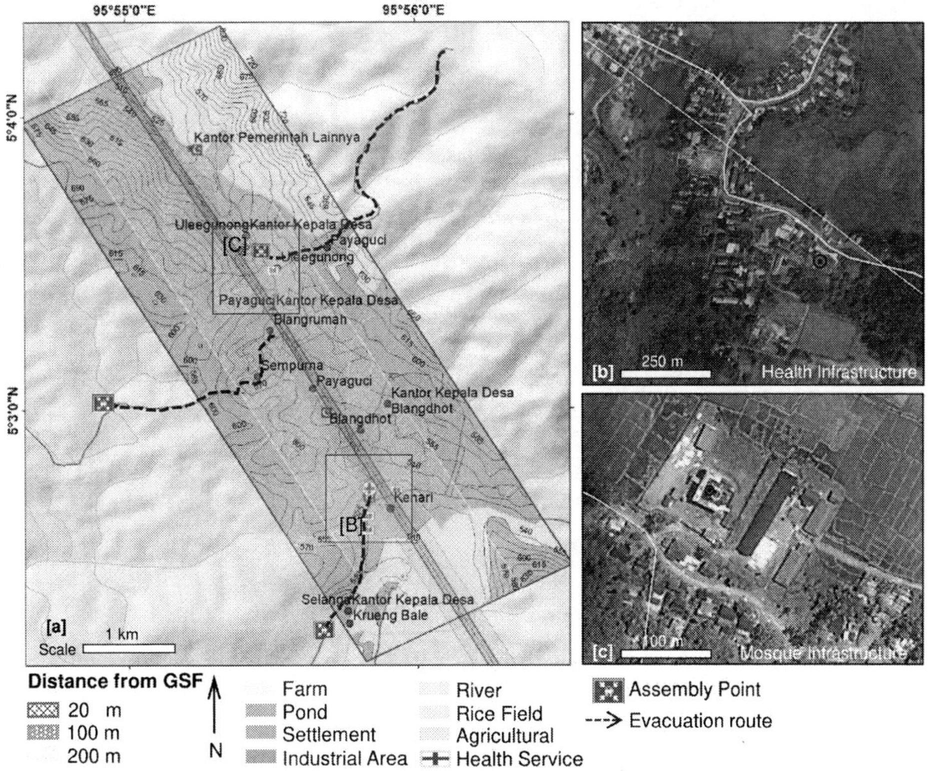

Figure 4. is the distance of the Seulimum Segment fault area to public activities and public infrastructure in Tangse Village, Pidie Regency, while (b) is infrastructure such as hospitals, settlements, and village administrative offices, and (c) is the infrastructure for places of worship that are close to the Seulimeum Segment.

Therefore, several recommendations are formulated as an effort to reduce the risk of infrastructures that are built close to the Seulimeum Segment, for example, recommendations for establishing evacuation routes and building assembly points. The assembly point at Lamreh, Krueng Raya, is recommended to be in the east along the coastline and relatively close to places of worship. Meanwhile, in the Tangse area, it is recommended that the evacuation route and assembly point be built on the northeast and southeast, it caused to a safe distance from the fault (as shown by the blue dotted line in Figure 4). In addition, other recommendations are to limit infrastructure development in areas that intersect the fault or adjacent to the fault line, conduct settlement relocation, relocate buildings, prohibit reconstruction, prohibit tenancy, and prohibit the construction of new buildings that are excluded based on recommendations from comprehensive geological studies.

Based on the analysis data, the feasibility study of the Regional Spatial Plan, residential areas, and infrastructure distribution conclude that the potential for geological disasters has not been considered both in the preparation of the Regional Spatial Plan and the development process in the area along the Seulimeum Segment. The recommendation that is proposed based on this research is to review the Regional Spatial Plans of areas that intersect with active faults. This is crucial to avoid disaster risks and formulate future disaster mitigation plans. In addition, a policy regarding structures and buildings that have been erected along active faults needs to be made. The policy is expected to be able to limit development and relocation and be able to provide knowledge to the public about the major risks to be faced in the future.

References

[1] Bellier O, Sebrier M, Pramumijoyo S, Beaudouin T, Harjono H, Bahar I, Forni O 1997 *J. Geodyn.*, **24** 169–183.
[2] Sieh K, Natawidjaja D 2000 *J. Geophy. Res.: Solid Earth,* **105** 28295–326.
[3] US Geological Survey. 2020. *Earthquake Catalog*.
[4] Ito T, Gunawan E, Kimata F, Tabei T, Simons M, Meilano I, Agustan, Ohta Y, Nurdin I, Sugiyanto D 2012 *J. Geophys. Res.: Solid Earth,* **117**.
[5] Muzli M, Muksin U, Nugraha A D, Bradley K E, Widiyantoro S, Erbas K, Jousset P, Rohadi S, Nurdin I, Wei S 2018 *Seismol. Res. Lett.,* **89** **(5)** 1761-72.
[6] Ismail N, Yanis M, Abdullah F, Irfansyam A, Atmojo B S W. 2018. *J. Phys.: Conf. Ser.,*. **1120** 012035.
[7] Yanis M, A Bakar M, Ismail N 2017 *23rd European Meeting of Environmental and Engineering Geophys.,* **1** 1-5.
[8] Muksin U, Bauer K, Muzli M, Ryberg T, Nurdin I, Masturiyono M, Weber M 2019 *J. Asian Earth Sci.,* **171** 20–27.
[9] Ismail N, Yanis M, Idris S, Abdullah F, Hanafiah B 2017 *J. Phys.: Conf. Ser.,* **846** **(1)** 012016.
[10] Marwan, Asrillah, Yanis M, Furumoto Y 2019 *Int. J. GEOMATE,* **17** 210–216.
[11] Rizal M, Ismail N, Yanis M, Muzakir, Surbakti M S 2019 *IOP Conf. Ser.: Earth and Environ. Sci.,* **364** 012036.
[12] Yanis M, Marwan M, Ismail N 2019 *Indonesian J. Applied Phys.,* **9** 61.
[13] Husni J 2010 *Kanun Jurnal Ilmu Hukum,* **51**. *(in Bahasa Indonesia).*
[14] Newcomb K R, McCann W R 1987 *J. Geophys. Res.: Solid Earth,* **92** 421–439.
[15] Adiyoso W, Kanegae H 2013 *Procedia Environ. Sci.*, **1** **(17)** 918-927.

A proposed method for calculating dimension of irrigation channel section

A Yulianur BC*, A Fauzi
Department of Civil Engineering, Faculty of Engineering Universitas Syiah Kuala, Banda Aceh, Indonesia

*E-mail: fian_7anur@unsyiah.ac.id

Abstract. In general, the cross section of irrigation channels in Indonesia is planned to be trapezoidal. The calculation of the dimensions of the trapezoidal section uses the quadratic equation when calculating the hydraulic depth in the channel. Calculations using quadratic equations can produce imaginary numbers if the discriminant value is negative, resulting in a failed calculation of channel dimension. This study proposes a method for calculating the dimensions of a trapezoidal cross-sectional irrigation channel, in which the discriminant value of quadratic equation is not negative. The analysis is carried out on all formulas that produce the independent variable for the discriminant value in the quadratic equation. The independent variable is the mean flow velocity in the channel or the channel base slope. This study produces a formula for calculating the mean flow velocity in a channel as well as a formula for calculating the channel base slope which results do not result in a negative discriminant value. Thus, the calculation of the hydraulic depth in the channel has not failed and the calculation of other channel dimensions can be continued.

1. Introduction

Irrigation channel planning requires that the dimensions of the planned channel must have the capacity or ability to drain a water discharge amount equal to or greater than the discharge of irrigation water needs [1, 2]. The channel dimensions that determine the channel capacity or channel discharge are the cross-sectional dimensions and the longitudinal sectional dimensions. From these two sectional dimensions, the area of the channel section and the mean flow velocity in the channel can be calculated. The multiplication of the area of channel section with the mean flow velocity gives the channel discharge value [3, 4].

Generally, in Indonesia, the cross section of irrigation channels that is planned is a trapezoidal section [5]. The calculation of the dimensions of the trapezoidal section uses the quadratic equation when calculating the hydraulic depth in the channel [3]. Calculations using this quadratic equation require conditioning of the discriminant value not to be negative [6], because the negative discriminant value produces imaginary numbers and results in failure in the planning of irrigation channel dimensions. Therefore, it is necessary to conduct a study to produce a method of calculating the dimensions of the irrigation channel that discriminant value is not negative when quadratic equation is used to calculate hydraulic depth.

In this study, an analysis of all independent variables in the formula was carried out to calculate the discriminant value. From this analysis, the formula for calculating the independent variable is obtained which results in a non-negative discriminant value. Then proposed a method for calculating the dimensions of a trapezoidal cross-sectional irrigation channel in the form of a calculation sequence and a calculation flow chart.

2. Literature review

2.1. Discharge of channel

For any flow, the discharge Q at a channel section is expressed by formula below [3, 4].

$$Q = AV \tag{1}$$

Where Q is discharge at a channel section, m^3/sec; A is the flow cross-sectional area normal to the direction flow, m^2; and V is the mean velocity, m/sec.

The mean velocity of flow in channel, V can be calculated using Manning Formula as below [3,4,7]

$$V = \frac{1}{n} R^{2/3} S^{1/2} \tag{2}$$

Where n is roughness coefficient of Manning; R is hydraulic radius, m; and S is base slope of channel.

2.2 Geometric elements of channel section

Artificial channel is usually designed with section of regular geometric shapes, such as trapezoid and rectangle. The formulas of geometric element of channel section, especially trapezoid section is shown in Figure 1 [3].

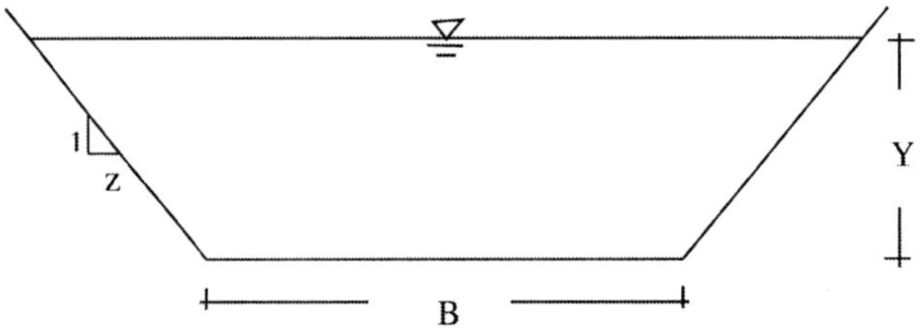

Figure 1. Trapezoid section of channel [3]

$$A = (B + zY)Y \tag{3}$$

$$P = B + 2Y\sqrt{1 + z^2} \tag{4}$$

$$R = \frac{A}{P} \tag{5}$$

Where A is area of channel section; B is width of channel; Y is hydraulic depth; P is wetted perimeter; R is hydraulic radius; and z is side slope.

3. Study methodology

In the planning of the dimensions of the irrigation channel, the amount of water that is flowed by the channel or channel discharge, Q must be greater than or equal to the amount of irrigation discharge, Q_{ir} so that irrigation water needs are met. This relationship is shown below [1, 2].

$$Q \geq Q_{ir} \tag{5}$$

The discharge of irrigation needs, Q_{ir} is obtained from calculating the need for irrigation water which has been known in advance. Based on formula 3.1, in calculating the dimensions of the channel, the amount of channel discharge, Q is equal to the amount of irrigation discharge, Q_{ir}.

Calculation of channel dimensions begins with analyzing the all formulas in the literature review section above. The analysis carried out is described in the following

1. From formula 4, formula 6 is obtained.

$$P = B + 2Y\sqrt{1+z^2}$$
$$B = P - \left(2\sqrt{1+z^2}\right)Y \quad (6)$$

2. Then, from the substitution of formula 6 into formula 3, formula 7 is obtained.

$$A = (B + zY)Y$$
$$A = \left\{P - \left(2\sqrt{1+z^2}\right)Y\right\}Y + zY^2$$
$$A = PY - \left(2\sqrt{1+z^2}\right)Y^2 + zY^2 \quad (7)$$

Formula 7 above is a quadratic equation which is then simplified to formula 8.

$$\left(2\sqrt{1+z^2} - z\right)Y^2 - PY + A = 0 \quad (8)$$

3. As it is known that the solving of quadratic equations is done with equations as shown in Formulas 9 and 10 [6].

$$Y_1 = \frac{-b + \sqrt{b^2 - 4ac}}{2a} \quad (9)$$
$$Y_2 = \frac{-b - \sqrt{b^2 - 4ac}}{2a} \quad (10)$$

The values a, b, and c in formulas 9 and 10 are obtained from formula 8, namely:

$$a = 2\sqrt{1+z^2} - z \quad (11)$$
$$b = -P \quad (12)$$
$$c = A \quad (13)$$

4. The requirement that the results of the quadratic equation calculation of formulas 9 and 10 above do not produce imaginary values is discriminant value, D cannot be negative [6].

$$D = b^2 - 4ac \geq 0 \text{ or}$$
$$b^2 \geq 4ac \quad (14)$$

5. To meet the requirements of D value cannot be negative, it is necessary to take steps to solve b^2, starting from formula 5 to be changed to formula 15

$$R = \frac{A}{P}$$
$$P = \frac{A}{R} \quad (15)$$

Then, by substituting formula 15 into formula 12, then formula 16 is obtained.

$$b = -P$$
$$b = -\frac{A}{R} \quad (16)$$

From formula 1, formula 17 is obtained.

$$Q = AV$$
$$A = \frac{Q}{V} \quad (17)$$

Then from formula 2, formula 18 is obtained.

$$V = \frac{1}{n} R^{2/3} S^{1/2}$$

$$R = \left(\frac{V\,n}{S^{1/2}} \right)^{3/2} \tag{18}$$

Furthermore, by substituting formula 17 and formula 18 into Formula 16, then the Formulas 19 and 20 are obtained.

$$b = -\frac{A}{R}$$

$$b = -\left\{ \frac{\frac{Q}{V}}{\left(\frac{V\,n}{S^{1/2}} \right)^{3/2}} \right\}$$

$$b = -\left(\frac{Q\,S^{3/4}}{V^{5/2}\,n^{3/2}} \right) \tag{19}$$

$$b^2 = \frac{Q^2\,S^{3/2}}{V^5\,n^3} \tag{20}$$

The Formula 20 above is the solution for b^2 in the Formula 14.

6. Next, the solution for *4ac* is in formula 14, starting by substituting formula 17 into Formula 13, then Formula 21 is obtained.

$$c = A$$

$$c = \frac{Q}{V} \tag{21}$$

Then based on the value of *a* in Formula 11 and the value of *c* in Formula 21, the solution for *4ac* is obtained as shown in Formula 22.

$$4ac = 4\left(2\sqrt{1+z^2} - z \right)\frac{Q}{V} \tag{22}$$

The formula 21 above is the solution for *4ac* in the formula 10.

7. Then b^2 in formula 20 and *4ac* in formula 21 are substituted into formula 14, then formula 23 is obtained.

$$b^2 \geq 4ac$$

$$\frac{Q^2\,S^{3/2}}{V^5\,n^3} \geq 4\left(2\sqrt{1+z^2} - z \right)\frac{Q}{V}$$

$$V^4 \leq \frac{Q\,S^{3/2}}{4\left(2\sqrt{1+z^2} - z \right)n^3} \qquad \text{if} \quad \alpha = 4\left(2\sqrt{1+z^2} - z \right)$$

$$V \leq \left(\frac{Q\,S^{3/2}}{\alpha\,n^3} \right)^{1/4} \tag{23}$$

The formula 23 above is the formula for calculating the mean flow velocity in the channel, *V* with the condition of the channel base slope, *S* is determined. If it turns out that the value of *V* is less than the mínimum velocity allowed or greater than the máximum velocity allowed, then the calculation is carried out to find the channel base slope, *S* with the value of mean flow velocity in the channel, *V* is determined. The formula is as follows.

$$S \geq \left(\frac{V^4 \, \alpha \, n^3}{Q} \right)^{2/3} \tag{24}$$

4. Result and discussion

The methodology section above shows that the calculation of the dimensions of the irrigation canal must go through the quadratic equation, which requires discriminant value, D must not be negative in order to avoid imaginary numbers. Based on non-negative discriminant conditions and after analyzing the formulas in the literature review section, a formula is obtained to calculate the mean flow velocity in the channel, V with the channel base slope, S is determined as well as the formula is obtained to calculate the channel base slope, S with the mean flow velocity at channel, V is determined.

In calculating the dimensions of the irrigation channel, several variables have been determined in advance. The variable of channel discharge, Q is equal to irrigation discharge, Q_{ir} which is calculated based on irrigation water needs. Then the channel side slope, z and Manning's roughness coefficient, n values are both determined based on the type of channel material. Whereas the variable mean flow velocity in the channel, V and the channel base slope, S is selected one of thier value.

Based on the results of the analysis that has been carried out, the method of calculating the dimensions of the irrigation channel with a trapezoidal cross-section is obtained, namely the calculation of the hydarulic depth, Y and width of channel, B in the order described below.

1. Calculate the value of α with the side slope value of the channel, z has been determined as planned.

$$\alpha = 4\left(2\sqrt{1+z^2} - z \right)$$

2. Calculate the value of the mean flow velocity in the channel, V or the base slope the channel, S with the Manning roughness coefficient, n and the channel discharge, Q has been predetermined as planned.

$$V = \left(\frac{Q \, S^{3/2}}{\alpha \, n^3} \right)^{1/4} \text{ or } S = \left(\frac{V^4 \, \alpha \, n^3}{Q} \right)^{2/3}$$

3. Calculate the value of a

$$a = 2\sqrt{1+z^2} - z$$

4. Calculate the value of b

$$b = -\left(\frac{Q \, S^{3/4}}{V^{5/2} \, n^{3/2}} \right)$$

5. Calculate the value of c

$$c = \frac{Q}{V}$$

6. Calculate the value of the hydraulic depth in channel, Y_1 and Y_2

$$Y_1 = \frac{-b + \sqrt{b^2 - 4ac}}{2a}$$

$$Y_2 = \frac{-b - \sqrt{b^2 - 4ac}}{2a}$$

7. Determine the P value.

$$P = -b$$

8. Calculate width of channel, B_1 and B_2

$$B_1 = P - \left(2\sqrt{1+z^2} \right) Y_1$$

$$B_2 = P - \left(2\sqrt{1+z^2}\right)Y_2$$

9. Select a channel dimension pair Y_1 and B_1 or Y_2 and B_2
10. Perform the control calculations in the following order.

$$A = (B + zY)Y$$

$$P = B + 2Y\sqrt{1+z^2}$$

$$R = \frac{A}{P}$$

$$V = \frac{i}{n} R^{2/3} S^{1/2}$$

$$Q = AV \geq Q_{ir}$$

The flow chart of the method for calculating the dimensions of a trapezoidal cross-sectional irrigation channel is illustrated by Figure 2.

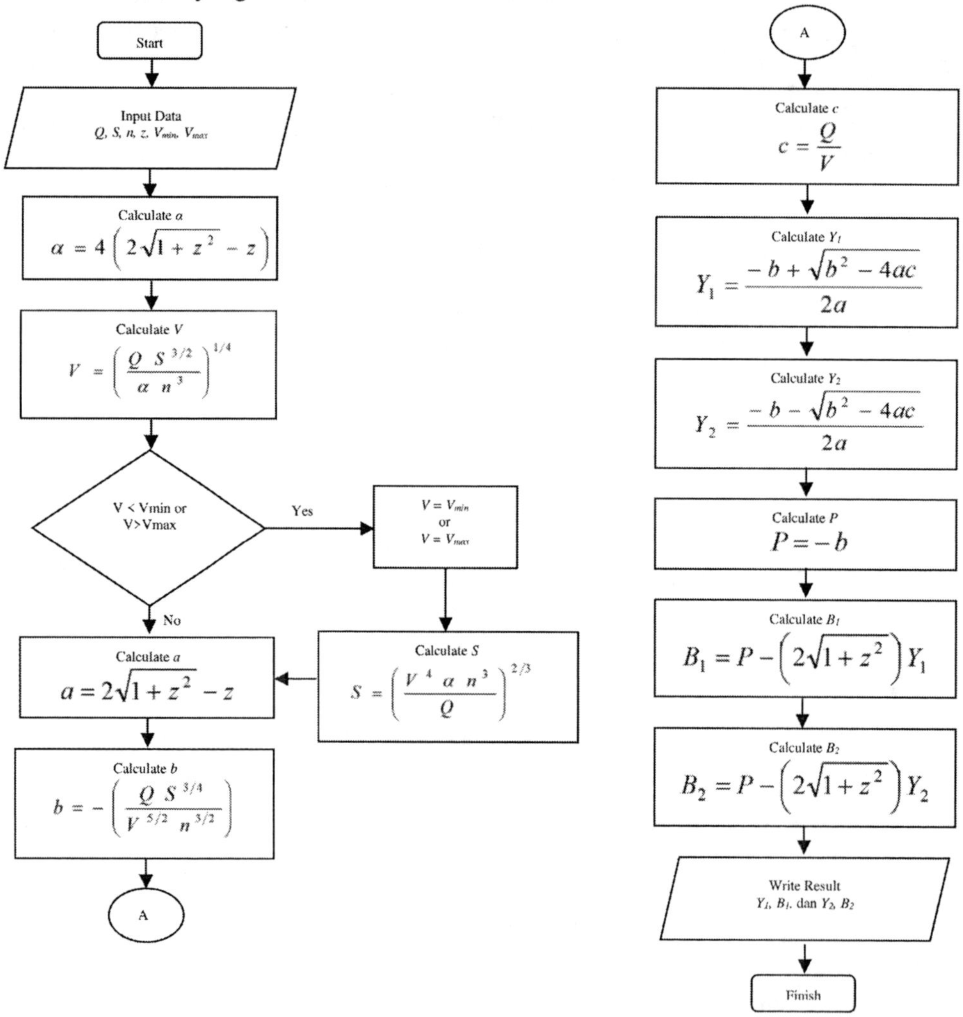

Figure 2. The flow chart of the method for calculating the dimensions of a trapezoidal cross-sectional irrigation channel.

As an example, the proposed method is applied to calculate the dimensions of the primary canal in the Krueng Jreu irrigation area located in Aceh Besar District, Aceh Province. According to [8] Krueng Jreu Irrigation has an area of 3,175 ha, the length of the primary channel is 1,839.80 m, the maximum need for irrigation water is 1.7 ltr/sec/ha, so that the irrigation discharge that is flowed by the primary channel is, Q = 5.4 m^3/sec. The primary channel is planned to be made of concrete, has a trapezoidal cross-section, and the slope of the base of the channel in the direction of the flow is the same as the slope of the irrigation area, S = 0.0016. According to [5] for concrete channels: roughness coefficient of Manning, n = 0.015, the side slope can be planned, z = 1.5 and the maximum permissible velocity, V_{max} = 3 m/sec.

At first, the calculation was tried by following the procedure for calculating the dimensions of the channel according to [3], whose calculation steps are as follows.

1. $V = \frac{1}{n} R^{2/3} S^{1/2}$

   $$R = \left(\frac{n\,V}{S^{1/2}}\right)^{3/2} = \left(\frac{0.015\times 3}{0.0016^{1/2}}\right)^{3/2} = 1.2\,m$$

2. $Q = A\,V$

   $$A = \frac{Q}{V} = \frac{5.4}{3} = 1.8\,m^2$$

3. $P = \frac{A}{R} = \frac{1.8}{1.2} = 1.5\,m$

4. $A = (B + z\,Y)Y = B\,Y + z\,Y^2$

   $$1.8 = B\,Y + 1.5\,Y^2$$

5. $P = B + 2\,Y\sqrt{1+z^2}$

   $$B = P - 2\,Y\sqrt{1+z^2}$$

   $$B = P - 2\,Y\sqrt{1+z^2} = 1.5 - 2\,Y\sqrt{1+1.5^2} = 1.5 - 3.61\,Y$$

6. Substitute 5 into 4, then the quadratic equation is obtained.

   $$1.8 = (1.5 - 3.61\,Y)\,Y + 1.5\,Y^2 = 1.5\,Y - 2.1\,Y^2$$

   $$2.1\,Y^2 - 1.5\,Y + 1.8 = 0$$

   The quadratic equation above has a = 2.1; b = -1.5; and c = 1.8

7. Solving the quadratic equation to calculate the value of hydraulic depth, Y.

   $$Y_{1.2} = \frac{-b \pm \sqrt{b^2 - 4ac}}{2a}$$

   $$Y_{1.2} = \frac{1.5 \pm \sqrt{1.5^2 - 4\times 2.1\times 1.8}}{2\times 1.8} = \frac{1.5 \pm \sqrt{-12.87}}{4.2}$$

   It turns out that the value of flow depth, Y cannot be obtained from the above calculation procedure because the discriminant value, $D = b^2 - 4ac$ is negative, resulting in an imaginary number.

In order for the discriminant value not to be negative, it is calculated using the proposed method which is the sequence of calculations as follows.

1. $\alpha = 4\left(2\sqrt{1+z^2} - z\right) = 4\left(2\sqrt{1+1.5^2} - 1.5\right) = 8.4$

2. $V=\left(\dfrac{Q\,S^{3/2}}{\alpha\,n^3}\right)^{1/4}=\left(\dfrac{5.4\times 0.0016^{3/2}}{8.4\times 0.015^3}\right)^{1/4}=1.8\;m/\text{sec}$

V < $Vmax$ *and* V > $Vmin$ = *0.76 m/sec* [3].

3. $a=2\sqrt{1+z^2}-z=2\sqrt{1+1.5^2}-1.5=2.1$

4. $b=-\left(\dfrac{Q\,S^{3/4}}{V^{5/2}\,n^{3/2}}\right)=-\left(\dfrac{5.4\times 0.0016^{3/4}}{1.8^{5/2}\times 0.015^{3/2}}\right)=-5.4$

5. $c=\dfrac{Q}{V}=\dfrac{5.4}{1.8}=3$

6. $Y_1=\dfrac{-b+\sqrt{b^2-4ac}}{2a}=\dfrac{5.4+\sqrt{5.4^2-4\times 2.1\times 3}}{2\times 2.1}=1.760\,m$

$Y_2=\dfrac{-b+\sqrt{b^2-4ac}}{2a}=\dfrac{5.4-\sqrt{5.4^2-4\times 2.1\times 3}}{2\times 2.1}=0.812\,m$

7. $P=-b=5.4$

8. $B_1=P-\left(2\sqrt{1+z^2}\right)Y_1=5.4-\left(2\sqrt{1+1.5^2}\right)1.8=-0.9\,m$

$B_2=P-\left(2\sqrt{1+z^2}\right)Y_2=5.4-\left(2\sqrt{1+1.5^2}\right)0.8=2.5\,m$

Selected: width of channel, B = 2.5 m and hydraulic depth, Y = 0.812 m, as displayed in Figure 3.

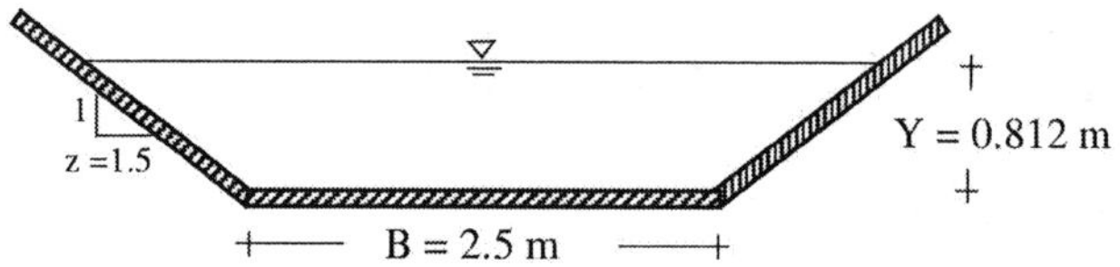

Figure 3. Dimensions of the cross section of the trapezoidal channel

Control:

$A=(B+z\,Y)Y=B\,Y+z\,Y^2=2.5\times 0.812+1.5\times 0.812^2=3.019\,m^2$

$P=B+2\,Y\sqrt{1+z^2}=2.5+2\times 0.812\times\sqrt{1+1.5^2}=5.428\,m$

$R=\dfrac{A}{P}=\dfrac{3.019}{5.428}=0.556\,m$

$V=\dfrac{1}{n}R^{2/3}S^{1/2}=\dfrac{1}{0.015}\times 0.556^{2/3}\times 0.0016^{1/2}=1.804\,m/\text{sec}$

$Q=A\,V=3.019\times 1.804=5.446\,m^3/\text{sec}$ > Q = 5.4 m^3/sec (fulfill the requirements).

5. Conclusion

The hydraulic depth in the irrigation channel with a trapezoidal section calculated by the quadratic equation, the discriminant value in the quadratic equation must not be negative so that the calculation results are not imaginary. The negative value can be avoided by analyzing the independent variable, namely the mean flow velocity or the channel base slope. Based on the non-negative discriminant value, it are obtained the formula for calculating the mean flow velocity with the channel base bottom slope is determined as well as the formula for calculating the channel base slope with the mean flow velocity is determined. The proposed method for calculating the dimensions of the irrigation channel with a trapezoidal cross-section is in the form of a calculation sequence and a calculation flow chart which, if followed, will avoid calculation failure due to imaginary numbers appearing in the calculation of the hydraulic depth in the channel calculated by the quadratic equation.

References

[1] Yulianur A 2004 *Jurnal Teknik Sipil,* **3** **(2)**. *(in Bahasa Indonesia).*

[2] Yulianur A 2008 *Jurnal Teknik Sipil,* **6** **(3)**. *(in Bahasa Indonesia).*

[3] Chow V T 1959 *Open-Channel Hydraulics*. Tokyo: McGraw-Hill Kogakusha LTD.

[4] Henderson F M 1966 *Open Channel Flow*. New York: MacMillan Publishing Inc.

[5] Departemen Pekerjaan Umum 1986 *Standar Perencanaan Irigasi Kriteria Perencanaan Bagian Saluran KP-03.* Bandung: C.V. Galang Persada. *(in Bahasa Indonesia).*

[6] Stroud K A 1990 *Engineering Mathematics Programmes and Problems Third Edition*. Hampshire: Macmillan Education Ltd.

[7] Te Chow V 1988 *Applied Hydrology*. Singapore: McGrow-Hill Company.

[8] Kementerian Pekerjaan Umum 2015 *Daerah Irigasi Krueng Jreu, Ranting Indrapuri*. Available at: https://rtg-indrapuri.blogspot.com/2015/05/daerah-irigasi-krueng-jreu.html. *(in Bahasa Indonesia).*

IOP Conf. Series: Materials Science and Engineering 1087 (2021) 012028 doi:10.1088/1757-899X/1087/1/012028

The comparison between the method of Bina Marga and the pavement condition index (PCI) in road damage condition evaluation (case study: Prof. Ali Hasyimi Street, Banda Aceh)

R Faisal[1,*], M Ahlan[1], C Mutiawati[1], M Rozi[2], Zulherri[2]
[1]Department of Civil Engineering, Universitas Syiah Kuala, Banda Aceh, Indonesia.
[2]Vocational education, Department of Civil Engineering, Universitas Syiah Kuala, Banda Aceh, Indonesia

*E-mail: ruhdi.faisal@unsyiah.ac.id

Abstract. Monitoring in-service road surface conditions is a part of road pavement maintenance. The most effective maintenance type will save costs and time. Road damage usually occurs on roads that have reached the design age, but some roads have been damaged before reaching the design age. To assess road damage, several methods can be used. The most widely used method in Indonesia is the Bina Marga method and the PCI (Pavement Condition Index) method. This study aims to compare both methods, in terms of accuracy and workability, in monitoring the surface condition of Prof. Ali Hasyimi street, which is located in the city of Banda Aceh. The data required for these two methods is the primary data, which are geometric, visual road damage, and traffic volume. Based on this study's results, the Bina Marga method is easier and faster to use, while the PCI method is more detailed to utilize. In terms of accuracy, both methods relatively have the same accuracy.

1. Introduction

Roads are one of the most important transportation infrastructures to support the economic growth of a region. However, the newly constructed roads will experience damage either due to natural factors or vehicle loads. Road damage causes enormous losses, especially for road users, such as long travel times, congestion, accidents.

The concept of the pavement management system is used to address the road damage problem. There are several steps in the pavement management system, including in-service monitoring and evaluation. The in-service monitoring and evaluation data are used to define the most appropriate interruption and schedule in maintaining the road structure [1]. Also, the data is used as the decision support system in infrastructure investment [2]. There are three most frequently used methods to monitor and evaluate the road surface in Indonesia: Bina Marga, PCI, and the Asphalt Institute method [3]. While PCI is also used in other countries [4, 5].

PCI method has been compared with other road-surface monitoring methods, such as the international roughness index (IRI) method and pavement structural strength index (PSSI) [6, 7]. However, only a few studies compare the Bina Marga method with the PCI method. This study aims to compare both methods in evaluating the surface condition of a secondary arterial road, a road that lies in built-up areas and serves much traffic.

2. Study area

This research was conducted on a segment of Prof. Ali Hasyimi street, which is between the BPKP intersection and the Pango Bridge. This road is located in the city of Banda Aceh, and it serves much traffic. Banda Aceh is the capital of Aceh Province with a population in 2019 of around 270,321 people. Since the 2004 tsunami, the city of Banda Aceh has experienced very rapid development. One aspect to support this development is the policy of the Aceh Regional Spatial Plan, which establishes the city of Banda Aceh City as the national activity center. According to the city's land-use planning, Prof. Ali Hasyimi street is planned as an office and trading area. This street has transportation facilities, such as the off-street parking facilities made to prevent on-street parking activities.

Based on the historical data, Prof. Ali Hasyimi street was completed in 2011. The street planned age is 10 years. As a result, it is estimated that by 2021 the road has reached the design age. Because of the short remaining service life—one year—the researcher wants to examine the condition of Prof. Ali Hasyimi's road by using two different methods: Bina Marga Method and the PCI method.

Prof. Ali Hasyimi street has a length of about 1.65 km, a width of 18 meters, and 3 meters median. The type of road is 6/2 D. It is classified as a class II road. The road functional status is secondary arterial. For more details, the location of the road position can be seen in Figure 1.

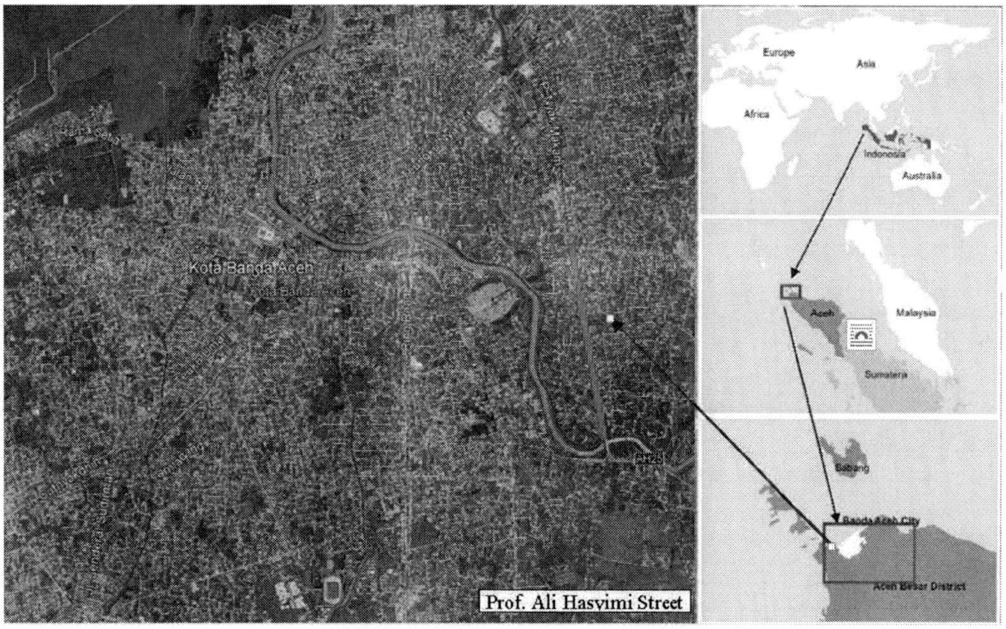

Figure 1. Map of the study area

3. Methodology

This research began with the formulation of the problem, background, and the research objective. Then, the literature study was conducted. After that, the primary and secondary data was collected. The primary data in this research is the road geometry obtained by measuring the dimensions of the road using a length measuring instrument. The Secondary data in this study are administrative maps, road status based on current regulations, and a sketch of the research location obtained from google maps. Then, data processing and data analysis were carried out using two types of methods, namely the Bina Marga method and the PCI method.

Prof. Ali Hasyimi street has a total width of 21 meters, a pavement width of 18 meters, and a median of 3 meters. The cross-section of the road can be seen in Figure 2. The length of the road is 1,650 meters.

Furthermore, it is divided into five segments, each of which is 330 m long. The image containing the segment divisions can be seen in Figure 3.

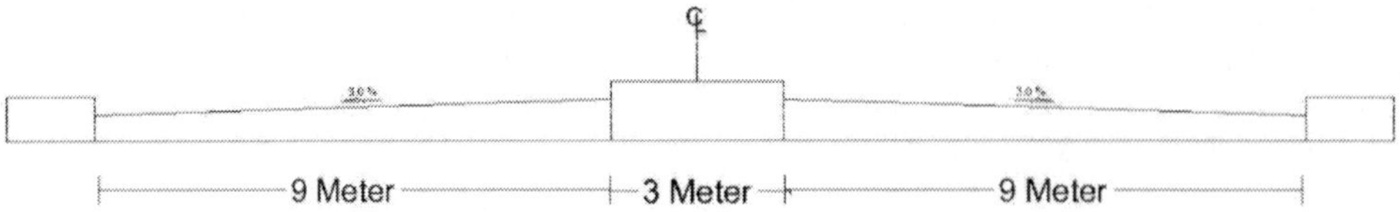

Figure 2. Cross section

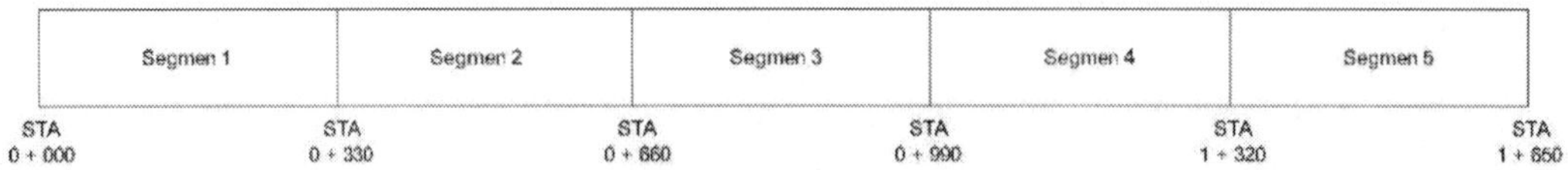

Figure 3. Segment division

The road damage survey was conducted from March 10 to March 11, 2020. Road damage data can be seen in Figure 3. The traffic survey was carried out in three days, namely Tuesday, Wednesday, and Saturday, which are 16, 17, and 20 June 2020, respectively. The data collected through the traffic survey can be seen in Figure 4.

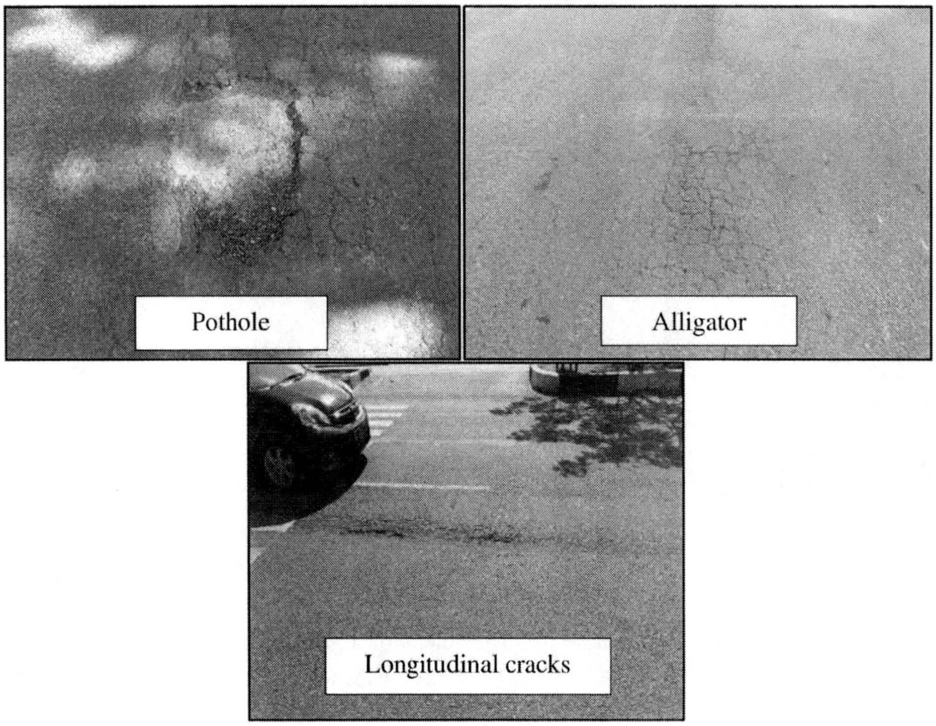

Figure 4. Photos and types of road damage

Regarding the step for the Bina Marga method, first, the road segments were determined [8]. Next. the areas of road damage and the level of damage were calculated. Road damage data can be seen in the figure. Then, the value of the annual average daily traffic was collected. Furthermore, the value of the road classes was determined. Then, the results of the survey were determined, and the data was classified according to the type of damage. After that, the parameters for each type of damage was calculated, and an assessment of each type of damage was conducted. Then, each number for all types of damage was added up, and a road condition value to the total damage figure was assigned. The value of road conditions based on the total number of damage was obtained from each segment.

The Bina Marga method decides the priority order of the maintenance types. For the priority sequence of 0 to 3, it indicates that the road should be included in the road improvement program. For the order priority of 4 to 6, it indicates that the roads need to be included in a regular maintenance program. While for the order of priority of more than 7, the road is sufficiently included in the routine maintenance program.

As mentioned above, the Bina Marga method requires some data including traffic volume. The traffic volume is the number of motorized vehicles passing a certain point at a specific time. Traffic volume is expressed in vehicles per hour (pcu / hour) [9]. The formula for traffic volume can be seen in Formula 1 and 2. Formula 2 is for calculating traffic volume, from peak hours to average daily traffic flow. The K-factor for outer-city roads is 0.11 and the K-factor for inner-city roads is 0.09. The formula for determining the road priority value can be seen in formula 3.

$$Q_{pcu} = PCE_{LV}\ x\ LV + PCE_{HV}\ x\ HV + PCE_{MC}\ x\ MC \quad (1)$$

Information:
Qpcu: Total traffic flow (pcu / hour)
LV: Light vehicle, pce value = 1
HV: Heavy vehicle, pce value = 1.3
MC: Motorcycle, pce value = 0.5
Pcu: private car unit
Pce: private car equivalent

$$\text{ADT}= \frac{peak\ hour}{k} \quad (2)$$

$$\text{Priority Value} = 17 - (\text{ADT Class} + \text{Road Condition Value}) \quad (3)$$

Where:
ADT: Average daily traffic

The results of the calculation of the Bina Marga method are shown in Table 1. The average priority value is 9, which is entered into the periodic maintenance handler.

Table 1. Recapitulation of research results

No.	Segment	Priority value
1	1	10
2	2	8
3	3	9
4	4	8
5	5	10
Average		9

Regarding the steps in the PCI method, first, the visual-observation based data of damage was recorded. Then, the level of damage was decided according to the formula [3]. The recapitulation of the percentage value of road damage can be seen in Table 2. Next, the severity of the pavement damage according to the damage conditions was decided. The severity was divided into low (L), medium (M), and High (H). Then, the deduct value—the reduction value for each type of damage that is slotted into the curve—was decided. Next, the maximum allowable reduction amount was determined. After that, the maximum corrected reduction value was determined based on the graph of Figure 5. The qualification of pavement quality according to PCI can be seen in Figure 6.

$$(density)\ (\%) = \frac{Ad}{As} \text{ x } 100\% \tag{4}$$

Information :
A^d = the total area of damage types for each level of damage (m^2),
A^s = the total area of the segment unit (m^2),

Table 2. Recapitulation of 1.65 km road damage rate

No.	Type of Damage	L	M	H	Total Damage Rate	Average PCI Value (%)
1	Pothole	1.27	0.44	0.53	2.23	3.01
2	Alligator	34.54	35.76	0.00	70.30	94.88
3	Longitudinal cracks	0.06	1.45	0.06	1.56	0.15
	Total				74.09	100.00

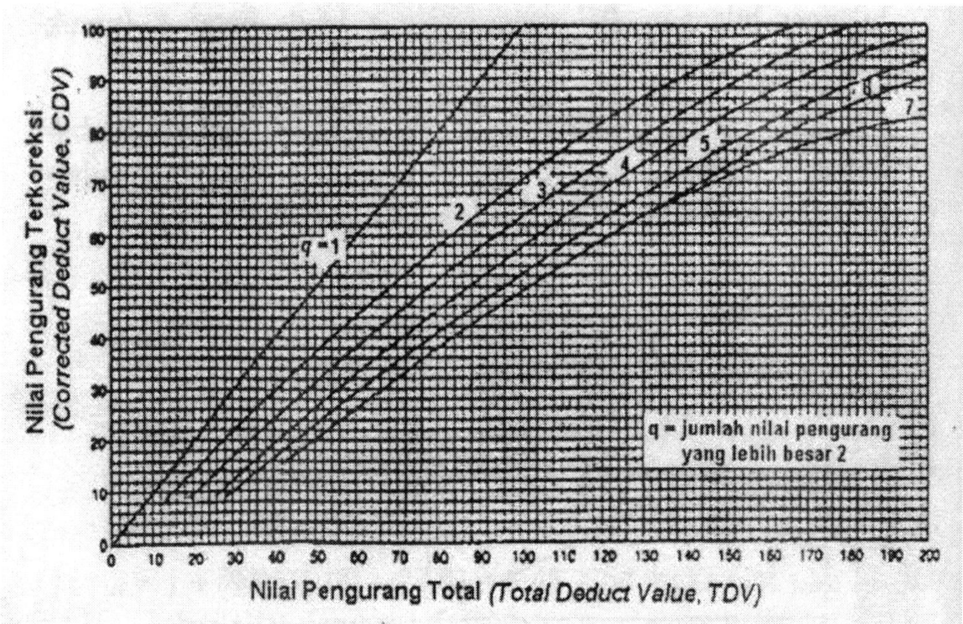

Figure 5. Corrected Deduct Value (CDV) graph

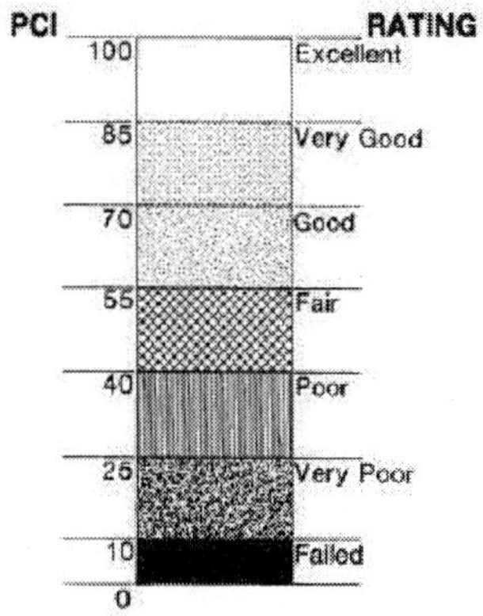

Figure 6. Qualification of pavement quality according to PCI

The results of the calculation of the PCI method can be seen in Table 3. From the above values, it can be concluded that the pavement value in the segment of Prof. Ali Hasyimi street is categorized as perfect (excellent). The average PCI value is 94.4.

Table 3. PCI Value

No.	Segment	Total PCI Value	Condition
1	Segment I	100	Excellent
2	Segment II	100	Excellent
3	Segment III	84	Very good
4	Segment IV	88	Excellent
5	Segment V	100	Excellent
	∑	475	

4. Result and discussion

Based on the analysis using the Bina Marga Method, the obtained order of priority is 9, which means the road is included in the routine maintenance program. For the analysis based on the PCI method, the obtained PCI value is 9.46, which means the road is in perfect condition. From bot methods, it can be concluded that the two methods have the same treatment recommendation results. However, the PCI and Bina Marga method have several differences in determining the assessment of road damage conditions. The differences can be seen in Table 4.

Table 4. Comparison of the Bina Marga Method and the PCI Method

Bina Marga	PCI
1. Conducting average daily traffic survey	1. Not conducting the average daily traffic survey
2. The analysis includes the number of damage condition and the average daily traffic class value	2. The analysis uses a graph according to the type of damage
3. The final result is in the form of the order of priority for road damage handling	3. The final result is in the form of the level of road pavement damage

There are several advantages and disadvantages of both Bina Marga and PCI methods. For the Bina Marga method, the advantage is that the method involves the volume of traffic passing through a road section. Also, the Bina Marga method process is faster because the calculation does not use many graphs with much data needed. However, the limitation of this method is that the method lacks detail in the assessment because it only includes the type of available damage to the road damage value. Furthermore, for potholes damage, this method compares with the percentage area of the road under review. This type of pothole damage calculation could underestimate small damages in the road divided into many segments.

In terms of the PCI method, the advantage is that the method has a more detailed damage analysis. This is because the graph for each type of damage, which is different one by one, is used in this method. The disadvantage of this method is that the process takes a longer time than those of the Bina Marga method because the method requires the input of each type of damage, which should be conducted one by one into the graph. Furthermore, the PCI method does not include the actual traffic volume while the actual traffic volume should have a high impact on road damage.

5. Conclusion

The conclusion that can be drawn from this research is that, in terms of workability, the Bina Marga method is easy to use compared to the PCI Method that involves many graphics and requires several repetitions to get the best value. Furthermore, in terms of accuracy, the Bina Marga method lacks detail in assessing road damage because it only includes the type of road damage without paying attention to the number in a segment. To address this issue, we encourage to use smaller segmentation in the Bina Marga method. On the other hand, the PCI method analyzes in more detail so that any type of road damage can be analyzed using this method. However, the PCI method does not include the traffic volume factor, which should have an impact on road damage. From the results of the assessment, these two methods have relatively the same final value.

References

[1] Uddin W, Hudson W R, Haas R 2013 *Public Infrastructure Asset Management, Second Edition*. New York: McGraw-Hill Education.

[2] Kelly G, Delaney D, Chai G, Mohamed S 2016 *J. Traffic Transp. Eng.*, **3** **(5)** 465–474.

[3] Hardiyatmo H C 2007 *Pemeliharaan Jalan Raya*. Yogyakarta: Gadjah Mada University Press. *(in Bahasa Indonesia).*

[4] Werkineh A B, Demissie B A 2019 *Am. J. Sci., Eng. Technol.,* **4** **(4)** 73–79.

[5] Shah Y U, Jain S S, Tiwari D, Jain M K. 2013. *Procedia Social and Behavioral Sci.,* **104** 332–341.

[6] Suryoto, Siswoyo D P, Setyawan A 2017 *Procedia Eng.,* **171** 1435–42.

[7] Qian G, Hu C, Yu H, Gong X 2020 *J. Mater. Civil Eng.,* **32** **(7)** 1–9.

[8] Direktorat Pembinaan Jalan Kota 1990 *Tata Cara Penyusunan Pemeliharaan Jalan Kota (No.108/T/BNKT/1990)*. Jakarta: Direktorat Jendral Bina Marga Departemen PU. *(in Bahasa Indonesia).*

[9] Direktorat Jenderal Bina Marga Republik Indonesia 1997 *Manual Kapasitas Jalan Indonesia. (in Bahasa Indonesia).* Jakarta: Direktorat Jenderal Bina Marga Republik Indonesia.

Effect of enforcement operations during festive seasons called Ops Sikap on road traffic accidents of Malaysia

Y Darma[1,2,*], M R Karim[1], S Abdullah[3]
[1]Transportation Science Society of Malaysia.
[2]Department of Civil Engineering, Universitas Syiah Kuala, Banda Aceh, Indonesia
[3]Accredited Road Safety Auditor, Malaysia.

*E-mail: yusriad@gmail.com

Abstract. The road safety statistics have shown a remarkable increase in number of injuries during the festive seasons in Malaysia. To anticipate the burden of the injuries, the police has carried out an advanced enforcement operations called "Ops Sikap" since 2001. Evaluations on the effectiveness of the operations are, however, found inconclusive. This study employed ARIMA and intervention model to further evaluate the effect of the Ops Sikap. The effect of an intervention on the number of accidents is measured using the pulse function. ARIMA model is used to forecast the series without such intervention and become the basis for the ARIMA(p,d,q) process of the intervention model. Forecasted accidents from the intervention model are then compared with the ARIMA one to attain how far the effect. The results show that the operations have reduced the number of accidents in many years. Out of twenty five Ops Sikap carried out, fourteen times (56%) have successfully brought the accidents down. The operations have provided the reduction of the monthly accidents up to 0.7% as compared with the accidents if there were no such operations taken.

1. Introduction

Malaysia targeted to have the rate of road traffic fatalities of 2.0 per 10,000 vehicle registrations. To achieve the rate, the government implements various strategies, interventions and safety measures. The rules relating to traffic safety have been imposed since the early 1970s. As such, the 1973 Motorcycles (Safety Helmet) Rules, the 1978 Motor Vehicles (Seat belt) Rules, Road Transport Act 1987, and the 1989 Motor Vehicles (Speed Limit) Rules. Besides the rules, the authorities released the standards for vehicle safety, technical guidelines for road safety, launched the national road safety plans and programmes. The government has formed the agencies to manage the road safety programmes of the country as well. The country has also joined the Decade of Action for Road Safety 2011–2020 initiated by the United Nations.

The enactment of the rules and regulations are, however, not promptly reduce the number of injuries in Malaysia. Even, the trend of fatalities continues to increase over the years. The rates of fatalities per vehicle registrations, however, continue to decline. The remarkable decline (8.9%) on the number of fatalities occurred in the period 1997–1998 [1, 2]. This may be owing to the effects of the National Road Safety Plan launched in 1996 that consists of various strategies and efforts, including the more intense campaigns and enforcements to improve the road safety. Nevertheless, another explanation can also be related to the major decline i.e. the financial crisis occurred in 1997–1998. It is believed that traffic exposure to the risk on the roads decreased at that time due to the substantial reduction (6.7%) of the road sector energy consumption in the period 1997–1998. This indicates that more efforts to improve the safety are still required.

The government may realise that with the less intense campaigns and enforcements, the rules and regulations shall not provide a remarkable improvement on road safety. Many studies also have shown that enforcement is essential. For instance, the use of speed camera shall reduce the injuries and efficient [3-6]. Thus, in the recent years, the more intense safety campaigns and followed by enforcements have been conducted. Moreover, the road safety statistics have shown that higher number of injuries occurs during the festive seasons, specifically during the holydays for the *Eid* festival, the Chinese New Year (CNY) and the Deepavali respectively. To anticipate the burden of the injuries during the festive seasons, the road safety enforcement operations called "Ops Sikap" has been taken by the Royal Malaysian Police (PDRM) since 2001. The operation is intended to increase the compliance on the road traffic safety rules and regulations. Later, some other government agencies as such the Road Transport Department (JPJ), the Public Works Department (JKR) and the Road Safety Department (JKJR) joined the enforcement operations in accordance with their responsibility. This integrated operations aim to intensify the enforcement during the festive seasons and called "Ops Bersepadu" and have been taken place since 2006. The police conducted the *Ops Sikap* until the early 2012 and since the mid 2012 onwards they named the operations as "Ops Selamat."

The police claimed that the enforcement operations has generally provided the better road safety situations during the festive seasons over the years, especially in reducing the injuries. Several studies have also been conducted to evaluate the effectiveness of the enforcement operations, especially under the Malaysian Institute of Road Safety Research (MIROS) [7-10]. The study by Jamilah et al. [10] urged that enforcement activities should be improved in the future to further reduce the number of crashes and fatalities in the country. Another study by Yaacob, Husin, Abd. Aziz, and Nordin [11] revealed that out of fifteen Ops Sikap implemented (for the period 2001–2007), only Ops Sikap II conducted in February 2002, VI (January & February 2003), XIII (January 2004), XII (October 2006) and XIV (October & November 2007) show a reduction in the number of road accidents occurrence though the reduction is not statistically significant.

We undertook this study to further evaluate the effect of the Ops Sikap as there are various conclusions and opinions on the achievement of the enforcement operations. The more monthly accidents data are used i.e. from January 1998 to January 2012. So, the impact of the twenty five Ops Sikap on the number of accidents are made possible to be analysed.

2. Materials and methods

2.1 Data

This study prefers to analyse the impact of the enforcement operations on the number of fatalities. The unavailability of the data, however, become the limitation. In Malaysia, the road traffic safety related data are collected, recorded and published by the PDRM. At the moment, only number of accidents are published in monthly basis, the rest types of injuries are provided annually. Hence, the series of the monthly number of accidents are used in the analysis. Other data that are needed to know are the months when the enforcement operations conducted. Figure 1 shows the series of the accidents and the months when the Ops Sikap were carried out.

2.2 Methods to attain the impact of an intervention

In general, there are two patterns of the intervention effect i.e. step and pulse function. The step function represents an intervention occurring at time T that remains in effect thereafter. Whilst, the pulse function represents an intervention taking place at only one time period [12]. On the basis of the time plot graph (Figure 1), the pulse function is appropriate to be used to attain the impact of the safety operations on the number of accidents. Therefore, the effect of an intervention on the number of accidents is measured using the pulse function of the intervention model. This model is the combination of the intervention function and the noise process. Hence, the autoregressive integrated moving average (ARIMA) model for the series is developed first. This model is also used to forecast the series without such intervention and become the basis for the ARIMA (p, d, q) process of the intervention model. Forecasted accidents from the intervention model are then compared with the ARIMA one to attain how far the effect.

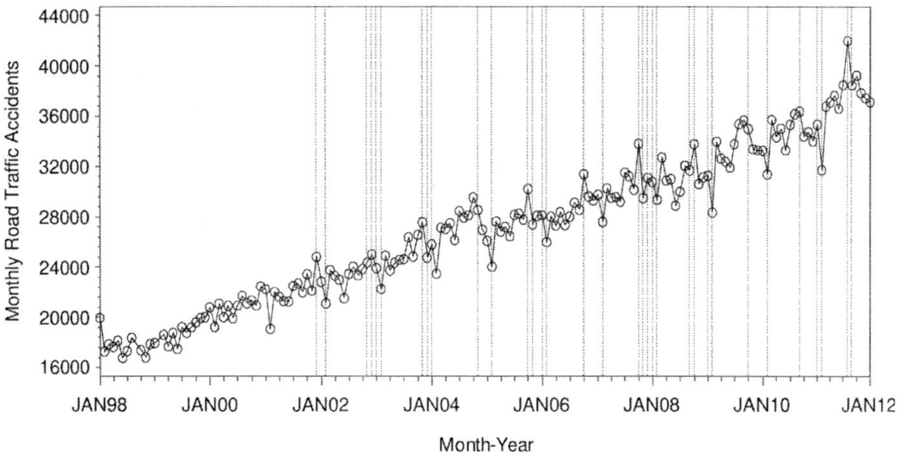

Figure 1. Number of road traffic accidents and the month when the Ops sikap carried out

2.3 ARIMA model

A combination of the autoregressive and moving average models is called the ARMA model. Typically, this model is denoted by ARMA (p, q), where p is the order of the autoregressive part and q is the order of the moving average part. An ARMA (p, q) model has the general form as follows [13, 14].

$$Y_t = \phi_0 + \phi_1 Y_{t-1} + \phi_2 Y_{t-2} + \cdots + \phi_p Y_{t-p} + \varepsilon_t - \omega_1 \varepsilon_{t-1} - \omega_2 \varepsilon_{t-2} - \cdots - \omega_q \varepsilon_{t-q} \quad (1)$$

where

Y_t	=	response (dependent) variable at time t
$Y_{t-1}, Y_{t-2}, \ldots, Y_{t-p}$	=	response variable at time lags t - 1, t - 2, ..., t – p respectively. These Y's play the role of independent variables.
$\phi_0, \phi_1, \phi_2, \ldots, \phi_p$ $\omega_1, \omega_2, \ldots, \omega_q$	=	coefficients to be estimated
ε_t	=	error term at time t that represents the effects of variables not explained by the model. The assumptions about the error term are the same as those for the standard regression model
$\varepsilon_{t-1}, \varepsilon_{t-2}, \ldots, \varepsilon_{t-q}$	=	errors in previous time periods that, at the time are incorporated in the response Y_t.

Time series needs to be stationary (white noise) to apply the ARMA (p, q) models. When $q = 0$, the model change to a pure autoregressive model of order p (ARMA (p, 0)). Similarly, when $p = 0$, the model change to a pure moving average model of order p (ARMA (p, 0)). An ARMA (p, q) model forecasts will depend on current and past values of the response Y as well as current and past values of the errors (residuals). The orders p and q in an ARMA model are determined from the patterns of the sample autocorrelation function (ACF) and partial autocorrelation function (PACF). For an AR(p) model, the pattern of the ACF is die out and PACF is cut off after the order p of the process. Whilst for an MA(q) model, the ACF pattern shows a cut off after the order q of the process and PACF shows a die out pattern. If both ACF and PACF show the die out pattern, the model is an ARMA (p, q) model.

The steps in Box-Jenkins (ARIMA) approach consist of initial model identification, coefficient estimation, and model checking (residuals analysis). Usually, an iterative process is conducted until the model residuals indicate no further modification is necessary. Then, the fitted model can be used for forecasting.

2.4 Model identification

As the time series data should be stationary i.e. appears to vary about a fixed level, the first task taken is to determine whether the series is stationary. The plot of the series along with the sample ACF and

PACF may be used to check whether the time series is stationary or not. A nonstationary time series is indicated if the series appears to grow or decline over time and the sample autocorrelations fail to die out rapidly. Many time series are not stationary, but they can be converted to a stationary series. Typically, the Box-Cox transformation followed by a differencing technique are employed. Thus, the original series is replaced by a series of differences. An ARMA (*p, q*) model is then defined for the differenced series and this is called ARIMA(*p, d, q*) model. Consequently, modeling changes are taken rather than modeling the levels.

The original time series with order $p = 1$ and $q = 1$ has the following form.

$$Y_t - Y_{t-1} = \phi_1(Y_{t-1} - Y_{t-2}) + \varepsilon_t - \omega_1\varepsilon_{t-1} \quad (2)$$

If $Y_t - Y_{t-1} = \Delta Y_t$ and $Y_{t-1} - Y_{t-2} = \Delta Y_{t-1}$, thus the equation above can be written

$$\Delta Y_t = \phi_1\Delta Y_{t-1} + \varepsilon_t - \omega_1\varepsilon_{t-1} \quad (3)$$

Sometimes the first differencing is not adequate to provide the stationary data. Subsequently, the second differencing is conducted. This process can be achieved through the following equation

$$\Delta^2 Y_t = \Delta(\Delta Y_t) = \Delta(Y_{t-1} - Y_{t-2}) = Y_t - 2Y_{t-1} + Y_{t-2} \quad (4)$$

The level of differencing given i.e. 1, 2 ... to achieve stationarity is denoted by *d*. The differencing technique, however, aims to transfer the series to be stationary. Forecasts for the original series can always be computed directly from the fitted model.

After the stationary time series attained, the next tasks are to compute and plot the ACF and PACF. This is to compare the ACF and PACF computed from the data to the theoretical ACF and PACF for the various ARIMA models. Various theoretical correlations for some of the more common ARIMA models can be referred from literatures as the guidance. To simply the process, usually, both the sample ACF and PACF are compared with $\pm 2/\sqrt{n}$, where n is the number of observations in the time series. From these tasks the tentative ARIMA models are attained. The judgment on which model is adequate will be done in the next steps as some measurements need to be conducted. Nevertheless, it was kept in mind that simple models are preferred to complex models (the principle of parsimony).

2.5 Model estimation

The first task in this part is to estimate the coefficients (parameters) of the tentative models. This can be done through minimising the sum of squares of the fitting errors. The least squares or maximum likelihood technique is used to specify the parameters to be estimated. Once the parameters and their standard errors are determined, the t-value of each parameter can be constructed and interpreted. Parameters that are significantly different from zero (p-value < 0.05) are included in the fitted model. Conversely, parameters that are not significant are excluded from the model.

2.6 Model checking

The model should be checked for its adequacy before using for forecasting. The adequate model is obtained when the residuals cannot be used to improve the forecasts. Thus, the residuals should be random. Some criteria in checking the model include: the individual residual autocorrelations rk(e) should be small and generally within $\pm 2/\sqrt{n}$ of zero. The residual autocorrelations as a group should be consistent with those produced by random errors. The checks on adequacy can be done through a Chi-square (X2) test based on the Ljung-Box Q statistic. The residual autocorrelations as a group is examined using this test. The Q statistic is

$$Q = n(n+2)\sum_{k=1}^{m}\frac{r_k^2(e)}{n-k} \quad (5)$$

where

n = the number of residuals

k = the time lag
m = the number of time lags included in the test
$r_k(e)$ = sample autocorrelation function of the residuals lagged k time periods.

It is approximately that the Q statistic distributed as a Chi-square random variable with m - r degrees of freedom where r is the total number of parameters estimated in the ARIMA model. If the p-value associated with the Q statistic is small ($p < 0.01$ for this study), the model is considered inadequate. A new or modified model need to be developed until a satisfactory model has been determined.

2.7 Forecasting with the model
Forecasts are usually made using the satisfactory or might be the one that classified as the best model. Associating the confidence level to determine the upper and lower bound is also useful to see the variation that might be occurred. Ensuring the model is still possible in this step, as such, by monitoring the forecast errors. If the magnitudes of the most recent errors tend to be consistently larger than previous errors, reevaluation of the model is needed. Model correction is also required if the recent forecast errors tend to be consistently positive (underpredicting) or negative (overpredicting).

2.8 Box-Cox transformation
Box-Cox transformation is the most commonly used variance-stabilising transformation. This transformation changes the variance of residuals becomes constant. The equation below illustrates the Box-Cox transformation

$$y = \begin{cases} \dfrac{(Y^{\lambda} - 1)}{\lambda}, & \lambda \neq 0 \\ Log\, Y, & \lambda = 0 \end{cases} \quad (6)$$

where λ is the shape parameter and a real number. The λ of 0 means that the suggested transformation is using a logarithmic equation ($y = \log Y$ or $\ln Y$). If the λ equal to -0.5, the $y = 1/\sqrt{Y}$ transformation is suggested to be used, while the $y = 1/Y$ transformation is for the λ equal to -0.1, and so forth.

2.9 Model selection criteria
Identifying the ARIMA models by examining the plot of the series and matching their sample ACF and PACF patterns is the initial approach. Further analysis is, however, required when selecting the satisfactory model have been developed. The scientists have developed the approaches for selecting the best model. These involve the use of the Akaike information criterion (AIC) and the Schwarz's Bayesian criterion (SBC) respectively. Often, the two criteria to produce the same result. Nevertheless, as the SBC imposes a greater penalty for a number of parameters than in the AIC criterion, the decision based on the minimum SBC will result in a simpler model. Hence, the SBC is employed in this study. The criterion is obtained using the following equation. The second term of the equations is called the penalty factor for including additional parameters in the model. The minimum value of the SBC will show the best model.

$$BIC = ln\,\hat{\sigma}^2 + \frac{\ln n}{n} r \quad (7)$$

where
ln = the natural log
$\hat{\sigma}^2$ = the residual sum of squares divided by the number of observations
n = the number of observations (residuals)
r = the total number of parameters (including the constant term) in the model.

2.10 Intervention analysis with pulse function pattern

The model for the single intervention input with pulse function and the ARIMA (*p, d, q*) as the noise is given by

$$Y_t = \mu + \frac{\omega_0(B)}{1 - \delta_1(B)} I_t + \frac{(1 - \theta_q B)}{(1 - \emptyset_p B)(1 - B)^d} a_t \quad (8)$$

where μ is the constant mean, while the second part is the intervention function that consist of:

I_t is the input time series at time t and in this study the input is considered as a pulse function that lead to an abrupt and temporary effect.

$\omega_0(B)$ is the numerator polynomial of the intervention function for the input series.

$\delta_1(B)$ is the denominator polynomial of the intervention function for the input series.

(see e.g. [14-16] for the detailed).

3. Results

The ARIMA and the intervention modeling processes have provided a total of twenty five paired models. This is in accordance with the number of Ops Sikap. Initially, the series of the monthly number of accidents was not stationary. Thus, the pre-whitening process was carried out. The transformation using the equation of y = 1/Y and the second order of differencing has successfully provided the stationary series. Table 1 shows the parameters of the twenty five ARIMA and the intervention models developed in this study. All the parameters of the models have the p-value < 0.01, while the Chi-square of the Q statistic at most of lags are greater than 0.01. Thus, all the models developed are white noise series and sufficient for forecasting purposes.

Table 1. The parameters of the ARIMA and the intervention models.

Ops Sikap	ARIMA models	Intervention models	Ops Sikap	ARIMA models	Intervention models
I	ϕ_2 = -0.391	ω_0 = 1.662E-7 δ_1 = 0.100	XIV	ϕ_2 = -0.379	ω_0 = 5.420E-8 δ_1 = 0.100
	θ_{12} = -0.250	ϕ_2 = -0.376 θ_{12} = -0.229		θ_{12} = -0.468	ϕ_2 = -0.367 θ_{12} = -0.379
II	ϕ_2 = -0.379	ω_0 = 1.339E-7 δ_1 = 0.100	XV	ϕ_2 = -0.373	ω_0 = 5.314E-8 δ_1 = 0.100
	θ_{12} = -0.266	ϕ_2 = -0.392 θ_{12} = -0.297		θ_{12} = -0.464	ϕ_2 = -0.374 θ_{12} = -0.378
III	ϕ_2 = -0.425	ω_0 = 1.122E-7 δ_1 = 0.100	XVI	ϕ_2 = -0.380	ω_0 = 5.253E-8 δ_1 = 0.100
	θ_{12} = -0.366	ϕ_2 = -0.428 θ_{12} = -0.305		θ_{12} = -0.462	ϕ_2 = -0.367 θ_{12} = -0.382
IV	ϕ_2 = -0.426	ω_0 = 1.064E-7 δ_1 = 0.100	XVII	ϕ_2 = -0.374	ω_0 = 4.969E-8 δ_1 = 0.100
	θ_{12} = -0.376	ϕ_2 = -0.437 θ_{12} = -0.312		θ_{12} = -0.464	ϕ_2 = -0.374 θ_{12} = -0.379
V	ϕ_2 = -0.411	ω_0 = 9.232E-8 δ_1 = 0.100	XVIII	ϕ_2 = -0.380	ω_0 = 4.938E-8 δ_1 = 0.100
	θ_{12} = -0.372	ϕ_2 = -0.412 θ_{12} = -0.317		θ_{12} = -0.462	ϕ_2 = -0.365 θ_{12} = -0.383
VI	ϕ_2 = -0.413	ω_0 = 9.038E-8 δ_1 = 0.100	XIX	ϕ_2 = -0.372	ω_0 = 4.743E-8 δ_1 = 0.100
	θ_{12} = -0.381	ϕ_2 = -0.417 θ_{12} = -0.317		θ_{12} = -0.467	ϕ_2 = -0.357 θ_{12} = -0.387
VII	ϕ_2 = -0.405	ω_0 = 8.000E-8 δ_1 = 0.100	XX	ϕ_2 = -0.364	ω_0 = 4.512E-8 δ_1 = 0.100
	θ_{12} = -0.418	ϕ_2 = -0.402 θ_{12} = -0.334		θ_{12} = -0.465	ϕ_2 = -0.358 θ_{12} = -0.388
VIII	ϕ_2 = -0.403	ω_0 = 7.526E-8 δ_1 = 0.100	XXI	ϕ_2 = -0.362	ω_0 = 4.352E-8 δ_1 = 0.100
	θ_{12} = -0.424	ϕ_2 = -0.382 θ_{12} = -0.352		θ_{12} = -0.464	ϕ_2 = -0.356 θ_{12} = -0.390
IX	ϕ_2 = -0.387	ω_0 = 6.874E-8 δ_1 = 0.100	XXII	ϕ_2 = -0.364	ω_0 = 4.169E-8 δ_1 = 0.100
	θ_{12} = -0.435	ϕ_2 = -0.379 θ_{12} = -0.363		θ_{12} = -0.467	ϕ_2 = -0.359 θ_{12} = -0.393
X	ϕ_2 = -0.384	ω_0 = 6.593E-8 δ_1 = 0.100	XXIII	ϕ_2 = -0.365	ω_0 = 4.008E-8 δ_1 = 0.100
	θ_{12} = -0.442	ϕ_2 = -0.377 θ_{12} = -0.368		θ_{12} = -0.469	ϕ_2 = -0.360 θ_{12} = -0.393
XI	ϕ_2 = -0.383	ω_0 = 6.120E-8 δ_1 = 0.100	XXIV	ϕ_2 = -0.363	ω_0 = 3.864E-8 δ_1 = 0.100
	θ_{12} = -0.454	ϕ_2 = -0.372 θ_{12} = -0.375		θ_{12} = -0.459	ϕ_2 = -0.358 θ_{12} = -0.391
XII	ϕ_2 = -0.377	ω_0 = 5.851E-8 δ_1 = 0.100	XXV	ϕ_2 = -0.360	ω_0 = 3.752E-8 δ_1 = 0.100
	θ_{12} = -0.463	ϕ_2 = -0.367 θ_{12} = -0.377		θ_{12} = -0.461	ϕ_2 = -0.356 θ_{12} = -0.391
XIII	ϕ_2 = -0.385	ω_0 = 5.476E-8 δ_1 = 0.100			
	θ_{12} = -0.462	ϕ_2 = -0.373 θ_{12} = -0.381			

Note: all the parameters have p-value < 0.01

All the models are then used to produce the monthly number of accidents. The forecasted accident based on the ARIMA model is considered as the number of accident without the integrated enforcement of the safety rules and regulations or the "business as usual scenario." Whilst, the forecasted accident produced by the intervention model is with the intervention (enforcements). In addition, some of the enforcement operations are conducted in days of two subsequent months. This study only measured the impact of the Ops Sikap in the second month when the impact is usually greater. Table 2 shows the forecasted number of accidents from December 2001 to January 2012 in the month when the Ops Sikap was conducted. The difference in the number of accidents between with and without the enforcements are also highlighted in the table to show whether there is a decrease or an increase, respectively.

Table 2. Forecasted number of accidents with and without the *Ops Sikap*

Month-Year	Number of accidents		The difference		Ops Sikap
	With Ops Sikap	Without Ops Sikap	Number	%	
Dec 2001	23370	23375	5	0.02%	I
Feb 2002	23245	23105	-140	-0.60%	II
Dec 2002	24474	24396	-78	-0.32%	III
Feb 2003	23753	23787	34	0.14%	IV
Dec 2003	26731	26724	-7	-0.03%	V
Jan 2004	26309	26316	7	0.03%	VI
Nov 2004	29446	29240	-207	-0.70%	VII
Feb 2005	27285	27315	30	0.11%	VIII
Nov 2005	27563	27678	114	0.42%	IX
Feb 2006	27174	27412	238	0.88%	X
Oct 2006	29240	29146	-94	-0.32%	XI
Feb 2007	29429	29403	-26	-0.09%	XII
Oct 2007	31437	31319	-118	-0.38%	XIII
Nov 2007	31368	31240	-127	-0.41%	XIV
Jan 2008	29621	29735	115	0.39%	XV
Feb 2008	31056	31211	155	0.50%	XVI
Sep 2008	30012	30093	81	0.27%	XVII
Oct 2008	31797	31716	-81	-0.25%	XVIII
Feb 2009	31250	31289	39	0.13%	XIX
Oct 2009	35075	34977	-98	-0.28%	XX
Feb 2010	32268	32489	220	0.68%	XXI
Sep 2010	35945	35920	-26	-0.07%	XXII
Feb 2011	33727	33693	-34	-0.10%	XXIII
Sep 2011	38447	38432	-15	-0.04%	XXIV
Jan 2012	38314	38270	-44	-0.11%	XXV

Note: " - " indicates a decrease and " + " for an increase

4. Discussion

Since the early years of road safety study, researchers have suggested the three E's ways i.e. education, enforcement and engineering to improve the safety on the roads. In the review on enforcement effect road safety, Mohan in [4] summarised: (a) Most attempts at enforcing road traffic legislation will not have any lasting effects, either on road user behavior or on crashes unless the enforcement is continuous and widespread. This level of enforcement is expensive and difficult to sustain in most situations. (b) Imposing stricter penalties (in the form of higher fines or longer prison sentences) does not affect road-user behavior significantly and imposing stricter penalties also reduces the level of enforcement. (c) Increased normal, stationary speed enforcement is in most cases cost-effective. Automatic speed enforcement with cameras seems to be even more efficient. (d) There is no evidence proving mobile traffic enforcement with patrol cars is cost-effective.

An enforcement action corresponding to a policy implemented in Korea is believed to have taken place at the policy onset, suggesting an abrupt and instant (i.e. no time delay) effect [17]. More than 70% fatality reduction can be achievable in the U.S. within one decade by intensified enforcement and more appropriate safety device and limit laws [18]. Soole et al. in [5] investigated the impact of police speed enforcement methods on self-reported speeding behavior. Results indicate that marked patrol vehicles parked on the side of the road were reported as the most effective on freeways (62.7%) and school zones (56.3%), but only somewhat effective on urban roads (49.4%). A literature review by [19] showed that by augmenting the enforcement, meaning more severe penalties and more frequent controls, in most cases the number of traffic accidents and violations will decrease, which will have a positive effect on increasing traffic safety levels.

In this study, the effect of the integrated enforcement operations was examined. Based on the forecasted number of accidents produced by the ARIMA and the intervention models developed, it can be seen that out of twenty five Ops Sikap carried out, fourteen Ops Sikap (56%) have successfully reduced the accidents. Ops Sikap VII that was conducted in November 2004 for the Eid festive has provided the highest reduction of the monthly accidents (0.7%). This is followed by Ops Sikap II in February 2002 that designated for CNY festive (0.6%). Ops Sikap XIV in November 2007 for the Deepavali festive ranked third in reducing the number of accidents (0.4%). Nevertheless, whether the reductions are significant or not, it cannot be provided as the data series used are the absolute number of accidents. Using the paired t-test shall provide a less accurate result since there are other variables that contribute to the change of number of accidents during the festive seasons i.e. the traffic exposure and major modes shift. If the exposure to the risks on the roads, for instance, the vehicle-kilometres travelled during the festive season increase substantially, the reduction forecasted by the models can be considered significant and vice-versa. At the moment, the comprehensive vehicle-kilometres travelled are still developed for the country. The modes shift is also critical as the risk of the modes is not similar from each other. Long distance travels are common during the festive seasons and thus the use of motorcycle decline, whereas this mode of transport is the predominant in Malaysia and considered as unsafe.

Previous studies, as such Abdul Rahmat et al. in [7] looked into the effectiveness of the Ops Bersepadu CNY 2007 in reducing the accident and fatality rates. They specified that for overall fatalities, the daily average death for 15 days before the enforcement operations was 15.1 but dropped to 13.8 during the operations. According to the report, the daily traffic volumes were counted at before and during the festive season on several federal road segments. The locations can be classified as suburb and rural areas. From the traffic data, there were significant increase of the traffic volumes, especially for car daily volumes, but not for motorcycle and lorry. Indeed, the percentage of motorcycles was 25% before the operations, dropping to 22% during the operations. Whilst, in Malaysia, motorcycle fatalities accounted for 60% of the total fatalities. Hence, the decline of fatalities during the enforcement operations may not be owing to the integrated operations, but effect of the lower volumes of motorcycle that lead to lower risk on the roads.

In addition, Mohd Fauzi et al. in [8] have conducted an evaluation on the enforcement operations during the Eid festive in 2007. The study has included the survey on the vehicle-kilometres travelled (vkt) to provide the rate of fatalities per vkt. Although the number of samples may be relatively few, it is a better technique to attain the effectiveness of the operations. They found that the rate of fatalities declined during the festive i.e. from 16.26 to 13.97 billion vkt. The vehicle-kilometres travelled during the festive season was higher about 4%. Nevertheless, they also noticed that the volume of motorcycles dropped about 3.6% during the festive season as compared with the working days just before the festive. Moreover, Jamilah et al. in [10] have further analysed on the Ops Bersepadu and focused on the effectiveness of the operations conducted for Eid festive in 2011. Among the main results, they found that the increase in the total number of accidents during the operations for Eid festive in 2011 is significant when compared to those of the operations for Eid 2010 and 2009. The higher numbers in the total accidents and total fatalities during operations for Eid 2011 are also significant as compared with the operations for CNY 2011. In short, there was a significant increase on the number of accidents during the festive season in 2011 as compared with the festive season of the previous years.

5. Conclusion

The models have shown that integrated enforcement operations called Ops Sikap conducted since in December 2001 have reduced the number of accidents during the festive seasons in many years. Out of twenty five Ops Sikap carried out, fourteen times (56%) have successfully reduced the accidents. The highest achievement was attained during the implementation of Ops Sikap VII that was conducted in November 2004 for the Eid festive. The operations have provided the reduction of the monthly accidents up to 0.7% as compared with the accidents if there were no such operations taken. With the assumption that there was a substantial increase on the traffic volumes during the festive seasons, the reductions can be classified as significant. The recent Ops Sikap that conducted from 2010 to 2012 also provided a positive impact of road traffic safety as they have successfully brought down the number of accidents if compared with the "business as usual scenario."

References

[1] Umar R S R 1998 *Pertanika J. Sci. Technol.,* **6** **(2)** 1–13.
[2] Umar R S R 2007 *7th Malaysian Road Conference, Sunway, Malaysia.*
[3] Hess S, Polak J 2003 *Transp. Res. Record.,* **1830** **(1)** 25-33.
[4] Mohan D 2004 *The road ahead: Traffic injuries and fatalities in India*. New Delhi: Indian Institute of Technology Delhi.
[5] Soole D, Watson B, Alexia L 2009 *Australasian Road Safety Research, Policing and Education Conference: Smarter, Safer Directions, 10–12 November 2009 Sydney.*
[6] De Pauw E, Daniels S, Thierie M, Brijs T 2014 *Accident Anal. Prevention,* **62** 426–431.
[7] Rahmat A M A, Azad A R A, Alvin P W H, Allyana S M R S, Zulhaidi M J, Umar R S R 2007 *OPS Bersepadu conducted over the Chinese New Year period from 11–25 February 2007.* Kuala Lumpur: Malaysian Institute of Road Safety Research.
[8] Faudzi M Y M, Sumangala P, Umar R S R 2007 *The effectiveness of Ops Bersepadu conducted over the Hari Raya Period from 7–21 October 2007.* Kuala Lumpur: Malaysian Institute of Road Safety Research.
[9] Rasid O M, Fuad A, Faradila P N, Suffian A M, Hafiz Z Z, Wahida A B, Ilhamah O, Atiqah M F S 2011 *Evaluation of the effectiveness of Ops Bersepadu Chinese New Year 2009 with respect to selected SHE practices* (MRR 04/2011). Kuala Lumpur: Malaysian Institute of Road Safety Research.
[10] Jamilah M M, Norlen M, Rasid O M, Fuad A, Faudzi M Y M, Wong S V 2012 *Evaluation of the effectiveness of Ops Bersepadu Hari Raya 2011: Conducted over the Hari Raya Period from 23 August 2011 to 6 September 2011* (MRR 03/2012). Kuala Lumpur: Malaysian Institute of Road Safety Research.
[11] Yaacob W F W, Husin W Z W, Aziz N A, Nordin N I 2011 *J. Applied Sci.,* **11** **(7)** 1105–12.
[12] Wei W W S. 2006. *Time series analysis: univariate and multivariate methods 2nd Ed.* New York: Pearson Education Inc.
[13] Hanke J E, Wichern D W 2005 *Business forecasting 8th Ed.* New Jersey: Pearson Education Inc.
[14] Box G E P, Jenkins G M, Reinsel G C 1994 *Time series analysis forecasting and control 3rd Ed.* New Jersey: Prentice Hall Inc.
[15] Brocklebank J C, David A D 2003 *SAS for Forecasting Time Series 2nd Ed.* Cary: SAS Institute Inc.
[16] Montgomery D C, Jennings C L, Kulahci M 2008 *Introduction to Time Series Analysis and Forecasting*. New Jersey: John Wiley & Sons Inc.
[17] Kim K S, Myeong M H, Kweon Y 2006 *Transport Rev.,* **26** **(3)** 293–304.
[18] Koornstra M J 2007 *Sadhana,* **32** **(4)** 365–395.
[19] Antić B, Vujanić M, Jovanović D, Pešić D 2011 *Scientific Res. Essays,* **6** **(29)** 6176–84.

More accurate study of seismicity effect on Sumatra-Fault for liquefaction potential in Banda Aceh by using cyclic loading mobility data

Munirwansyah, R P Munirwan*, Irhami
Department of Civil Engineering, Faculty of Engineering, Universitas Syiah Kuala, Banda Aceh, Indonesia, 23111.

*E-mail: r.munirwan@unsyiah.ac.id

Abstract. Banda Aceh is a prone city, predictions from earthquake experts with a potential for a tsunami in the return period of 500 years there will still be a large earthquake that threatens development in Aceh, especially Banda Aceh. A great earthquake has occurred in Aceh on December 26, 2004, which destroyed government infrastructures and property and human life in Banda Aceh. The high number of casualties during the natural disaster occurred due to low government preparedness and public understanding of earthquake disasters caused by several earthquake sources surrounding the city of Banda Aceh, such as; Aceh-Sumatra great fault, the subduction of the Indo-Australian plate with the Euro Asia plate which move side by side along the side of Aceh province - Andaman Island. This research will provide important information regarding the orientation of the depth of the strong layer and the distribution (horizon) of the subsoil of the city surface in several densely populated zones and surrounded areas. The study method was carried out by using a cross-section of the sub-surface of the earth's plate from the coast - transition - land using Surfer software. Quantitative and qualitative analysis results obtained 3D image surface sand shading relief; soft layers, moderately layers and hard layers, each of which the thickness and depth of the soil was used to perform an analysis of the phenomenon of liquefaction potential; soft layers less than 10 kg/cm^2 thickness 0.2-3.6 m depth, moderately layers 10-50 kg/cm^2 thick 1.0 -7.0 m depth 1.2-10.6 m, hard layers > 120 kg/cm^2 thickness 2.0-18.6 m depth 5.0-46.8 m, respectively

1. Introduction
In earthquake-prone areas, including Aceh Province, are areas prone to the threat of earthquakes and tsunamis. From combination data of the NEIC-USGS and BMKG catalogues for the period 1963-2013, showing the distribution of seismicity points from shallow to medium depths was already so tightly covering almost the entire land and ocean area of Aceh. A great earthquake measuring 9.3 MW that triggered tsunami waves up to an altitude of 24 M, it occurred on December 26, 2004.

Munirwansyah et al. [1] stated that the effect of the earthquake that was felt in the city of Banda Aceh in the earthquake was so great that it caused many monumental buildings and multi-story buildings built from reinforced concrete structures to collapse due to the shock of the earthquake. Since then, studies on geotechnical aspects, seismic load effects [2], and sub-surface investigations on the distribution of bearing layers for substructure building planning purposes such as deep foundations have been increasingly focused and have become the focus of attention of researchers at Universitas Syiah Kuala, Banda Aceh. in detailed engineering design planning

Muniwarsyah et al. [1] have data on earthquake distribution that resembles a saw blade as shown in Figure 1, that the magnitude of the earthquake appears to vary and fluctuate in a function of time and return period. Munirwansyah et al. [2] found a layer of appearance bearing stratum between soil profile of bearing layers in the form of a layer of sand in the depth under the saturated surface of the groundwater which is vulnerable in soil character change will occur liquefaction under Banda Aceh. This layer has the potential to experience liquefaction when receiving vibration loads such as cyclic loading mobility. This article shows the distribution of soil layers under the city of Banda Aceh which is still limited. More accurate data were obtained with more accurate measurement results using the GP3D (3D Global Positioning System) measuring instrument, which was compared again using the 3D Globe GIS.

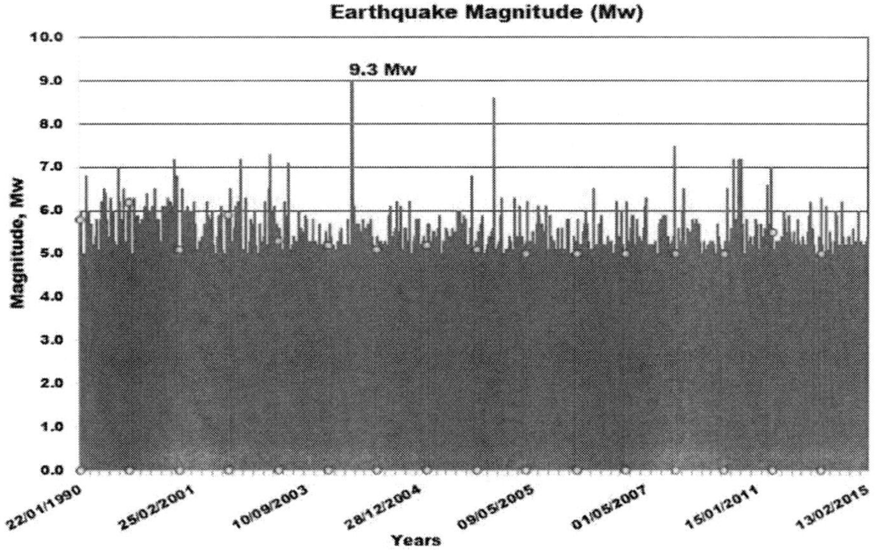

Figure 1. Earthquake magnitude variation (Mw) from 1990-2016 recorded at BMKG in Aceh Province

Google Earth used to make it easier to find out the location of development planning, bring up the geographical conditions of a location, show the locations of remote areas, see coordinates and surface elevations, and can make survey trajectories from a planning location. Fajrianto [3] suggests a device used as a satellite signal receiver in general use is called a GPS Tracker or GPS Tracking. This GPS uses a tool that allows a user to track the position of the place and the location of the point planning. Moreover, Fajrianto [3] explains that google earth is a program to map the earth from the superimposition of images collected from satellite mapping, aerial photography, and 3D GIS globe. Fajrianto [3] also explains that Global Positioning System (GPS) is a navigation system using satellites designed to provide instant position, speed, and time information in almost any place on earth, any time and in any weather conditions.

Cone Penetration Test (CPT) data was used to determine the profile of soil or rock at a certain depth in the site locations. CPT data can be used to plan the foundation and the location of the hard soil layer. Surfer is an application or device that can be used in making contour maps and the contour can be made into three dimensions by basing it on the data grid. Aprilyawan [4] explains that this software plots irregular XYZ tabular data into a regular grid of rectangular dots. Detailed contour analysis by surfer software allows the creation of three-dimensional maps of tabular data compiled using a worksheet. The surfer also facilitates the formation of contours involving topography with the help of google earth and shows the results of the soil profile in the zone under study.

Munirwan et al. [5] suggests that in determining the bearing stratum, for simple buildings, soil layers with a soil cone resistance value qc = 0-10 kg/cm² (soft soils) can be used for buildings with moderate loads up to three-story building can use a layer of soil with a hardness of qc = 10-50 kg/cm² (medium soil), for buildings with heavy loads a High-Rise-Building can use a layer of soil with a hardness of qc = 50-120 kg/cm² (hard soil) and for Super-Tall-Buildings can be piled in a layer of soil with qc hardness ≥ 120 kg/cm² (stiff soil).

2. Methodology

Over the past period, several significant changes have been made to the assessment framework for interpreting the hazards, structural responses, damage, and losses. This research is not based on the analytical method, the initial implementation for searching data coordinate (XYZ) and topographic surface analysis mapping was carried out using Google Earth sand continued with re-measurement using a more accurate GPS-3D tool that able to show the height of the position of the point on the earth's surface on above sea level, which is then converted to qc-CPT data to obtain the depth and thickness of each type of soil layer at the measurement location. The data obtained from measurement methods then combined with the GPS coordinate grid system and the SURFER-3D software to process and digitize the distribution of layers and thicknesses in each zone studied, namely the coastal zone which is a zone coastal city and inner land zone of Banda Aceh. The distribution of CPT-GPS measurement points in this study, as shown in Figure 2.

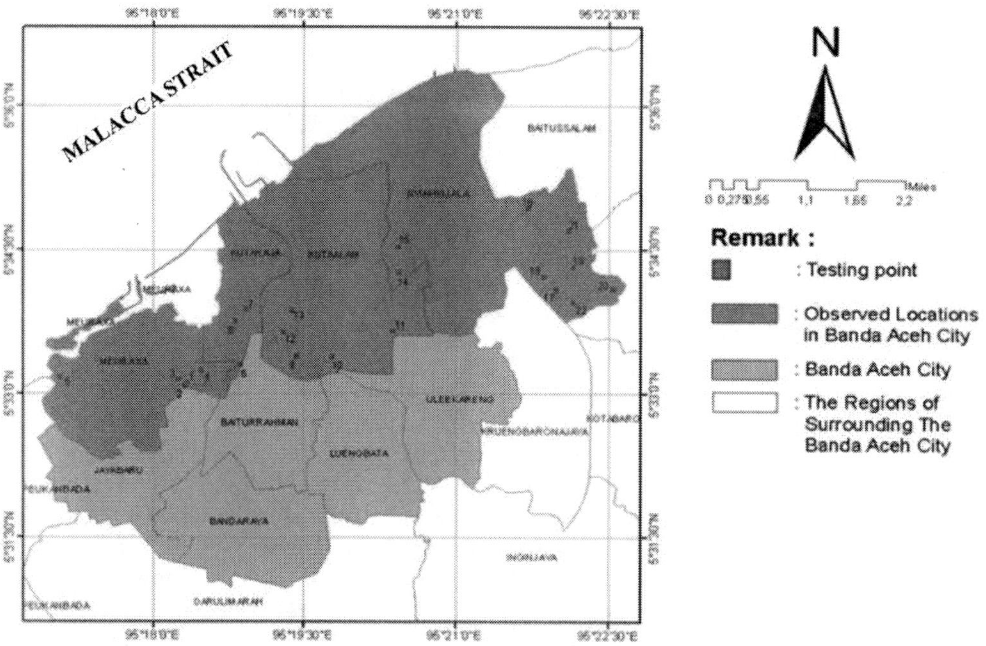

Figure 2. Distribution of measurement points and the accuracy of the Pole Measurement Device within the zone observed locations in Banda Aceh City

The coordinate measurement method is carried out by placing the GPS at the prepared point in a more stable position and using the principle of the benchmarking measurement pole device (BM-MPD) point mapping system as shown in Figure 3. The measurement data is finally processed into each map contour layer. which spreads below the location of the research area or zone with the Surface Mapping

System (Surfer) application. The number of data collection points in the coastal area of Banda Aceh City consists of 22 points. These points are scattered along the coast of Banda Aceh, including 5 points in Meuraxa District, 3 points in Kuta Raja District, 5 points in Kuta Alam District, and 9 points in Syiah Kuala District.

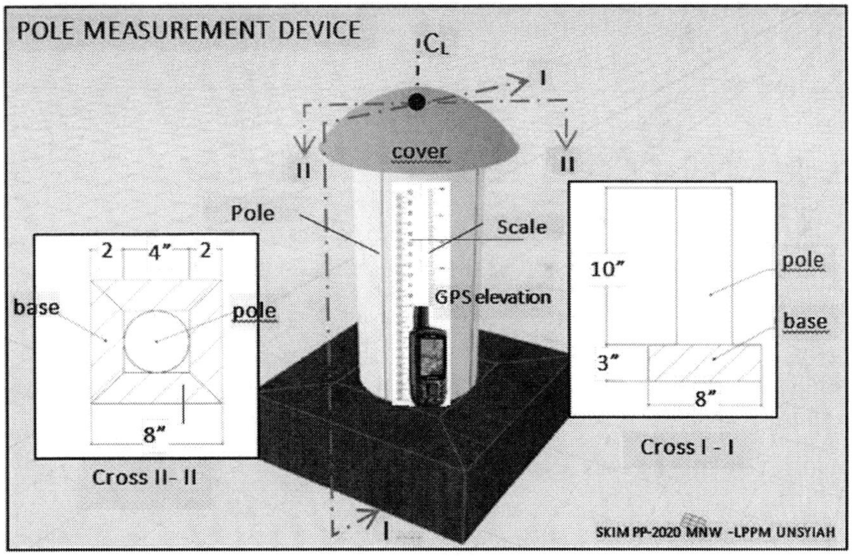

Figure 3. Pole Measurement Device and GPS positioning to calculate coordinate (x,y,z) by Google Earth and Global Positioning System for 3D-Surfer analysis

Primary and secondary data that have been obtained are then processed using Surfer Software. Surfer has a working principle of spreading X-Y-Z tabular data. The irregular X-Y-Z data is spread out into a sheet of regular rectangular dots called a grid. Before the map was formed, the data from the contour height measurements in the field were entered into a part of the surfer called a worksheet. Keckler [6] said that surfer software can function as an exact interpolator or as a data smoother, depending on the parameters used so that the data is more detailed.

3. Results and discussion

Parameters that were used for data analysis are Google Earth coordinates, Global Positioning System (GPS) coordinates, depth of CPT data, and qc values in CPT data. The analysis results are presented in the form of maps, tables, 3D images obtained by using calculations using surfer software. The research results obtained from data processing in the form of a research map points and a bearing layer contour analysis model that can be used for infrastructure planning as earthquake mitigation and liquefaction.

The distribution of research location points is needed for the accuracy of the data plotted into the Surfer worksheet so that the accuracy of the results obtained can be used to mitigate liquefaction disasters, earthquakes, and an initial view of the construction site conditions [7]. The map of the distribution of research points is described based on the coordinate data obtained from Google Earth.

Based on data analysis and processing of primary and secondary data at the research location, a table is obtained that can be used as complete information and helps to create a 3D-mapping layer distribution map. The table is presented in detail, including the coordinates, the depth of each layer, and the strength of the soil layers at the research location. The contours of the bearing layer in the X-Y and Z directions of the Cross-section in the map illustrate the distribution of location points whose coordinates are measured, while the Z direction represents the soil profile of the soil layer depth and strength, respectively. The resulting contours of each soil layer can be seen as shown in Figure 4.

The contours of the surface layer of soil surface are as shown in Figure 4(a). From the results of the analysis of each area, the data is obtained that; Meuraxa District is at an altitude of + 2.0-3.0 M above sea level, Kuta Raja District is + 2.0 M, Kuta Alam District is + 2.0-4.0 M, while the height of Syiah Kuala District is + 2.0-5.0 M above sea level. It can be seen that the soft soil layer qc value is at 0-10 kg / cm2. obtained at an altitude of 0.2 m to 2.0 M above sea level. The contour distribution of the soft layer can be seen in Figure 4(b).

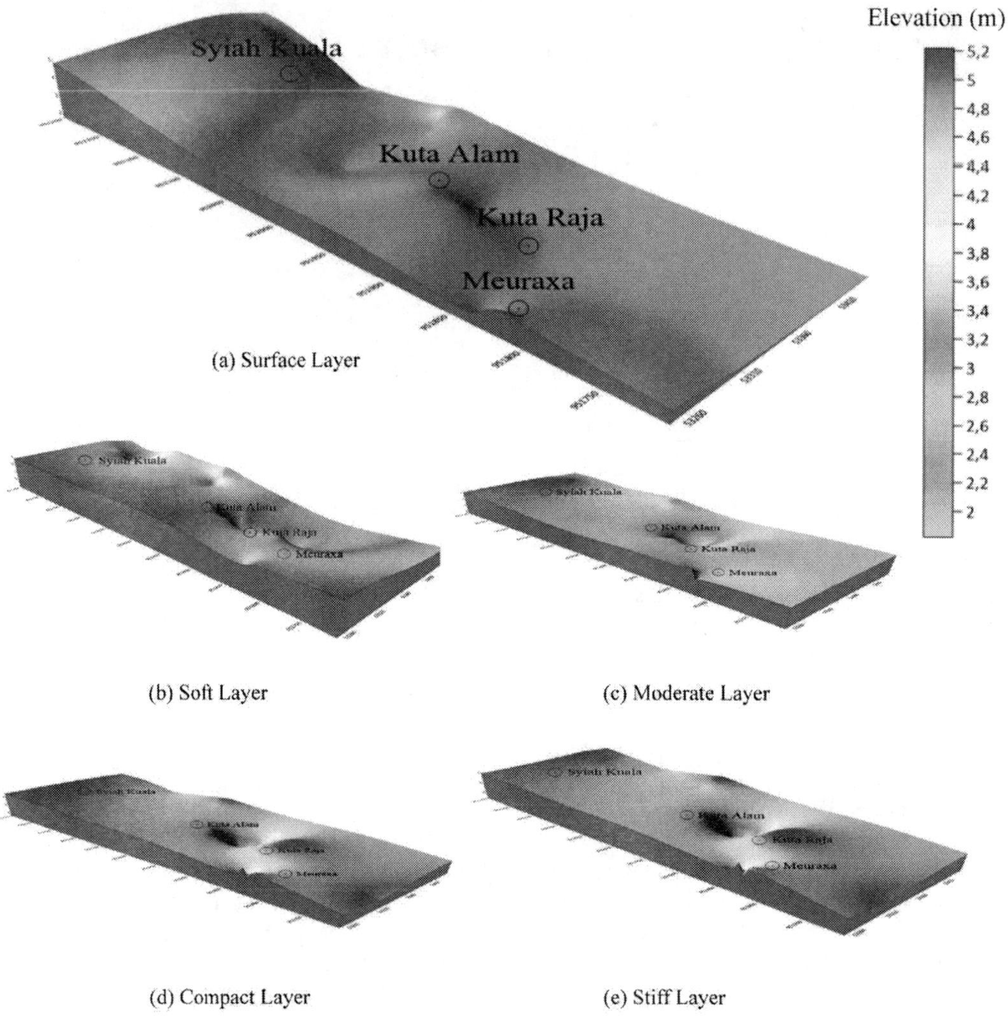

Figure 4. Distribution of layer under Banda Aceh. (a) surface layer, (b) soft layer, (c) moderate layer, (d) compact layer, and (e) stiff layer

Based on the value of qc = 10-50 kg/cm^2, the distribution of the moderate layer ranges from 1.2 m to 7.0 m above sea level, as shown in Figure 4(c). For medium dense layers, the qc value is at 50-120 kg/cm^2. Based on this category, it can be determined that the depth of the soft soil layer is spread in a thickness ranging from 2.0 - 17.6 m. This layer can be seen in Figure 4(d). Furthermore, for soil layers, the value of qc> 120 kg/cm^2 is illustrated by the distribution and thicknesses ranging from 3.4 to 18.6 M. This layer can be seen in Figure 4(e). In brief, Figure 5 shows the cross-section distribution of the 3D Surface Sand Shading Relief profile, which is a very useful result in geotechnical aspects in planning the substructure as a foundation and other civil engineering structures.

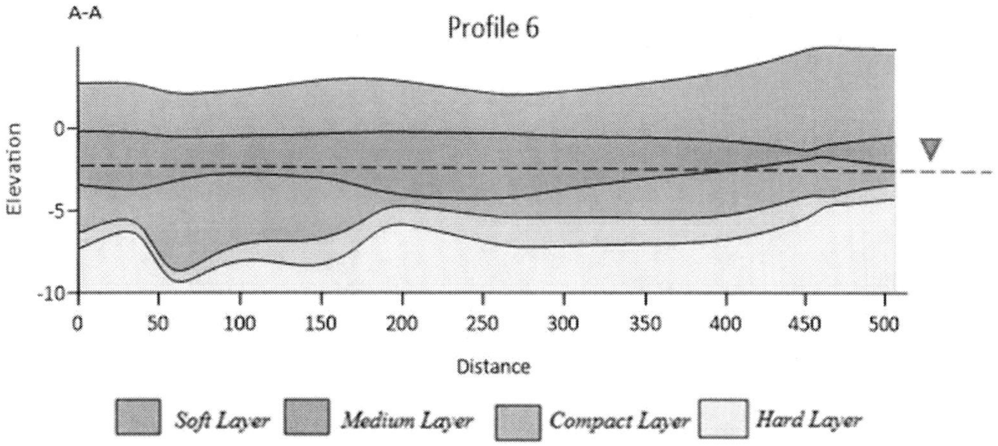

Figure 5. Cross-section of distribution of layer under Banda Aceh

4. Conclusions

Based on research data, it can be concluded that the distribution of soil layers under the city of Banda Aceh in the coastal zone is as follows:

- Soft soil layers with a value of qc = 0-10 kg/cm^2 obtained a thickness from 0.2 M to 3.6 M
- Medium density soil layer with a value qc = 10-50 kg/cm^2 thickness from 1.2 to 7.0 M.
- Solid soil layer with a value of qc = 50-120 kg/cm^2 thickness 2.0 to 17.6 M.
- Hard soil layer with a value of qc = 50-120 kg/cm^2 thickness 3.4 to 18.6 M

In the coastal zone, almost all areas have a layer of saturated sand with apparent strength that is susceptible to the phenomenon of liquefaction. Districts that have the liquefaction potential include Meuraxa, Kuta Alam, and Syiah Kuala Districts.

5. Acknowledgement

Special gratefully to (LPPM) Universitas Syiah Kuala, which has funded this research through the 2020 Professor Research Scheme, with contract No. 268/UN11/SPK/PNBP/2020. Thank you also given to the research team and research assistants and others who have helped directly and suggested opinions that have been given so that the research has been carried out and helped publish the output of this research.

References

[1] Munirwansyah, Munirwan R P, Yunita H 2018 *Int. J. Advanced Sci., Eng. Information Technol.*, **8** (**3**) 870-875.
[2] Munirwansyah, Munirwan R P, Sungkar M, Melinda Z 2019 *IOP Conf. Ser.: Mater. Sci. Eng.*, **523**.
[3] Fajrianto 2009 *Rekayasa: Jurnal Ilmiah Fakultas Teknik, Universitas Lampung*. **13** (**2**). *(in Bahasa Indonesia)*.
[4] Aprilyawan D 2015 *Software Untuk Pemetaan. (in Bahasa Indonesia)*.
[5] Munirwan R P, Munirwansyah 2019 *IOP Conf. Ser.: Earth Environ. Sci.*, **273(1)**.

[6] Keckler D 1994 *Surfer for Windows, Contouring and 3d Surface Mapping*. Golden Software.
[7] Munirwansyah M, Fulazzaky M A, Yunita H, Munirwan R P, Jonbi J, Semeru K 2020 *Geodesy and Geodynamics,* **11 (6)** 455-467.

IOP Conf. Series: Materials Science and Engineering 1087 (2021) 012031 doi:10.1088/1757-899X/1087/1/012031

The vulnerability of housing on the north coast of Banda Aceh to disasters from the sea

C Nursaniah[1,*], M Irwansyah[1], L Qadri[1], Mariana[2]
[1]Departement of Architecture, Engineering Faculty, University Syiah Kuala, Banda Aceh, Indonesia
[2]Departement of Chemical Engineering, Engineering Faculty, University Syiah Kuala, Banda Aceh, Indonesia

*E-mail: cutnursaniah@unsyiah.ac.id

Abstract. Alue Naga is an example of a coastal village in the northern city of Banda Aceh which was severely damaged during the earthquake and tsunami in 2004, but until now there has been no maximum environmental improvement and mitigation for this coastal settlement. The ideal initial mitigation study to be carried out for coastal settlements is to analyze the physical vulnerability of housing as a physical structure characteristic that determines the potential damage to disasters. This study uses a qualitative descriptive method with data collection through observation, a global positioning system, interviews and questionnaires. Data were analyzed using geographic information system techniques to determine the spatial condition of the area. Assessment of the potential vulnerability of housing in each cluster of Alue Naga settlements is obtained by scoring and weighting methods for variables that contribute to the level of disaster vulnerability, namely the type of building construction, the shape of the ground floor of the building, and the orientation of the building to the coast. The results showed that the potential vulnerability in each cluster was different for each variable being reviewed, there were clusters that were vulnerable, more vulnerable, and very vulnerable to the threat of disasters from the sea. It takes serious action from stakeholders to further improve protection and mitigation against disasters from the sea in the northern coastal settlements of Banda Aceh City with structural mitigation and non-structural mitigation.

1. Introduction

Coastal areas are very vulnerable to damage caused by various natural phenomena from their geographical conditions. The tsunami disaster has reminded us of the importance of urban spatial planning that considers disaster aspects. According to [1], the devastating damage that hit the city of Banda Aceh in 2004 was due to the lack of barriers to the breakthrough of sea waves entering the city, due to the vertical road structure to the coastline. Even a ship weighing hundreds of tons is capable of "landing" into a city up to five kilometers from its place of origin by the sea. The road structure on the northern coast of Banda Aceh is indeed dominant with its position perpendicular to the coastline, while the function of vegetation as a barrier is very important in these coastal environmental conditions. This phenomenon happened in Alue Naga Village (Figure 1), causing the village to suffer very serious damage when the tsunami disaster occurred in 2004. Before the tsunami, Alue Naga was very arid, due to exploitation of mangrove forests for ponds. According to [2] coastal vegetation is effective in dissipating incoming wave energy, offering valuable protection to coastal communities.

Figure 1. Due to the lack of coastal vegetation to block the breakthrough of sea waves, the collector road and river that divides Alue Naga Village functions like an ocean flow tunnel in the 2004 tsunami disaster.

Currently, the damage to the settlement of Alue Naga Village has not been fully reconstructed and mitigated, so there is still a risk of further environmental damage. The barren and non-ecological conditions of the coast as seen in Figure 2 will not be able to protect the environment if a disaster comes from the direction of the sea. Coastal areas without mitigation are prone to disasters that can instantly destroy all facilities for houses, boats, coastal resources, etc. as a source of daily livelihoods. [3] concluded that disaster risk reduction actions on the Banda Aceh coast are still sporadic and there is no commitment to more detailed institutional policies.

Figure 2. The phenomenon of Alue Naga Village as an example of the northern coast of Banda Aceh City has not been optimally mitigated.

Ideally, mitigation studies on coastal settlements begin by looking at the physical vulnerability of the settlement. [4] mentions vulnerability as a specific condition that results in increased damage, losses and losses due to a disaster depending on characteristics such as types of building materials, demographics, and geographic location. Furthermore, [5] stated that physical vulnerability is defined as a characteristic of the physical structure that determines the potential damage caused by a disaster (type of material and quality of the building). Assessing the potential vulnerability of housing in coastal areas is very important to determine how resilient these settlements are when a disaster occurs. This is one of the mitigation efforts to minimize and reduce the impact of disasters on society. Special considerations must be made when selecting materials for coastal buildings. [6] said harsh environments require more substantial building materials to be used and more caution when using these materials to ensure durability, resistance to hazards, and less maintenance. [7] stated that buildings with stilt is safe against extreme ocean waves and buildings with an orientation position perpendicular to the coastline are relatively safe against these disasters.

2. Methods

This research according to its purpose is a qualitative descriptive study by combining quantitative data and qualitative data.

2.1 Collection of data

Primary data collection is carried out in two ways:

- Field observations to observe the housing characteristics of each cluster in the study location and determine the housing location using the Global Positioning System. The collected data is overlaid with google earth imagery to get a spatial picture with attribute information, then various maps are made according to the condition of Alue Naga Village.
- Interview and distribution of questionnaires to the people of Alue Naga Village to obtain information and perceptions related to the characteristics of residential housing. The data from the questionnaire was then processed in descriptive form to support the results of research based on community perceptions regarding housing and disasters in the area.

2.2 Data analysis

- Geographical information system technique with ArcGIS software is used as a spatial analysis technique / georeference. To determine the potential for disaster vulnerability, scoring and weighting methods are used for each variable.
- Assessment of building vulnerability is obtained by measuring the physical element variables of the building that contribute to the level of vulnerability to earthquakes, tsunamis, tidal waves, and strong winds. The element variables used are 1 The quality of the building is indicated by the materials that compose the building based on three types of materials including reinforced concrete and plastered bricks (permanent), concrete and wood construction (semi-permanent), and wood construction; 2). The form of the ground floor includes, houses constructed on stilts are considered more capable of escaping water runoff compared to houses built on the ground with openings and closed ones; 3). Building orientation is the tendency of the shape of the building towards the shoreline divided into three orientation classifications (to the coastline), namely the long side of the building tends to be perpendicular, the long side of the building tends to tilt, and the long side of the building tends to be parallel.
- Each element has a different weight to the physical vulnerability of the building. It is given the numbers 5 (very good), 4 (good, 3 (not good), 2 (not good), 1 (very bad).

2.3 Study location

The research location is in Alue Naga Village and divided into four clusters based on the hamlets in the village, namely Beunot, Musafir, Kutaran, and Po Diamat. Each cluster has a different distance from the coastline, can be seen in Figure 3.

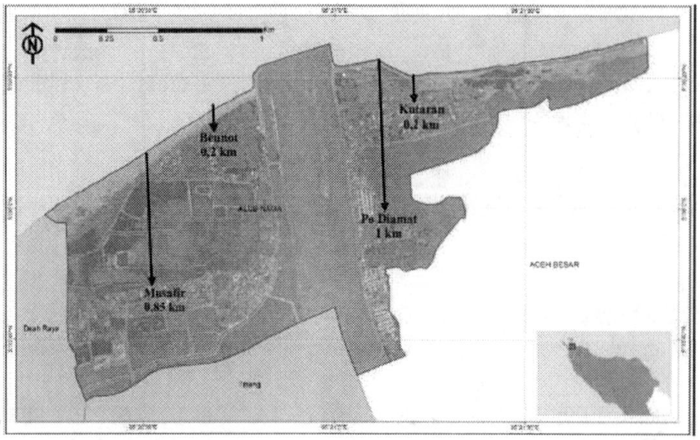

Figure 3. Study location and clusters distance from the shoreline

3. Results and discussion

3.1 Identification of disaster prone conditions

The northern coastal area of Banda Aceh, especially Alue Naga village, is a flat area where around 89% is an area with a slope of 0 to 2%. In this area, development can still be carried out but must pay attention to the safety factor which is closely related to the type of geological hazard. Geological protected areas, particularly geological disaster prone areas in this region include:

a. Earthquake-prone areas
To anticipate earthquake disasters it is necessary to have a building code (Building Code), such as residential development needs to use earthquake resistant buildings that are traditional in nature, while for heavy buildings it is necessary to consider earthquake acceleration.

b. Tsunami prone areas
To minimize the impact of damage from the tsunami threat, development in the Alue Naga coastal area needs to pay attention to the height of the tsunami run-up, namely elevation up to <5 m above sea level, coastal morphology (the shape of the bay, river estuary, and the topography of the plains parallel to the coastline). It is necessary to build a representative escape building for the evacuation of the population, because in this area which is very close to the sea, there are no buildings for rescue. Alue Naga Village is located at an altitude of 1 meter above sea level.

c. Areas with High Liquefaction Potential
The threat of liquefaction is high because in this area there are lanauan sand deposits with a carrying capacity of 0.2-0.5 kg/cm^2, which are currently filled with trade and service activities. Therefore, for buildings on this type of soil, it is necessary to choose a building construction that is liquefaction-resistant, to build according to civil engineering standards and following the characteristics of the threat.

The people who have lived in the study locations are very familiar with the various kinds of disasters that often hit their villages and understand the causes of disasters. According to the community, the main causes of disasters are heavy rain and strong winds, beach / wave abrasion, natural / weather factors and due to "Allah's Power". Even the community already knows that disasters occur because they do not protect the environment, global warming, changing seasons, shifting plates and there is no greening.

Based on various types of threats, the most frequent ones are strong winds, tidal waves and coastal abrasion. Meanwhile, the threat of earthquake and tsunami disasters are also very familiar to these coastal communities. They are very aware that the potential for a tsunami disaster someday threatens the area where they live again. Potential disasters that threaten the coast of Alue Naga can be seen in

Table 1 below. Table 1 shows the types of disasters that have hit the study location, and how much threat the community feels from each of these disasters. The "total column" shows the ranking of the most feared disasters according to the community, namely earthquakes, then tsunamis, high winds, tidal waves, coastal abrasion, and finally floods. Meanwhile, the "Rank column for disaster threat in the cluster" shows the position of the cluster against the threat, that is, the community feels the threat of disaster in the Beunot cluster is the biggest of the four clusters, while the safest is in the Musafir cluster.

Table 1. Alue Naga coastal community by type of disaster.

No.	Kind of disaster	Cluster Po Diamat (1)	Cluster Kutaran (2)	Cluster Musafir (3)	Cluster Bunot (4)	Total	Disaster Threat Ranking
1	Strong wind	2	3	1	3	**9**	**3**
2	Tidal wave	3	1	1	3	**8**	**4**
3	earthquake	3	3	3	3	**12**	**1**
4	Beach abrasion	1	2	1	3	**7**	**5**
5	Tsunami threat	3	3	2	3	**11**	**2**
6	Flood	2	1	1	1	**5**	**6**
	Disaster threat rating on the cluster	**15**	**14**	**10**	**17**		

3.2 Housing quality parameters

The majority of housing in Alue Naga Village are houses of type 36 plus square meters which were reconstructed by the Rehabilitation and Reconstruction Agency (BRR) NAD-NIAS and several countries that donated housing after the tsunami disaster. Generally, the residential buildings in this area are constructed from brick and cement materials totaling 521 houses with a total percentage of 78%. Meanwhile, the number of buildings that are not permanent is 150 with a percentage of 22%.

The Australian Government built housing in Beunot and Musafir with a green zinc roof. The Catholic Relief Services (CRS) Institute built housing in Kutaran and Podiamat with a blue zinc roof. Furthermore, housing assistance from Caritas Germany in Podiamat characterized by red zinc roofs and housing construction in Kutaran characterized by white zinc roofs. Meanwhile, housing development assistance from BRR in Musafir and Beunot was characterized by red zinc roofs and green walls. Table 2 shows the housing model and types of housing construction materials in Alue Naga.

Table 2. Housing models and types of building materials.

No.	Cluster	Housing Models	Donor	Materials
1.	Bunot and Musafir		Australian government	- Concrete - Zinc roof - Mountain rock foundation - Wooden door - Glasses window - Concrete truss
2.	Kutaran and Po Diamat		Catholic Relief Sevices	- Concrete - Zinc roof - Mountain rock foundation - Wooden door - Glasses window - Concrete truss

3.	Po Diamat		Caritas Jerman	- Concrete - Zinc roof - Tread foundation - Wooden door - Glasses window - Concrete truss
4.	Musafir and Bunot		BRR	- Concrete - Zinc roof - Mountain rock foundation - Wooden door - Glasses window - Concrete truss
5.	Kutaran		Caritas Jerman	- Concrete - Zinc roof - Foundation wells - Wooden door - Glasses window - Concrete truss

The community considered that the construction and materials of their houses were adequate for protection and safe from disasters. About 26% rated their house construction as very good, 33% of the people rated their house construction as good, 23% considered it not good, 14% thought it was not good, and 1% considered their house construction very bad. The results of the translation of the hamlets in the Alue Naga village are presented in Figure 4.

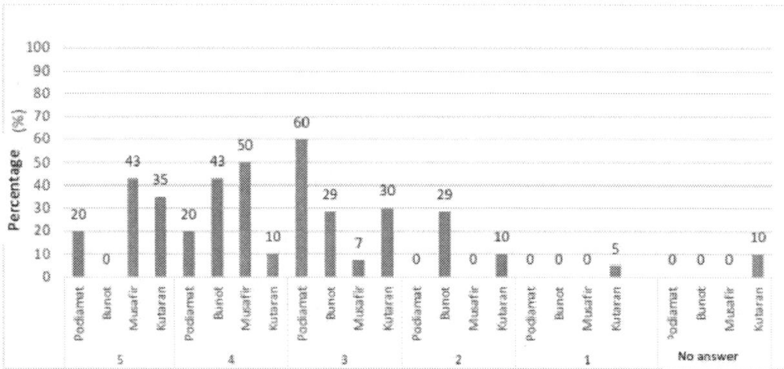

Figure 4. Graph of community assessment of housing construction related to disasters.

For residential buildings, in terms of structural strength, they have met the strength standard for housing in the earthquake area, namely a one-story concrete building using reinforced concrete construction. using concrete truss and stucco brick.

The buildings in the study location that are not permanent (constructed with wood) are the community shops in the conditions shown in Figure 5. These wooden stalls are built in the coastal location and its surroundings on silt clay sediment types that are prone to liquefaction. The building is very vulnerable to disasters, however, the wooden building is not used by the community as a shelter.

Figure 5. Sample of non-permanent buildings in Alue Naga village.

3.3 Ground floor condition parameters

The types of buildings in the study location are dominated by residential buildings. The characteristics of the buildings in the area show that the ground floor is more closed than the open ground floor / stilted. The floor height of a dwelling that is constructed on the ground is too low, causing the dwelling to be at risk of becoming damp and accessible to tidal waves because in this area the ground water level is high. It is better if the houses are made on stilts such as those built by Caritas Germany in Kutaran cluster and Po Diamat cluster.

Figure 6 shows that the stilt-floored residential building in Kutaran has a floor height of half a meter from the ground level, and the stilt floor dwelling in Po Diamat has a height of 1 (one) meter from the ground. The open ground floor conditions can allow water to pass and reduce the pressure load of the tsunami. However, the resistance to tsunamis with the construction of houses on stilts has no guarantee of strength and is highly dependent on the wave height, because the height of the stage is relatively low compared to the height of the tsunami waves that occur in coastal settlements. In this case the role of vegetation as a barrier to wave speed, as well as protection for beaches and settlements is very decisive. The complete relationship between construction and vulnerability can be seen in Table 3.

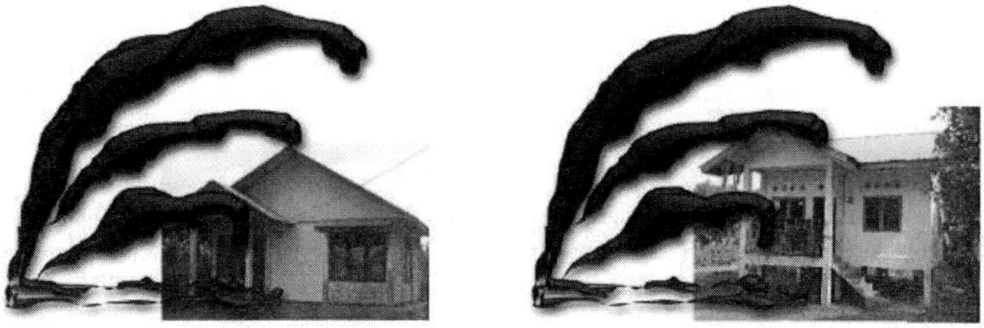

Figure 6. Illustration of resistance on the ground construction and stilted residential construction to tsunami

Building structures that respond to tsunamis are structures that can break and are strong enough to withstand the impact of the waves that hit them. As well as having an area that is higher than the tsunami wave front as a means of saving the building users themselves, see Figure 7.

Table 3. Housing Models and Vulnerability

No.	Housing Models	Cluster	Construction	Vulnerability
1.		Bunot and Musafir	On the ground	- Damp - Tsunami threat
2.		Kutaran and Po Diamat	On the ground	- Damp - Tidal wave - Tsunami threat
3.		Po Diamat	Stilted 100 cm	- Tsunami threat
4.		Musafir and Bunot	On the ground	- Damp - Tidal wave - Tsunami threat
5.		Kutaran	Stilted 50 cm	- Tsunami threat

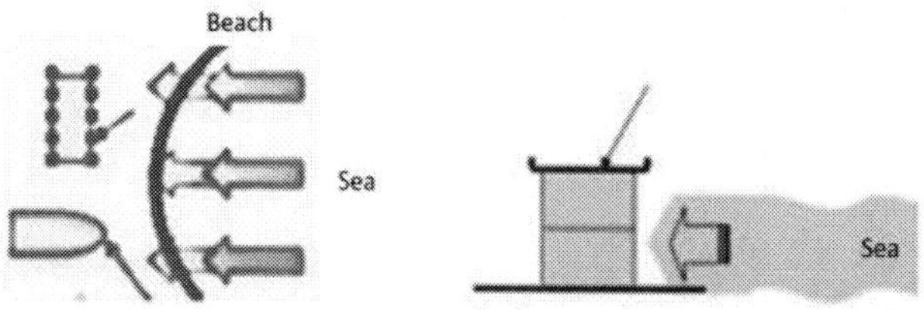

Figure 7. Good structure to respond to tsunamis (Source: [8])

3.4 Building orientation parameters

The observation result of building orientation at the study location shows the tendency of buildings that are perpendicular to the coast resulting in a low vulnerability of buildings to tsunamis and strong winds, related to the area of impact received by the buildings. The impact of the tsunami caused the building to be destroyed, while the impact of the strong winds coming from the direction of the sea resulted in the tightening force of the roof area, as shown in Figure 8.

The greater the angle of the roof, the greater the area of the roof that is in direct contact with the incoming wind (high risk of wind damage). As stated by [9], the amount of air pressure on the roof depends on the slope of the roof itself. The gable forms a large angle / ramps and the ramps are larger than other roofs. Therefore, it is better if in windy areas to avoid the shape of the roof with large angles / ramps. Houses in windy areas can use a roof with a small angle or a roof with a curved model to make it more aerodynamic.

The group of buildings with the orientation perpendicular to the coast indicates that in this area the buildings tend to be more dominant to be built with their long sides perpendicular to the coastline. This condition is considered good enough in terms of building orientation to reduce the level of vulnerability to tsunamis. Table 4 shows the vulnerability between the orientation of the buildings towards the coast in each cluster in Alue Naga Village.

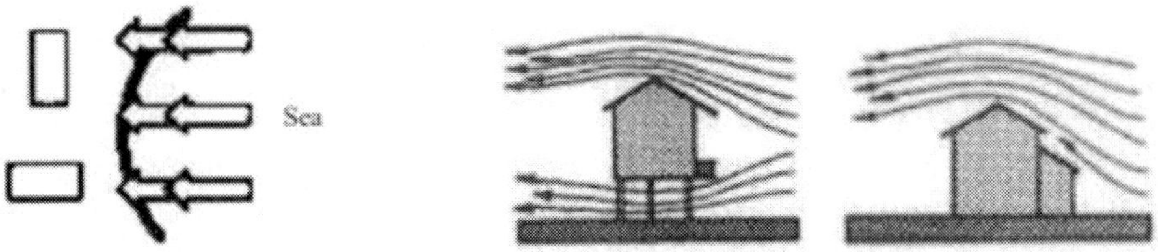

a b

Figure 8. a). Wave behavior towards building orientation and b). Wind behavior towards roof slope (Source: [8])

Table 4. The orientation of the building towards the coast in relation to its vulnerability.

No.	Cluster	Description	Vulnerability
1.	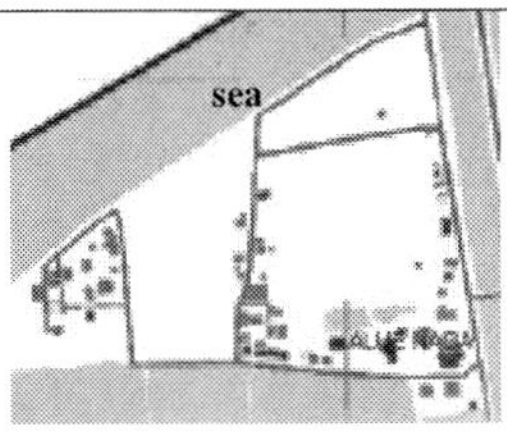 Beunot	The orientation of the buildings and dwellings in this cluster is irregular. 70% of the buildings are perpendicular to the coast and 30% of the buildings are parallel to the coast.	Vulnerable to Tsunamis and strong winds
2.	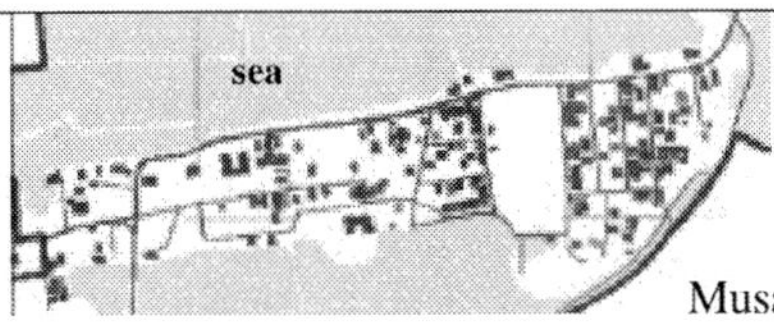 Musafir	The orientation of the buildings and dwellings in this cluster is irregular. 50% of the buildings are perpendicular to the coast and 50% of the buildings are parallel to the coast.	very vulnerable to Tsunamis and strong winds
3.	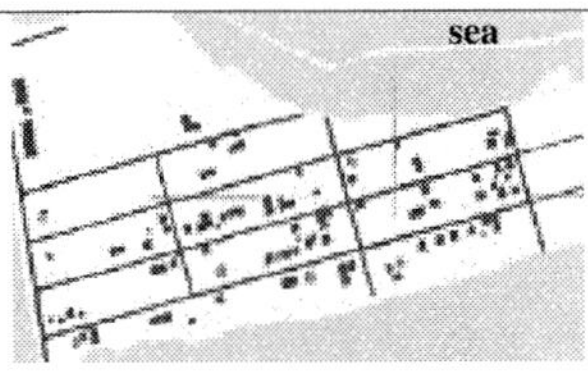 Kutaran	The orientation of the buildings and settlements in this cluster is somewhat irregular. 90% of the buildings are perpendicular to the coast and 10% of the buildings are parallel to the coast.	somewhat prone to Tsunamis and strong winds
4.	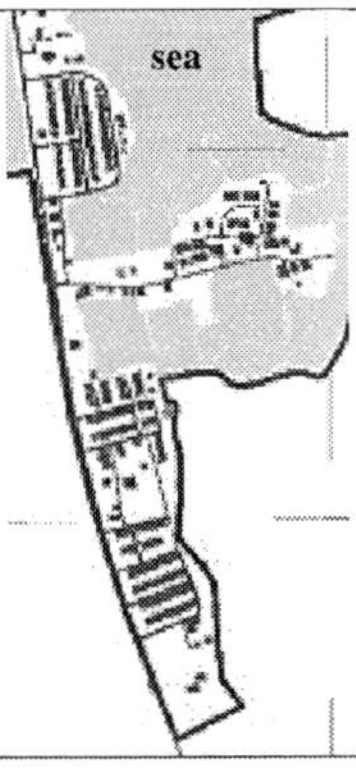 Po Diamat	The orientation of the buildings and dwellings in this cluster is irregular. 90% of the buildings are perpendicular to the coast and 10% of the buildings are parallel to the coast.	somewhat prone to Tsunamis and strong winds

4. Conclusions

From the discussion above, several conclusions can be drawn, including:

1. Housing in each cluster in Alue Naga village has used reinforced concrete construction and building materials that comply with civil engineering standards to respond to earthquakes;
2. Construction of buildings on the ground applies a very low peil floor height to the beunot and Musafir clusters which will cause the building to become damp, because the water level of the swampy / brackish soil in the area is high. Brackish water seeping into building walls also has the potential to make plaster and brick walls porous;
3. The ideal housing form for Alue Naga Village and the northern coastal area of Banda Aceh is one which constructs a stage to respond to humidity and tidal waves. However, to respond to tidal waves and tsunamis, there must be environmental protection in the form of soft protection and hard protection, which the government strives to help empower local communities. The ideal stilt construction to respond to tsunamis is a reinforced concrete frame construction structure;
4. The stage construction in the Kutaran and Po Diamat clusters, if seen from the height of the floor, the stage has not responded optimally to the tsunami. It is necessary to build a rescue building for this coastal village area because the achievement of evacuation must go through a fairly long road;
5. The orientation of the building towards the coast in order to respond to strong winds and the threat of tsunami has not been carried out optimally (house orientation is irregular), the vulnerability of each cluster varies, vulnerable clusters (Kutaran and Po Diamat), more vulnerable clusters (Beunot), and highly vulnerable clusters (Musafir).

5. Acknowledgement

Special thanks to Direktorat Riset dan Pengabdian Masyarakat (DRPM) for supporting this research fund, so that this research can be done well. Thanks also to the Universitas Syiah Kuala Research Institute and Community Service which have facilitated and coordinated this research.

References

[1] Edyanto C B, Herman 2011 *Jurnal Teknologi Lingkungan*, **12 (3)** 309 – 318. *(in Bahasa Indonesia).*

[2] Contestabile, Vicinanza 2020 *Water*, **12 (9)** 2594.

[3] Irwansyah M, Nursaniah C, Qadri L, Mariana 2019 *The European Proceedings of Multidisciplinary Sciences EpMS.*

[4] Noson L 2002 *Proceding of The Regional Workshop on Best in Disaster Management.*

[5] Ebert A, Kerle N, Stein A 2009 *Nat. Hazards,* **48 (2)** 275-294.

[6] Ajrawat. A, Kumar S A 2017 *Int. J. Emerging Technol.*, **8 (1)** 285-290.

[7] Diposaptono. 2014 *Materi Paparan Direktur Tata Ruang Laut, Pesisir, dan Pulau-pulau Kecil Kementerian Kelautan dan Perikanan. Yogyakarta, 25 September.*

[8] Muhari A 2010 *J. Earthq. Tsunami,* **4 (4)** 341-368.

[9] Lechner N 2007 *Heating, Cooling, Lighting: Metode Desain untuk Arsitektur.* Jakarta: Rajawali. *(in Bahasa Indonesia).*

Priority of provincial road maintenance in Kabupaten Aceh Besar based on analytic hierarchy process method

Risdiawati[1,*], S M Saleh[2], M Isya[2]
[1]Master Student, Department of Civil Engineering, Universitas Syiah Kuala, Banda Aceh, Indonesia
[2]Department of Civil Engineering, Universitas Syiah Kuala, Banda Aceh, Indonesia.

*E-mail: risdiawati88@gmail.com

Abstract. Road damage that has occurred on several roads has caused enormous losses, especially for road users, such as long travel times, congestion, accidents, and others. It is necessary to pay attention to this so that there is no deterioration in road quality due to road surface damage so that it can affect safety, comfort, and smoothness of traffic. Therefore, researchers damaged road surface conditions, which can affect the travel time of the vehicle. The purpose of this research is to identify the type and level of road damage and to determine road-handling priorities. The results of calculating the condition of each provincial road section in the Aceh Besar District using the PCI method are included in the very good, good and moderate categories, so that the type of handling can be determined is routine maintenance. From the results of the analysis of respondents by using Analytic Hierarchy Process (AHP), said that the priority road-handling by using several criteria, namely: the order of 1 is the type of treatment with a weight of 0.47, then the second is the condition of the road with a weight of 0.37 , then the third-order is accessibility with a weight of 0.12, and the fourth place is a land use with a weight of 0.04. As for the alternative road-handling priority obtained a score's greatest at road Keutapang Dua - Mata Ie with the acquisition of a score of 0.34, then alternate both at road Batas Kota Banda Aceh - Sp. Lam Ateuk 0.32, the third alternative on road Ir. M. Thaher 0.26, for the fourth alternative on road Sp. Lamreueng-Sp. Cot Iri 0.06, the fifth alternative on road Kota Jantho-Alue Glong 0.011, the sixth alternative on road Bts. Banda Aceh-Blang Bintang 0.010, and for the seventh alternative on road Krueng Raya - Batas Pidie with a score of 0.003. The first priority for road handling based on an assessment of road conditions using the PCI method is road Krueng Raya - Batas Pidie section.

1. Introduction

Good transportation is a major supporting factor for determining the advancement of the economic growth of a region or country. Transportation is the lifeblood of the economy. The availability of good road transportation will provide services for vehicles that transport goods and can pass quickly, safely and comfortably to their destination. In addition to the construction of new roads, supervision and maintenance of existing roads must be carried out continuously so that they do not suffer damage before the calculated plan age has been reached [1].

Generally the cause of damage to the road there are many reasons that a design life on the road that has been passed, a puddle of water on the surface of the road that cannot flow due to drainage is not good, the traffic load recurring loads (overloaded) which causes the shelf life shorter paths of planning. Improper planning, inadequate supervision and implementation that is not in accordance with existing plans. Apart from that, minimal maintenance costs, delays in budget expenditures and inaccurate priorities for handling are also the causes [2].

Road maintenance should be carried out regularly or periodically to keep pavement performance in good condition and avoid premature damage. The main objective of road maintenance is to maintain the infrastructure that has been built so that it is always close to its original condition (steady condition, i.e. good and moderate) in order to be able to provide optimal services to support regional social and economic activities [3].

As the condition of provincial roads in the Aceh Besar District, from year to year, it is always faced with the quality of road services. In general, the provincial roads in the Aceh Besar District have almost reached good condition, but in certain segments there are still road conditions that experience damage due to the increase in traffic volume and the large load of vehicles crossing the road, resulting in road damage that can be damaged. Interfere with the activities of road users.

Based on these conditions, this study assesses the existing conditions of the Provincial roads in the Aceh Besar District and can determine road handling priorities based on the level of importance of the criteria by using the decision-making method. This is due to the large number of roads that must be handled, both for road improvement and maintenance, while road management funds are very limited, it is necessary to set road handling priorities so that the allocation and use of limited funds becomes effective and efficient. The research method used to determine the priority scale of determination is the AHP (Analytic Hierarchy Process) method. The method is obtained from a ranking of the assessments given by respondents who are competent in the field of road management.

2. Research methodology

The location of this research is in the Aceh Besar District, the map of the Aceh Besar District can be seen in the Figure 1 to be precise on the provincial road section in the Aceh Besar District.

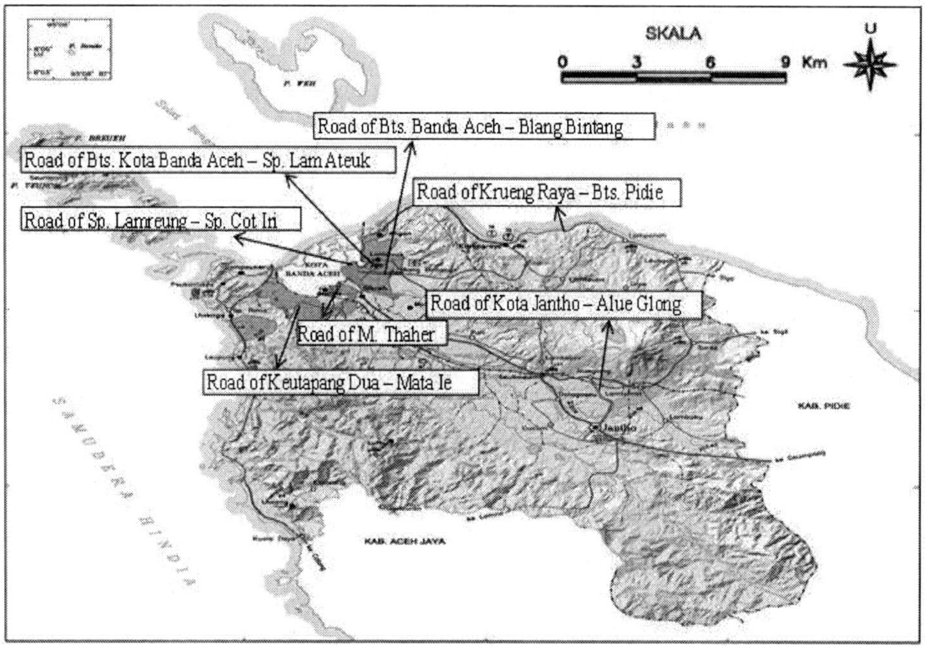

Figure 1. Map of the research location

On provincial roads in the Aceh Besar District, there are several roads that are handled by the Aceh Public Works and Spatial Planning Agency with the budget being financed from the Aceh Revenue and Expenditure Budget (APBA). The road sections under study are the main road and connecting roads used for various accesses to locations both to government centers, education centers, shopping centers and tourist attractions. These roads can be seen in Table 1.

Table 1. Determination of roads

No.	Road Section Name	Road Length (Km)
1	Road of Batas Kota Banda Aceh - Simpang Lam Ateuk	4.99
2	Road of Ir. M. Thaher	3.32
3	Road of Keutapang Dua - Mata Ie	3.17
4	Road of Simpang Lamreueng – Simpang Cot Iri	4.42
5	Road of Batas Banda Aceh –Blang Bintang	7.99
6	Road of Kota Jantho - Alue Glong	7.04
7	Road of Krueng Raya – Batas Pidie	33.44

In this study, the method used is the Pavement Condition Index (PCI) method to obtain road conditions to be handled in order to obtain the type of handling that will be carried out. Road conditions are obtained from the results of the Road Condition Survey on the road sections under review. After obtaining the results of road conditions and handling, then determining the type of road handling based on 4 (four) criteria using the Analytic Hierarchy Process (AHP) method. For data processing, the Microsoft Excel program is used for the Road Condition Survey data and data from distributed questionnaires. The steps for determining the value of conditions and types of handling can be described as follows:

a. The value of damage in the Road Condition Survey form that has been obtained is then carried out by calculating the Pavement Condition Index (PCI) for the road segment for each road damage with the help of the Microsoft Excel program where the assessment of the pavement condition is carried out according to the following formula:

$$\text{Density} = \frac{Ad}{As} x100\% \quad (1)$$

b. Deduct Value of each sample unit, enter the density percentage on the graph of each type of damage, then draw a vertical line to cut the severity of the damage (low, medium, high). Then a horizontal line is drawn and the deducts value (DV) is obtained.
c. Total Deduct Value (TDV) is the total amount of the reduction value of each sample unit or the total value of the individual deduct value of each type of damage and the level of damage that exists in a segment unit.
d. Corrected Deduct Value is obtained from the curve of the relationship between the TDV value and the CDV value by selecting the curve according to the number of individual deduct values that have a value greater than 2. The graph shows the relationship between the corrected deduct value (CDV) and the total deduct value (TDV).
e. The PCI value of each sample unit is calculated using the following formula:

$$PCI\ (s) = 100 - CDV \quad (2)$$

$$PCI = \frac{\sum PCI(s)}{N} \quad (3)$$

As for solving the problems faced by using the AHP method, the following steps are necessary:
(a) Defining the problem and determining the desired solution;

(b) Creating a hierarchical structure starting with a general objective, followed by the criteria and alternatives;
(c) Create a pairwise comparison matrix that describes the relative contribution or influence of each element to each goal and criterion level above it. Comparisons are made by assessing the importance of an element with other elements;
(d) Perform pairwise comparisons to obtain a total judgment of nx ((n-1) / 2), with the number of elements being compared;
(e) Calculating the eigenvalues and testing the consistency, otherwise the data collection must be repeated;

$$Wi = {}^{n}\sqrt{\text{Number of Rows}} \tag{4}$$

$$\text{Vector Eigen (Xi)}=\frac{Wi}{\Sigma\, Wi} \tag{5}$$

In the pairwise comparison assessment, there is often inconsistency in the references provided by the decision-maker. [4] has proven that the consistency index of matrices of order n can be obtained by formulas:

$$C.I = \frac{\lambda\ \text{maksimum - n}}{\text{n-1}} \tag{6}$$

$$C.R = \frac{CI}{RI} \tag{7}$$

If the CI is zero, it means that the matrix is consistent. The inconsistency limit is measured using a random generator value (RI), as summarized in Table 2.

Table 2. Random index (RI)

n	1	2	3	4	5	6	7	8	9	10
RI	0	0	0,58	0,9	1,12	1,24	1,32	1,41	1,45	1,58

The value of the *Consistent Ratio (CR)* is smaller than 0.1, which means that it is smaller than 10%, then the value is in accordance with the consistency requirements, which must be less than 0.1 or 10%.
(f) Determine road handling priorities based on the value of alternative performance.

3. Results and discussion

Assessment of road conditions can be done by looking at the damage to the road surface. This assessment is carried out by calculating the percentage of damage along the road sections under review on provincial roads in the Aceh Besar District. The damage that is mined is crack, potholes, depression, edge cracking, and corrugation. The road conditions for each road segment can be seen in Table 3.

Provincial road in the district of Aceh Besar is more than 50% have a category of good condition, from the results of field observations, roads in good condition is a road Batas Banda Aceh - Simpang Lam Ateuk with PCI value 66, Road Ir. M. Thaher with a PCI score of 66, Road Keutapang Dua - Mata Ie with a PCI value of 58, Road Simpang Lamreueng - Simpang Cot Iri with a PCI value of 67, and Road Kota Jantho - Alue Glong with a PCI score of 68. However, there are still road conditions with the medium category which is on the Krueng Raya - Batas Pidie road section with a PCI value of 53. The road of Banda Aceh - Blang Bintang is in very good condition with a PCI value of 75.

Table 3. PCI values

No.	Road Section Name	PCI Value	Road Condition
1	Road of Batas Kota Banda Aceh - Simpang Lam Ateuk	66	Good
2	Road of Ir. M. Thaher	66	Good
3	Road of Keutapang Dua - Mata Ie	58	Good
4	Road of Simpang Lamreueng – Simpang Cot Iri	67	Good
5	Road of Batas Banda Aceh –Blang Bintang	75	Very good
6	Road of Kota Jantho - Alue Glong	68	Good
7	Road of Krueng Raya – Batas Pidie	53	Fair

In the Analytic Hierarchy Process (AHP) method, it begins with distributing questionnaires to several respondents, in this case it has been carried out to 8 respondents. The data collected from these respondents are primary data from questionnaires. The number of questionnaires distributed to 8 selected respondents. The questionnaire data were collected from 8 respondents who were directly involved in the maintenance of provincial roads in the Aceh Besar District.

The largest sample, 67% has a Masters in Engineering education, it causes a reliable mindset and understanding of the work. The distribution of questionnaires as many as 8 copies were carried out by giving directly to respondents. The figures given to the respondents' perceptions are a scale of comparison of each of the criteria and alternative factors.

Answers to the assessment on the "criteria", from the results of interviews with respondents by conducting questions in the questionnaire in determining the level of importance of the criteria, answers were obtained based on the scale / rating range given on the questionnaire sheet. The answers to each respondent's perceptions of the "criteria" are tabulated as shown in Table 4.

Table 4. Recapitulation of respondents answers to "criteria"

Respondents	RESPONDENT PERCEPTION					
	A: B	A:C	A: D	B: C	B: D	CD
R1	2	3	3	2	4	3
R2	3	2	4	3	2	3
R3	2	2	4	3	2	2
R4	2	4	3	2	2	3
R5	2	3	2	3	3	2
R6	3	3	3	3	5	2
R7	2	3	2	3	2	3
R8	2	2	3	2	2	3

There are several criteria that influence the decision-making policy on road maintenance assessments. As for the criteria factors that influence the basis for selecting road maintenance are road condition, type of handling, accessibility and land zone. Weight of Assessment Criteria of each criterion were analyzed using the Analytic Hierarchy Process (AHP) method with the following steps:

- Step 1 Calculation of the initial matrix for the assessment "criteria"

Beginning with analyzing the data in Table 4, it is analyzed by inverse calculation according to the pairwise comparison matrix. The data is as complete as it is shown in Table 5.

Table 5. Pairwise comparison

Respondents	RESPONDENT PERCEPTION					
	A: B	A:C	A:D	B: C	B: D	CD
R1	2.00	2.00	2.00	0.50	4.00	3.00
R2	3.00	0.25	3.00	3.00	2.00	3.00
R3	2.00	2.00	0.25	4.00	0.50	0.50
R4	2.00	2.00	0.33	2.00	0.50	3.00
R5	0.50	2.00	0.20	3.00	3.00	2.00
R6	3.00	0.25	0.25	3.00	5.00	0.50
R7	2.00	0.33	2.00	0.33	2.00	0.33
R8	0.50	0.33	0.33	2.00	2.00	3.00
∑R	15.00	9,17	8.37	17.83	19.00	15.33
R / 8	1.88	1.15	1.05	2.23	2.38	1.92

Information:
∑R = the cumulative number of assessment comparison scales
R / 8 = Average comparison of ratings

Table 6. Initial "Criteria" Matrix

Criteria	A	B	C	D
A	1.00	1.88	1.15	1.05
B	0.53	1.00	2.23	2.38
C	0.87	0.45	1.00	1.92
D	0.96	0.42	0.52	1.00
∑	3.36	3.74	4.90	6.34

- Step 2. Calculation of the Eigenvectors.

Table 7. Value Eigenvectors for determining the scale of the "Criteria"

Criteria	A	B	C	D	total	Wi	E-Vector
A	1.00	1.88	1.15	1.05	2.25	0.56	0.37
B	0.53	1.00	2.23	2.38	2.82	0.71	0.47
C	0.87	0.45	1.00	1.92	0.75	0.19	0.12
D	0.96	0.42	0.52	1.00	0.21	0.05	0.04
∑	3.36	3.74	4.90	6.34	6.03	1.51	1.00

- Step 3. Calculation of the "Criteria" Maximum Eigenvalues is obtained from the Initial Matrix from Table 6, which is multiplied by the E-Vector of each matrix indicated by Table 7, and then the product is added. This is shown in Table 8 below.

Table 8. "Criteria" Maximum Eigen Vector Matrix

	A	B	C	D		E-Vector		
A	1.00	1.88	1.15	1.05		0.37		1.89
B	0.53	1.00	2.23	2.38	X	0.47	=	1.60
C	0.87	0.45	1.00	1.92		0.12		0.53
D	0.96	0.42	0.52	1.00		0.04		0.12
∑	3.36	3.74	4.90	6.34				4.13

- Step 4. Control of the Consistency Index (CI)
  Consistency Index (CI) = (4.13 - 4) / (4-1) = 0.038
  Consistency Ratio (CR) = 0.043 / 0.9 = 0.048 <0.1 (consistent)

- Step 5. Weighting Criteria. The weight of the element is obtained from the *E-Vector value* which is expressed as a percentage as shown in Table 9.

Table 9. Weights of Criteria for "Priority Scale Handling of Road"

Criteria	Weight
Road Condition	0.37
Type of Handling	0.47
Accessibility	0.12
Land Zone	0.04
Total	1.00

From Table 9 above, it can be seen that the respondent's assessment of several criteria shows that the criteria for road conditions have an influence on the level of importance namely with a weight of 0.37 (37%) then followed by the type of handling factor with a weight of 0.47 (47%), the accessibility factor with a weight of 0.12 (12%), and finally the land zone factor with a weight of 0.04 (4%).

The selected alternative is obtained by multiplying the total criterion weight with an alternative assessment or scoring. Table 10 displays a summary of the results:

Table 10. Multiply the weight of the criteria by the alternative score

Criteria	Weight	Alternative						
		x1	x2	x3	x4	x5	x6	x7
Road Condition	0,373	0,119	0,097	0,125	0,023	0,004	0,004	0,001
Type of Handling	0,468	0,149	0,122	0,157	0,029	0,005	0,005	0,002
Accessibility	0,124	0,040	0,033	0,042	0,008	0,001	0,001	0,000
Land Zone	0,035	0,011	0,009	0,012	0,002	0,000	0,000	0,000
Total		0,319	0,261	0,335	0,061	0,010	0,011	0,003

By analyzing the criteria for the implementation of routine road maintenance using the Analytic Hierarchy Process (AHP) method , there are weights that indicate the ranking order of the criteria given by 8 respondents. The following is a weighting order of criteria which can be seen in Table 11 below.

Table 11. Weights of criteria: handling of road priority scale

Criteria	Weight	Weights
Road Condition	0.37	2
Type of Handling	0.47	1
Accessibility	0.12	3
Land Zone	0.04	4

Table 11 shows that the criteria considered for the implementation of routine road maintenance are based on 8 respondents. These respondents are respondents who are directly involved in handling routine road maintenance. From the results of the analysis using the Analytic Hierarchy Process (AHP) method, it is known that the order that is considered in the selection of priority road handling is the 1st order is the type of handling with a weight of 0.47, then the second is the road conditions with a weight of 0.37, then the sequence third is accessibility with a weight of 0.12, and fourth is land zone with a weight of 0.04.

The results of the weighting of the criteria using AHP on 4 (four) criteria are carried out to determine the priority of road handling based on the results of a survey of respondents' perceptions showing that the type of handling criteria has the most influence in making road handling decisions. This is based on the calculation results shown in Figure 2. which shows that the handling specific gravity value has the greatest weight.

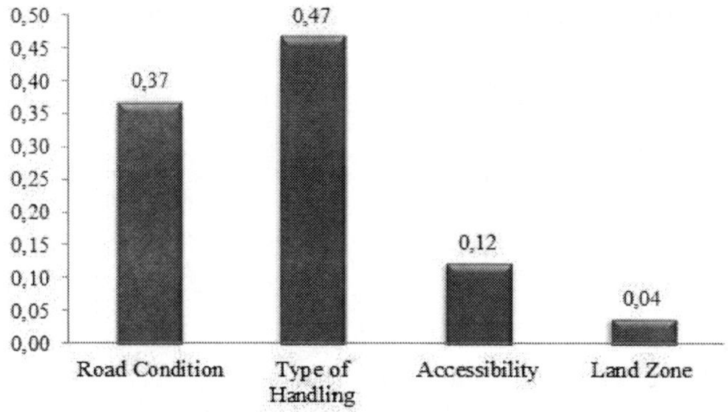

Figure 2. Weight criteria

Based on the calculation results in Table 11, the results of the assessment of the alternative options analyzed using AHP can be explained as follows on road of Keutapang Dua - Mata Ie the weight gain is 0.34, then on road Batas Kota Banda Aceh - Sp. Lam Ateuk obtained a weight of 0.32, then road of Ir. M. Thaher 0.26, on road of Sp. Lamreueng-Sp. Cot Iri 0.06, road of Kota Jantho-Alue Glong 0.011, road of Batas Banda Aceh - Blang Bintang 0.010, and for road of Krueng Raya - Batas Pidie with a weight gain of 0.003. The results of the calculation of the assessment based on the respondents' perceptions indicate that the priority for handling of roads with the greatest weight value is on the road Keutapang Dua - Mata Ie. From the results of alternative analysis using the Analytic Hierarchy Process (AHP) method in selecting road handling priorities, as shown in Table 12.

Table 12. Sequence of Alternative Weights of 8 Respondents

Road Section Name	Weight	Weights
Road of Batas Kota Banda Aceh - Simpang Lam Ateuk	0.319	2
Road of Ir. M. Thaher	0.261	3
Road of Keutapang Dua - Mata Ie	0.335	1
Road of Simpang Lamreueng – Simpang Cot Iri	0.061	4
Road of Batas Banda Aceh –Blang Bintang	0.010	6
Road of Kota Jantho - Alue Glong	0.011	5
Road of Krueng Raya – Batas Pidie	0.003	7

Based on the results of the above analysis using the Analytic Hierarchy Process (AHP) method, it shows that of the 8 respondents who presented the opinion that the priority of road handling with this type of routine maintenance is better done especially on road Keutapang Dua - Mata Ie. This is because the respondent's choice of weight is the largest choosing road Keutapang Dua - Mata Ie with a weight of 0.34. Figure 3 shows that road of Keutapang Dua - Mata Ie has the highest weight compared to other provincial roads in Aceh Besar District.

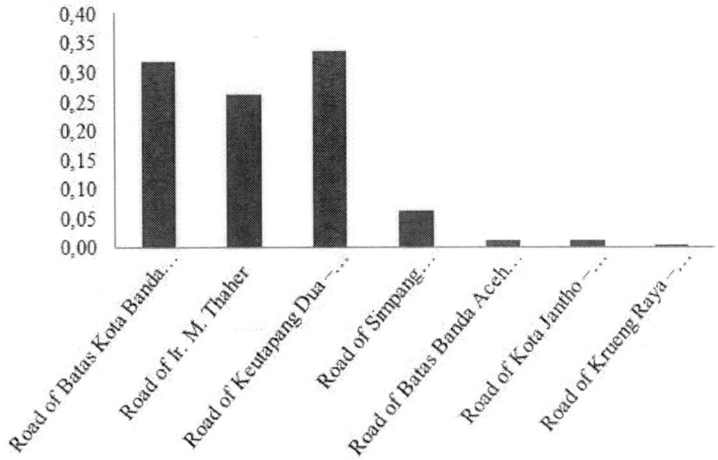

Figure 3. Alternative weight sequence

4. Conclusion

On provincial road in the district of Aceh Besar based on observations by the method of Pavement Condition Index (PCI), the obtained results for all the roads under review is more than 50% have a good condition, but there are roads with category Area on the Krueng Raya - Batas Pidie road. The road of Batas Banda Aceh - Blang Bintang is in very good condition. Types of handling for provincial roads in Aceh Besar District are at the level of routine maintenance where the PCI values of each observed road segment are in the very good, good and moderate categories with PCI values ranging from 51 - 75. The results of the analysis of 8 respondents using the Analytic Hierarchy Process (AHP) method show that for the greatest weight of the criterion assessment, namely the type of handling with a weight of 0.47, then road conditions with a weight of 0.37, followed by accessibility of 0.12, and land zone of 0.04. Whereas for the alternative priority of road handling the highest weight was obtained on road Keutapang Dua - Mata Ie with a weight of 0.34, then the second alternative was on road Batas Banda Aceh -Sp. Lam Ateuk 0.32, the third alternative on road Ir. M. Thaher 0.26, for the fourth alternative on road Sp. Lamreueng-Sp. Cot Iri 0.06, the fifth alternative on road Kota Jantho - Alue Glong 0.011, the sixth alternative on road Batas Banda Aceh - Blang Bintang 0.010, and for the seventh alternative on road Krueng Raya - Batas Pidie with a weight gain of 0.003.

References

[1] Irvan K 2014 *Jurnal Teknik Sipil,* **2** ISSN 2339-0271. *(in Bahasa Indonesia).*
[2] Suswandi A, Sartono W, Hardiatmo H C 2008 *Forum Teknik Sipil.* **XVIII** 934-946. *(in Bahasa Indonesia).*
[3] Widodo A D 2018 *Evaluation Of Pavement Conditions And Prediction Of Before Life With The Pavement Condition Index, Fostering And Mechanistic-Empiric Methods With The Kenpave Program.* Magister Thesis. Islamic University of Indonesia. *(in Bahasa Indonesia).*
[4] Saaty T L 1980 *The Analytic Hierarchi Process.* New York: McGraw-Hill International.

The influence in selection of pile type and foundation piling method

O L Sari*, A A I Saputra, A Subhakti, A P Putri, C C S Khala

Civil Engineering Study Program, Kalimantan Institute of Technology, Balikpapan, Indonesia

*E-mail: oryza@lecturer.itk.ac.id

Abstract. One of the jobs that have a high complexity in civil works is foundation pile work. Practitioners are often faced with many choices and various conditions in the field in making decisions about the types of piles and methods of piling. The selection of the type of piles and the method of piling will have environmental, economic and social impacts. Many conditions need to be considered when making this selection. This paper presents a literature review in the study of influence in the selection of pile type and foundation piling method in problem identification and approaches. The results of this study support the influence factor used by case studies of decision-making taking into triple constraint in project management.

1. Introduction

In general, the structure of the building is divided into two basic parts, namely the substructure and the upper structure. The foundation is the most stable structure and functions as a load support as a whole on the part of the structure that is above the ground [1].

In choosing the type of foundation, there are aspects that need to be considering because it will affect the method that should be used, such as soil bearing capacity, hard soil depth, the load of the upper structure, cost, and length work available [2]. In order to get the best cost and processing time, it is very important to pay attention to several existing criteria because they are influential in choosing the type of foundation and the type of foundation that there is also very much [3]. In general, the process of implementing piles usually has an impact on the environment that is detrimental to human comfort or physical damage to buildings. So that a distance of 200 m from the construction site is considered an ideal distance that needs to be done to pay attention to the limitations of the given environment [1].

It can be concluded that foundation work is a job that has many criteria that need to be considered and has many alternative types of foundation piles and pile methods that can be used. Therefore, in selecting the type of foundation piles and the method of piles, unique knowledge is required with strict regulations and must be followed so that the required costs and execution time are in accordance with the economic conditions and can avoid environmental and social damage [4].

A lot of research on influence in decision-making types of foundation poles and piling methods. This research presents literature review in the study of the influence of decision making in determining the type of foundation pole and piling method. The main problem in researching the influence of the selection of foundation pole types and piling methods is to review what influence affects in the retrieval of the foundation. This research was conducted by grouping research on the influence of the selection of the type of foundation pole and the method of piling along with its approaches. Each study group was analyzed by combining and comparing emerging issues. This was developed using a methodology map, in which the results of the analysis are prepared to find the direction of future research.

The foundation consists of two types, namely shallow foundation and deep foundation. Shallow foundations do not require deep soil digging, as hard soil is at a fairly shallow depth and is commonly used in simple house buildings. While deep foundations are used on hard soil layers that have a deep enough depth, it is commonly used in multi-story buildings, bridges, coastal or offshore buildings, and so on. The foundation of the stake is a type of foundation [5].

In the selection of the type of stake to be used, there are three main factors that must be reviewed, namely the location and type of the building, the durability of the pole, and the characteristics of the soil obtained from soil testing at the point of the piling site [6]. The limitations of the construction site or construction site are an obstacle that must be overcome by regulating the type of pole to be used, piling equipment, materials, and labor. The characteristic factor of the soil in the selection of the foundation design is determined by the testing of soil samples at the site. Depth of hard soil layer by getting qc conus value and fs value. In selecting the type of foundation pile practitioners use parameters related to information on soil characteristics to determine decision making [7].

In the selection of foundation piling methods, it is worth noting several factors such as environmental impact, cost, location. The need to pay attention is the demand for tranquility from the environment due to the implementation of development such as noise and vibration disturbances. The vibration sparked by the piling activity is quite difficult because it is very disruptive and can damage existing buildings in the vicinity of the construction is carried out in a densely populated location.

2. Method

This paper presents a literature review in case studies on the selection of foundation pole types and piling methods from previous studies. It is necessary to clarify the issues raised in this study in order to be useful as a basis for thinking for the development of further research [8]. The method of literature review used is to collect related literature, analysis, synthesis and results.

The first stage of the literature review is to collect and read some previous research related to the issues raised [9]. In collecting related research, it is done by determining relevant keywords and then searching for literature using relevant keywords. The research that has been obtained is filtered by reading the title and abstract of the research, then followed by the analysis process by mapping the previous research grouped based on the research method used. The results of the mapping were synthesized by linking the differences and similarities of each previous study. From each of these studies, it can be analyzed the similarities and differences between each study aimed at finding supporters of influence in the selection of foundation pole types and piling methods.

3. Result and discussion

3.1. Research map

The results of the study influence the selection of the type of foundation pole and the method of piling presented in Figure 1. Based on the results of the analysis, it was obtained that there are two research groups used in literature review this time, namely primary quantitative and qualitative primary. Almost all of the studies reviewed came from the primary quantitative group. It has been explained that the research used determines the criteria for the selection of foundation types with survey questionnaires and continued with AHP study [10] [11] [12]. Another research analyzes the effect of foundation piling vibrations and demolishing buildings on the residential environment [1] [13]. And several research projects comparing foundation shape and design against the cost, soil structure, and soil characteristics [3][4] [6] [7]. For more detail, the research methods in literature review are shown in the Table 1.

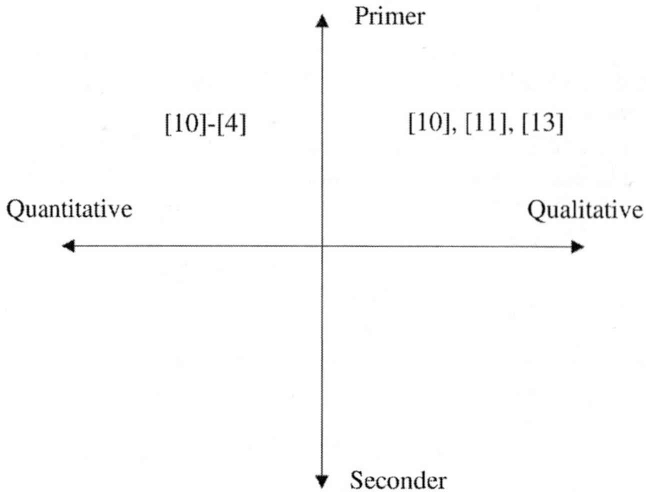

Figure 1. Map of the research methodology.

Table 1. Research methods.

Authors	Year	Method
R. Sutjipto Tantyonimpuro dan Agustina Dwi Retnaningtias [10]	2006	Qualitative quantitative research, AHP
Mohamad Ridwan [2]	2008	Data analysis, quantitative
Mahendra Cipta A.N., Guntur Panji Wijaya, Hermawan, M. Agung Wibowo [11]	2010	Qualitative quantitative research, interviews, AHP
Juniada Pagehgiri [3]	2015	Comparative research, survey and interview
Joko Yulianto [12]	2017	Observation and interviews, AHP
Oryza Lhara Sari [13]	2019	Qualitative and quantitative research
Kukuh Prayogo and Hasriyasti Saptowati [7]	2016	Data Analysis, quantitative
Lilies Widojoko [6]	2015	Data Analysis, quantitative
Taufan H. Saputra, Habir and Achmad Munajir [4]	2016	Comparative, quantitative

3.2. Research position

Based on the results of previous literature studies, previous research on decision-making models of foundation pole types and piling methods, then obtained research positions grouped based on variables discussed in research and research models. The four group variables used as a comparison include the criteria of piling, the type of foundation pole, the method and the impact of the piling, as indicated in Table 2. Slices of the variables of each literature reviewed that are illustrated using venn diagrams can be seen in Figure 2.

Table 2. Research position.

Authors	Piling Pile Criteria	Pile Type	Piling Method	Impact	Research Model
R. Sutjipto Tantyonimpuro dan Agustina Dwi Retnaningtias [10]	√	√		√	AHP
Mohamad Ridwan [2]			√	√	Experimental Research
Mahendra Cipta A.N., Guntur Panji Wijaya, Hermawan, M. Agung Wibowo [11]	√	√			AHP
Juniada Pagehgiri [3]		√		√	Comparative Research
Joko Yulianto [12]	√		√	√	AHP
Oryza Lhara Sari [13]				√	Modeling
Kukuh Prayogo and Hasriyasti Saptowati [7]	√	√			Experimental Research
Lilies Widojoko [6]	√	√			Experimental Research
Taufan H. Saputra, Habir and Achmad Munajir [4]	√	√		√	Experimental Research

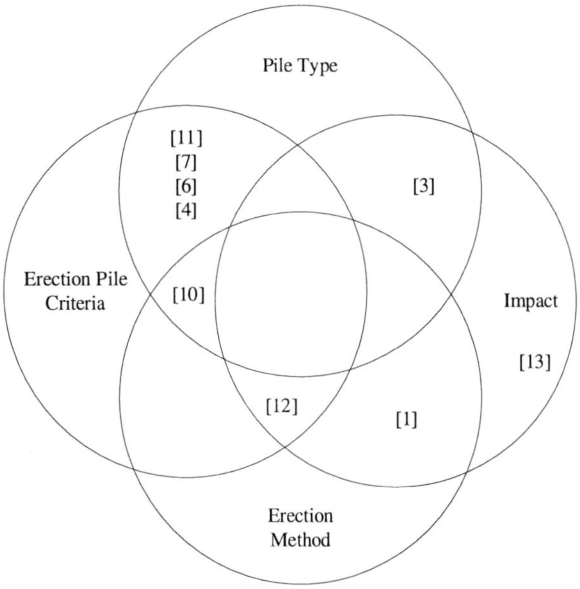

Figure 2. Venn diagram of variables.

4. Conclusion

Based on the literature review in the selection of the type of foundation pole and piling method, it can be concluded that the influence of soil characteristics, cost, characteristics of buildings and impact on the environment greatly affects the type of poles and piling methods used. Such influence should be considered as a basis in research on the selection of foundation pole types and future piling methods. The results of this study support the influence factor used by case studies of decision-making taking into triple constraint in project management.

5. Acknowledgement

The writes appreciate the recognition and awards in the form of research grants from the "Hibah Penelitian dan Pengabdian Masyarakat Institut Teknologi Kalimantan 2020, Grant Number:1134/IT10/PPM.05/2020". All concepts and results presented in this paper are part of research project entitled "Model Pengambilan Keputusan Pada Pemilihan Jenis Driven Pile dan Metode Pemancangan Pondasi Berbasi Multiclass Multilabel Classification". Authors would like to thank ITK for funding the research.

References

[1] Ridwan M 2008 *Journal Permanente,* **3 (1)** 44-54.
[2] Fauzi L A and Ikhya I 2016 *RekaRacana: Jurnal Teknik Sipil* **2 (2)** 36-46.
[3] Pagehgiri J 2015 *Extrapolasi Jurnal Teknik Sipil Untag Surabaya,* **8 (1)** 121-136.
[4] Saputra T H, Habir and Munajir A 2016 *Jurnal Universitas 17 Agustus 1945 Samarinda*, 1 **(1)** 1-15.
[5] Yusti A and Fahriani F 2014 *Jurnal Forum Profesional Teknik Sipil,* **2 (1)** 19-31.
[6] Widojoko L 2015 *Jurnal Teknik Sipil UBL,* **6 (2)** 818-842.
[7] Prayogo K and Saptowati H 2016 *Jurnal Perangkat Nuklir,* **10 (1)** 30-49.
[8] Boote D N and Beile P 2005 *Educ. Res.*, **34 (6)** 3-15.
[9] Bryman A 2011 *Social Research Method 4th Ed* (New York: Oxford University Press).
[10] Tantyonimpuro R S and Retnaningitas A D 2006 *Jurnal Teknik Sipil Unika Soegijapranata,* **3 (2)** 77-87.
[11] Cipta M A N, Wijaya G P and Wibowo M A 2010 *Konferensi Nasional Teknik Sipil,* **4** 17.
[12] Yulianto J 2017 *Jurnal Riset Rekayasa Sipil* **1** 50-58.
[13] Sari O L 2019 *Proceed. Int. Conf. Sustain. Innovation – Technol. Eng. Series: Advances in Eng. Res.* **187** 58-61.

Territorial characteristic through residential boundaries on Indo-Tiongkok relocation housing in Neuheun, Aceh Besar

A Shah[1], E Wulandari[2*], Izziah[2]

[1]Graduate Student on the Department of Architecture and Planning, Faculty of Engineering, Universitas Syiah Kuala, Darussalam, Banda Aceh 23111, Indonesia
[2]Department of Architecture and Planning, Faculty of Engineering, Universitas Syiah Kuala, Darussalam, Banda Aceh 23111, Indonesia

*E-mail: elysa_wulandari@unsyiah.ac.id

Abstract. Residents create boundaries through territories control according to what they feel, need, and their abilities in technical and technology. That is why territorial boundaries will be able to reflect the characteristics of the particular residents. This research is conducted in Indo-Tiongkok relocation housing which is located in a rural hilly area. Those aided houses were uniformly shaped given to the tsunami victims with very diverse socio-cultural conditions. This research uses survey methodology, observes the architectural object of plot boundaries, and in-depth interviews to find out the reasons for the creation of the boundary pattern. The result shows the characteristic in general, the pattern of the fence as a boundary appears based on social needs (self-actualization) and external disturbances, namely: a) disturbance that appears in limiting the activities of foreigners to the residential area; b) disturbance from animal pests to protect gardens and houses.. This research concludes that the development of residential boundaries for relocation housing is influenced not only from the internal aspects of the residents but also from the external aspects of the environment. In determining the relocation of post-disaster relief homes in rural areas, the threat of wild animals needs to be considered in planning housing areas.

1. Introduction

The 2004 tsunami devastated most of the coastal settlements so that many villages were relocated by providing assistance in the form of special housing areas for tsunami victims [1]. Relocation housing assistance after the 2004 tsunami in Aceh is generally built-in suburban rural areas that are safe from the threat of the tsunami hazard, such as in the Neuheun Village, Mesjid Raya District, Aceh Besar Regency. The housing development is generally designed in a certain pattern with a uniform type of house and the same footprint area, so that housing assistance does not differ between occupants and looks uniform.

The uniformity of the character of the residents' dwellings is gradually changing, especially in the development of plot boundaries. Plots boundaries as architectural works become important to mark the boundaries of the controlled areas and even become identities of the residents [2]. The boundaries of the residential lot will form the architectural space of the residence. A dwelling has an area boundary which means ownership of a residence. Control of the coverage area means the ability to cover the area, limiting entry from outside interference. This is a way of human instinct in understanding the built environment.

The territory has the meaning of the occupancy where humans live and limits what comes in and out of the area.

Figure 1. Location of Indo-Tiongkok housing in Aceh Besar regency. Contour map, interval: 2 m.

The Indo-Tiongkok relocation household area, which consists of 600 lots with a house type of 42 m^2, is built in a grid following the contour pattern. The house is placed in the middle of the plot of the lot, where each plot does not have a clear land boundary visually, but the owner of the house knows the boundaries of each plot. The drawing of the location can be seen in Figure 1, while the environmental design, and the placement of the house in the lot can be seen in Figure 2.

Figure 2. Housing condition in the initial development without fences after two years of inhabitations.

Also from Figure 2, it can be seen that the community in their early years of occupation, the developer does not provide a fence directly. Homeowners start to build their fences with their characteristics according to their needs. The character of territorial boundaries also arises from the social right to defend oneself against outside interference. It is revealed that the territory in the architectural environment and behavior is defined as the extent to which a living organism determines its demands, marking, and defend it, especially from the possibility of intervention from other parties [3]. This concept was originally developed for a living organism is not human, but later used for human and environmental concerns are also perceived as an imaginary environment. Meaning for humans, the concept of territory is more than the demand for a regional spatial and physical, but also emotional and cultural needs.

The selection of Indo-Tiongkok as a subject is the reason why this research is important. Currently, the boundary is built using various kinds of types, from wooden fences, iron, bricks which can be questioned for the specific reasons toward the construction of these types of boundaries. The more diverse the character of the community, the more varied the pattern of control of the occupancy site will be. The research was conducted in an Indo-Tiongkok relocation housing located in a rural hilly area, because the aid houses were uniformly shaped, given to tsunami victims with very diverse socio-cultural conditions. This study intends to understand how each person develops boundaries that describe their identity. This research is important to do concerning post-disaster relief evaluation of relocation housing, how the process of assimilation of a heterogeneous society, through the architectural creation of its residential environment. We can also observe how creativity in character building of a society, that views the forms of freedom of expression and which positive values will be transmitted to other residents. That is why the selection of Indo-Tiongkok relocation housing is an option where based on the data obtained, there is a diversity of ethnicities and work backgrounds of the residents of the housing. Besides that, the disturbance of wild animals from the hills outside the housing area has become a major consideration for how people protect their living space.

2. Theoretical review

Territoriality is related to human behavior toward their environment [4]. Territoriality is one behavior attributes that have a relationship between individuals, groups/organizations with the physical setting. Accordingly, it is explained that individuals in a physical setting related to behavior, activities, places, and times [3]. While, a physical setting has physical element and activity [5]. From some of these points of view, it can be concluded that there are three main territoriality elements, those are physical setting (territory), actors (individual/group), and a variety of activities. The need to cultivate crops makes the need for physical boundaries to protect residential areas from the environment. For example, garden predators such as disturbance of cows, goats, and wild boar. The three elements in the area boundary are land for cultivation, users as well as the farming activities.

2.1 Territory and boundary characteristic pattern

The territory is defined as occupying the behavior of humans. The physical formation might not yet be included here. The boundary that creates the territory is not mandatory. But more on the control of space. Personal space is a universal phenomenon, although a special dimension is determined by the culture of society. Territorial boundaries are not maintained by body language. However, it must be communicated with a marker. Other people must understand the boundaries of the territory being communicated so that these boundaries are not violated. The most basic way can be seen in the form of stones to pillars that indicate the borderline.

Although the boundary does not always mean territory consistently. The livestock fence is to restrict livestock from leaving, not to prevent others from entering. Walls and fences divide space, it may or may not mean a boundary. Gates as a housing boundary are constantly being challenged. If no one prohibits these restrictions from being breached, then the gate does not mean an area limitation that prohibits entry for everyone. And this will be repeated over and over again until there is a direct ban. Any vacant land where there is a certain owner when there is no marker can be violated until real control is taken.

2.2 Claim on boundary

Claims on the boundaries of a residential area can be realized either in real or abstract ways. Fences, walls, or plants usually mark the coverage of a residence. This shape becomes a marker of territory for property owners, and a marker for its surroundings. Even without a fence, the property can still appear to be a private area with manicured grass that claims the boundaries of the residential area.

Meanwhile, the abstract form with differences in the ground material or using landscapes as a form of social limitation from the owner. Even though the doors and terraces of the house are boundary forms, they do not necessarily mark the boundaries of the house area. It's the garden of the house where the house is built. Doors and terraces can be boundaries to higher hierarchies. The built-up formation can be a claim to territorial boundaries, continuously from various outside environment disturbances.

2.3 Territorial hierarchy

The boundaries of a residential area have a hierarchy, housing has a hierarchy to even reach the city scope. In each hierarchy, different permission is required to enter within the boundaries of the region. In architecture, usually, the term of public, semi-public to private spaces are used as a guideline for the hierarchy of boundaries in a residential area. The fence plays the most crucial role in what can enter into a deeper room. Control from the top of the hierarchy plays the most important role in what is allowed within the boundaries of the area. The fence clarifies the boundaries between the dwellings in this discussion.

3. Research method

This study uses a survey methodology to observe the footprint of the housings and in-depth interviews to know the reason how the creation of the boundaries occurs. Direct observation is to see more clearly the house fences and boundaries between neighbors. This makes it easier to understand the boundaries between occupants that occur in the social community. Where the fence becomes a clear boundary of the residential area between spaces resulting from area boundaries. Data is taken in the form of random sample house sites, which shows a significant difference in architecture.

Environmental studies require the development of a methodology that includes inspection of the physical area of observation in the form of photos and floor plans, user observations of the environment, and direct interviews with users, which focus on use and reactions to the environment and are carried out by investigators with qualified skills. After seeing each house firsthand, we can compare the boundaries of each dwelling. These boundaries will greatly depend on the culture and perceptions of the people with different hierarchies of boundaries. The finding can be successfully spotted based on previously researched theories.

Furthermore, in our observations, we can compare the physical boundaries of a dwelling with the perception of area boundaries directly depending on the differences in people's perceptions of the boundaries of occupancy and local culture. The discussion in this study will emphasize the comparison between settlements and how culture can influence the definition of boundaries seen from the Indo-Tiongkok housing complex. Especially in finding the defensive character patterns of the local community.

4. Result and discussion

The territory is the formation of an area to achieve optimal privacy that is sought by developing physical settings [6]. Types and levels of privacy are influenced by behavior patterns in the cultural context, personality, and aspirations of individuals. Walls, screens, symbolic boundaries, and tangible boundaries, as well as distance, are mechanisms for showing privacy. The character of defense can be seen from the behavior patterns of the local community in a cultural and emotional context. A house that is a symbolic residence requires a marker at a higher level to give clearer boundaries. Further varied needs of residents to complement the psychological and aesthetic needs of the building.

From direct observation and discussion with residents of the houses in block C of the Indo-Tiongkok relocation housing. It was found that approximately 50% of the residents do not make the dwelling there as the main house for the owner. Some leave their houses empty, some are renting out, and some are using them as visiting houses. The difference in the atmosphere of a house that is occupied permanently and that which is not permanent can be seen in the management of the house barriers (Figure 3).

If we compare the coupling house as in Figure 3. It is clear that the character of the different boundaries between the occupants who make the main house and the house that is not inhabited by residents. Fences appear as markers that help restrict outsiders from entering private areas. Not only outsiders, but many animals are predators of plants in gardens. Moreover, residents like to plant plants in the form of chilies, pineapples, jackfruit, bananas, and so on. Fences become territorial boundaries that prevent plant predators such as wild boar, goats, cows from entering their territory. The need for gardening of the occupants creates physical boundaries that directly protect their territory from outside threats.

(a) (b)

Figure 3. Coupling house: the difference in condition on occupied houses and vacant.

4.1 Emotional and cultural needs

The need for residents of the house to create clear boundaries to present a better aesthetic as in Figure 3. Besides that, it is also able to provide emotional comfort so that residents can feel safer and protected when in their homes.

In Figure 3b, the aesthetic that appears shows a high control over the function of the fence, the appearance of modern materials with contemporary colors shows that the residents come from educated urban people, have a high profile and high individual impression. Unlike the fence in Figure 3a, whose aesthetic function comes from natural materials, it is simple but has good natural knowledge. Low profile and met the purpose to protect themselves.

Territoriality is an individual or group relationship with the physical setting, which is illustrated with a sense of belonging and efforts to control the use of space [6]. In concrete terms, characterized by territoriality placement (occupancy) symbolically with attachment points (attachment). The territory is a space that can be defined and controlled by an individual or group through the use of physical space, ownership, defense, use exclusively, or signs of identity-oriented spatial access [7]. The identity of residential ownership is a factor in the existence of a marker of a physical residence. Furthermore, a part of the dwelling becomes a temporary residence for the settlers and further boundaries become a solution.

The residents of the housing complex who were victims of the earthquake and tsunami in 2004 consisted of various cultural, professional, ethnic, linguistic, and religious backgrounds. Table 1 indicates the types of work for residents of the housing complex.

Table 1. Types of occupation for residents of the Indo-Tiongkok.

No	Occupation	Number of people
1	Farmers	73
2	Fishermen	59
3	Entrepreneur	34
4	Construction workers	90
5	Traders	47
6	Government employees	30
7	Retired	7
8	Others	150

The residents of the housing complex live blended between employees, fishermen, farmers, motorcycle taxi drivers, civil servants, traders, and entrepreneurs. In terms of ethnicity, there are indigenous Acehnese, Aceh-Javanese, Padang, Batak, Chinese, and other ethnic groups. There are so many differences between fellow residents who live in the housing complex. However, of the many differences that exist, the most highlighted is the existence of religious differences. The housing complex is inhabited by some non-Muslim Chinese residents.

A visiting house needs clearer boundaries to avoid outside disturbances. This is also supported by the owner's higher economic level. Every need is realized optimally, not only based on the need for regional boundaries, but also aesthetics and culture. Territoriality has four main characters; ownership or rights from somewhere, personalization or marking of a particular area, the right to defend itself from outside interference, and control of multiple functions, ranging from meeting basic needs to the satisfaction of psychological cognitive and aesthetic needs [8]. The boundary has four main characters. Where in Indo-Tiongkok relocation housing, the character of the territorial boundaries arises from the type of fence that is presented by the occupants of the house. Because the house built from the developer does not provide a fence initially. Homeowners build their fences with their characters according to their needs. The character of territorial boundaries also arises from the social right to defend oneself against outside interference. From the data collected, socially the people in the Indo-Tiongkok relocation housing consists of various ethnicities and professional backgrounds, with different interests.

So it can be concluded that the community does not integrate completely with fellow residents of the housing. From the observation, the fence looks stronger for people who have sufficient funds to build according to their needs. The control that emerges from Lang is further explained by Robert Sommer following the needs of the homeowner. A shelter that becomes a symbolic temporary residence requires a marker at a higher level to provide clearer boundaries. Further varied needs of residents to complement the psychological and aesthetic needs of the building appear from several buildings.

4.2 Securing activities and natural predators

There are four types of territoriality that are useful in environmental design; attached territory, central territory, supporting territory, and peripheral territory [8]. We can identify three distinctive levels of territoriality, namely: personal space, home base (a space that is well maintained), and home range (is a behavior set that is built by somebody's life) [9]. From the author's observations on the habits of the residents as seen from one end unit of block G, the occupants of the house create a wider area boundary (home range) from their daily work activities. The inhabitants raise chickens across from their house. Road boundaries are no longer public places that can be crossed by many people. However, it is claimed to be the boundary of the user's territory. Moreover, the last house beside them was no longer inhabited for a long time. It appears that the claim for boundaries includes roads that are positive for the improvement of the community's economy (Figure 4a).

(a) (b)

Figure 4. Territorial claim for land on the outskirts of a living neighborhood.

Sommer [10] stated that a territory is an area controlled by a person, family, or other face-to-face collectivity. Control is reflected in actual or potential possession rather than evidence of physical combat or aggression - at least at the human level. Robert Sommer emphasized that possession/ownership in the terms of the territory is more important than a desire to defend the territory from encroachment. Territories are areas controlled based on ownership and exclusiveness of use. 'This is Mine' or 'You keep off' [10]. From what Robert Sommer said, ownership is far more important than a concern to defend territorial boundaries from disturbance. In Indo-Tiongkok housing, there is a lot of disturbance from livestock and wild animals. So that active behavior to make clear boundaries is very important. This is done to protect the territory from disturbance because animals do not use the same social assumptions as humans.

From the shelter in block G, there is a case from a garden who is own by Batak, a similar statement has been made stronger but the boundaries are strengthened by the presence of dogs to guard the garden (Figure 4b). Large gardens filled with organic vegetables require assistance from guard dogs to protect the garden from predatory animals. The differences in the culture of the residential owners who came from outside Aceh provided a choice of types guarding in various ways.

4.3 Boundary transparency

The same thing can be understood from public spaces such as SDN 3 Neuheun which has a transparent fence. This is done to provide an understanding that the area is open to admitting other people with interests such as school employees and students studying at the school.

We can distinguish the territory into four types; personal territory, territorial communities, society territory, and free territory [8]. The four types of these territories are classified based on: 1) Control degree that is used by other people, 2) A person or group of people who do the counterpart, and 3) the existence of a signal which has the function as a control point. If one of these elements do not exist or do not work effectively, the ability of territoriality formation territory will decrease. Area boundary formation is determined based on these three things for optimal boundary clarity. It is also seen that the boundaries of social facilities are more open than those of private areas [11]. This can be seen from the Village Secretariat facility in the area. It can therefore be concluded that the character that emerges from a boundary is determined by the need to limit external interference. From this theory, we can also see the hierarchy of the territorial boundaries in the Indo-Tiongkok relocation housing and how the strength level of the territorial boundaries is present in this housing.

The degree to which the individual feels inaccessible to the family (or family to the neighborhood) may be expressed in the thickness of the boundary, its degree of imperviousness, the distance from family spaces. The degree to which the individual feels accessible to the family (or family to the individual) may be expressed in these physical boundaries by the degree to which the boundary is broken by doors, windows, its proximity to family spaces, etc. The thickness, resistance of the fence, and the distance from the family room can be a transparent hierarchy of boundaries of a dwelling. A residence that uses a high and sturdy fence will have the meaning of keeping a distance from other people who live around it. Meanwhile, a house with a fence that is more transparent and simple means that the residential approach is more open to other people, as shown in Figure 5.

(a) (b) (c)

Figure 5. Fence transparency which accommodate occupant personal character.

The character of the fence used in the Indo-Tiongkok relocation housing represents the character of the territorial boundaries. From observations, it can be seen that the use of housing in the form of permanent fences and semi-permanent fences in the majority of dwellings. However, some houses do not have a fence, wherefrom the observation, the houses are mostly uninhabited or rented out to other people. In other words, it is not the house that is occupied by the owner of the residence. The character of the fence is almost always raised by the owner. The looks of the fence types vary. This represents the fact that residents want to show their territorial boundaries from the residential space they are occupying.

5. Conclusion

The boundary has four main characters. Where in Indo-Tiongkok relocation housing, the character of the territorial boundaries arises from the type of fence that is presented by the occupants of the house. Because the house results from developer to the owner does not provide a fence directly. Homeowners build their fences with their occupant characteristics according to their needs. The character of territorial boundaries also arises from the social right to defend oneself against outside interference.

From the fact that the Indo-Tiongkok relocation housing consists of various ethnicities and work backgrounds, it appears that the interests of each occupant make the fence as a barrier appear very diverse, both in quantity and quality. This can be concluded that the community does not integrate completely with fellow residents of the housing. The difference in the economy of the people is visible in the appearance of the fence. However, in general, the pattern of the character of the fence as a regional boundary defense appears based on social needs (self-actualization) and external disturbances, namely: a) disturbance that appears in limiting the activities of foreigners to the residential area; b) disturbance from animal pests to protect gardens and houses.

This research concludes that the development of residential boundaries for relocation housing is influenced not only from the internal aspects of the residents but also from the external aspects of the environment where the housing location is located. Rural hilly areas are dangerous for residential gardening architecture from disturbance of wild animals. Therefore, it is necessary to pay attention to the existence of a buffer environment outside the housing environment, so that wild animals do not interfere. In determining the relocation of post-disaster relief homes in rural areas, the threat of wild animals needs to be considered in planning housing areas.

6. Acknowledgments

This paper is an initial review of the Master of Architecture thesis plan at the Faculty of Engineering Unsyiah and at the same time being involved in the research of the head lecturer Dr. Ir. Elysa Wulandari, M. T. entitled: Changes in the spatial housing area for tsunami relocation and its impact on the area's sustainability system, Case study: Neuheun Village, Mesjid Raya District, Aceh Besar Regency.

References

[1] Elysa W, Fahmi A, Evalina Z, Myna A 2020 *IOP Conf. Ser.: Earth Environ. Sci.*, **452 (1)** 012117.
[2] Habraken N J 1998 *The Structure of the Ordinary: Form and Control in the Built Environment* (Cambridge: MIT Press).
[3] Haryadi and Setiawan B 2014 *Arsitektur Lingkungan dan Perilaku* (Yogyakarta: Gadjah Mada University Press).
[4] Weissman G D 1981 *J. Man Environ. Relation*, **1 (2)** 315-346.
[5] Rapoport A 1977 *Human Aspect of Urban Form* (New York: Pergamon Press).
[6] Altman I 1975 *The Environment and Social Behavior* (Monterey, CA: Wadsworth).
[7] Gifford R 1987 *Environmental Psychology: Principle and Practice* (Boston: Allyn and Bacon. Inc.).
[8] Lang J 1987 *Creating Architectural Theory: the Role of the Behavioral Sciences in Environmental Design* (New York: Van Nostrand Reinhold).
[9] Porteous J D 1977 *Environment And Behavior* (England: Addison Wesley Publishing Company).
[10] Sommer R 1983 *Social Design: Creating Building with People in Mind* (New Jersey: Prentice-Hall Inc.).
[11] Brower S N 1976 *Human Behavior and Environment* (New York: Plenary Press).

Analysis of the effect of land use on flood height in Balikpapan city (case study: Posindo sub-basin)

R M Kadaryanti*, A N Dewanti, A Ghozali

Department of Civil Engineering and Planning, Institut Teknologi Kalimantan, Indonesia

*E-mail: rossa.margareth@lecturer.itk.ac.id

Abstract. One of the problems with infrastructure in the City of Balikpapan is the inadequate drainage system and is one of the causes of flooding in the City of Balikpapan. Drainage problems in the city of Balikpapan concentrated in the central area of the city, one of which system of drainage that is located in the city center is the Ampal/Klandasan Besar Basin. In that drainage system, there is one of the sub-basin where the majority of its land use is built-up area, namely the Posindo Sub-basin. This study aims to analyze the effect of land use on high floods that occurred in the Posindo Sub-basin. Flood discharge due to rain was analyzed using the HEC-HMS program. The results of the hydrological discharge of the HEC-HMS model are compared with the cross-sectional capacity of the channel so that the flood height above the channel is obtained. Based on the analysis of channel capacity and flood discharge, the results show that the drainage channel in the Posindo Sub-basin is unable to drain the hydrological discharge due to rain without overflow.

1. Introduction

Balikpapan City is one of the big cities in eastern Indonesia and is included in the province of East Kalimantan. This condition makes the City of Balikpapan the potential as a national sea transportation node, national air transportation node, a mainstay area and an integrated economic development area [1]. This has an impact on the development of the City of Balikpapan which is moving rapidly in various sectors. The development of the city should also be supported by the ability of the provision of infrastructure and urban facilities, one of the problems that occur is the unavailability of urban drainage infrastructure and facilities by their needs, resulting in flooding in some locations. This condition is due to land clearing for settlement resulting in reduced water catchment areas, thus increasing rainwater runoff, and increasing sediment in rivers and channels.

Following the drainage master plan for the City of Balikpapan, the drainage system for the City of Balikpapan is divided into 16 drainage systems, each of which consists of primary, secondary, and tertiary channels. Based on the City of Balikpapan Mid-Term Plan (RPJMD) 2016-2021, the flood control program is one of the priority programs of the Balikpapan City Government with a target of reducing flood points from 51 points in 2016 to 35 points in 2021. Until the end of 2018, point There are still 47 flood points where most of them are concentrated in 3 drainage systems, namely the Ampal/Klandasan Besar system, the Sepinggan system, and the Klandasan Kecil system [2]. The Balikpapan City Government has made efforts to compile a flood control plan in the intended drainage system by referring to the drainage master plan. Various efforts have been made to overcome the problem of flooding, the majority of which only overcome short-term problems in the form of cleaning sedimentation and garbage, as well as carrying out maintenance and repair of drainage construction [3].

One of the problems that occur in the Posindo Sub-watershed is flooding due to drainage channels that are unable to drain the flood discharge without overflow. This condition is influenced by land use and heavy rainfall intensity. Based on these conditions, it is necessary to carry out an analysis related to the effect of land use in the Posindo Sub-watershed on the flood height in the channel. This can be a reference or reference related to further strategic steps that can be taken by the Balikpapan City government. This study aims to analyze the effect of land use on the flood height in Balikpapan City (Case study: Sub- watershed Posindo). The objectives of this research are:

1. Analyzing the intensification of land use in the study area
2. Analyze the flood discharge in the study area
3. Analyzing the flood height in the study area

2. Methods

2.1 Land use intensification analysis

At this stage, land use identification and an analysis of land use suitability were carried out in the study area with the City of Balikpapan Spatial Planning (RTRW).

2.2 Analysis of rainfall distribution

At this stage, identification of the type of distribution that best suits the conditions of the rain distribution in the research area is carried out. Statistical parameter analysis includes the calculation of the following variables: 1) Average value ($\overline{X}$); 2) Standard deviation (S); 3) Coefficient of variation (Cv); 4) The slope coefficient (Cs); 5) Sharpness coefficient (Ck).

2.3 Return period rainfall analysis

The calculation of the return period rainfall uses the Gumbel Type I distribution equation as follows:

$$X = \overline{X} + (k.S) \tag{1}$$

with:

X = estimated value expected to occur in a certain return period
S = standard deviation
$\overline{X}$ = the average value of events
k = frequency factor k for the extreme price of Gumbel

2.4 Rain Distribution Analysis

In the analysis of rainfall distribution, the rainfall calculation is done at the 0th hour to the t-hour by using the following formula:

$$RT = xRt - [(t-1)xR(t-1)] \tag{2}$$

with:

RT = t-hour rainfall (mm)
Rt = Average rainfall at t-hour (mm)
T = Rainy time from the beginning to t-hour (hour)
R(t-1) = Average rainfall from the beginning to (t-1) hour (mm)

2.5 Analysis of flood discharge with the HEC-HMS program

In this study, the HEC-HMS program was used for hydrological analysis of the Posindo Sub-watershed using a hydrological approach in the field. The data used in the HEC-HMS modelling are area effective rainfall data, channel length, and the area of the Posindo Sub-watershed. The result of model simulation is the number of synthesis unit hydrograph [4]. Calculations with the HEC-HMS auxiliary program are carried out by making a watershed model, which is a drawing of a basin model and inputting data on rainfall, land use, and so on.

2.6 Hydraulic discharge analysis with the HEC-RAS program

The calculations are carried out using the HEC-HMS auxiliary program. Modelling of channel flow profile conditions will be carried out using the help of the HEC-RAS program. In this modelling is done by applying the flood hydrograph resulting from the HECHMS hydrological modelling. The results of the hydraulic analysis using the HEC-RAS assistance program are water surface elevation data in the channel section [5].

3. Result and discussion

3.1 Description of the research area

Posindo Sub-basin is a small part of the Ampal Basin. The Posindo basin consists of 6 secondary channel segments which lead to the Inhutani secondary channel. Table 1 is the location and schematic of Posindo Sub-basin.

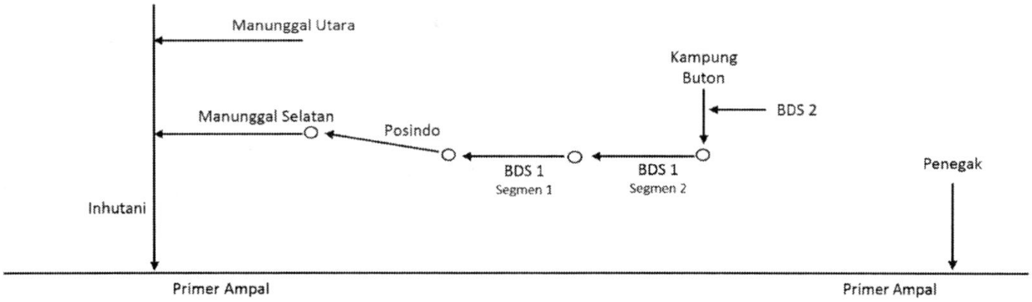

Figure 1. Posindo sub-basin schematic map.

The Posindo sub-basin has an area of 113,156 hectares with Inhutani main channel. This sub-basin has 6 catchments with different areas as shown in Table 1. Each catchment area has a channel that runs runoff water directly to the main river.

Table 1. Catchment area of Posindo sub-basin.

No.	Sub-basin	Area of sub-basin		Average of land slope
		[Ha]	[Km^2]	[%]
1	BDS 2	26.049	0.260	3.610
2	BDS 1 Segmen 1	15.931	0.159	10.224
3	BDS 1 Segmen 2	15.721	0.157	9.872
4	Kampung Buton	5.605	0.056	9.016
5	Posindo	14.954	0.150	8.354
6	Manunggal Selatan	3.201	0.032	7.310

3.2 Analysis of land use intensification

Based on the results of the primary survey it was found that the land use in the study areas varied in each of the handling areas. In this part, land use analysis is carried out within the scope of the sub-basin area to be handled so that it is known the percentage of built up and non-built up area as well as the suitability of the City Spatial Plan for Balikpapan in 2012-2032.

The Posindo sub-basin area is included in the secondary channel of Jalan Penegak and Posindo resident. The total area of this sub-basin is around 113 ha. Based on the results of the field survey it was found that land use in the study area was dominated by residential and settlement land use with a percentage of 55.05%. While the use of green open space is 18%. Then, if seen from the classification of built up area and non-built up area, this region still has non-built up area at 29% of the total area. This condition allows this region to still have good water absorption, although runoff is not evenly distributed.

Based on the review of the spatial plan of the Balikpapan City Spatial Plan for 2012-2032, it is known that there are 3 main designations in this area, namely settlements, urban forests, and trade and services. Although from the existing land use this area still has a large amount of non-built up area, the development of the built up area was found to be incompatible with the designation of the spatial plan. The results of this land suitability analysis found 1.31% of land use is not following the intended designation. This is a result of the allotment of urban forest land that has been built up by housing and trade and services. In this region, land suitability was also found to be quite suitable because the existing land use did not contradict the designation of the spatial plan such as the allotment of housing used for trade and services. These land uses are commonly found along Jalan Kampung Buton and Jalan Manunggal. Picture of built up and non-built up area (open field and vegetation) of the Posindo Sub-basin can be seen in Figure 2.

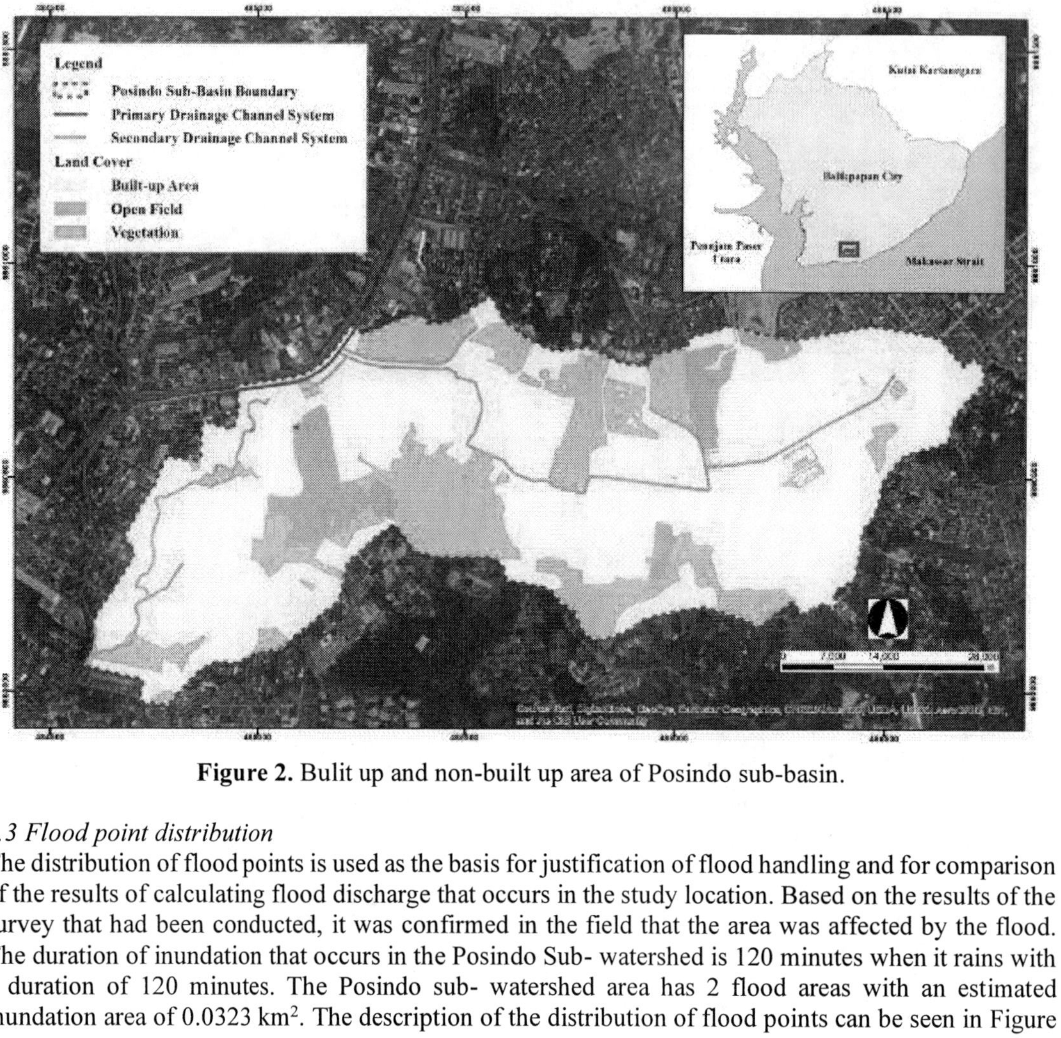

Figure 2. Bulit up and non-built up area of Posindo sub-basin.

3.3 Flood point distribution

The distribution of flood points is used as the basis for justification of flood handling and for comparison of the results of calculating flood discharge that occurs in the study location. Based on the results of the survey that had been conducted, it was confirmed in the field that the area was affected by the flood. The duration of inundation that occurs in the Posindo Sub- watershed is 120 minutes when it rains with a duration of 120 minutes. The Posindo sub- watershed area has 2 flood areas with an estimated inundation area of 0.0323 km^2. The description of the distribution of flood points can be seen in Figure 3.

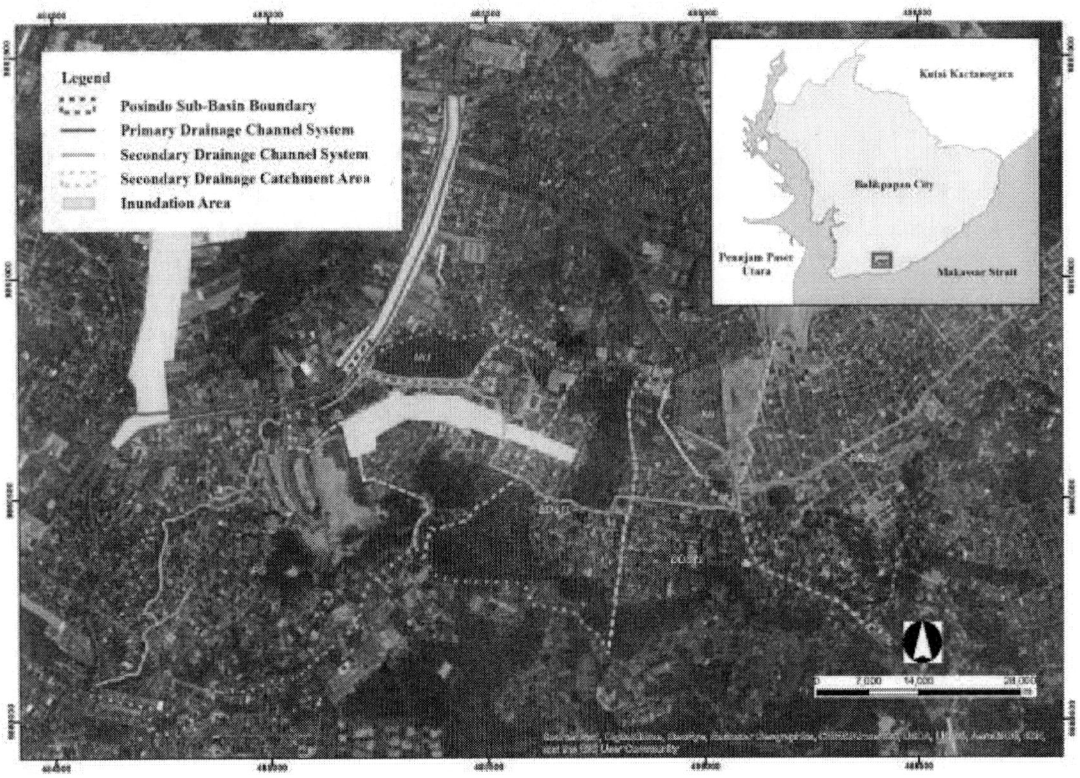

Figure 3. Inundation areas of Posindo sub-basin.

3.4 Flood discharge analysis
Hydrological discharge analysis was carried out to obtain the planned flood discharge size for the 5 year return period. Flood discharge analysis aims to determine the amount of discharge due to rain that overflows the Posindo watershed. In the analysis process, the flood discharge is obtained from the calculation of rainfall plans for a return period of 5 years, the calculation of the flow coefficient, and the characteristics of the watershed covering the area, length of the river, etc. From this analysis, it is obtained the hydrological discharge value/flood discharge that flows in the channels in the Posindo sub-basin.

3.4.1 Analysis of precipitation rain return period. The return period rainfall calculation aims to determine the rain intensity of the return period plan of 2 to 100 years. Based on the rain data obtained, the following calculations are made:

3.4.2. Mean value:

$$\bar{X} = \frac{\Sigma X}{n} = 136.99$$

3.4.3. Standard deviation:

$$S = \sqrt{\frac{(X \quad \overline{X})^2}{n \quad 1}} = 39.86$$

Calculation of rainfall period is re-written in Table 2.

Table 2. Gumbel distribution return period rainfall.

Return period [year]	Value of variant correction [Yt]	Frequency factor [K]	Xmax [mm]
2	0.37	-0.15	131.09
5	1.50	0.92	173.60
10	2.25	1.62	201.74

For the analysis of the flood reduction study in the Posindo Sub-watershed, a 5 year return period was used. This is by the initial planning of the secondary channel in Balikpapan City. The 5 year return period was also chosen according to the channel plan in the Posindo sub-basin.

3.4.2 Flood discharge analysis with HEC-HMS assistance program. The distribution of rain based on rainfall data is only obtained daily data, while to get the discharge using the HEC-HMS program, data is needed in seconds and hours. Converting the daily rainfall data to hourly rainfall requires an existing hourly rainfall data model.

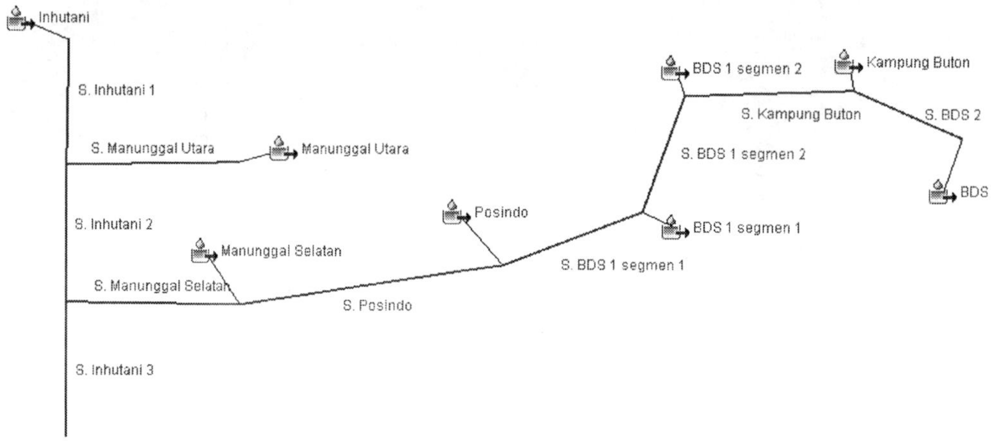

Figure 4. Catchment area model of Posindo sub-basin.

In this study, the HEC-HMS program was used for hydrological analysis of the Posindo Sub-watershed using a hydrological approach in the field. Calculations with the HEC-HMS auxiliary program are carried out by making a watershed model, which is a drawing of a basin model and inputting data on rainfall, land use, and so on. The depiction of the water catchment area model and schematic map of the direction of flow in the Posindo Sub-watershed can be seen in Figure 4.

Figure 4 illustrates the Posindo watershed model in the HEC-HMS. This modeling consists of 6 sub-watersheds and 6 channels which are part of the DAS. Each channel segment has a different catchment area as presented in Table 3. In this analysis, the SCS method (surface runoff estimation method) is used as the loss rate method (water loss process). Loss rate method is a model for calculating water loss that occurs due to the infiltration process and reduction of storage. SCS developed an empirical curve number parameter that assumes various factors from soil layer, land use, and porosity to calculate the total rainfall runoff [6]. In contrast to the hydrograph transformation of runoff units, a synthetic unit hydrograph from SCS (soil conservation service) is used. The main parameter required is the time lag, namely the time lag between the weight of the effective rain and the gravity of the hydrograph.

One of the factors that influence the amount of runoff discharge is the value of the coefficient of flow (CN). This coefficient aims to determine the ratio of the percentage of the amount of water that overtakes the soil surface and absorbs it into the soil. This coefficient value can also change according to the conditions of the influencing factors, namely the type of land and land use. The value of the CN (curve number) varies from 100 (for waterlogged surfaces) to about 30 (for non-waterproof surfaces with high infiltration values). CN values are obtained from the results of land classification based on their use. To find out the CN value in each sub-watershed, it is necessary to calculate the combined CN and the results can be seen in Table 3.

Table 3. Curve number sub-basin.

Sub-basin	Area [m^2]	Combined CN
BDS 2	260577.52	84.52
BDS 1 Segmen 1	159319.05	82.09
BDS 1 Segmen 2	157209.59	82.27
Kampung Buton	56049.71	80.95
Posindo	149265.53	82.98
Manunggal Selatan	32009.60	82.46

Table 3 is the result of calculating the value of the flow coefficient of each of the Posindo Sub-watersheds. From the Table, it is known that the average CN value in the sub-watershed is 82.60. A CN of 82.60 means that 82.60% of the rainwater that falls on the soil surface will flow into runoff and the remaining 17.40% will seep into the soil through the infiltration process.

3.4.3. Modeling of the Posindo sub-basin. In this research, the HEC-HMS program is used for hydrological analysis of Posindo sub-basin using a hydrological approach in the field. The results of the running process in the form of output data consist of hydrographic graphs at each control point that is equipped with the amount of discharge at each point and the discharge data at the Posindo sub-basin. In the previous discussion, it was explained that the Posindo sub-basin has six sub-basin with different channel lengths and lengths. The area and length of the channels for each sub-basin to be included in the HEC-HMS model are written in Table 4.

Table 4. Area and dan river length of sub-basin.

No.	Sub-basin	Area [km^2]	River	River length (L) [m]
1	BDS 2	0.260	S.BDS 2	513.53
2	BDS 1 Segmen 1	0.159	S. BDS 1 Segmen 1	314.78
3	BDS 1 Segmen 2	0.157	S. BDS 1 Segmen 2	247.68
4	Kampung Buton	0.056	S. Kampung Buton	266.45
5	Posindo	0.150	S. Posindo	584.76
6	Manunggal Selatan	0.032	S. Manunggal Utara	363.09

Table 4 is the data that will be entered into the HEC-HMS modeling to obtain runoff discharge values. This data is a compiler of the watershed model in HEC-HMS. This data will be converted into hydrological elements that represent the Posindo River Basin physical. There are two hydrological elements used in this modeling, these elements are the channel (river) and sub-basin.

3.4.4 Modeling Results of Posindo sub-basin. From the HEC-HMS modeling using the SCS method, the runoff (hydrological) discharge value is obtained by Table 5. Based on Table 5, it is obtained the value of runoff discharge (hydrology) that flows in each channel in the Posindo Sub basin. In the channel section, namely upstream discharge channel obtained BDS 2 1.00 m^3/sec and increased significantly in the downstream channel that is Manunggal Selatan at 3:20 m^3/sec. The results of this analysis were carried out without considering the base flow. This is following the conditions of the field, where the entire channel is a city drainage channel that does not have the baseflow.

Table 5. Maximum hydrological discharge of sub-basin.

No.	River	Hydrological discharge (maximum) [m^3/sec]
1	S.BDS 2	1.00
2	S. BDS 1 Segmen 2	1.30
3	S. BDS 1 Segmen 1	2.60
4	S. Kampung Buton	1.30
5	S. Posindo	3.20
6	S. Manunggal Selatan	3.20

3.5 Drainage channel cross section capacity analysis

Analysis of the cross-sectional capacity of the channel aims to determine the channel capacity compared to the flood discharge that flows in the channel. This comparison aims to determine the ability of the channel to drain the flood discharge that occurs. If the channel capacity is greater than the hydrological discharge, the river is free from overflow or flooding [7]. If the cross-sectional capacity value of the channel is smaller than the value of the flood discharge, it means that the channel is unable to drain the flood discharge due to rainwater without overflow. This condition can also be said to be a flood condition. The Posindo sub - watershed has flood problems caused by increased built-up areas, silting and narrowing of the channels due to sedimentation. The results of the analysis of the existing conditions based on the dimensions of each channel and the proportion of current land use show an overflow of water as in Table 6.

Table 6. Drainage channel capacity.

No.	River	Channel Dimensions		Water speed	Channel capacity
		wide	high		
		(m)		(m/sec)	(m^3/sec)
1	S.BDS 2	1.00	0.50	0.23	0.12
2	S. BDS 1 Segmen 2	1.60	1.00	0.17	0.27
3	S. BDS 1 Segmen 1	2.00	1.50	0.32	0.96
4	S. Kampung Buton	1.30	0.70	0.48	0.44
5	S. Posindo	2.00	1.50	0.82	2.46
6	S. Manunggal Selatan	2.00	1.50	0.94	2.82

Based on Table 6, the results of the analysis of existing conditions, the canal dimensions in the Posindo Sub-watershed have dimensions varying from 1 x 0.5 m to 2 x 1.5 m. Sub-basin channel capacity is between 0:12 m^3/sec up to 2.82 m^3/sec. Channel capacity is related to the channel's ability to drain hydrological discharge due to rain. When compared with the value of hydrological discharge is 1.00 m^3/s up to 3.2 m^3/s, it can be concluded that the drainage channels on Posindo Sub-basin not able to drain discharge hydrology due to rain without overflowing. Table 7 obtains an overview of the height and discharge of overflowing water.

Table 7 shows the six channels that became the observation points experiencing overflows because the channel capacity is smaller than the hydrological discharge flowing into the channel. The height of the overflow or flooding of the channel varies between 3.79 m -1.40 m.

Table 7. Height and overflow discharge.

No.	River	Channel Dimensions		Channel capacity	Overflow discharge	Water level	Freeboard	Flood level	Status
		Wide	High						
		(m)		(m^3/sec)	(m3/sec)		(m)		
1	S.BDS 2	1.00	0.50	0.12	0.89	4.29	0.30	3.79	Flood
2	S. BDS 1 Segmen 2	1.60	1.00	0.27	1.03	4.25	0.30	3.25	Flood
3	S. BDS 1 Segmen 1	2.00	1.50	0.96	1.64	4.22	0.30	2.72	Flood
4	S. Kampung Buton	1.30	0.70	0.44	0.86	4.17	0.30	3.47	Flood
5	S. Posindo	2.00	1.50	2.46	0.74	4.02	0.30	2.52	Flood
6	S. Manunggal Selatan	2.00	1.50	2.82	0.38	2.90	0.30	1.40	Flood

4. Conclusion

Based on studies that have been carried out in the Posindo sub-basin area, it is known that the land use intensification in the Posindo Sub-basin by 71% is a built-up area with the dominance of land use are housing and settlements, and land use mismatch is obtained by 1.31%. Based on the land use condition, it is obtained that the planned flood discharge for a 5-year return period flows in the upstream Posindo sub-basin channel of 1.00 m^3/s and downstream of 3.20 m^3/sec. Floods in the Posindo sub-basin are caused by a lack of channel capacity to drain the flood discharge. The flood height that occurs in the channel varies between 3.79 m - 1.40 m.

5. Acknowledgments

Thank you to Republic Indonesia through Ministry of Education and Culture for funding this research by PDP program. With this program, the research has been successfully finished with expected result.

References

[1] Pemerintah Kota Balikpapan 2012 *Materi Teknis Penyusunan Rencana Tata Ruang Wilayah Kota Balikpapan Tahun 2012-2032* **1** 2.
[2] Pemerintah Kota Balikpapan 2016 *Dokumen Informasi Kinerja Pengelolaan Lingkungan Hidup Daerah Tahun 2016* **1** 20.
[3] Wong T and Brown R R 2008 *Transitioning to Water Sensitive Cities: Ensuring Resillience through a new Hydro-Social Contract* (Edinburgh: IWA Publishing).
[4] Hydrlogic Engineering Center 2002 *HEC-HMS Hydrolic Modelling System (User's Manual)* **1** 2.
[5] Hydrlogic Engineering Center 2002 *HEC-RAS River Analysis Sistem (User's Manual)* **1** 2.
[6] Ponce V M and Hawkins R H 1996 *J. Hydro. Eng.*, **1 (1)** 11-19.
[7] Suripin 2007 *Sistem Drainase Kota yang Berkelanjutan* **1** 139.

IOP Conf. Series: Materials Science and Engineering 1087 (2021) 012036 doi:10.1088/1757-899X/1087/1/012036

Architecture in between: connecting between traditional Acehnese house and the current condition of modern Acehnese community

Izziah*, E Meutia, Z Sahputra, M Irwansyah
Department of Architecture and Planning, Faculty of Engineering, Universitas Syiah Kuala, Banda Aceh, Indonesia

*E-mail: izziah.hasan@unsyiah.ac.id

Abstract. The formation of traditional architecture cannot be separated from the local wisdom of the community which is very friendly to environment. In this regard, the notion of 'locality' (such as local material, technology and culture) and 'climate' are used as a guide to the construction of buildings. The traditional Aceh house, the so-called Acehnese house (*rumoh Aceh*), that referred to as a type of stilted house is earthquake resistant. The house was also designed to anticipate flood and ocean tide and could provide comfort to the dwellers and be friendly to the local environment. The existence of Acehnese houses, however, has been extinguished. Houses owners prefer to demolish or sell the houses and to reconstruct new buildings with different styles. This study, thus, aims to explore the development of traditional architectural concept that suit to present context. It assesses three representative traditional houses found in Lambenot village. In order to accomplish the study, primary data collection was carried out through field observation, documentation and interviews with several informants including house owners. In additions, literature review is needed to support primary data. In doing this study, the paper shows the form of conservation efforts carried out by the Lambenot village community in order to maintain their collective memory of life journey. The benefit of this study is to give alternative toward preservation effort on how Acehnese houses can be used as an adaptive housing model for contemporary needs both architectural and structural aspects.

1. Introduction

Aceh traditional house, the so-called Aceh House (*Rumoh Aceh*) is an architecture that was built through the local knowledge and has been passed down from generation to generation. The ancestors built their houses based on knowledge and intuition that concern on how the built structure fits with its environment. In additions, the formation of traditional architecture is inseparable from the local wisdom of the community [1]. The involvement of local wisdom in creating traditional architecture, among others, are local material, climate, surrounding environment, site and topography, economic condition, local technology, daily life necessity, symbolism and meaning [2]. Those aspects become a guide to the construction of buildings. Generally, the traditional method of house construction had not been changed radically until the nineteenth century when the Dutch came to Aceh [3]. The stilted construction was designed to protect the dwellers from flood and ocean tide. Moreover, the house construction is earthquake resistant as well as addresses climatic needs.

This study is part of ongoing research on traditional architecture and conservation, exploring the concept of traditional architecture, which the involvement of local knowledge value. It is expected that the research will provide guidelines on how to preserve traditional Acehnese House that suits to the current condition of the local community. The location of this research takes place in Lambeunot village,

Aceh Besar. The village is located about 35 km from the capital city of Banda Aceh. During recent years, the existence of traditional houses in the traditional Lambeunot village starts to become extinct. Although the construction of Aceh House, as mentioned above, offers some advantages to the dweller in terms of thermal comfort and natural disasters, a number of house owners have been ruined their traditional houses and reconstruct with modern domestic architecture. This is the main concern of this study. There are two main reasons for such a matter. First is because the owners' financial conditions that caused them un-able to build a new house. Second is because the house owners want to keep the memories passed down from their ancestors. Having identified such problems, the main questions rise; In what extend does most Lambeunot community prefer to build modern brick houses rather than maintain their traditional house? How to maintain the Aceh house so that its function as a living space in the modern period fit with the conditions of the cultural development of the local community? In view of this, the study aims to examine the traditional houses chosen as samples of this study and explore the development of the traditional of Aceh houses concept that suits to present context.

2. Methods

In doing this study, the qualitative descriptive method was carried out as a research approach. This study focuses on three Acehnese houses that considered the oldest houses found in Lanbeunot village. In collecting data, primary sources are mostly used. Its data was collected through field observation and documentation to the houses, investigating some changes that occur in rooms function and building materials. In additions, the data collection is carried out through interview with several Lambeunot villagers, especially the house owners. In order to support primary data, A few key sources related to traditional Acehnese architecture are collected and reviewed. Such sources provide detailed and illustrated explanations of the constructions of the Aceh House in relation to its structure, material and function [4,5]. Another source used in this study is Hasan's writing that searches on rethinking architectural identity in modern life [1]. The paper shows the form of conservation efforts carried out by the Lambenot village community in order to maintain their collective memory of life journey. The benefit of this study is to give alternative toward preservation effort on how the Aceh House can be used as an adaptive housing model for contemporary needs both architectural and structural aspects. It enriches the variation of Acehnese architecture in modern time.

3. Result and discussion

3.1 Local wisdom and traditional architecture

In the beginning of the formation of Lambeunot village, the village dwellers were comprised of one generation. Parents build houses for married daughters around their land. Then, the married daughters with the husband provided houses for their married daughter. In other words, the women who own houses are descendants of the same female progenitor. It is therefore, in Aceh, a house belong to women. During these days, a group of houses found in one plot of land are dwelled by several family members of three to four generations. The settlement pattern of the Lambeunot village at this time still follows the traditional village pattern. As shown in Figure 1 those houses are set in groups along the village road. Several houses occupy a large plot of land. Dwellings in Lambeunot village are situated in a fenced courtyard that, planted with variety of plants. The dwellers share to use the open space found between houses in one plot of land.

Orientation of house becomes an important aspect of design in Traditional life of Acehnese. Like in other village settlement in Aceh, dwellings found in Lambeunot village are set in an east – west orientation especially those houses located along the village streets. According to local belief, this orientation had a ritual function, within which this has been passed from generation to generation [5]. In his writing on "The Traditional Acehnese House" Dall claims that such an orientation is applied to all dwellings in the village streets throughout Aceh region [5]. Moreover, Dall adds that before Islam came to Aceh, the east – west orientation of the Aceh house roof derived from Hindu's belief. In the Hindu's tradition, this orientation was used in order to avoid the house entrance facing the sunset. They believe that west or sunset symbolizes death, while east or sunrise signifies life. When Islam came to

Aceh, the east–west orientation continued to be used by Muslims for religious reasons. This orientation identifies the ritual direction for daily prayers for Muslims. It is therefore now days people claims that the Acehnese house reflects Islamic values through the roof on an east-west axis as Figure 2 portrays.

Figure 1. Map of Lambeunot village.

Figure 2. Houses that are set in east – west orientation and situated in fenced courtyard.

In connection with local geographical conditions, an east–west orientation was also necessary to prevent house damage – especially the roof – by the strong force of what the local people called as *angina barat* (west wind) and now days known as the northern and eastern winds. There are two gables (*tulak angen*) are placed right under the roof on east and west sides of the house with the gable screens sloping outward. The reason for such installment is to prevent damage by heavy winds. Besides, the gable screens that are perforated are constructed under the roof to allow the wind from outside of the house to pass through the gable screen on the roof part so that the wind cools of the rooms inside.

Rumoh Aceh itself is a stilted house type with its floor is elevated 2 – 2.5 meters from the ground. The house floors are made of timber slats are arranged at certain intervals. The stilted construction aimed to protect the residents from wild animals. Moreover, in terms of climate, such a house construction addresses climatic needs. Elevating the house floor up to 2.5 m allows the cool moist air from the space underneath the house to be drawn up to the rooms inside the house through the timber floor arranged with certain intervals. This local wisdom on how to deal with climate has been approved by previous study conducted by Sari at all [6]. The study found that inside air temperature of the traditional house is closer to the comfort air temperature, ranging from 26.6 to 29.7°C.

The construction method of Acehnese house uses the knock down construction system. Thus, all joints that are found in the construction of traditional Acehnese house do not use nails. This can be seen in Figure 3 that shows that the joints between posts and beams are supported by wooden pegs and wedges and the wood joins on the roof construction are held together and lashed with *ijuk* (palm fibre) twine. As a result, such a flexible system shows that structural elements are able to safe themselves from earthquake shakes [7].

Figure 3. Houses that are set in east – west orientation and situated in fenced courtyard.

In terms of space zoning, Acehnese house is divided into three zones. They are male zone/verandah (*seramoe agam*) located at the front of the house and used as public space and praying room; bedrooms zone (*dalam*) at the middle, and female zone/verandah (*seuramo inong*) at the back of the house that is used for house hold activities. Since the space inside Acehne house is used to perform daily prayer, the Acehnese considered the interior of Acehnese house as a clean space. It is therefore water vessel was provided near the base of the stair so that people had to clean their feet before entering the house.

3.2 Traditional architecture and the current local conditions

Most Acehnese houses found in Lambeunot village are around 135-150 years old. The three houses taken as samples in this study belonged to Mrs. Wahyuni, Mrs. Nursiah and Mrs. Putri. To some extent, there are some traditional aspects in the design of Acehese house that are still both appropriate and not appropriate to be adapted in the current condition. This circumstance is consistent with the writings of Hasan [1] which referred to an architectural theorist, Curtis [8] a well-known western architectural theorist who emphasized the concept of architectural regionalism. In referring to Curtis's writing, Hasan [1] emphasizes that some of local traditions are still fit to be adapted in the era of globalization and this should reflect to architectural works where the local inhabitants who are the user of its architecture are able to feel at home in their own environment.

In line with that, this study observes the existence of the three houses and found that the appearance of houses exterior, in general, has not extremely changed. However, there are some changes in the use of space as well as building materials. The space underneath the house is no longer used as a place to keep pets, instead, the dwellers change its function and used it for various activities. Couples rooms that are highly used every day such as bedrooms, kitchens and living room are moved down to the ground by adding new structure in it. Meanwhile, spaces that are not always used, such as guest rooms, storage, and children's bedrooms are remained in the Acehnese house. The occurrence of the changes mentioned above is due to economic and practical reasons. The additional structures are built on the ground either attached to or side by side with Acehnese house itself. This new structure is known as *rumoh yub* (the additional structure(s) built on the ground). For the owner of the houses, removing those rooms is for practical reason. The dwellers do not have to climb up and down to travel between rooms within the house. Concerning this, Households more often do activities in the new structure under the house.

The open space underneath the house, thus, seems to be the most desirable space. It becomes the main occupied space during the day, especially for women's activities such as feeding their children, lullabying the baby, making handy crafts and even having relaxed with family members and friends as shown in Figure 4. In certain times, this space is also used as communal kitchen where the women in this village work hand in hand to prepare food for celebrations such as weddings, commemorating the Prophet's birthday, and other celebrations. In view of this, *rumoh yup* is more widely used now days than the Acehnese house itself. As result, the use of Rumoh Aceh becoming less and less.

Figure 4. The multi-function space underneath the Aceh house.

Thus, there are very few rooms up stair in the Aceh House needed in order to accommodate dwellers' activities. Concerning this, the interior space has been rearranged by the house owners. The space mostly used as a storage room. Besides that, there are a public space (*seramoe agam*) that can be used as a sitting or praying area as well as reading quran and a children's bedroom. The dwellers create a bigger space for *seramo agam* by taking the bedroom into part of *seramo agam*. In view of this, the wall that separated between these two rooms is taken off. The reason for combining these two spaces is to create a brighter and wider space, as Figure 5 illustrates. In additions, the fresh air and wind can pass to the room through windows freely without being blocked by insulating walls. Now days the dwellers prefer to have a room that has sufficient light and fresh air. The children bedroom available in each three existing Acehnese house is often left empty since the room owner often uses *seramo agam* as a space to do his or her activities, especially studying, and sleeping. Instead, the bedroom tends to become a private storage.

Concerning new material used in Rumoh Aceh, the house owners used variety of woods in order to replace rotten or damaged wood material. The damage materials mostly found in houses stairs, floor, and walls. Concerning the new structure built underneath the Acehnese house, the owner used brick as the main building material, glass for windows and timber for window frames and doors. Difficulties in obtaining wood material due to its quality as well as its price has resulted the house owners to obtain alternative material for their houses. The striking sight of the new house material can be seen from the use of zinc material for the roof covering and the use of glass material for the windows. The dwellers have been replacing the roof material from thatch to zinc as this material has been easily obtained in the building material stores nearby and with affordable price. Meanwhile, to have thatch leaf material, one needs to order it and wait for 3 to 4 months. Moreover, its price is much higher than the price of zinc. The characteristics zinc, however, is a very good material type of heat conductor. As a result, the zinc creates the extreme heat to the space under the zinc roof. To avoid being in such a high temperature room, the house holders prefer to be in the open space or in additional rooms underneath the house in order to carry out daily household activities.

Figure 5. The open space in the interior, function as a public space

4. Conclusion

The local wisdom is a form of local people' expression in carrying out their activities based on their cultural tradition and knowledge. Directly or indirectly, culture influences the formation of local wisdom. It is therefore, the formation of traditional architecture is inseparable from the local wisdom of the community. The ancestors built their houses based on knowledge and intuition that concern on how the built structure fits with its environment. They have paid great attention to geographical condition of the region, emphasizing 'locality' and 'climate' as a guide to the construction of traditional Acehnese House.

This study has observed how the dwellers in Lambunot village treat their Aceh House to become sustainable house based on their knowledge. This study shows that traditional architecture that has local wisdom values can still be maintained by considering the cultural development of today's society. Moreover, the dwellers have been carried out some efforts of traditional building preservation. Due to practical and economic reasons, the house owners change the function of few existing rooms and the use of building materials. In view of this, the dwellers have been creating architecture as a living space that is able to make its dwellers feel at home in their environment. Thus, the Acehnese house can be used as an adaptive housing model for contemporary needs both architectural and structural aspects.

5. Acknowledgment

This research is funded by Universitas Syiah Kuala through PNBP scheme 2020.

References

[1] Hasan I 2010 *Architecture and the Politics of Identity in Indonesia: A Study of Cultural History of Aceh* (Doctoral Dissertation: University of Adelaide).
[2] Sudigdo D, Harysakti A and Usop T B 2014 *Perspektif Arsitektur,* **9** (**1**) 37-47.
[3] Nas P J 1999 *Ethnic Identity in Urban Architecture. Generation of Architects in Banda Aceh* (Leiden: Leiden University).
[4] Leigh B 1989 *Tangan-tangan Terampil* (Jakarta: Penerbit Djambatan).
[5] Dall G 1982 *The Malay-Islamic World of Sumatra* (Clayton, Victoria: Monash University).
[6] Sari L H, Izziah, Irwansyah M and Meutia E 2017 *Tesa Arsitektur,* **15** 1-12.
[7] Meutia E 2011 *Proceedings of the International Conference on SENVAR 11*, (Malang: Universita Brawijaya).
[8] Curtis W J R 1986 *Mimar: Arc. Dev.*, **19** 24-31.

IOP Conf. Series: Materials Science and Engineering 1087 (2021) 012037 doi:10.1088/1757-899X/1087/1/012037

Risk identification at the pre-construction stage in post-earthquake community based housing reconstruction project in Pidie Jaya Regency

N Malahayati[1,2]*, M Munirwansyah[2], M Afifuddin[2], S Syamsidik[2]
[1]Engineering Doctoral Program, Universitas Syiah Kuala, Banda Aceh, Indonesia
[2]Civil Engineering Department, Universitas Syiah Kuala, Banda Aceh, Indonesia

*E-mail: malahayatijj@unsyiah.ac.id

Abstract. The post-earthquake housing reconstruction project is inseparable from being exposed to various risks. Risk is the possibility for an adverse event to occur due to uncertainty. The reconstruction in Pidie Jaya Regency after the 2016 earthquake is still ongoing and uses community-based methods. The purpose of this paper is to identify risks during pre-construction stage in the implementation of community houses reconstruction after the earthquake disaster in Pidie Jaya Regency. Stakeholders who were directly involved in the reconstruction phase were the Management Consultants, Community Group (Pokmas), and the Regional Disaster Management Agency (BPBD) and each had its own potential risks. Data were obtained using 58 respondents from these stakeholders by distributing questionnaires in the research location, namely Pidie Jaya Regency. A total of 29 risk variables were identified, of which there are 8 variables that had high risk. The results of the analysis shown that when viewed from risk factors, the Pokmas formation factor is the most potential risk factor compared to other risk factors because these risk variables are owned by the three groups of respondents with the highest ranking.

1. Introduction

On Sunday, December 7th, 2016 an earthquake of 6.5 Mw occurred in the northern part of the Indonesian island of Sumatra. The Earthquake Center is located in Pidie Jaya Regency, Aceh Province. This destructive earthquake also affected the districts closest to Pidie Jaya, namely Pidie and Bireun Regencies, but the worst impact was in Pidie Jaya Regency. This disaster is included in the status of a regional disaster based on the number of casualties, property losses, damage to infrastructure and land, the wide coverage of the affected areas and the socio-economic impacts it causes. The community housing sector is the worst affected sector where 17,673 houses were damaged, including 2,202 severely damaged and 4,542 moderately damaged and 10,929 slightly damaged [1]. The house reconstruction phase in Pidie Jaya Regency used a community-based method and involved Management Consultants and the Regional Disaster Management Agency (BPBD). Initial research was conducted and it was found that there were delays in the community housing reconstruction project, namely in the pre-construction and construction phases [2,3]. There is no construction project that does not have risks in its implementation, especially in post-disaster reconstruction projects with a very complex situation compared to normal and community-based situations in the construction process [4]. If the risk is not managed, there will be chances of not achieving project goals such as delays, lack of costs and performance or reputation. Project risk management includes the process of implementing risk management planning, identification, analysis, response planning, response implementation, and risk monitoring on a project. The objective of project risk management is to increase the probability and / or impact of positive risks and to reduce the probability and / or negative impacts [5].

Risk management has not been widely practiced in post-disaster housing reconstruction projects [6]. The pre-construction phase is an important phase that must receive great attention because this phase has been identified as giving a very large contribution to the success of reconstruction projects using community-based methods [4]. There are several regulations issued by the Indonesian government regarding the rehabilitation and reconstruction of houses using community-based methods [7,8,9], and this method was born and has been directly practiced in the reconstruction of houses after the earthquake disaster and tsunami 2004 in Aceh Province. The community not only received benefits from housing assistance but they participated in building their own houses, controlled and were accountable for the assistance funds that had been provided by the government assisted by the facilitators. Several countries such as BAM, Iran, Gujarat, and Indonesia apply community-based methods and have garnered many successes [4], however other studies suggest that the failure of reconstruction projects was due to one of them being a lack of community participation [10].

The identification of risks in the pre-construction phase of the housing reconstruction project has been carried out by several researchers and found the factors that have the potential to cause the average risk on the part of the government [4, 11]. There were two activities that took a lot of time in the pre-construction phase, namely the formation of Community Groups (Pokmas) and design activities. It took 5 months in the process of forming Pokmas for post-disaster reconstruction of the 2004 earthquake and tsunami in Aceh Province [12]. There are five main factors that cause the delay in carrying out the reconstruction of the house at the pre-construction stage, namely idea activities, identification of damaged houses, recruitment of facilitators, training programs for the community, formation of Community Groups (Pokmas), program socialization and coordination, inventory, verification and data validation. damage, and design [2, 4, 11]. These factors in the pre-construction phase need to be identified again to get the risks that are the responsibility of each party, namely the Community Group (Pokmas), Management Consultants (KM) and the government (in this research is BPBD). The purpose of this paper is to identify risks during pre-construction taking place in the implementation of reconstruction of community houses after the earthquake disaster in Pidie Jaya District. The benefits of this research can be used as input in decision making for related parties to overcome the negative consequences that occur in the handling of post-earthquake community-based housing reconstruction projects in the future.

2. Methods

The research location that was used as a place for distributing questionnaires, interviews and observations was Pidie Jaya Regency. The risk identification used in this study is based on the risk management method contained in the PMBOK Guide Sixth edition [5]. The inputs, tools / techniques and outputs used are described below:

- The inputs used are: (1) the document for the implementation of reconstruction from the Regional Disaster Management Agency (BPBD) of Pidie Jaya Regency which consists of the 2016 Pidie Jaya Earthquake Action Plan, Technical Guidelines for Post-Earthquake Rehabilitation and Reconstruction of Pidie Jaya Regency which contains data on damage to people's homes, estimated duration and list of stakeholders; (2) business environment factors, namely the results of similar studies that have been published on risk identification.
- The tools and techniques used are data collection by: (1) Brainstorming, namely identifying a list of risk ideas from stakeholders, as many as 5 people who are directly involved in the post-disaster housing reconstruction process in Pidie Jaya Regency consisting of CM, Pokmas, and BPBD. In this risk identification, the risk owner will choose the risk that is in accordance with what has been experienced and state the response actions they have taken. Data collection and analysis is carried out for 2 months and the results will be compiled into a questionnaire containing risks and risk owners; (2) A checklist containing risks that have occurred before and which are relevant to the

scope of this study. A total of 5 journals / proceedings that examine risk identification during the pre-construction phase of post-disaster housing reconstruction projects [2,3,4,11,12]; (3) Interviewing and distributing questionnaires to risk owners by visiting directly to the research location and also by means of cell phones and e-mails. The questionnaire contains the characteristics of the respondent and the risk probability in the pre-construction phase with a Likert measurement scale. The assessment categories used were very frequent (5), frequent (4), moderate (3), rare (2) and never (1). Data collection was conducted from May 2020 to July 2020. Snowball sampling technique was used to get respondents. A total of 58 respondents were used in this study consisting of 10 people from BPBD Pidie Jaya Regency, 20 people from Management Consultants and 28 people from Pokmas.

- The output is a risk register containing a list of risks, risk owners and potential risk responses that have an impact on the implementation of housing reconstruction after the earthquake disaster in Pidie Jaya Regency.

To test the quality of the research instrument, the validity and reliability tests were used. The validity test used the product moment correlation with a significant level of 5% and the reliability test used the Cornbach alpha (c-alpha) with a value of ≥ 0.6 which was considered feasible. The risk proportions were analyzed using the Frequency Index (FI) method and from the FI results the potential risk factors that occurred in housing construction after the earthquake in Pidie Jaya Regency were obtained. The FI formula is as follows:

$$FI = \frac{\sum_{i=1}^{5} \alpha_i . n_i}{5N} \quad (1)$$

Where i = Response category index, α_i = the weight associated with the i- response value, n_i = frequency of the i-th response as a percentage of the total respondents for each factor, dan N = The total number of respondents.

Then after getting the FI value for risk probability, it is continued by categorizing the risk based on the value of the FI value shown in Table 1.

Table 1. Classification of the FI rating scale.

Score	Weight	Risk Assessment
1	$0,000 \leq FI/SI \leq 0,125$	Very low
2	$0,125 \leq FI/SI \leq 0,375$	Low
3	$0,375 \leq FI/SI \leq 0,625$	Moderate
4	$0,625 \leq FI/SI \leq 0,875$	High
5	$0,875 \leq FI/SI \leq 1,000$	Very High

Analysis of variance (ANOVA) was used in this study to test the comparison hypothesis of more than two samples and each sample consisted of two or more types simultaneously. similarity mean of three or more populations. The purpose of a two-way analysis is to examine the differences between two or more groups where there are two or more factors being considered, namely the possibility of risk variables occurring in the three groups of risk owners in the pre-construction phase. The initial hypothesis is stated that there is no significant effect between the risk variable and the risk owner on the likelihood / opportunity of risk. Two-way ANOVA summary Table as described in the Table 2.

Table 2. Two-way Anova summary.

Source of Diversity (SK)	Sum of Squares (JK)	degrees of freedom (db)	Middle Square (KT)	Fcount
Row Average	JKB	db numer1 = r-1	s^2B = KTB = $\frac{JKB}{r-1}$	f count = $\frac{KTB}{KTG}$
Colum Average	JKK	db numer2 = k-1	s^2K = KTK = $\frac{JKK}{k-1}$	f count = $\frac{KTK}{KTG}$
Error	JKG	db denum = (r-1)(k-1)	s^2G = KTG = $\frac{JKG}{(r-1)(k-1)}$	
Total	JKT	r.k -1		

3. Analysis and discussion

The data obtained are in the form of respondents' answers from questionnaires that have been distributed. The questionnaire was addressed to 58 respondents involved in the implementation of housing reconstruction rehabilitation. Characteristic data, validity tests, reliability tests, and Frequency Index (FI) analysis are described below.

3.1 Characteristics of respondents

Respondent characteristics are divided into three groups according to the object of the respondent, namely the Management Consultant (CM), Community Group (Pokmas) and Government (BPBD). Respondent characteristics consist of the respondent's position, age, latest education, and experience in construction. It is summarized in Table 3.

3.2 *Risk identification*

Risk identification starts with the activities carried out during the pre-construction phase of the community housing reconstruction project, in this study only the pre-construction scope of socialization activities, validation of house damage assessment, identification of beneficiaries, establishment of Pokmas and Pokmas training.

Table 4 explained that risk identification starts from a check list of the risks of pre-construction construction of houses in the post-earthquake reconstruction phase that has occurred and is adjusted to the conditions that occurred in this study by giving 5 respondents who were considered experts and were directly involved in the reconstruction process, namely from the CM including the facilitator, Pokmas and BPBD. Respondents will choose the risks they face and explain the causes of the risks and respond to the risks they have done. The risk response put forward by the expert is how the strategy they have implemented to overcome the risk in order to minimize the threat and reduce the overall risk exposure. In this study, data on risk response is only at the risk identification stage, while in more detail the risk response will be discussed in the next research after the quantitative risk analysis stage is carried out by following the stages in risk management based on the PMBOK Guide Sixth edition.

It can be explained that the socialization factor with variable V1 is the lack of facilitators in program socialization, the 5 experts gave their opinion that the risk owners are CM, PK and BPBD and on average choose the risk response to accept. Accepting means accepting these risks and acknowledging the threats that will occur, namely that time and costs will increase in outreach activities. There is compensation from the BPBD for the extension of time in this outreach activity so that it means that this acceptance strategy is active. The risk responses for the other variables can be explained in the Table 4. Then, this checklist was compiled into a questionnaire and was distributed to respondents using a likert rating scale.

3.3 The result of validity and reliability tests

Data from the distribution of questionnaires will be tested first whether the questionnaire is valid and reliable. The validity test with a significant level of 5% was used in this study. Testing by taking 20 respondents from the Management Consultant with $r_{table} = 0.444$, Pokmas as many as 28 respondents with $r_{table} = 0.378$ and BPBD as many as 10 respondents with $r_{table} = 0.632$. The test criterion is if $r_{count} \geq r_{table}$, then the instrument is declared valid and vice versa declared invalid. Table 5 describes the results of data processing for testing the validity of the instrument. All questionnaires were declared valid so that they could be carried out for the next stage of testing, namely the reliability test.

Table 3. Respondent characteristics.

Respondent Data	CM		Pokmas		BPBD	
	Freq.	Percentage	Freq.	Percentage	Freq.	Percentage
Respondent Age						
< 20 Year	0	0%	0	0.0%	0	0%
20 - 40 Year	11	55%	1	3.6%	5	50%
>40 Year	9	45%	27	96.4%	5	50%
Gender						
Male	18	90%	27	96.4%	5	50%
Female	2	10%	1	3.6%	5	50%
Position / Occupation						
Leader Management Consultant	1	5%	-	-	-	-
Senior Facilitator	8	40%	-	-	-	-
Technical Facilitator	11	55%	-	-	-	-
Farmers / Fishermen	-	-	13	46.5%	-	-
Entrepreneur / Gov. Employee	-	-	5	17.8%	-	-
Builder	-	-	9	32.1%	-	-
Housewife			1	3.6%		
BPBD Rehab / Recon Officer	-	-	-	-	2	20%
BPBD Recon Rehab Staff	-	-	-	-	8	80%
Level of Education						
Elementary School	0	0%	3	10.7%	0	0%
Junior High School	0	0%	9	32.1%	0	0%
Senior High School	0	0%	12	42.9%	0	0%
Diploma	5	25%	1	3.6%	1	10%
Undergraduate Degree (S1)	15	75%	3	10.7%	7	70%
Graduate Degree (S2)	0	0%	-	-	2	20%
Work Experience In Construction						
0 Year	0	0%	8	28.6%	0	0%
1-4 Year	2	10%	12	42.9%	5	50%
4-8 Year	10	50%	3	10.7%	2	20%
>8 Year	8	40%	5	17.9%	3	30%
Experience as a companion to the community self-management program						
1 Times	3	15%	-	-	-	-
2-5 Time	5	25%	-	-	-	-
5-8 Time	6	30%	-	-	-	-

Table 4. Risk identification.

Factor	Variable Code	Variable	Owner of Risk			Risk response
			CM	PK	BPBD	
Socializationsi (F1)	V1	Lack of facilitators in program socialization	√	√	√	Accept
	V2	Inexperienced facilitator	√	√	√	Accept
	V3	Schedule tight / short	√	√		Mitigate
Validation of damage assessment (F2)	V4	Lack of a housing database	√		√	Accept
	V5	Too many parties are involved			√	Mitigate
	V6	Non-uniform damage assessment methods	√			Mitigate
	V7	Coordination problems in assessment	√	√	√	Mitigate
	V8	The number of surveyors / insufficient no. assessment facilitators			√	Transfer
	V9	Surveyor / facilitator inexperienced		√	√	Mitigate
	V10	Weak assessment by the government		√		Transfer
	V11	Local authority pressure on damage assessments			√	Transfer
	V12	Lack of involvement of the entire community in identifying damage to houses			√	Mitigate
Recipient identification (F3)	V13	Transportation access is difficult to identify and land ownership			√	Accept
	V14	Collusion in determining beneficiaries			√	Mitigate
	V15	Problems with land ownership / rights	√	√	√	Mitigate
	V16	Validation issues identify ownership	√		√	Transfer
	V17	The results of the assessment by the government are inaccurate		√		Transfer
Pokmas formation (F4)	V18	Facilitator who has no experience in forming community organizations (Pokmas)		√		Mitigate
	V19	Failure to build community organizations		√	√	Mitigate
	V20	Community resistance to the process of forming community organizations	√		√	Transfer
	V21	Society manipulated by other parties	√	√	√	Mitigate
	V22	Disagreement with community contracts / agreements		√		Mitigate
	V23	Brief schedule for the stages of forming community organizations	√	√		Mitigate
Pokmas / Worker training (F5)	V24	Lack of facilitators in community / worker training	√	√		Accept
	V25	Facilitator who is not experienced in training		√		Accept
	V26	Limited knowledge about how to build earthquake resistant houses	√			Mitigate
	V27	The training material is lacking and beyond the expected results		√		Accept
	V28	Lacking Government support for training	√	√		Mitigate
	V29	The low role of the community	√			Avoid

Table 5. Validity test results.

Variable Code	Number of Variables	Value Range r_{Count} Risk Probability	r_{tabel}	Value Range r_{Count} Risk Probability	r_{tabel}	Value Range r_{Count} Risk Probability	r_{tabel}	Explanation
		Management Consultant		Pokmas		BPBD		
V1-V3	3	0.636 – 0.706	*0.444*	0.618-0,934	0.374			Valid
V1,V2	2					0.740-0.777	0.632	Valid
V4, V6, V7	3	0.576 – 0.669	*0.444*					Valid
V7, V9, V10	3			0.558-0.952	0.374			Valid
V4-V12	7					0.636-0.757	0.632	Valid
V13, V15,V16	3	0.514 – 0.918	*0.444*					Valid
V15, V17	2			0.920-0.969	0.374			Valid
V13-V16	4					0.662-0.717	0.632	Valid
V20, V21, V23	3	0.673 – 0.905	*0.444*					Valid
V18, V19, V21, V22, V23	5			0.434-0.897	0.374			Valid
V19, V20, V21	3					0.666-0.762	0.632	Valid
V24, V26, V28, V29	4	0.556 – 0.777	*0.444*					Valid
V24, V25, V27, V28	4			0.582-0.822	0.374			Valid

The results of the reliability test which aims to determine the level of reliability of the instrument using the Cronbach Alpha value exceeding the value of 0.6. The reliability test is carried out from the results of the questionnaire from each risk owner. The test results show that from the Management Consultant there is 1 factor that has a Cronbach Alpha value of less than 0.6, namely identification of beneficiaries so that these factors are eliminated and not used in further analysis, as indicated by Table 6.

Table 6. Management consultant reliability test results summary.

Questioner Code	Number of Variables	C-Alpha Management Consultant	C-Alpha Pokmas	C-Alpha BPBD
V1-V3	3	0.7	0.63	
V1,V2	2			0.84
V4, V6, V7	3	0.67		
V7, V9, V10	3		0.62	
V4-V12	7			0.83
V13, V15,V16	3	0.52		
V15, V17	2		0.83	
V13-V16	4			0.65
V20, V21, V23	3	0.76		
V18, V19, V21, V22, V23	5		0.83	
V19, V20, V21	3			0.71
V24, V26, V28, V29	4	0.67		
V24, V25, V27, V28	4		0.61	

3.4 Result of the frequency index (FI) analysis of risk variables

The pre-construction phase in the implementation of post-disaster housing reconstruction, where each risk owner has a number of different factors and variables. Management Consultant has 4 factors and 13 variables, Pokmas has 5 factors and 17 variables, while BPBD has 4 factors and 16 variables. Potential factors and variables are determined based on the results of the FI analysis. The results of this analysis provide an overview of the indexes of each variable ranging from 0 to 1. The higher the index value, the greater the frequency or chance of risk arising in the implementation of activities in the pre-construction phase. Table 7 below describes the highest FI rating for each risk owner.

IOP Conf. Series: Materials Science and Engineering 1087 (2021) 012037 doi:10.1088/1757-899X/1087/1/012037

Table 7. Frequency Index (FI).

Factor	Variable	Owner of Risk					
		CM		PK		BPBD	
		FI	Rank	FI	Rank	FI	Rank
F1	V1	0.280	13	0.314	14	0.440	7
	V2	0.490	8	0.429	7	0.420	12
	V3	0.470	9	0.264	17		
F2	V4	0.710	1			0.440	8
	V5					0.760	1
	V6	0.430	10				
	V7	0.530	6	0.389	12	0.560	4
	V8					0.360	14
	V9			0.421	10	0.360	15
	V10			0.443	5		
	V11					0.720	2
	V12					0.420	11
F3	V13					0.320	16
	V14					0.420	10
	V15			0.264	16	0.400	13
	V16					0.440	9
	V17			0.629	3		
F4	V18			0.386	13		
	V19			0.493	4	0.460	5
	V20	0.600	4			0.640	3
	V21	0.610	3	0.679	1	0.460	6
	V22			0.671	2		
	V23	0.320	12	0.400	11		
F5	V24	0.330	11	0.421	9		
	V25			0.429	8		
	V26	0.600	5				
	V27			0.443	6		
	V28	0.670	2	0.286	15		
	V29	0.520	7				

Table 7 describes the ranking of each pre-construction risk variable. where each risk owner has different potential risk variables. Management Consultant has 2 high risk variables and Pokmas and BPBD have 3 high risk variables. When viewed from the risk factors. the Pokmas formation factor (F4) is the most potential risk factor compared to other risk factors in the pre-construction phase because there is a potential risk variable with the highest rank held by the three parties. This is closely related to the initial research [2]. which was carried out due to the increase in the number of damaged house data after re-validation so that the number of Pokmas automatically increased. resulting in Pokmas formation and training activities not running effectively. The involvement of the community with a very complex background and the absence of a risk management plan from the government triggers this risk. These potential factors and variables should be of concern to the government because if the identified risks are not immediately planned for a resolution strategy. it is possible that these risks will likely reoccur in the future.

3.5 Analysis of variance (Anova)
Based on the results of the analysis. F count > F table and P value $<\alpha$ (0.05). then the initial hypothesis is rejected. It can be concluded that there is a significant influence between risk variables and risk owners on the likelihood / likelihood of risk occurring in the pre-construction phase.

4. Conclusion
Based on the results of the research and discussions that have been carried out. it can be concluded that the pre-construction stage activities in the post-earthquake housing reconstruction process in Pidie Jaya Regency have the potential for risks that can impact time. cost and quality. The government, in this case the BPBD, as the coordinator in the rehabilitation and reconstruction of houses after an earthquake

disaster has the responsibility to reduce the risks that have the opportunity to arise again in the future if risk management is not planned. The list of potential risk identification that has been obtained from this study can be used as input in planning risk management for the pre-construction stage of post-disaster housing reconstruction when using community-based methods.

5. Acknowledgments

Thanks to the research team and respondents for their contribution. This paper is part of the Head Lecturer Research Program funded by the Institute for Research and Community Service at the Universitas Syiah Kuala for the 2020 fiscal year and also as research from the Engineering Doctoral Program of Universitas Syiah Kuala.

References

[1] Pemerintah Aceh 2017 *Pedoman Umum Penyelenggaraan Rehabilitasi dan Rekonstruksi Sektor Permukiman Berbasis Masyarakat (Insitu) Pascabencana Gempa Bumi Kabupaten Pidie Jaya, Pidie dan Bireuen* (Banda Aceh: Pemerintah Aceh).

[2] Malahayati N and Afifuddin M 2019 *IOP Conf. Ser.: Earth Environ. Sci.,* **273** **(1)** 012037.

[3] Malahayati N, Munirwansyah, Afifuddin M and Syamsidik 2019 *Prosiding Konferensi Nasional Teknik Sipil Ke 13 19-20 September 2019.*

[4] Ophiyandri T, Amaratunga D, Pathirage C and Keraminiyage K 2013 *Int. J. Disaster Resil. Built Environ.* **4** (2) 236-249.

[5] Project Management Institute 2017 *A Guide to the Project Management Body of Knowledge (PMBOK® Guide)* Sixth Edition.

[6] Da Silva J 2010 *Key considerations in post-disaster reconstruction* (UK: Practical Action Publishing).

[7] Pemerintah Indonesia 2007 *Undang-Undang Republik Indonesia No. 24 Tahun 2007 Tentang Penanggulangan Bencana* (Jakarta: Pemerintah Indonesia).

[8] Badan Nasional Penanggulangan Bencana (BNPB) 2008 *Peraturan Kepala Badan Nasional Penanggulangan Bencana No 11 Tahun 2008 Tentang Pedoman Rehabilitasi dan Rekonstruksi Pasca Bencana* (Jakarta: BNPB).

[9] Badan Nasional Penanggulangan Bencana (BNPB) 2013 *Peraturan Kepala Badan Nasional Penanggulangan Bencana Nomor 3 Tahun 2013 Tentang Petunjuk Teknis Pelaksanaan Anggaran Kegiatan Rehabilitasi dan Rekonstruksi Wilayah Pasca Bencana* (Jakarta: BNPB).

[10] Coffey V and Trigunarsyah B 2012 *Proceedings of 8th Annual International Conference of the International Institute for Infrastructure, Renewal and Reconstruction* (IIIRR) 2012 Apr 23 292-300.

[11] Boy W and Suripin W M 2017 *Int. J. Civil Eng. Technol.* (IJCIET) **8** (10) 1220-36.

[12] Ophiyandri T, Amaratunga D and Keraminiyage K 2016 *Int. J. Disaster Resil. Built Environ.* **7** (4) 420-431.

IOP Conf. Series: Materials Science and Engineering 1087 (2021) 012038 doi:10.1088/1757-899X/1087/1/012038

Lightweight concrete in cement-treated subbase construction

F M Suryani[1]*, S Rizal[2], Abdullah[1], M Isya[1]
[1]Civil Engineering Department, Engineering Faculty, Universitas Syiah Kuala, Darussalam, Banda Aceh 23111, Indonesia
[2]Mechanical Engineering Department, Engineering Faculty, Universitas Syiah Kuala, Darussalam, Banda Aceh 23111, Indonesia

*E-mail: fitrika_mitasuryani@unsyiah.ac.id

Abstract. Following the development of transportation technology, steady road infrastructure is needed. The choice of construction type is a crucial point for the expected quality. Road construction consists of flexible and rigid pavement construction. Rigid pavement has higher durability and is produced from concrete, a strong, flexible, and fire-resistant material. Lightweight concrete, containing high amounts of pores, can be produced by using lightweight aggregate or a mixture of lightweight-coarse aggregate fulfilling the requirements for compressive and flexural strength. This study evaluated the compressive and flexural strength of lightweight concrete without coarse aggregate as a rigid pavement material. This study used specimens from type I Portland cement for compressive and flexural strength test, in 0.5 water-cement ratios and the specific gravity of 1.0. The result revealed a compressive strength value of 38.91 kg/cm^2 and a flexural strength value of 1.32 MPa. Based on the results obtained, it was concluded that lightweight concrete material fulfilled the general specification criteria for the compressive strength requirements of the cement-treated subbase (CTSB).

1. Introduction

Transportation is an activity to move people or goods from one place to another and is mainly purposed to connect humans to land use. Transportation has characteristics and attributes indicating its specific meaning and function. In this 20th century, the growth of transportation developed rapidly in line with the latest technological advances [1]. In accordance with the development of transportation technology, a steady road infrastructure is needed [2]. The development of road planning and construction is in line with the increasing need for speed, comfort, safety, and ease of traveling on land [3].

Road construction consists of flexible and rigid pavement construction. Flexible pavement is composed of asphalt concrete which is placed on top of a base or subbase layer supported by compacted subgrade. Rigid pavement generally consists of two layers, namely a concrete layer and a subbase [4]. In Indonesia, the use of flexible pavement structures is more dominant due to the lower cost. However, in terms of maintenance, flexible pavement structures require higher maintenance costs than rigid pavement. On the other hand, rigid pavement structures have higher durability compared to flexible pavements, so that high frequency loads, even of heavy vehicles, do not become a problem [5].

Based on this background, we conducted research on rigid pavement construction. Concrete, a construction material that has been commonly used, is the key substance required in the construction of rigid pavement. Various building constructions have used concrete as materials due to its high compressive strength, flexibility, fire resistance, affordability, and cost-effectiveness [6]. Lightweight concrete is one of the various types of concrete. It is a mixture of cement, water, and foam produced by the foam agents. Also being known as cellular or aerated concrete, this type of concrete is produced through certain techniques leading to the large amounts of pores contained. The pores, becoming a specific characteristic of lightweight concrete, are intentionally formed to reduce the weight of the concrete [7].

2. Literature review

2.1 Rigid pavement

The pavement type of choice varies based on traffic volume, design life, and condition of road foundation. Engineers should consider the lowest costs throughout the design life, in addition to the implementation effectivity and limitations. For light to moderate traffic loads, flexible pavement is more suitable than rigid pavement. This can be seen especially in certain rural or urban areas where road construction is less disruptive to the traffic. Rigid pavements can be a cheaper option for urban roads with limited access for heavy vehicles. In a limited area, the implementation of rigid pavement will be easier and faster than that of flexible pavement [8].

2.2 Rigid pavement with cement-treated base (CTB)

For roads serving medium and heavy traffic, a CTB foundation layer is preferred than a grained foundation layer due to its cost-effectiveness. The CTB can economize the use of asphalt and grained materials and is less sensitive to water than grained foundation layers. It can be substituted by Lean Mix Concrete (LMC) that is easier to implement in narrow working areas such as pavement widening work or work in urban areas [8].

Overloaded vehicles are a real condition that must be anticipated. Such loads can cause premature cracking of the CTB layer. Therefore, CTB design is only based on the CTB stiffness modulus at the post-fatigue cracking stage without considering the age of pre-fatigue cracking [8]. The unconfined compressive strength requirements of the Cement Treated Base (CTB) and Cement Treated Subbase (CTSB) at the age of 7 (seven) days are 45 - 55 kg/cm^2 and 35 - 45 kg/cm^2, respectively [9]. The CTB construction requires a competent contractor with adequate equipment resources, as CTB pavements are only selected if the required resources are available [8].

2.3 Lightweight concrete

Based on SNI 03-3449-1994, lightweight structural concrete is a concrete that uses lightweight aggregate, or a mixture of lightweight-coarse aggregate and natural sand as a substitute for lightweight fine aggregate and meets the requirements for compressive and flexural strength with the maximum weight of 1850 kg/m^3. The flexural strength of air-dry lightweight concrete ranges from 70% to 90%, compared to the general concrete of the same compressive strength. Nevertheless, if being exposed continuously to a certain level of moisture, these two types would have the same level of flexural strength [10].

The development of lightweight concrete at Universitas Syiah Kuala (USK) began in 2006. USK has been able to produce lightweight concrete with structural quality (compressive strength value > 17 mPa). Various qualified products using lightweight concrete as a material have been invented [11].

Lightweight concrete is classified into low density concrete (content weight less than 50 pcf (800 kg/m^3)), medium strength concrete (content weight ranging from 60 - 80 pcf (960 - 1360 kg/m^3) and compressive strength between 1000 - 2500 psi (6.89 - 17.23 MPa)) and structural concrete (content weight ranging from 90 - 120 pcf (1440 - 1920 kg/m^3) and equal compressive strength to the general concrete) [12]. In general, concrete is obtained from a mixture of Portland cement, water and aggregate, while for lightweight concrete, the constituent materials depend on its type. Lightweight concrete is a mixture of cement, water, air, and foam produced from foam agents. Production of lightweight concretes is generally done by adding air bubbles into the concrete mixture or by using lightweight aggregates, such as clay and pumice [13].

2.4 Compressive and flexural strength

Compressive strength is an illustration of the concrete quality. According to SNI 03-1974-1990, concrete compressive strength is the amount of load per unit area that causes the concrete specimens to collapse when loaded with a certain compressive force generated by a compressive testing machine [14]. Flexural strength is the ability of concrete placed on two positions to withstand the force given in the direction perpendicular to the axis of the specimen until collapsed which is stated in Mega Pascal (MPa) force per unit area. The concrete flexural strength affects its ability to overcome the initial cracks before being loaded [15].

3. Materials and methods

3.1 Materials

The manufacture of the cylindrical specimens with a diameter of 15 cm and a height of 30 cm was intended for compressive strength test, while another specimen in the form of blocks with a size of 15 cm x 15 cm x 60 cm was used for the flexural strength test. The water-cement ratio was 0.5, and the specific gravity was 1.0. The cement used was type I Portland cement, as shown in Figure 1. The equipment used included a press loading machine with a capacity of 100 Tf, balance, filter, concrete mixer, and foam generator.

Figure 1. Type I Portland cement.

3.2 Methods

3.2.1. Specimen manufacture and maintenance. The specimen manufacture, listed in Table 1, was carried out at the Construction and Building Materials Laboratory of the Civil Engineering Department of Unsyiah in the following procedures:

- Preparation and measure of the materials according to the predetermined proportion of the mixture.
- Addition of water and cement into the foam produced by the generator in the concrete mixer until well blended.
- Adjustment of the concrete mixture on the mold and formation of the concrete cast.
- Treatment of the concrete was carried out after hardened.
- Removal of the concrete cast from the mold to be put under immersion.

Table 1. The lists of specimen manufacture for each test.

Test	Specimen	Quantity
Compressive Strength	Cylinder 15 cm x 30 cm	3
Flexural Strength	Block 15 cm x 15 cm x 60 cm	3
	Total	6

3.2.2. Compressive strength test. The cylinder specimens were tested at the age of 7 (seven) days of concrete after immersion. Before doing the tests, it was confirmed that the specimens were dried, weighed and measured. The test was carried out by giving loads until the specimens were destroyed. The compressive strength test equipment is displayed by Figure 2.

Figure 2. Compressive strength test equipment.

3.2.3. Flexural strength test. The block specimens were tested at the age of 7 (seven) days of concrete after immersion. Before doing the tests, it was confirmed that the specimens were dried, weighed and measured. The test was carried out by giving loads until the specimens were destroyed. The flexural strength test equipment is shown in Figure 3.

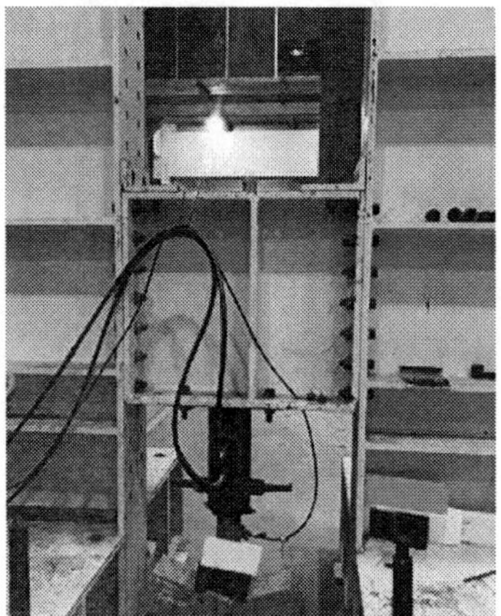

Figure 3. Flexural strength test equipment.

4. Results

The compression and flexural strength value obtained are shown in Table 2 and Table 3. Table 2 shows the average value of the compressive strength of lightweight concrete at the age of 7 (seven) days was 38.91 kg/cm^2, and the average value of flexural strength obtained at the age of 7 (seven) days was 1.32 MPa. Based on the literature, the criteria for the unconfined compressive strength of CTSB at the age of 7 (seven) days is 35 - 45 kg/cm^2 [9]. These results indicate that the quality of lightweight concrete tested at the age of 7 (seven) days has met the compressive strength requirements. In normal and lightweight concrete, higher concrete age increases the concrete quality. Likewise, the flexural strength value is also directly proportional to the concrete age [16]. These results were considered interesting because the specific gravity of lightweight concrete tested for use in a pavement layer was very small compared to the specific gravity of normal concrete which reaches 2.4. This was very significant to the reduction of concrete own weight in the road foundation.

Table 2. The compression strength values.

Specimen	Weight (kg)	Dimension		Area (mm²)	Force			Compression Strength (kg/cm²)	
		Height (mm)	Diameter (mm)		(Ton)	(kN)	(N/mm²)	Unit	Average*
1	5.58	301.00	149.40	17521.48	6.00	58.86	3.36	34.26	38.91 ± 4.37
2	5.38	302.00	150.20	17709.63	7.00	68.67	3.88	39.56	
3	5.32	305.00	149.20	17474.60	7.50	73.58	4.21	42.93	

* The data is presented as mean ± SD.

Table 3. The flexural strength values.

Specimen	Weight (kg)	Dimension				Maximum Load		Flexural Strength (MPa)	
		Length (mm)	Width (mm)	Height (mm)	l (mm)	(tf)	(kN)	Unit	Average*
1	11.84	522.00	150.00	150.00	450.00	0.75	7.36	0.98	1.32 ± 0.58
2	13.96	601.00	150.00	149.00	450.00	0.75	7.36	0.99	
3	14.16	600.00	148,00	150.00	450.00	0.82	8.04	1.09	

* The data is presented as mean ± SD.

5. Conclusions

Based on the results obtained, it was concluded that lightweight concrete material fulfilled the general specification criteria for the compressive strength requirements of CTSB. As in normal concrete, the quality of lightweight concrete without coarse aggregate is directly proportional to the concrete age. Therefore, lightweight concrete can be used as a material in rigid pavement. The utilization of this material in pavement construction can be considered to improve the pavement stability. We propose that further research is required to identify the effect of substituting lightweight concrete with alternative materials in order to reduce the initial cost of rigid pavement.

References

[1] Salim A 2004 *Manajemen Transportasi* (Jakarta: PT. Raja Grafindo Persada).
[2] Saodang H 2009 *Kontruksi Jalan Raya* (Bandung: Nova).
[3] Sulaksono S 2011 *Rekayasa Jalan* (Bandung: Penerbit Institut Teknologi Bandung).
[4] Papagiannakis A and Masad E 2016 *Pavement Design and Materials* (Hoboken: John Wiley & Sons, Inc.).
[5] Mohod M and Kadam K 2016 *IOSR J. Mech. Civil Eng.* **13** 84–88.
[6] Tjokrodimuljo K 2007 *Teknologi Beton* (Yogyakarta: Biro Penerbit KMTS Fakultas Teknik

Universitas Gadjah Mada).
[7] Harjanto S, Prima S, Suharno B and Ashadi H 2008 *Jurnal Sains Materi Indonesia* **9** 139–144.
[8] Direktorat Jenderal Bina Marga 2017 *Manual Desain Perkerasan Jalan* (Jakarta: Direktorat Jenderal Bina Marga).
[9] Direktorat Jenderal Bina Marga 2018 *Spesifikasi Umum 2018 untuk Pekerjaan Konstruksi Jalan dan Jembatan* (Jakarta: Direktorat Jenderal Bina Marga).
[10] Direktorat Pekerjaan Umum 1994 *Tata Cara Rencana Pembuatan Campuran Beton Ringan dengan Agregat Ringan (SNI 03-3449-1994)* (Bandung: Direktorat Pekerjaan Umum).
[11] Abdullah and Bustari Fonna S 2007 *Pengembangan Beton Ringan Mutu Tinggi dengan dan Tanpa Serat* (Banda Aceh: Universitas Syiah Kuala).
[12] Winter G and Nilson A 1993 *Perencanaan Beton Bertulang* (Jakarta: PT. Pradnya Paramita).
[13] Gunawan P, Prayitno S and Warsino W 2015 *Matriks Teknik Sipil* **1** 679–687.
[14] Badan Standarisasi Nasional 1990 *Metode Pengujian Kuat Tekan Beton (SNI 03-1974-1990)* (Jakarta: Badan Standarisasi Nasional).
[15] Badan Standarisasi Nasional 1997 *Metode Pengujian Kuat Lentur Normal dengan Dua Titik Pembebanan (SNI 03-4431-1997)* (Jakarta: Badan Standarisasi Nasional).
[16] Badan Standarisasi Nasional 2002 *Metode Pengujian untuk Mengukur Nilai Kuat Tekan Beton pada Umur Awal dan Memproyeksikan Kekuatan pada Umur Berikutnya (SNI 03-6805-2002)* (Jakarta: Badan Standarisasi Nasional).

Reventing brittle hybrid high-strength reinforced concrete slab collapse due to punching shear using coal flyash substitution, tie wire fiber and polypropylene fiber

T B Aulia[1]*, Y Hayati[1], Y Idris[1], Z Amalia[1], E Handoko[2]

[1]Department of Civil Engineering, Univeritas Syiah Kuala, Banda Aceh 23111, Indonesia
[2]Department of Physics, Universitas Negeri Jakarta, Jakarta 13220, Indonesia

*E-mail: aulia@unsyiah.ac.id

Abstract. Heavy vehicles crossing a bridge could lead in punching shear failure at reinforced concrete deck slab due to highly concentrated axial compressive load from their wheels. This may occur in a very brittle manner and difficult to fix. Effort in improving punching shear capacity of deck slab is with using high-strength concrete and adding fibers into concrete. However, high-strength concretes are not environmentally friendly because using a lot of cement contributing to large CO_2 emissions to the atmosphere during its production, therefore coal flyash is used as a part of cement substitution. Based on these facts, this research aims to analyze punching shear capacity of hybrid high-strength reinforced concrete slabs using coal flyash as cement substitution which added with tie wire fiber and polypropylene fiber. The amount of coal flyash was taken 15% from cement weight as cement substitution, tie wire fiber 2% from concrete volume having aspect ratio l/d of 40 and straight shaped fiber geometry, as well as polypropylene fiber 0.2% from concrete volume. Three hybrid high-strength reinforced concrete slabs of 60 x 60 cm^2 (two-way-slab) with a thickness 12 cm, w/c-ratio of 0.30 and concrete cylinder compressive strength of f'_c = 60 MPa were made with 15% coal flyash and 2% tie wire fiber (CFA-TWF), 15% coal flyash and 0.2% polypropylene fiber (CFA-PPF), as well as 15% coal flyash and combination of 2% tie wire fiber and 0.2% polypropylene fiber (CFA-TWF-PPF). One plain slab without flyash and fibers (PSHSC) was tested as a comparison. To guarantee the punching shear failure mechanism, the slabs were reinforced with tensile reinforcement of 10 Ø 10.5 mm which was distributed to each direction and tested then by giving a concentrated punching load. The results showed that using of coal flyash substitution and addition of fibers could significantly increase the punching shear capacity and structural ductility of hybrid high-strength concrete slabs with the highest enhance both achieved in CFA-TWF-PPF, i.e., 67.91% for punching shear capacity and 92.82% for structural ductility compared with plain slab PSHSC.

1. Introduction

Punching shear is a failure mechanism occurring in slab structures that are subjected to high concentration of axial load. Such failure generally causes dangerous brittle damage and difficult to repair. However, the rules of reinforced concrete slab design in many countries are more focused on bending capacity neglecting safety due to punching shear failure in the structure. One of the structures to be prone to punching shear failure is the bridge deck slab resulting from heavy vehicles traffic, which transfers a large concentrated wheel load to the deck slab. Other than that, the floor flat slabs riveting on the columns in the building construction are also at risk for punching shear failure.

Due to difficulties in installation, reinforced concrete slab is often designed without using stirrups as shear reinforcement. As the result, punching shear capacity is fully determined by slab thickness and concrete shear strength without contribution of shear reinforcement.

Investigation of punching shear capacity of reinforced concrete slabs without shear reinforcement had been carried out by [1] with variations in concrete strength of 30 MPa and 60 MPa, slab thickness of 10 cm and 12 cm, using tie wire fiber with fraction volume (V_f) of 2.0% from the concrete volume. The findings showed that punching shear capacity was predominantly determined by concrete strength and slab thickness, while tie wire fibers enhanced the punching shear capacity and structural ductility of reinforced concrete slab significantly through its contribution as micro-shear reinforcement. Without using synthetic fibers, in case of the concrete compressive strength was increased from about 36 MPa to 130 MPa, the punching capacity of reinforced concrete slabs rose 42% with longitudinal reinforcement ratios of 1.25%, but when reinforcement ratio increased from 0.94% to 1.48%, the punching capacity increased of about 13%, as reported by [2].

Research on the use of synthetic fibers that are added individually to the concrete mixtures on the punching capacity of reinforced concrete slabs has been widely carried out. Study by [3] using steel fibers with fiber volume ratio 0.5%, 1%, 1.5% in the area d, 2d, 3d from the edge of the column and reinforcement ratio 1.2% reported that slabs having 1.5% steel fiber ratio resulted in highest punching shear capacity. The results showed that the use of steel fibers only in a part of slab which was equal to slab thickness *d* from column face was adequate to provide optimum increase in both punching strength and ductility behavior. In [4], it was mentioned that variations in concrete strength of 56 MPa and 123 MPa, fiber content of 0, 1, 2, 3%, and types of fibers, i.e., steel fiber, polypropylene fiber and Aramid fiber glass could increase the punching shear strength of 22-66% depending on fiber content, in comparison with similar slabs without fibers. It was concluded that the fibers used had a significant effect on the punching shear capacity and ductility behavior of reinforced concrete slabs. Moreover, with using hooked ended steel fibers of 1.2% and 1.4% on high-strength reinforced concrete slabs of 64.5 MPa, it was reported that steel fibers contributed to the enhance in punching shear capacity of 34% and ductility improvement in cracking behavior [5]. Further, by varying slab thicknesses of 12 mm, 22 mm, 32 mm and 42 mm using 3% steel fibers, it was reported that punching shear capacity of high-strength reinforced concrete slabs was significantly increased with the slab thickness [6].

In addition to steel fibers, polypropylene fibers are widely used to study the punching shear capacity of reinforced concrete slabs. Punching shear capacity of reinforced concrete slabs made with different concrete compressive strength of 20, 30, 40 MPa and different ratios of micro-polypropylene fibers (i.e., 6, 7.5 and 9% from cement weight) was investigated by [7]. It was concluded that micro-polypropylene fibers enhanced punching shear capacity with concrete strength of 20 MPa and 30 MPa by 15% and reinforced concrete slab with concrete strength of 40 MPa by 12% with using 9% micro-polypropylene fibers and showed a significant effect on deformability. Then, an increase in punching shear capacity of reinforced concrete slabs incorporated with 0.5% polypropylene fibers amounting to 15.28% and displacement of 20.14% compared with normal plain slab was reported by [8]. Besides being able to enhance the punching shear capacity, the use of tie wire fibers and polypropylene fibers can significantly increase the bending capacity of high-strength reinforced concrete beams, mainly in terms of flexural strength, ductility and bending behavior as reported by [9].

High-strength concrete made with a lot of cement having a low water to binder ratio, which is added with additives and admixtures. On the contrary, production of cement at factory is not environmentally friendly because it needs a lot of calcium based raw material and releases CO_2 emission to the atmosphere during the combustion of clinker at temperatures up to 1500°C. Production of 1 ton cement requires 1.7 tons of natural raw material and on the other, releases 1 ton CO_2 into the atmosphere [10]. To reduce the usage of cement, additives containing some chemical compositions those found in cement can be used. One of such additives is coal flyash.

The research using partial substitution of cement with coal flyash, i.e., 15% from cement weight on the mechanical capacity of high-strength reinforced concrete beams had been carried out by [11] on bending capacity, by [10] on shear capacity, as well as by [12] on torsion capacity. The results showed that substitution of cement with 15% coal flyash could generate mechanical properties and mechanical behavior of high-strength reinforced concrete beams that were closer to or even better than entirely cement-based high-strength reinforced concrete beams.

Based on these facts, punching shear performance of high-strength reinforced concrete slab without shear reinforcement by combining geopolymer flyash and hybrid synthetic fibers has not been sufficiently investigated. Therefore, the aim of this study is to find potential solution in preventing sudden failure in structures due to punching shear phenomenon. The use of tie wire fiber and polypropylene fiber as well as coal flyash are on the basis of efficiency, environmental conservation and availability of materials which are relatively inexpensive, in which case tie wire fiber is construction project waste and coal flyash is electric steam power plant waste.

2. Experimental program

2.1 Research method

This research was carried out to determine punching shear resistance of hybrid high-strength reinforced concrete slabs using 15% geopolymer coal flyash as cement substitution which added with 2% tie wire fiber and 0.2% polypropylene fiber both individually and jointly. The resulting punching shear behavior was then compared with plain high-strength reinforced concrete slab without flyash substitution and fibers.

Critical punching shear failure occurs at the perimeter plane of concentrated axial compressive load which is described by the rupture of slab along the surface of axial load to the lowest side of slab surface. The stresses formed by punching shear in a flat slab can be analogous as a two-dimensional shear stress occurring in a beam causing more critical stresses compared with the shear stress of beam. Perimeter of critical area is considered to be at a half distance from column face to the failure tangent with the lower slab surface as shown in Figure 1(a) [13]. Several rules that govern the critical punching shear limits are described in Figure 1 (b) [13].

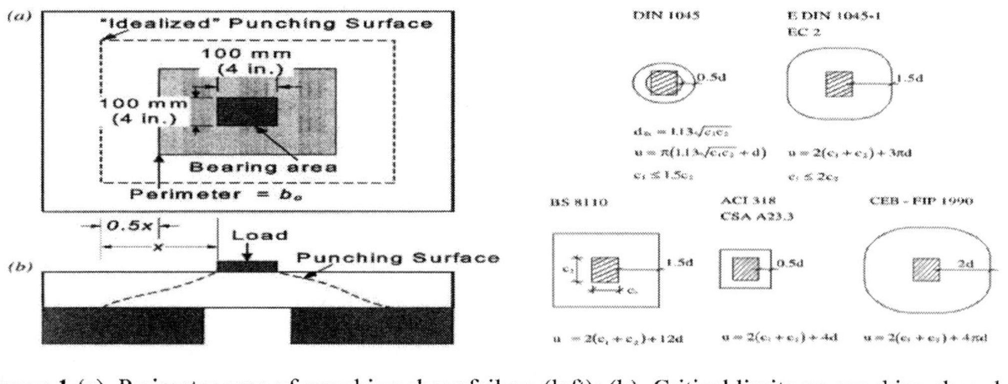

Figure 1 (a). Perimeter area of punching shear failure (left); (b). Critical limits on punching shear by several Codes (right).

The specimens used were high-strength reinforced concrete slabs with a concrete compressive strength of 60 MPa, dimension of 60 x 60 cm2 and slab thickness of 12 cm. To gain the main mechanical properties of concrete for controlling the characteristics of slabs, concrete compressive strength test using cylinder specimens with a diameter of 15 cm and height of 30 cm and concrete flexural tensile strength test with beam specimens of 15 x 15 x 60 cm were conducted.

The width of slab was taken to be 60 cm according to [13], which must be greater $4d + c_0 = (4 \times 12) + 10 = 58$ cm, where d is the slab width and c_0 is width of the steel plate pressing block surface (footing) having dimension of 10 x 10 cm with a thickness of 10 cm. It was taken into account to avoid the initial rupture occurred in supporting area.

To prevent bending failure, the slabs were reinforced with tensile reinforcement of 10 Ø 10.5 mm with the yield strength of 240 MPa, which was distributed to each direction (two-way-slab) as shown in Figure 2 (a). Amount of tensile reinforcement was obtained from theoretical calculations taking into account the dimension of slab, concrete compressive strength and yield stress of flexural reinforcement. The results were then controlled to be safe against bending ($P_y \geq 1.5V_c$) and ensured the occurrence of punching shear in the slab structure ($V_n \geq V_c$), where Py is flexural strength of slab, V_c is shear strength of concrete, V_n is nominal shear strength occurring in slab structure.

Strain gauges were used to monitor deformations on tensile reinforcement and concrete, each 4 units for reinforcement and 2 units for concrete. Deformations were recorded and read using Portable Data Logger TDS-302. The placement of strain gauges on tensile reinforcement and concrete is given in Figure 2 (b).

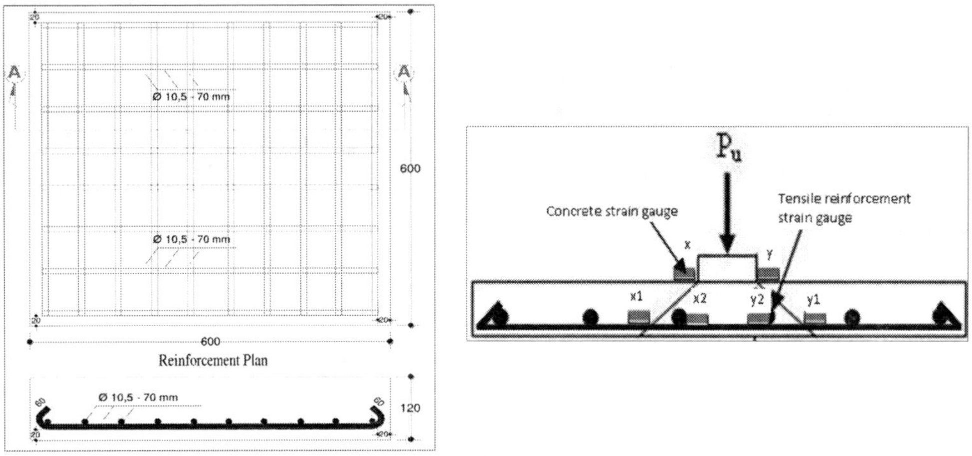

Figure 2 (a). High-strength reinforced concrete slab specimen (left); (b). Placement of strain gauges on tensile reinforcement and concrete (right).

To determine deformability of the slab, ductility index which describes the ability of structure in carrying inelastic deformation prior to collapse was calculated with the following equation $\mu = \Delta_u/\Delta_y$, where Δ_u is the ultimate deformation and Δ_y is deformation at first yield. Ductility constitutes structural property which plays an important role in designing earthquake-resistant structures. It is widely used to control brittle collapse of structure.

2.2 Materials

The coal flyash used originated from electric steam power plant in Nagan Raya Regency, Aceh, which possessed chemical composition as follows SiO_2 = 26.65%, Al_2O_3 = 9.6%, Fe_2O_3 = 17.56%, and SO_3 = 2.51% [10]. Amount of SiO_2, Al_2O_3, and Fe_2O_3 content is 53.81% which is categorized into Class C flyash referring to ASTM C.618, that has a minimum SiO_2, Al_2O_3, and Fe_2O_3 content about 50% [14]. Dosage of coal flyash used was 15% from cement weight.

The fibers employed were tie wire fiber and polypropylene fiber which were used separately and combined into a hybrid composite. The tie wire fibers were cut from the construction reinforcing binding-wire having a diameter of 0.75 mm and a length of 30 mm with tensile strength of 38.5 N/mm^2 and specific gravity of 6.60 gr/cm^3. The fiber fraction volume (V_f) was 2.0% from the volume of concrete.

Polypropylene fibers are plastic-based fibers from the polyolefin family in the form of bundles that made in small bands (fibrils). The use of polypropylene fibers can increase the ductility of concrete by employing the higher tensile strength and elastic modulus of the fiber than concrete, polypropylene fibers can bridge cracks formed in the concrete matrix thus slowing crack propagation. The pullout process of polypropylene fibers from the concrete can provide high fracture energy and fracture toughness preventing the sudden collapse of concrete. Besides, polypropylene fiber can function as micro reinforcement for cracking control which is more cost effective than other synthetic fibers. The polypropylene fibers used in this research owned a diameter of 18 microns and a length of 12 mm, thus having the aspect ratio of 66.7.

High-strength concretes were produced using the weight ratio method by trial and error referring to concrete mixture volume of 1,000 liters. Portland cement type I in amount of 550 kg/m^3 with a water cement ratio of 0.3, silica fume 8% from cement weight as additive and superplasticizer Sikament NN with a dosage of 2% by cement weight were used to make plain high-strength concrete without using coal flyash and fibers. Aggregates used consisting of fine aggregates from river sand with a maximum diameter of 4.76 mm and coarse aggregates in form of split with a maximum diameter of 15.9 mm.

2.3 Instrumentation and testing

The slabs were tested at the age 28 days. During testing, data about loading, tensile reinforcement strain and concrete strain, deflection, as well as crack pattern of slabs must be collected in detail. The specimen was first placed on the rigid metal frame support mounted rigidly on a steel plate with dimension of 80 x 80 x 10 cm as strong floor, which was connected with the bottom steel frame beam. The testing instruments were installed in a H-beam steel frame of 200 x 100 x 5 mm which attached as columns and beams. Testing was carried out by giving punching load on the pressing block (footing) of 10 x 10 x 10 cm through a hydraulic jack attached to the load cell. Load was continuously increased until the formation of initial cracks and ended until the collapse of slab.

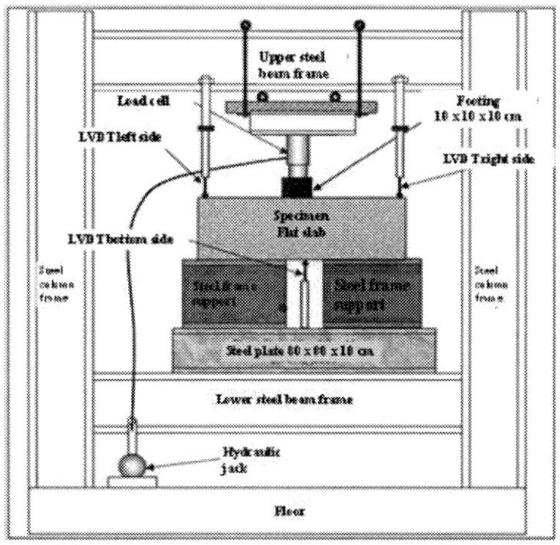

Figure 3. Set-up of punching shear testing.

Three Linear Variable Displacement Transducers (LVDTs) were vertically installed on slab, 2 units on the top side to record rotation at the slab ends and 1 unit at the bottom side to record the deflection at the middle of slab. Deformations on tensile reinforcement and concrete were monitored by strain gauges installed each 4 units for reinforcement and 2 units for concrete. All deformations and deflections occurring during testing were recorded using Portable Data Logger TDS-302 and printed automatically on the print paper P-60 installed on the Portable Data Logger. A space with a width of 15 cm between the rigid metal frame supports was provided for the placement of LVDT under the mid span of slab and to put camera for observing crack propagation on the bottom surface of slab during loading. Set-up of punching shear test for flat slab specimens on the steel frame is presented in Figure

3. After testing, slab specimens were moved to observe crack propagation and developing failure pattern visually from both surfaces, bottom side and top side.

3. Experimental results and discussion

3.1 Compressive and flexural tensile strength of concrete

The compressive test results show that all specimens yield higher compressive strength compared with plain high-strength concrete without using coal flyash and fibers (PSHSC) as demonstrated in Figure 4. Hybrid high-strength concrete made with 15% coal flyash and combination of 2% tie wire fiber and 0.2% polypropylene fiber (CFA-TWF-PPF) resulted in the highest compressive strength of 68.59 MPa followed then by high-strength concrete with 15% coal flyash and 2% tie wire fiber (CFA-TWF) of 67.68 MPa, as well as with 15% coal flyash and 0.2% polypropylene fiber (CFA-PPF) amounting to 61.94 MPa respectively. PSHSC itself had a compressive strength of 55.00 MPa.

The same findings were obtained from flexural tensile strength tests as shown in Figure 4, where CFA-TWF-PPF produced the best result amounting to 8.97 MPa, followed by CFA-TWF of 8.40 MPa and CFA-PPF of 7.36 MPa. High-strength concrete without using coal flyash and fibers (PSHSC) earned the lowest flexural tensile strength, i.e., 5.01 MPa.

Increase in compressive strength and flexural tensile strength on hybrid high-strength concretes with coal flyash and fibers, both tie wire fiber and polypropylene fiber, was generated by pozzolanic and filling effect of coal flyash, which composed of high cementitious minerals in form of SiO_2, Al_2O_3 and Fe_2O_3, as well as contributed by the higher modulus of elasticity and tensile strength of synthetic fibers in comparison with the cement paste in concrete matrix.

Through the pozzolanic effect, the cementitiuos material on the coal flyash will bind $Ca(OH)_2$ which is formed from the primary hydration process of cement to form extra Calcium Silicate Hydrate (C-S-H) which plays an important role in increasing the compressive strength of concrete. Due to the more finer particle of coal flyash compared with Portland cement particles size, thus a denser matrix with a lower porosity will be created in concrete.

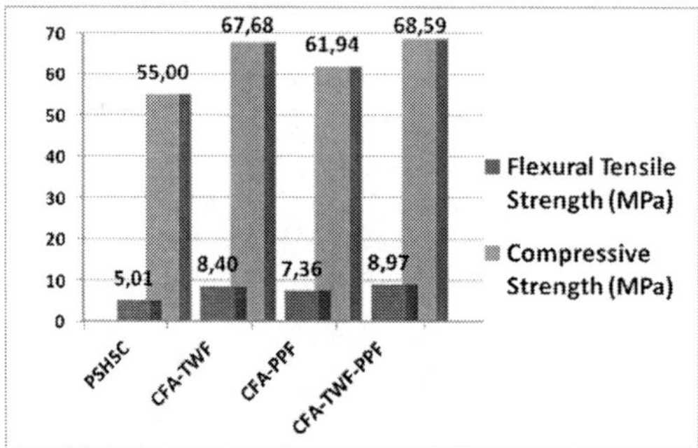

Figure 4. Compressive and flexural tensile strength of concrete slabs

3.2 Punching shear testing of high-strength reinforced concrete slabs

3.2.1. Load and deflection. The load and deflection relationships at the midst of slab for all investigated high-strength reinforced concrete slabs are described in Figure 5. From these relationships it could be revealed the loads at initial crack, at yield and at ultimate condition of tested slabs during loading history.

Referring to the graphs, hybrid high-strength concrete slab CFA-TWF-PPF yielded the highest punching shear strength, i.e., 34.08 ton resulting in a deflection of 12.98 mm followed by CFA-TWF which earned punching shear capacity of 31.48 ton with deflection of 7.34 mm and CFA-PPF which resulted in maximum punching shear strength of 28.46 ton with deflection of 6.16 mm respectively. Plain high-strength reinforced concrete slab without flyash and fibers (PSHSC) had a lowest punching shear capacity, that was 20.3 ton corresponding with a deflection of 5.17 mm.

Based on the graphs, it could also be observed that the slabs of CFA-TWF-PPF, CFA-TWF as well as CFA-PPF developed the load-deflection characteristics in a close manner each other indicating a higher stiffness and much better ductility in comparison with PSHSC slab which generated a sloping and flatter load-deflection curve. Thus, it could be obviously observed that the use of coal flyash and such synthetic fibers, i.e. tie wire fiber and polypropylene fibers both added individually and jointly could significantly enhance punching shear capacity of high-strength reinforced concrete slabs.

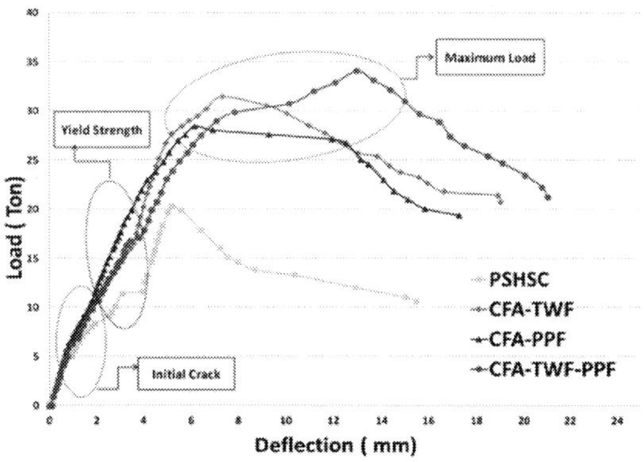

Figure 5. Load and deflection at the midst of slab

3.2.2. Load and deformation of flexural tensile reinforcement. The load and deformation behavior during testing that developed on the flexural tensile reinforcement for all punching shear tested slabs is given in Figure 6. In accordance with those relationships, it could be analyzed that the flexural tensile reinforcements installed in all high-strength reinforced concrete slabs had not reached its yield strain of 0.002 (2000με) when the slabs reached its ultimate load. This condition justified the assumption of slab reinforcement design where the slab must be safe against bending but will collapse due to punching shear ($P_y \geq 1.5V_c$).

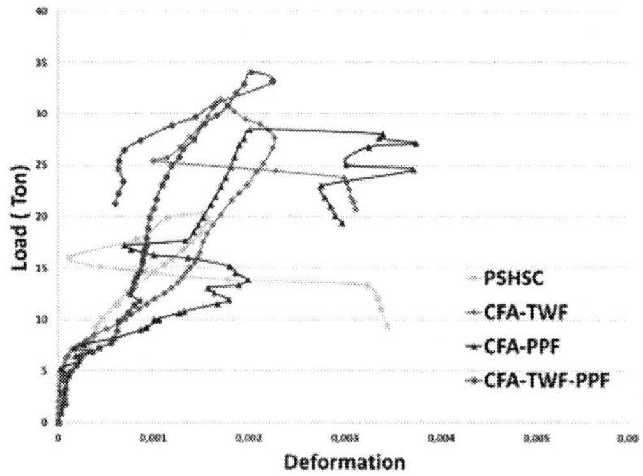

Figure 6. Load and strain of flexural tensile reinforcement of slab

The maximum flexural tensile reinforcement strain at ultimate load recorded was 0.0021 (2100με) yielded by CFA-PPF slab, followed by CFA-TWF-PPF slab of 0.00202 (2022με), then CFA-TWF slab of 0.00171 (1710 με) and the lowest was reached by PSHSC slab of 0.00153 (1530 με) respectively. Thus, it could be noticed that the slabs made with coal flyash and both fiber types added individually and in combination altogether demonstrated a better post-yield deformation in comparison with plain slab PSHSC, signifying a better ductile behavior. In other words, it means that the use of those coal flyash and fibers could delay the attainment of tensile reinforcement yield strain at a higher load.

3.2.3. Load and strain of concrete. The load and concrete strain relationships of all high-strength reinforced concrete slabs investigated in this study were compared each other in Figure 7. Hybrid high-strength reinforced concrete slab CFA-TWF-PPF generated the maximum concrete strain at ultimate load in amount of 0.00265 (2650με), followed by CFA-TWF slab of 0.00201 (2010με), then CFA-PPF slab of 0.00092 (920 με), and PSHSC slab resulted in the lowest concrete strain of 0.003 (300 με) respectively. According to Figure 7, it could be noticed that high-strength reinforced concrete slabs containing coal flyash and fibers that had higher punching shear capacity generated larger concrete strain due to punching load, explaining the concrete rupture at the perimeter plane of the given concentrated axial compressive load, which became critical punching shear failure.

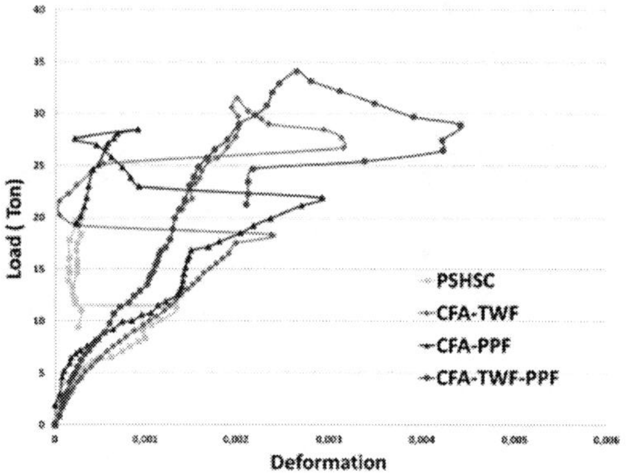

Figure 7. Load and strain of concrete.

3.3 Ductility and failure pattern of high-strength reinforced concrete slab

Structural ductility of high-strength reinforced concrete slabs is calculated with the formula $\mu = \Delta_u/\Delta_y$, where Δ_u is the ultimate deflection and Δ_y is deflection at yield point of structure. Hybrid high-strength reinforced concrete slab CFA-TWF-PPF generated the highest ductility, that was 3.222 followed by CFA-TWF slab of 2.324, after that CFA-PPF slab of 2.038 and PSHSC slab of 1.671 respectively. In comparison to plain slab PSHSC, there was a very significant increase in ductility of hybrid high-strength reinforced concrete slabs containing coal flyash and fibers, i.e., 92.82% for CFA-TWF-PPF, 39.08% for CFA-TWF as well as 21.96% for CFA-PPF.

Failure pattern on the bottom surface of punching shear tested slabs is given altogether in Figure 8. Punching shear failures on the bottom slab surface shown in Figure 8 revealed almost the identical crack pattern for all tested slabs, unless it differed only in the number of cracks developed. Plain high-strength reinforced slab PSHSC showed less number of cracks in comparison with slabs containing coal flyash added with fibers. Crack propagation was firstly initiated by micro cracks occurred due to flexural behavior arising in the midst of slab. These initial cracks developed after the concrete flexural tensile strength was exceeded by load increment. Then the cracks propagated further with a diagonal pattern accompanied by emerging new cracks which formed branches and spread continuously to the entire slab surface until the emergence of spalling on concrete at peak load.

Cracking propagation on the top face of slab was initiated by penetration of steel plate pressing block as load carrier to the concrete in the footing vicinity area of about 2 cm deep. In the high-strength reinforced concrete slabs with coal flyash and fibers, punching shear failure occurring on the top surface emerged after ultimate concrete strain was exceeded. The cracks started from the perimeter of footing face but did not develop to the outside of footing area at all. Conversely, at the plain high-strength reinforced concrete slab, after occurring in the perimeter of footing, the crack propagated further to the outside of footing area with a diagonal propagation direction as shown in Figure 9.

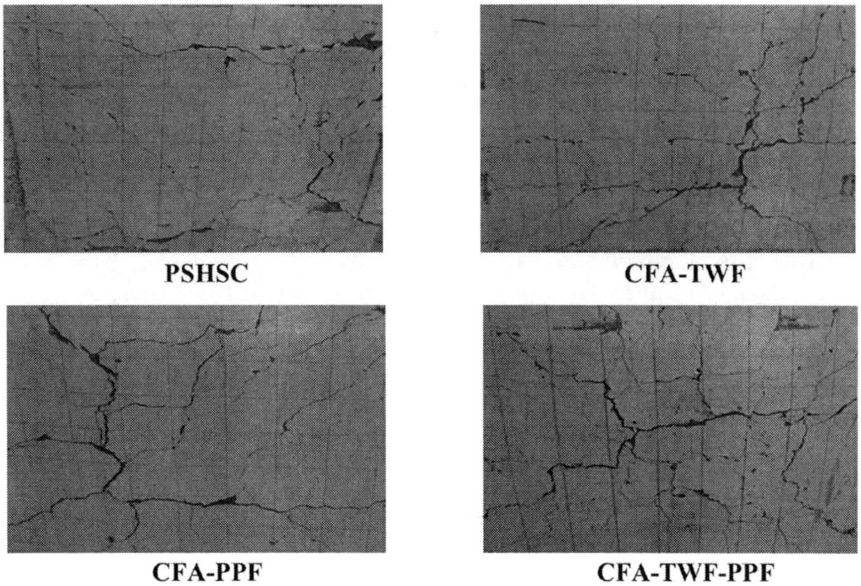

Figure 8. Failure pattern on the bottom surface of slab

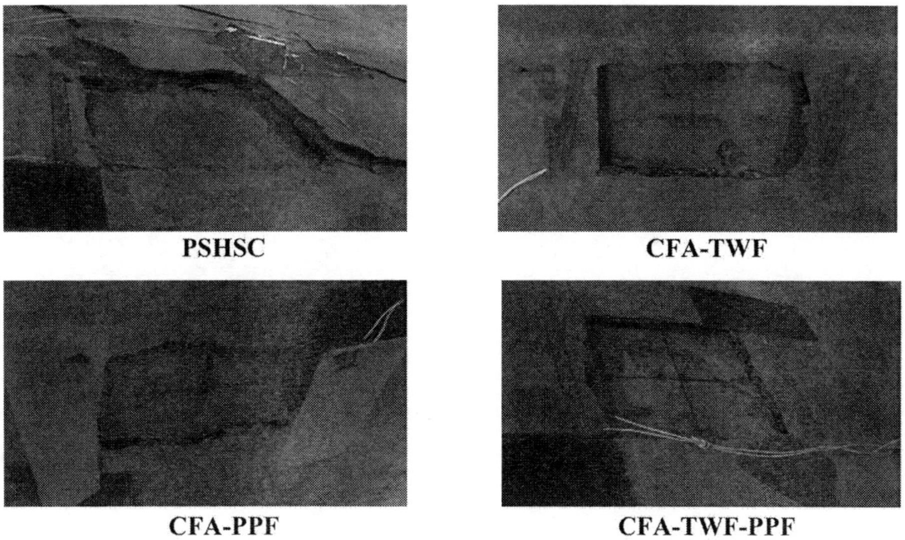

Figure 9. Failure pattern on the top face of slab

4. Conclusions

The geopolymer coal flyash used as cement substitution and tie wire fiber as well as polypropylene fiber that could contribute as micro-shear reinforcement added both individually and in combination altogether could increase significantly the punching shear behavior of hybrid high-strength reinforced concrete slabs, better than plain high-strength concrete slab. The highest enhance in punching shear capacity was given by the hybrid high-strength reinforced concrete slab added with coal flyash as cement substitution and both tie wire fiber and polypropylene fiber altogether. Thus, the use of such materials could provide inexpensive and environmental friendly high-strength reinforced concrete slabs contributing to the environmental preservation, because coal flyash and tie wire fiber are industrial and construction project waste.

References

[1] Aulia T B 2013 *Proceedings of The 8th International Conference on Numerical Analysis in Engineering,* Pekanbaru, Indonesia.
[2] Shembade K T 2017 *Int. J. Eng. Res. Technol.,* **6 (8)** 296–309.
[3] Abdel-Rahman A M, Hassan N Z and Soliman A M 2018 *Hous. Build. Nation. Res. Center J.,* **14** 272–281.
[4] Yassin M E, Shoukry A M and Tarabia A M 2020 *Proceed. Inst. Civil Eng. Struct Build.*
[5] Abadel A, Almusallam T, Al-Salloum Y and Abbas H 2014 *Proced. the Second Int. Conf. on Adv. Civil, Struct. Environ. Eng. (ACSEE).*
[6] Lampropoulos A P, Duncan J N and Tsioulou O T 2019 *IABSE Congress – The Evolving Metropolis.*
[7] Abdel-Jaber M, Shatnawi A and Bsisu K A 2016 *Advances in Intelligent Systems Research 4th International Conference on Sensors, Mechatronics and Automation (ICSMA).*
[8] Sermet F and Ozdemir A 2016 *Procedia Eng.,* **161** 458 – 465.
[9] Aulia T B and Rinaldi 2015 *Procedia Eng.,* **125** 1121-28.
[10] Aulia T B, Muttaqin, Afifuddin M, Zaki M and Merriza S 2018 *IOP Conf. Ser.: Mater. Sci. Eng.,* **434** 012199.
[11] Aulia T B, Afifuddin M and Opirina L 2016 *Proceed. The 7th Civil Eng. Conf. in the Asian Region (CECAR7).*
[12] Aulia T B, Muttaqin, Afifuddin M, Zaki M and Nastiti G 2019 *J. Physics: Conf. Ser.,* **1402** 022014.
[13] Naaman, E A, Likhitruangsilp V and Parra-Montesinos G 2007 *ACI Structural J.,* **104 (2)** 170-179.
[14] ASTM C618-17 2013 *Standard specification for coal fly ash and raw or calcined natural pozzolan for use in concrete. Annual Book of ASTM Standard.* (West Conshohocken: ASTM International).

Analysis of flood discharge using Snyder and Melchior methods at Krueng Meureubo watershed in Pasi Mesjid village

M Safriani*, I Y Salena, C S Silvia
Civil Engineering Department, Teuku Umar University, Jl. Alue Peunyareng, Kecamatan Meureubo, Aceh Barat, Indonesia

*E-mail: meylissafriani@utu.ac.id

Abstract. Pasi Mesjid villages in Meureubo district are vulnerable to floods in the rainy season. Every year, the village was flooded and submerges many houses, rice fields, and plantation land owned by residents. The phenomenon of annual flooding is a concern for the local community because it causes material losses. Research on the flood discharge in the Meureubo watershed, Pasi Mesjid Village, was carried out to determine the amount of flood discharge with various return periods. The flood discharge calculation methods are using Melchior and Snyder. The scope of this study includes frequency analysis, rainfall design, flood discharge design by the Melchior and Snyder methods with return periods of 2, 5, 10, 25, and 50 years. The hydrological analysis results showed that the flood discharge values of the Snyder and Melchior methods were not much different, namely in the 2-year return period of 1607.77 m^3/s. With the Melchior method, a 2-year return period obtained the plan flood discharge of 1813.286 m^3/s, a 5-year return period of 2791.62 m^3/s, a 10-year return period of 2938.14 m^3/s, a 25-year return period of 3153.19 m^3 / s, and a 50-year return period of 3336.5 m^3/s. With the Snyder method, the 5 (five) year return period is 2374.710 m^3/s, the 10 (ten) year return period is 2762.09 m^3/s, the 25 year return period is 3284.96 m^3/s, and for the 50 year return period, the value is 3699.18 m^3/s. Based on this analysis results, the Snyder and Melchior methods can be used in analyzing the flood discharge in the Meureubo watershed area.

1. Introduction

The overflow event of river water is caused by an increase in water volume in the river called a flood. Overflowing river water can flow inland to both residential and non-residential areas. A flood is a natural disaster that occurs when the land is inundated by excessive water flow. Floods often occur, especially when it rains. Based on information from the West Aceh Regional Disaster Management Agency (BPBD), floods occurred in nine sub-districts in West Aceh Regency. The nine sub-districts that were flooded were Sungai Mas, Woyla Induk, West Woyla, East Woyla, Pante Ceureu¬meen, Johan Pahlawan, Meureubo, Arongan Lambalek, and Kaway XVI. In these nine sub-districts, at least 46 villages hit by floods with an altitude between 30-100 cm. One of the youth leaders in Pasi Mesjid Village, Meureubo District, and Indra stated that the flood had entered people's homes since 04.00 AM. At least 70 houses of residents were flooded with an altitude of 30-50 cm [1].

The impact of the flooding caused hundreds of families to flee to safe locations because the water that had inundated their settlements continued to rise. Household goods, electronic equipment, plantations, rice fields, livestock were also flooded, and based on data from BPBD [2], there are no reports of casualties due to flooding from 2015 to 2016. One of the villages that are prone to flooding in the rainy season in Meureubo District is Pasi Mesjid Village. This village almost every year experiences flooding. If rain duration with a high capacity is only a few days, the water has spread to the residential areas with inundation levels reaching 50-100 cm. The number of flood events in 2016 in Pasi Mesjid Village was very high because, in the past year, there have been three flood inundations with an average height of 70 cm above ground level. The flood inundated many houses and rice fields, and plantation land belonging to residents in Pasi Mesjid Village, and become concerned in the local community about the annual flood phenomenon because it cause material losses. The floods have caused material losses for residents, such as household items damaged by the flood, rice fields that have failed to harvest, and dead livestock. The floods that have occurred in Pasi Mesjid Village showed in Figure 1.

Figure 1. Flood conditions in Pasi Mesjid village

The causes of flooding are the amount of surface flow / direct flow, saturated soil, the lack of water absorption in the soil, significant changes in land use, and the capacity of the river can no longer withstand the intensity of rainfall. Research on the flood discharge in the Meureubo watershed in Pasi Mesjid Village is necessary to determine the value of the flood discharge that passes through the village. With this information, flood risk countermeasures made either through structural or non-structural developments. The scope of this study includes frequency analysis, rainfall design, and flood discharge design by the Melchior and Snyder methods.

Several researchers have analyzed the flood discharge in the Meureubo watershed. Analyzed the flood discharge in the Meureubo watershed with HSS Nakayasu with a minimum return period of 2 years is 2436.24 m^3/s, and a maximum return period of 100 years is 3950.53 m^3/s. For the value of the Krueng Meureubo watershed using the Nakayasu Synthetic Unit Hydrograph (HSS) method, the time from rainfall (Tr) = 7.12 hours, the time from the start of the flood to the peak of the flood (Tp) = 15.21 hours, the time from the peak of the flood to 0.3 times the peak of the flood (T0.3) = 19.021 hours, the peak discharge (Qp) = 23.11 m^3/s [3]. Analyzed the flood discharge that occurred in the Krueng Meureubo River using the Gama method. The results obtained are the flood discharge in the Q2 year return period of 71.16 m^3/s, Q5 years of 817.57 m^3/s, Q10 years of 969.43 m^3/s, Q25 years of 1176.18 m^3/s, Q50 years amounting to 1341.57 m^3/s, and Q100 for the year 1516.93 m^3/s [4].

In this study, flood discharge analysis used a different method from previous researchers, namely the Snyder and Melchior methods. For the Snyder method, [5] have analyzed the flood discharge in Citarik Rancaekek Majalaya River, Bandung Regency. Data collection includes causes of flooding, analysis of flood discharge, and flood management planning. By using the HSS Snyder method, a planned discharge is obtained which is then input into the HEC RAS software. River normalization is an effort made to improve the cross-section of the river, namely by widening the river and deepening the river (dredging) so that the river capacity increases so that it can accommodate flood discharge. The design flood is the amount of discharge planned to pass through a river cross-section with a certain return period. The amount of flood discharge is determined based on rainfall and river flow, including the amount of rain intensity, and the area of the watershed. In this study, the method used to calculate the flood discharge design analysis is the Melchior and Der Weduwen methods [6].

1. Melchior Method

One way to calculate the flood discharge design for the watershed with an area > 100 km^2 and less than 1000 km^2 is by using the Melchior method [7].

$$Q_T = \alpha \times \beta \times q \times A \times \frac{R_T}{200}$$

$$\beta = \frac{180+0,75A}{150+A} \quad (1)$$

where :

Q_T = T year return period flood (m^3/s);
α = rainwater runoff coefficient (0.6);
β = regional reduction coefficient for watershed rainfall
q = maximum rainfall intensity (m^3/km^2/s)
A = watershed area (km^2);
R_T = rainfall design for the annual return period T (mm).

2. Snyder Method

The design flood discharge is the maximum planned discharge in a river or natural channel with a certain return period that can be flowed without endangering the surrounding environment and river stability. Design flood discharge analysis can be done by observing and measuring the river directly at the location of the river or by analyzing the maximum rainfall data at rainfall measurement stations in the river basin. One method in determining the design flood discharge is a synthetic unit hydrograph. The synthetic unit hydrograph used in this research is the Snyder synthetic unit hydrograph.

Four parameters (Equation 2), namely time of lag, peak flow, base time, and standard duration of effective rain for unit hydrographs, are associated with the watershed's physical geometry with the following relationship [8].

$$Tp = Ct\ (L\ Lc)\ 0.3$$
$$Qp = CP\ A\ /\ tp$$
$$T = 3 + (tP\ /\ 8)$$
$$TD = tP\ /\ 5.5 \quad (2)$$

If the effective rain duration, tr is not the same as the standard duration tD, then *TpR* and *QpR* are calculated (Equation 3).

$$TpR = tp + 0,25\ (tr - tD)$$
$$QpR = Qp\ tp\ /\ tpR \quad (3)$$

Where:

tD = standard duration of effective rain (hours);
tr = effective rain duration (hours)
tp = the grace period from the beginning of the rain to the peak of the flood (hours)

tpR = time from the center of gravity tr rain duration to the top of the unit hydrograph (hours)
T = hydrograph base time unit (days)
Qp = peak discharge for duration tD QpR: peak discharge for duration tr;
L = the length of the main river to the control point under review (km);
Lc = the length of the river is measured tothe point closest to the watershed (km);
A = watershed area (km^2);
Ct = coefficients that depend on the slope ofthe watershed, which vary from 1.4 to 1.7;
Cp = coefficients that depend on thecharacteristics of the watershed, which vary from 0.15 to 0.19.

By using the formulas above, the unit hydrograph can be drawn. Some formulas (Equation 4 and 5) are calculated to make it easier to describe.

$$W_{50} = \frac{0{,}23.\ A^{1,08}}{Q_{pR}^{\ 1,08}} \quad (4)$$

$$W_{75} = \frac{0{,}13.\ A^{1,08}}{Q_{pR}^{\ 1,08}} \quad (5)$$

W50 and W75 are hydrograph unit width at 50% and 75% of the peak discharge, expressed in hours. As a reference, the widths of W50 and W75 made with a ratio of 1: 2, with the short side to the left of the unit hydrograph, as in Figure 2 shown.

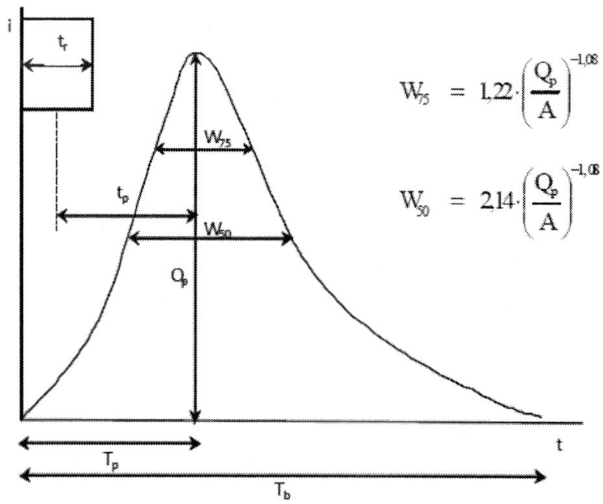

Figure 2. General form of HSS *Snyder*
Source: [8]

2. Research methods

The data used in this research is secondary data. The secondary data in this study is the rainfall data obtained from the Meteorology, Climatology, and Geophysics Agency (BMKG) Cut Nyak Dhien, West Aceh Regency.

The rainfall data used for the calculation is the maximum annual daily rainfall data with a record length of 12 years from 2005 to 2018. This data will be used to calculate the rainfall design and flood discharge design. Meanwhile, the Meureubo watershed map and topographic map were obtained from BP DAS Banda Aceh. The DAS map is used to obtain the watershed area and length of the river in calculating the design of flood discharge, while the topographic map serves as a location map to determine the elevation of the river area and the Pasi Mesjid Village. This research's location is a river that passes through the river in the village of Pasi Mesjid. Figure 3 below shows the study locations.

Figure 3. the study locations in Pasi Mesjid Village

Data analysis is the activity of processing input data to obtain output data. This activity includes activities to study the watershed characteristics, analysis of rainfall distribution, rainfall design, and design of flood discharge. Analysis of watershed characteristics includes area and land use in the Krueng Meureubo watershed. The watershed area and land use were analyzed using GIS software. Hydrological analysis is needed to determine the Krueng Meureubo watershed hydrological characteristics, especially in Pasi Mesjid Village, Meureubo District, West Aceh Regency. Hydrological analysis used to determine the magnitude of the flood discharge plan in a water construction plan. The data for determining the planned flood discharge in this study is rainfall data, where rainfall is one of several data that be used to estimate the amount of planned flood discharge.

2.1 Rainfall distribution analysis
A frequency distribution analysis carried out to find a distribution in accordance with the data available from the station. The selection of rain distribution calculated using statistical data parameters, namely by using the frequency analysis method Normal, Log Normal method, Log Pearson III, and Gumbel. Then the value obtained adjusted to the terms of each distribution.

2.2 Rainfall design analysis
The rainfall design is determined by analyzing the available rainfall data following the Normal, Log Normal, Gumbel distribution. Based on the type of rainfall distribution calculation, the rainfall design is calculated according to the selected distribution type. The rainfall design is then calculated to determine the rainfall design that occurs in the return period of 2 years, 5 years, 10 years, 25 years, 50 years and 100 years. The results of this calculation are used to calculate the flood discharge design.

2.3 Flood discharge design analysis
The flood discharge design in this study will be analyzed using the Melchior and Snyder methods. The method in this study was determined because the area of rain storage at the location for this study was above >1000 km^2. The planned flood discharge calculation is done to determine the design of flood discharge that occurs in the return period of 2 years, 5 years, 10 years, 25 years, 50 years and 100 years in Krueng Meureubo River in Pasi Mesjid Village.

3. Result and discussion

The results of the calculation and discussion include frequency analysis, rainfall design, and flood discharge design.

3.1 Maximum rainfall

The data used for the calculation of the planned rainfall is the maximum monthly rainfall data. Table 1 shows the maximum monthly rainfall data.

Table 1. Monthly maximum rainfall data.

No	Year	Month											
		Jan	Feb	Mar	Apr	May	Jun	Jul	Aug	Sep	Oct	Nov	Dec
1	2005	43	73	106	83	68	37	48	46	85	53	72	56
2	2006	50	70	52	42	19	54	66	65	88	107	60	31
3	2007	39	41	31	37	50	37	60	101	35	135	42	94
4	2008	94	100	31	95	94	40	96	90	96	75	77	69
5	2009	25	63	96	88	100	66	45	59	95	45	75	26
6	2010	91	56	69	96	72	50	30	40.5	100.5	58	65	80
7	2011	71,5	50	90.5	105	55	40	90	90	50.5	90.5	90	90
8	2012	90	100	100	74.5	45.5	75	93	48	51	106.5	76	90
9	2013	78.5	80.5	60.5	72.5	70.5	45.5	15.5	65.5	25	15	85.5	80.5
10	2014	84	96	125	65	51	82	58	146	71	98	96	59
11	2015	49.1	47.3	41	119.9	53.3	115.8	53.3	59.7	76.7	99.1	172.7	90.8
12	2016	187.9	86.6	99.9	106	193.8	61.6	54.3	187.4	38	115.4	93.6	59.6
13	2017	127.8	63.1	77.2	110	44.7	29.5	41.5	58.3	50.5	203.99	75.4	186.8
14	2018	41.3	51.5	87.1	89.8	21.3	12.9	95.8	179.7	71.7	72.7	137.9	50.9

3.2 Frequency analysis

After obtaining the maximum monthly rainfall value, the calculation of the frequency analysis is carried out, as Table 2 summarizes. The value of this frequency analysis can be seen which distribution is suitable for use in this study.

Table 2. Frequency analysis with various types of distribution

Distribution Type	Requirement	Calculation Result	conclusion
Normal	$Cs \approx 0$ $Ck = 3$	$Cs = 1.29$ $Ck = 3.80$	Not eligible
Log Normal	$Cs \approx 3\ Cv + Cv3$	$Cs = 1.14$ $Ck = 1.14$	Not eligible
Gumbel	$Cs = 1.1396$ $Ck = 5.4002$	$Cs = 1.29$ $Ck = 3.80$	Not eligible
Log Pearson III	Apart from the above values	$Cs = 1.087$	Eligible

From these results, the distribution that meets the requirements is the Log Person type III distribution with the condition $Cs \neq 0$, and the results obtained are $Cs = 1.087$.

3.3 Chi Square test analysis

The four distributions results indicate that the distribution meets the requirements of the Pearson III Log Distribution distribution. After obtaining the distribution to be used, the next step is to calculate the value for Chi-Square. The chi-square test calculation results for the Log Pearson III distribution can be seen in Table 3.

Table 3. Calculation of the value of X^2 for the Log Pearson III distribution

Class	Interval	Ef	Of	Of - Ef	$(Of - Ef)^2 / Ef$
1	>135.24	2.8	4	1.2	0.514
2	132.11-135.24	2.8	3	0.2	0.0142
3	129.89-132.11	2.8	3	0.2	0.0142
4	127.76-129.89	2.8	1	-1.8	1.157
5	<127.76	2.8	3	0.2	0.0142
	Σ	14	14		1.714

The calculation results obtained the degree of confidence (DK) = 1.2 $\alpha = 5\%$. The critical value for the Chi-Square distribution $\alpha = 5\% = 5.991$, so that the chi square analysis is fulfilled.

3.4 *Flood discharge analysis design*

The calculation of the flood discharge design on K = Krueng Meureubo River in Pasi Mesjid Village calculated by Melchior method. This method selected because the areas of the watershed in the planning study location was 490.13 km^2 and to calculate the flood discharge design with a watershed area of more than 100 km^2 and less than 1000 km^2 was the Melchior method. The planned flood discharge for the return period T year is calculated by entering the rainwater runoff's coefficient, the reduction coefficient of the area for watershed rainfall, the maximum rainfall intensity, the area of the watershed, and the rainfall design for the T year return period. It is summarized by Table 4.

Table 4. The calculation result of the flood discharge design of the Melchior Method

Parameter	Return Period (Year)				
	2	5	10	25	50
Tc	11.9	11.9	11.9	11.9	11.9
A	0.6	0.6	0.6	0.6	0.6
Q	4.27	4.27	4.27	4.27	4.27
B	0.855	0.855	0.855	0.855	0.855
A	490.12	490.12	490.12	490.12	490.12
Rt	113.13	147.36	174.63	214.66	248.8
Qt	1607.77	2791.62	2938.14	3153.19	3336.5

The flood discharge design calculated using the Snyder method with a return period of Q2, Q5, Q10, Q25, and Q50. The planned flood discharge is the amount of discharge that is considered capable of meeting the extreme conditions in an area and is used to calculate the amount of flooding in an area. The results of the calculation of the planned flood discharge can be seen in the following table.

Table 5. The calculation results of the flood discharge design with the Snyder Method

No	Return Period (year)	Flood Discharge (m^3/s)
1	2	1813.29
2	5	2374.71
3	10	2762.09
4	25	3284.97
5	50	3699.18

The obtained value in the table above shows the magnitude of the flood value for each return period (T). The 2-year return period obtained a planned flood discharge of 1813.286 m^3/s, a 5 (five) year return period of 2374.71 m^3/s, a 10 (ten) year return period of 2762.09 m^3/s, a 25 year return period of 3284.97 m^3/s and for the 50 year return period, the value is 3699.18 m^3/s. The maximum discharge of the unit hydrograph obtained 4.38 m^3/s. The graphical depiction of the basic time of the Snyder hydrograph showed in Figure 4.

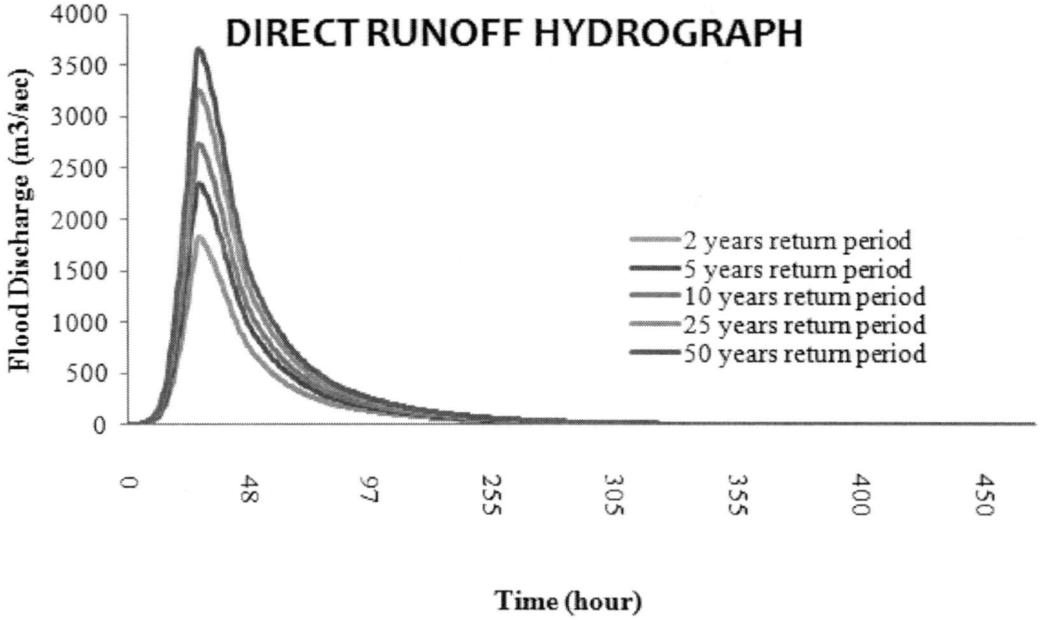

Figure 4. Hydrograph Snyder on flood discharge in the Meureubo watershed

4. Conclusion

The hydrological analysis results showed that the flood discharge values of the Snyder and Melchior methods were not much different, namely in the 2-year return period of 1607.77 m^3/s. With the Melchior method, the 2 years return period obtained a planned flood discharge of 1813.29 m^3/s, a 5 years return period of 2791.62 m^3/s, a 10 years return period of 2938.14 m^3/s, 25 years return period of 3153.19 m^3/s, and a 50-year return period of 3336.5 m^3/s. With the Snyder method, the 5 years return period is 2374.71 m^3/s, the 10 years return period is 2762.09 m^3/s, the 25 years return period is 3284.97 m^3/s, and for the 50 years return period, the value is 3699.18 m^3/s. Based on this analysis results, the

Snyder and Melchior methods can be used in analyzing the flood discharge in the Meureubo watershed area.

References

[1] Harian Analisa 2016 *Aceh Barat Banjir, Ratusan KK Mengungsi.*
[2] Badan Penanggulangan Bencana Daaerah (BPBD) 2016 *Jumlah Korban Jiwa Banjir di Kabupaten Aceh Barat.* Aceh Barat: Badan Penanggulangan Bencana Daerah (BPBD) Kabupaten Aceh Barat.
[3] Arifianto Y D 2017 *Jurnal Infrastruktur,* **3** 112.
[4] Zikra A, Masimin and Fatimah E 2018 *Jurnal Arsip Rekayasa Sipil dan Perencanaan,* **1** 172-182.
[5] Ramadhan I R and Muspitasari M 2017 *Jurnal Teknik Lingkungan,* **18** 43-53.
[6] Harto S 2000 *Hidrologi, Teori Masalah dan Penyelesaian* (Yogyakarta: Nafiri Offset).
[7] Noor M A and Utomo B 2013 *Jurnal Info Teknik,* **14** 81-91.
[8] Triatmodjo B 2016 *Hidraulika II* (Yogyakarta: Beta Offset).

Pavement condition assessment using SDI and PCI method on Geumpang road – West Aceh boundary

M Melyar[1]*, M Isya[2], S M Saleh[2]
[1]Master Student, Department of Civil Engineering, Universitas Syiah Kuala, Banda Aceh, Indonesia
[2]Department of Civil Engineering, Universitas Syiah Kuala, Banda Aceh, Indonesia.

*E-mail: monamelyar0@gmail.com

Abstract. Construction failure can occur in every project activity. This has been included in Law No.19 of 1990 concerning construction services which require that building failures be included in the contract in order to maintain the principle of justice. The factors causing this construction failure basically can occur due to various reasons depending on the aspects seen. It is necessary to pay attention to this so that there is no deterioration in road quality due to damage to the road surface so that it can affect safety, comfort, and smoothness of traffic. The purpose of this study is to evaluate the pavement functionally and compare the value of pavement damage conditions based on two methods, namely the Surface Distress Index (SDI) and Pavement Condition Index (PCI) methods which are used as the basis for determining the type of road maintenance. This research took place on the road Geumpang Road - West Aceh boundary. The results of the calculation obtained the type and level of damage to the road surface on the Guempang road - the boundary of West Aceh which became the object of research, there were several types of damage such as holes, 8.70%, cracks 4.10%, subsidence 3.16%, patches 0.83%, edge damage 3.33%, and 0.51% peeling off. Overall, the average IRI value obtained from the PUPR Office in 2019 is 3.45% in very good conditions, for the average SDI value is 15.02% in good conditions, while the average PCI value is 79,63% were in very good condition. The type of maintenance on the Geumpang Road - West Aceh boundary, Pidie Districts as a whole based on the Surface Distress Index (SDI) method and the Pavement Condition Index (PCI) method is routine maintenance. The main factors causing road damage are standing water, running water, and landslides.

1. Introduction

Road is a land transportation infrastructure covering all parts of the road, including complementary buildings and equipment intended for traffic, which are on the ground level, above the ground surface, below the ground or water surface, as well as above the water surface, except railroad roads, lorries and cable roads [1]. It is not uncommon for road damage to occur before the end of the road life plan and in general, the damage that occurs is not caused by just one factor, but is a combination of interrelated causal factors. Therefore, negligence of a minor damage would inflict much greater damage. If the road is still in good service, then it should be evaluated periodically to find trends that will affect its condition in the future. Such a program provides continuous inventory data, so that problems can be detected and adequate corrective action can be taken effectively and efficiently [2-3].

This research is reviewed by conducting surveys directly into the field and classifying road pavement damage based on the type and degree of damage. This study used two methods, namely Surface Distress Index (SDI) and Pavement Condition Index (PCI).
In the context of the transportation system, roads are alternative infrastructure used by the community as a means of moving traffic to carry out activities or moving from one area to another. According to its function, the road network system in Indonesia can be distinguished into a primary road network system and a secondary road network system [4].

The objectives of this study are, Knowing the types and levels of damage and the comparison of road condition on the road segment Geumpang Road - West Aceh boundary, Pidie Regency using two methods, namely the Surface Distress Index (SDI) and the Pavement Condition Index (PCI). Knowing the type of maintenance of maintenance on Geumpang Road - West Aceh boundary district of Pidie by using methods Surface Distress Index (SDI) and Pavement Condition Index (PCI). Knowing the main factors causing road pavement damage to the Geumpang Road - West Aceh boundary road section.

2. Research method

The area of research is the province with the study of the case: Geumpang Road - West Aceh boundary (P.024.11) which is a segment of road Simpang Bangkeh - Border Pidie /Aceh Barat along 16.38 km. The location of research can be seen on Figure 1.

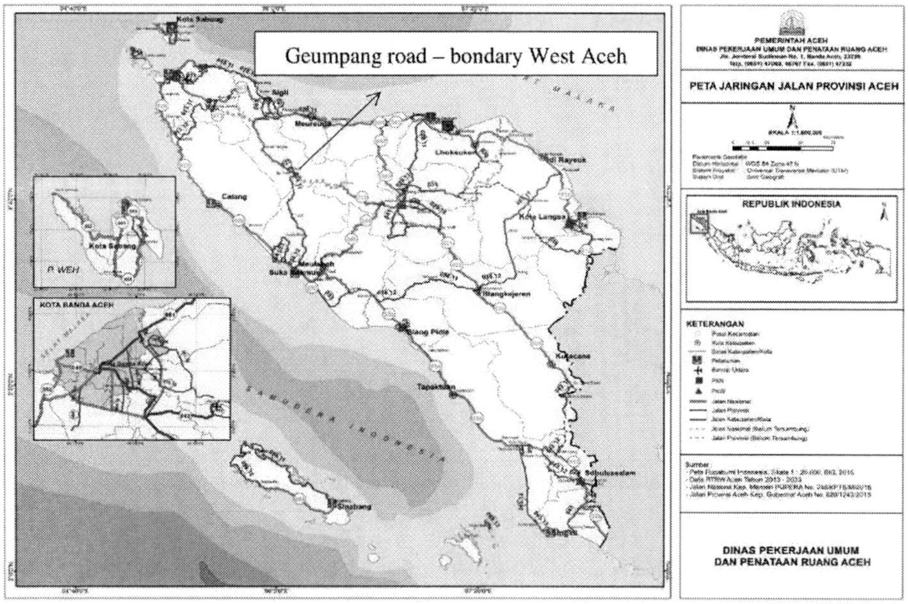

Figure 1. Map of the study area.

Road conditions are obtained from the results of the Road Condition Survey (RCS) on the road sections under review. In this research, the method used is the Surface Distress Index (SDI) method and the Pavement Condition Index (PCI) method to obtain the road conditions to be handled to obtain the type of handling to be performed. Determining the sample unit with a road length of 16.38 km and a survey area length per segment of 100 m, then N = 164 segments.

The steps for determining the value of road conditions and the type of handling using the Surface Distress Index (SDI) method can be described as follows:

a. The value of damage on the Road Condition Survey form that has been obtained is then calculated using the Surface Distress Index method with a road length of 16.38 km. Road

IOP Conf. Series: Materials Science and Engineering 1087 (2021) 012041 doi:10.1088/1757-899X/1087/1/012041

Condition Survey by manual method is to record the condition of easy-to-change road sections for asphalt roads and the purpose of developing plans and programs for road network development (Figure 2).

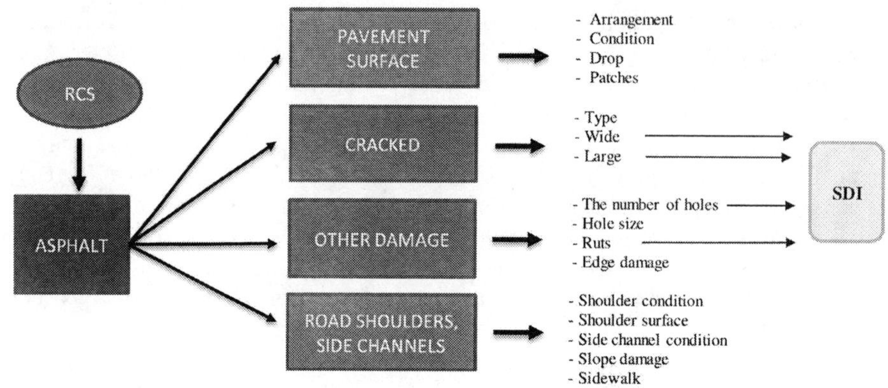

Figure 2. Overview of the asphalt road surface.

b. The SDI value for the assessment of each segment of the observation then the entire length of the road is accumulated and then the average value is taken as the SDI value for the observed road segment. Calculation of Surface Distress Index (SDI) to assess the condition of the road can be seen in Figure 3.

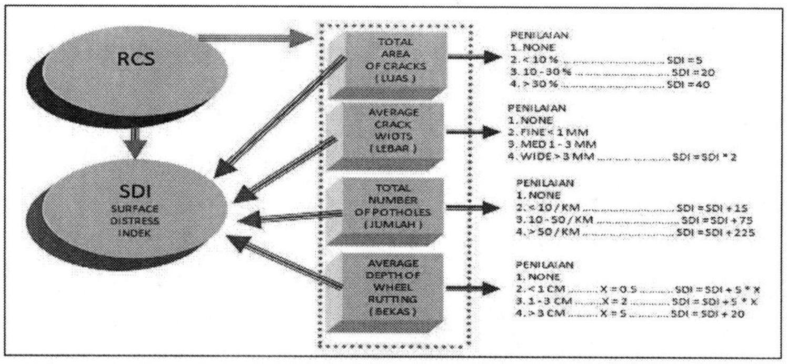

Figure 3. Calculation of SDI for Asphalt Roads Results from RCS.

In assessing and determining segment of the road is obtained from the results of assessment of each type of damage pavements with seeing the value of Surface Distress Index (SDI) based on the condition of the road as in Table 1.

Table 1. Surface Distress Index (SDI) technique

Road Condition	SDI
Well	<50
Moderate	50 - 100
Light Damage	100 - 150
Heavy Damage	> 150

c. The International Roughness Index (IRI) value is secondary data obtained from the Aceh PUPR Office, where the data is the value of each observation segment. The IRI value obtained is then accumulated as a whole. The accumulated IRI values are then averaged to obtain an average IRI value for roads, as shown in Table 2.

Table 2. Surface type and International Roughness Index (IRI)

No.	IRI	Surface Type	Information
1	< 4	Asphalt	Very Good
2	4 - 8	Asphalt	Good - Fair
3	8 - 12	Asphalt	Fair - Poor
4	12 - 16	Asphalt	Poor - Bad
5	16 - 20	Asphalt	Bad
6	≥ 20	Asphalt	Very Bad
7	Any	Unsealed	Unsealed

d. Based on the value of the Surface Distress Index (SDI) and the value of the International Roughness Index (IRI), then the road conditions and types of handling are determined, as Table 3 shows,

Table 3. Determination of road conditions

IRI (m/Km)	SDI			
	< 50	50 - 100	100 - 150	> 150
< 4	Good	Moderate	Moderate	Light Damage
4 - 8	Moderate	Moderate	Light Damage	Light Damage
8 - 12	Light Damage	Light Damage	Heavy Damage	Heavy Damage
> 12	Heavy Damage	Heavy Damage	Heavy Damage	Heavy Damage

The activity stages of determining the value of conditions and types of handling using the Pavement Condition Index (PCI) method can be described as follows:

a. Density
Density value is a type of damage that is differentiated based on the level of damage. Density can be calculated using the following equation.

$$Density = \frac{Ad}{As} \, x \; 100\% \quad (1)$$

b. Adeduct Value (DV)
To determine the deduct value of each sample unit, enter the density percentage on the graph of each type of damage then draw a vertical line to cut the severity of the damage (low, medium, high). Then a horizontal line is drawn and the deduct value (DV) will be obtained.

c. Total Deduct Value (TDV)
The total reduction value is the total amount of the reduction value in each sample unit or the total value of the individual deduct value for each type of damage and the level of damage that exists in a segment unit.

d. Corrected Deduct Value (CDV)
Corrected deduct value is obtained from the curve of the relationship between the TDV value and the CDV value by selecting the curve according to the number of individual deduct values that have a value greater than 2. The graph shows the relationship between the corrected deduct value and the total deduct value.

e. The PCI value for each sample unit is calculated using the following formula:

$$PCI\ (s) = 100 - CDV \quad (2)$$

$$PCI = \frac{\sum PCI(s)}{N} \quad (3)$$

PCI is an index numerical whose value ranges between 0-100 as in the table following this:

Table 4. PCI (Pavement Condition Index) value

PCI Value	Condition
0 – 10	Failed
11-25	Very Bad
25 - 40	Bad
41 - 55	Medium
56 - 70	Good
71 - 85	Very Good
86 - 100	Perfect (Excellent)

3. Results and discussion

After processing the data that has been done, then proceed with the analysis of the results. The results of the survey are in the form of the data type of damage to the road, the level of damage to roads was analyzed by two methods, namely methods Surface Distress Index (SDI) and Pavement Condition Index (PCI) to be able to determine the type of maintenance.

Geumpang Road - West Aceh boundary (P.024.11) has a length of 16.38 km, with starting Point to be at the point of coordinates 04°50'.07" South Latitude - 96°07'.48" East Longitude and the last Point are at coordinates 04°44'.31" South Latitude - 96°07'.05" East Longitude. Geumpang Road - West Aceh boundary is the Primary Collector (K1) road.

3.1 Road damage using the Surface Distress Index (SDI) method

Damage to the surface of the road is obtained from the results of a survey of visual or observation directly in the field with a distance interval per 100 m length of road that is done on the date of 13 - 15 January 2020. The survey was conducted by dividing the Geumpang Road - West Aceh boundary where the research was conducted into 8 segments.

From the level of damage to the road surface in I - VIII Geumpang road section - the upper border of West Aceh then the percentage of road surface damage is obtained which can be seen in Figure 4.

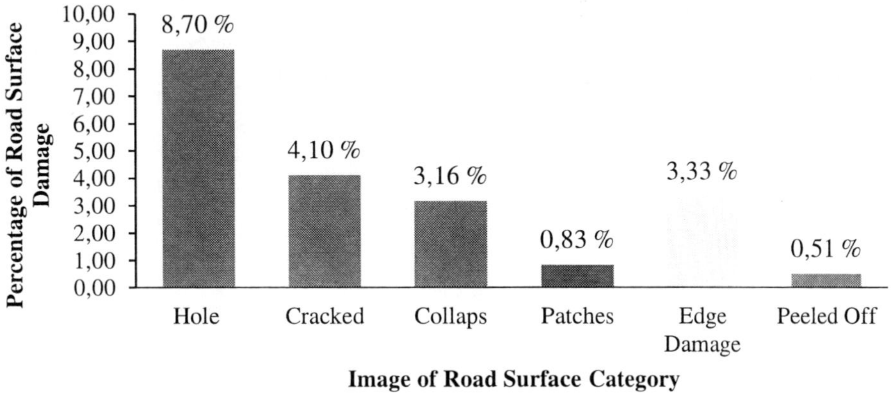

Figure 4. Total percentage of damage to road surfaces on Geumpang Road - West Aceh boundary, using the Surface Distress Index (SDI) method

3.2 Damage to the road using the Pavement Condition Index (PCI) method

The following is an example of a survey sheet for the Geumpang road - West Aceh Boundary at Point 0+000 to 0+100:

- Density
  The density calculation for Point 0+000 to 0+100 can be seen as follows:

Edge cracking $L = \frac{1,25}{100} \text{ x } 100\ \% = 1,25\ \%$

Longitudinal crack $L = \frac{8,50}{100} \text{ x } 100\ \% = 8,50\ \%$

Raveling $L = \frac{2,80}{100} \text{ x } 100\ \% = 2,80\ \%$

- Calculating the Reduction Value, as Figure 5 indicates.

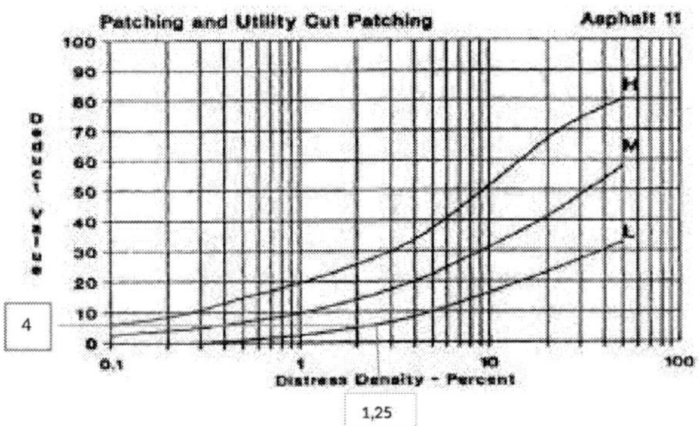

Figure 5. Graph of deduct values for crack edge damage

- Calculating the Total Deduct Value
  All the deduct values that have been obtained are then added up to obtain the total deduct value or total deduct value. The data is then presented in Table 5.

Table 5. The following table is the Deduct Value for Point 0 + 000 to 0 + 100

Type of Damage	Density (%)	Severity Level	Deduct Value
Edge Cracking	1.25	L	4
Longitudinal crack	8.5	L	17
Raveling	1.86	L	3
Total Deduct Value (TDV)			**24**

- Corrected Deduct Value (CDV)
  From the data obtained the number q = 3 and subsequently plotted to the graph CDV as in Figure 6 of the graph of the obtained value of CDV (TDV = 24) is 12.

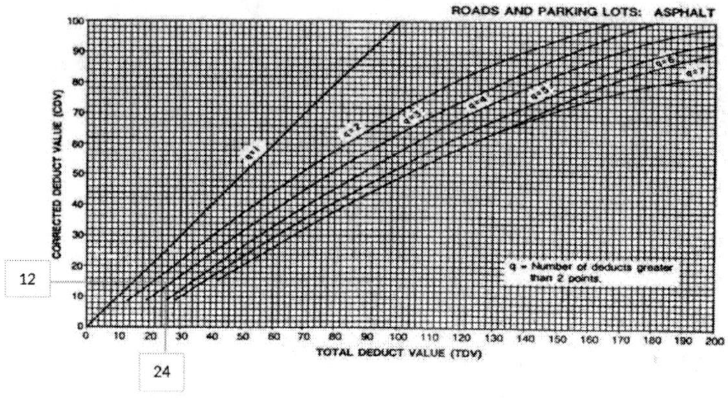

Figure 6. Graph of the corrected deduct value (CDV)

- Pavement Condition Index (PCI)
  The PCI values are obtained using the following equation:

$$PCI = 100 - CDV$$
$$= 100 - 12$$
$$= 88$$

The pavement condition level for Point 0+000 to 0+100 on the Geumpang road with the PCI value = 88 is Excellent. The calculations for the other Sta are done in the same way as for Point 0+000 to 0+100.

3.3 Main factors causing road pavement damage

From the data obtained from the PUPR Office for Fiscal Year 2019 it is known that several factors causing road damage at the Geumpang Road - West Aceh boundary were caused by inundated springs, so that they overflowed into the road body and filled the road. While when it rains there is a puddle that caused by running water because the land is covered by roads. At present, there are several untreated locations where inundation occurs.

Landslides that occur on the path of Geumpang Road - West Aceh boundary which are hilly and mountain topography are generally triggered by cutting road cliffs with slope angles ranging from 35° to > 60°, likewise, this forest area has quite high rainfall. During the rainy season, the weathering of slope constituents containing relatively open clay fractions will absorb (absorb) rainwater that falls on the soil surface, eventually decreasing the soil adhesiveness resulting in landslides.

IOP Conf. Series: Materials Science and Engineering 1087 (2021) 012041 doi:10.1088/1757-899X/1087/1/012041

4. Conclusion

From the results of data analysis, several conclusions can be drawn, including: Based on the results obtained, the type and level of damage to the road surface on the Geumpang road - the boundary of West Aceh which was the object of the study, there were several types of damage such as holes, 8.70%, cracks 4.10%, subsidence 3.16%, patches 0.83%, edge damage 3.33%, and 0.51% peeling off. The level of road surface damage for the total of several types of damage is 20.64% and roads that are not damaged are 79.36% of the total length of the road, namely 16.38 km.

In overall value of IRI average obtained in the segment Geumpang Road - West Aceh boundary border is at 3,45% with the condition is very good, for the value of SDI average was at 15.02% with the conditions well, while the value of PCI average is amounting to 79.63% with the conditions very well. Based on the comparison of the value of IRI, SDI, and PCI level of damage to roads on average in each Sta already reached the level is very good, but they are also the level of state road that reach was the kind of maintenance periodically while on the level of state roads that reach the level of bad needs to do improvement reconstruction. The type of maintenance on the Geumpang Road - West Aceh boundary, Pidie District road as a whole is based on the Surface Distress Index (SDI) method and the Pavement Condition Index (PCI) method is routine maintenance. The main factors causing road damage are standing water, running water, and landslides.

References

[1] Bina Marga 2006 *Government Regulation No. 34 of 2006 concerning Roads* (State Secretariat of the Republic of Indonesia, Jakarta).

[2] Suswandi A, Sartono W and Hardiatmo H C 2008 *Forum Teknik Sipil*, **XVIII** 934-946.

[3] Hardiatmo H C 2007 *Highway Maintenance* First Edition (Gadjah Mada University Press, Yogyakarta).

[4] Sukirman S 1999 *Road Bending Powers* (Bandung: Penerbit Nova).

Hydraulic analysis and ecological and physical study of hydraulic Alas-Singkil river Southeast Aceh, Indonesia

Ziana*, Azmeri, A Yulianur, E Meilianda

Civil Engineering Department, Faculty of Engineering, Universitas Syiah Kuala, Jl. Tgk. Syeh Abdul Rauf No. 7, Darussalam – Banda Aceh 23111, Indonesia

*E-mail: ziana@unsyiah.ac.id

Abstract. The problem of flooding that occurs around the Lawe Alas-Singkil, Aceh Province is the conservation function of the area from upstream to downstream needs serious attention. The area around the watershed is degraded due to several factors. This factor is closely related to illegal logging, exploitation of nature with mineral mining to the conversion of area functions. This study aims to identify the existing conditions of the riverbanks, how the relationship between community social activities and the banks of the Alas-Singkil River, and to study the physical conditions of the river from upstream to downstream. The study method in this research includes a literature study, hydrological conditions of the Alas-Singkil Watershed, field surveys, identification of existing conditions, and study of the ecology and physicality of the river. Survey the condition of flora and fauna in the study area and the economic and social activities of the community. The results of the study on the hydrological conditions of the Alas-Singkil watershed with an area of 10,090.13 km2 and a river length of 316.26 km, in the upstream form of mountains and natural forests where deforestation has occurred due to forest conversion to cocoa plantations, planting of corn on the hills and so on, The maximum annual daily rainfall is high (151-300 mm), so the potential for flood disaster is even greater. In the rainy season, large amounts of water cause mountains or hills in Southeast Aceh to melt and become a flash flood disaster. Many of the rivers in the middle and downstream have turned into plantations and rice fields and some are still swamps. The condition of the river in the Alas-Singkil watershed generally occurs silting or sedimentation in the river body, the existing condition of the Alas-Singkil riverbank, where there are still many settlements on the riverbanks. The downstream Krueng Singkil is experiencing more frequent flooding, either due to the overflow of Krueng Singkil in the upstream part and the blocked river flow and rising tides. The frequency of inundation and flood disasters in Singkil City reaches 3 to 4 times each year with an inundation height of 0.5-3.0 meters. It is necessary to repair the riverbanks by using vegetation components found on the banks of the Alas-Singkil River and overcoming floods in the upstream and downstream rivers to protect the ecosystem and people living on the riverbanks. The solution that can be done is to restore the conservation function of the area around the Lawe Alas watershed. and built channel diversion flood from the Lae Soraya river in the village of Gelombang directly to the sea.

1. Introduction

The Alas-Singkil River has an important role for the people who live along the riverbanks. Residents who live on the banks of the river consider the river to be able to meet their needs and sustain their lives, both from the aspects of transportation and mobility, economic, social, cultural and political. From the concentration of the population and their interaction with the river, the river culture was born. River culture is characterized by the presence of riverside settlements, floating markets, and jukung, as well as social interactions that occur in them. The people who come from the majority of traders make this culture egalitarian, cosmopolitan and open.

Based on Presidential Decree No. 12/2012 concerning the Designation of River Basins, Aceh has nine major river basins. One of them is the Alas-Singkil River area which covers an area of 10,090.13 km^2. It is along the riverbanks that the people of Singkil (Aceh Singkil-Subulussalam city since they started civilization, but now almost 80% of the local population has moved to areas some distance from the watershed due to natural disasters, especially to avoid flash floods and frequent overflow floods. The longest river in Aceh is also a historical witness about billions of cubic of good quality wood from the Singkil forest being assembled, pulled by tugboats through the estuary to the sea and finally shipped to various countries.

According to The National Disaster Management Agency the problem of flooding that occurs around the Lawe Alas watershed (Catchment area), Singkil, Aceh Province where the conservation function of the area from upstream to downstream needs serious attention. The area around the watershed is degraded due to several factors. This factor is closely related to illegal logging, exploitation of nature with mineral mining to the conversion of area functions. The solution that can be done is to restore the conservation function of the area around the Lawe Alas watershed.

Gayo Lues to Aceh Tenggara changes in vegetation, where the area is dominated by agriculture and plantations, and the Alas River undergoes physical changes, sedimentation occurs, so if the potential for a flood disaster can occur because the area's absorption capacity is reduced when the rain intensity reaches its highest point. Human activities (atithropogenic activities) in the Alas-Singkil River are very important factors in morphological, ecological, and hydraulic changes. These changes will disrupt the balance of the river in question and lead to unpredictable destabilization of the river. In the development of river functions, the concept of pure hydraulic development cannot be used because it does not consider ecological aspects and impacts that will occur after construction.

Efforts to restore the conservation function may experience resistance from residents who are comfortable with farming and gardening. Solutions with social impacts also need to be considered together. Restoring the conservation function needs to be accompanied by efforts of economic value; on the other hand, so that the ecological function is not neglected.

Vegetation has a very vital hydraulic function, namely as a component of flood resistance and erosion resistance of river beds and cliffs. Ecological and hydraulic factors support each other. During a flood, the vegetation along the river will function as a retention factor that will inhibit the speed of river flow to the downstream area. Because the speed of the water is slowed down, the water level will rise and inundate the riverbank areas where vegetation grows. This dynamic inundation of the riverbanks with natural duration is very much needed by the river flora and fauna for their survival.

The flood is affected by geomorphometric of watershed [2]. The relationship between hydrometeorology and social science is considered important to improve our ability to cope with flash flood [9]. In the context of risk management, the perception of flood risk from the local population is the first step towards achieving community and community resilience, so it is necessary to take watershed management measures as an effort to control flash floods [4].

Structural river management or river normalization aims to drain water as quickly as possible downstream, build a sheet pile, restore river width, increase river capacity by dredging sedimentation, protect the area around the river from flooding and optimal water utilization. The construction of physical structures such as embankments or cliff concreting and diversion channels and pumping water into the sea is expected to protect rivers from flooding, this is usually done downstream of the river. River management and flood prevention in a non-structural manner (naturalization) are usually carried out upstream of the river. The naturalization process is restoring the river to its natural condition and restoring the degraded river so that it will increase the flow capacity, add large rocks as natural reservoirs for the river and the ecological restoration process with trees, shrubs and flowers. This will restore the river's biodiversity, and create new open spaces along the river banks which are greened, but the constraints require more time and more open land.

The purpose of this study was to analyze the hydrological and ecological conditions, how the relationship between community social activities and the Alas-Singkil riverbank, as well as to examine the hydraulics of the river from upstream to downstream. The location can be seen in Figure 1.

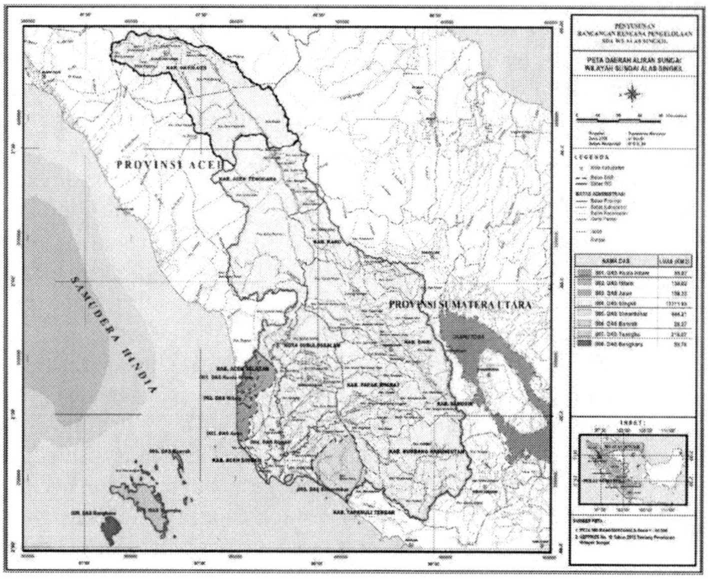

Figure 1. Map of the Lawe Alas-Singkil watershed.

Eco-hydraulics have existed from the results of the integration of hydraulics and ecology in the management of aquatic ecosystem assessment [10]. Much history has demonstrated a hydraulic integration model with biological functional response to assess the ecology of regime flows in river systems [5]. According to [3] provide the best analysis of methods and concepts in ecohydraulic applications that form the basis of river restoration. According [6] explains very well how previous research on eco-hydraulics, the state of the art in the development of methodology, is very clearly illustrated from case studies.

According to [11] stated that the vegetation on the riverbanks affects the deposition process and prevents erosion. According [7], flood control is generally carried out with a purely hydraulic concept, namely increasing physical or structural development, for example making embankments, making river diversions, dredging riverbeds, making the slope of river cliff, and others. Realize that this is an incorrect step in the application of the eco-hydraulic concept. Flood control by physical development causes a decrease in river retention, an enlarging river slope, resulting in increased flow discharge, peak flow and decreased time to peak discharge which results in increased flood retention in the downstream area of the river.

According to [12], the bad result of structural development is that the watershed will lose the flora and fauna habitat around the river environment and change the characteristics of the river shape. This is included in pure hydraulics. With the development rate of environmental awareness and the discovery of the enormous negative impact of pure hydraulic engineering as mentioned above, a new interdisciplinary engineering pattern was developed by incorporating ecological considerations in every water problem solving. This interdisciplinary ecohydraulics approach is seen as a new approach pattern that can be accepted by both ecologists and hydro civil engineering people and has high sustainability. Figure 2 presents an illustration of pure hydraulic engineering of a river. The method of calculating the influence of vegetation in the riverbank area on water velocity and discharge is based on the Merten equation [8].

1. Coefficient of resistance

$$\frac{1}{\sqrt{\lambda}} = -2{,}03\log\left(12{,}27\frac{R}{K_s}\right) \tag{1}$$

2. Roughness Value

$$Ks = Cb + 1{,}5dp \tag{2}$$

with :
C = vegetation composition coefficient
b = riverbank width (m)
dp = vegetation diameter (m)

3. Vegetation composition coefficient

$$C = 1{,}2 - 0{,}3(B/1000) + 0{,}06(B/1000)^{1{,}5} \tag{3}$$

$$B = \left[\frac{\alpha_x}{d_p} - 1\right]^2 \frac{\alpha_y}{d_p} \tag{4}$$

with :
B = vegetation parameter
αx = transverse vegetation distance (m)
αy = distance between vegetation in longitudinal direction (m)

4. Water velocity according to Darcy - Weisbach

$$V = (\frac{1}{\lambda}8gRI)^{0{,}5} \tag{5}$$

with:
λ = resistance coefficient
g = gravity speed (m/s^2)
R = hydraulic radius (m)
I = river slope

Calculation of the influence of the width of the riverbank on the water discharge for all three types of vegetation. The water discharge is calculated by the equation:

$$Q = (A1 \times V1) + (A2 \times V2) + (A3 \times V3) \tag{6}$$

where:
A1, A2, A3 = cross-sectional area (m^2)
V1, V2, V3 = velocity of water (m/s)
Q = discharge (m^3/s)

Q1 and Q3 are the capacity of the riverbanks to accommodate water overflow from the riverbed, while Q2 shows the capacity of the river.

2. Data analysis

The first location of the study was in the area of Badar District, Kutacane City, Aceh Province, with the main research location being the upstream area of the Lawe Alas River. The research was conducted from September 2018 to November 2018. In the study, the method of calculating flood discharge used the Melchior method because it has a watershed area of more than 100 km^2.

The data required are river hydrometry data, daily rainfall data and topographic maps. River hydraulic data is obtained from measurements for 7 days in the field such as measuring river depth, river width. Rainfall data are analyzed using statistical parameters which can then be determined the appropriate distribution type between Normal, Normal Log, Gumbel and Pearson Log distributions. Then the distribution test was carried out with the Chi Square test. After that, the flood discharge was calculated using the Melchior Method.

The purpose of the eco-hidraulics analysis is to determine the eco-hidraulics plan, which is a riverbank design. The stages to be carried out in the eco-hidraulics analysis in this study are as follows:

1. To plan the width of the bank, from the measurement it is obtained that the left bank width is 50 m and the right bank is 20 m, so that in this study the width of the bank is 20 m;
2. Planning the diameter of the vegetation, which consists of 10 cm, 20 cm, 30 cm, 40 cm, and 50 cm;
3. Designing the distance between vegetation along the transverse and longitudinal direction of 1 m;2 m; 3 m; 4m; and 5 m.
4. Calculating the channel roughness using equation (2) based on various vegetation diameters;
5. Calculating the value of water velocity for various vegetation diameters using equation (5);
6. Calculating the flood discharge after the effect of vegetation planting; and
7. Comparing flood discharge before and after vegetation planting.

3. Result and discussion

3.1 Lawe Alas river

The research results obtained data from the Lawe Alas River. Watershed area = 312.05 km^2, river length = 89.7 km and river slope = 0.0245. The table of discharge and velocity measurements can be seen in Table 1.

Table 1. Results of measurement of velocity and discharge for a week.

Tanggal	Velocity (V)	Discharge (Q)
	(m/s)	(m^3/s)
09-Nov 2018	0.76	514.36
10-Nov 2018	0.67	441.55
11-Nov 2018	0.61	388.13
12-Nov 2018	0.56	346.02
13-Nov 2018	0.50	297.70
14-Nov 2018	0.46	260.48
15-Nov 2018	0.43	236.75

Based on the Rating curve shows the maximum discharge that can be accommodated by the Lawe Alas river is 514.360 m^3/s. Table 2 calculation the roughness parameters, drag coefficient, velocity and discharge for various vegetation diameters in the Lawe Alas river. Then, a graph of the relationship between the Diameter of the Vegetation and the Roughness and the Coefficient of inhibition can be seen in Figure 2. Based on Figure 2, it can be seen that the larger the diameter of the vegetation, the greater the roughness, and likewise the drag coefficient the greater the effect of the diameter of the vegetation is getting bigger. The graph of the relationship between vegetation diameter and velocity and discharge can be seen in Figure 3.

Based on Figure 3, it can be seen that the greater the diameter of the vegetation, the smaller the velocity, and likewise the discharge decreases the greater the diameter of the vegetation. For the design of the banks with a width of 20 m, a vegetation diameter of 10-50 cm, the discharge is 356.91 - 338.26 m^3/s.

The discharge can be reduced by 32.5-36% from before the arrangement of the riverbank. The velocity value with 10 - 50 cm vegetation planting is 1.164 m/s - 1.104 m/s. The velocity obtained before the arrangement of the banks is 1.725 m/s, the velocity that can be reduced is 32.5-36 %.

Table 2. Calculation results of roughness parameters, drag coefficient, velocity and discharge for various vegetation diameters in the Lawe Alas river.

Diameter Vegetation	Horizontal Distance	Transverse Distance	Parameter Vegetation	Coefficient Vegetation	Rouhgness	Cross Section Area	Radius of hidraulis (R)	Roughness Coefficient	Velocity (V)	$Q_{eco\text{-}hydraulic}$
(m)	(m)	(m)	(B)	(C)	(Ks)	(m^2)	(m)	(λ)	(m/s)	(m^3/s)
0	0	0	0	0	0	306.5	2.881	0	1.725	528.67
0.1	1	1	810	1.001	20.165	306.5	2.881	4.085	1.164	356.91
0.2	2	2	810	1.001	20.315	306.5	2.881	4.195	1.149	352.2
0.3	3	3	810	1.001	20.465	306.5	2.881	4.308	1.134	347.52
0.4	4	4	810	1.001	20.615	306.5	2.881	4.426	1.119	342.87
0.5	5	5	810	1.001	20.765	306.5	2.881	4.547	1.104	338.26

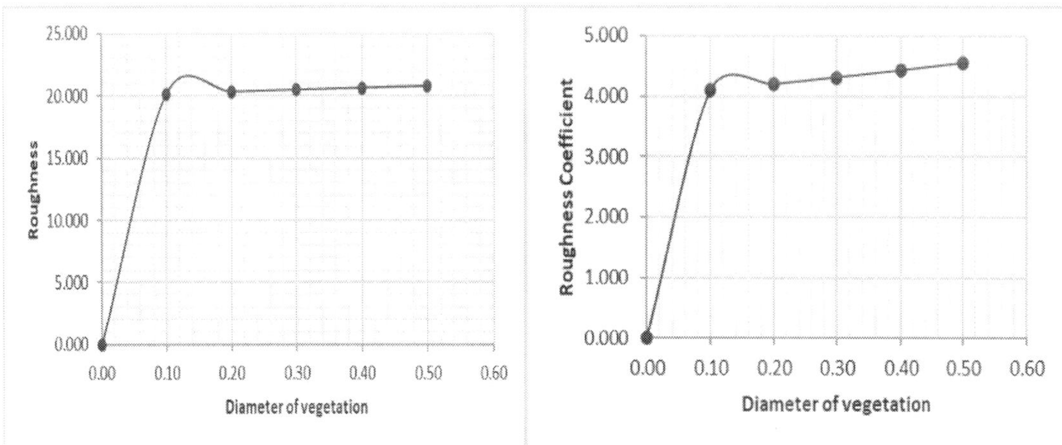

Figure 2. Graph of the relationship between diameter and roughness and the rouhgness coefficient.

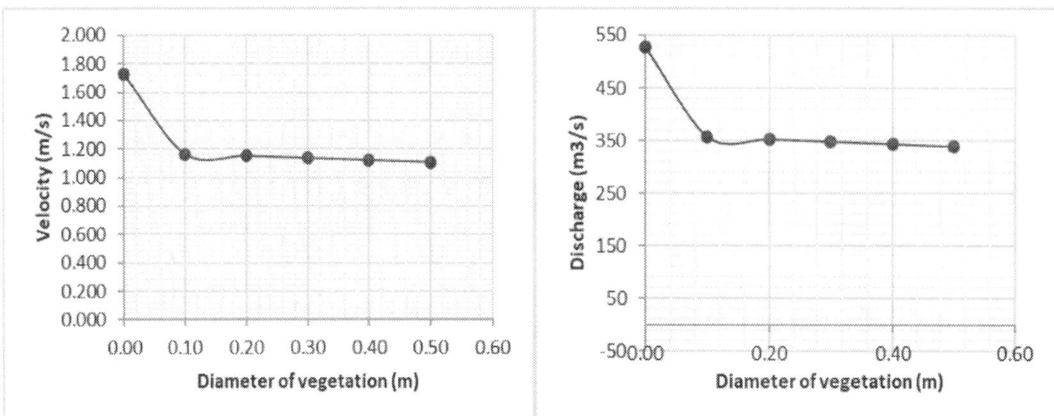

Figure 3. Graph of the relationship of diameter of vegetation with velocity and discharge.

3.2 Lae Soraya river

The second research location is in the district of Rundeng, Subulussalam municipality, Aceh Province, with the main research location being the Lae Soraya River area. The research was conducted from February 2017 to June 2017. River hydraulic data was obtained from measurements for 7 days in the field such as measurements of river depth, river width, velocity, discharge and river water level, as shown in Table 3. Table 4 summarizes ecohydraulic analysis and the relationship between velocity and discharge.

Table 3. Results of measurement of discharge and velocity for 7 days in the Lae Soraya river.

Date	Flow Area (m^2)	Velocity (V) (m/s)	Discharge (Q) (m^3/s)
23-Mei 2017	826.73	0.909	751.45
24-Mei 2017	802.49	0.790	634.31
25-Mei 2017	780.67	0.739	577.24
26-Mei 2017	761.28	0.664	505.80
27-Mei 2017	741.90	0.602	446.72
28-Mei 2017	720.08	0.567	408.14
29-Mei 2017	698.27	0.517	360.77

Table 4. Ecohydraulic analysis and the relationship between velocity and discharge.

Vegetation Diameter (m)	Roughness (Ks)	Velocity (m/s)	Discharge (m^3/s)
0	0	1.672	1406.946
0.07	68.483	0.915	792.690
0.10	100.224	1.391	1293.484
0.15	114.397	1.556	1467.417
0.20	118.036	1.595	1508.593
0.25	119.336	1.609	1522.999
0.50	120.691	1.623	1522.999

The maximum discharge that can be accommodated by the Lae Soraya river, when the water level reaches 3.66 m, is 1090.734 m^3/s. Furthermore, the calculation of the eco-hydraulics analysis with the Merten equation [8]. The relationship between vegetation on the riverbank effects on changes in flow velocity, which can be seen in Figure 4.

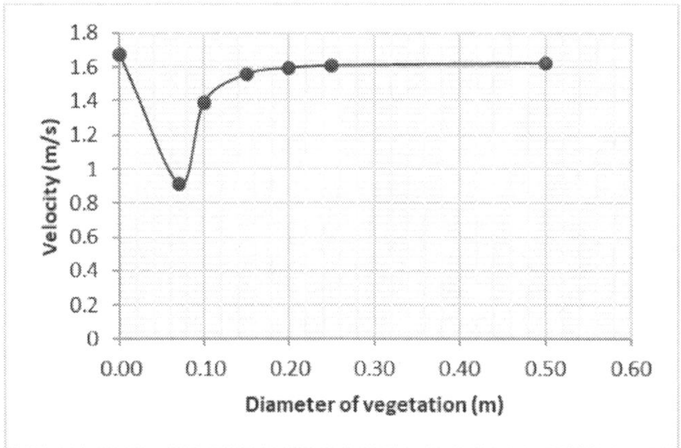

Figure 4. Relationship between vegetation diameter and velocity.

Figure 4 is a graph showing that the initial water velocity before the riverbank arrangement was 1.672 m/s, while in the presence of 0.07 m vegetation, the water velocity on the banks decreased to 0.952 m/s or the water velocity could be reduced by 43%. The velocity on a larger vegetation diameter results in a greater velocity, but at a lower value compared to conditions without vegetation. The smallest speed reduction with the arrangement of the banks is 2.57% on a vegetation diameter of 0.5 meters.

3.3 Hydrological analysis

3.3.1 DAS (Watershed). Based on Environmental Impact Analysis, data by the Aceh Transportation Service, in 2019, the hilly and mountainous topography has resulted in many rivers that have quite heavy flows. Hydrology in Southeast Aceh District is characterized by a long river, namely the Lawe Alas River and its tributaries (hundreds in number) which originate from many mountains, including Mount Leuser, Mount Kemiri, Mount Bendahara and Mount Perkison. Generally, the tributaries run water all year round.

The water catchment areas in Southeast Aceh Regency, including Mount Perkison and Mount Bendahara, are a source of drinking water for rivers flowing to the east coast. Apart from irrigation and transportation facilities, rivers are also used as drains for draining dirty water and rainwater. The Alas -Singkil watershed in the upper reaches of the mountains and natural forests are still in pretty good condition. In the middle and downstream parts, many people have used it to become cultivated land in the form of plantations and luxury and some are still swamps. The condition of the Alas-Singkil river generally occurs silting or sedimentation in the river body.

The topographical conditions of the Sungai Gelombang and Singkil watersheds are classified as lowland areas (altitude 0-200 MDPL), with an area of the Waves and the Singkil River Basin each covering an area of 5752.68 km^2 with river lengths respectively 219.84 km and 316.28 km.

3.3.2 Surface water quality. Based on Environmental Impact Analysis, data by the Aceh Transportation Agency, in 2019, the results of the river water quality tests observed were Alas River, Soraya River and Singkil River, all of the parameters tested were below the established quality standards. Specifically, for the DO parameter, the value can be declared good if the values are below the quality standard, except at the points of Singkil and Sigrun Villages.

3.3.3 Groundwater quality. Based on the Environmental Impact Analysis, data by the Aceh Transportation Agency in 2019, the results of testing the quality of clean water for all the parameters tested were below the established quality standards, the quality standards used to compare the results of the measurement / analysis were the Clean Water Quality Standards PerMenKes No 416 / MenKes / Per / IX / 1990. Even so, overall clean water around the location of the activity is still good enough to be utilized, but we must pay more attention to the procedures for using clean water, especially if the water is used for consumption.

3.4 Ecological analysis

3.4.1. Vegetation. Based on data from the 2019 Transportation Service Environmental Impact Analysis, the vegetation cultivated at the Salim Pipit, Wave, Rundeng, Panji and Singkil locations is in the form of mixed gardens such as oil palm, coffee, cocoa, cloves and rubber, community garden plants such as bananas and cassava. The dominant commodities are areca nut and palm oil.

The natural forest vegetation in the study area includes Alstonia Scholaris, Terminalia sp, Bridelia Pustulata, Macaranga sp, Delonix sp, Intsia Bijuga, Senna Alata, Duabanga Moluccana, Endiandra Maingayi, Litsea Grandis, Aglaia Forbessi.

Based on data from the Aceh Province Natural Resources Conservation Center, 2018, the protected vegetation types are found in the forests of Mount Leuser National Park, namely Rafflesia, Semar Bags, Carrion Flowers, Shoe Orchids and Umbrella Palm.

3.4.2. Animal. The types of mammals in the study area are wild boar, squirrel, squirrel, bat, bat, rat, field rat and ground squirrel. There are also 13 bird species in the study area. Types of amphibians and reptiles include monitor lizards, rice field snakes, pythons, estuarine crocodiles as well as lizards and frogs.

3.4.3. Aquatic Biota. In general, the aquatic biota in the study area is grouped into plankton, benthos, and nekton. The study of marine biota is to determine the richness of marine biota species found in the area.

3.5 *Environmental, social, economic analysis*

The study of community perceptions in this study is focused on 3 things, namely the community's perception of the plan to improve river banks in an ecohydraulically manner, the community's perception of land acquisition and the community's perception of land management with vegetation. The number of respondents studied was 100 households consisting of 86 male and 14 female heads. Respondents' ages ranged from 20 to 69 years. The average number of dependents is 2 people per family, and two of them work. Judging from the education level, it is known that 18 people have not graduated from elementary school, 39 people have graduated from elementary school, 20 people have graduated from junior high school and 17 people have graduated from high school.

The results of interviews with communities in the study locations indicated that there had never been any conflicts over land ownership or conflicts between ethnic groups and religions. Security and order were well maintained. Based on the results of a survey on community perceptions of land acquisition, 98% agreed and 2% disagreed with reasons of concern that the value of compensation was not what they wanted [1].

4. Conclusion

1. Flood Management in Lae Soraya River to Singkil, starting from the village of Gelombang a diversion channel must be made to drain some of the flow into the sea.
2. The government immediately looks for a location to move the slum settlements of residents who live along the river. Based on interviews with the community, they all wish to move as long as they are provided with a place by the local government because most of them are low income earners.
3. Flood management with eco-hydraulics, namely the arrangement of river banks with local vegetation in the upper reaches of the Lawe Alas River. Q50 = 528,670 m3 / s, the riverbanks L= 20 m, D = 10-50 cm, the discharge is 356.91 - 338.26 m3 / s. The discharge can be reduced by 32.5-36% from before the arrangement of the riverbank.
4. The value of velocity with vegetation planting 10 - 50 cm is 1.164 m/s - 1.104 m/s. The velocity obtained before the arrangement of the banks is 1.725 m/s, the velocity that can be reduced is 32.5-36%.
5. The results of research in the upper reaches of the Alas river, 50 cm diameter vegetation is best planted on the riverbanks with a width of 20 m. The results showed that all vegetation diameters could reduce river flow and velocity, so planting vegetation along the riverbanks was highly recommended to immediately reduce flood discharge downstream of the river.
6. The results of the study in the middle of the Lae Soraya river show that the maximum storage capacity of the Lae Soraya River is 1090.734 m^3/s. The appropriate eco-hydraulic design for 5-year to 10-year flood discharge is 100 m wide and the use of vegetation with a diameter of 7 cm to 10 cm with a spacing between vegetation is 100 cm.
7. This is a preliminary study conducted by the author in further analysis of the conditions in the Alas Singkil river area.

References

[1] Aceh Transportation Agency 2019 *Environmental Impact Analysis in Aceh Tenggara District, Subulussalam Municipality and Aceh Singkil District*. Banda Aceh: Aceh Transportation Agency. *(in Bahasa Indonesia)*.
[2] Azmeri A, Isa A I 2018 *J. IJDRS,* **10 (1)** 1-6.
[3] Clifford N J, Harmar O P, Harvey G, Petts G E 2006 *Aquatic Conservation: Marine and Freshwater Ecosystems,* **16** 389-408.
[4] Guardiola-Albert C, Díez-Herrero A, Cuervo-Arango M A, Bodoque J M, García J A, Naranjo-Fernández N, Aroca-Jiménez E 2019 *J. FRM,* **13 (1)** 01-16.
[5] Leclerc M. 2005 *J. JHR,* **43** 63-64.
[6] Maddock I, Harby A, Kemp P, Wood P 2013 *Ecohydraulics: An Introduction*. In: Maddock I, Harby A, Kemp P, Wood P, editors, Ecohydraulics: an integrated approach (Chicester: Wiley).

[7] Maryono A 2014 *Dealing with Floods, Drought, and the Environment* (Yogyakarta: Gadjah Mada University Press).
[8] Maryono A 2017 *Eco-Hydraulics of Eco-Friendly River Management* (Yogyakarta: Gadjah Mada University Press).
[9] Montz B E, Eve Gruntfest 2002 *J. JEH,* 4(1), 15-22.
[10] Palmer M A, Bernhardt E S, Allan J D, Lake P S, Alexander G, Brooks S, Carr J, Clayton S, Dahm C N, Follstad Shah J 2005 *J. JAE,* **42** (2) 208-217.
[11] Sandercock P J, Hooke J M, Mant J M 2007 *Progress in Phys. Geo.*, **31** (2) 107-129.
[12] Sari J 2015 *Ecohydraulic Design for Flood Control in Symmetric River Morphology* (Bogor: Bogor Agricultural Institute).

Environmental adaptation and hazard reduction strategies in relocation housing development by its occupants case: Indo-Tiongkok housing, Neuheun Aceh Besar

E Wulandari, Zahriah*, Z Fuadi
Department of Architecture and Planning, Faculty of Engineering, Universitas Syiah Kuala, Kopelma Darussalam, Banda Aceh 23111, Indonesia

*E-mail: zahriah@unsyiah.ac.id

Abstract. The construction of relocation housing is a form of housing assistance for victims of the 2004 tsunami, with various environmental designs. Indo-Tiongkok housing (Jacky Chan housing) is one of the relocation housing in the Neuheun area, designed with an approach adjusting the hilly topography. The disadvantage of building on contoured land is that it is susceptible to local erosion, which can damage buildings on the land (roads and buildings). This study reveals the phenomenon of environmental adaptation strategies by the community to reduce the risk of hazards in their home environment, and this is important to disclose to find out people's attitudes and actions in maintaining the sustainability of their living space. The research method was in-depth interviews with the community to understand the understanding of environmental hazards and community strategies in reducing the risk of dangers that might occur. The information required relates to 2 things: 1) types of environmental hazards, risks, and patterns of threats; 2) attitudes and strategies of society to overcome these dangers. The results showed that the community understands the risk of hazards logically, by looking at signs of natural elements such as traces of erosion, and the impact on artificial elements (houses, soil, yard boundaries). The community strategy to reduce the risk of hazards is to combine various factors from engineering aspects (building materials and techniques), economic aspects, and suitability aspects of plant selection. Results showed that the strategy for adaptation patterns to the environment would appear in each occupant according to their problems and abilities so that the results of the appearance of the residential environment design would vary. This is so unique in the culture of living in each place, which will distinguish it from other places.

1. Introduction

The 2004 tsunami natural disaster that hit most of the coast of Aceh has damaged, destroyed, and even eliminated some of the coastal settlements in the area [1]. The Aceh rehabilitation and reconstruction program is a program initiated by the central government to plan and manage various types of assistance for victims of the tsunami disaster. One of its programs in the housing and settlement sector is housing relocation assistance, one of which is in Neuheun, Aceh Besar District, whose construction began in 2006 [2].

The Neuheun area, located in Aceh Besar district, is the largest residential relocation area in Aceh. There are five relocation housing that was built and scattered on the edge of the hills of Ujong Batee. Indo-Tiongkok relocation housing (known as Jacky Chan housing) is one of the housings constructed in the hills. The housing development was financed by the Tiongkok government, which was inaugurated in July 2007. The housing is located at an altitude of 300 meters above sea level on an area of 22.4 hectares consisting of 606 housing units with type 42 [3].

The hills of Ujong Batee are less fertile, rocky stony, and grow shrubs. The potential of mountainous rock is used as a C-excavation mining location [4]. Based on the explanation of the village head (Keuchik) Neuheun, the land was used by the local community for traditional grazing.

Based on observations, the construction of the relocation housing used the cut and fill method and directed the hilly walls so that terracing land was formed according to the needs and planning of the housing environment. This matters a part of the hilly land surface open and prone to erosion during the rainy season. Based on the Government Regulation of the Republic of Indonesia Number 14 of 2016 concerning the implementation of housing and residential areas, one of which is mandated that the housing environment must be safe from danger and not cause environmental damage that reduces the quality of the human environment.

This paper aims to see how far the community has succeeded in adapting to the hilly environment and what strategies the community has taken to prevent erosion so that the environment is disaster-safe. The quality of the housing environment increases. This study is essential to conduct as an evaluation material for the construction of relocation housing for disaster victims regarding the adaptation pattern of the relocation community in a new place. It turns out that the location is not free of other natural hazards after they have been removed from the natural dangers of the tsunami.

2. Theoretical studies

2.1. Housing environment and hill topography

Understanding the residential environment can be approached with existing theory, which explains five essential elements, namely: nature, man, society, shell, and network [5]. Nature (Nature) is a place of life that is chosen by considering one of them is the aspect of being safe and safe from the threat of natural hazards. Humans who will come from nature will carry out adaptation processes with their knowledge and engineering capabilities to create artificial buildings that make sense of sustainable security. The community will learn in communal challenges so that the pattern of adaptation to the natural environment will become a typical character of the community and, in the long run, will become local wisdom. Houses and other environmental facilities as Shell and Network are designed to follow the character of the natural environment so that the artificial environment is safe from the dangers of the natural environment.

Hilly areas are vulnerable due to sloping topography, and certain geological conditions can be unstable due to developments on them. In selecting the site of a residential neighborhood, it is justified to choose a location with a particular slope and relatively stable geological conditions, by engineering the site, by making terrasering to get flat land for the house site in it [6]. Related to the movement and people movement and vehicles in environmental space are reflected in the determination of the pattern of the housing environment and the form of environmental facilities and infrastructure. Therefore, based on geographical considerations, various concepts of the provision of facilities can be applied integratively.

2.2. Natural hazards and adaptation strategies

Natural hazards will become natural disasters when natural disasters threaten human life and destroy the natural environment itself [7]. In the Law of the Government of the Republic of Indonesia No.24 of 2007 concerning disaster management, it explains about the standard of rehabilitation and reconstruction of natural disasters, the scale of handling various approaches, namely; a) time aspect (immediate, medium-term, and long term); b) Scope of the problem (partial; holistic/integrative); c) Treatment objectives: (emergency, medium-term, sustainable). The design of the housing environment is required to follow geographical conditions related to geological, soil, air, and climate conditions so that the artificial building is adaptive to withstand damage from natural dangers in the long term and sustainable.

Sustainable development is following the concept of resilience, which is defined as a system problem that is disrupted and rebuilt when it changes to maintain the same function, identity structure, and feedback [8]. In handling for the long term, holistic/integrative and sustainable, an approach is needed in all aspects of life for disaster-affected communities for the sustainability of their lives. It is necessary to know the level of vulnerability of a place and the character of the community that will occupy it. The concept of vulnerability of a place, which is related to the form of its threat, vulnerability to vulnerability. Concerning the formula, if an increase in the vulnerability of a regional community and the community will be reduced. Therefore, knowledge of disaster forms and patterns, and understanding personal vulnerability, humans can create the necessary adaptation patterns.

To develop adaptation patterns, it requires public openness to knowledge about disasters or transparency to previous experiences, which made the conditions mentioned above vulnerable. In the next life process, a culture close to disaster will be formed (cultural disaster). In a cultural disaster, people know and understand several things related to disasters, namely: a) types of threats; b) frequency of threats; c) the magnitude of the danger; and d) the aftermath of the risk. Thus, they can create adaptation strategies for their resident culture.

Adaptation patterns can be developed in three models according to the problems at hand [10], namely:

a) Limitation, avoidance, related to the threat and risk of known disasters, then take avoidance actions.
b) Stabilization/protection, related to reducing threats and simultaneously increasing the capacity factor to prevent disasters.
c) Mitigation, related to efforts to reduce or eliminate the impact of a disaster in a combination of limitation and stabilization.

Natural and environmental hazards that can be recognized from an area can be converted into potentials through engineering on land. It is managed by hilly areas that are prone to runoff from rainwater and minimal groundwater reserves, often applying the concept of an air collecting system to prevent air shortages and function from capturing flowing water [9]. Also, an area landscape can be engineered to obtain the desired landscape quality so that it can improve the quality of human living space, such as outdoor plants, which are following geographical conditions [10]. Thus it is hoped that, with a formally planned housing environment, the quality of the hilly landscape will be better and can improve the land surface in the housing environment. The experience in several places where the artificial environment of housing can enhance the natural qualities of nature itself.

3. Research methods

The research approach is qualitative-phenomenological-interpretive research on the spatial adaptation of the physical environment, to see the tendency of changes in people's behaviour in reading patterns of natural environmental hazards and survival strategies through the creation of responsive artificial environments [11]. The data needed is related to community perceptions about the form of dangers and threats, and survival strategies and patterns of community capacity building. Data obtained by in-depth interviews, reconstructing hazards, and what plans the community is using to reduce the risk of harm. Interviews were conducted with community leaders and selected community members because they can know about people's lives in general. The data is processed and analysed using the dynamic system method to compile causal events so that the optimal strategy can be found for the survival of life in the future. The validation by looking at the structure and logic of thinking.

4. Results and discussion

4.1. Geographical characteristics of Neuheun area and natural environmental hazards

The geographical characteristics of the Indo-Tiongkok relocation housing area cannot be separated from the geographical characteristics of the Neuheun area, which is an integral part of the northern tip of Sumatra Island. Natural hazards come from hydrometeorological and geological threats. The Neuheun area is on the eastern side of the hills around the Banda Aceh Delta, which is influenced by the monsoon climate. These climatic dynamics affect the waters of the Andaman Sea and the Indian Ocean in the West and the Asian Continent in the North, which cause relatively sudden, high-rainfall West winds, hurricanes, and hot and dry air. Meanwhile, the wind from the east from the Malacca Strait causes eastern winds with the character of the air a lot of rain, and the wind is relatively soft [12]. The difference between the two seasons affects the geographical characteristics of the hills around the Banda Aceh Delta, including the conditions of the Ujong Batee hills. See a map of the geographic location of the Neuheun Region against the Banda Aceh Delta and the monsoons as Figure 1 illlustrates.

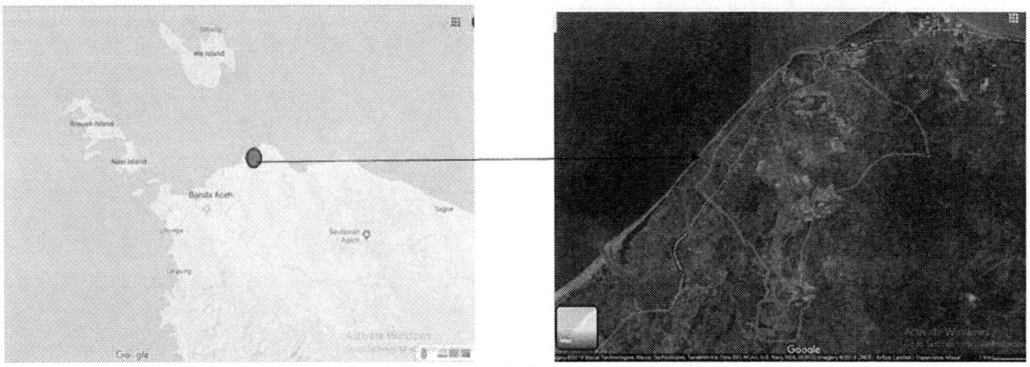

Figure 1. Location of Neuheun area in the hills of Ujong Batee to the geographic delta of Banda Aceh City [13].

By looking at the monsoon wind pattern map in the Delta area of Banda Aceh City, the condition of the hilly area around the city of Banda Aceh, during the East wind season (the rainy season), will experience the danger of flooding and landslides. In contrast to the West (dry season), hot winds have occurred, which have caused fires in the hills around the city of Banda Aceh, including the hills of Ujong Batee [14].

The geographical characteristics of the Neuheun area are rocky hills mixed with relatively infertile soils. The large stones appear on the surface scattered all over the surface of the hills, making it difficult to grow large plants, which grow in the form of shrubs and grass. The flow of rainwater easily peels off the ground surface. The land is suitable for livestock grazing, which the local community has cultivated for a long time and has become the traditional way of life of local people as livestock herders. Reconstruction of the Neuheun area after the tsunami, including the C mine excavation site, many parts of the area are now open.

4.2. Indo-Tiongkok relocation housing development

The Aceh provincial government determined Indo-Tiongkok housing built in the Neuheun hill area. The distance from the city of Banda Aceh is about 20 km. This hilly area has a height of 300m above sea level, making it very safe from the reach of the tsunami, making it ideal for coastal housing. Figure 2 shows the location of the Neuheun hill area and the placement of the relocation housing.

From Figure 2, it can be seen that the geographic character of Neuheun is a relatively open landscape. At the time after development in 2007, the hills appear rocky and relatively minimal plants, so that it seems less fertile so that development planning must be careful about natural hazards if the extreme weather is rainy or dry. In 2020, there will be environmental conditions that are relatively greener than before, and this indicates that there is an environmental potential that can be empowered and managed.

Based on the explanation of the village head (keuchik) Neuheun, the construction of an Indo-Tiongkok relocation house was carried out in 2005 and then occupied in 2006. That is what was built with the concept of "building Indo-Tiongkok brotherhood". Therefore, the residents of the Indo-Tiongkok Housing who were victims of the 2004 earthquake and tsunami were relatively heterogeneous, with diverse cultural, professional, ethnic, linguistic, and religious backgrounds. Based on the explanation of the chairman of Indo-Tiongkok housing, initially, this housing consisted of 600 households according to the number of housing units and 100 ethnic Chinese families who came from

Chinese residents in Peunayong Banda Aceh as victims of the tsunami, but currently, the number of households is only around 490 households.

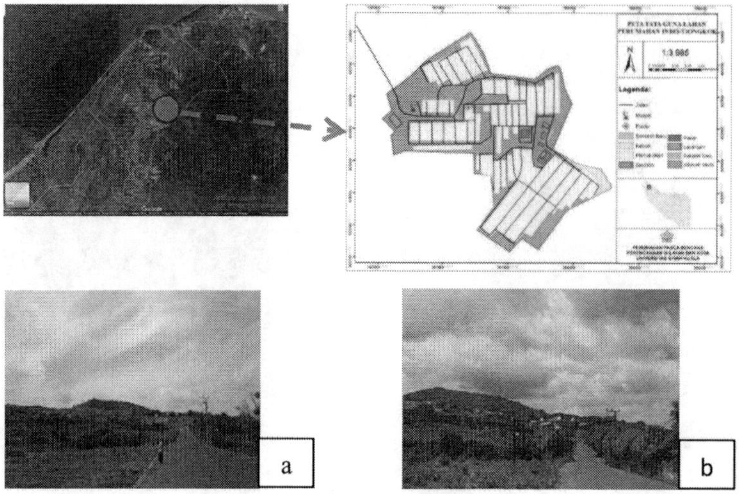

Figure 2. Image of relocation housing location, Indo-Tiongkok housing environment map and conditions a) 2007, and b) 2020.

Various types of occupations for residents of the Indo-Tiongkok housing are farmers, fishermen, businessmen, construction workers, traders, civil servants, retirees, and others. With a variety of jobs that are expected to be a positive aspect of the sustainability of housing, people will learn from each other from experiences of various aspects of life and culture. The residents of the housing consist of various ethnicities, ethnic groups, indigenous tribes, Aceh-Java, Padang, Batak, Chinese and other tribes; they live mingled together.

4.3. The development pattern of Indo-Tiongkok relocation housing environment

The Indo-Tiongkok relocation housing was designed to follow the contours of the hills with the concept of cut and fill, cutting cliffs for residential sites and neighborhood roads, so that an environmental pattern was formed in the form of a semi-circular and grid compound. The Indo-Tiongkok relocation housing is designed with a residential area of around 22.4 ha, with 606 units divided into seven blocks and equipped with public and social facilities.

The houses are built with type 42 M2 on plots where the land has been leveled as needed; meanwhile, the ground in the yard is open, with varied topography, so that the soil feels narrow, prone to local erosion, and unsafe for daily activities occupants.

Figure 3. Condition of housing environment at the beginning of construction (a) and currently (b, c).

In Figure 3, it can be seen that the condition of the land at the beginning of development in 2009 (a) in the residential area located on the hills looks arid and lacks trees and the current condition (b) (2020) also looks arid in the hilly part outside the housing (c) so that it is vulnerable to natural hazards, both geological and hydrometeorological factors.

4.4. Characteristics of environmental hazards in Indo-Tiongkok *relocation housing*
Several natural events threaten the existence of relocation housing, including a) flash floods due to extreme rain and soil erosion, weakening the construction of houses and roads; b) rock slide, prone to hit people's homes; and c) extreme heat often accompanied by strong winds, prone to fires.

Flash floods due to the extreme rainy season and soil erosion have occurred in several places around the hilly areas. According to the residents, this was due to the existence of a C-dug mine on the hill. Currently, the community is anxious about the condition of the hilly environment because the land cover is already open; there are no large trees; the exposed mountain rocks cannot absorb rainwater. During the rainy season, the water will flow swiftly towards the housing. Security in the initial housing planning by building a housing ring road and a relatively wide road border so that the road functions as a buffer. However, the road is also used for access to hills, for the illegal opening of the C excavation business. This effort was unsettling for the community because of the threat of flash floods and landslides, so that in early 2019, the road to the hills was closed [5].

Mountain rock avalanches had occurred in the Cinta Kasih housing nearby and traumatized the community so that the Indo-Tiongkok housing community prevented any construction on the hilly land that was spread across the housing. The effort is being made to control the road to the hills outside the housing.

Extreme heat and strong winds very often occur because of the geographical location in the hills and facing West, which is the direction of the West wind flow or the transition period that often causes drought. People are worried about forest fires that can spread to housing. Therefore people want the area around the housing to be green. However, the community cannot do much to improve the surrounding environment. The effort can be made at this time is only limited to handling in the territory of his residence.

4.5. Environmental adaptation strategies by occupants of Indo-Tiongkok housing on occupancy area
The Indo-Tiongkok housing community, in general, understands the threat posed by contoured landforms. They know the characteristics of the land. Based on their experience while living in Neuheun for more than ten years, in general, this area is safe from floods and landslides, but the community is aware of the threat of extreme rainy season that occurs every year.

Extreme weather conditions, such as heavy rains, will damage the forested yard due to local erosion. Therefore the community generally controls rainwater through technical engineering of the land in the yard. However, land engineering can also be beneficial for residents both in the economic aspects of daily needs and socio-culture.

4.5.1. Engineering front yard. Land improvement in the front yard by changing the ground level parallel to the outside of the building, as seen in Figure 4.

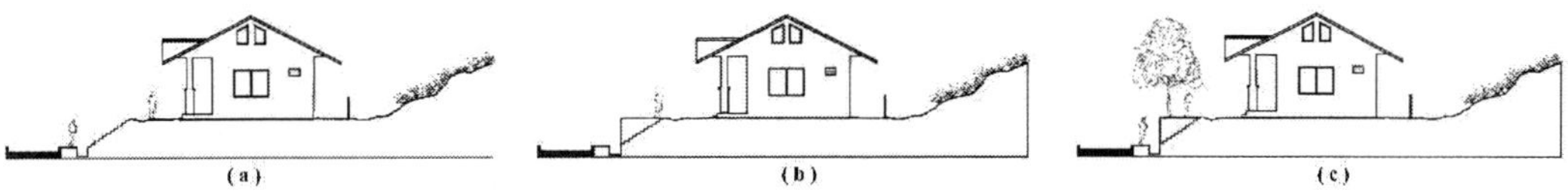

Figure 4. The pattern of land change in front of the house lots a) initial house b) front land c) in the yard area.

Changing the front yard land (by hoarding and elevating the land) makes the yard area broader and more qualified. In general, people who grow plants are useful in terms of daily kitchen needs and to get a relaxed home environment. The distance between the house and the neighborhood road as an imaginary separate public space, this allows residents to sit on chairs with the large courtyard in front of the house, the distance between the home and the neighborhood road as an imaginary separate public space, this allows residents to sit on chairs. Another benefit, a large yard is useful for expanding social areas, such as receiving a larger number of guests.

The engineering on the land demonstrates a compromise between disaster risk and the need for adequate housing. Thus, the archetype is based on disaster mitigation adaptation.

4.5.2. Backyard engineering. The condition of the land behind the house is divided into two types, namely: the ground behind the house on the inner plot of the house and the land behind the house on the field on the outskirts of the housing, which borders the land outside the housing.

4.5.2.1. Rear plot in housing. The land behind the house is generally contoured as high as the house building and borders the neighborhood road, above which is fenced by hard plants such as horse trees. Thus, the limit of land management is back to the boundary of the environmental road above it. Changes in the land behind the house occur when there is a need for house expansion due to the demands of the home owner's needs. More space is added to the back or side of the house by cutting the ground close to the wall of the house. This makes the house wall unite the earth wall so that the house wall also functions as a follower of the land or a separation between the house wall and the cliff, as shown in Figure 5.

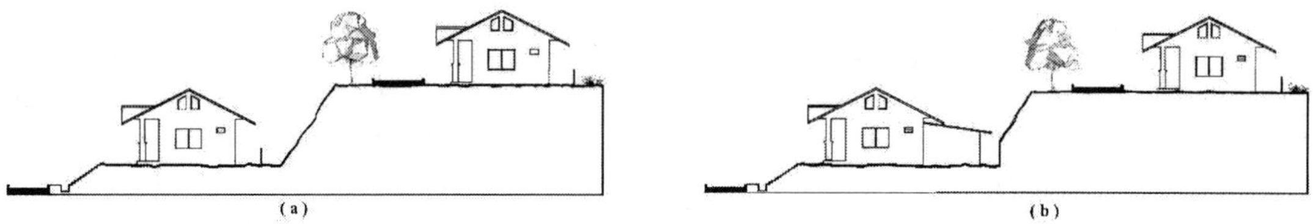

Figure 5. Techniques to build a wall of earth behind the house.

According to the community represented by Mrs. Ani, these plants can prevent land erosion. Based on the above conditions, the danger of erosion always threatens the existence of the house. However, the community realizes this condition, trying to make the back land green by planting lush trees such as soursop, jemblang, and srikaya trees.

The community has built drainage channels around the back and sides of the house, so is no puddle around the house. Rainwater is stuck beside the house, which means used to fertilize the plants around it, and excess is then channeled into the ditch in the front house. Community action to manipulate the site is a form of strategy to reduce the risk of natural disasters in the way of disaster mitigation measures.

4.5.2.2. Rear lots at the edge of the housing. The backyard is relatively long sloping contours. It is possible to function as a border for roads outside the environment. It is hoped that the effects of activities and natural dynamics outside of housing will not affect nearby housing. The backyard is used for orchards while maintaining the land cover with grass. The community is aware of the dangers of erosion that may occur as Figure 6 displays.

Figure 6. Management of residential periphery land which also functions as a neighborhood road border outside the housing.

IOP Conf. Series: Materials Science and Engineering 1087 (2021) 012043 doi:10.1088/1757-899X/1087/1/012043

Many of the back yards are a divider for residential areas to the hilly environment outside, which is very suitable for the management that will be given to the owners of the plots concerned. The sense of ownership of the land will encourage efforts to care for an environment where the threat is high. If the community has adequate management capacity, then disaster risk will be reduced. Besides, it will increase the food security of the plot owner.

5. Conclusion

Changing the front yard land is a strategy for adapting the interests of residents and improving small-scale ecology, namely:

i. from the aspect of the residents 'interests, a flat yard will expand the occupants' activity space, both for social, cultural, and daily activities of residents, such as sitting to fill their spare time or a place to receive large numbers of guests.
ii. From an ecological aspect, the landfill is relatively fertile so that it can be planted with tree crops such as fruit trees. Changing the yardland into a cool garden and producing microclimate changes that will prevent strong winds from entering the house at the same time and cool the atmosphere of the place during the dry season.

Indo-Tiongkok housing shows that the people have succeeded in improving their occupancy environment through technical engineering in site housing. It is beneficial in improving the residential landscape as well as hills, which can enhance the micro-climate environment. However, there is very little attention to public facilities, so that the existence of these facilities is threatened by natural hazards, such as erosion of land and building quality that is not affected due to extreme climate change.

In general, the housing environment has improved the character of the hilly area, but what is worrying is the dangers that come from outside the housing environment that cannot be controlled. Adaptations that are carried out, in general, are disaster mitigation adaptations in the form of compromising between human and environmental interests.

Concerning the land outside the housing which belongs to the local community and individuals, part of it has been cultivated as C excavation (mountain rock quarry), which is also used as pasture land. Therefore, it is necessary to understand the community and carry out dialogue that results in an agreement for the benefit of immigrant communities as well as local communities.

6. Acknowledgments

This paper is part of the research report of the Head Lecturer Research Scheme, PNBP notes for 2020. Thank you to the Syiah Kuala University and the Indo-Tiongkok relocation housing community for the information, as well as the students participating in the coastal area planning course Urban and Regional Planning Study Program Department of Architecture and Planning, Faculty of Engineering, Universitas Syiah Kuala, even semester 2019-2020, which has helped study data activities and make the necessary maps.

References

[1] Elysa W, Fahmi A, Evalina Z, Myna A 2020 *IOP Conf. Ser.: Earth Environ. Sci.,* **452** 012117.
[2] Badan Rehabilitasi dan Rekonstruksi (BRR) NAD-Nias 2008 *Enriching the construction of recovery, Annual Report BRR* (Jakarta: Badan Rehabilitasi dan Rekonstruksi (BRR) NAD-Nias).
[3] British Broadcasting Corporation (BBC) 2019 Tsunami Aceh, 15 Tahun kemudian: Bantuan hibah asing beralih fungsi. Available at: https://www.bbc.com/indonesia/indonesia-50682475.
[4] Pemerintah Kabupaten Aceh Besar 2008 *Rencana Tata Ruang dan Wilayah Kabupaten Aceh Besar Tahun 2009-2029.* Jantho: Pemerintah Kabupaten Aceh Besar.
[5] Doxiadis C. 1968. *Ekistics: An Introduction to The Science of Human Settlements* (New York: Oxford University Press).
[6] Strom S, Nathan K 1997 *Site Engineering for Landscape Architects* (USA: John Wiley & Son, Inc.).

[7] Smith K. Petley D 2009 *Env. Hazards: Asseng Risk and Reducing Disaster (Fifth ed)* (London: Routledge).
[8] Pickett S T A, Cadenasso M L, Grove J M 2004 *Lands. Urban Plan.,* **69** 369–384.
[9] Zakaria Z 2004 Stabilisasi dan Rancang Bangun Lereng Terpadu, salah satu usulan untuk penanganan masalah longsor. *in Permasalahan, kebijakan dan Penanggulangan Bencana Longsor di Indonesia* ed Naryanto H S (P3-TPSLK BPPT dan HSF).
[10] Williams D E 2007 *Sustainble Design: Ecology, Architecture and Planning* (USA: John Wiley & Son, Inc.).
[11] Yunus H S 2010 *Metoda Penelitian Wilayah Kontemporer* (Yogyakarta: Pustaka Pelajar).
[12] Pemerintah Kota Banda Aceh 2008 *Rencana Tata Ruang dan Wilayah Kabupaten Aceh Besar Tahun 2009-2029* (Banda Aceh: Pemerintah Kota Banda Aceh).
[13] Google 2020 Location of Neuheun Area in the Hills of Ujung Batee to the geographic Delta of Banda Aceh.
[14] Analisa Aceh Besar Rawan Bencana 2017 Available at: https://analisadaily.com/berita/arsip/2017/9/27/421788/aceh-besar-rawan-bencana.

Characterization and utilization of coconut coir and bentonite-based adsorbents for removal of lead metal ion from hazardous liquid waste

Mariana[1,2]*, F Mulana[1], I Perkasa B[3], T Muhamad A F[3]

[1]Department of Chemical Engineering, Universitas Syiah Kuala, Banda Aceh 23111, Indonesia
[2]Research Center for Environmental and Natural Resources, Universitas Syiah Kuala, Banda Aceh 23111, Indonesia
[3]Student of Department of Chemical Engineering, Universitas Syiah Kuala, Banda Aceh 23111, Indonesia

*E-mail: mariana_hasyim@yahoo.com

Abstract. Adsorbent is one of the most commonly used materials because of its high adsorption effectiveness. However, adsorption with one type of adsorbent causes a lower adsorption efficiency. This study aims to determine the characterization of adsorbents from coconut coir and bentonite-based adsorbents and the combined adsorbent from coconut coir charcoal and bentonite and to examine the adsorption capacity of lead metal ion on these adsorbents. The types of adsorbents used in this study were coconut coir charcoal without activation, coconut coir charcoal with activation, pure bentonite, bentonite with activation, and combination of activated coconut coir charcoal: activated bentonite with mass ratio of 70:30. The research was initiated with pyrolysis of coconut coir at temperature of 400°C, then activated chemically with 0.3 M HCl solution. The resulting adsorbents were characterized using Scanning Electron Microscopy (SEM) to determine the morphological structure, Fourier Transform Infra-Red Spectrophotometer (FTIR) to determine the functional groups, X-Ray Diffraction (XRD) to determine the crystalline structure. Then the adsorption of lead metal ions was carried out by using these adsorbents. The results showed that the best adsorption efficiency is by adsorbent combination of activated coconut coir charcoal: activated bentonite with mass ratio of 70:30. The adsorption efficiency of lead metal ions using activated coir adsorbents: activated bentonite (70:30) is 35.0%. The activated coconut coir charcoal: activated bentonite with mass ratio of 70:30 has the largest adsorption capacity of lead metal ions which is 182.71 mg/gr.

1. Introduction

Coconut coir that is a waste from coconut plantations is available in large quantities and is easily found in Indonesia [1], can be used as sorbent because it contains quite high lignin, cellulose and hemicellulose and has a porous structure.According to research by Etim et al, the results of testing coconut coir ash (CCD) against methylene blue in aqueous solution were very effective to reduce the levels of methylene blue and the material was also inexpensive compare to other materials [2]. Furthermore, the results of research conducted by Tan et al stated the results of KOH adsorption: by impregnated charcoal can adsorb 191.73 mg / g from a solution of 2,4,6 TCP [3].Bentonite has a layer structure that has the ability to swell and has cations that can be exchanged with others. Although bentonite clay functions as adsorption, its adsorption capacity is still limited. These deficiencies can be regulated through the activation process by using acid solutions such as HNO_3, HCl and H_2SO_4 [4].

The use of bentonite together with coconut coir to produce a novel adsorbent would be an alternative way to form adsorbent with higher sorbent surface reactivity. In this study the use of adsorbents from coconut coir and bentonite-based adsorbents and the combination between activated charcoal from coconut coir and activated bentonite was performed to have the combined adsorbent with the best adsorption capacity. The adsorbent obtained was tested for adsorption capacity by using lead metal ion solution.

2. Method

2.1. Materials and tools

The tools used in this study are Fourier Transform InfraRed Spectroscopy (FTIR), Scanning Electron Microscopy (SEM), X-Ray Diffraction (XRD), hot plates and stirrers, tube furnaces, muffle furnaces, digital scales, mechanical sieves 100-200 meshand desiccator. While the materials used are coconut coir, bentonite, HCl 0.3 M, Pb(NO3)2solution and distilled water.

2.2. Carbonization process of coconut coir

Coconut coir was put into the tube furnace and then carbonized at 400 °C for 2 hours [6]. Then the coconut charcoal was cooled for 24 hours. Then it was ground and sieved using a 100-200 mesh sieve [5,6].

2.3. Bentonite activation process

Bentonite originating from BlangDalam Village, Nisam Sub-District, Aceh Utara was washed with water to remove soil and dirt, then dried, milled and sieved to particle size of 100-200 mesh. Subsequently bentonite was put into an erlenmeyer and activated by the addition of HCl at temperature of 110 °C and an activation time of 30 minutes.

2.4. The mixing and carbonization process

Activated charcoal from coconut coir and activated bentonite was mixed with mass ratio of 70:30. The adsorbentswere then dissolved in distilled water with a ratio of 10: 1 (liquid:solid) and stirred using a stirrer for 2 hours. After stirring, the sorbents were put in oven for 4 hours at temperature of 110 °C to remove moisture content. Next, the dried materials were put in a muffle furnace and heated at temperature of 800 °C for 2 hours [7]. After the carbonization process the material was stored in a desiccator.

2.5. Adsorben characterization

The produced adsorbent was characterized using Scanning Electron Microscopy (SEM) to determine the morphological structure, Fourier Transform Infra-Red Spectrophotometer (FTIR) to determine the functional group and X-Ray Diffraction (XRD) to determine the crystalline structure.

2.5.1. Water content. Two grams of combined adsorbents were put into a weighted porcelain cup. The cup was put into the oven at temperature of 105°C for 3 hours then cooled in a desiccator and weighed. The formula of moisture content can be seen on Equation 1.

$$\text{Moisture Content (\%)} = \frac{a-b}{a} \text{ x } 100\% \quad (1)$$

Where :
a = initial carbon weight (g)
b = dry carbon weight (g)

2.5.2. Ash content. Two grams of combined adsorbents were put into a weighted porcelain cup. The cup was put into the muffle furnace and heated at 600°C for 6 hours then cooled in a desiccator and weighed. The formula of ash content can be seen on Equation 2.

$$\text{Ash Content (\%)} = \frac{a}{b} \text{ x } 100\% \quad (2)$$

Where:
a = ash weight (g)
b = initial dry carbon weight (g)

2.5.3. Adsorption of lead metal ion. One hundred ml of 1000 mg/L artificial $Pb(NO_3)^2$ waste solution was contacted with 0.5 gram of adsorbent with variation of contact time 30, 60, 90, 120 and 150 minutes while stirring at speed of 250 rpm. After being contacted with adsorbents,the concentration of artificial $Pb(NO_3)^2$ waste solution was measured using AAS. The formula of percentage of adsorption and adsorption capacity can be seen on Equation 3 and 4 respectively.

$$\text{adsorption (\%)} = \frac{\text{(initial concentration of adsorbate} - \text{final concentration of adsorbate)}}{\text{initial concentration}} \, x\ 100\% \quad (3)$$

$$\text{adsorption Capacity} = \frac{\text{(initial concentration of adsorbate} - \text{final concentration of adsorbate}}{\text{the amount of adsorbent}} \, x\ solution\ vol \quad (4)$$

3. Result and discussion

3.1. Characterization on of prepared adsorbents

The analysis showed that the activated carbon from coconut coir was activated HCl 0.3 M and without activator has met the quality standard of activated carbon according to SNI No. 06-3730-1995. Initial characterization of prepared adsorbents to see the water content and ash content of each adsorbent can be seen in Table 1.

Table 1. Characterization of prepared adsorbents.

No.	Type of Adsorbent	Water content	Quality Standard SNI No. 06-3730-1995	Ash Content	Quality Standard SNI No. 06-3730-1995
1.	Coconut Coir	5 %	Max 15%	1.84%	Max10%
2.	Coconut Coir Charcoal	6.5%	Max 15%	1.73%	Max10%
3.	Activated Coconut Coir Charcoal	7 %	Max 15%	2.09%	Max10%
4.	Purified Bentonite	4.3%	Max 15%	0.53%	Max10%
5.	Activated Bentonite	7.5%	Max 15%	2.32%	Max10%
6.	Activated Coconut Coir Charcoal: Activated Bentonite (70:30)	3.5%	Max 15%	1.84%	Max10%

Table 1 shows the analysis results of prepared adsorbents, showing that the water content and adsorption of iodine met the quality standards of activated charcoal based on SNI No. 06-3730-1995. The highest water content was found in activated bentonite by 7.5%, while the lowest water content was found in activated coconut coir charcoal: bentonite (70:30) was by 3.5%. This is due to the chemical activation of the bentonite and the washing process with water. Water in the activation and washing solution enters the bentonite, causing the water content in the bentonite to increase. Whereas, the activated coconut coir charcoal adsorbent: activated bentonite (70:30) that is physically activated at a temperature of 800 °C, the water in the adsorbent would evaporate causing a decrease of water content compared to previous adsorbent. According to research by Abdul Ghani et al, the water content is influenced by various processes such as stirring, size reduction and combustion [8], a low water content would improve carbon quality because the adsorption of activated carbon against gases or liquids is higher. Then the highest ash content was found in sample of activated bentonite that was

2.32%, while the lowest ash content was found in activated coconut coir charcoal: activated bentonite (70:30) as much as 0.53%. This is because in activated bentonites the use of chemicals as activators often results in contamination of the activated charcoal produced. Generally, activators leave unwanted remnants, for example substances that are not soluble in water at the time of washing.

3.2. Scanning Electron Microscope (SEM) analysis

The SEM (Scanning Electron Microscope) characterization was carried out to see the surface morphology of the adsorbent. The results of SEM analysis can be seen in Figure 1.

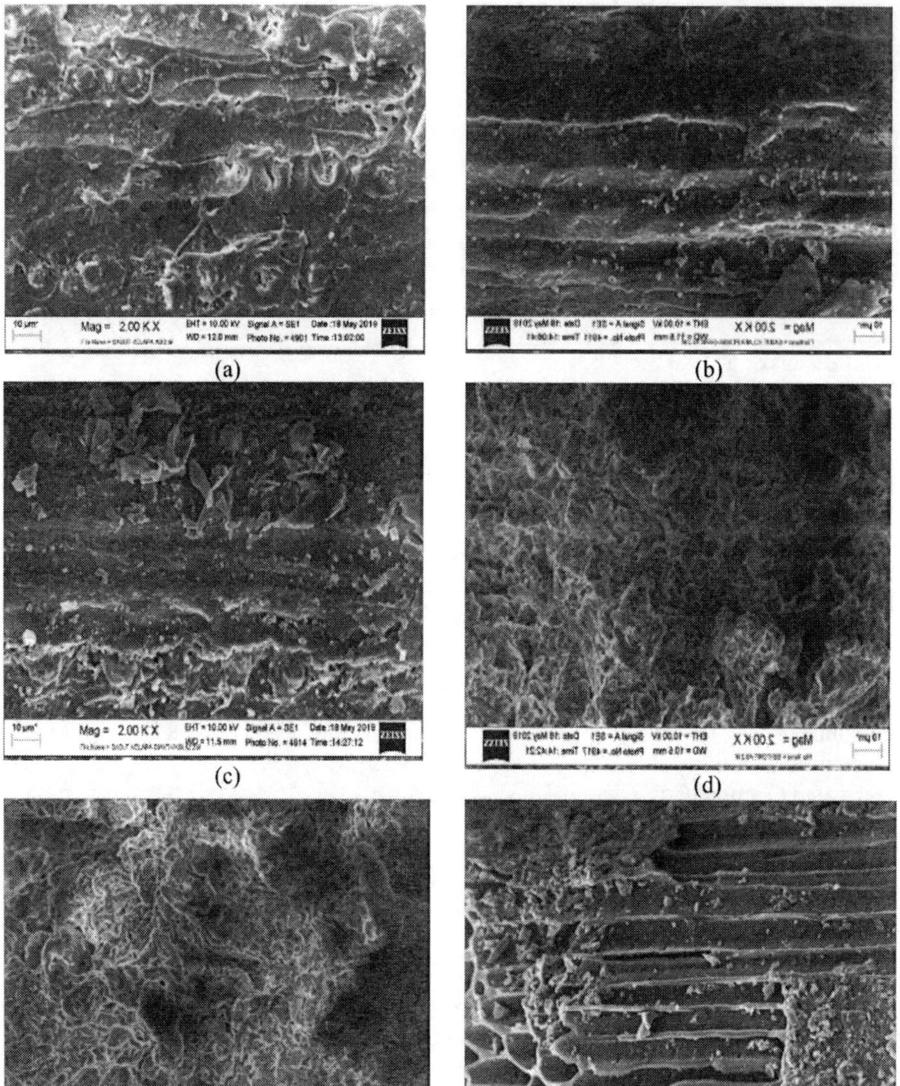

Figure 1. SEM photographs of (a) pure coconut coir, (b) coconut coir charcoal without activation, (c) coconut coir charcoal with activation, (d) pure bentonite (e) bentonite with activation and (f) activated coconut coir charcoal: activated bentonite (70:30) by magnification 2000 times.

Coconut coir charcoal with chemical activation of HCl has a more regular pore distribution with a greater number of pores than without chemical activation. In addition, in coconut coir charcoal with HCl activation has much higher visible surface pores compared to coconut coir charcoal without activation. This is because activation with HCl is more able to dissolve impurities so that more pores are formed, and the adsorbate adsorption process becomes more higher than without activation which has fewer number of pores. Carbon pores without activation, was still covered by some impurities. Coconut coir charcoal activated by HCl solution shows a change with appearance of the number of open pores [9].

In Figure 1 (d) pure bentonite shows that the pore morphology is still closed, in Figure 1 (e) bentonite with activation begins to appear open pores, and in Figure 1 (f) activated coconut coir charcoal: activated bentonite (70:30) the open pores became larger. This shows that chemical activation in bentonite can open pores bigger than in pure bentonite. The pore morphology of activated coconut coir charcoal: activated bentonite (70:30) is greater than coconut coir adsorbent types and bentonite so that the distribution of activated coconut coir : activated silica (70:30) is more regular with more pores than the type of adsorbent coconut coir and bentonite [8].

3.3. Fourier Transform Infra-Red (FTIR)

Characterization of Fourier Transform Infra-Red (FTIR) was conducted to determine the functional groups or components contained in the adsorbent of activated coconut coir charcoal: activated bentonite. The results of the FTIR analysis can be seen in Figure 2.

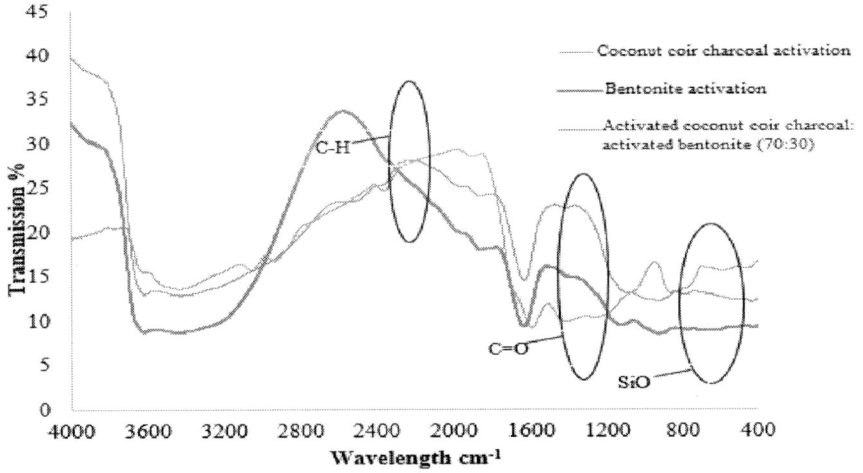

Figure 2. FTIR spectrum for several adsorbents.

Table 2. IR Interpretation of several prepared adsorbents.

Functional Groups	Bond	*Wavenumbers Range* (cm^{-1})			
		Reference	Activated Coconut Coir Charcoal	Acivated Bentonite	Activated Coconut Coir Charcoal: Activated Bentonite (70:30)
CH_4	C–H	3000-2700	-	-	-
Aldehide, Keton, Carboxyl acid	C=O	1900-1650	1653	1859,37	1840,04
SiO_2	Si-O-Si	1130-1000	1016,48	1082,06	1100,37

Source: Ma et al. (2015).

Based on Table 2 the FTIR spectrum of activated coconut coir charcoal: activated bentonite shows the shifting of wave numbers. This shows that the activation process affects the absorption intensity at the wavelength. Coconut coir adsorbent has several functional groups which are suitable for the structure of activated carbon in general which contains C-H, C = O, and Si-O-Si which are used as active groups to absorb adsorbate.

3.4. X-rays diffraction (XRD) analysis

The characterization of XRD aims to determine the crystalline structure of the coconut coir and bentonite-based adsorbents at an angle of 2θ. The following analysis test results using XRD for 5 types of adsorbents in Figure 3.

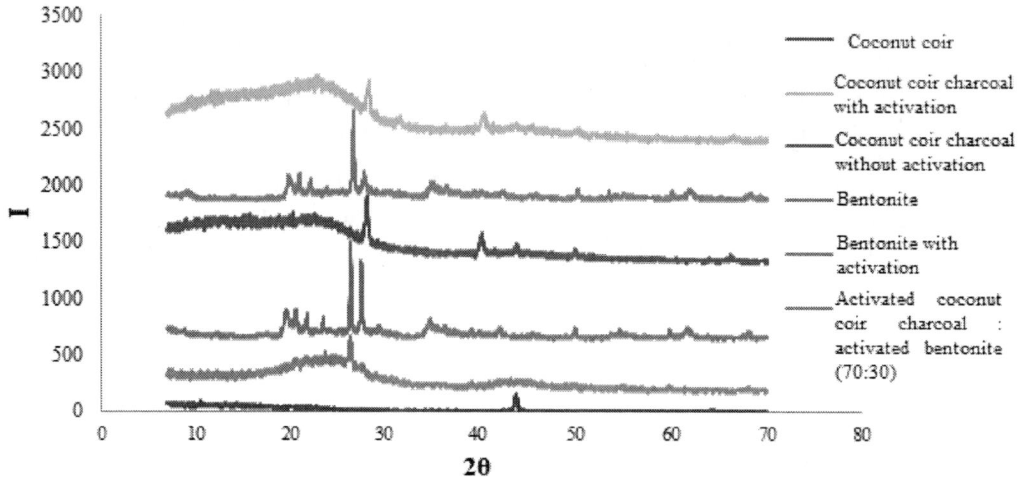

Figure 3. XRD profile for several prepared adsorbents.

Figure 3 shows that the sharp peak region is a characteristic of silicon oxide (SiO_2) and the reflection of XRD results also shows crystalline form. The success of chemical activation in coconut coir and bentonite adsorbents can be studied through testing of coconut coir charcoal samples before activation and after chemical activation, pure bentonite and bentonite after chemical activation, and the combination of coconut coir and bentonite adsorbent by using X-Ray Diffraction (XRD) to see the composition level that affects the adsorbent. Coconut coir charcoal without activation shows a peak at 2θ namely: 22.46° (d = 3.95537 Å) indicating the present of SiO_2, then coconut coir charcoal with activation shows a peak at 2θ namely: 23,16° (d = 3,90052 Å) which shows the present of SiO_2. On sample of pure bentonite shows peak at 2θ namely: 19.68° which shows SiO_2, on bentonite with activation shows peak at 2θ namely: 20.01° (d = 4.43469 Å) which shows SiO_2 and on coconut coir: bentonite shows a peak at 2θ namely: 23.82° (d = 3.73253 Å) which shows SiO_2. Activation of coconut coir charcoal decreases a peak intensity of coconut coir charcoal compared to coconut coir charcoal before activation at the diffraction angle from 2θ namely: 22.46° (d = 3.95537 Å) to 2θ namely: 23.16° (d = 3,90052 Å). The new diffraction angle shows that the activation of coconut coir charcoal causes coconut coir charcoal to be more amorphous compared to before activation. The activation process which involves chemical activation and high temperature heating causes a change in the structure of the carbon aromatic compound bonds and the partial breakdown of the double bonds between carbon atoms into a single bond. This is largely due to the decomposition of oxygen-containing bonds as well as the removal of carbon atoms due to the effect of pressure which makes the existing layers fragmented.

Table 3. The composition of each adsorbents.

Type of Adsorbent	Composition (%)		
	SiO_2	Al_2O_3	Fe_2O_3
Coconut coir charcoal	51.3	34.5	14.3
Activated coconut coir charcoal	61.3	28.9	9.8
Pure bentonite	43.5	34	22.5
Activated bentonite	81.4	14.3	4.3
Activated coconut coir charcoal: activated bentonite (70:30)	73.5	20.6	5.8

From these compositions, it can be seen that after activation process, the composition change of SiO_2, Al_2O_3 and Fe_2O_3 occurred, the amount of silica composition increased for both samples of activated coconut coir charcoal and activated bentonite. The activators are useful for increasing pore volume on the surface of the adsorbent. The amount of silica composition for samples of combination activated coconut coir charcoal and activated bentonite is between the value of activated coir charcoal composition and activated bentonite.

3.5. The effect of adsorbent types on adsorption efficiency and adsorption capacity of lead metal ion

In this study the adsorption process of lead metal ion solution by prepared adsorbents was measured by using an AAS tool. Figure 4 shows that adsorption efficiency increase with increasing of the contact time until reaching optimum contact time. After the optimum contact time is achieved, the adsorption efficiency value decreases significantly. However, for sample of activated coconut coir charcoal: activated bentonite (70:30) after the optimum contact time is reached the adsorption efficiency value decreases and after that tends to be constant. The highest adsorption efficiency value for lead metal ion as much as 35 % was obtained after 90 minutes contact time for sample of activated coconut coir charcoal: activated bentonite (70:30). Table 3 also show that the SiO_2 content found in sample of activated coconut coir charcoal: activated bentonite (70:30) is higher than in sample of activated coconut coir charcoal and it is slightly below the activated bentonite. From the XRD results, the silica content found in sample of activated coconut coir charcoal: activated bentonite (70:30) is as high as 73.5%. Silica plays an important role in the adsorption process because it has good adsorption and ion exchange capabilities [9]. Besides, the presence of greater Al_2O_3content in sample of activated coconut coir charcoal: activated bentonite (70:30) makes the ability of this adsorbent better than other adsorbents.

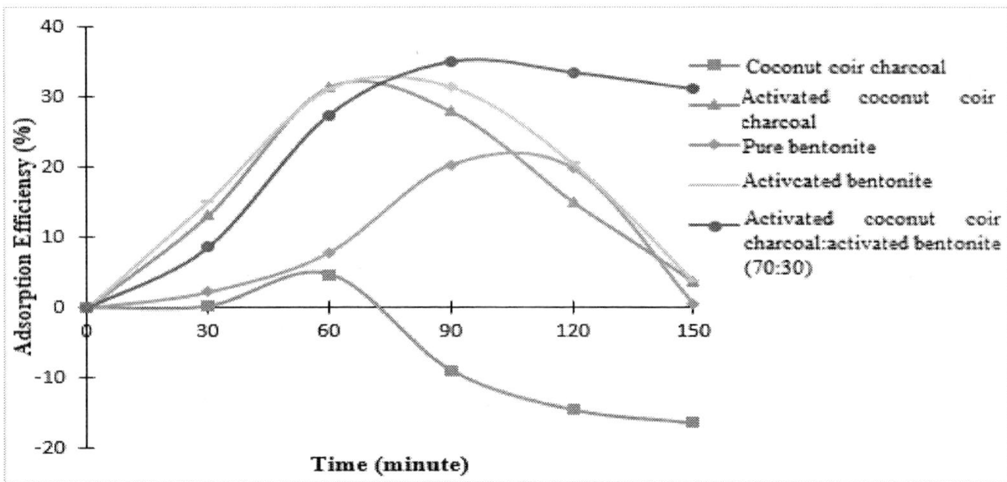

Figure 4. Relationships among contact time (minute), types of adsorbent and adsorption efficiency.

Adsorption capacity shows the ratio of the amount of adsorbate that be able be adsorbed by the adsorbent. The relationship between the type of adsorbent and the adsorption capacity can be seen in Figure 5.

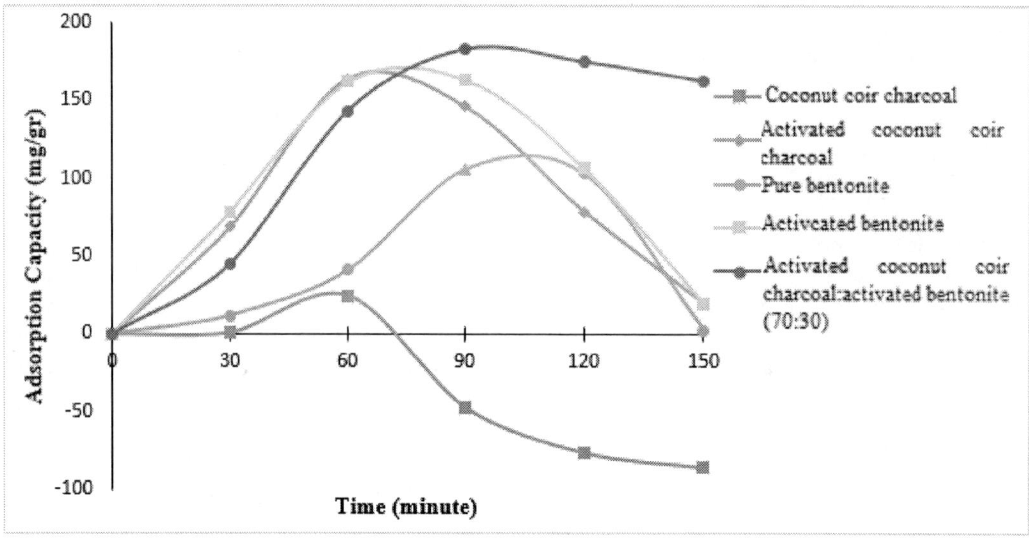

Figure 5. Relationships among contact time (minute), type of adsorbents and adsorption capacity.

Figure 5 shows that adsorption capacity will increase with increasing of contact time between adsorbent and lead metal ion, and then after reaching the optimum adsorption time there is a decrease in lead adsorption and tends to decrease. The optimum adsorption time of lead metal ion by the adsorbent is 90 minutes. The highest adsorption capacity was 182.71 mg/g. Figure 5 show that the type of adsorbent and contact time affects significantly the adsorption capacity, and for the effect of contact time, the adsorption capacity would decrease after 90 minute of adsorption process.

4. Conclusion

Characterization of coconut coir, coconut coir charcoal, pure bentonite, activated bentonite, and activated coconut coir charcoal: activated bentonite (70:30) indicated that the water content and ash content have met the activated carbon quality standard based on SNI No. 06-3730-1995.SEM characterization indicated that activated coconut coir charcoal: activated bentonite (70:30) has the largest pore surface compared to coconut coir and bentonite typesadsorbent.FTIR characterization showed that the absorption of the C − H, C = O, and Si-O-Si groups dominates the five adsorbents. From the XRD test results, the silica content in sample of activated coconut coir charcoal: activated bentonite (70:30) was 73.5%.

References

[1] Abdullah A, Saleh A and Novianti I 2013 *Jurnal Al Kimia,* **1** 32-43.
[2] Etim U J, Umoren S A and Eduok U M 2016 *J. Saudi Chem. Society,* **20** 67-76.
[3] Tan I A W, Ahmad A L and Hameed B H 2008 *J. Hazard. Mater.,* **153** 709-717.
[4] Idrus R, Lapanporo B P and Putra Y S 2013 *Prisma Fisika,* **1** 50-55.
[5] Okafor P C, Okon P U, Daniel E F and Ebenso E E 2012 *Int. J. Electrochem. Sci.,* **7** 12354-69.
[6] Rahayu L H, Purnavita S and Sriyana H Y 2014 *Majalah Ilmiah Momentum,* **10** 47-53.
[7] Mariana and Oobuchi Y 2004 *J. Chinese Inst. Chem. Eng.,* **35** 491-494.
[8] Haji H G, Pari G, Nazar M and Habibati 2013 *Int. J. Eng. Sci.,* **5** 89-94.
[9] Sureshkhumar A and Mishra S 2018 *Brazilian J. Chem. Eng.,* **35** 1105-16.
[10] Hardyanti N and Suparni S R 2007 *Jurnal Presipitasi,* **2** 28-33.

IOP Conf. Series: Materials Science and Engineering 1087 (2021) 012045 doi:10.1088/1757-899X/1087/1/012045

The effect of particle size and dosage on the performance of Papaya seeds (*Carica papaya*) as biocoagulant on wastewater treatment of batik industry

R S Putra[1,2*], N H Airun[1,2]

[1]Department of Chemistry, Faculty of Mathematics and Natural Sciences, Universitas Islam Indonesia, Jl. Kaliurang km. 14, Yogyakarta 55584, Indonesia
[2]Environmental Remediation Research Group, Faculty of Mathematics and Natural Sciences, Universitas Islam Indonesia, Jl. Kaliurang km. 14, Yogyakarta 55584, Indonesia

*E-mail: rudy.syahputra@uii.ac.id

Abstract. In this study, the performance of papaya seeds (*Carica papaya*) as biocoagulant was evaluated using Batik wastewater which referred to the water sample. The effectiveness of particle size and dosage of papaya seeds were evaluated based on turbidity removal (%), total dissolved solids (TDS), and electrical conductivity (EC). The Tyndall effect of clear solution was also evaluated in this study. The results showed that the turbidity removal of water sample achieved as much as 92.2%, and had followed by the decreasing of TDS and EC. The optimum results from the decrease in turbidity, TDS, and EC were obtained at the particle size of 250 mesh and the biocoagulant dosage of 1000 mg. The measurement of the Tyndall effect has concluded that the water sample has to remain the colloidal particle even though the treatment water as clear as the quality standard of freshwater.

1. Introduction

The Sustainable Development Goals are 17 goals with 169 world targets with a deadline set by the United Nations as the world development agenda for the benefit of humans and planet Earth. A total of 169 targets, all of which cover health, energy, sanitation, social and environmental aspects. Environmental issues are still one of the obstacles in the achievement of these targets [1]. One of the causes of serious environmental problems today is related to the textile industry, for example, wastewater from the batik industry. Approximately 80% of water is used in the batik process including coloring which has turned into polluted water [2]. Some methods that can be used to deal with textile wastewater are coagulation-flocculation, precipitation, electrodialysis, electroflotation, and adsorption [3]. Coagulation-flocculation method is a method of treating wastewater that aims to reduce pollutants and colloidal particles in liquid waste. The mechanism of floc formation in the coagulation-flocculation process consists of three stages, namely the destabilization of colloidal particles, the stage of microphilic formation, and the stage of macrophilic formation [4]. The substances used in water treatment can be either synthetic or natural coagulants. However, the continued use of synthetic coagulants has trigger Alzheimer's disease and neurotoxicity [5-6]. Therefore, one way to reduce the adverse effects of synthetic coagulants is to use natural coagulants (biocoagulants). One of the natural ingredients that make a natural material can be used as a coagulant is a biocoagulant based on the protein [7].

Papaya fruit contains a large number of small black seeds. Papaya fruit and seeds have a large protein content. Papaya seeds contained 25.1% protein, 8.2% ash, and 45.6% crude fiber [8]. Papaya seeds can be used as a biocoagulant due to the presence of positively charged proteins that bind to negatively charged particles (silt, clay, bacteria, and toxins, etc.), then it can produce flocks that will precipitate and produce clear water (adsorption & charge neutralization). Papaya seed powder can bond with solid particles in water and then settle to the bottom [9]. The use of biocoagulant has various advantages including being environmentally friendly, easily obtainable raw materials, affordable prices, simple synthesis processes, and biodegradable. Therefore in this study, the performance of papaya seed (*Carica papaya*) as biocoagulant based on the protein was evaluated in terms of turbidity removal, TDS, EC, and Tyndall effect from the treated water sample.

2. Material and methods

2.1. Materials and instruments

Papaya seeds (*Carica papaya*) were obtained from fruit traders around Universitas Islam Indonesia. The seed was washed and then dried at 110 °C for 10 h before characterized by FT-IR spectroscopy (PerkinElmer Spectrum Version 10.5.1, USA). The dried seeds were then crushed and sieved to get fine powder with a size of 150; 200 and 250 mesh before it was stored in a plastic container. The evaluation of biocoagulant was carried out using wastewater from the batik industry based on turbidity (Waterproof Portable Turbidimeter TN100, Eutech, Netherland), total dissolved solids (TDS), and electrical conductivity (EC) (Multi Checkers Hanna HI 9813-5, Romania). The wastewater sample was obtained from the Sekar Idaman Batik Industry, Yogyakarta, Indonesia (GPS points, 7°40'56.1"S 110°22'04.1"E).

2.2. Evaluation of biocoagulant

The treatment of batik wastewater was carried out by the coagulation-flocculation process. As much as 250 mL of water sample has been added to four glass beakers, and then 150 mesh papaya seed powder was added with the dosage of 250, 500, 750, and 1000 mg. Furthermore, the coagulation and flocculation were carried out with rapid stirring at 100 rpm for 5 mins, followed by a lower speed at 40 rpm for 24 mins, then left for 60 mins in Mohr tube before filtered by Whatman 42 filter paper (GE Healthcare Life Sciences, USA). After that, the final characterization of the water sample was carried out based on turbidity, electrical conductivity, and Tyndall effect. The turbidity removal was calculated according to Equation 1. The Tyndall effect was evaluated by measuring the remaining light passing through the solution using a light meter (Lutron LX-103, USA) after the sample cell at a distance of 16.5 cm as shown in Figure 1. A similar evaluation was carried out for the coagulant size of 200 and 250 mesh.

$$\text{Turbidity removal (\%)} = [(T_i - T_f) / T_i] \times 100 \quad (1)$$

Where T_i is the initial turbidity (NTU) of the water sample and T_f is the turbidity (NTU) after the process.

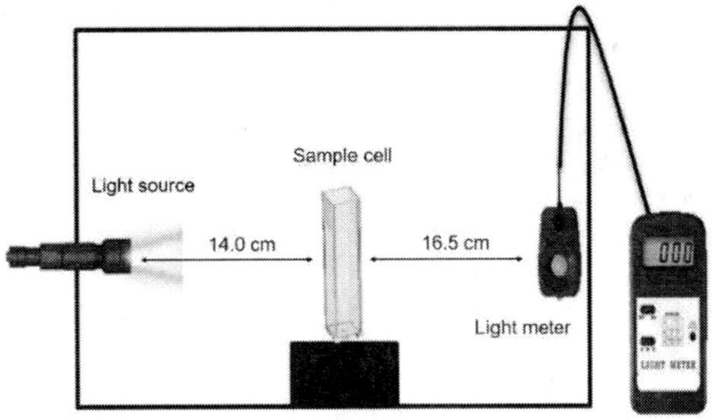

Figure 1. Schematic diagram of Tyndall effect measurement on the sample solution.

3. Result and discussion

3.1. Characterization of biocoagulant

The functional groups in the biocoagulant were measured to identify -OH, -COOH, and NH groups that infer possible intramolecular interaction between polymers biocoagulant and particles dissolved in a solution [10]. Figure 2 shows the FTIR peaks of functional groups in papaya seed (*Carica papaya*) as biocoagulant. The result showed that the molecule had a broad peak at 3280.41 cm^{-1} due to the O-H band, 1709,31 cm^{-1} due to C=O, 1640.11 cm^{-1} due to N-H, and C-O at 1035.45 cm^{-1}. The group of -OH, -CO and -NH in the spectral peak affirm the presence of hydroxyl, carboxyl acid, and amina groups that causes papaya seed to have more effect on the coagulation abilities [11].

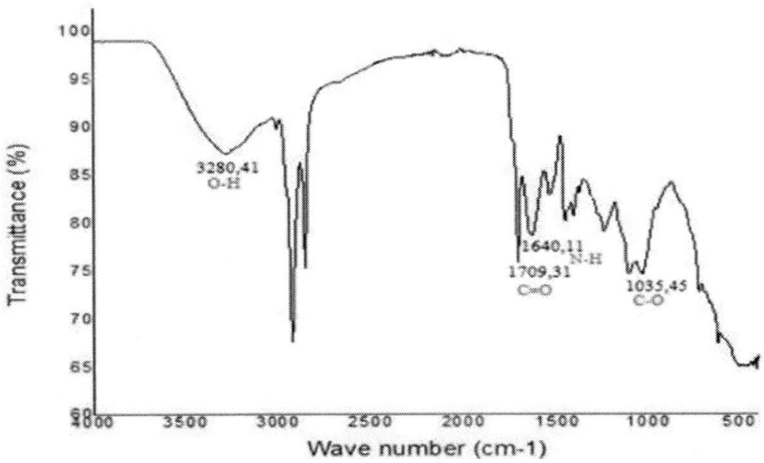

Figure 2. IR peaks of active groups in papaya seed (*Carica papaya* L.).

3.2. Effect of particle size and biocoagulant dosages on the turbidity removal

Figure 3 shows the effect of particle size and biocoagulant dosage on the removal of turbidity from the water sample. In general, the turbidity removal of water samples increased with the increase of particle size and biocoagulant dosage. In this regard, the biocoagulant with 150, 200, and 250 mesh of particle size obtained as much as 89.3%, 86.2%, and 92.2% of turbidity removal in 1000 mg/250 mL biocoagulant dosage. Based on these results, it can be concluded that the particle size of 250 mesh and the biocoagulant dosage of 1000 mg was the optimum process on the turbidity removal from the water sample. Particle size and coagulant dose are two parameters that play an important role in determining the efficiency of coagulation-flocculation. Particle size and coagulant dose are two parameters that play an important role in determining the efficiency of coagulation-flocculation [12]. The particle size affects the performance of the coagulant. The smaller the size of the coagulant particles, the larger the surface area of the coagulant. This can make the aggregation process between the coagulant and the particles in the solution was easier to be occurred, so that the potential for the aggregation process between solid particles and coagulants will be greater. It causes floc to be more easily formed and then separated from the wastewater easily. In addition, the higher the number of coagulant doses the greater the aggregation occurred between the particles. then it causes the particle destabilization. When the dose reaches the optimum concentration causes a greater number of particles in the solution to gather and settle so that flocs can form, and turbidity can decrease.

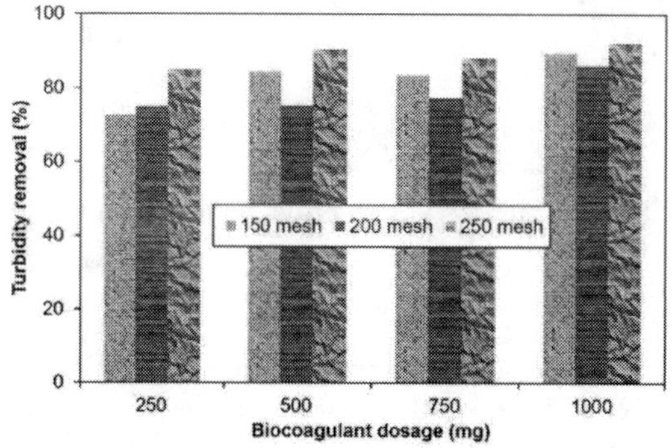

Figure 3. Effect of particle size and dosage on decreasing the turbidity of wastewater.

3.3. Effect of particle size and biocoagulant dosages on the TDS and EC

Figure 4 shows the effect of particle size and biocoagulant dosage on the TDS and EC measurements. Total dissolved solids (TDS) and electrical conductivity (EC) have indicated how much the presence of mineral elements, charged macromolecules, and other ionic compounds may be dissolved or dissociated into the solutions [13]. In general, the TDS and EC of water samples decreased by the decrease of particle size. In addition, the lowest TDS and EC were obtained when the particle size of 250 mesh was used for all biocoagulant dosage applied for. However, in this regard, the TDS would not be always decreased by the increase of biocoagulant dosage at the particle size of 250 mesh (see Figure 3a). In the optimum dose of biocoagulant, the colloidal particle has been neutralized and precipitated. When the biocoagulant dosage was added excessively in the solution, the water becomes turbid again because the biocoagulant cannot interact with the colloidal particles of different charge [14].

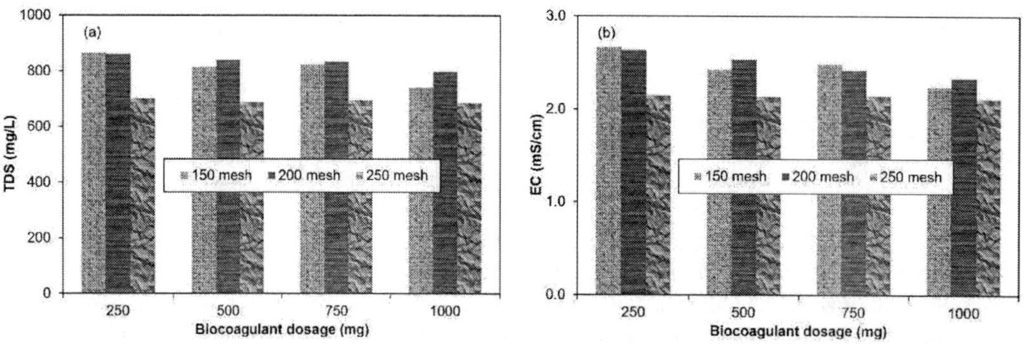

Figure 4. Effect of biocoagulant particle sizes and dosage on the TDS (a) and EC (b).

3.4. Effect of particle size and biocoagulant dosage on Tyndall effect

The Tyndall effect was evaluated based on light scattering by colloidal particles. The light beam passed through the solution and then the remaining light intensity was measured to indicate how much the colloidal particle remaining in the solution even though the solution as clear as the quality standard of freshwater. If the solution contains a high colloidal particle, the low measuring of light intensity after the sample cell was detected by the light meter in (see Figure 1). In general, the increasing biocoagulant dosage used in the treatment has resulted in the high intensity of light measurement as shown in Table 1. This means that the treatment solution has contained a low concentration of colloidal particles [15]. Also, the increase of particle size has decreased the colloidal particle concentration in the solution, so that the treatment has decreased the turbidity of the solution. The biocoagulant dosage of 1000 mg and a particle size of 250 mesh hash obtained a clear water treatment as well as freshwater.

Table 1. Tyndall effect (light measurement) on the wastewater quality after treatment

Biocoagulant dosage (mg)	Initial	250	500	750	1000	Tap water
	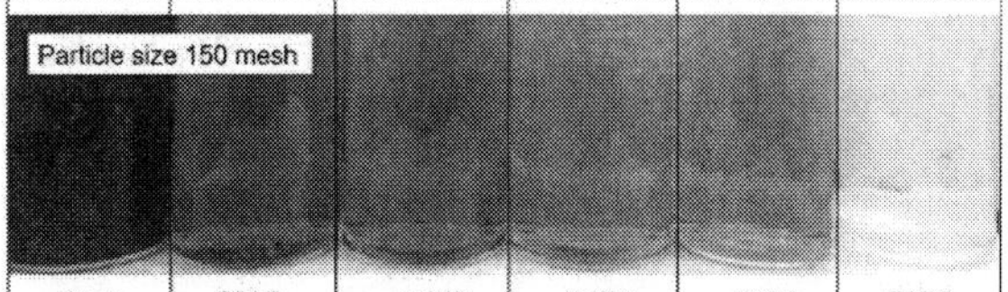					
Light measurement (Lux)	78.3	231.3	318.7	313.3	405.0	743.0
	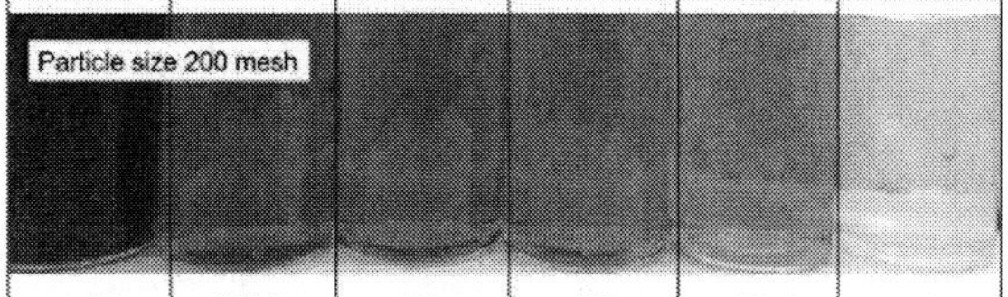					
Light measurement (Lux)	78.3	224.3	300.0	353.3	401.3	743.0
	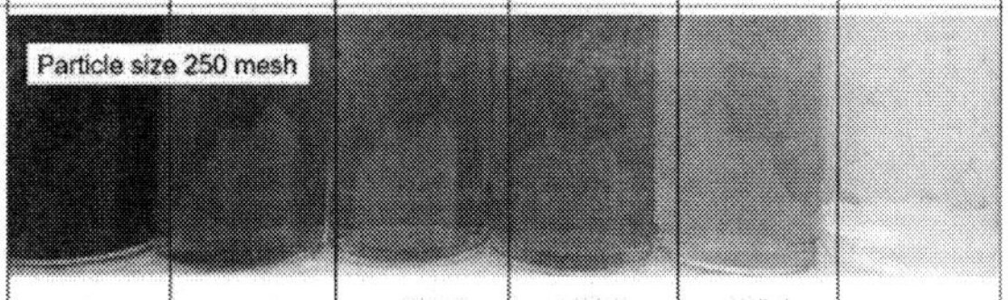					
Light measurement (Lux)	78.3	306.7	449.7	398.7	538.0	743.0

4. Conclusions

The result showed that the biocoagulant of papaya seed (*Carica papaya*) had a good performance in the coagulation-flocculation process in the treatment of Batik wastewater. This can be seen from the decreasing of turbidity removal (%), TDS, and EC from the sample solution. Based on the results showed that the particle size and biocoagulant dosage has greatly affected the performance of the biocoagulant. It was found that the smaller the particle size and the higher by coagulant dosage, the increasing of turbidity removal, and the decreasing TDS, and EC. The results showed that the particle size of 250 mesh and 1000 mg of biocoagulant dosage was the highest turbidity removal for as much as 92.2%, respectively. In this regard, the Tyndall effect of the treatment solution has concluded that the addition of a high dose (1000 mg) and particle size (250 mesh) of biocoagulant decreased the remaining colloidal particle in the wastewater to produce clear water as well as freshwater.

5. Acknowledgments

The authors would like to thank the Ministry of Education and Culture and Ministry of Research and Technology of the Republic of Indonesia for their financial support through Excellence Higher Education Basic Research (PDUPT) 2019-2020 (Research contract: 227/SP2H/LT/DRPM/2019; LLDIKTI and UII contract: 1627.1/LL5/PG/2020).

References

[1] Utama Y J, Ambariyanto A, Zainuri M, Darsono D, Setyono B, Widowati and Putro S P 2018 *J. Physics: Conf. Ser.,* **1025**

[2] Yulianto A, Hakim L, Purwaningsih I and Pravitasari V A 2009 *Jurnal Teknologi Lingkungan,* **5 (1)** 6-11.

[3] Ali Rasool M, Tavakoli B, Chaibakhsh N, Pendashteh A R and Mirroshandel A S 2016 *Ecol. Eng.,* **90** 431-437

[4] Teh C Y, Budiman P M, Shak K P Y and Wu T Y 2016 *Ind. Eng. Chem. Res.,* **55 (16)** 4363-89.

[5] Campbell A 2002 *Nephrol. Dial. Transplant.,* **17 (2)** 17-20.

[6] Mirza A, King A, Troakes C and Exeley C 2017 *J. Trace Elem. Med. Biol.,* **40** 30-36.

[7] Prihatiningtyas E and Effendi A J 2012 *Jurnal Teknik Lingkungan* **18** 97-105.

[8] Maisarah A M, Asmah R and Fauziah O 2014 *J. Nutr. Food Sci.,* **4 (2)** 1000267.

[9] Arya Chandran J and George D 2018 *Int. J. Sci. Eng. Res.* **6 (3)** 41-46.

[10] Kumar V, Othman N and Asharuddin S M 2017 *MATEC Web of Conferences* **103** 06016.

[11] Verma S, Mehraj I, Jain A and Ray A P 2015 *Int. J. Innov. Sci. Eng. Technol.,* **2 (5)** 693-696.

[12] Ngadi N and Ida N 2013 *Jurnal Teknologi* **1** 1-7.

[13] Magaji U F, Sahabi D M, Abubakar M K and Muhammad A B 2015 *Indonesian J. Edu. Stud.* **4 (2)** 19-26.

[14] Mudenur C, Sorokhaiban L G, Bhandari V, Raja S and Ranade V V 2016 *ACS Sust. Chem. Eng.* **4 (5)** 2495-07.

[15] Jerlov N G and Kullenberg B 1953 *Tellus* **5 (3)** 306-307.

The simultaneous of electroflotation and biocoagulation on the treatment of peat water using mung bean (*Vigna radiata*) as natural coagulant

R S Putra[1, 2*], N S Tyagustin[1], C I Putri[1]

[1]Department of Chemistry, Faculty of Mathematics and Naturals Sciences, Universitas Islam Indonesia, Jl. Klaiurang km. 14, Yogyakarta 55584, Indonesia
[2]Environmental Remediation Research Group, Department of Chemistry, Faculty of Mathematics and Natural Sciences, Universitas Islam Indonesia, Jl. Kaliurang km. 14, Yogyakarta 55584, Indonesia

*E-mail: rudy.syahputra@uii.ac.id

Abstract. The simultaneous of electroflotation and biocoagulation performance using mung bean flour (*Vigna radiata*) with 250 mesh as a natural coagulant on the treatment of peat water has been carried out. The electroflotation was carried out using graphite as anode and stainless steel as a cathode electrode with a DC constant voltage of 20 V for 30 mins operation. The mung bean in the doses of 1.0, 1.5, and 2.0 g/500 mL was used to treat the peat water. The effectiveness of the simultaneous of electrofotation and biocoagulation was in term of turbidity, total dissolved solids (TDS), and pH. Initially, the peat water sample had the parameters as much as 224.3 NTU, 1076 mg/L, and 2.7 respectively for turbidity, total dissolved solids (TDS), and pH. The results showed that the simultaneous of electroflotation and biocoagulation (EB) process had decreased those parameters as much as 84.5%, 70.2%, and changes the pH to be 6.8.

1. Introduction

The peat water is mostly obtained in lowland peatland areas. The characteristics found in peat water are a high degree of acidity (pH), a concentrated color intensity, and a high concentration of suspended particles [1]. One of the main compounds contained in peat water is humic acid which is known to damage endothelial cells when accumulated in the human body [2]. Humic acid in peat water can form a complex compound with metals through the adsorption of complex bonds and chelates [3]. Heavy metals such as iron (Fe) have also potential as toxicant causes for organisms, so it will be very dangerous if it accumulates in the human body and causes pollution in the water system [4]. The conventional methods commonly used in water remediation such as peat water are coagulation and flocculation [5]. However, the process of both methods still has difficulties caused by the complexity of the process and its high-cost rate [6]. Electrofotation is a method of purification by forming hydrogen gas and oxygen through electrolysis of water which causes particles dispersed in water to rise to the surface [7]. Coagulation is a chemical additive process commonly used in water treatment [8]. The coagulants that are often used in the coagulation process are aluminum, ferric chloride, and poly aluminum chloride (PAC) [9]. However, chemical coagulants based on aluminum compound can cause damage to the nervous system if it accumulates in the human body as in senile dementia (Alzheimer's) [10].

The use of plant-based coagulants (biocoagulant) is an alternative to be applied in clean water processing because the biocoagulant is non-toxic and had a high biodegradable level [8]. In this study, the coagulation process using a protein-based biocoagulants interaction with peat water samples is evaluated. Mung bean flour (*Vigna radiata*) is evaluated as biocoagulant as it contains a protein that can used to decrease the turbidity [11]. The enhancement of electroflotation using mung beans flour in the peat water treatment is also evaluated.

2. Materials and methods

2.1. Designed electroflotation reactors and voltage optimization

The electroflotation reactor was made from a glass plate with the dimensions of 10 cm (l) x 10 cm (w) x 15 cm (h) and can accommodate a 500 mL of peat water samples. The electrodes were used graphite with dimensions of 5 cm (l) x 7 cm (w) as an anode and stainless-steel net wire with dimensions of 5 cm (l) x 7 cm (w) as a cathode. Voltage optimization in the electroflotation process was carried out using 0.001 M NaCl as electrolyte solutions in the DC constant voltage of 10, 20, and 30 V for 30 mins. The bubbles were generated at the electrode surface and the image was captured by digital camera (Nikon Coolpix 3200, Japan)-assisted with 8× magnification of magnifying glass (Figure 1). Direct counting of bubbles from photo image was classified based on size categories namely <0.5, 0.5, and 0.5> mm. The bubble distribution was also verified by ImageJ software.

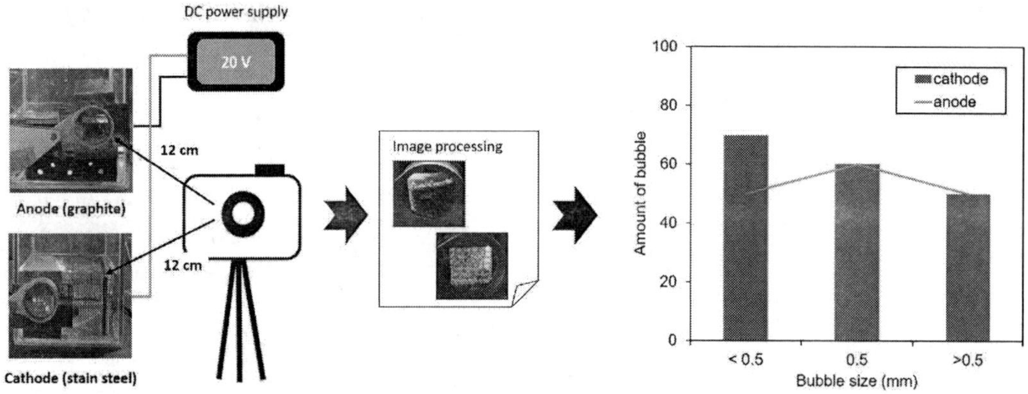

Figure 1. Bubble distribution was counting using a camera and image processor.

2.2. Electroflotation and biocoagulation (EB) process

Peat water sample was prepared by dissolving 0.05 g of humic acid into 1 L dionized water to be 50 mg/L of peat concentration according to elswhere publication [12] (Sudoh , 2015) The biocoagulant used in this study was mung bean flour (*Vigna radiata*). Dried mung bean seeds were mashed and sieved to obtain 250 mesh particle size. The biocoagulant dose used in the study was 1.0, 1.5, and 2.0 g in 200 mL of peat water. Biocoagulation of peat water was carried out using a magnetic stirrer with 30 secs. of rapid stirring and 30 mins. of slow stirring. The peat water was then allowed to stand for 30 mins. until mass precipitate was formed. The clarified solution was then measured the turbidity (Turbidimeter, TN100 Eutech, Netherland), total dissolved solids (TDS), and pH (pH/TDS meter, H19813-5, Romania). Furthermore, the optimum dose of biocoagulant obtained in this step further was used in the electroflotation of peat water treatment. The simultaneous of electroflotation and biocoagulation (EB) process of peat water treatment was carried out in 500 mL at constant voltage of 20 V for 30 min. The performance of EB process was evaluated based on the percentage removal of turbidity, and total dissolved solids (TDS) as shown in Eq-1 and pH solution.

$$\text{Removal (\%)} = [(\text{Initila score} - \text{Final score}) / \text{Initial score}] \times 100 \quad (1)$$

3. Results and discussion

3.1. Voltage optimization in electroflotation

Voltage optimization was conducted using an electrolyte solution to determine the optimum voltage applied in peat water treatment based on the size of the generated bubbles as described above. The results of bubble distribution were shown in Table 1. Bubble gases (i.e. O_2 and H_2) were formed in the electrolysis of water as shown in Eq-2 and Eq-3.

$$\text{Anode: } 2H_2O_{(aq)} \rightarrow O_{2\,(g)} + 4\,H^+_{(aq)} + 4e^- \tag{2}$$
$$\text{Cathode: } 2H_2O_{(aq)} + 2e^- \rightarrow H_{2\,(g)} + 2\,OH^-_{(aq)} \tag{3}$$

Table 1. Influence of constant voltage on the bubble distribution

<table>
<tr><th>Anode</th><th>Cathode</th><th>Bubble distribution</th></tr>
<tr><td>10 V
</td><td>10 V
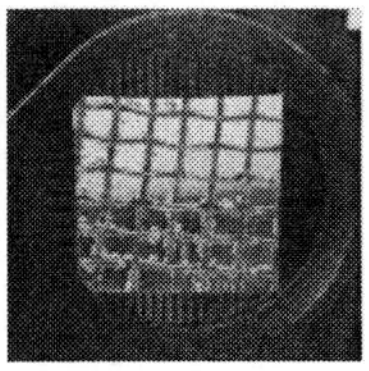</td><td>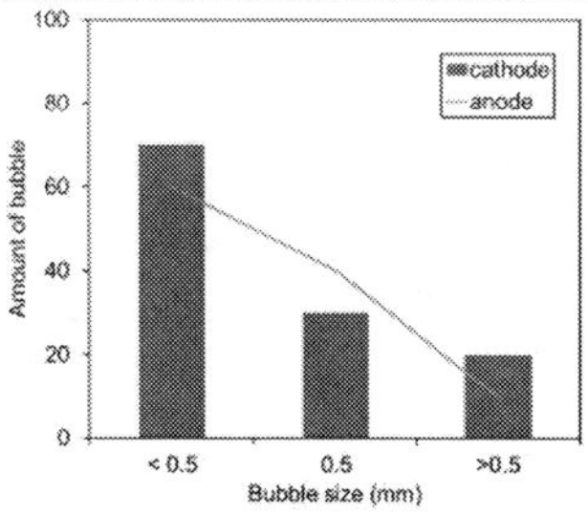
</td></tr>
<tr><td>20 V
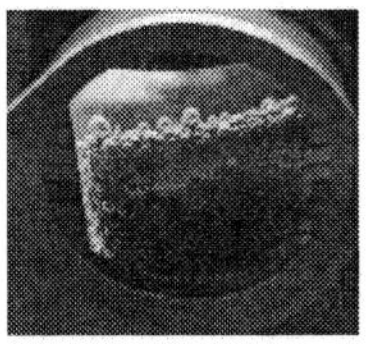</td><td>20 V
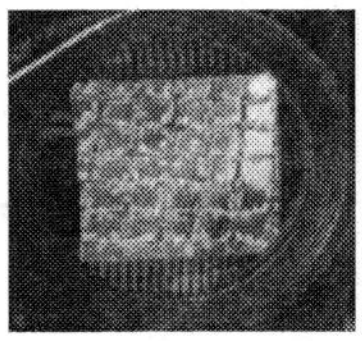</td><td>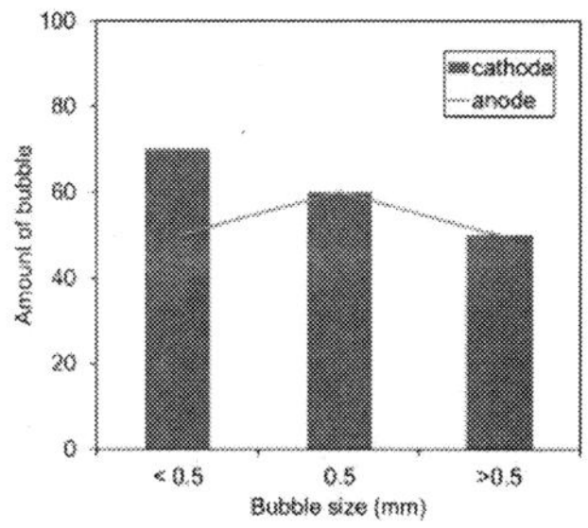
</td></tr>
<tr><td>30 V
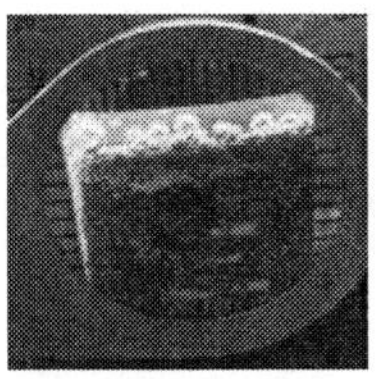</td><td>30 V
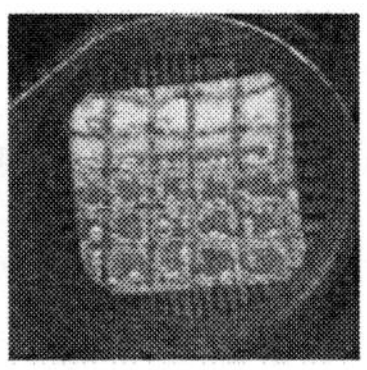</td><td>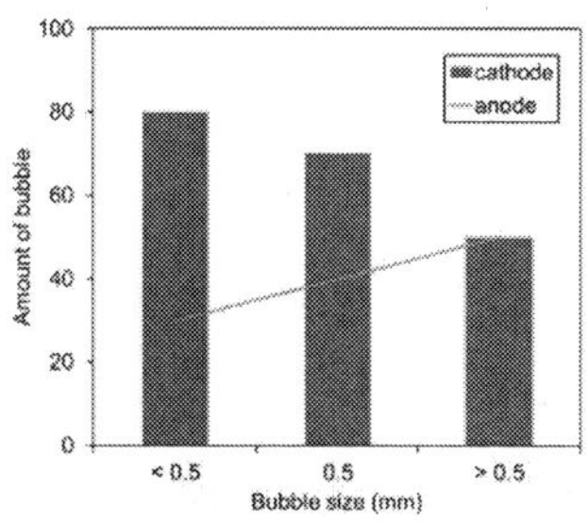
</td></tr>
</table>

The small number of bubbles were generated from the anode and cathode by the size of 0.5 mm and > 0.5 mm at a constant voltage of 10 V. While contrary results showed that small number of bubbles at the anode formed when 30 V of constant voltage was applied in the electroflotation. Finally, the constant voltage of 20 V was used on the treatment of peat water since it gave a uniformity in the bubbles size and continue bubble production at the anode and cathode which provide optimum flotation ability [13]. Figure 2 shows the comparison results of peat water treatment before and after the electroflotation.

Figure 2. Electroflotation of peat water (a) and the peat water treatment (b) before electrofotation (left) and after (right).

3.2. Optimization of biocoagulant dose

The biocoagulation of peat water was prepared using mung bean as biocoagulants. This fact could be based on the hydrophilic interaction between the protein contained in mung bean and peat water, so that facilitate the formation of the flock, which reduce the level of turbidity in the peat water [14]. Biocoagulant dosage were intended to determine the optimum dose to be used in peat water treatment using a simultaneous electroflotation and biocoagulation. Table 2 shows the optimum dosage of biocoagulant. A dose with 1.5 g of biocoagulant was the optimum concentration on the biocoagulation process, which used further on the simultaneous EB process.

Table 2. The optimum dosage of biocoagulant used in the biocoagulation process.

Dosage (g/500 mL)	Parameter				
	Turbidity (NTU)	TDS (mg/L)	pH	Sludge mass (%)	Light intensity (Lux)
Initial peat water	224.3	1076	2.7		152.3
Mung bean 1	111.6 (50.2%)	481 (55.2%)	6.1	61.7	311
Mung bean 1.5	77.7 (65.3%)	266 (75.2%)	6.8	72.2	323
Mung bean 2	208.6 (6.9%)	566 (47.3%	6.2	78	316

The percentage in parentheses was calculated using Eq-1 and pH was adjusted by the addition of 1.0 M NaOH.

3.3. Electroflotation-biocoagulation performance

The simultaneous electroflotation and biocoagulation (EB) process was carried out using 1.5 g of mung bean as biocoagulant in 200 mL peat water at a constant voltage of 20 V for 30 mins. Table 3 shows the results obtained from the simultaneous EB process. Biocoagulant was able to form micro-sized flocks in turbid water by van der Waals tensile forces, as well as neutralize the electric charge. The addition of biocoagulant formed the micro flock formed and then converted into macro flock, which uplifting to the surface by the bubble in the EB process [15]. This process had decreased the turbidity as much as 84.5% and TDS as much as 70.2%,. Also, pH of peat water changed from 2.7 to 6.8. Figure 3 shows the light intensity in the peat water, indicating the clearness of solution before and after treatment.

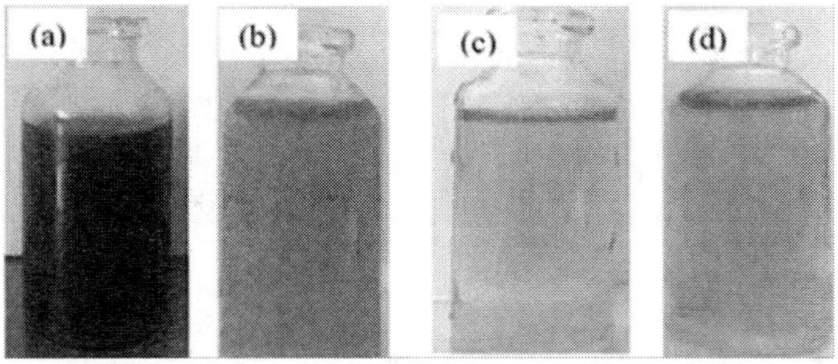

Figure 3. Initial peat water, 152.3 Lux (a), biocoagulation 1.5 g, 323 Lux (b), electroflotation 20 V, 361.6 Lux (c), and the simultaneous electroflotation-biocoagulation 1.5 g, 355 Lux (d).

Table 3. The results of the electroflotation-bicoagulation method using the optimum dose and voltage.

Peat water treatment	Parameter				
	Turbidity (NTU)	TDS (mg/L)	pH	Sludge Mass (%)	Light intensity (Lux)
Initial	224.3	1076	2.7		152.3
Electroflotation	30.2 (86 %)	305 (71.6 %)	5.0	57.1	361.6
EB 1.5 g	34.6 (84.5 %)	320 (70.2 %)	6.8	72.7	355

The percentage in parentheses was calculated using Eq-1

4. Conclusions

Based on the results presented in this study, it can be concluded that the mung bean as biocoagulant can improve the performance of electroflotation on peat water treatment on the decreasing of turbidity as much as 84.5% and TDS as much as 70.2%, and changed the pH 2.7 to 6.8 (closer to neutral freshwater).

5. Acknowledgments

The authors would like to thank the Ministry of Education and Culture and Ministry of Research and Technology of the Republic of Indonesia for their financial support through Excellence Higher Education Basic Research (PDUPT) 2019-2020 (Research contract: 26/E1/KPT/2020 and 227/SP2H/LT/DRPM/2020).

References

[1] Schwarzel K, Renger M, Sauerbrey R and Wessolek G 2002 *J. Soil Sci. Plant Nutr.,* **165** 479-486.
[2] Hseu Y C, Huang H W, Wang S Y, Chen H Y, Lu F J, Gau R J and Yang H L 2002 *Toxic. Applied Pharm.,* **182** 34-43.
[3] Kihara Y, Yustiawati, Tanaka M, Gumiri S, Ardianor, Hosokawa T, Tanaka S, Saito T and Kurasaki M 2012 *Environ. Toxic.,* **29** pp 916-925
[4] Zhuang Y, Han B, Chen R and Shi B 2019 *Water Res.,* **165** 1-8.
[5] Gone D L, Seidel J L, Batiot C, Bamorya K, Ligbana R, and Biemi J 2009 *J. Hazard. Mater.,* **172** 693-699.
[6] Henderson R, Sharp E, Jarvis P, Parsons S, and Jefferson B 2006 *Water Supply,* **6** 31-38.

[7] Yanqing X, Shang J Q, Yono F W, Gary G, Coleman D P, Sioshansi M and Sullivan S 2009 *Water Qual. Res. J. Canada,* **44** 189-200.
[8] Song Z, Williams C J, and Edyvean R G J 2004 *Desalination,* **164** 249-259.
[9] Vishali S and Karthikeyan R 2015 *Desalin. Water Treat.,* **56** 1489-97.
[10] Flaten T P 2001 *Brain Res. Bull.,* **55** 187-196.
[11] Wang X, Chen S, Gu X and Wang K 2009 *J. Hazard. Mater.,* **164** 81-86.
[12] Sudoh R, Islam M S, Sazawa K, Okazaki T and Hata N 2015 *J. Environ. Chem. Eng.*, **3** 770-774
[13] Sarkar M S K A, Evans G M and Donne S W 2010 *Minerals Eng.,* **23** 1058-65.
[14] Ahmed N and Jameson G J 1985 *Int. J. Mineral Proces.,* **14** 195-215.
[15] Putra R S, Ayu M, and Amri R Y 2020 *Key Eng. Materials* **840** 29-34.

The combination of electroflotation-biocoagulation process using *Aloe vera* for river water treatment

R S Putra[1,2*], C I Putri[1], N S Tyagustin[1]

[1]Department of Chemistry, Faculty of Mathematics and Naturals Sciences, Universitas Islam Indonesia, Jl. Kaliurang km. 14, Yogyakarta 55584, Indonesia
[2]Environmental Remediation Research Group. Department of Chemistry, Faculty of Mathematics and Natural Sciences, Universitas Islam Indonesia, Jl. Kaliurang km. 14, Yogyakarta 55584, Indonesia

*E-mail: rudy.syahputra@uii.ac.id

Abstract. An evaluation of the combination of electroflotation-biocoagulation performance using Aloe vera as biocoagulant in the treatment of river water has been conducted. Titanium electrodes were used as anode and stainless as cathode in the electroflotation reactor, with a 20 V DC of constant voltage for 30 minutes. The parameter dose of Aloe vera was 1.0, 1.5, and 2 g/200 mL water. The effectiveness of the electroflotation-biocoagulation process was reflected by the measurement of turbidity, total dissolved solids (TDS), pH. The results showed that the electroflotation-biocoagulation process decreased the turbidity as much as 97%, TDS as much as 82.64%, and changes in the pH value to 6.6 from the initial concentration of water.

1. Introduction

Water is a universal need. Most living things need water to support their lives. Water itself has an important role for humans, ranging from the needs of bathing, washing and toileting, household to industrial needs. The need for clean water increases with population growth. Clean water conditions are strongly influenced by human activities, both on a household and industrial scale. Water sources commonly used by the water treatment plant to process raw water into clean water are springs, river water, shallow wells, and deep wells [1]. The water quality status might be considered for water resources management [2].

The conventional physicochemical methods commonly used for water treatment in water treatment plants are coagulation/flocculation, and sedimentation. Coagulation is a treatment process by adding and mixing a coagulant. The most common coagulants are aluminum salts such as poly aluminum chloride (PAC), iron salts, and including inorganic polymers with large molecular weight [3]. Several studies report that aluminum compounds can trigger Alzheimer's disease [4]. However, the use of natural ingredients as a coagulant is currently being developed because it has several advantages, including being biodegradable, safe for human health, and free of toxins [5].

The available natural coagulants include extracts of microorganisms, animals, and plants that are safe for human health [6]. *Aloe vera* has the potential to be a natural coagulant because it contains complex carbohydrates and sugars that can bind to particles in water to reduce water turbidity. *Aloe vera* was used as a coagulant to clear well water [7]. The results showed that the turbidity of water decreased from 18 NTU to 5 NTU while hardness decreased from 314.28 mg /L to 114.77 mg /L, which was obtained at an optimum ratio of 0.3 mL in 500 mL of sample water. The color decreased from 187 TCU to 58.67 TCU, which was obtained at an optimum ratio of 0.5 mL in 500 mL of sample water. However, the coagulation process is not very efficient if the effluent contains a low suspended density [8].

Electroflotation is the process of separating pollutants in liquids by floating the substances or particles of pollutants dispersed in the water to the surface using the lifting force of bubbles of oxygen and hydrogen gas that is formed from the electrolysis reaction [9]. Electroflotation can be an alternative as it is efficient and can be used for small, medium, or even large-scale processing of suspended solids [10]. A study in the combination of electroflotation and coagulation has conducted by optimization of coagulant concentration, pH, processing time, and flow strength [8]. The experiment showed that at the coagulant concentration of 639 mg/L, the optimal condition with the results of COD reduction from 684 to 204 mgO_2/L and BOD_5 reduction from 100 to 85 mgO_2/L.

In this study, river water was prepared to produce clean water by the combination process of electroflotation and coagulation using *Aloe vera* as biocoagulant. The obtained result from electrpflotation and biocoagulation were compared with the result of the combination process of elctroflotation-biocoagulation.

2. Materials and methods

2.1. Preparing river water samples

The water sample used in this study was collected from Selokan Mataram river, Yogyakarta, Indonesia. Water samples were collected using a plastic bottle dipped in water until fully loaded. The water sample was then measured by the turbidity level using a turbidimeter (TN-100 Eutech, Netherland), TDS/pH using a pH/TDS meter (H19813-5, Romania).

2.2. Designed electroflotation reactors and voltage optimization

In the electroflotation process, a glass plate reactor was used with the dimensions of 10 (L) x 10 (W) x 15 (H). The electrodes were stainless steel wires (5 cm (L) x 7 cm (W) for the cathode and titanium (Ø 4.5 cm, 0.3 cm thick) for the anode. Voltage optimization (i.e. 10, 20, 30 V) in the electroflotation process was carried out by using a 0.001 NaCl electrolyte solution for 30 mins. The bubbles distribution was divided into three categories namely < 5, 5, and 5> mm. The number of bubbles was counted manually using a millimeter scale that was assembled with an 8x magnifying glass and an image of the bubble on the surface of the electrode was photographed by a digital camera (Nikon, Coolpix 3200, Japan) as shown in Figure 1. The bubble distribution was also verified by ImageJ software. The optimum voltage obtained which produced high bubble density was then used in the combination of electroflotation-biocoagulation process.

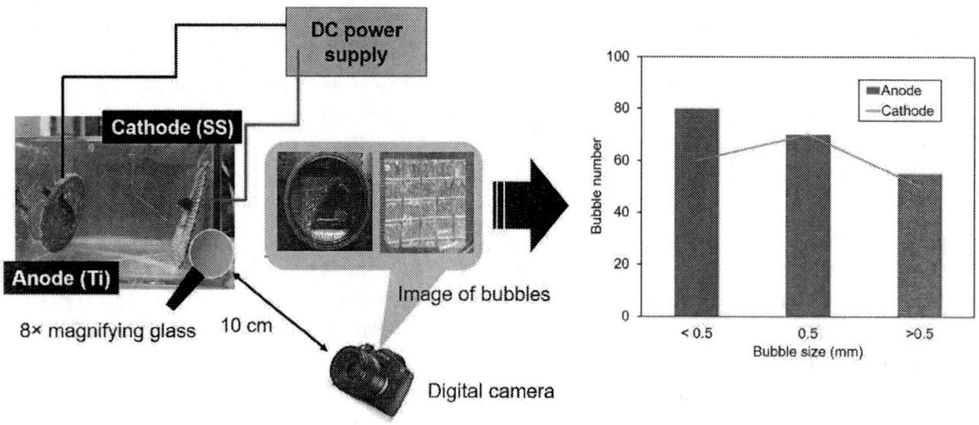

Figure 1. Image processing on the bubble counting.

2.3. Treatment of river water using Aloe vera as biocoagulant
The biocoagulant used in this study was Aloe vera which had been cleaned and the gel part was separated from remaining external skins. The dosage of Aloe vera used in the experiment as much as 1, 1.5, and 2 g in a 200 mL of river water. The coagulation was obtained by rapid stirring for 30 secs. and slow stirring for 30 mins. using a magnetic stirrer [11]. Directly after the coagulation ended then the solution was allowed to stand until the sediment formed before the clarified solution was tested for turbidity, TDS, and pH. The optimum dose was then further used in the combination of electroflotation-biocoagulation process.

2.4. Performance evaluation of electroflotation-biocoagulation
A river water of 500 mL was treated by the combination of electroflotation-biocoagulation at constant voltage of 20 V for 60 mins. The performance was evaluated by measuring the turbidity, TDS, and pH. The turbidity and TDS reduction were calculated according to Eq- and Eq-2.

$$\text{Turbidity reduction (\%)} = [(T_i - T_f) / T_i] \times 100 \quad (1)$$
$$\text{Total dissoved solids reduction (\%)} = [(TDS_i - TDS_f) / TDS_i] \times 100 \quad (2)$$

Where:
T_i = initial turbidity (NTU) of the water sample
T_f = turbidity (NTU) after the process
TDS_i = initial of total dissolved solids (mg/L) of the water sample
TDS_f = dissolved solids (mg/L) after the process.

3. Result and discussion
3.1. River water characteristic
At the initial evaluation, it was found that the turbidity, TDS, and pH of river water sample from the Selokan Mataram was as much as 866 NTU, 125 mg/L, and 6.7 respectively. Turbidity is derived from suspended materials such as mud, sand, organic and inorganic materials, plankton and other microscopic organisms [12]. High turbidity level in river water affects the aquatic life [13]. The initial parameter analysis aimed to evaluate the effectiveness of the combination of electroflotation-biocoagulation using *Aloe vera* as biocoagulant in the river water treatment.

3.2. Voltage optimization
Table 1 shows the number of bubbles formed in the voltage optimization process. In general, the number of bubbles produced on the surface of the cathode was higher than the anode. Assuming equal temperature and pressure for both gases, the produced hydrogen gas has, therefore, twice the volume of the produced oxygen gas because the mole ratio of hydrogen to oxygen is 2/1 (two atoms of H per 1 atom of O) in the water molecule. As the volume of gases depends on the number of moles of its components, the volume of hydrogen is twice as that of oxygen according to Eq-3 and Eq-4.

$$\text{Anode: } 2H_2O_{(aq)} \rightarrow O_{2\,(g)} + 4H^+_{(aq)} + 4e^- \quad (3)$$
$$\text{Cathode: } 2H_2O_{(aq)} + 2e^- \rightarrow H_{2\,(g)} + 2OH^-_{(aq)} \quad (4)$$

The number of bubbles produced at a voltage of 20 V was higher and more uniform than at voltage of 10 V and 30 V. Therefore, the constant voltage of 20 V was applied in the combination of electroflotation-biocoagulation process.

Table 1. Bubble distribution in determining the voltage optimization.

Anode	Cathode	Bubble size distribution
10V 	10V 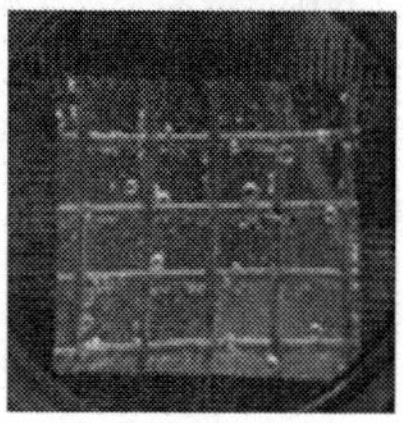	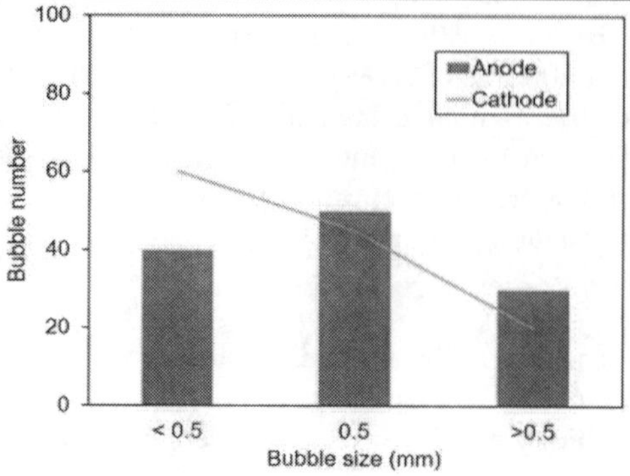
20V 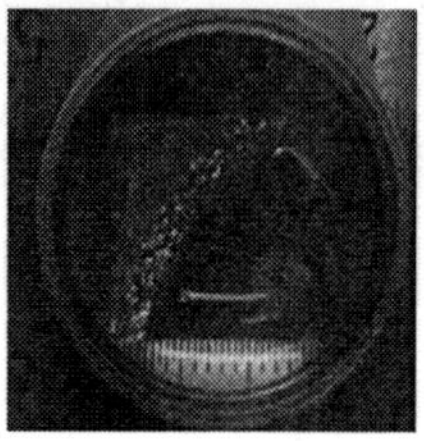	20V 	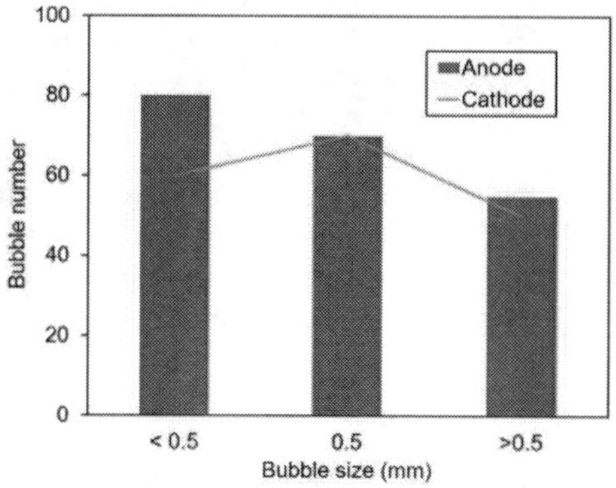
30V 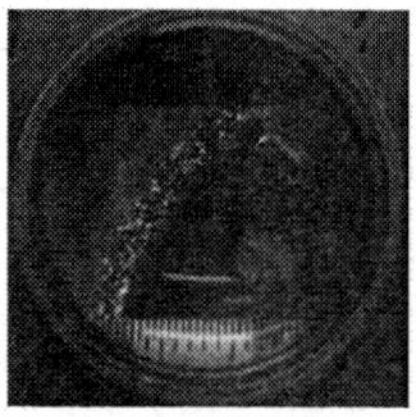	30V 	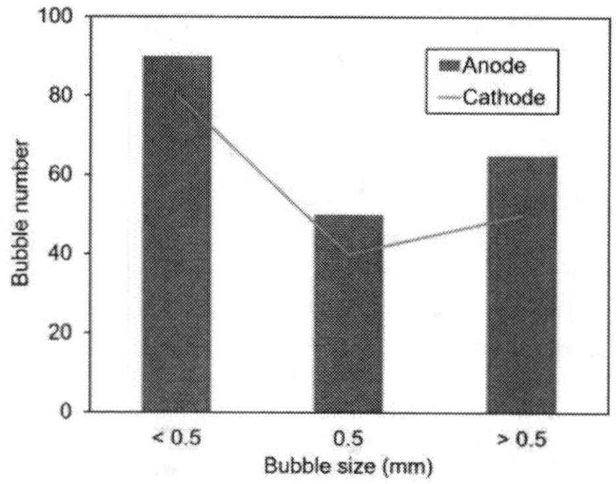

3.3. Effect of Aloe vera on biocoagulation

Table 2 shows the effect of *Aloe vera* doses on the biocoagulation. The pH was adjusted under an alkaline pH (i.e. ~8 to 10) before the evaluation, so that the initial physico-chemical of river water had been changed then the natural one. The addition of 1g of *Aloe vera* in river water decreased the turbidity level as much as 91.4%, and TDS as much as 61.9% from initial concentration. At the optimum dosage of biocoagulant, the colloids have been neutralized and precipitated, but excessive biocoagulant dosage will cause the water to become turbid again because since it cannot interact with another colloidal particles of different charge [14]. *Aloe vera* has the potential to decrease the turbidity in water since the material contains complex carbohydrates and sugar which can bind with pollutant particles in water [9].

Table 2. Effect of *Aloe vera* doses on biocoagulation.

Dosage (g/500 mL)	Parameter		
	Turbidity (NTU)	TDS (mg/L)	pH
Initial (after adjusted pH)	912	205	8.5
Aloe vera 1	78.1 (91.4%)	78.2 (61.9%)	7.1
Aloe vera 1,5	186 (79.6%)	185 (9.8%)	7.5
Aloe vera 2	191 (79.1%)	190 (7.3%)	7.7

Percentage in the bracket was the turbidity and TDS reduction

3.4. Combination of electroflotation-biocoagulation

Table 3 shows the effectiveness of the combination of electroflotation-biocoagulation process with *Aloe vera.* In this regard, the pH of the river water was not adjusted because under these conditions the hydrogen and oxygen bubbles were formed optimally [10]. The results showed that, in general, the combination of electroflotation-biocoagulation process was effective in treating water samples into clean water. The combination of electroflotation-biocoagulation process can reduce turbidity as much as 97% and TDS as much as 82.6%. The sludge characteristic was given by sludge mass (%) which was measured according to Eq-5.

$$\text{Sludge mass (\%)} = [(WS_{wet} - WS_{dried})/WS_{wet}] \times 100 \quad (5)$$

Where WS_{wet} is the weight of sludge in wet (g) and WS_{dried} is the weight of sludge in dried (g).

The formation of sludge was evaluated in each treatment process by calculating the sludge mass to determine the effectiveness of the combination of electroflotation-biocoagulation process. In the biocoagulation, the sludge formation was the lowest (38.32%) than EB (45.76%), and electroflotation. Lower the sludge mass, better the sludge settling [15]. However, the turbidity and TDS level in electroflotation and biocoagulation was higher than EB so that the process has competitiveness on the water treatment.

The results showed that the biocoagulant has high in the combination of electroflotation-biocoagulation process. This was due to the assistant of electroflotation on flocculation by electron transfer from water electrolysis. While the biocoagulation process was binding with colloidal particles only rely on the stirring process.

Table 3. Evaluation of electroflotation-biocoagulation process.

Treatments	Parameter			
	Turbidity (NTU)	TDS (mg/L)	pH	Sludge mass (%)
Initial (without adjusted pH)	866	125	6.7	
Electroflotation	250 (71.1%)	108 (13.6%)	6.6	54.34
Biocoagulant (*Aloe vera* 1g/500 mL)	78.1 (94.6%)	78.2 (57.0%)	7.1	38.32
Electroflotaion-biocoagulation (EB) (*Aloe vera* 1g/500 mL)	21.7 (97.0%)	6.6 (82.6%)	6.6	45.76

4. Conclusion

From this study, it can be concluded that the addition of biocoagulant (*Aloe vera*) can enhance the performance of electroflotation in the treatment of the water river (i.e. the combination of electroflotation-biocoagulation). This was indicated by the reduction of turbidity and TDS level in the combination of electroflotation-biocoagulation process respectively as much as 97% and 82.6% compared to the electroflotation and biocoagulation process alone.

5. Acknowledgements
The authors would like to thank Ministry of Education and Culture and Ministry of Research and Technology of the Republic of Indonesia for their financial support through Excellence Higher Education Basic Research (PDUPT) 2019-2020 (Research contract: 26/E1/KPT/2020 and 227/SP2H/LT/DRPM/2020).

Reference
[1] Asean Development Bank (ADB) 2016 Indonesia Country Water Assessment 978-92-9257-361-4 (e-ISBN) Mandaluyong City, Philippines: Asian Development Bank, pp. 11-24.
[2] Najah A, Eishafie A, Karim O A, Jaffar O 2009 *Eur. J. Sci. Res.,* **28 (3)** 422-35.
[3] Siregar S A 2005 Instalasi Pengolahan Air Limbah (Kanisius: Yogyakarta, Indonesia), pp. 10-20.
[4] Campbell A 2002 *Nephr. Dial. Transplant.,* **2** 17-20.
[5] Choy S Y, Prasad K M N and Wu T Y 2014 *J. Environ. Sci.,* **26** 2178-89.
[6] Kihampa C, Mwegoha W J S, Kaseva M E and Marobhe N 2011 *African J. Environ. Sci. Technol.,* **5** 867-872.
[7] Mujariah, Abram P H and Jura M R 2016 *Jurnal Akademika Kimia,* **5** 16-22.
[8] Mansour L B, Ksentini 1 and Elleuch B 2007 *Desalin.,* **208** 34-41.
[9] Yanqing X, Shang J Q, Yono F W, Gary G, Coleman D P, Sioshansi M and Sullivan S 2009 *Water Qual. Res. J. Canada* **44** 189-200.
[10] Kurniawan T A, Chan G Y S, Lo W H and Babel S 2006 *Chem. Eng. J.* **118** 83–98.
[11] Bin Ahmeda S, Ayoub G, Al-Hindi M, Azizi F 2015 *Desalin. Water Treat.*, **53** 3388–3396.
[12] Effendi H, Romanto, Wardiatno Y 2015 The 1st International Symposium on LAPAN-IPB Satellite for Food Security and Environmental Monitoring *Procedia Environ. Sci.,* **24** 228-237.
[13] Swer S, Singh O P 2004 Status of Water Quality in Coal Mining Areas of Meghalaya, India (IPHE: Bangladesh).
[14] Mudenur C, Sorokhaibam L G, Bhandari V, Raja S and Ranade V V 2016 *ACS Sustain. Chem. Eng.,* **4** 2495-07.
[15] Bramantyo A E and Mustofa A 2014 Pembuatan Biokoagulan dari Biji Kelor (Moringa Oleifera) sebagai Pengganti Koagulan Sintetis (Tawas). Diploma Thesis Institut Teknologi Sepuluh Nopember, Surabaya, Indonesia.

Enhancement of electroflotation using *Musa acuminate* peel as biocoagulant on the wastewater treatment of chemical laboratory

R S Putra[1,2]*, F R F Firdaus[1]

[1]Department of Chemistry, Faculty of Mathematics and Natural Sciences, Universitas Islam Indonesia, Jl. Kaliurang km. 14, Yogyakarta 55584, Indonesia
[2]Environmental Remediation Research Group, Department of Chemistry, Faculty of Mathematics and Natural Sciences, Universitas Islam Indonesia, Jl. Kaliurang km. 14, Yogyakarta 55584, Indonesia

*E-mail: rudy.syahputra@uii.ac.id

Abstract. In this study, the combined biocoagulation electroflotation process using *Musa acuminate* peel coagulant was evaluated using chemical laboratory wastewater as water sample. The effectiveness of the combined biocoagulation electroflotation process was seen based on the proportion of turbidity removal (%) and the proportion of TDS removal (%). The results showed that the proportion of turbidity reduction in chemical laboratory waste was 98,23% and the proportion of TDS disposal was 28,85% using an optimum coagulant dose of 0,5g/500 mL of chemical laboratory wastewater.

1. Introduction

Laboratory activities that use chemicals will inevitably produce both solid and liquid waste such as laboratory wastewater. There are two types of waste, solid waste and liquid waste. Liquid waste is more dangerous than solid waste [1]. Liquid waste must be treated before being discharged into the environment. Coagulation is the often method used in wastewater treatment by adding the coagulants such as aluminium salts to stabilize the colloidal material [2]. However, the chemical coagulants can cause Alzheimer's disease [3], and pre-senile dementia [4]. To avoid this, natural coagulants are used to replace the chemical one, whichs extracted from microorganisms, animals, and plants[5].

Banana is a plant that is given by Southeast Asia and spread widely to Africa, South America, and Central America and spread throughout the world. Bananas can be consumed in the form of fresh fruit and the form of processed and consists of various types. Banana peels are a lot of waste from bananas and contain 1/3 part of bananas. Bananas contains 10-21% of pectin [6-7].

Electroflotation focuses on the formation of bubbles based on the redox reaction of water. The size of the bubbles directly affects the effectiveness of the process. Electrodes material is the most important part of electroflotation [7]. Graphite and lead oxide are the electrodes commonly used in electroflotation[8]. Ketkar et al (1991) described graphite and PbO_2 electrodes have high potential to produce large O_2 bubbles [9]. Therefore, this study will combine the electroflotation process with biocoagulants so that it will improve the performance of the electroflotation process in treating chemical laboratory wastewater.

2. Experimental methods

2.1. Electroflotation reactor design

In the electroflotation process, the reactor design is made of glass with a design of 10cm (lenght) x 10cm (width) x 15cm (height). This electroflotation reactor can be used up to 1 L. Graphite electrodes are as anodes and cathodes. Figure 1 shows the dimensions of the electroflotation reactor used in this study.

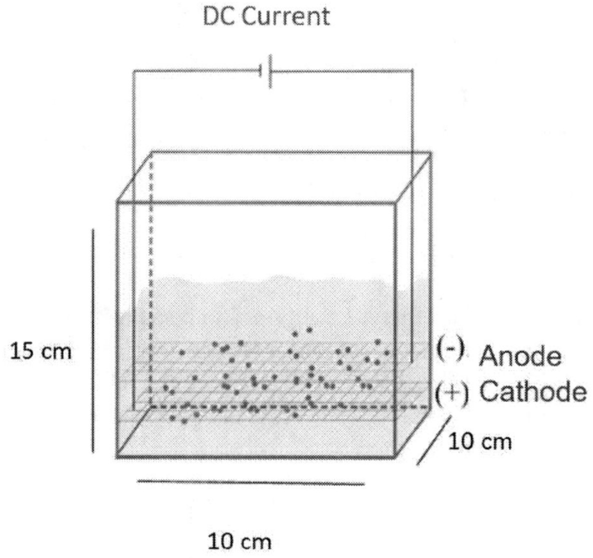

Figure 1. The dimension of electroflotation reactor used in this study.

2.2. Preparation of biocoagulant

Musa acuminate peels were cleaned with distilled water and then dried at room temperature for 2 hours. Furthermore, all peels were dried in oven at 100 °C for 8 hours. Dried peel was then mashed and sieved using 150 mesh to obtain a uniform size.

2.3. Voltage optimization

The optimization of the electroflotation process was carried out using 0,001 M $MgSO_4$ as electrolyte solution in several of DC voltage of 10, 14, 18, 22, 26 and 30 V for 30 minutes. The optimum results were concluded by direct measuring the distribution and size of bubbles at each voltage. Bubble image was captured by using 8 times of magnifying glass and digital camera in the distance between the camera and reactor was 17 cm. This measurement was then calibrated by Image J software.

2.4. Dosage optimization

As much as 500 mL of wastewater was put into the beaker and then added coagulant *Musa acuminate* peels coagulant as much as 0,25; 0,5; 1; 3; and 5 g then stirred for 1 hour. The determination of the optimal dose of coagulant is done by looking at the lowest turbidity number at each dose.

2.5. Electroflotation processes without biocoagulant

The optimum voltage was applied in the electroflotation of wastewater for 60 minutes at 500 mL solution. This process was carried out without adding a coagulant to compare with the electroflotation with biocoagulant.

2.6. Electroflotation processes with biocoagulant

As much as 500 mL of liquid waste is put into the reactor then coagulant is added according to the optimum dose with the optimum dc current for 30 minutes.

3. Result and discussion

3.1. Characterization of biocoagulant

Functional groups contained in the coagulant are identified to determine the presence of functional groups that may interact with dissolved particles in solution[10]. Pectin is a macromolecule containing carboxyl and methoxyl groups which can function as coagulants and flocculants agent [11]. Figure 2 shows FTIR peaks of functional groups in*Musa acuminate* peel coagulant. Peak observations were made in the range 1730-1760 cm^{-1} and 1600–1630 cm^{-1} to see the presence of a carbonyl group (C=O) and a carboxyl group (COO^-)[12]. In Figure 2 there are absorptions peak at 1603 cm^{-1} which indicates the presence of a carboxylate group (COO^-) and a peak of 1736 cm^{-1}which indicates the presence of a carbonyl group (C=O). It can be concluded that *Musa acuminate*peel coagulant contain biopolymers which can be used as coagulants.

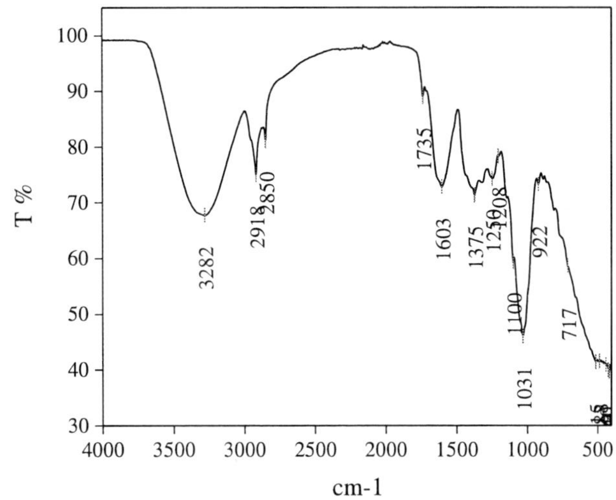

Figure 2. IR peaks of active groups in *Musa acuminate* peel coagulant.

3.2. Voltage optimization

Table 1 shows the plot results of bubble sizes distribution was generated at different DC currents in the cathode and anode electrode. The linear regression value (R^2) was used to evaluate the optimal voltage in the electroflotation. Direct measurement of bubble size distribution using camera image was plotted with software measurement with ImageJ software to get R^2 value. In the electroflotation process, the amount of current is directly proportional to the formation of gas bubbles. The greater the current used the faster the formation of gas bubbles to affect the electroporation performance due to shrinkage of the bubble nucleation time. When nucleation shrinkage occurs, the bubbles produced will have a smaller size and cause the collision process between bubbles and particles to be greater so that the efficiency of the electroflotation process will increase [13]. From the 6 voltage variations, the voltage of 26 V has a linear regression that is close to one at both the anode and cathode. This shows the closeness of the calculation results between manual calculations with ImageJ software in determining the size of the bubbles. So it can be concluded that the optimum voltage for the electroflotation process is 26 V.

Table 1. Results of manual measurement vs ImageJ plots on determining bubble size.

DC Voltage	Cathode	Anode
10 V	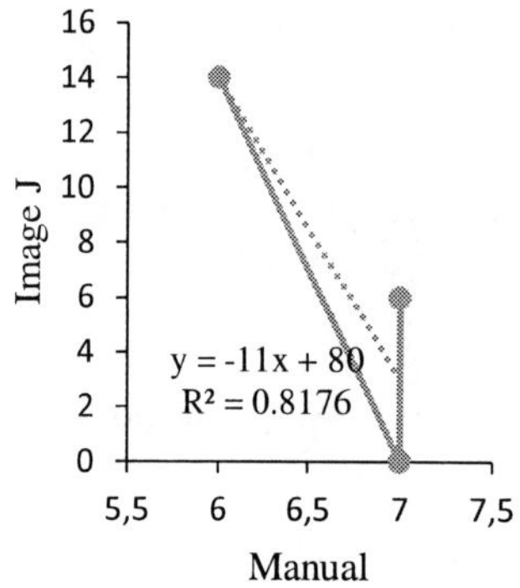	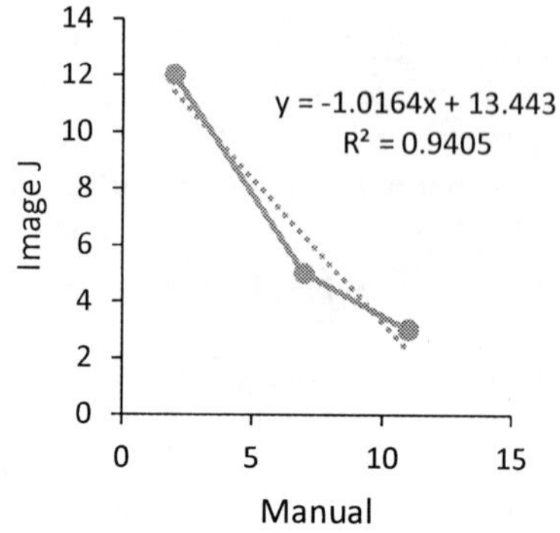
14 V	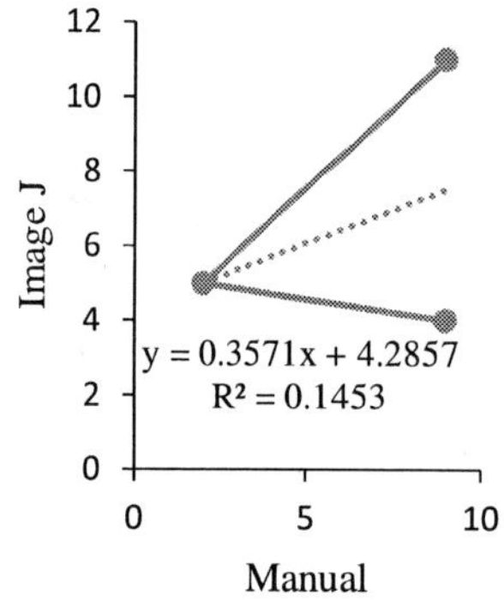	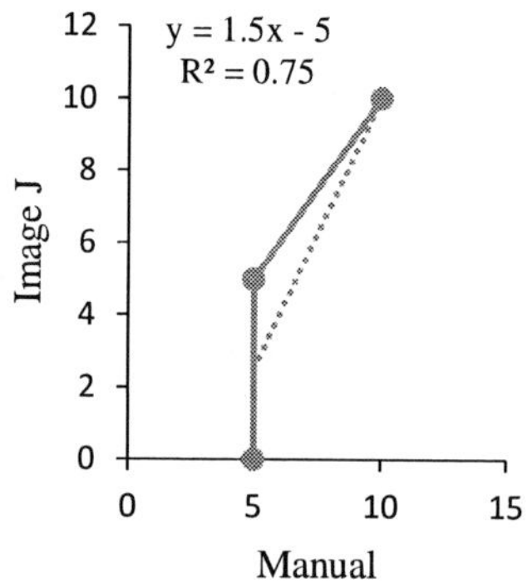
18 V	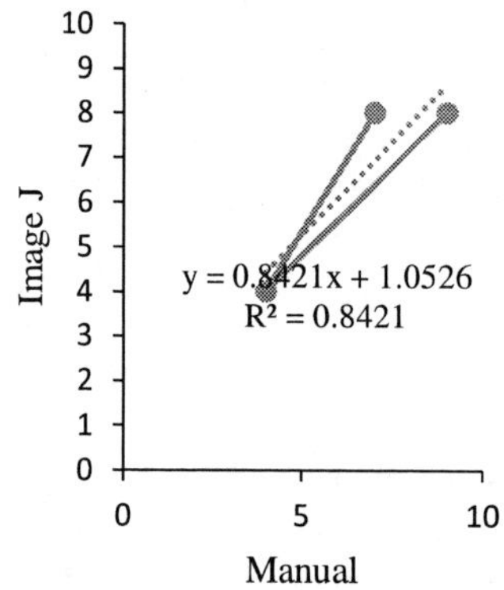	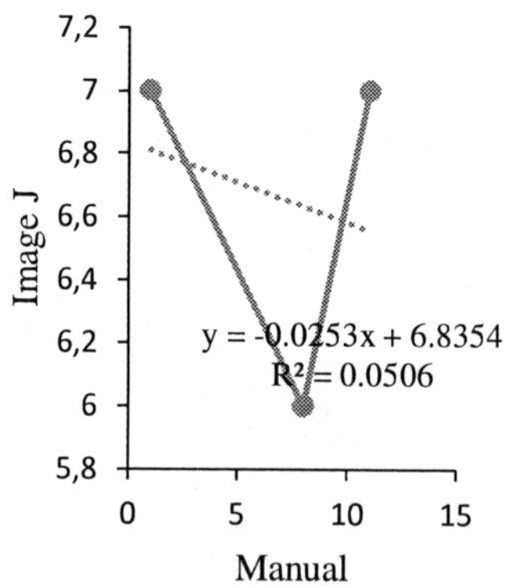

22 V

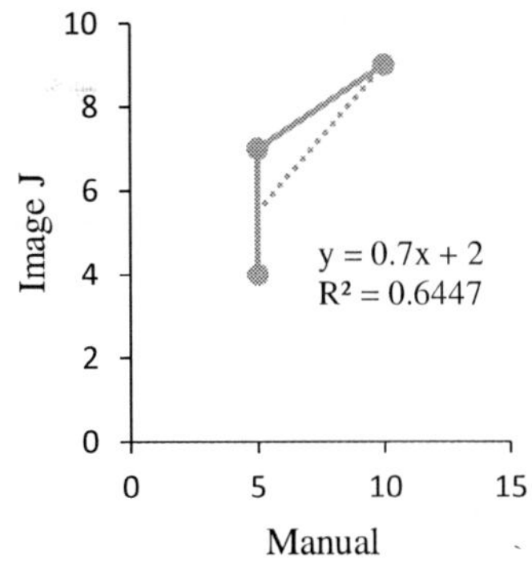
Image J
Manual
y = 0.7x + 2
R² = 0.6447

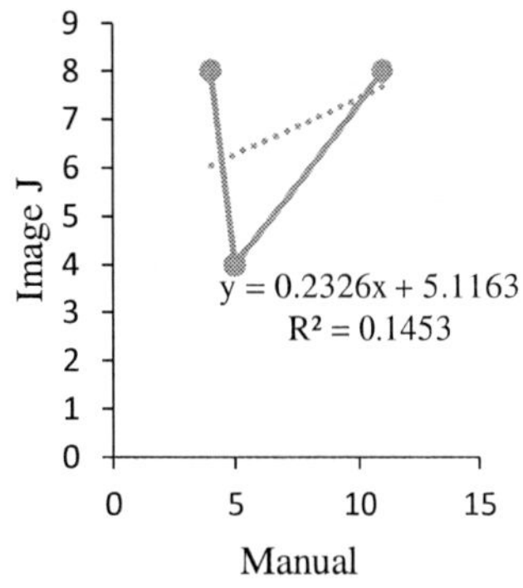
Image J
Manual
y = 0.2326x + 5.1163
R² = 0.1453

26 V

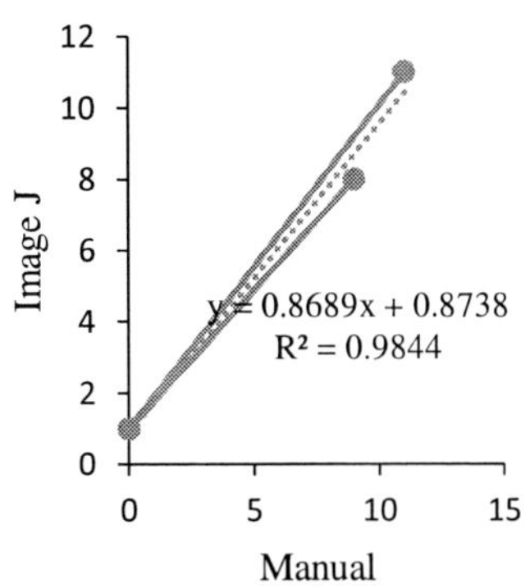
Image J
Manual
y = 0.8689x + 0.8738
R² = 0.9844

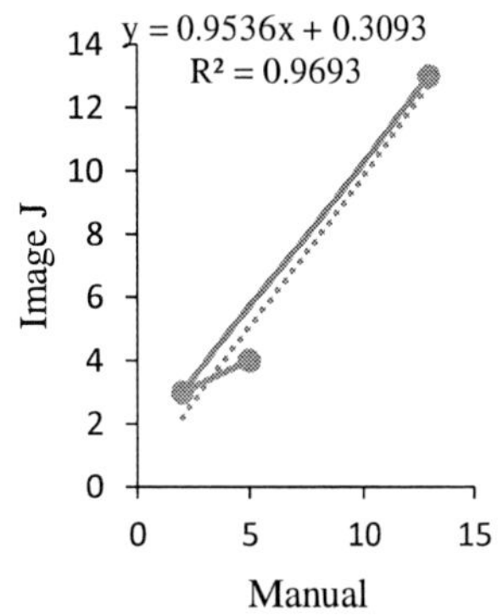
Image J
Manual
y = 0.9536x + 0.3093
R² = 0.9693

30 V

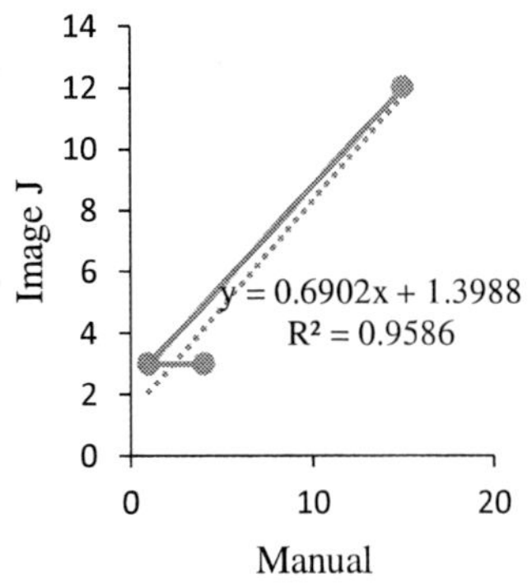
Image J
Manual
y = 0.6902x + 1.3988
R² = 0.9586

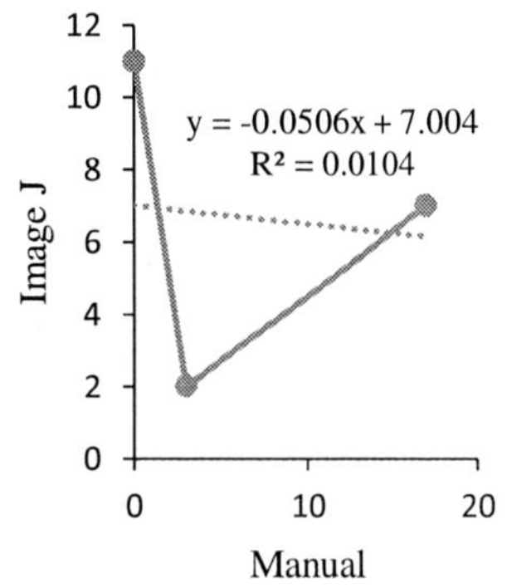
Image J
Manual
y = -0.0506x + 7.004
R² = 0.0104

IOP Conf. Series: Materials Science and Engineering 1087 (2021) 012048 doi:10.1088/1757-899X/1087/1/012048

3.3. Dosage optimization

Some important parameters in determining water quality are turbidity and TDS. Figure 3 shows the effect of biocoagulants on decreasing the wastewater turbidity. The optimum dose of coagulant is 0.5 g/500 mL with a turbidity of 1.26 NTU with percent removal 99.16%. This shows that the pectin in *Musa acuminate* peel coagulant that has a COO^- functional group will bind to positive particles in wastewater [11]. However, there is an increase in the value of turbidity if the coagulant dose exceeds 0.5 g due to the repulsion between particles with the same charge so that flock deflocculation occurs as a result, the turbidity of the wastewater increases [14]. On the other hand, *Musa acuminate* peel coagulant did not affect the decrease in TDS which can be seen in Figure 4 where the more coagulant doses were added, the TDS value did not decrease. It can be concluded that *Musa acuminate* peel coagulant can reduce turbidity due to the presence of the COO^- functional group but does not have an impact on decreasing TDS in waste. The optimal dose of coagulant is 0.5 g.

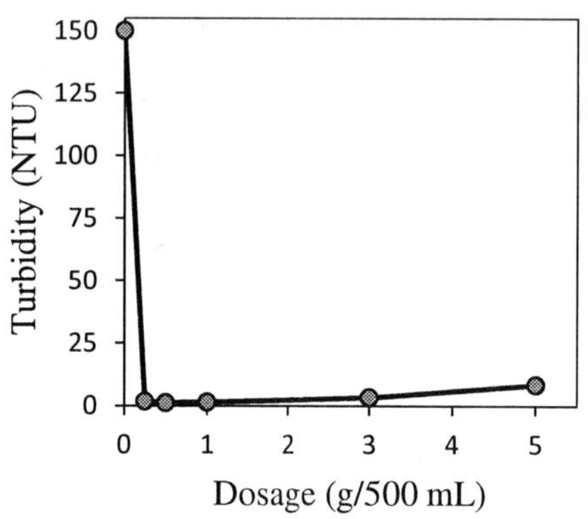

Figure 3. Effect of biocoagulant dosage on turbidity.

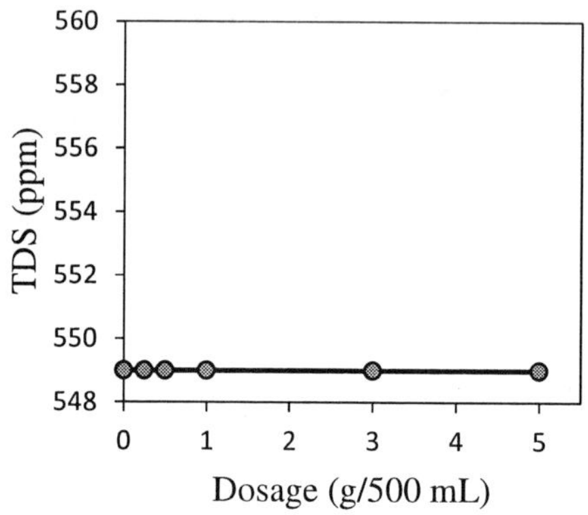

Figure 4. Effect of biocoagulant dosage on TDS.

3.4. Chemical laboratory wastewater treatment using electroflotation

In this study, an electrode in the form of graphite was used both as anode and cathode. The use of graphite electrodes is due to its high efficiency in the production of hydrogen. This is due to the presence of a layer in the graphite electrode and the porous nature of the material so that it can increase water diffusion which causes the production of hydrogen to increase [15]. Table 2 shows the comparison of the results between the electroflotation process and the electroflotation-biocoagulation. In the electroflotation process, the turbidity decreased by 97.02% and the TDS decreased by 32.08%. In the biocoagulation electroflotation process, the decrease in turbidity was greater, 98.23%, but the decrease in TDS was only 28.85%. It can be concluded that the combined electroflotation-biocoagulation process is better at reducing wastewater turbidity compared to electroflotation.

Table 2. Comparison between electroflotation and electroflotation-biocoagulation process.

Treatment	TDS (ppm)	Turbidity (NTU)
Wastewater	5580	150
Electroflotation	3790	4.47
Combine process (electroflotation biocoagulation)	3970	2.65

4. Conclussion

Based on the research results, it can be seen that *Musa acuminate* peel biocoagulants can improve electroflotation performance when compared to electroflotation without using biocoagulants. In electricity without using a coagulant, the percentage of turbidity reduction and TDS reduction percentage was 97.02% and 32.08%. Whereas in the combined process the decrease in the percentage of turbidity increased to 98.23% but the percentage decrease in TDS was only 28.85%. It can be concluded that the combined biocoagulation electroflotation process using *Musa acuminate* peel coagulants is considered good because it can reduce turbidity beyond the usual electroflotation process.

5. Acknowledgements

The authors would like to thank Ministry of Education and Culture and Ministry of Research and Technology of the Republic of Indonesia for their financial support through Excellence Higher Education Basic Research (PDUPT) 2019-2020 (Research contract: 227/SP2H/LT/DRPM/2019; LLDIKTI and UII contract: 1627.1/LL5/PG/2020).

References

[1] AlMadury S, Fakhrunnisa F, Panjaitan R, Ni'mah S and Chandra T P 2014 *Khazanah,* **6** 55-66.
[2] Domínguez J R, González T, García H M, Sánchez-Lavado F and Beltrán de Heredia J 2007 *J. Hazard. Mater.,* **148** 15–21.
[3] Campbell A 2002 *Nephr. Dialysis Transplant.,* **2** 17-20.
[4] Niquette P, Monette F, Azzouz A and Hausler R 2004 *Water Qual. Res. J.,* **3** 303–310.
[5] Kihampa C, Mwegoha W J S, Kaseva M E, and Marobhe N 2011 *African J. Env. Sci Technol.,* **5** 867–872.
[6] Mohapatra D, Mishra S and Sutar N 2010 *J. Sci. Ind. Res.,* **69** 323–329.
[7] Chen X, Chen G and Yue P L 2002 *Environ. Sci. Technol.,* **36** 778–783.
[8] Hosny A Y 1996 *Separations Technol.,* **6** 9–17.
[9] Ketkar D R, Mallikarjunan R and Venkatachalam S 1991 *Int. J. Mineral Proces.,* **31** 127–138.
[10] Kumar V, Othman N and Asharuddin S 2017 *MATEC Web Conf.,* **103** 9.
[11] Ibarra-Rodríguez D, Lizardi-Mendoza J, López-Maldonado E A and Oropeza-Guzmán M T 2017 *Chem. Eng. J.,* **323** 19–28.
[12] Singthong J, Ningsanond S, Cui S W and Douglas Goff H 2005 *Food Hydrocoll.,* **19** 793–801.
[13] Da Cruz S G, Dutra A J B and Monte M B M 2016 *J. Env. Chem. Eng.,* **4** 3681–87.
[14] Nisa N I F and Aminudin A 2019 *Jurnal Riset Sains dan Teknologi,* **3** 61.
[15] Yuvaraj A L and Santhanaraj D 2014 *Mater. Res.,* **17** 83–87.

Filtration performance of polyethersulfone (PES) composite membrane incorporated with organic and inorganic additives

A Fathurrahman[1], R Arisandi[1], A Fahrina[1,2], N Arahman[1,3,4,5,*], F Razi[1]
[1]Department of Chemical Engineering, Universitas Syiah Kuala, Jl. Syeh A. Rauf, No. 7. Banda Aceh, 23111 Indonesia
[2]Doctoral Program, School of Engineering, Universitas Syiah Kuala, Jl. Syeh A. Rauf, No. 7. Banda Aceh, 23111 Indonesia
[3]Graduate School of Environmental Management, Universitas Syiah Kuala, Jl. Tgk Chik Pante Kulu No. 5, Banda Aceh, 23111, Indonesia
[4]Research Center for Environmental and Natural Resources, Universitas Syiah Kuala, Jl. Hamzah Fansuri, No. 4, Darussalam, Banda Aceh, 23111Indonesia
[5]Atsiri Research Center, Universitas Syiah Kuala, Jl. Syeh A. Rauf, No. 7. Banda Aceh, 23111 Indonesia

*E-mail : nasrular@unsyiah.ac.id

Abstract. Membrane separation technology has been extensively applied in separation processes. Improving membrane hydrophilicity plays a significant role in mitigating fouling phenomena and maintain membrane filtration performance. Fouling is a pore blockage of the membrane due to the interaction between hydrophobic molecules and membrane surface that cause membrane flux declining, shorter membranedurability, and increasing energy consumption. Incorporating inorganic particles to enhance membrane hydrophilicity is very attractive nowadays. Inorganic particles have unique characters due to their small size, excellent mechanical properties, and hydrophilic nature. This research aims to improve PES membrane filtration performances by employing nanocarbon and Fe_2O_3 as an inorganic additive. The membrane manufacturing process was carried out using a phaseinversionmethod., especially non-solvent induced phase separation (NIPS). The resulted membranes were analyzed by filtration soy protein solution to evaluate protein permeation flux and flux recovery ratio (FRR). The result experiments showed that PES/Nanocarbon could provide the highest FRR of up to 76.46%.

1. Introduction

Among a large number of separation technology, membrane filtration is one of the most frequently applied technology for water purification, wastewater treatment, and protein concentration. The advantages of using membrane are high solid retention, high volumetric rate, flexible in process, environmentally friendly, and small footprints [1,2]. Membrane technology classified in different types, including micro-filtration (MF), ultra-filtration (UF), nano-filtration (NF), and reverse osmosis (RO) [2,3]. Ultra-filtration (UF) is currently used due to its performance in separating colloids, macromolecules, and particles. The UF membrane should provide better anti-fouling property to maintain permeation stability and energy consumption.

Polyethersulfone (PES) is a polymer that extensively used as membrane materials due to its mechanical and environmental endurance, chemical resistance, and secure processing [4][5]. However, the character of PES significantly affects on the performance of the resulted membrane. Membrane fouling such as protein adsorption, denaturation, and aggregation occurs due to the hydrophobic nature of the PES membrane surface that caused water flux decline, increase energy consumption, and high maintenance required. Therefore, an anti-fouling should be added into membrane matrices to increase hydrophilicity and reduce membrane fouling. Some methods to modify membrane surfaces are coating, surface grafting, and blending. Among this modification, the blending technique is the most straightforward way to improve membrane hydrophilicity.

Nowadays, inorganic particles are extensively used in membrane preparation through the blending method. The advantages of inorganic application materials provide uniform pores in the membrane surface due to their small sizes, have mechanical and chemical stable properties, and excellent hydrophilicity [6]. Furthermore, the material could enhance the mass transfer, increase selectivity in separation applications, and mitigate fouling phenomena [7,8]. The most common inorganic particles applied in the membrane are TiO_2, SiO_2, CNT, HNTs, and Al_2O_3[6,7]. The increasing membrane hydrophilicity can be evaluated by the calculated flux recovery percentage of filtration. In this study, nanocarbon and Fe_2O_3 was employed to enhance membrane hydrophilicity analyzed by calculating membrane flux recovery of soy protein concentrate.

2. Materials and methods

2.1. Materials

PES with an average molecular weight of 65,000 Da (BASF, Ludwigshafen, Germany) was used as the primary polymer to prepare a flat sheet membrane. Dimethylformamide purchased from Merck KgaA, Germany, was supplied as a solvent to dissolve the polymer. Fe_2O_3 and nanocarbon were kindly provided by the Physic Science Laboratory and Energy laboratory of Syiah Kuala University, respectively.

2.2. Membrane preparation

For the original membrane, the dope solution is prepared by mixing 16wt% of PES in DMF. In other dope solution, nanocarbon and Fe_2O_3 of 0.05 wt% was added separately to prepare the modified PES membrane. The detailed composition of the membrane solution is summarized in Table 1. All dope sample is stirred for about 24 hours to obtain the homogeneous solution. After releasing the air bubble, the solution is poured onto a glass plate using an applicator knife with a thickness of 300 μm. The casted solution and the glass plate then dipped in a coagulation bath containing pure water for solidification of the membrane.

Table 1. Membrane composition.

Concentration (wt%)				Code
PES	DMF	Fe_2O_3	Nanocarbon	
16	84	-	-	M1
16	83.95	0.05	-	M2
16	83.95	-	0.05	M3

2.3. Membrane filtration performance

Filtration performance of the fabricated membrane was analyzed in terms of water flux and flux recovery ratio (FRR). Soy protein solution with 5000 ppm concentration was filtrated by the formed membranes using a cross-flow module for 190 minutes. The filtration experiment was carried out at constant pressure (1 bar), and the permeate was collected every 10 minutes. Every 60 minutes of filtration, the membrane inside the module was washed using distilled water for 5 mins, then soy protein filtration was continued, and the permeation was measured again. Soy protein flux was calculated by measuring the volume of permeate that passes through the active area of the membrane during a certain time, as shown in Equation (1). Flux recovery ratio (FRR) of the membrane was determined by using Equation (2).

$$J_p = \frac{V_p}{A \times \Delta t} \quad (1)$$

$$FRR\ (\%) = \frac{J_{w1}}{J_{w2}} \times 100 \quad (2)$$

Where:

J_p = flux of soy protein (L/m^2.hour)
V_p = permeate volume (L)
A = membrane effective surface area (m^2)
Δt = permeationTime (hour)
FRR = Flux Recovery Ratio (%)
Jw_1 = protein flux before washing (L/m2.hour)
Jw_2 = protein flux after washing (L/m2.hour)

3. Result and discussion

3.1. Soy protein permeability

The soy protein flux test was carried out to determine the rate of separation of soy protein using the PES membrane. Flux is a standard that can be used in evaluating membrane performance before and after use. The membrane is tested on the cross-flow module for 190 minutes, and the permeate rate is recorded every 10 minutes then backwashing every 60 minutes. The PES membranes flux rate of soy protein is presented in Figure 1.

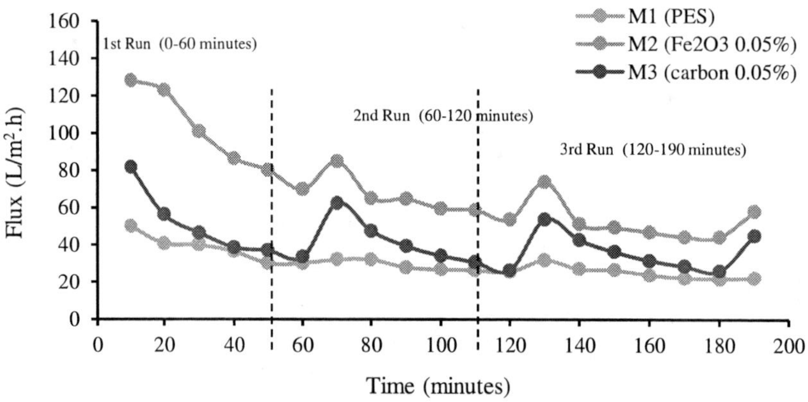

Figure 1. Soy protein permeation flux of all membranes.

Based on Figure 1, soy protein flux tends to decrease significantly in the first cycle (0-60 minutes) and falls stably in the second and third cycles. It is due to membrane fouling phenomena in the initial filtration caused by adsorption of protein onto the membrane surface. Moreover, the protein molecules blockage membrane pores and consequently decline membrane flux. Backwashing helps to decrease membrane fouling and increase membrane flux. As shown in Figure 1, the pristine membrane (M1) without the addition of additives have a more stable flux, but also have the lowest flux compared to the other membranes while the PES membrane with the presence of inorganic particles; nanocarbon and Fe_2O_3 particles showed higher flux. Inorganic particles are containing oxygen-functional groups promising hydrophilic properties because of high interaction with water molecules [9]–[11]. The addition of hydrophilic particles in membrane solution leads to rapid solidification in the coagulation bath resulting in higher porosity [7], [12]–[15]. As confirmed in our previous research, PES/Fe_2O_3membrane has longer finger-like structure in membrane sub-layer due to its hydrophilic nature compared with the nascent PES and PES/nanocarbon membranes [16]. Therefore, in this study M2 and M3 membranes produce higher protein flux than the M1 as a pristine membrane.

3.2. Flux recovery ratio (FRR)

The determination of the flux recovery ratio aims to find out the effectiveness of the reused membrane after filtration. The backwash process is carried out to normalize membrane performance from reversible fouling [16]. In this study, fouling was occurred due to the interaction between soy protein and PES membrane surface that has the same hydrophobic nature. Therefore, flux recovery ratio (FRR) should be evaluated to determine the impact of anti-fouling additives, nanocarbon, and Fe_2O_3 used in this experiment.

Figure 2 displays the flux recovery ratio of all membranes. As shown in Figure 2, the PES membrane, with the addition of 0.05% carbon (M3), has the highest FRR value of 76 %, followed by the PES membrane with the addition of 0.05% Fe_2O_3 (M2) of 66 % while the PES pristine membrane showed the lowest value of FRR. From this study, nanocarbon and Fe_2O_3 given the positive impact to increase membrane anti-fouling property. More hydrophilic surfaces of membrane lowered the interaction between protein molecules and membrane surface, resulting in reversible fouling that easier to clean up by physical cleaning [17,18].

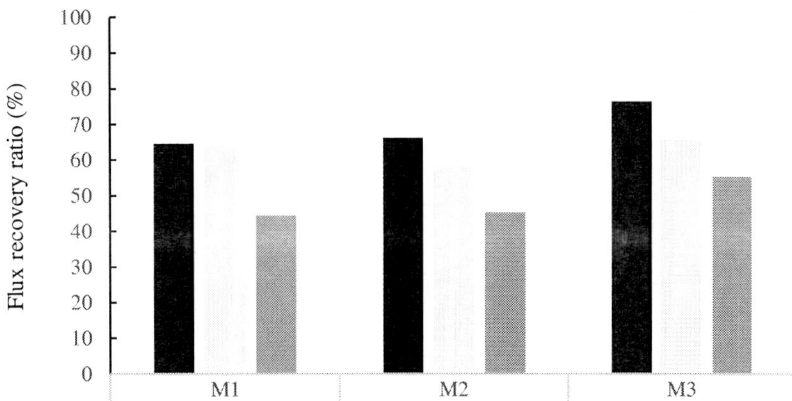

	M1	M2	M3
■ 60 minute	65	66	76
120 minute	64	58	66
▒ 180 minute	44	45	55

Figure 2. The flux recovery ratio of PES membranes.

4. Conclusion

Based on the research, the addition of nanocarbon and Fe_2O_3 increases the anti-fouling property of the PES membrane indicated by producing higher protein permeation and flux recovery ratio (FRR). PES/nanocarbon reach the highest value FRR of 76 %; PES/ Fe_2O_3 provides FRR of 66 %, while PES pristine membrane showed the lowest FRR of 55.39%.

References

[1] Saleem M, Alibardi L, Cossu R, Cristina M, Spagni A 2017 *Chem. Eng. J.,* **312** 136–143.
[2] Saraswathi M S A, Kausalya R, Kaleekkal J N, Rana D, Nagendran A 2017 *J. Environ. Chem. Eng.,* **5** 2937–43.
[3] Lin J, Ye W, Zhong K, Shen J, Jullok N, Sotto A, Van der Bruggen B 2016 *Chem. Eng. Proc.,* **107** 194-205.
[4] Wang Z, Wang H, Liu J, Zhang Y 2014 *Desalination,* **344** 313–320.
[5] Liu F, Hashim N A, Liu Y, Abed M R M, Li K 2011 *J. Membr. Sci.,* **375** 1–27.

[6] Wang X, Wang X, Xiao P, Li J, Tian E, Zhao Y, Ren Y 2016 *Colloids Surfaces A: Physicochem. Eng. Asp.,* **508** 327–335.
[7] Arahman N, Mulyati S, Fahrina A, Muchtar S. 2019 *Molecules,* **24** 1-13.
[8] Arahman N, Mukramah, Syawaliah, Maimun T, Bilad M R 2018 *Int. J. GEOMATE,* **15** 50-57.
[9] Sali S and Mackey H R. 2019. *Nanomaterials,* **9** 1-16.
[10] Shi H, He Y, Pan Y, Haihui D, Zeng G, Zhang L, Zhang C 2016 *J. Membr. Sci.,* **506** 60-70.
[11] Rahimi Z, Zinatizadeh A A L, Zinadini S 2015 *J. Ind. Eng. Chem.,* **29** 366–374.
[12] Arahman N, Mulyati S, Fahrina A 2019. *Mater. Res. Express,* **6** (**66419**) 1-8.
[13] Mavukkandy M O, Bilad M R, Kujawa J, Al-Gharabli S, Arafat H A. 2017. *Sep.Purif. Technol.,* **187** 365–373.
[14] Chang J, Zuo J, Zhang L, O'Brien G S O, Chung T S. 2017 *J. Membr. Sci.,* **539** 295–304.
[15] Yeow M L, Liu Y T, Li K 2004 *J. Appl. Polym. Sci.* **92** 1782–89.
[16] Arisandi R, Fathurrahman A, Fahrina A, Razi F, Jalil Z, Arahman N. 2020. *IOP Conf. Series: Mater. Sci. Eng.,* **845** 1-6.
[17] Zuthi F M R, Guo W, Ngo H H, Nghiem D L, Xia S, Li J, Li J, Liu Y 2017. *Bioresour. Technol.,* **238** 86–94.
[18] Jamal S, Chang S, Zhou H 2014 *Membranes,* **4** 319–332.

Characterization of polyethersulfone (PES) membrane entrapping with ginger extract (GE) as a green additive

A Fahrina[1,2], N Arahman[1,3,4,5*], S Mulyati[1], S Aprilia[1,3,4], M R Bilad[6]
[1]Department of Chemical Engineering, Universitas Syiah Kuala, Jl. Syeh A. Rauf, No. 7 Banda Aceh, 23111 Indonesia
[2]Doctoral Program, School of Engineering, Universitas Syiah Kuala, Jl. Syeh A. Rauf, No. 7 Banda Aceh, 23111 Indonesia
[3]Graduate School of Environmental Management, Universitas Syiah Kuala, Jl. Tgk Chik Pante Kulu No. 5, Banda Aceh, 23111, Indonesia
[4]Research Center for Environmental and Natural Resources, Universitas Syiah Kuala, Jl. Hamzah Fansuri, No. 4, Darussalam, Banda Aceh, 23111 Indonesia
[5]Atsiri Research Center, Universitas Syiah Kuala, Jl. Syeh A. Rauf, No. 7. Banda Aceh, 23111Indonesia
[6]Chemical Engineering Department, Universiti Teknologi PETRONAS, Seri Iskandar, Perak 32610, Malaysia

*Email: nasrular@unsyiah.ac.id

Abstract. The hydrophilicity of the membrane surface plays an essential role in improving membrane filtration performance. Better hydrophilicity could facilitate higher water permeation and prevent fouling phenomena. In this research, a green additive from ginger extract (GE) was employed to improve the polyethersulfone (PES) membrane hydrophilicity. The PES/GE membrane was fabricated via the phase inversion method by dispersing GE powder in PES/NMP solution using sonication followed by mechanical stirred. The influence of ginger extract (GE) on the PES membrane was characterized in terms of hydrophilicity, morphological structure, porosity, water permeation, chemical composition, and mechanical property. Water contact angle measurements showed that PES/GE membrane with 0.1% GE has the highest hydrophilicity. Scanning Electron Microscopy (SEM) indicated a porous finger-like structure on the cross-section membrane after the addition of GE.

1. Introduction

Polymeric ultrafiltration (UF) membrane is widely applied in water purification, wastewater treatment, and chemical engineering process [1]. However, the hydrophobic nature of pristine membranes is one of the significant obstacles that usually hinders membrane applications [2,3]. Fouling phenomena due to the interaction between membrane surface and organic molecules causes a substantial decline of the water permeation, increase in energy consumption, and shorten membrane lifetime [4]. Therefore, the improvement of membrane hydrophilicity is highly necessary.

Hydrophilic groups, such as hydroxyl, carboxyl, and sulfonate groups, are introduced to improve membrane surface hydrophilicity [4,5]. Some plant extracts from Piper betel, *Vanilla planifolia,* and natural products such as 2(5H)-Furanone that contains aldehyde, hydroxyl, and ether functional groups have been explored to increase membrane hydrophilicity [6]. Ginger (*Zingiber Officinale Roscoe*) is another plant derivative with active substances in the form of zingerone, shogaols, gingerols, and volatile oils (essential) [7,8]. The hydroxyl groups are found in the chemical structures of Ginger extract (GE) and potentially improved membrane hydrophilicity. However, the use of ginger extract for enhancement membrane hydrophilicity has not been reported.

In the present study, polyethersulfone (PES) was utilized as the primary membrane polymer because it owes mechanical stability and chemical resistance [9]–[11]. Herein, ginger extract (GE) was explored to enhance membrane hydrophilicity by blending technique. The effect of ginger extract on the morphology, hydrophilic, and mechanical properties of the PES membrane was reported. Besides, the employing plant extract as membrane additive material is a green alternative to produce an appropriate membrane from an environmental point of view.

2. Materials and experimental set up

2.1. Material

Polyethersulfone (PES, Ultrason E6020P) as the main material waspurchased from BASF Co. (Ludwigshafen, Germany). N-Methyl Pyrrolidone (NMP) (Wako Pure Chemical Industries, Japan) was directly used as a solvent. *Zingiber Officinale Roscoe* extract (purity 100%) (Sciyu bio. Tech, China) was utilized as an antibacterial additive. Distilled water (DI) was used as a non-solvent for the solidification process.

2.2. Membrane fabrication

The phase inversion technique was used to prepare flat sheet membranes. Ginger extract (GE) was initially dissolved in the N-Methyl Pyrrolidone (NMP) by sonication for 10 minutes, followed by 5 minutes stirring. Then, polyethersulfone (PES) was added into the suspension, and the solution was kept to continue stirring until homogeneous. The composition of membrane solutions in this study is shown in Table 1. The membrane solutions were then cast on flat glass with an air gap of 200 μm using a casting knife (Yoshimitsu, YBA-3, Japan). Next, the glass plate is dipped in a water bath to induced the solidification process. The formed membranes are stored in containers containing distilled water to remove any remaining solvents.

Table 1. PES membrane composition.

Membrane	PES (wt%)	GE (wt%)	NMP (wt%)
PG1	16	0	84
PG2	16	0.01	83.99
PG3	16	0.03	83.97
PG4	16	0.05	83.95

2.3. Characterization

Membrane chemical group composition was identified by attenuated total reflectance (ATR) instrument (Thermo Scientific iD5 ATR–Nicolet iS5 FTIR spectrophotometer, Japan). Hydrophilicity on the membrane surface was measured by contact angle meter (Drop Master 300, Kyowa Interface Science Co., Japan). The membrane morphological structures on surface and cross-sections were visualized using scanning electron microscopy (SEM; JSF-7500F, JEOL Co., Ltd., Japan) at 5.0 kV. The mechanical property was evaluated with a tensile test instrument (Autograph AGS-J, Shimadzu Co., Japan), by following ASTM D 638-14.

3. Result and discussion

3.1. Hydrophilicity

Figure 1 displayed the water contact angle of the formed membranes as a function of hydrophilicity. The water contact angle is evaluated by placing a water droplet on the membrane surface, then the angle between the droplet and membrane surface is calculated [12]. It means the smaller contact angle facilitates better hydrophilicity.

As shown in Figure 1, the increment of the loading of ginger extract (GE) particles increases PES membrane hydrophilicity. PG4 membrane has the highest hydrophilicity that the other membranes and decreases the PES membrane water contact angle from 68.50° to 58.31. In this case, the hydroxyl groups (−OH) in GE particles play a significant role in improving the hydrophilicity of the membrane surface. IR-spectra confirmed the presence of hydroxyl groups (−OH) in the ginger extract in Figure 2.

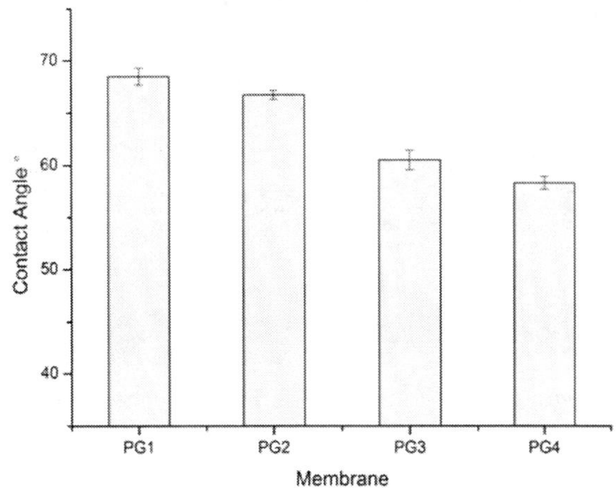

Figure 1. The water contact angle of the PES membrane.

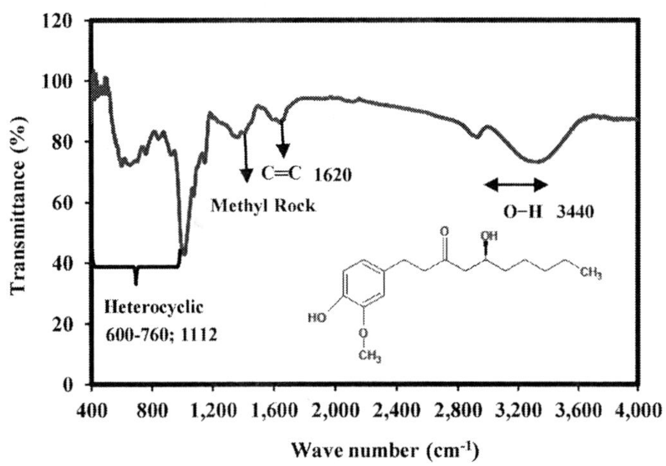

Figure 2. IR-Spectra of ginger extract.

3.2.Morphological structure

Figure 3 showed the cross-section of the formed membranes with a completely finger-like structure. As shown in Figure 3, the increasing concentration of GE particles resulted in higher porosity of membrane indicated by more extensive finger formation. Since GE additive has hydroxyl groups and soluble in water, it can leach out from the membrane during the solidification process leading the creation of porous [13,14]. Furthermore, the increasing finger size in the cross-section membrane is also induced by the fast exchange of solvent and non-solvent due to a closer affinity between membrane solution and water as non-solvent caused by higher hydrophilicity of membrane solution [13–16].

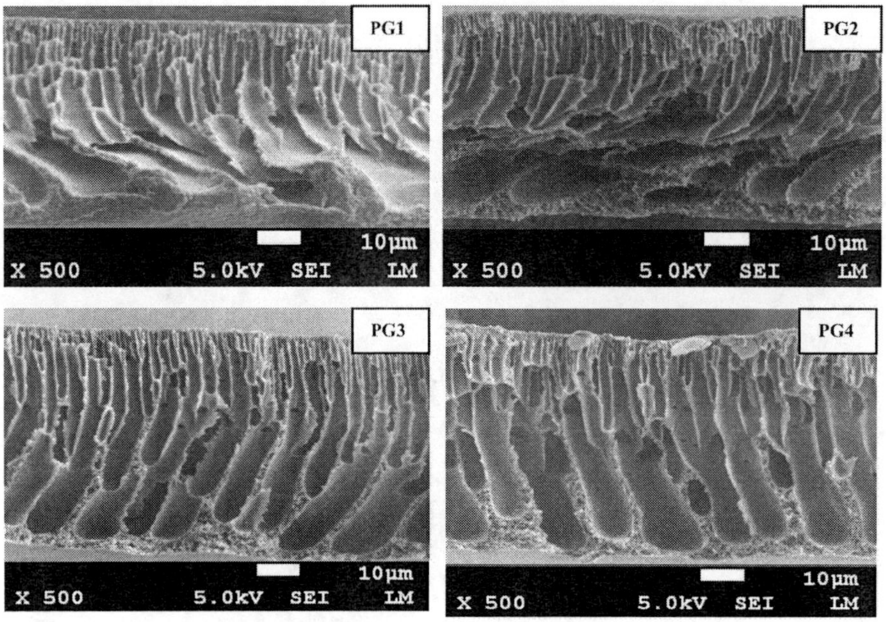

Figure 3. SEM image of PES cross-section membrane.

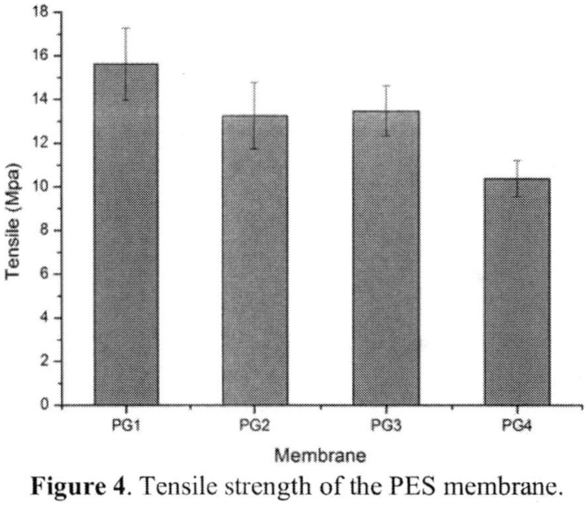

Figure 4. Tensile strength of the PES membrane.

3.3. Mechanical property

The mechanical property of the membranes was investigated using the tensile strength test. As presented in Figure 4, the pristine membrane (PG1) has the most durable mechanical property. The loading of GE particles leads to the decreasing of PES membrane tensile strength from 15.6 to 10.4 Mpa. From the results, the formation of macrovoid in membrane structure affected the tensile strength of the membrane. It weakens the density of the membrane and products in a more fragile form [17].

4. Conclusion

Based on the results, ginger extract (GE) showed a positive impact on enhancing the hydrophilicity of the membrane. The addition of GE particles into the PES membrane matrix also induced the formation of finger-like structures that indicated a more porous membrane. However, the decrease of tensile strength due to the addition of GE still provides excellent performance of mechanical property; it is above 10 Mpa.

5. Acknowledgments

This research was funded and supported by Universitas Syiah Kuala, Indonesia, through PRUUPDScholarship Program.

6. References

[1] Hu M, Cui Z, Li J, Zhang L, Mo Y, Dlamini D S, He B, Li J, Matsuyama H. 2019 Journal of Colloid and Interface Science. **552** 319–31.
[2] Kumar S R, Arthanareeswaran G, Thuyavan L Y. 2017 *J. Taiwan Inst. Chem. Eng.,* **72** 200–12.
[3] Chen M, Zhang X, Wang Z, Wang L, Wu Z. 2017 *Wat. Res.,* **120** 256–64.
[4] Wang S, Fang L, Cheng L, Jeon S, Kato N. 2018. *J. Membr. Sci.,* **564** 672–81.
[5] Liu F, Hashim N A, Liu Y, Abed M R M. Li K. 2011 *J. Membr. Sci.,* **375** 1–27.
[6] Kumar R S, Arthanareeswaran G. 2019 *Mater. Sci. Eng.,* **94** 258–69.
[7] Dhanik J, Arya N, Nand V. 2017 *J. Pharmacogn. Phytochem.,* **6** 174–84.
[8] Park M, Bae J, Lee D. 2008 *Phytotherapy Research.* **1449** 1446–49.
[9] Wang W, Zhu L, Shan B, Xie C, Liu C, Cui F, Li G. 2018 *J. Membr. Sci.,* **548** 459–69.
[10] Lin J, YeW, Zhong K, Shen J, Jullok N, SottoA, Van der Bruggen B. 2016 *Chem Eng Process,* **107** 194–205.
[11] Rahimi Z, Zinatizadeh A A L, Zinadini S. 2015 *J Ind Eng Chem.,* **29** 366–74.
[12] Arahman N, Mukramah, Syawaliah, Maimun T, Bilad M R. 2018 *Int. J. GEOMATE,* **15** 51–7.
[13] Vatanpour V, Siavash S, Moradian R, Zinadini S, Astinchap B. 2012 *Sep. Purif. Technol.,* **90** 69–82.
[14] Nikooe N, Saljoughi E. 2017 *Appl. Surf. Sci.,* **413** 41–9.
[15] Fahrina A, Maimun T, Humaira S, Rosnelly C M, Lubis M R, Bahrina I, Sunarya R, Ghufran A, Arahman N. 2018 *MATEC Web Conference.* **197** 1–4.
[16] Chang J, Zuo J, Zhang L, O'Brien G S, Chung T S. 2017 *J. Membr. Sci.,* **539** 295–304.
[17] Arahman N, Mulyati S, Fahrina A. 2019 *Mater. Res. Express,* **6** 1–10.

IOP Conf. Series: Materials Science and Engineering 1087 (2021) 012051 doi:10.1088/1757-899X/1087/1/012051

The implementation of K-Means clustering in kovats retention index on gas chromatography

T R Noviandy[1], A Maulana[1], N R Sasmita[2], R Suhendra[1], Muslem[3], G M Idroes[5], M Paristiowati[6], Z Helwani[7],E Yandri[8], S Rahimah[9], Muhammad[10], Irvanizam[1], R Idroes[3,4,*]

[1]Department of Informatics, Faculty of Mathematics and Natural Sciences, Universitas Syiah Kuala, Kopelma Darussalam, Banda Aceh 23111, Indonesia
[2]Computational and Applied Statistics Research Group, Department of Statistics, Faculty of Mathematics and Natural Sciences, Universitas Syiah Kuala, Kopelma Darussalam, Banda Aceh 23111, Indonesia
[3]Department of Chemistry, Faculty of Mathematics and Natural Sciences, Universitas Syiah Kuala, Kopelma Darussalam, Banda Aceh 23111, Indonesia
[4]Department of Pharmacy, Faculty of Mathematics and Natural Sciences, Universitas Syiah Kuala, Kopelma Darussalam, Banda Aceh 23111, Indonesia
[5]Department of Chemical Engineering, Faculty of Engineering, Universitas Syiah Kuala, Kopelma Darussalam, Banda Aceh 23111, Indonesia
[6]Department of Chemistry, Faculty of Mathematics and Sciences, Universitas Negeri Jakarta, Jl. Rawamangun Muka, Jakarta 13220, Indonesia
[7]Department of Chemical Engineering, Universitas Riau, Pekanbaru 28293, Indonesia
[8]Graduate School of Renewable Energy, Darma Persada University, Jl. Radin Inten 2, Pondok Kelapa, East Jakarta 13450, Indonesia
[9]Department of Food Industrial Technology, Faculty of Agroindustrial Technology Universitas Padjadjaran, Jatinangor 54363, Indonesia
[10]Faculty of Teacher Training and Education, Abulyatama University, Lampoh Keudee, Aceh Besar 23372, Indonesia

*E-mail: rinaldi.idroes@unsyiah.ac.id

Abstract. In this study, the retention index data of 146 compounds that are found in coal and petroleum-derived liquid fuels were grouped using the K-means clustering method, and the similarities between each cluster were analyzed. The psycho-chemical properties of each compound in the cluster were identified and compared with other clusters. Each compound's retention index is grouped based on the similarity between the column polarity and heating rate of one compound to another. Based on the results of tests carried out on nine different*k* values, it is known that the grouping with the value of $k = 3$ is the best determined from the obtained silhouette score = 0.568, where this score is higher than the score obtained on the other k values. The results of clustering with $k = 3$ obtained three clusters, namely cluster C1, cluster C2, and cluster C3. Cluster C1 and cluster C2 consist of chemical compounds that have a relatively low carbon number and molecular mass, but in cluster C2 the molecular mass of the compound is lower than in cluster C1. In contrast, the C3 cluster consists of chemical compounds that have a relatively high carbon number and molecular mass.

IOP Conf. Series: Materials Science and Engineering 1087 (2021) 012051 doi:10.1088/1757-899X/1087/1/012051

1. Introduction

The Kovats Retention Index was introduced by Ervin Kovats in 1958 as a second parameter that is fixed and resistant to the change ofexperimental conditions [1]. The Kovats retention index has been extensively developed for various compounds and is applied to gas chromatography [2] and liquid chromatography [3]. The retention index's essence is to obtain the psycho-chemical [4] and thermodynamic information [5] from a chromatographic analysis that the retention time cannot provide. The nature of the retention time that is dependent on changing the analysis conditions causes the retention time to be limited as a marker of separation. The identification of separate compounds must be made using spectroscopic methods as a tandem such as FTIR [6], NMR [7], or MS [8–11].

In general, the retention index is calculated using data of retention time and dead time with an algorithm's support. The dead time value can be determined by measuring the inert gas or homologous series method [12]. In comparison, the retention time is measured from the compound to be determined by its retention index. However, this method significantly depends on the accuracy of the dead time determination because it will affect the accuracy of the retention index calculation [13,14]. The retention index value can also be predicted based on its chemical structure. Idroes et al. have succeeded in predicting the Kovats retention index of flavors and fragrances [15,16] and essential oils [17] compounds using the genetic algorithm (GA) [18] combined with multiple linear regression (MLR).

As a representation of chemical information, retention index values can have a relationship and pattern to these chemical properties. These relationships can be analyzed using statistical methods. Some statistical methods commonly used to analyze the relationship between chemical parameter values and their properties are PCA [19,20] and clustering. Clustering is a method of partitioning the dataset into groups called clusters. Clustering is done by separating data in such a way that points in the same cluster are similar, and points in other clusters are different [21]. The purpose of clustering is to form a compact representation of the data (a subset selection), assessment of the diversity, and detect redundant variables [22]. Husna et al. have grouped the retention index of compounds in liquid fuels derived from coal and petroleum using the hierarchical cluster analysis (HCA) method [23], and Idroes et al. have grouped flavor and fragrance retention index compounds using the biplot method [24].

K-means clustering is a clustering method that looks for cluster centers that represent certain regions of the data. K-means clustering has numerous advantages, such as not requiring a distance matrix as required by hierarchical clustering and having fast computation time[25]. K-means clustering has been applied to the field of chemometrics, such as cluster analysis on Chinese medicine samples and Chinese tea [26] and pattern recognition analysis of chromatographic fingerprints [27].

In this study, the retention index data of 146 compounds that are found in coal and petroleum-derived liquid fuels were grouped using the K-means clustering method, and the similarities between each cluster were analyzed. Coal and petroleum-derived liquid fuels compounds are used because they are difficult to identify using only mass spectra, so this method is combined with retention indices. This combination can be used to identify individual compounds in complex mixtures with high confidence [28]. The purpose of clustering in this study is to classify similar compounds based on their psycho-chemical properties. The psycho-chemical properties of each compound in a cluster were identified and compared with other clusters.

2. Data and methods

2.1. Data collection

The dataset used in this study was the retention index data of 146 compounds that are found in coal and petroleum-derived liquid fuels obtained from Lai & Song literature [28]. The columns used are intermediately polar capillary column coated with 50% phenyl-50% methyl polysiloxane (Rtx-50), and a slightly polar column coated with 5% phenyl-95% methyl polysiloxane (DB-5). The column temperature is programmed linearly at heating rates of 2, 4, and 6 °C min^{-1}.

2.2. Data Clustering

K-means clustering is performed using Orange Data Mining software [29]. To determine the best k value, tests were carried out on nine different values of k, ranging from 2 to 10. Grouping for each value of k was done by ten times repetitions and the number of iterations of 300 iterations for each repetition. The best k value will be determined based on the silhouette score obtained. The equation used to calculate the silhouette value for each entity y_i can be seen in equation 1.

$$s(y_i) = \frac{b(y_i) - a(y_i)}{max\{a(y_i) - b(y_i)\}} \quad (1)$$

Where $b(y_i)$ ais the minimum dissimilarity to all clusters and$a(y_i)$is the average dissimilarity of y_i to all points in the same cluster [30].

3. Results and discussions

Retention index data of 146 compounds found in coal and petroleum-derived liquid fuelsweregrouped using K-means clustering based on the similarity between column polarity and each compound's heating rate. The results of the test conducted on nine different k values are available in Table 1. It can be seen that the $k = 3$ is the best determined from the obtained silhouette score = 0.568, where this score is higher than the scores obtained on the other k values. This means that at $k = 3$, the level of similarity between a compound to another compound in the same cluster and the level of dissimilarity between a compound against other compounds in a different cluster is higher than the other k values.

Table 1. Silhouette score for each *k*.

k	Silhouette Score
2	0.547
3	**0.568**
4	0.508
5	0.506
6	0.506
7	0.506
8	0.462
9	0.447
10	0.445

The results of clustering with $k = 3$ obtained three clusters, namely cluster C1, cluster C2, and cluster C3. The results of grouping for each compound from 146 compounds found in coal and petroleum-derived liquid fuels can be seen in Table 2. Cluster C1 consists of 64 compounds, cluster C2 consists of 66 compounds, and cluster C3 consists of 16 compounds. The visualization of the clustering results is seen in Figure 1.

Table 2. Results of clustering.

No	Compound	Cluster
1	n-Tridecane	C1
2	1-Tridecene	C1
3	1,3,5_Triethylbenzene	C1
4	n-Hexylbenzene	C1
5	Bicyclohexyl	C1
6	Naphthalene	C1

IOP Conf. Series: Materials Science and Engineering 1087 (2021) 012051 doi:10.1088/1757-899X/1087/1/012051

No	Compound	Cluster
7	2,4,6-Trimethylphenol	C1
8	n-Tetradecane	C1
9	1-Tetradecene	C1
10	Benzothiophene	C1
11	2,3-Dihydroindole	C1
12	Cyclohexylbenzene	C1
13	1-Indanol	C1
14	n-Octylcyclohexane	C1
15	Quinoline	C1
16	n-Pentadecane	C1
17	2-Methylnaphthalene	C1
18	1-Pentadecene	C1
19	1-Methylnaphthalene	C1
20	8-Methylquinoline	C1
21	1,2-Dicyclohexylethane	C1
22	5,6,7,8-Tetrahydro-3-methylquinoline	C1
23	n-Octylbenzene	C1
24	n-Hexadecane	C1
25	1,2,3,4-Tetrahydroquinoline	C1
26	2,6-Di-tert-butylphenol	C1
27	2-Ethylnaphthalene	C1
28	1-Hexadecene	C1
29	2,7_Dimethylnaphthalene	C1
30	2,6_Dimethylnaphthalene	C1
31	l,l'-Biphenyl (diphenyl)	C1
32	2-Methylbiphenyl	C1
33	1-Ethylnaphthalene	C1
34	1,3_Dimethylnaphthalene	C1
35	1,6_Dimethylnaphthalene	C1
36	2-lsopropylnaphthalene	C1
37	1,4_Dimethylnaphthalene	C1
38	1,5_Dimethylnaphthalene	C1
39	1-Isopropylnaphthalene	C1
40	n-Decylcyclohexane	C1
41	1,2_Dimethylnaphthalene	C1
42	n-Heptadecane	C1
43	5,6,7,8-Tetrahydro-I-naphthol	C1
44	3-Methylbiphenyl	C1
45	1,8_Dimethylnaphthalene	C1
46	4-Methylbiphenyl	C1
47	Acenaphthene	C1
48	1,2_Diphenylethane (bibenzyl)	C1
49	Dibenzofuran	C1
50	n-Decylbenzene	C1
51	n-Octadecane	C1
52	1-Octadecene	C1
53	3,3'-Dimethylbiphenyl	C1
54	2-Naphthol	C1
55	4,4'-Dimethylbiphenyl	C1
56	Fluorene	C1

No	Compound	Cluster
57	n-Nonadecane	C1
58	2,6-Diisopropylnaphthalene	C1
59	Dibenzyl ether	C1
60	9,10-Dihydroanthracene	C1
61	9,10-Dihydrophenanthrene	C1
62	1,2,3,4,5,6,7,8-Octahydroacridine	C1
63	Phenanthrene	C1
64	Anthracene	C1
65	n-Pentane	C2
66	n-Hexane	C2
67	1-Hexene	C2
68	n-Heptane	C2
69	Cyclohexane	C2
70	1-Heptene	C2
71	Cyclohexene	C2
72	Methylcyclohexane	C2
73	Benzene	C2
74	n-Octane	C2
75	trans-1,4-Dimethylcyclohexane	C2
76	c&1,3-Dimethylcyclohexane	C2
77	1-Octene	C2
78	trns-1,2-Dimethylcyclohexane	C2
79	1-Methylcyclohexene	C2
80	cis-1,4-Dimethylcyclohexane	C2
81	rruns-1,3-Dimethylcyclohexane	C2
82	cis-1,2-Dimethylcyclohexane	C2
83	Ethylcyclohexane	C2
84	Toluene	C2
85	n-Nonane	C2
86	Pyridine	C2
87	1-Nonene	C2
88	n-Propylcyclohexane	C2
89	Ethylbenzene	C2
90	1,4-Dimethylbenzene (p-xylene)	C2
91	1,3_Dimethylbenzene (m-xylene)	C2
92	n-Decane	C2
93	1,2-Dimethylbenzene (o-xylene)	C2
94	1-Decene	C2
95	rrans-Octahydro-lH-indene	C2
96	tert-Butylcyclohexane	C2
97	Isopropylbenzene (cumene)	C2
98	cis-Octahydro-1 H-indene	C2
99	n-Propylbenzene	C2
100	n-Butylcyclohexane	C2
101	1-Ethyl-3-methylbenzene	C2
102	1-Ethyl-4-methylbenzene	C2
103	1,3,5_Trimethylbenzene (mesitylene)	C2
104	tert-Butylbenzene	C2
105	4,5,6,7_Tetrahydroindan	C2
106	1-Ethyl-2-methylbenzene	C2

No	Compound	Cluster
107	n-Undecane	C2
108	1,2,4-Trimethylbenzene	C2
109	1-Undecene	C2
110	set-Butylbenzene	C2
111	truns-Decalin	C2
112	1,I'-Bicyclopentyl	C2
113	1,2,3_Trimethylbenzene	C2
114	Phenyl ethyl ether	C2
115	n-Butylbenzene	C2
116	n-Pentylcyclohexane	C2
117	cis-Decalin	C2
118	Indan	C2
119	n-Dodecane	C2
120	1-Dodecene	C2
121	2-Methylindan	C2
122	l-Methylindan	C2
123	1,2,4,5_Tetramethylbenzene	C2
124	n-Hexylcyclohexane	C2
125	1,4-Diisopropylbenzene	C2
126	5-Methylindan	C2
127	1,2,3,4-Tetramethylbenzene	C2
128	2,6-Dimethylphenol	C2
129	4-Methylindan	C2
130	Tetralin	C2
131	n-Eicosane	C3
132	n-Henicosane	C3
133	1,2,3,4-Tetrahydrocarbazole	C3
134	n-Docosane	C3
135	1-Phenylnaphthalene	C3
136	n-Tricosane	C3
137	n-Tetracosane	C3
138	Fluoranthene	C3
139	n-Pentacosane	C3
140	Pyrene	C3
141	9,10-Dimethylanthracene	C3
142	n-Hexacosane	C3
143	p-Terphenyl	C3
144	n-Triacontane	C3
145	Chrysene	C3
146	n-Hentriacontane	C3

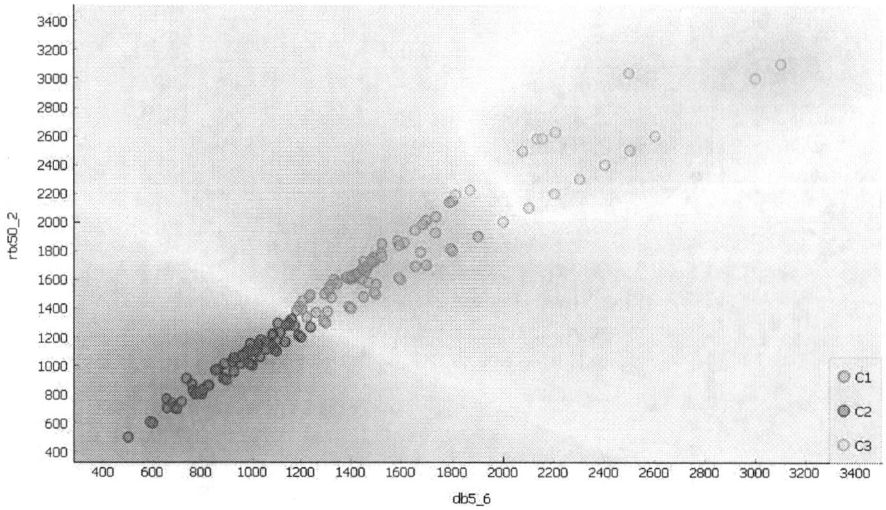

Figure 1. K-means clustering results visualization.

The results indicate that each cluster has its characteristics that are formed based on their psycho-chemical properties. There are two chemical parameters that show the different characteristics between the clusters. The C3 cluster consists of chemical compounds that have a relatively high carbon number and molecular mass. Cluster C1 and cluster C2 consist of chemical compounds that have a relatively low carbon number and molecular mass. However, the molecular mass of compounds in cluster C2 is lower than compounds in cluster C1.

4. Conclusion

Based on the results of the clustering of 146 compounds that are found in coal and petroleum-derived liquid fuels with several k ranging from 2 to 10, the best cluster results were obtained at $k = 3$ with a silhouette score of 0.568. From the three divided clusters, it can be seen that cluster C1 and cluster C2 consist of chemical compounds that have a relatively low carbon number and molecular mass, but in cluster C2 the molecular mass of the compound is lower than in cluster C1. In contrast, the C3 cluster consists of chemical compounds that have a relatively high carbon number and molecular mass.

Reference

[1] Kovats von E. 1958 *Helv. Chim. Acta.,* **41** 1915–32.
[2] Pratiwi S U T, Lagendijk E L, Weert S de, Idroes R, Hertiani T, Hondel C Van den 2015 *Int. J. Appl. Res. Nat. Prod.,* **8** 1–13.
[3] Idroes R. 2009 *Indones. J. Pharm.,* 133–140.
[4] Nowotnik D P, Narra R K. 1993 *Liq. Chromatogr.,* **16** 3919–32.
[5] Didaoui L, Touabet A, Badjah Hadj Ahmed A Y, Meklati B Y, Engewald W. 1999 *J. High Resolut. Chromatogr.,* **22** 559–564.
[6] Berdeaux O, Fontagné S, Sémon E, Velasco J, Sébédio J L, Dobarganes C. 2012 *Chem. Phys. Lipids,* **165** 338–347.
[7] Albert K. 2014. *Handbook of Spectroscopy Second Ed.* 1679–1716.
[8] Earlia N, Rahmad R, Amin M, Prakoeswa C, Khairan K, Idroes R. 2019 *Sains Malaysiana.* **48** 1019–24.
[9] Helwani Z, Ramli M, Saputra E, Bahruddin B, Yolanda D, Fatra W, Idroes G M, Muslem M, Mahlia T M I, Idroes R. 2020 *Catalysts.,* **10** **(164)** 1-13.
[10] Estevam E C, Griffin S, Nasim M J, Zieliński D, Aszyk J, Osowicka M, Dawidowska N, Idroes R, Bartoszek A, Jacob C. 2015 *Nat. Prod. Commun.,* **10** 1934578X1501001025.
[11] Earlia N, Suhendra R, Amin M, Prakoeswa C R S, Idroes R. 2019 *Sci World J., (***2019)** 1-7.

[12] Idroes R, Muslem, Saiful, Mahmudi, Idroes G M, Suhendra R, Irvanizam, Zamzami M P. 2019 *IOP Conf. Series: Mater. Sci. Eng.,* **364** 1-7

[13] Idroes R, Husna I, Muslem, Mahmudi, Rusyana A, Helwani Z, Idroes G M, Suhendra R, Yandri E, Rahimah S, Sasmita N R. 2019 *IOP Conf.Series: Earth Environ. Sci.,* **364** 1-7.

[14] Idroes R, Muslem, Mahmudi, Saiful, Idroes G M, Suhendra R, Irvanizam. 2020. *Heliyon.* **6** 1-8.

[15] Idroes R, Noviandy T R, Maulana A, Suhendra R, Sasmita N R, Muslem M, Idroes G M, Irvanizam I. 2019 *Int. Rev. Model. Simulations,* **12** 373-382.

[16] Maulana A, Noviandy T R, Idroes R, Sasmita N R, Suhendra R, Irvanizam I 2020 *International Conference on Electrical Engineering and Informatics (ICELTICs) Banda Aceh.*

[17] Noviandy T R, Maulana A, Sasmita N R, Suhendra R, Irvanizam I, Muslem M, Idroes G M, Yusuf M, Sofyan H, Abidin T F, Idroes R. 2020 *J. Eng. Sci. Technol., xx-xx*

[18] Idroes R, Maulana A, Noviandy T R, Suhendra R, Sasmita N R, Lala A, Irvanizam. 2020. *IOP Conf.Series: Earth Environ.,* **796** 012033.

[19] Nadia Y, Ramli M, Muslem, Japnur A F, Rusyana A, Idroes G M, Suhendra R, Muhammad, Sasmita N R, Tallei T E, Idroes R. 2019 *IOP Conf.Series: Earth Environ.,* **364** 012040.

[20] Wahidah S, Khairi, Lelifajri, Idroes R, Rahmadi, Lala A, Mahmudi, Muslem, Japnur A F. 2019 *IOP Conf.Series: Earth Environ.,* **364** 012016.

[21] Müller A C, Guido S. 2016 *Introduction to machine learning with Python: a guide for data scientists.* California: O'Reilly Media, Inc.

[22] Drab K, Daszykowski M. 2014 *J. AOAC Int* **97** 29–38.

[23] Husna I, Rusyana A, Muslem, Idroes G M, Suhendraand R, Idroes R. 2020 *IOP Conf. Series: Mater. Sci. Eng.,* **796** 012064.

[24] Idroes R, Japnur A F, Suhendra R, Rusyana A. 2019 *IOP Conf. Series: Mater. Sci. Eng.,* **523** 012007.

[25] Kundu M, Kundu P K, Damarla S K. 2017 *Chemometric Monitoring: Product Quality Assessment, Process Fault Detection, and Applications.* Florida: CRC Press.

[26] Sun L-X, Xu F, Liang Y-Z, Xie Y-L, Yu R-Q. 1994. *Chemometr Intell Lab Syst.,* **25** 51–60.

[27] Aliakbarzadeh G, Sereshti H, Parastar H. 2016. *Anal. Bioanal. Chem.,* **408** 3295–3307.

[28] Lai W. 1995. *Fuel,* **74** 1436–51.

[29] Demšar J, Curk T, Erjavec A, Gorup C, Hocevar T, Milutinovic M, Možina M, Polajnar M, Toplak M, Staric A, Stajdohar M, Umek L, Žagar L, Žbontar J, Žitnik M, Zupan B. 2013. *J. Mach. Learn. Res.,* **14** 2349–2353.

[30] de Amorim R C, Hennig C. 2015. *Inf. Sci.,* **324** 126–145.

Study of linearity and stability of Pb(II)-1,10-phenanthroline complex with the presence of Fe (II) dan Mg (II) matrix ions using UV-Vis spectrophotometry

K Nisah[1,3], M Ramli[2], Marlina[2], R Idroes[2,4] E Safitri[2,*]
[1]Graduate School of Mathematics and Applied Science, Universitas Syiah Kuala, Banda Aceh 23111, Indonesia
[2]Chemistry Department Faculty of Science, Universitas Syiah Kuala,Banda Aceh 23111, Indonesia
[3]Chemistry Department Faculty of Science and Technology, Ar -raniry State Islamic University, Banda Aceh 23111, Indonesia
[4]Pharmacy Department Faculty of Science, Universitas Syiah Kuala, Banda Aceh 23111, Indonesia

*E-mail : e.safitri@unsyiah.ac.id

Abstract. The effect of Fe (II) and Mg (II) as matrix ions in the determination of Pb (II) based on 1,10-phenanthroline complexing agent at pH 7 had been investigated.The objective of this study is to determine the effect of Fe (II) and Mg (II) on the determination of Pb (II) using the complexing agent 1,10-phenanthroline. The absorbance determination using a spectrophotometer was conducted at a wavelength of 302.5 nm for Pb (II)-1,10-Phenanthroline. Linear correlation between the investigated absorbance on Pb concentration was represented by a linear equation of $y = 0.0131\,x + 0.8693$ ($R^2 = 0.9613$). From the linear equation, it was obtained the increase in sensitivity and linearity for Mg (II), as well as the decrease in sensitivity and linearity for Fe (II). It revealed that Fe (II) and Mg (II) affects the analysis of Pb (II) using 1,10-phenanthroline complexing agent. In order to achieve stability of Pb (II)-1,10-Phenanthroline, it was compared with Mg (II)-1,10-Phenanthroline and Fe (II)-1,10-Phenanthroline. It suggested that Pb (II), Mg (II) and Fe (II) had an equivalent binding strength against 1,10-Phenanthroline.

1. Introduction

The contamination of heavy metal in the environment still become a major problem. Besides it present naturally by geological or geochemical activities [1–5], the intense uses by the construction [6]and industry [7,8]make the metal pollution has caused exposure in many cases[9–11] This can be more dangerous because the heavy metal can transport from an ecosystem to others [12]and poisoning the organism[13].

Lead (Pb) is one of the hazardous heavy metals which threatens the life of humans and animals. Pb is also a component that is commonly used as an additive in water pipe [14], paint [15], and octane number reducing agent in premium gasoline [16]. The use of Pb in those materials is due to its specific function and low price [17].

Pb can be accumulated in bone tissue [18,19], which further damages the nervous system and enzymatic function stem from its neurotoxicity [17]. This case often occurs on children [20], including brain damage. The toxicity of Pb is a big problem due to its broad use. Therefore, the presence of Pb essentially needs to be analyzed in a periodical manner to observe the level of pollution. It further helps the decision-making on the countermeasures. Based on the stated problems, Pb analysis has to be carried out.

Several methods had been applied to analyze Pb, either conventional or instrumental methods. Determination of Pb for both methods requires specific criteria in term of quantity and types of samples. The instrumental determination usually is more sensitive due to its ability to measure Pb up to the smallest concentration unit. Some methods used for Pb determination includes Atomic Absorbance Spectroscopy (AAS), Graphite Furnace AAS, chemiluminescence, potentiometry, Anodic Stripping Voltammetry and UV-Vis spectrophotometry [21]. Concentration determination of Pb through UV-Vis spectrophotometry had been used widely due to its quick, easy, and cheap operation. Pb (either in the form of Pb (II) or Pb (IV)) is firstly required to be priorlycomplexed using Pb ions complexing agents. Some Pbcomplexing agents include dithizone[22], 5-Bromo-2-hydroxy-3-methoxy benzaldehyde-hydroxybenzoic hydrazine[23], diphenylcarbazone and 1,10-phenanthroline [24].

Spectrophotometry methods with Pb complexing agent, dithizone, has been studied by dithizone [22]. On the stated method, acidic medium (concentrated HNO_3) was used within the presence of 50 matrix cations during the determination of Pb (II). Meanwhile, B. Saritha, A. Giri, and T. Sreenivasulu Reddy studied a similar method; however, they used different complexing agents (-Bromo-2-hydroxy-3-methoxy benzaldehyde-hydroxybenzoic hydrazine). At optimum pH (between 7-9) [23,24], it was reported that the quantity of metal ions and anions did not affect the determination of Pb (II). Mikuła et al., 2009 conducted research using phenanthroline complexing agents and Pb adsorbent. Complexing agent 1,10-phenanthroline is a simple reagent that can be used to complex aqueous Pb (II).

The references in introduction to prove the finding by author which used as the basis in this research, several references for the Pb existence both naturally and manmade. It also describe the case of Pb exposure on human life and the disease appeared.This description is the reason why Pb identification is important.Other references described about the recent various methods used for Pb.These are well methods in accuracy and reliability,but less ones in simplicity and cost.These descriptions provided the basis why the UV-Vis offered as a quick,easy and cheap methods.

Analysis of Pb spectrophotometrically using complexing agent 1,10-phenanthroline was affected by several metal ions such as nickel and cadmium [25]. Nickel and cadmium ions affect the determination of Pb ions by decreasing the absorbance. A study by B. Yusuf, A. and S. Nurliana, at pH 8.5, the linear equation was found to be $Y= 0.1596 X + 0.1932$, with $R^2 = 0.9991$[26]. Furthermore, M. Yayinie, M. Anza, and T. Shifaraw in 2015 conducted the same study at pH 10-11.5, where the determination yielded a sensitivity as high as 0.014 with $R^2 = 0.99781$, standard deviation = 0.01793 and recovery = 93.8% [21].

As discussed above, it appears that the studies of matrix ions were merely conducted against metal ions from the transition group due to its similar properties. Nonetheless, in this work, we present an investigation results on Mg (II) and Fe (II) ions. Both metals are commonly found in an aqueous sample, which motivates our work in finding how those metal ions affect the stability of 1,10-Phenanthroline Pb (II) complex.

2. Instruments and materials

UV-Vis Shimadzu 1800 Spectrophotometer, glassware (Pyrex), pH-meter Bran Orion Star.

2.1. Materials

Pb $(NO_3)_2$, $FeCl_2.6H_2O$, Mg $(NO_3)_2$, 1,10-Phenanthroline, Buffer Solution pH 7 NaOH and acetone were purchased from Merck.

2.2. Methods

Preparation of Pb stock solution. Pb (II) stock solution (100 ppm) was taken using pipette as much as 0.1 - 0.5 mL and added into a volumetric flask 10 mL to give Pb (II) solutions with 1-5 ppm.

Into each volumetric flask, 1.5 mL 1,10-Phenanthroline 1000 ppm was added, followed by 1.5 mL buffer solution NaOH pH 7, 5 mL acetone, and distilled water until the total volume reached 10 mL. The mixture was shaken, then left for 5 min. The mixture was measured for its absorbance at maximum wavelength using UV-Vis spectrophotometer. This procedure was repeated three times. Then, the calibration curve was constructed between the absorbance and the Pb (II) concentration. The linear equation and R^2 were obtained from the plot.

Absorbance Measurement In Determination of Fe (II) and Mg (II) on Pb (II) analysis at pH 7 Pb Pb(II) stock solution (1-5 ppm) was then respectively added with 0.01 mL and 0.05 mL Fe (II) solution 100 ppm, where it was followed by the addition of 1.5 mL 1,10-Phenantrolin 1000 ppm, 5 mL buffer solution NaOH pH 7 and 5 mL acetone. Next, the distilled water was poured until the volume reached 10 mL. The mixture was shaken and left for 5 min. It was then measured for the absorbance at maximum wavelength using UV-Vis spectrophotometer. This procedure was conducted with three times repetitions. The same procedure was also applied for Mg (II) solution.

Investigation of complex stability (K_c) of Pb (II), Fe (II) and Mg (II) on 1,10 phenanthroline complexing agent.

The investigation of complex stability(K_c) of Pb (II), Fe (II) and Mg (II) with 1,10 phenanthroline complexing agent was determined based on the concentration of Pb (II), Fe (II) dan Mg (II) with 1,10 phenanthroline ligand which had been affected by the used ions.

3. Results and discussion

In this research, we studied the comparative matrix effect of Fe (II) and Mg (II) against Pb (II) analysis using 1,10-Phenanthroline complexing agent at pH 7 with UV-Vis spectrophotometer. This work was initiated by complexing Pb (II) with 1,10-Phenanthroline forming Pb (II)-1,10-Phenanthroline complex, which appeared colourless. Pb (II)-1,10-Phenanthroline complex at pH 7 [23-24] was measured for its absorbance to determine the maximum wavelength using spectrophotometer UV-Vis. The effect of Fe (II) and Mg (II) against the maximum wavelength of Pb (II) would be discussed in this work.

The plot obtained revealed that the maximum absorbance was observed at 302.5 nm, which suggested that the maximum wavelength for Pb (II)-1,10-Phenantroline was 302.5 nm [15]. 1,10-Phenantrolin is a complexing agent that is capable of reacting with various types of metal to form complex compounds. 1,10-Phenantrolin could form a strong complex with Pb (II) and appeared colourless as shown by Figure 1.

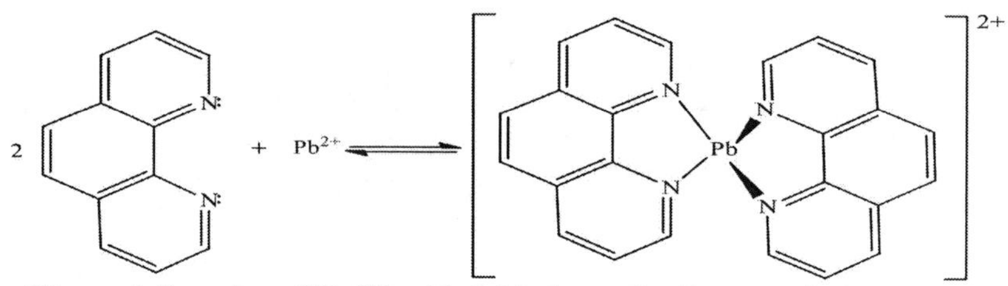

Figure 1. Reaction of Pb (II) with 1,10 phenanthroline complexing agent

Table 1 presents the effect of Fe (II) and Mg (II) against the linearity of Pb (II)-phenanthroline complex. In line with the sensitivity, the R^2 and experimental t [25-26], it can be observed that the presence of Fe (II) and Mg (II) in the Pb (II) measurement system using 1.10 phenanthroline complexing agent, affected the determination[24]. The value of sensitivity on Pb (II) determination (1-5 ppm) changed significantly by the presence of both ions. The difference was significantly based on the t-experimental which is far above the t-theoretical on the Pb (II) sensitivity before and after the presence of the matrix ions (Fe (II) and Mg (II)). The t-test was carried on the sensitivity for n = 3 at confidence level of 95% (t-theoretical = 2.353). Matrix effect is given by Fe (II) 0.1 ppm and Fe (II) 0.5 ppm decreased the measurement sensitivity as much as 8 x10^{-4} and 1.6 x 10^{-3}, respectively. Next, the effect of Mg (II) against the linearity of Pb (II) (1-5 ppm) had caused a significant increase in the sensitivity, which reached 9.1 x10^{-2} and 9.52 x 10^{-2} for Mg (II) 0.1 ppm and Mg (II) 0.5 ppm, respectively.

Table 1. Sensitivity and linearity of Pb (II)-1,10-phenanthroline complex with the presence of matrix ions Mg (II) and Fe (II).

	b (Sensitivity)	R^2	t experimental
[Pb] (1-5 ppm)	0.0131 ± 0.00012	0.9613	
[Pb] (1-5 ppm) + Fe 0.1ppm	0.0123 ± 0.00021	0.9544	8.32
[Pb] (1-5 ppm) +Fe 0.5 ppm	0.0115 ± 0.00015	0.9274	13.6
[Pb] (1-5 ppm) + Mg 0.1 ppm	0.1038 ± 0.00035	0.9867	30.9224
[Pb] (1-5 ppm) + Mg 0.5 ppm	0.1083 ± 0.00179	0.9908	7.24

Pb ion analysis using 1,10-Phenanthroline complexing agents could be affected by several matrix ions such as nickel and cadmium [25]. 1,10-Phenanthroline could form a complex with all bivalent metals [16]. In this work, it can be observed that the presence of Mg (II) and Fe (II) within Pb (II)-1,10-Phenantroline complex system affected the absorbance at each investigated Pb concentration. It suggested that Mg (II) and Fe (II) contributed in the uncertainty of Pb (II) analysis using 1,10-Phenanthroline complexing agent. It is owing to the reactivity of Mg (II), and Fe (II) is higher than Pb (II) in binding 1,10-Phenanthroline [15]. The proposed reactions occurred when Mg (II) and Fe (II) was complexed with 1,10-Phenanthroline are depicted by Figure 2.

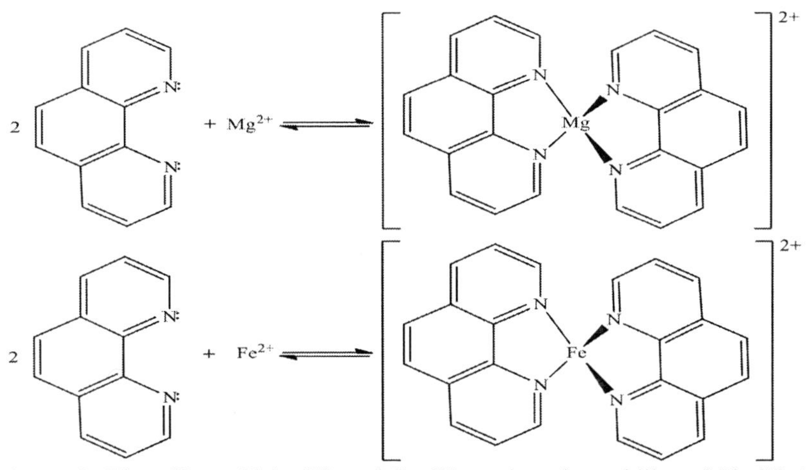

Figure 2. The effect of Mg (II) and Fe (II) against the stability of Pb (II)-1,10-Phenanthroline complex.

In this work, the linearity of Pb (II)-1,10-Phenanthroline complex for Pb (II) with a concentration ranged from 1-5 ppm was investigated for its complex equilibrium constant (Kc) profile for both Mg (II) and Fe (II). Tables 2 and 3. present the values of equilibrium constant (Kc) of Pb (II)-1,10-Phenanthroline with and without the presence of matrix ions (Mg (II) and Fe (II)) [18]. Kc values of Pb (II)-1,10-Phenanthroline without the presence of matrix ions were relatively the same, compared with the presence of Mg (II) and Fe (II) matrix ions, either for 0.1 ppm or 0.5 ppm.

Table 2. Kc Profile of Pb (II)-1,10-Phenanthroline with the addition of Mg (II) 0.1 and 0.5 ppm.

Concentration Pb^{2+} (ppm)	Concentration Mg^{2+} (ppm)	y (Absorbance Mg^{2+})	x (Concentrationof determined Pb^{2+} (ppm))	Kc
1	0	0.881	0.893	3.970×10^{-5}
2	0	0.898	2.191	19.474×10^{-5}
3	0	0.911	3.183	42.443×10^{-5}
4	0	0.915	3.489	62.019×10^{-5}
5	0	0.938	5.244	116.539×10^{-5}
1	0.1	0.881	0.857	3.810×10^{-5}
2	0.1	0.91	2.337	20.771×10^{-5}
3	0.1	0.923	3	40×10^{-5}
4	0.1	0.934	3.561	63.311×10^{-5}
5	0.1	0.967	5.245	116.553×10^{-5}
1	0.5	0.89	0.644	2.8630×10^{-5}
2	0.5	0.92	2.485	22.086×10^{-5}
3	0.5	0.93	3.098	41.309×10^{-5}
4	0.5	0.941	3.773	67.076×10^{-5}
5	0.5	0.977	5.9816	132.924×10^{-5}

Table 3. The effect of Fe (II) ions addition.

Concentration Pb^{2+}(ppm)	Concentration Fe^{2+} (ppm)	y (Absorbance Fe^{2+})	x (Concentration of determined Pb^{2+} (ppm))	Kc
1	0	0.881	0.893	3.970×10^{-5}
2	0	0.898	2.191	19.474×10^{-5}
3	0	0.911	3.183	42.443×10^{-5}
4	0	0.915	3.489	62.019×10^{-5}
5	0	0.938	5.244	116.540×10^{-5}
1	0.1	0.888	0.854	3.794×10^{-5}
2	0.1	0.905	2.236	19.874×10^{-5}
3	0.1	0.917	3.211	42.818×10^{-5}
4	0.1	0.92	3.455	61.427×10^{-5}
5	0.1	0.942	5.244	116.531×10^{-5}
1	0.5	0.898	1.667	7.407×10^{-5}
2	0.5	0.911	2.130	18.937×10^{-5}
3	0.5	0.923	3.174	42.319×10^{-5}
4	0.5	0.924	3.261	57.971×10^{-5}
5	0.5	0.949	5.435	120.773×10^{-5}

Therefore, it could be stated that the stability of Pb (II)-1,10-Phenanthroline was relatively the same with Mg (II) -1,10-Phenanthroline and Fe (II) -1,10-Phenanthroline complexes. It was ascribed to the surrounding reaction condition at pH 7, where at the stated pH, Pb could perform optimally [24]. It could also be associated with the atomic radius of Fe (II) cations, which is smaller than that of Pb (II) cation [17].

4. Conclusion

Mg (II) and Fe (II) could affect the analysis of Pb (II)-1,10-Phenantroline, due to the change of sensitivity and linearity of the formed complex. Therefore, a more sensitive method is required in analyzing the metal complex. The Kc values of Pb (II)-1,10-Phenantroline were relatively similar to that of Mg (II)- 1,10-Phenantroline and Fe (II)-1,10-Phenantroline. It implied that Pb (II)-1,10-Phenantroline has a similar strength in binding 1,10-Phenanthroline, in comparison with Mg (II)- 1,10-Phenantroline and Fe (II)- 1,10-Phenantroline.

References

[1] Duivenvoorden L J, Roberts D T,Tucker G M 2017 *Environ. Earth Sci.,* **76** 320.
[2] Idroes R, Yusuf M, Alatas M, Subhan, Lala A, Saiful, Suhendra R, Idroes G M, Marwan 2018 *IOP Conf. Ser. Mater. Sci. Eng.,* **334** 012002.
[3] Idroes R, Yusuf M, Saiful S, Alatas M, Subhan S, Lala A, Muslem M, Suhendra R, Idroes G M, Marwan M, Mahlia T M I 2019 *Energies,* **12** 4442.
[4] Idroes R, Yusuf M, Alatas M, Subhan, Lala A, Muhammad, Suhendra R, Idroes G M, Marwan. 2019 *IOP Conf. Ser. Mater. Sci. Eng.,* **523** 012012.
[5] Idroes R, Yusuf M, Alatas M, Subhan, Lala A, Muslem, Suhendra R, Idroes G M, Suhendrayatna, Marwan, Riza M. 2019 *IOP Conf. Ser. Mater. Sci. Eng.,* **523** 012010.
[6] Shi P, Xiao J, Wang Y, Chen L. 2014. *Int. J. Environ. Res. Public Health* **11** 2504–2520.
[7] Helwani Z, Ramli M, Saputra E, Bahruddin B, Yolanda D, Fatra W, Idroes G M, Muslem M, Mahlia T M I, Idroes R 2020 *Catalysts,* **10 (164)** 1-13.
[8] Helwani Z, Ramli M, Saputra E, Putra Y L, Simbolon D F, Othman M R, Idroes R. 2020 *Catalysts,* **10 (724)** 1-16.
[9] Chhabra D, Oda K, Jagannath P, Utsunomiya H, Takekoshi S, Nimura Y 2012 *Asian Pac. J. Cancer Prev.,* **13** 187–190.
[10] Wahidah S, Khairi, Lelifajri, Idroes R, Rahmadi, Lala A, Mahmudi, Muslem, Japnur A F. 2019 *IOP Conf. Series: Earth and Environ. Sci.,* **364** 12016.
[11] Irnawati I, Manaf S A, Suhartono E, Rusyana A, Idroes R. 2019 *IOP Conf. Ser. Earth Environ. Sci.* **348** 12068.
[12] Fang H, Huang L, Wang J, He G, Reible D. 2016 *J. Hazard. Mater.,* **302** 447–457.
[13] Suhartono E, Thalib I, Aflanie I, Noor Z, Idroes R. 2018 *IOP Conf. Ser. Mater. Sci. Eng.,* **350** 12008.
[14] Zweifel H, Maier R D, Schiller M, Amos S E. 2001 *Plastics additives handbook.*
[15] Völz H G, Kischkewitz J, Woditsch P, Westerhaus A, Griebler W-D, De Liedekerke M, Buxbaum G, Printzen H, Mansmann M, Räde D, Trenczek G, Wilhelm V, Schwarz S, Wienand H, Adel J, Adrian G, Brandt K, Cork W B, Winkeler H, Mayer W, Schneider K, Leitner L, Kathrein H, Schwab E, Jakusch H, Ohlinger M, Veitch R, Etzrodt G, Pfaff G, Franz K-D, Emmert R, Nitta K, Besold R, Gaedcke H. 2006. *Pigments, Inorganic in Ullmann's Encyclopedia of Industrial Chemistry*. Weinheim: Wiley-VCH Verlag GmbH & Co. KGaA.
[16] Oxtoby D W, Gillis H P, Nachtrieb N H, Suminar S A, Amalia S, Simarmata, Silvester L. 2003 *Prinsip-prinsip Kimia Modern*. Jakarta: Erlangga. *(in Bahasa Indonesia).*
[17] Boldyrev M. 2018. *Wiki J. Sci.,* **1** 1-7.
[18] Food and Drug Association. 2015. *Elemental Impurities Guidance for Industry* 1–85.
[19] Suhartono E, Noor Z, Edyson, Budianto W Y, Idroes R. 2019. Effect of chronic lead exposure on bone using ATR-FTIR spectroscopy. *AIP Conf. Proc.,* **2108** 020025.
[20] Sarfo D K, Izake E L, O'Mullane A P, Ayoko G A. 2018. *Sensors Actuators B Chem.,* **255** 1945–1952.
[21] Yayinie M, Anza M, Shifaraw T. 2015 *J. Phys. Chem. Sci.,* **3**.
[22] Khan H, Ahmed M J, and Bhanger M I. 2007 *Anal. Sci.,* **23** 193–199.

[23] Saritha B, Giri A, Reddy T S. 2014 *J. Chem. Pharm. Res.,* **6** 1571–1576.
[24] Mikuła B, Puzio B, Feist B. 2009 *Microchim. Acta* **166** 337–341.
[25] De Toni L, Tisato F, Seraglia R, Roverso M, Gandin V, Marzano C, Padrini R, Foresta C. 2017 *Toxicol. Reports,* **4** 234–239.
[26] Yusuf B, Nurliana S. 2015 *J. Kim. Mulawarman* **11** 1-3.

CaO from chicken eggshell supported on activated carbon and KOH (CaO/C/KOH) as catalyst for biodiesel production from off grade palm oil

Z Helwani[1], I Zahrina[1], S Z Amraini[1], R I Sianturi[1], G M Idroes[2], Muslem[5], R Idroes[3,4,*]

[1]Department of Chemical Engineering, Faculty of Engineering, Universitas Riau, Kampus Bina Widya Panam, Pekanbaru 28293, Indonesia

[2]Department of Chemical Engineering, Faculty of Engineering, Universitas Syiah Kuala, Banda Aceh 23111, Indonesia.

[3]Department of Chemistry, Faculty of Mathematics and Natural Sciences, Universitas Syiah Kuala, Banda Aceh 23111, Indonesia

[4]Department of Pharmacy, Faculty of Mathematics and Natural Sciences, Universitas Syiah Kuala, Banda Aceh 23111, Indonesia

[5]Department of Chemistry, Faculty of Science and Technology, Universitas Islam Negeri Ar-Raniry, Banda Aceh 23111, Indonesia

*E-mail: rinaldi.idroes@unsyiah.ac.id

Abstract. This study carried out the catalyst synthesis of calcium oxide (CaO) from chicken eggshells supported on activated carbon (C) and activated by KOH solution (CaO/C/KOH). The effects of variation of KOH concentration and the CaO:C mass ratio to catalyst performance were also investigated. The catalyst was used in the biodiesel production from off-grade palm oil. The catalyst with the highest catalytic activity was obtained at the CaO:C mass ratio of 12:6 and the KOH concentration of 25%. The transesterification that was performed under the conditions of the oil: methanol molar ratio of 1:10, the catalyst concentration of 1% -b oil, the reaction temperature of 65 °C, and reaction time of 3 hours produced a biodiesel yield of 85.74%. The catalyst characterization by using XRD, Hammet indicator, and Brunauer-Emmet-Teller (BET) showed the product was in the crystalline phase of CaO, K_2CO_3, K_2O $Ca(OH)_2$, and $CaCO_3$, the base strength of H_> 12.2, and the specific surface area of 17.890 m^2/g, respectively. The results showed that the KOH concentration in the catalyst activation process was directly proportional to the biodiesel yield. Several methyl esters such as methyl palmitate and methyl oleate were found in biodiesel through Gas Chromatography-Mass Spectrometry (GC-MS) analysis.

1. Introduction

After using fossil energy has caused many negative impacts, and its resource is decreasing,alternative energy becomes a popular subject in both scientific and engineering. Many countries explore and develop the new energy according to the available energy resources they have, such as wind, hydro, solar [1], geothermal [2–8], and even biomass [9,10]. Especially for the agricultural country, biomass conversion to be energy resources is highly interesting. Besides solving agricultural waste, it also increases the value of some less valuable agricultural products, such as off-grade palm oil.

Off-grade palm oil is a product from the residual sorting in the Palm Oil Mill that has not been fully utilized. The amount of available off-grade palm oil is quite large, around 7-10% of the product. This oil is usually sold 30-40% cheaper than CPO [11]. Off-grade palm oil can be used as an alternative raw material for biodiesel production. As a palm family, It contains high FFA content [12]. Transesterification is the most common method used to convert this content to biodiesel.

A homogeneous alkaline catalyst is one of the catalysts that is usually used in transesterification because it has high catalytic activity, high biodiesel conversion, and relatively short reaction time (± 1 hour) [13]. However, this catalyst has some disadvantages: it cannot be re-used or regenerated, difficult to separate from the product, require more equipment for separation and purification, so that production costs increased [14]. The heterogeneous solid base catalysts can be used as an alternative because it has advantages, including environmentally friendly, easy to separate, providing high selectivity, low cost, sustainable processes, and less waste [15].

CaO catalysts are one of a heterogeneous solid base catalyst. It is usually obtained from limestone, calcium carbonate, and calcium hydroxide. This catalyst is easily provided from chicken eggshell waste or synthesized from fine material. Chicken eggshells contain 94% $CaCO_3$, 1% $MgCO_3$, 1% $Ca_3(PO_4)_2$ and 4% organic ingredients. According to [16], the use of CaO directly as a catalyst causes oxygen ions on the CaO surface to form hydrogen bonds with glycerin. So that the viscosity of glycerin increases and forms a suspension. The formation of the suspension makes it difficult for CaO and glycerin to be separated from the product. To overcome this problem, CaO must be impregnated with catalyst support or metal oxide. Previous research has reported the use of CaO from $Ca(NO_3)_2.H_2O$ and eggshells as a source and provide well catalytic activities in biodiesel production [17–20].

As reported previously, the performance of CaO catalysts supported on activated carbon was influenced by the CaO: C mass ratio, temperature, and calcination time [21,22]. The more significant the amount of CaO usedthe more increase the biodiesel yield. Different things were obtained from the effect of calcination temperature, which indicates that the increase in calcination temperature reduced product yield [23]. In this study, CaO was impregnated on activated carbon and KOH (CaO/C/KOH). Activated carbon has the advantage of increasing the surface area. Its composition is higher and easily separated from the product [21]. It is expected that the synthesis of the CaO/C/KOH catalyst will increase the catalyst product's performance so that it increases the yield of biodiesel.

2. Materials and methods

The chicken eggshell wastes were collected from the local bakery, KOH, aquadest, off-grade palm fruits were obtained from PTPN V, H_2SO_4, methanol, and Hammet indicators were obtained from PT. Bratachem. The tools used were spindle hydraulic press, condenser, three-neck boiling flask, thermometer, magnetic stirrer, and heating mantle.

2.1. Synthesis of CaO/C/KOH Solid base catalyst

Chicken eggshells were cleaned and dried at 110 °C for 12 hours, mashed using a shaker mill and sieved with 100-200 mesh sizes to homogenize the size, and calcined at 900 °C for 3 hours to get calcium oxide (CaO) [21]. The activated carbon was uniformed in size using the 100-200 mesh sieve. Furthermore, CaO and activated carbon were weighed according to the variation of mass ratio, 12 gr CaO each with a variation of 6 gr, 8 gr, and 10 gr of activated carbon. Each of the mixtures was dissolved in KOH with the concentration variation of 15, 20, and 25% each as much as 100 ml. Then it was stirred until homogeneous to have the $Ca(OH)_2$ solution. The process condition was carried out at room temperature for 24 hours with a stirring speed of 400 rpm. The slurry formed was dried in the oven at 110 °C for 5 hours and calcined for 5 hours at 500 °C.

The study of the catalytic performance was studied on the transesterification process using the same method reported in the previous research by [19,20].

2.2. Characterization of catalysts and biodiesel product

The crystalline phase, basicity, and specific surface area of catalysts were characterized by using X-Ray Diffraction (XRD), Hammet Indicators, and Brunaeur-Emmet-Teller (BET) methods, respectively. The characteristics of biodiesel tested included the acid number, flash point, cetane number, and viscosity (kinematic viscosity). The composition of biodiesel was analyzed using GC-MS.

3. Result and discussion

Table 1. Physical properties and FFA of crude palm-oil off grade.

Parameter	Off grade palm oil	Standard CPO SNI 01-2901-2006
Colour	Orange	Yellow
Density 40°C (kg/m$^{3)}$	892.11	-
Viscosity 40°C (mm^2/s)	29.47	-
Water content (%)	3.5	Max 0.5
FFA (%)	9.28	Max 0.5

It can be seen in Table 1 that off-grade palm oils have high water content, which can cause the hydrolysis process to occur so that the FFA level increased [24]. Also, water can react with the catalyst, so that the catalytic activity will decrease [25,26]. Water also causesto form soap during the transesterification process[27]. Therefore, water needs to be removed by heating the oil above the boiling point of the water. High FFA levels are carried out through an esterification process to reduce FFA levels from 9.28% to 1.38 [28].

3.1. Hammet indicator titration of CaO/C/KOH catalyst

All CaO/C/KOH catalysts that made from the variation of CaO:C mass ratio, that are (12:6), (12:8), and (12:10) and the variation of KOH concentrations, that are 15, 20, and 25 (% w) calcined at a temperature of 500 °C were tested the basic strength using the Hammet indicator. The results showed that the catalyst alkalinity ranged from H_> 9.3 - 12.2. The basic nature of the catalyst increases with increasing KOH concentration and CaO:C ratio. Catalysts with an H_> 9.3 basicity are relatively strong catalysts for the transesterification process [29].

3.2. Characterization of CaO/C/KOH catalyst using x-ray diffaction (XRD)

The effect of calcination temperature and KOH concentration on the crystalline phase of the catalyst was studied by comparing each XRD pattern of the catalysts. The XRD pattern showed that the amount of CaO was very small because the percentage of CaO content has reacted with KOH and activated carbon in the catalyst in the impregnation process to form K_2CO_3. So that K_2CO_3 has a greater composition than CaO, that is 52.18% with a peak of 2θ: 34.74°, 28.18°, 41.44°, 58.15°, and 55.68°with the peak of 34.74°as the highest peak, as shown in Figure 1.

The carbonate compound formed on activated carbon after impregnation with KOH, which was thought of as K_2CO_3, where potassium binds to a carbonate group. When activated carbon was modified with KOH, acidity neutralization occurred at the carbon surface. KOH also reacted with carbon as follows:

$$CaO + C + 4KOH \rightarrow K_2CO_3 + Ca(OH)_2 + 2K + H_2 \quad (1)$$

$$4KOH + C \rightarrow K_2CO_3 + K_2O + 2H_2 \quad (2)$$

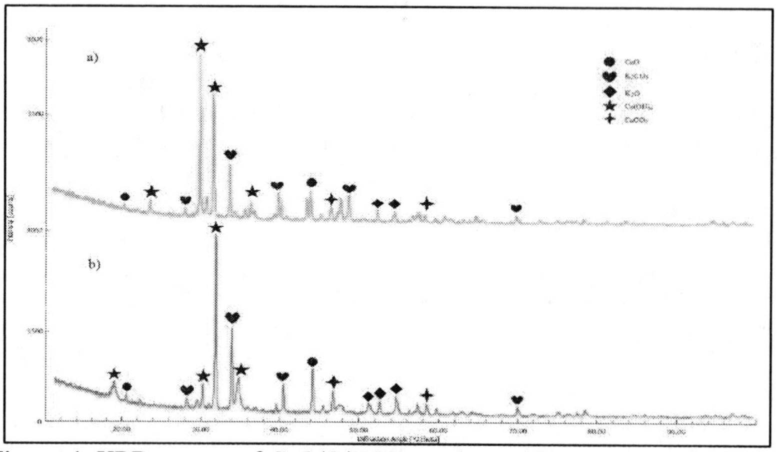

Figure 1. XRD pattern of CaO/C/KOH catalyst with CaO:C mass ratio of 12:6 and KOH concentration of: (a) 15% and (b) 25%.

Figure 2. XRD pattern of CaO/C/KOH catalyst with KOH concentration of 15% and CaO:C mass ratio of: (a) (12:6) and (b) (12:10).

In these reactions, there are K and H ions which will release due to the calcination process at high temperatures, so that the K_2CO_3 compound was produced. The three XRD patterns of the synthesized CaO/C/KOH catalysts, shown in Figure 2, reveals that there was still a few impurities of $Ca(OH)_2$ and $CaCO_3$. These were caused by the sample contamination with air to form $Ca(OH)_2$ through H_2O absorption from the atmosphere [30]. Besides, these impurities' presence indicated that $Ca(OH)_2$ and $CaCO_3$ have not been converted to CaO during the calcination. The presence of the $CaCO_3$ peaks was due to the imperfect decomposition of $CaCO_3$ into CaO and CO_2, so that it formed $CaCO_3$ in the calcite phase. The XRD pattern of the three samples of CaO/C/KOH catalyst also appeared to have a peak that tended to be the same. This indicated that the three samples have the same phase except for the peak intensity of the K_2CO_3 compound, which was slightly larger as the KOH concentration increased [22].

3.3. The surface area of CaO/C/KOH catalyst using Brunaeur-Emmet-Teller (BET)

The CaO/C/KOH catalyst with CaO:C mass ratio of 12: 6 and KOH concentration of 25%, which gave the highest yield, was characterized using BET. It resulted in the catalyst surface area of 17.980 m^2/g. There was an increase of surface area caused by activation by KOH on CaO/C/KOH catalyst compare to research by [23] that reported the surface area of the CaO/C catalyst without activation by KOH was 15.890 m^2/g. This suggests that the activation by KOH increased the outer surface of the catalyst.

3.4. Yield and biodiesel characterization

The effect of the variation of CaO mass ratio and KOH concentration on biodiesel yield is shown in Figure 3. The highest yield of 85.74% was obtained at the CaO mass ratio of 12: 6 and KOH concentration of 25%. Biodiesel yield decreased with increasing the amount of CaO, but it increased with increasing KOH concentration. This is caused by CaO accumulation to form clumps (agglomeration) on the surface of the catalyst so that it prevented contact between the active side of the catalyst and the reactants [16]. Also, agglomeration covered the micropore of the catalyst, so that the surface area became small, and the active base side is reduced [13,16,31]. Meanwhile, the KOH concentration is directly proportional to the biodiesel yield. The presence of KOH increased the alkaline catalyst that increased the activity of the catalyst [31]. However, The excess KOH concentration also reduced the activity of the catalyst. If the KOH concentration is too high, CaO will dissolve, thereby reducing the number of the active side of the catalysts [22].

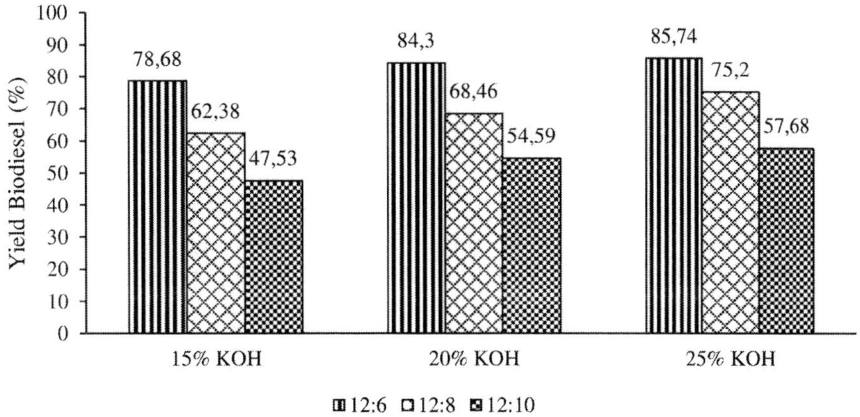

Figure 3. Effect of CaO:C mass ratio and KOH concentration on biodiesel yield.

The effect of activated carbon as the support for CaO catalyst was studied by comparing the biodiesel yield obtained from transesterification by using CaO/C, CaO/C/KOH, and without a catalyst. The higher biodiesel yield was obtained by using CaO/C/KOH (85.7%), followed by CaO/C (61.96%) and without catalyst (31.55%) as Figure 4 portrays. This is because the CaO/C/KOH catalyst has a larger surface area of 17.890 m^2/g than the surface area of the CaO catalyst without metal oxide of 1.963 m^2/g. According to [14], the loading of CaO into an alkaline or metal oxide increased the number of base sides and the surface area of the catalystsso that the catalytic activity of the CaO catalyst increased.

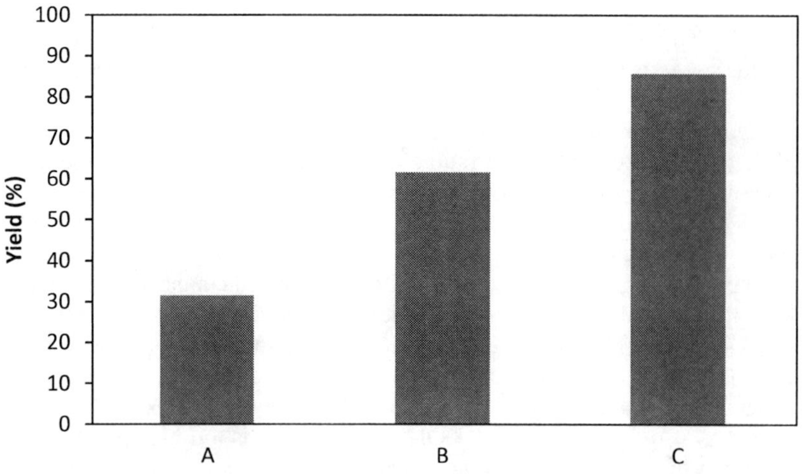

Figure 4. Biodiesel yield by using the various CaO catalyst(A) without catalyst; (B) CaO/C; (C) CaO/C/KOH.

In Figure 5, it can be seen the two highest peaks of methyl esters, which are methyl oleate with a biodiesel composition of 60.32% (peak number 4) and methyl palmitate with a biodiesel composition of 34.52% (peak number 3). The conversion result from carboxylic acid groups to methyl esters in the biodiesel production process reached 100%. This methyl ester content was included in the predetermined biodiesel quality standards (SNI 7182: 2015), which is 96.5%. This indicates that the biodiesel product produced was pure. Based on the results of the GC-MS analysis, CaO/C/KOH catalyst can be used as a catalyst in biodiesel production.

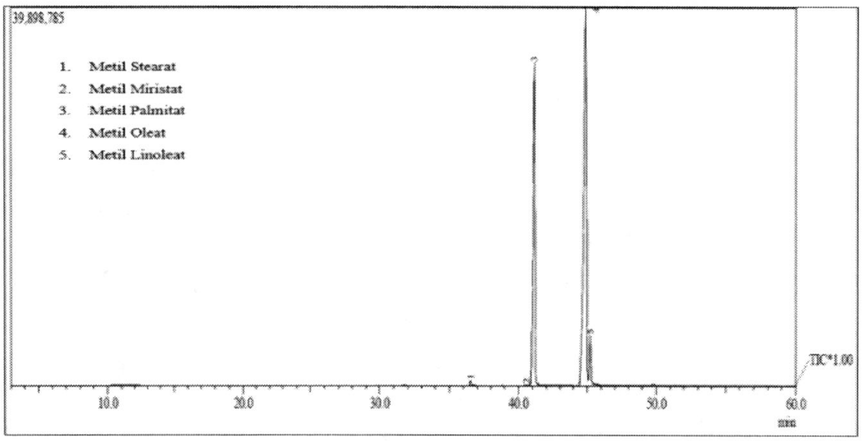

Figure 5. The GC chromatogram of the obtained biodiesel by using CaO/C/KOH calcined at a temperature of 500 °C for 5 hours, KOH concentration of 25% and CaO:C mass ratio of 12:6.

To ensure a machine has optimal work and strong durability, the biodiesel that is about to be used for the machine should first meet a certain quality standard. Hence, in this research, the obtained biodiesel was characterized, in which the characteristics were later compared with the Indonesian standard (SNI 7182: 2015), as presented in Table 2.

Table 2. Characteristics of the obtained biodiesel.

No	Characteristic	Unit	Result	SNI 7182:2015
1	Density	kg/m^3	860-880	850-890
2	Kinematic viscosity	mm^2/s	3.17-5.64	2.3-6.0
3	Acid number	mg-KOH/g	0.35	Max. 0.5
4	Cetane number	-	52	Min. 51

4. Conclution

The catalyst characterization showed that the CaO/C/KOH catalyst calcined at a temperature of 500 °C for 5 hours was in the crystalline phase of CaO, K_2CO_3, K_2O $Ca(OH)_2$, and $CaCO_3$, the base strength of $H\_ > 12.2$, and the specific surface area of 17.890 m^2/g, respectively. The best catalytic activity of the CaO/C/KOH catalyst was obtained at the addition of a KOH concentration of 25% and CaO:C mass ratio of 12:6 with a biodiesel yield of 85.74%.

References

[1] Agarkar B . and Barve S B 2011 *Int. J. Curr. Eng. Technol.,* **4** 188–91
[2] Idroes R, Yusuf M, Alatas M, Subhan, Lala A, Saiful, Suhendra R, Idroes G M and Marwan 2018 *IOP Conf. Ser. Mater. Sci. Eng.,* **334** 012002
[3] Idroes R, Yusuf M, Alatas M, Subhan, Lala A, Muslem, Suhendra R, Idroes G M, Suhendrayatna, Marwan and Riza M 2019 *IOP Conf. Ser. Mater. Sci. Eng.,* **523** 012010
[4] Idroes R, Yusuf M, Alatas M, Subhan, Lala A, Muhammad, Suhendra R, Idroes G M and Marwan 2019 *IOP Conf. Ser. Mater. Sci. Eng.,* **523** 012012
[5] Idroes R, Yusuf M, Saiful S, Alatas M, Subhan S, Lala A, Muslem M, Suhendra R, Idroes G M, Marwan M and Mahlia T M I 2019 *Energies,* **12** 4442
[6] Marwan, Syukri M, Idroes R and Ismail N 2019 *Int. J. GEOMATE,* **16** 141–7
[7] Marwan, Yanis M, Idroes R and Ismail N 2019 *Int. J. GEOMATE,* **17** 173–80
[8] Putri D R, Nanda M, Rizal S, Idroes R and Ismail N 2019 *IOP Conf. Ser. Earth Environ. Sci.,* **364** 012003
[9] Helwani Z, Zulfansyah, Fatra W, Fernando A Q, Idroes G M, Muslem and Idroes R 2020 *IOP Conf. Ser. Mater. Sci. Eng.,* **845** 012019
[10] Helwani Z, Ramli M, Rusyana A, Marlina M, Fatra W, Idroes G M, Suhendra R, Ashwie V, Mahlia T M I and Idroes R 2020 *Processes,* **8** 777
[11] Arifin J K 2009 (Universitas Sumatera Utara. Medan)
[12] Budiawan W.; Helwani, Z. R Z . F 2013 *IChESA Conf.,* 40–50
[13] Hu S, Guan Y, Wang Y and Han H 2011 *Appl. Energy* **88** 2685–90
[14] Marinković D M, Stanković M V., Veličković A V., Avramović J M, Miladinović M R, Stamenković O O, Veljković V B and Jovanović D M 2016 *Renew. Sustain. Energy Rev.,* **56** 1387–408
[15] Liu X, He H, Wang Y, Zhu S and Piao X 2008 *Fuel,* **87** 216–21
[16] Liu C, Lv P, Yuan Z, Yan F and Luo W 2010 *Renew. Energy,* **35** 1531–6
[17] Helwani Z, Fatra W, Saputra E and Herman S 2016 *Pros. Sem. Nas. Ind. Kim. & SDA* (Banjarmasin: Universitas Lambung Mangkurat) 13–8
[18] Helwani Z, Fatra W, Saputra E and Maulana R 2018 *IOP Conf. Ser. Mater. Sci. Eng.,* **334** 012077
[19] Helwani Z, Ramli M, Saputra E, Putra Y L, Simbolon D F, Othman M R and Idroes R 2020 *Catalysts,* **10** 724
[20] Helwani Z, Ramli M, Saputra E, Bahruddin B, Yolanda D, Fatra W, Idroes G M, Muslem M, *Catalysts,* **10** 164
[21] Hadiyanto H, Afianti A H, Navi'a U I, Adetya N P, *J. Environ. Chem. Eng.,* **5** 4559–63
[22] Zhang J and Meng Q 2014 *World J. Eng. Technol.,* **02** 184–91
[23] Rahayu R P, Prasetyo R A, Purwanto D A, Kresnoadi U, Iskandar R P D and Rubianto M 2018

Vet. World, **11** 765–70
[24] Pahan I 2006 (Jakarta: Penebar Swadaya)
[25] Ulfayana S and Helwani Z 2015 *Abstract Book: Regional Conference on Chemical Engineering* (Yogyakarta) 7:22
[26] Helwani Z, Aziz N, Kim J and Othman M R 2016 *Renew. Energy,* **86** 68–74
[27] Jin H, Kolar P, Peretti S W, Osborne J A and Cheng J J 2017 *Energies,* **10** 1920
[28] Farag H A, El-Maghraby A and Taha N A 2013 *Int. J. Chem. Biochem. Sci.,* **3** 1–8
[29] Ho W W S, Ng H K, Gan S and Tan S H 2014 *Energy Convers. Manage.,* **88** 1167–78
[30] Umdu E S 2008 (İzmir Institute of Technolog)
[31] Tang S, Wang L, Zhang Y, Li S, Tian S and Wang B 2012 *Fuel Process. Technol.,* **95** 84–9

Application of Near-Infrared spectroscopy and chemometric (PCA) in variety *holothuria atra and holothuria scabra* in Simeuleu, Aceh province

I Ikhsan[1], R Yusnaini[1], R Nasution[2], A A Munawar[3], R Idroes[2,4,*]
[1]Graduate School of Mathematics and Applied Sciences, Universitas Syiah Kuala, Banda Aceh 23111, Indonesia
[2]Department of Chemistry, Faculty of Mathematics and Natural Sciences, Universitas Syiah Kuala, Banda Aceh 23111, Indonesia
[3]Department of Agricultural Engineering, Universitas Syiah Kuala, Banda Aceh 23111, Indonesia
[4]Department of Pharmacy, Faculty of Mathematics and Natural Sciences, Universitas Syiah Kuala, Banda Aceh 23111, Indonesia

*E-mail: rinaldi.idroes@unsyiah.ac.id

Abstract. As the largest maritime country globally, Indonesia has a variety of potential marine biota that can be explored as a source of nutrition and raw materials for medicine. As one of the marine biota, Sea cucumbers have been widely recognized by the public as a medicine for treatment from generation to generation. With the breadth and variation of Indonesian waters' characteristics, it is possible to have differences in the nutritional and bioactive content of various types and sources of sea cucumber. This study analyzes the compounds in Holothuria atra, and Holothuria scabra dried sea cucumbers using Near-Infrared Spectroscopy (NIRS). Spectrum data of dried sea cucumber samples were obtained from the absorbance spectrum in the wavelength range of 780 nm - 2500 nm with a resolution of 0.02 nm and 32 scans. The spectrum data is then projected onto principal component analysis (PCA) to extract data and observe similarities between the two samples. The results showed that there was no difference between the Holothuria atra and Holothuria scabra dry samples in terms of the spectral data showing their molecular bonds. However, the essential chemical structure related to the content of sea cucumbers is vibrating in the Near-infrared region, especially in the waves 1147, 1310, 1452, 1667, 1819, 1920, and 2235 nm, indicates the presence of molecular bonds C-H, O-H, C-O, C-H-O, N-H, and S-H. Based on the results obtained, it can be concluded that NIRS has the potential to analyze and determine several quality attributes of sea cucumber samples with further calibration modeling.

Introduction

Sea cucumber has a high nutritional content and pharmacological effects, making it a potential food and medicinal source [1]. As a food source, sea cucumber has been shown to contain high protein, low fat, and high levels of essential minerals [2]. Various active compounds in this species are thought to be capable of being a therapeutic breakthrough in various diseases [3]. Also, sea cucumbers have high economic value on the world market and are a valuable export commodity. Therefore, as the largest archipelagic country, Indonesia has the potential to develop sea cucumber processing as food, medicinal ingredients, and export commodities.

Sea cucumbers have been known for a long time to treat hypertension, asthma, rheumatism, impotence, and constipation by the Chinese and other Asian nations [4].Various chemical constituents have been extracted from these marine biota ranging from simple ones, such as polypeptides, to complex ones such as polyether [5]. Various phenolic compounds found in sea cucumbers include chlorogenic acid, pyrogallol, coumaric acid, and catechins. Experts have noted that there are 1,400 species of sea cucumber scattered throughout the world, and 55 of them are found in Indonesian waters, so there may be differences in the content of nutrients and other bioactive compounds [6]. Research conducted in the Aceh Besar area and Simeuleu shows various types of sea cucumbers [7].

Sea cucumbers are usually processed into a dry product known as beche-de-mer. Beche-de-mer qualities are classified as high, medium, or low based on species, shape, odor, color, and body wall thickness. Classification based onphysical property has a weakness because it depends on panelist abilities that are possibly differentfrom one to others. Instrumental methods can be more accurate by providing more detailed information such as chemical compounds, property value, and even quantitative data.

In the application of organic material, the use of a spectroscopy instrument is a well choice. It has been widely and accurately applied on the identification of organic samples, such as identification of protein, lipid and fish meat with Uv-Vis [8–10], food and coal by laser [11,12], leaf tissue with image processing [13], bone and polysaccharide with FTIR[14,15], secondary metabolites and fatty acid with mass spectrometry [16–19], etc.

In the purpose of the cucumbers sample, fatty acid profile are identified as a basis in its quality determination. The use of FTIR and polymerase chain reaction (PCR) have shown a precise and reliable result. However, both techniques are complicated in the case of time-consuming and sample pre-treatment process. NIRS is known to work based on the theory of electromagnetic radiation interaction from the biological samples used. Many studies have used NIRS because of its ease in analyzing chemical properties or quality without special sample preparation. It is environmentally friendly because there is no need to use chemicals; the process is fast and does not damage the sample. Besides, NIRS also can predict and determine the desired quality or nutrient parameters by using spectral data simultaneously[20,21].

Near-infrared spectroscopy is an analytical technique that utilizes infrared light rays, which have a wavelength range of 780-2500 nm. In this area, vibrations of the -CH, -OH, -SH, and -NH bonds generally occur. Thus, near-infrared can be used to analyze the chemical content of the organic matter that has these bonds. NIR spectroscopy can guarantee the quality of both raw materials and finished products with easy sample preparation. Analysis using near-infrared also has several other advantages, including non-destructive, accurate, fast analysis time, and does not require chemicals[22]. So the use of near-infrared spectroscopy (NIRS) can be a good option to simplify the time and eliminate the treatment.

Regardless of a convenient analysis offered by this spectroscopy method, a spectra interpretation difficulty is still found. So, a statistical application, chemometric, is required to solve this problem. The combination of instrumental and chemometric measurements is a perfect choice for reliable and straightforward procedures [23–25]. One of the chemometric methods that have recently applied in several analytical works is PCA. This method can classify complex data based on the similarity or difference pattern one data to others[26–28]. This benefit is useful in the purpose to determine the cucumber samples into certain grade quality. Therefore, the use of NIRS combined with PCA in this research can potentially be methods of checking the contents and classification of various sea cucumbers quickly and easily.Based on the above reasons, the authors intend to assess whether NIRS can be used to distinguish the chemical content between Holothuria atra and Holothuriascabra.

IOP Conf. Series: Materials Science and Engineering 1087 (2021) 012054 doi:10.1088/1757-899X/1087/1/012054

Materials and methods

1.1. Sample source

The sea cucumbers used in this study were dried types of Holothuria atra and Holothuria scabra obtained in Simeuleu waters, Aceh, Indonesia. Holothuria atra and Holothuria scabra were identified based on morphology to determine the type of species in the Marine Biology Laboratory, Faculty of Marine and Fisheries. On the other hand, near-infrared spectra data of studied samples were acquired and measured at the Laboratory of Instrumentation, Dept. of Agricultural Engineering, Universitas Syiah Kuala.

1.2. Infrared spectrum data

Equipment used: Portable sensing device Near-infrared spectroscopy (PSD-NIRS), i16 coupled with photodiode array sensors and programmable logic controller. Moreover, Spectra data were analyzed using principal component analysis using the CAMO Unscrambler X 10.3 (NC license, Olso Norway).

Spectrum data of dried sea cucumber samples were obtained from the absorbance spectrum in the wavelength range of 780 nm - 2500 nm with a resolution of 0.02 nm and 32 scans. Optical amplification and correction are also used to ensure smooth acquisition and eliminate spectrum interference during acquisition. The absorbance spectrum data were then projected onto principal component analysis (PCA) as a chemometric method for qualitative analysis to extract data and observe the similarities between the two types of samples. PCA classification method works by extracting spectrum data by entering the absorption data in a new projection matrix called main components (PCs). To get a strong classification, cross-validation is applied by using the non-iterative partial least square.

Result and discussion

The NIR spectra of Holothuria atra and Holothuria scabra are shown in Figure 1. From the results of the NIRS absorbance examination in the two species, visualization of the NIR spectrum did not show any differences between Holothuria atra and Holothuria scabra, where the two spectra appeared to coincide (not separate). This is in line with some of the previous studies which state that sea cucumbers have a lot of nutritional content such as various types of vitamins, high in minerals, especially calcium. Many research has proven that the main components of various sea cucumbers are collagen and polysaccharides, which have been studied to have various biological activities such as anti-inflammatory, anti-thrombotic, anti-tumor, and wound healing properties for various types of sea cucumbers. The secondary metabolites of sea cucumbers are associated with the presence of saponins, chondroitin sulfates, glycosaminoglycan polysaccharides, and essential fatty acids [29,30].

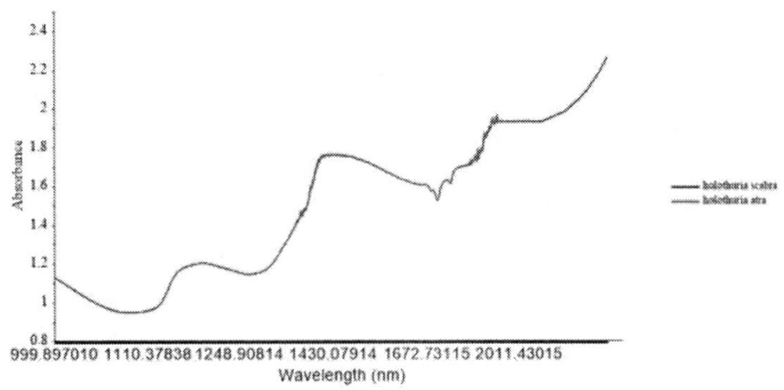

Figure 1. The NIR spectra of holothuria atra and holothuria scabra.

The two species optimum wavelengths in Figure 2 show the presence of absorption and strain over several wave ranges. The NIR spectrum in the 780 nm - 2500 nm wave range shows vibrations at 1147nm, 1310 nm, 1452 nm, 1667 nm, 1819 nm, 1920 nm, and 2235 nm wavelengths. This vibration indicates that NIRS noted the presence of an electromagnetic spectrum as a reaction to the presence of

O-H, C-H, C-O, N-H, C-H-O, and C-H-O-O-S groups and correlates with the specific chemical compound in sea cucumbers such as protein, fatty acids, and sulfates. This indicates the possible amino acid content, fatty acids, and aromatic compounds at overtones 1 and 2. Vibration in the 1819 nm region indicates sulfate groups' presence, while carboxylate groups are seen at 2235 nm [31].

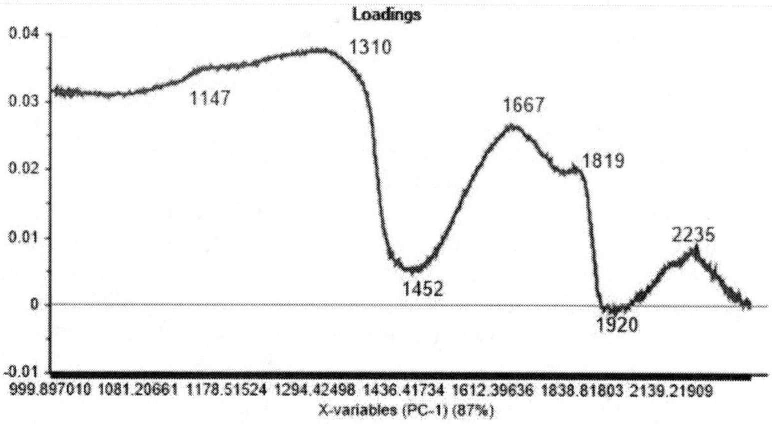

Figure 2. Loading plot derived from PCA analysis to determine optimum wavelength vibration of the group compound in holothuria atra and holothuria scabra.

The molecular bonds found in Holothuria atra and Holothuria scabra and adjusted for the distribution of organic bonds in NIRS electromagnetic waves can be seen in Table 1.

Table 1. Chemical bonds of holothuria atra and holothuria scabra in 1000-2500 nm waves [10].

Wavelength (nm)	Chemical Bonds
1147	CH, CH_2,CH_3
1310	CH, CH_2,CH_3
1452	H_2O, ROH
1667	ArCH, CH, CH_2,CH_3
1819	SH
1920	H_2O
2245	NH2, CHO, CH, CH_2, CH_3

These results align with other studies that stated that sea cucumbers contain lots of fatty acids, collagen-rich protein, and amino acids [32]. Glycosaminoglycan molecules such as chondroitin sulfate are allegedly present in sea cucumbers as components of the extracellular matrix found in various connective tissues such as cartilage, bone, skin, ligaments, and tendons. Giving chondroitin sulfate to patients with osteoarthritis can reduce pain and also improve joint function [33,34]. Meanwhile, the presence of CH, CH2, and CH3 groups indicates the possibility that triperpenoid compounds such as sequelants have been extensively studied as anticancer agents because they exhibit cytotoxic activity in certain types of cancer cells [35].

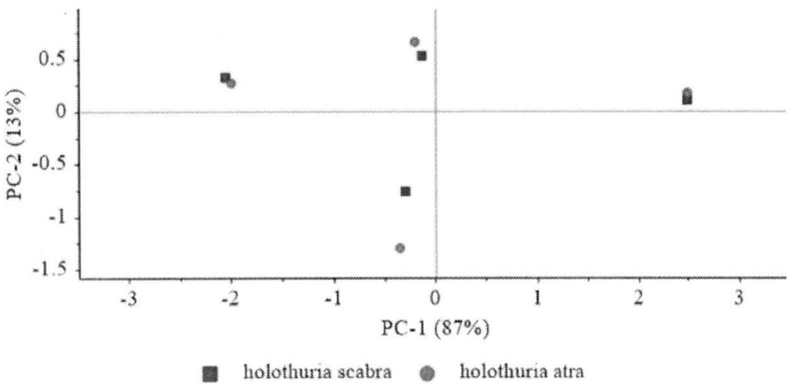

Figure 3. PCA analysis of holothuria atra and holothuria scabra.

From Figure 3, the main information derived from the PCA analysis results, there are no differences between Holothuria atra and Holothuria scabra. This means each sample can be replaced for their advantage purposes. Moreover, It also showed that chemometrics, utilizing the PCA method, can distinguish samples based on their origins. Loading plot, the PCA also mentioned that relevant and essential wavelengths are closely associated with several respective chemical constituents of sea cucumber samples.

Furthermore, to find out whether the content of chemical compounds on all sides of each sample's body is the same or different, the authors tried to illuminate 4 different sides of the sample, from the upper and lower ends and two points from the midsection. The results obtained were analyzed using principal component analysis (PCA), which shows the euclidian distance away from each radiation position. This indicates the possibility of certain types of compound content in the upper, lower, and middle ends of the sea cucumber's body, but the amount cannot be determined.

However, as a whole, based on the PCA results in Figure 3, it can be seen that there are similarities in the content of compounds between Holothuria atra and Holothuria scabra both in the upper, middle, and lower parts of the body of the two species.Thus, the main information obtained from the PCA analysis results, there is no difference between Holothuria atra and Holothuria scabra.PCA can also be employed to discover leverage and outlier data of spectra before spectra enhancement and correction. Consequently, it can provide more robust classification overview. This will be performed later on a further step in combination with discriminant analysis to classify samples

Conclusions

Based on principal component analysis (PCA) results, there are relatively no differences in chemical compounds between the two samples of sea cucumbers. The PCA method attempted to extract and classify spectral data based on their similarities and characteristics of sea cucumber samples. However, the important chemical structures associated with the molecular bonds of C-H, O-H, C-O, C-H-O, N-H, and S-H produce vibrations at 1147, 1310, 1452, 1667, 1819, 1920, and 2235 nm. The obtained results indicate that NIRS, combined with the PCA data analyses, were able to extract buried information inside spectra data. The respective information is that there are no differences between Holothuria atra and Holothuria scabra.This means each sample can be replaced for their advantage purposes. Moreover, It also showed that chemometrics, utilizing the PCA method, can distinguish samples based on their origins. Loading plot, the PCA also mentioned that relevant and important wavelengths are closely associated with several respective chemical constituents of sea cucumber samples.

References

[1] Khotimchenko Y 2018 *Int. J. Mol. Sci.,* **19** 1342
[2] Sroyraya M, Hanna P J, Siangcham T, Tinikul R, Jattujan P, Poomtong T and Sobhon P 2017 *Funct. Foods Heal. Dis.,* **7** 168
[3] Zhang J-J and Zhu K-Q 2017 *Exp. Ther. Med.,* **14** 1653–8
[4] Pangestuti R and Arifin Z 2018 *J. Tradit. Complement. Med.,* **8** 341–51
[5] Xu C, Zhang R and Wen Z 2018 *J. Funct. Foods* **49** 73–84
[6] Setyastuti A, Wirawati I and Iswari M Y 2018 *Biodiversitas J. Biol. Divers.,* **19** 676–82
[7] Fadli N, Muchlisin Z A, Soraya I, Dewiyanti I, Purnawan S, El-Rahimi S A, Sofyan H, Affan M and Siti-Azizah M N 2019 *IOP Conf. Ser. Earth Environ. Sci.,* **348** 012076
[8] Suhartono E, Thalib I, Aflanie I, Noor Z and Idroes R 2018 *IOP Conf. Ser. Mater. Sci. Eng.,* **350** 12008
[9] Hasanah U, Setyowati M, Efendi R, Muslem M, Md Sani N D, Safitri E, Yook Heng L and Idroes R 2019 *Biosensors,* **9** 60
[10] Hasanah U, Sani N D M, Heng L Y, Idroes R and Safitri E 2019 *Biosensors,* **9** 135
[11] Hedwig R, Lahna K, Idroes R, Karnadi I, Tanra I, Iqbal J, Kwaria D, Kurniawan D P, Kurniawan K H, Tjia M O and Kagawa K 2019 *Microchem. J.,* **147** 356–64
[12] Lahna K, Idroes R, Idris N, Abdulmadjid S N, Kurniawan K H, Tjia M O, Pardede M and Kagawa K 2016 *Appl. Opt.,* **55** 1731
[13] Suhartono E, Setiawan B, Santosa P B, Idroes R and Indrawan M S 2019 *J. Phys. Conf. Ser.,* **1374** 012057
[14] Suhartono E, Noor Z, Edyson, Budianto W Y and Idroes R 2019 *AIP Conf. Proceedings* 2108 020025
[15] Hasanah U, Setyowati M, Edwarsyah, Efendi R, Safitri E, Idroes R, Heng L Y and Sani N D 2019 *IOP Conf. Ser. Mater. Sci. Eng.,* **523** 12014
[16] Estevam E, Griffin S, Nasim J, Zieliński D, Aszyk J, Osowicka M, Dawidowska N, Idroes R, Bartoszek A and Jacob C 2015 *Nat. Prod. Commun.,* **10** 1733–8
[17] Pratiwi S U T, Lagendijk E L, de Weert S, Idroes R, Hertiani T and den Hondel C Van 2015 *Int. J. Appl. Res. Nat. Prod.,* **8** 1–13
[18] Earlia N, Muslem, Suhendra R, Amin M, Prakoeswa C R S, Khairan and Idroes R 2019 *Sci. World J.,* **2019** 1–7
[19] Earlia N, Rahmad R, Amin M, Prakoeswa C, Khairan K and Idroes R 2019 *Sains Malaysiana* **48** 1019–24
[20] Xu Q, Xu Q, Zhang X, Peng Q and Yang H 2016 *J. Mar. Biol. Assoc. United Kingdom* **96** 197–204
[21] Zhong M, Chen T, Hu C and Ren C 2015 *J. Food Sci.,* **80** C671–9
[22] Poji M, Mastilovi J M 2012. Rijeka(CR): InTech.
[23] Nadia Y, Ramli M, Muslem, Japnur A F, Rusyana A, Idroes G M, Suhendra R, Muhammad, Sasmita N R, Tallei T E and Idroes R 2019 *IOP Conf. Ser. Earth Environ. Sci.* **364** 012040
[24] Idroes R, Husna I, Muslem, Mahmudi, Rusyana A, Helwani Z, Idroes G M, Suhendra R, Yandri E, Rahimah S and Sasmita N R 2019 *IOP Conf. Series: Earth and Environ. Sci.,* **364** 12020
[25] Husna I, Rusyana A, Muslem, Idroes G M, Suhendraand R and Idroes R 2020 *IOP Conf. Ser. Mater. Sci. Eng.,* **796** 012064
[26] Idroes R, Japnur A F, Suhendra R and Rusyana A 2019 *IOP Conf. Ser. Mater. Sci. Eng.,* **523** 12007
[27] Idroes R, Noviandy T R, Maulana A, Suhendra R, Sasmita N R, Muslem M, Idroes G M and Irvanizam I 2019 *Int. Rev. Model. Simulations,* **12** 373
[28] Wahidah S, Khairi, Lelifajri, Idroes R, Rahmadi, Lala A, Mahmudi, Muslem and Japnur A F 2019 *IOP Conf. Ser. Earth Environ. Sci.,* **364** 12016
[29] Bordbar S, Anwar F and Saari N 2011 *Mar. Drugs,* **9** 1761–805
[30] Chen S, Xue C, Yin L, Tang Q, Yu G and Chai W 2011 *Carbohydr. Polym.,* **83** 688–96

[31] Cen H and He Y 2007 *Trends Food Sci. Technol.,* **18** 72–83
[32] Ridhowati S and Asnani A 2015 *J. Mat. Sains dan Teknol.,* **16** 20–7
[33] Jerosch J 2011 *Int. J. Rheumatol.* **2011** 1–17
[34] Henrotin Y, Mathy M, Sanchez C and Lambert C 2010 *Ther. Adv. Musculoskelet. Dis.,* **2** 335–48
[35] Prakash V 2018 *Pharmacogn. Rev.,* **12** 166

The influences of operating conditions on the removal of Fe(II) in water by adsorption using bentonite as a low-cost adsorbent

M Jannah[1], S Mulyati[2,*], C M Rosnelly[2], S Muchtar[2]
[1]Graduate School of Chemical Engineering, Universitas Syiah Kuala University , Jl. Tgk Chik Pante Kulu No. 5, Banda Aceh, 23111, Indonesia
[2]Department of Chemical Engineering, Syiah Kuala University, Jl. Syeh A. Rauf, No. 7 Banda Aceh, 23111 Indonesia

*E-mail: sri.mulyati@unsyiah.ac.id

Abstract. The contamination of Fe^{2+} in groundwater is a problem that needs to be handled seriously as it can cause health problems for humans. In this study, the adsorption process method is conducted by use of natural bentonite from North Aceh. This process is expected to reduce Fe^{2+} as effectively as possible. The adsorption process is carried out by varying the contact time, adsorbent dose, and pH of the sample. The most effective contact time is obtained at 160 minutes, the adsorbent dose is 1 g/L and the pH of the sample is in neutral condition. With the initial concentration of metal in the sample solution of 20 mg/L, the adsorption process is able to remove 85.5% of Fe with the remaining Fe in the water of only 2.82 g/L.

1. Introduction

Water pollution caused by toxic compounds is a major problem because it affects environmental quality such as plants, animals and human health. Heavy metals are toxic and dangerous substances even at very low concentrations[1].The term heavy metals is described as pollutants that are discharged into the environment in the form of air, water, or soil [2]. Heavy metals are defined as elements with a density higher than 5 g/cm^3[3]. Heavy metals are dangerous because they are not biodegradable, accumulated in the food chain, and are persistent [4]. One of the heavy metals that are often present in water sources is ferrous (Fe). Although Fe is an essential mineral that helps transport oxygen in the blood, its presence in groundwater at a certain level makes water unusable due to change in taste, smell, and cloudiness[5].Besides water hardness, the presence of Fe is the most common problem faced by consumers and water providers [6]. Fe which is deposited in the distribution system can increase the growth of microorganisms so that their concentration increases in the water. Fe is commonly found in groundwater supplies which are used as a source of raw water for clean water providers. If the content of this compound exceeds the maximum allowed, the water will visually change in color, which will result in poor piping and water quality. Therefore, the World Health Organization (WHO) has determined that the maximum permissible level of iron is 0.3 mg/L [7][8]. Measurement of iron content is particularly important in boreholes and other surface water sources. If the iron content exceeds 0.3 mg/L it is necessary to consider the most effective treatment that can be used.

Several methods have been used to remove iron compounds in water, including through the aeration process[9], oxidation-coagulation [10], chlorination[11], electrochemistry [12] and adsorption [13]. Of all the methods, the adsorption process is an attractive option. This is due to the cost, flexibility, simplicity of design, ease of operation, insensitivity to toxic pollutants and better removal efficiency [4]. Activated carbon has been found to be the most effective adsorbent for removing metals from solution due to its high adsorption capacity. However, the costs involved are high, limiting their widespread use in most industries and developing countries. Many researchers have studied the use of low-cost adsorbents to remove metal ions from solutions, thereby minimizing the high cost problem. Some of the low-cost adsorbents that have been used include agricultural waste and biomass materials, clay, zeolite, silica, fly ash and bentonite. In this research, the natural bentonite acquired from local source (North Aceh Regency) is used. Besides its abundant availability, bentonite has also been reported to have a high adsorption capacity for heavy metals due to its high specific surface area, small particle size, high porosity and high cation exchange capacity [14]. Bentonite is used without chemical modification or maintenance to keep process costs low. The effects of pH, initial metal ion concentration, contact time, adsorbent dose.

2. Materials and methods

2.1. Materials

Natural bentonite acquired from North Aceh Regency was employed as the adsorbent material. Fe_2O_3 (Wako, Japan) was used as a model contaminant feed solution. HCl (Merck, Germany) and NaOH (Sigma Aldrich, Germany) were used as pH controller. Distilled water was used for solvent.

2.2. Preparation of bentonite adsorbent

The adsorbent used in this work was based of natural local bentonite. Natural bentonite in the form of rocks was cleaned of impurities. Natural bentonite was pounded with a stainless steel hammer to reduce the size and on the ballmill to obtain a size of 80 mesh.

2.3. Batch adsorption experiments

The adsorption experiments were carried out in batch reactor at a room temperature using bentonite adsorbent with particle size of 80 mesh. The experiment was conducted by studying various parameters such as contact time, adsorbent dose, and sample pH on the uptake of Fe (II) as a contaminant model solution. The Fe-containing feed solution was prepared in a concentration of 20 mg/L by dissolving 20mg of pure Fe in 1L of distilled water. A total of 100 ml feed solution was mixed with certain amount of adsorbent with continuous stirring. The pH was also continuously adjusted to the expected value using HCl and NaOH solutions. The concentration of the filtrate was analyzed using a UV-VIS spectrophotometer.

To study the effect of contact time on the adsorption of Fe(II), the experiment was carried out at a pH of 7, adsorbent dose of 1 g/L, and the stirring rateof 100 rpm. The effect of the adsorbent dose on theabsorption of Fe (II)was carried out by adding 0.25 g/L, 0.5 g/L, 0.75 g/L and 1.0 g/L in 100 mL of feed solution. Fe (II) with a concentration of 20 g / L.

To determine the effect of pH on Fe (II) removal, the pH of the sample was set to the range 2,3,4,5,6,7,8, and 9. The dose of adsorbent used was 1 g/L, the sample concentration was 20 gr/L and stirring at 100 rpm.

3. Result and discussion

3.1. Influence of contact time on the removal of Fe (II)

The effect of contact time on the adsorption of Fe (II) by bentonite adsorbent was carried out by adding 0.1 gram of adsorbent into a beaker glass which filled with 100 ml of Fe (II) sample with a concentration of 20 mg/L. The mixture was then stirred at a speed of 100 rpm. Figure 1 shows the effect of contact time on Fe (II) metal removal. From Figure1, it can be seen that the effect of contact time on Fe (II) removal efficiency ranged from 0-20 minutes, the efficiency of the increase in metal removal was quite significant. In the next time interval, the elimination efficiency continued to increase, although it was not like in the 0-20 minutes interval range. The equilibrium condition is reached when the sample contact time with the adsorbent is 180 minutes. At the beginning of the adsorption process, the surface area of the contact on the adsorbent surface is still sufficiently available so that there is no competition between metal ions on the adsorbent surface, but over time, the available space for ion absorption others become limited. Due to the repulsion between the metal ions absorbed on the adsorbent surface and the ions in the solution, further ion absorption becomes more difficult [14, 15]. Repulsion is when ions are separated having their core electron wavefunctions cannot overlap and have the similar quantum numbers. As the ions get closer, some of the ions might be forced into having higher state of energy due to the overlap of wavefunctions. Because the electrons cannot be in the same energy state some of the electrons are pushed into higher states and preventing the ions from approaching closer to each other [14, 15]. When the entire adsorbent surface is saturated, there is an equilibrium (steady state) condition, where the absorption efficiency will be constant even though the absorption time is longer. In this process the optimal efficiency is obtained at 160 minutes contact time, with an absorption efficiency of 84.35 percent.

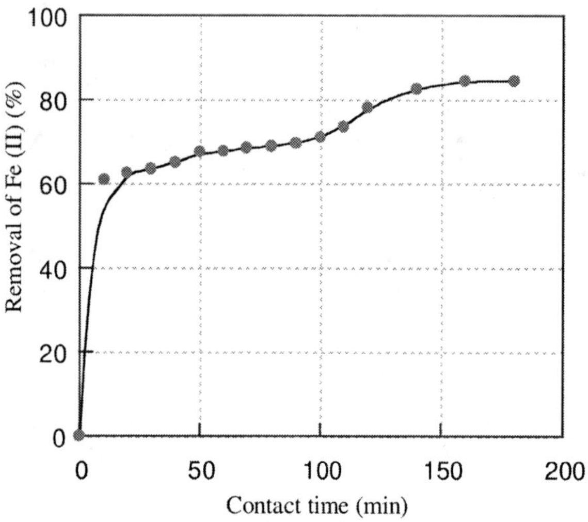

Figure 1. Influence of contact time on Fe (II) removal [[adsorbent] = 1 g/L; particle size 80 msh; [Fe (II)]$_0$ = 20 mg/L; pH =7.

3.2. Influence adsorbent dosage on the removal of Fe (II)

The effect of adsorbent loading on Fe^{2+} adsorption by bentonite adsorbent is illustrated in Figure 2. From the results obtained, it can be concluded that the adsorbent dose greatly affects the removal performance. Increasing the adsorbent concentration to 1 g/L resulted in a large increase in the percentage of Fe uptake (0–80% Fe adsorption). The increase in the percentage of Fe removed by increasing the adsorbent dose can be associated with an increase in the surface area of the adsorbent, which can increase the number of available adsorption sites. A further increase in the concentration of the adsorbent is not significantly affecting the percentage of metal removal, probably because almost all ions of Fe have bonded to the adsorbent and on the formation of an equilibrium between the ions bound to the solid phase and those remaining in solution. The optimal adsorbent dose used in this study is 1 g/L.

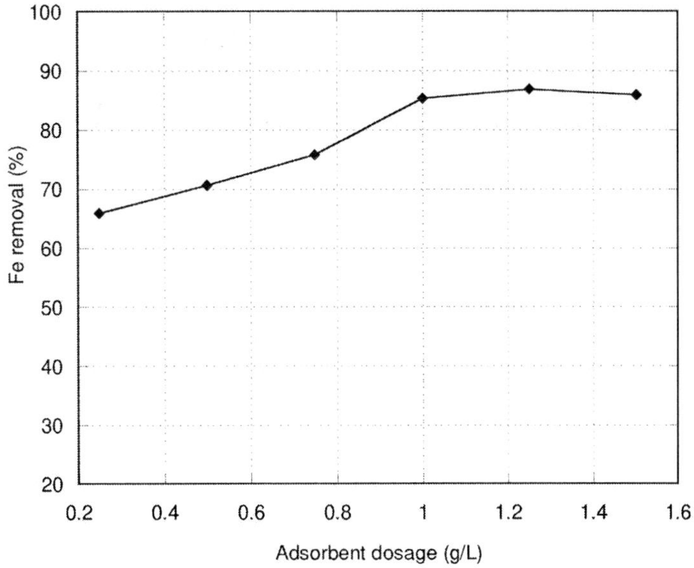

Figure 2. Influence of contact time on Fe (II) removal.The contact time = 140 min; particle size 80 mesh; $[Fe\ (II)]_0 = 20$ mg/L; pH =7.

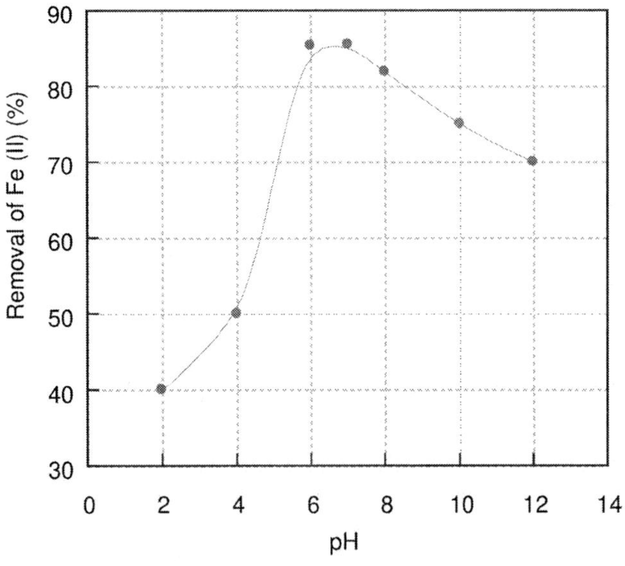

Figure 3. Influence of pH of sample on Fe (II) removal. The contact time = 140 min; particle size 80 mesh; $[Fe\ (II)]_0 = 20$ mg/L; dosage adsorbent =1 gr/L

3.3. Influence of pH on the removal of Fe (II)

In the adsorption process, the pH of the solution is one of the main controllingparameters. The absorption capacity of metal ions and the adsorption mechanism are influenced by the pH of the solution [16].The acidity or basicity of the solution affects the charge on the adsorbent surface as well as the chemical equilibrium, both on the adsorbate and on the adsorbent. The effect of pH of the feed solution on the Fe^{2+} absorption by bentonite adsorbent is illustrated in Fig. 3. From Figure 3, it can be seen that in acidic conditions the efficiency of Fe^{2+}absorption is low.This is because in this condition the surface of the active side of the adsorbent is positively charged thus there is competition between hydronium ions and metal ions.[17]. The absorption efficiency increased at pH 7 and reached a maximum value of 85.5%. Increasing pH to pH 7, the active surface of the adsorbent becomes negative causing Fe^{2+} ions to be adsorbed more easily because of the electrostatic force between the surface of the adsorbent and the metal. In the pH range of 8-12, the efficiency of metal absorption decreases which is due to the precipitation of metal ions as hydroxide compounds.

4. Conclusion

Based on the results of this study, it is concluded that the use of bentonite as an adsorbent is capable to remove Fe metal in the water sample. The optimal operating conditions were obtained at 160 minutes of contact time with an adsorbent dose of 1 g/L and the pH of the sample was at a neutral condition. The best removal efficiency was obtained at 85.5%. The final concentration of Fe in the effluent is 2.82 mg/L. This condition has met the quality standard for sanitary water that is permitted to be used for community.

5. Acknowledgement

This study was funded by Thesis Magister 2020 research grant.

References

[1] Liu C, Bai R, San L Q. 2008 *Wat. Res.,* **42** 1511-1522.
[2] Ghaedi M, Mosallanejad N. 2013 *J. Chem. Health Risks,* **3** 7-22.
[3] Ntwampe O I, Moothi K. 2018 *Heavy Metals.* **33**.
[4] Akpomie K G, Dawodu F A. 2015 *Beni-Suef Univ. J. Basic Appl. Sci.,* **4** 1-13.
[5] El-Naggar H M. 2010 *J. Egypt. Public Health Assoc.,* **85** 169-88.
[6] Vasudevan S, Lakshmi J, Sozhan G. 2009 *Clean–Soil, Air, Wat.,* **37** 45-51.
[7] Al-Shahrani S. 2013. *Int. J. Technol.,* **13** 1-12.
[8] Pascu D-E, Alina T G, Nechifor A C, Raluca M A. 2016 *J. Electrochem. Sci. Eng.,* **6** 47-55.
[9] Sanusi M A, Adlan M N, Rozainy M R, Jamil R. 2016 *Mater. Sci. Forum,* **6** 509-513.
[10] Bordoloi S, Nath S K, Gogoi S, Dutta R K. 2013 *J. Hazard. Mate.,r* **260** 618-626.
[11] Khadse G, Patni P, Labhasetwar P. 2015 *Sustain. Wat. Resources Manage.,* **1** 157-165.
[12] Tong M, Yuan S, Zhang P, Liao P, Alshawabkeh A N, Xie X, Wang Y. 2014 *Env. Sci. Technol.,* **48** 5145-53.
[13] Ujile A, Joel O. 2013 *Int. J. Eng. Sci. Technol.,* **5** **(6)** 979-988
[14] Sadanand P. 2017 *J. Mol. Liq.,* **241** 1091-1113.
[15] Jalali M, Aboulghazi F. 2013 *J. J. Mater. Cycles Waste Manag.,* **15** 548-555.
[16] Heidari A, Younesi H, Mehraban Z, Heikkinen H. 2013 *Int. J. Biol. Macromol.,* **61** 251-263.
[17] Ozdes D, Duran C, Senturk H B. 2011 *Environ. Manage.,* **92** 3082-3090.

Modification of polyetersulfone (PES) membrane by using jernang (*daemonorops draco blume*) as a natural additive on humic acid fouling

A C Ambarita[1], S Mulyati[2,*], N Arahman[2], Suhendrayatna[2], C M Rosnelly[2]
[1]Magister of Chemical Engineering, Syiah Kuala University, Banda Aceh, Indonesia, 23111
[2]Department of Chemical Engineering, Syiah Kuala University, Banda Aceh, Indonesia, 23111

*E-mail: sri.mulyati@unsyiah.ac.id

Abstract. This paper reports on the effect of adding Jernang additives on the performance of PES membranes. The membrane was made by mixing PES 16.5%, Jernang 1%, and NMP 82.5%, using the Non-solvent Induced Phase Separation (NIPS) method. The chemical group of Jernang has been studied in this paper. The addition of Jernang additives has succeeded in improving the performance of PES membranes. Pure water flux increases to 18.77 LMH from the initial flux of 1.66 LMH. Humic acid removal showed reasonable results up to 77.52%. Moreover, this membrane has good antifouling properties, but less stable under strong acid and alkaline solution.

1. Introduction

Global water crisis is the biggest problem in the 21st century, especially in developing countries. Although water is a renewable source, only 1% of 3% of fresh water can be used by humans for consumption (97% is sea water). According to many experts even 1% is sufficient because the supply of fresh water is not limited through the pure water cycle, but its availability is not evenly distributed around the world [1].Various water treatment methods are used to reduce contaminants in water, including coagulation-sedimentation [2], disinfectants [3], chemical oxidation [4], and aerobic and anaerobic digestion processes [5]. Among these methods, membrane technology has attracted many researchers and industrial sectors because of several advantages, including low energy consumption, simple operation process, easy to combine with other processes, optimizing membrane performance, and does not require a large area [6].

Polyethersulfone (PES) is the polymer most often used for ultrafiltration membranes. This is due to the PES has good thermal, chemical resistance and mechanical strength. However, this polymer is hydrophobic, which causes membrane fouling easily when applied to water treatment. Fouling can cause a decrease in membrane performance; therefore, modification of the membrane surface is important to produce a membrane that has anti-fouling properties [7]. The most effective modification is by adding additives to the polymer solution. This modification not only increases the hydrophilicity of the membrane, but is also useful for changing the interaction between microparticles and the membrane surface [8].There have been many additives used in membrane modification. Generally obtained from chemicals which are relatively expensive. Therefore, several researchers have developed additives from nature and have done it successfully. Various natural additives that have been used include nanosilica from rice huskabd sugarcane bagasse [9], nanocarbon from palm shell waste [10], and activated carbon from *Jatropha* seed shell [11]. Besides these additives, there is a natural additive that has never been studied before, that is Jernang.

Jernang or globally known as dragon's blood is the result of red rattan secretions. Dragon's blood is generally used for all types of resins obtained from four genus, including *Croton* (*Eeuphorbiaccac)*, *Dracaena* (*Dracaenaceac*), *Daemonorops* (*Palmaccae*), and *Precarpus* (*Fabaceac*). The type of dragon's blood that grows in Indonesia comes from the genus *Daemonoropsspp*, which is known by residents as jernang. Previous studies have reported that this resin contains terpenoids, flavonoids, phenols and dracorodine [12]. The amount of flavonoid content varies depending on the type of jernang and its quality. Flavonoids can increase the hydrophilicity of the membrane by substituting a hydroxyl group (-OH) into the polymer structure [13]. This research will use jernang from Aceh (*Daemonoropsdracoblume*) as an additive. This research studied the effect of adding additives to the performance of PES membranes. These experiments include filtration tests, humic acid rejection, antifouling and membrane stability.

2. Experimental

2.1. Material

Polyethersulfone (PES, Ultrason E6020P) used as the main polymer which was purchased from BASF Co. (Ludwigshafen, Germany). 1-N-methyl-2-Pirrolidone (NMP, 99.5%) used as the polymer solvent which was purchased from Merck (Hohenbrun, Germany). Jernang (Daemonoropsdraco Blume) used as the membrane additive which was obtained from farmers in Central Aceh. Humic acid (sodium salt, technical grade 50-60%) purchased from Sigma-Aldrich which was used as the model organic foulant to test the rejection and antifouling ability membrane.

2.2. Preparation and characterization of Jernang

Jernang was minimized using a mortar and pestle until powdery. Then the Jenang powder was filtered using a 180 μm sieve. This characterization consists of analyzing the Jernang functional groups using Fourier Transform Infrared (FTIR). FTIR spectra were analyzed by Shimadzu Prestige FT-IR 6400 in the range 4000 cm^{-1} to 400 cm^{-1} at room temperature. Before analysis, Jernang was mixed with KBr with a ratio of 1: 9 (Jernang: KBr).

2.3. Membrane fabrication

In this study two flat sheet membranes were fabricated. The first membrane (M0) was composed of 17.5% PES and 82.5% NMP. Then the second membrane (M1) was composed of PES 16.5%, Jernang 1%, and NMP 82.5%. The membrane was produced using the Non-solvent Induced Phase Separation (NIPS) method. Dope solution was made by dissolving PES and Jernang into NMP for 24 hours until homogeneous. Then leave for 6 hours at room temperature until the air bubbles completely disappear. Each dope solution was molded onto a lip at room temperature using an automatic applicator (YBA-3, Japan) with a thickness of 300 μm. Then the lisp was dipped into a bath containing non-solvent distilled water, during the immersion process a thin layer will be formed which is separated from the lips. Finally, the membrane was washed and stored as shown in Figure 1.

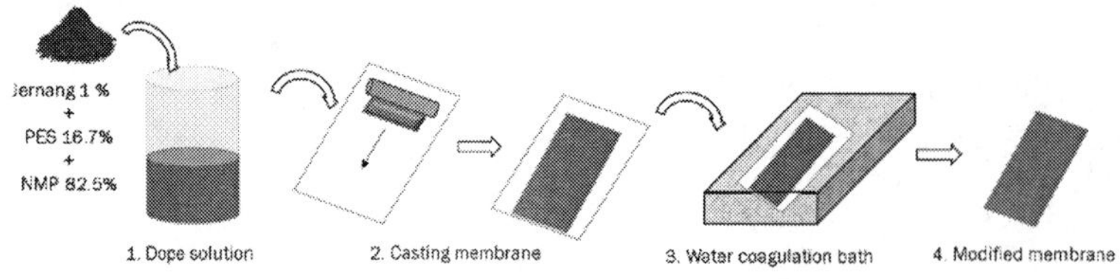

Figure 1. Illustration of membrane fabrication.

2.4 Evaluation of permeation and separation performance

Membrane permeation was evaluated by filtering the membrane with distilled water. The membrane was filtered at a trans membrane pressure of 3 bar. This test is useful for knowing how much permeate rate can be obtained. The dead-end module was used with a batch system as Figure 2 portrays. Permeate was collected in a beaker and its total weight was recorded every 10 minutes until the permeate weight was constant. Pure water flux was calculated using Equation 1

$$J = \frac{Q}{A.t} \tag{1}$$

Where J is the flux of pure water ($L/m^2.h$), Q is the permeate volume (L). t is the filtration time (hours), and A is the membrane surface area, which in this study was $5{,}667 x 10^{-4} m^2$.Separation performance was evaluated by filtering the membrane with an artificial solution of humus acid with an initial concentration of 50 ppm. The permeate concentration was analyzed using a UV-VIS spectrometer. The rejection coefficient (R) was calculated using equation 2.

$$R = \frac{Cf - Cp}{Cf} \, x\, 100\% \tag{2}$$

Where R is the rejection coefficient (%), C_f is the feed concentration (mg/L), C_p is the permeate concentration (mg/L).

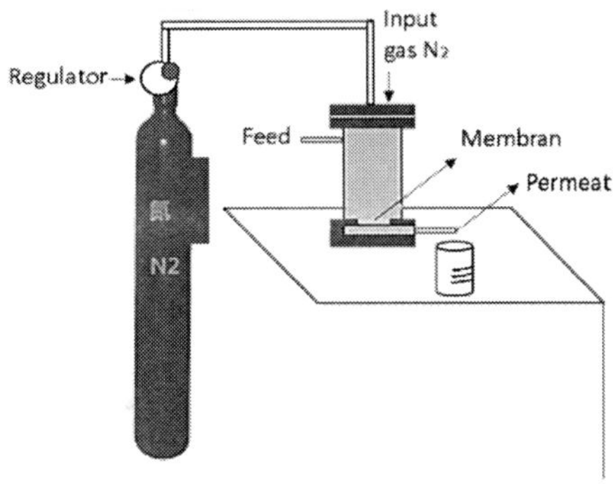

Figure 2. Set up filtration process.

2.5. Antifouling performance assessment

The antifouling test was useful for studying the resistance of membranes to fouling. This test was carried out in several stages, referring to the research of Jiang et al, 2014 [19]. The membrane was filtered with distilled water to obtain the initial pure water flux (J_{w1}). Then the membrane was filtered with humus acid solution (J_{HA}). After that, the membrane was washed using the backwash technique for 10 minutes with a pressure of 0.5 bar. In the final stage, the cleaned membrane was re-filtered with distilled water (J_{w2}). Data from this entire test series can be used to calculate the quantity of water flux that can be recovered after backwash, which is commonly known as flux ratio recovery (FRR). FRR can be calculated using Equation 3.

$$FRR = \frac{J_{W1}}{J_{W2}} \, x\, 100\% \tag{3}$$

Besides FRR, the membrane fouling behavior can also be studied from the flux loss caused by total fouling. Total fouling (R_t) is a combination of reversible (R_r) and irreversible (R_{ir}) fouling. These three behaviors are expressed mathematically in Equation 4 – 6.

$$R_T = \frac{J_{W1} - J_{HA}}{J_{w1}} \, x \, 100\% \quad (4)$$

$$R_r = \frac{J_{W2} - J_{HA}}{J_{w1}} \, x \, 100\% \quad (5)$$

$$R_{ir} = \frac{J_{W1} - J_{HA}}{J_{w1}} \, x \, 100\% \quad (6)$$

2.6. Membrane stability test

This test is to study the stability of the membrane when used in strong alkaline or strong acid feed conditions. This test was carried out by immersing the membrane in 1 N HCl solution (pH: 1) and 1 N NaOH (pH: 14) for a week. Then the membrane was re-tested for pure water flux and compared with its initial flux, so that the stability of the membrane could be evaluated.

3. Result and discussion

3.1. FTIR characterization of Jernang

FTIR analysis was carried out to identify chemical groups in Jernang. The FTIR spectrum is presented in Figure 3. The results of this analysis show a similar trend to the study reported by Nunes et al [15]. There are a dominant chemical group in Jernang, including the wave number 3267 cm^{-1} O-H stretch, 2846; 2939; 2999 cm^{-1} C-H stretch alkanes, 1730 cm^{-1} C = O stretch, 1602; 1647; 1500 cm^{-1} N-H bend, 1423; 1448; 1500 cm^{-1} C-C aromatic, and 1112 cm^{-1} C-N aliphatic amines.

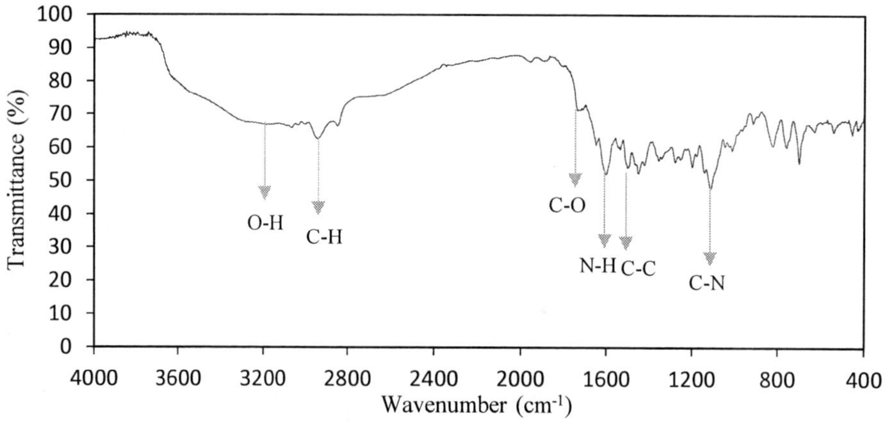

Figure 3. FTIR spectrum of Jernang.

3.2. Pure water flux

Figure 4.a shows that the addition of 1% Jernang to the PES membrane can increase the pure water flux up to 11 times greater than the pure PES membrane, which is 18.77 LMH. This proves that Jernang can increase the hydrophilicity of PES membranes. The hydrophilicity of a membrane increased the flux of pure water because it can reduce hydraulic resistance and increase pore size and membrane porosity [9].

3.3. Humic acid removal

In Figure 4.b humic acid flux shows the same tendency as the pure water flux, the difference in the amount of permeate. The humic acid flux is less than the pure water flux due to concentration polarization throughout the filtration time. Based on Figure 4, it can be seen that the addition of 1% Jernang causes rejection to decrease by 3.77% percent of the pure PES membrane. This is because the pore size and porosity of the membrane are increased which allows more foulant to be released during filtration.

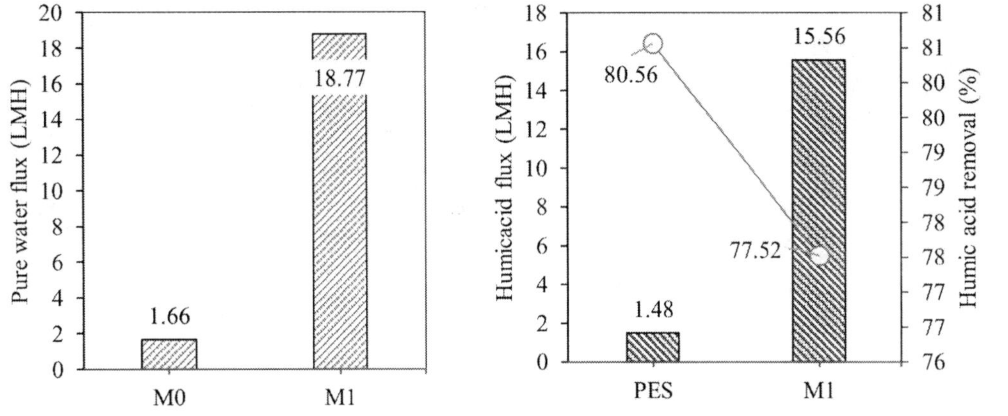

(a) (b)

Figure 4. (a) Effect of Jernang additive on pure water flux, (b) Effect of adding jernang additives on humic acid removal.

3.4. Anti-fouling performance

Fouling is caused by deposits of organic impurities on the surface or inside the pores of the membrane. Membrane fouling can be reversible and irreversible. The concentration polarization is reversible, so it can be solved by backwashing. However, in irreversible fouling, the foulant is in the pore of the membrane which causes pore constriction, so at this stage the foulant cannot be physically removed [14]. The effect of impurities on the membrane can be observed from the performance of the membrane, which is shown by the loss of flux during the filtration process, as in Figure 5.a.

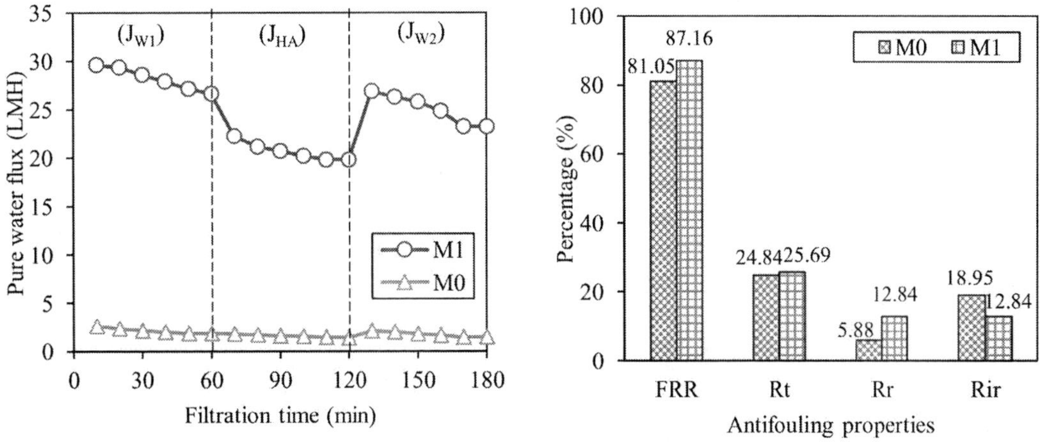

(a) (b)

Figure 5. (a) Reduction of flux and membrane recovery profiles during filtration with water and humic acid solution,(b) anti-fouling performance of pure PES membrane and PES + Jernang 1%.

This whole series of processes is carried out in order to measure the quantity of water flux that can be recovered after backwash, which is commonly known as flux ratio recovery (FRR). Apart from FRR, the fouling behavior can also be studied from the flux loss due to total fouling, reversible, and irreversible, which are summarized in Figure 5.b The results of the anti-fouling study showed that adding 1% Jernang to PES membrane solution increased the antifouling properties of the membrane. This was indicated by the FRR value on the M1 membrane reaching 87.16% and irreversible fouling of 12.84%. The test results prove that this modification is able to produce a membrane that has excellent fouling resistance.

3.5. Stability of membranes

The results of pure water flux testing before and after immersion can be seen in Figures 6. The immersion of the membrane in strong acids and alkaline solution increases the pure water flux of the modified PES membrane (M1). The flux increases 3 times more than the initial flux before the membrane is immersed. This significant value indicates that the PES membrane with added 1% Jernang is not stable against strong acid and alkaline conditions. Besides, the neat PES (M0) presents a great result. The pure water flux after immersion in strong acids and alkaline shows an amount similar to the initial flux. So, it can be concluded that neat PES has very good chemical stability.

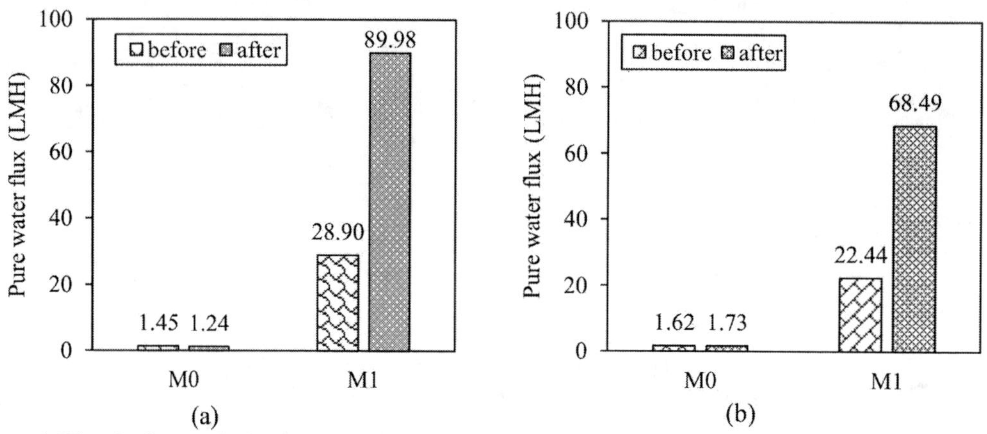

Figure 6. Flux before and after immersing the membrane in (a) strong acid solution and (b) strong alkaline solution.

4. Conclusion

The conclusion of this research is that modification of PES membrane with the addition of 1% jernang additive has been able to increase the hydrophilicity of the membrane followed by an increase in pure water flux. Humic acid removal showed reasonable results up to 77.52% and this membrane has good antifouling properties. However, PES membrane modification causes the membrane to be unstable under strong acid and alkaline conditions. It is necessary to learn more about the characteristics of this membrane and the effect of Jernang concentration on the performance of PES membranes.

5. Acknowledgment

This research was funded by the Institute for Research and Community Service (LPPM) Syiah Kuala University, in the scheme of Program Riset Unggulan Unsyiah Percepatan Doktor (PRUUPD) 2020.

References

[1] Ranjidar S R 2014 *Membrane Technology and Engineering for Water Purification: Application, Systems Design and Operation.* Oxford: Butterworth-Heinemann.
[2] Kato R, Asami T, Utagawa E , Furumai H, Katayama H 2018 *Water. Res.,* **132** 61–70.
[3] Chen Y, Lin T, Chen W 2019 *J. Environ. Sci.,* **78** 315–327.
[4] Chakraborty B, Indra A, Menezes P V, Driess M, Menezes P W 2020 *Mater. Today Chem.,* **15** 852–859.

[5] Aziz A, Basheer F, Sengar A, Irfanullah, Khan S U, Farooqi I H 2019 *Sci. Total Environ.,* **686** 681–708.

[6] Cui Z F, Muralidhara H S 2010 *Membrane Technology: A Practical Guide to Membrane Technology and Applications in Food and Bioprocessing* Amsterdam: Elsevier Science.

[7] Baker R W 2004 *Membrane Technology and Applications* Hoboken: Wiley.

[8] Das C, Gebru K A 2018 *Polymeric Membrane Synthesis, Modification, and Applications: Electro-Spun and Phase Inverted Membranes* Florida: CRC Press.

[9] Mulyati S, Muchtar S, Yusuf M, Arahman N, Sofyana S, Rosnelly C, Fathanah U, Takagi R, Matsuyama H, Shamsuddin N, Bilad M 2020 *J. Membr. Sci.,* **10** 1-12.

[10] Arahman N, Yusuf M, Muchtar S, Maimun T, Bilad M 2018 *Int. J. GEOMATE,* **15** 51–57.

[11] Aprilia S 2018 *J. Environ. Chem. Eng.,* **13** 87–92. *(in Bahasa Indonesia).*

[12] Gupta D, Bleakley B, Gupta R K 2008 *J. Ethnopharmacol.,* **115** 361–380.

[13] Escobar J D, Prieto C, Pardo-Figuerez M, Lagaron J M 2018 *Mol.,* **23** 26-41.

[14] Jiang J H, Li P, Zhang H T, Zhu, B K, Xu Y Y 2014 *J. Membr. Sci.,* **457** 73–81.

[15] Nunes O R, Cordeiro M M, Oliveira F, Passos T, Quilty B, Thiré R, Mcguinness G 2016 *Rev. Mater.,* **21** 767–779.

The effect of amino acids extracted from tofu wastewater on chitosan performance for methylene blue removal from water

H Fathana[1], Rahmi[1,*], S Lubis[1], Adlim[2]
[1]Department of Chemistry, Faculty of Mathematics and Natural Sciences, Universitas Syiah Kuala, Banda Aceh, Indonesia
[2]Department of Chemistry Education, Faculty of Teacher Training and Education, Universitas Syiah Kuala, Banda Aceh, Indonesia

*E-mail: rahmi@fmipa.unsyiah.ac.id

Abstract. In this study, we have modified chitosan using amino acids to form a chitosan-amino acid film. Amino acids were extracted from the tofu wastewater of the tofu producer, Timbul Jaya in Banda Aceh. The amino acid loaded on chitosan was varied (1, 0.5, 0.25, and 0.1 mL). The best chitosan-amino acid films were obtained at 0.5 mL amino acid loading. Chitosan-amino acid films were characterized using XRD (X-Ray Diffraction), TGA (Thermogravimetric Analysis), DSC (Differential Scanning Calorimetry), and FT-IR (Fourier Transform Red Spectroscopy). XRD data showed that amino acid loading did not affect the crystallinity of chitosan. Based on TGA and DSC data, the addition of amino acids slightly decreased the thermal properties of chitosan. FTIR data confirmed the amount of amino acids loaded can increase the amine group in chitosan. Adding amino acids to chitosan film was able to increase the adsorption capacity of chitosan.

1. Introduction

Chitosan is a biopolymer that is widely used when removing contaminants from water. The most common application of chitosan is in the membrane preparation process and the adsorption process [1]. This is due to the presence of functional groups amine and hydroxyl as these functional groups play an important role in the adsorption process of a substance. In addition, chitosan is also environmentally friendly, biodegradable, and biocompatible [2]. Although chitosan is a useful adsorbent, it has a relatively low adsorption capacity. Previous scholars have conducted research about how to improve the adsorption performance of modified chitosan [3].

Chitosan can be modified through several methods, such as: adding functional groups, adding photocatalysts, and adding crosslinked agents. For instance, Ren [4] modified chitosan through amino acid addition to increase the adsorption of Pb^{2+} ions in water. The research results indicated a 91% increase in adsorption efficiency at an adsorption temperature of 30°C, pH 4, and contact time of 20 hours. Boggione et al. [5] modified chitosan with amino acids and copper to increase endoglucanase adsorption. The results showed a 40% increase in endoglucanase adsorption compared with unmodified chitosan. Based on the previous research, the addition of amino acids during chitosan modification improved the adsorption capacity of chitosan. However, most researchers who have modified chitosan with amino acids did so using commercial amino acids. No one has yet modified chitosan with amino acids from tofu wastewater.

Currently, Aceh's tofu manufacturers dump tofu wastewater into the river which pollutes the water. Tofu wastewater contains organic substances that can cause rapid microbial growth in water, which causes the oxygen level in the water to drop dramatically. Additionally, tofu wastewater contains suspended particles and substances which make the water dirty or turbid [6]. Chua and Liu [7] stated that tofu wastewater contains high concentrations of protein, fat, and carbohydrates. If the organic compounds decompose within the water, methane gas, carbon dioxide, and other gases will be produced. The high levels of protein and other organic compounds cause BOD (Biochemical Oxygen Demand) and COD (Chemical Oxygen Demand) levels in the water to increase. To produce 1 ton of tofu or tempeh, 3,000-5,000 liters of wastewater are produced [6]. One of the potential strategies to decrease the impact of water pollution due to tofu wastewater is by repurposing the wastewater as an amino acid source. Therefore in this research, tofu wastewater from the Tahu Sumedang Timbul Jaya manufacturing company in Banda Aceh was used to produce amino acids. The resulting amino acids were used to modify chitosan and applied as methylene blue adsorbent.

2. Materials and methods

2.1. Materials

The tofu wastewater (TW, pH 5.6) was obtained from Tahu Sumedang Timbul Jaya (T. Umar street, Bandar Raya, Banda Aceh, Aceh, Indonesia). The chitosan was purchased from Tokyo Chemical Industry Co., Ltd. in Japan and was manufactured from shrimp shell with 75.0-85.0% deacetylation degree. The HCl 37% used during analysis, NaOH and Glacial Acetic Acid were obtained from Merck, Germany. Methylene Blue and Distilled Water were also used in the process.

2.2. Extraction of amino acids from tofu wastewater

In order to extract the Amino acids from the wastewater, 200 mL of tofu wastewater was combined with NaOH 2 N. Thus, the solution yielded a pH of 10 and was heated at a temperature of 50°C for 1 hour. It was filtered and then the resulting filtrate was combined with HCl 2 N to generate a pH of 4.5 (isoelectric pH) and stored at a temperature of 10°C for 1 night. The next day, it was centrifuged at -4°C for 20 minutes. The pellets obtained from centrifugation were washed using distilled water and dried at 50°C for 5 hours using an oven. The proteins were then extracted and weighed [8]. The proteins extracted from the tofu wastewater were combined with HCl 6 N with an excess volume (0.5 mL/mg) [9]. The solution was heated at a temperature of 70°C for 4 hours in closed conditions and stirred using a magnetic stirrer. Finally, The solution was exposed to a temperature of 55°C and a vacuum of 45 mbar for 20 minutes so the excess HCl would evaporate. The resulting amino acids were stored in a freezer.

2.3. Preparation of chitosan-amino acid film

In order to prepare the chitosan-amino acid film, 1 gram of chitosan was dissolved into 99 mL of 2% acetic acid for 2 hours using a magnetic stirrer. Next, 1 mL of amino acid was added and stirred for 1 hour. Finally, the homogeneous mixture was poured into a 17x12 cm film mold and dried at 40°C in an oven for 48 hours. The procedure was repeated with various volumes of acetic acid and amino acids of (99.9: 0.1), (99.75: 0.25), and (99.5: 0.5).

2.4. Methylene blue adsorption study

0.025 grams of chitosan-amino acid film that had been cut 1x1 cm was added into Erlenmeyer containing 20 mL of methylene blue. The erlenmeyer flask was put on the shaker for 30 minutes. Methylene blue was filtered and analyzed its absorbance using a spectrophotometer UV-Vis at the maximum wavelength.

3. Result and discussion

3.1. Amino acids extracted from tofu waste water

The first stage of this research was protein extraction from tofu wastewater by adding NaOH 2N to obtain a pH of 10. Alkaline conditions during the protein extraction process can cause most amino acids negatively charged at pH above the isoelectric point. Amino acids with the same charge, typically repel each other, resulting inminimal interaction between the acid residues and amino acids. This means that there is an enhancement of purity in protein [8].

In order to reach an isoelectric point, HCl 2N was added until the solution reached a pH of 4.5 during protein extraction. At the isoelectric point, the total charge on each amino acid in the protein is equal to zero [8]. The extracted protein in this research weighed 1 gram, had a white colour as shown in Figure 1.

The most important aspect of protein hydrolysis in various substances is obtaining a hydrolysate solution with high levels of amino acids or according to the levels of amino acids in the sample. Thus, in the protein hydrolysis process carried out on materials containing proteins, it can completely break down the amino acids and reduce the amino acid degradation during protein hydrolysis.

In general, protein hydrolysis using HCl 6N is carried out at a temperature of 110°C for 24 hours [10]. However, based on the results obtained in this research, this method cannot be used for protein hydrolysis with tofu wastewater, because the protein hydrolysate was black and contained black powders such as ash that was estimated to cause amino acid damage in the protein hydrolysis process. This can be seen in Figure 2.

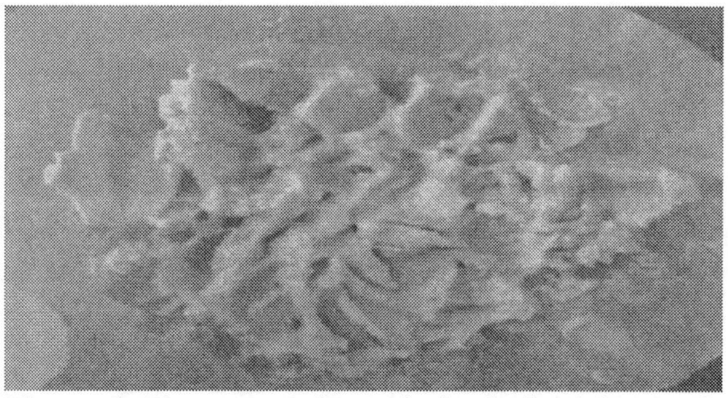

Figure 1. The extracted protein is obtained from tofu wastewater.

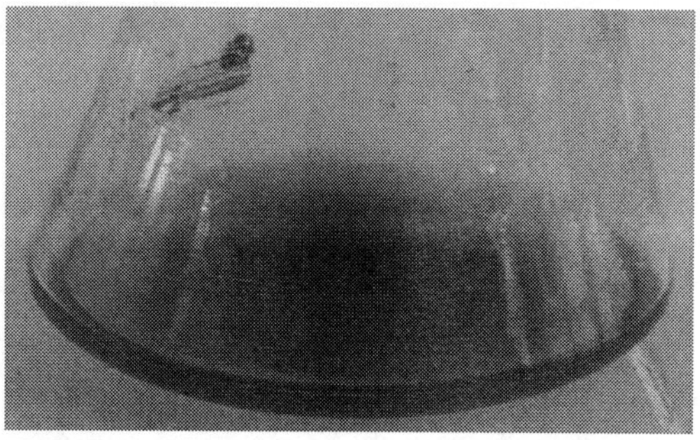

Figure 2. The results of protein hydrolysis of tofu wastewater using general protein hydrolysis (110°C, 24 hours).

The important parameters in protein hydrolysis are temperature, hydrolysis time, and composition of the hydrolysis mixture (volume of HCl and amount of protein) [11]. In this case, we tried to hydrolyze the proteins created from tofu wastewater at a temperature of 70°C. According to the research of Lamp et al. [10], protein hydrolysis was carried out between 2-140 hours. In this research, we used a protein hydrolysis time of 4 hours.

Protein hydrolosis from tofu wastewater using HCl 6 N with an excess volume of HCl 6N aims to reduce the concentration of carbohydrate components in the hydrolysis medium. This is because the tofu wastewater sample also contains much carbohydrates. The carbohydrates will decompose during the hydrolysis process to form humin compounds in the form of particles that can absorb certain amino acids. The protein hydrolysis process is also carried out in a closed state because oxygen from the air can oxidize amino acids [9]. The visual results of protein hydrolysate can be seen in Figure 3.

Figure 3. Protein hydrolysate.

According to the FT-IR spectrum,the protein hydrolysate (Figure 4) that was hydrolyzed at a temperature of 70°C showed a distinctive absorption band with amide I at wavenumbers 1654 cm^{-1}. Wavenumbers 1543 cm^{-1} shows a distinctive absorption band with amide II and wavenumbers 1462 cm^{-1} shows a unique absorption band with amide III [13]. In addition, the appearance of absorption bands at wavenumbers 1165 cm^{-1} indicates the presence of polysaccharides [12].

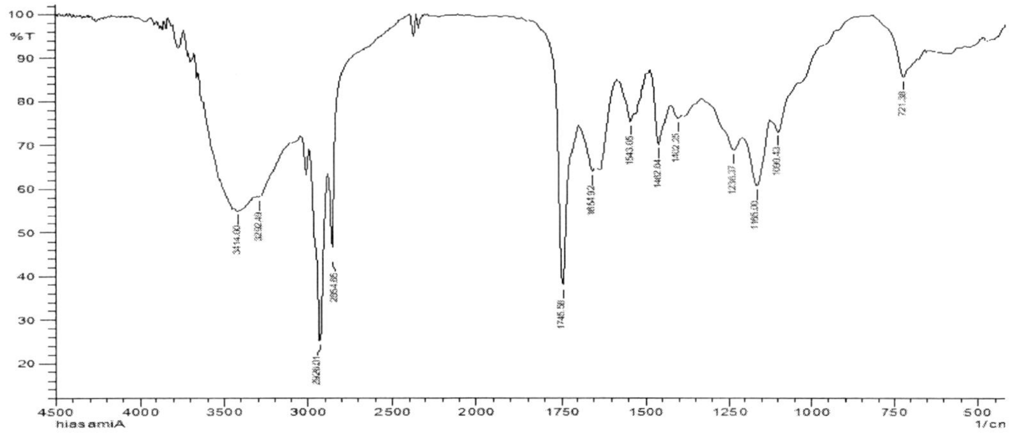

Figure 4. FT-IR Spectrum of protein hydrolysate.

3.2. Adsorption study

Chitosan that has been modified with amino acids has a better adsorption capacity than chitosan without amino acid modification. This proves that amino acids have been successfully added as new active groups in chitosan. The adsorption capacities obtained from the methylene blue adsorption experiments using chitosan and chitosan-amino acids film are shown in Figure 5.

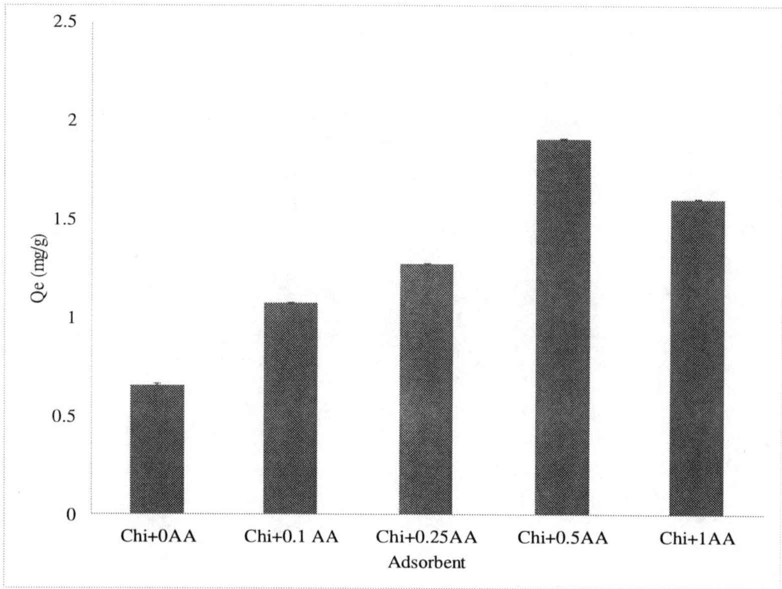

Figure 5. The adsorption capacity of chitosan and chitosan-amino acids film with various volume of amino acids.

Based on Figure 5, the best adsorption capacity was obtained using chitosan-amino acids film prepared with 0.5 mL of amino acids, which was 1.908 mg/g. The chitosan-amino acids film containing 1 mL of amino acid in its preparation shows the lowest adsorption capacity because there were too many active groups in the chitosan that had bound with the groups on the amino acids. Therefore, the ability of chitosan to adsorb methylene blue slightly reduced.

3.3. The characterization of chitosan-amino acids film

The chitosan-amino acid films were characterized by XRD (X-Ray Diffraction), TGA (Thermogravimetric analysis), DSC (Differential Scanning Calorimetry) and FT-IR (Fourier Transform-Red Spectroscopy).

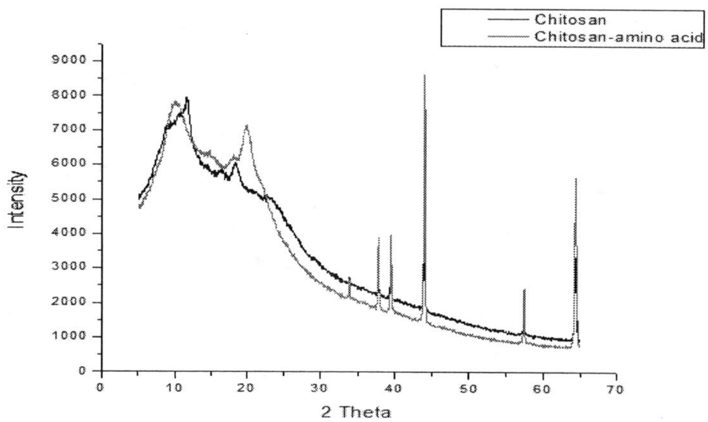

Figure 6. The XRD patterns of chitosan and chitosan-amino acids film.

Based on the XRD data, the typical diffractogram peak of chitosan appears at 2 θ of 19.5° as shown in Figure 6. The sharp peaks of the diffractogram in both samples showed high crystallinity. The addition of amino acids did not significantly affect the crystallinity of chitosan.

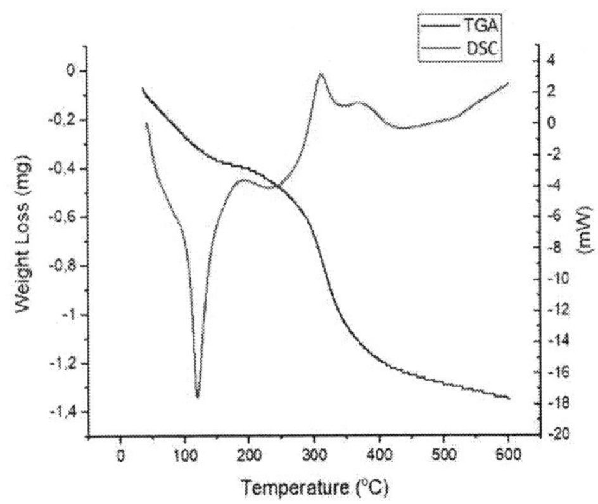

Figure 7. DSC and TGA curves of chitosan film.

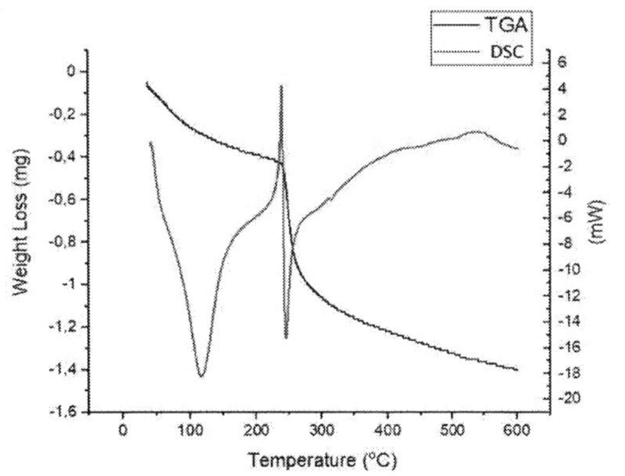

Figure 8. DSC and TGA curves of chitosan-amino acids film.

The DSC curves are consistent with the results of the TGA analysis. The thermogram of the chitosan film (Figure 7) and chitosan-amino acid film (Figure 8) show a two-stage thermal event. The first stage occurs between 50-100°C, and is associated with the loss of water within the sample. The second stage, starting at 279°C for chitosan film and 238.78°C for chitosan-amino acid film, was associated with the degradation of the sample. Endothermic peaks approaching 100°C can be attributed to water loss. At the exothermic peak around 315.30°C for the chitosan film and 238.35°C for the chitosan-amino acid films are related to the decomposition of chitosan [5]. The introduction of amino acids to chitosan film slightly reduces the thermal stability of chitosan due to the loss of hydrogen bonds.

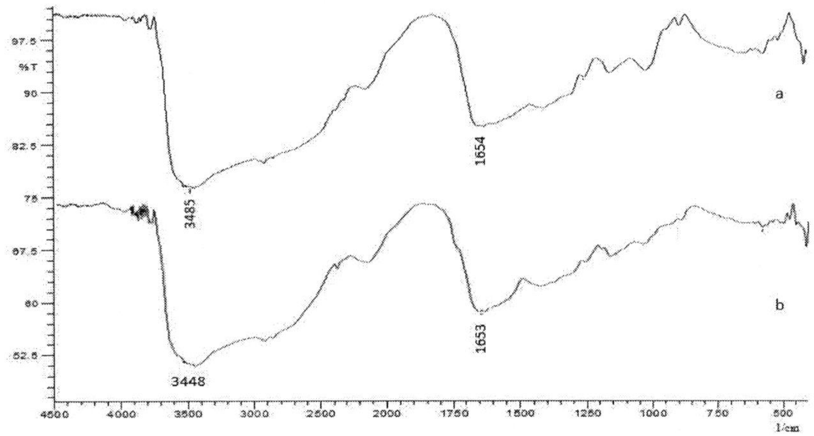

Figure 9. FTIR Spectra of chitosan film (a) and chitosan-amino acids film (b).

Based on the FT-IR data in Figure 9, chitosan modified with amino acids had no significant change compared to the functional groups of chitosan without modification such as the O-H and N-H groups. In unmodified chitosan, the IR spectrum showing the O-H group and N-H stretching vibrations were around 3485 cm^{-1}, while in the chitosan modified with amino acids, the stretching vibrations were around 3448 cm^{-1}. The wider peak about 3400 cm^{-1} of chitosan-amino acids film indicated the formation of $-NH_3^+$. The peak increase of about 1100 cm^{-1} of chitosan-amino acids film, indicated that loading amino acids can increase the amine group in chitosan [14].

4. Conclusion

Based on the research, it was determined that Tofu wastewater is able to become an amino acid source in chitosan modification. Chitosan that has been modified with amino acid extracted from tofu wastewater has a greater adsorption capacity than unmodified chitosan. However, the addition of amino acids to chitosan slightly decreased the thermal properties of chitosan.

5. References

[1] Crini G, Badot Badot P M. 2008 *Prog. Polym. Sci.*, **33** 399-447.
[2] Zhao, Weifeng, Xuelian H, Yilin W, Shudong S, Changsheng Z. 2016 *Carbohydr. Polym.*, **150** 201–208.
[3] Liu, Yi, Yian Z, Aiqin W. 2010 *J Environ Sci.*, **22** 486–493.
[4] Ren H, Li B, Neckenig M, Wu D, Li Y, Ma Y, Li X, Zhang N. 2018 *Carbohydr. Polym.,*. **207** pp. 737-746.
[5] Boggione M J, Mahl C R A, Beppu M M, Farruggia B. 2017 *Powder Technol.,* **315** 250–257.
[6] Azmi M., Edward HS, Andrio D. 2016. *J. Faculty Eng.,* **3** 1-5.
[7] Jian Y C, Shao Q L. 2019 *Trends Food Sci Technol.,* **91** 24-32.
[8] Nurdjannah N, Usmiatia S. 2006. *Center for Agricultural Postharvest Research and Development.*
[9] Sumardi. 1995. *Center for Applied Chemistry Research and Development-LIPI.* **5** 46-54.
[10] Lamp A, Kaltschmitt M, Lüdtke O. 2018. *Anal. Biochem.*, **543** 140-145.
[11] Fountoulakis M, Lahm H W. 1998. *J. Chromatogr. A.*, **826** 109-134.
[12] Rocha M, Delgadillo I, Marcela C, Jana C. 2003 *Carbohydr. Polym.*, **51** 383–389.
[13] Ossa J S H, Wagner J R, Palazolo G G. 2020. *Food Res. Int.,* **128**.
[14] Beppu M M, Arruda E J, Vieira R S, Santos N N. 2004. *J. Membr., Sci.,* **240** 227-235.

Molecular docking study of fatty acids from *Pliek U Oil* in the inhibition of SARS-CoV-2 protein and enzymes

K Khairan[1], R Idroes[1,2,*], S G Tumilaar[3], T E Tallei[4], G M Idroes[5], F Rahmadhany[1], M U Futri[1], N M Dinura[1], S Mauliza[1], M Diana[1], C P Maisarah[1], A Maulana[6], T R Noviandy[6], R Suhendra[6], Muslem[7], N Earlia[8]
[1]Department of Pharmacy, Universitas Syiah Kuala, Banda Aceh 23111, Indonesia
[2]Department of Chemistry, Universitas Syiah Kuala, Banda Aceh 23111, Indonesia
[3] Pharmacy Study Program, Universitas Sam Ratulangi, Manado 95115, Indonesia
[4]Department of Biology, Universitas Sam Ratulangi, Manado 95115, Indonesia
[5]Department of Chemical Engineering, UniversitasSyiah Kuala, Banda Aceh 23111, Indonesia
[6]Department of Informatics, Universitas Syiah Kuala, Banda Aceh 23111, Indonesia
[7]Department of Chemistry, Universitas Islam Negeri Ar-Raniry, Banda Aceh 23111, Indonesia
[8]General Hospital ZainoelAbidin, Aceh Province, Indonesia, 24415

*E-mail: rinaldi.idroes@unsyiah.ac.id

Abstract. This study aimed to analyze the fatty acid in*Pliek U* oil andevaluate its inhibitor potential based on the interaction to several proteins and enzymes in SARS-CoV-2 using the in silico approach. *Pliek U* oil containing capric acid, caprylic acid, lauric acid, linoleic acid, myristic acid, oleic acid, and palmitic acids, with oleic acid as a dominant substance. Molecular docking analysis showed that linoleic acid has the best interaction to the receptors with the lowest binding affinity to 3CLpro (6LU7), Spike protein (6VXX), PLpro (6WX4), RdRp (6M71), E protein (5X29), and Spike Ectodomain Structure (6VYB) of -4.9, -5.8, -4.7, -4.3, -5.3 and -5.5 kcal/mol, respectively. The finding suggests that the binding of linoleic acid to the SARS-CoV-2 protein and enzyme may cause impairment of viral attachments to host cells, thus reducing infectivity in COVID-19 patients.

1. Introduction

The end of 2019 saw the rise of a new pneumonia case in Wuhan, China. A few days later, China's health authorities confirmed that this case was related to coronavirus disease infection, hereinafter known as COVID-19 [1]. The first clinical report showed that 13 out of 14 (92%) COVID-19 patients had comorbidities [2], such as cardiovascular, diabetes, hypertension, and chronic obstructive pulmonary disease. Subsequent findings reported that 64 of the 138 cases (46.4%) of COVID-19 patients had known comorbidity. Other findings also prove that COVID-19 patients admitted to the ICU have high comorbidity of approximately 72.2%. This fact shows that comorbidity is a risk factor for patients of COVID-19 [3]. Therefore, a systematic risk evaluation or management is needed, not only for comorbidity but also risk management for the underlying disease in COVID-19 patients. Globally, at least 7900 deaths have been directly linked to COVID-19 and will continue to increase every day, as a vaccine has yet to be found to eradicate this virus.

The search for an effective cure against this disease continues to be conducted, through finding or synthesizing new active compounds. This work is handling by the expert whilst the pandemic is still ongoing by the pathogens. The alternative way to find a cure is by testing compounds that already have drug activity against the pathogens. It has been known that natural products have been widely reported having that activity to be antimicrobial [4,5], antibacterial [6–8], antibiofilm [9], antioxidant [10,11], and even antiviral [12]. The natural products are mostly found in plants as metabolites in almost plant parts [13]. It produces by itself through metabolism [14], or by endophyte [15,16]. The natural product is also preferred to be drugs in offering a safer, low side effect and green chemistry materials[17,18].

Fatty acid can be obtained naturally from animal and plant [19], coconut (*Cocos nucifera*) is one of a natural product as sources of fatty acid. In Aceh Province, coconut meat is traditionally used to produce fermented-coconut namely *Pliek U*. This process also produce a secondary product called *Pliek U* oil [20]. Earlia et al., reported that *Pliek U* oil contained capric acid, caprylic acid, lauric acid, linoleic acid, myristic acid, oleic acid, and palmitic acids, with oleic acid as a dominant substance [21,22], whereas the fatty acid content in *Pliek U* oil has a potential to be used as a pharmacy resource. Several studies reported that fatty acid affecting to bacterial and fungi, but limited reports focus on antiviral agents. Oleic acid reported has moderate antiviral activity against tobacco mosaic virus [23]. Fatty acids are one of the compounds reported to have reasonably good antiviral activity [24].

There are several essential protein targets for the COVID-19 therapy includingPL^{pro}, $3CL^{pro}$, E protein, Spike (S) glycoprotein, C-terminal RNA binding domain (CRBD), N-terminal RNA binding domain (NRBD), helicase and RdRp[25]. Some bioactive compounds have been plainly predicted as therapeutic agent candidates for COVID-19 infection. The potential of fatty acid in case of therapeutic agent for SARS-CoV-2 protein and enzyme are still limited reports. Recent in silico studies have shown that fatty acids have very stable interactions with receptor proteins from bond energy and structural orientation [22]. These are well potential for fatty acid to be tested as the antiviral against SARS-CoV-2. The resources of fatty acid are widely found on palm type such as coconut. As a product from fermented coconut, Pliek U oil should be contained much fatty acid.

The interaction of a chemical compound with receptor proteins is generally carried out through a spectroscopic approach, such as UV-Vis, sensors, lasers, and electromagnetics [26,27]. However, this method requires a lot of time and money. Therefore, a fast, inexpensive approach through computational methods, both virtual and in silico or molecular docking, is needed for an alternative in regard to the urgency related to COVID-19 research. In this study, a molecular docking study was carried out between the fatty acids contained in *Pliek U* against the SARS-CoV-2 protein and enzyme.

2. Material and methods

2.1. Molecular docking

2.1.1. Ligand determination. The compounds (ligands) used in the molecular docking study are fatty acid compounds contained in PUO and EPUOE in the form of capric acid, caprylic acid, lauric acid, linoleic acid, myristic acid, oleic acid, and palmitic acid.

2.1.2. Receptor determination. The nine SARS-CoV-2 proteins selected as receptors were: Main protease (Mpro) / 3CLpro (Protein Data Bank (PDB) code, PDB: 6LU7); Spike glycoprotein (S) (PDB code: 6VXX); and PLpro (PDB code: 6WX4); RdRp (PDB code: 6M71); Protein E (PDB code: 5X29); Nsp15 (PDB code: 2H85); Complex Nsp7,8,12 (PDB code: 6NUR); Receptor-binding domain, RBD (PDB code: 6YLA) and Spike ectodomain structure, SES (PDB code: 6VYB).

2.1.3. Preparation of ligands and receptors. The three-dimensional (3D) structure of the nine SARS-CoV-2 proteins as receptors was taken from the Protein Data Bank, PDB (http://www.rcsb.org//pdb) downloaded as files in pdb format. Files are stored in special folders for docking treatment. The file was opened using BIOVIA Discovery Studio Visualizer 2020. Water molecules and ligands attached to the receptors were removed, and the receptors were stored in pdb (name: receptor) format. Furthermore,

Autodock Tools [28] was then used, and polar hydrogen atoms were added to the receptors. The file was then saved in the pdbqt format (name: receptor.pdbqt).

The ligand structure was obtained from the PubChem website (http://pubchem.ncbi.nlm.nih.gov). The search was performed by entering the name of the ligand in the search option. The file of each ligand was then downloaded and saved. Files were downloaded in sdf format and then stored in a particular folder for docking treatment. Files are converted to pdb format using Open Babel [29] and saved in pdb format (name: ligand). The pdb-format ligands were opened using Autodock Tools. Torque adjustment was made by detecting the root and adjusting as desired. The file was then saved in the pdbqt format (name: ligand.pdbqt).

2.1.4. Active site determination. The location of amino acids as active sites at the receptors where the ligands will interact can be determined using the Autodock Tools. By arranging it on a three-dimensional map, a grid box was then created in the receptor regions where the ligands interact. The determination of this map can be based on the type of docking used. The three-dimensional map was made as wide as the receptor size, ensuring ligands to be docking to all parts of the receptors (blind docking). In 3CLpro docking, for example, the binding site is known to be the protein, so the three-dimensional map is only made as wide as the docked area (targeted docking).

2.1.5. Receptor-ligand docking. Docking was done using Autodock Vina [30]. The ligands and receptors saved in pdbqt format were copied to the Autodock Vina file to the docking folder. Then vina configuration file was typed into notepad and saved with the name 'conf.txt.' The Vina program was then run through the command prompt which was directed to the to-be-docked folder with the formula below. The docking results were then obtained from the output in a notepad format.

vina –config conf.txt –log log.txt. (1)

2.1.6. Analysis and visualization. The docking calculation result was shown from the output in notepad format. The docking ligand confirmation was then determined by selecting the pose with the highest affinity (the most negative Gibbs free energy). The results of the visualization of three-dimensional (3D) and two-dimensional (2D) structures from the results of docking calculations were carried out using BIOVIA Discovery Studio Visualizer 2020. Previously, the highest affinity selection was carried out using the vina program which was run through the command prompt which was directed to the docking calculation results folder, with the formula:

vina_split –input out.pdbqt (2)

Furthermore, the respective affinity values of the ligands were obtained in the pdbqt format. The files of receptors and ligands with the highest affinity in the pdbqt format were opened using the BIOVIA Discovery Studio Visualizer 2020. This enables the authors to see the interaction between ligands and receptors in the form of three-dimensional (3D) structures and show 2D diagrams for two-dimensional (2D) structures. Both files are saved in image files format.

3. Result

The content of fatty acid compounds from the analysis of *Pliek U*oil using GC-MS can be seen in Table 1 [21]. The results of the molecular docking analysis in the form of binding affinity between ligands (fatty acid compounds) with nine protein and enzyme receptors from SARS-CoV-2 can be seen in Table 2. The results of the binding affinity analysis show that several fatty acids from *Pliek U* oil have various binding values against some SARS-CoV-2 enzymes and proteins.

The image of code A shows the results of the docking analysis showing the interaction between ligands (fatty acids) and amino acid residues from the SARS-CoV-2 protein in the form of a three-dimensional (3D) structure. The image of the code B shows the results of the docking analysis showing the types of interactions that occur between ligands (fatty acids) and amino acid residues of the SARS-CoV-2 protein in two-dimensional (2D) form. Visualization results in 3D and 2D are performed based on the selection of poses with the highest affinity (most negative Gibbs free energy) of each fatty acid against the nine enzymes and SARS-CoV-2 protein.

Table 1. The results of the analysis of *Pliek U* oil using GC-MS [21].

No	Common Name	Chemical Structure	C:D
1	Caprylic acid	$CH_3(CH_2)_6COOH$	8:0
2	Capric acid	$CH_3(CH_2)_8COOH$	10:0
3	Lauric acid	$CH_3(CH_2)_{10}COOH$	12:0
4	Myristic acid	$CH_3(CH_2)_{12}COOH$	14:0
5	Palmitic acid	$CH_3(CH_2)_{14}COOH$	16:0
6	Oleic acid	$CH_3(CH_2)_7CH{=}CH(CH_2)_7COOH$	18:1
7	Linoleic acid	$CH_3(CH_2)_4CH{=}CHCH_2CH{=}CH(CH_2)_7COOH$	18:2

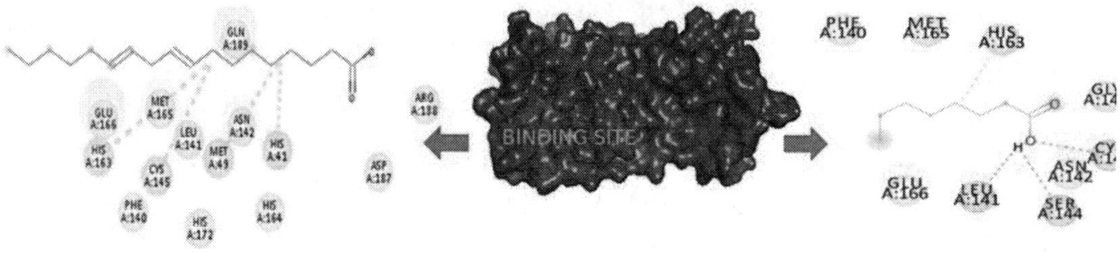

Figure 1. Molecular docking positions: Linoleic acid (A) and Palmitic acid (B) to the $3CL^{pro}$ receptor.

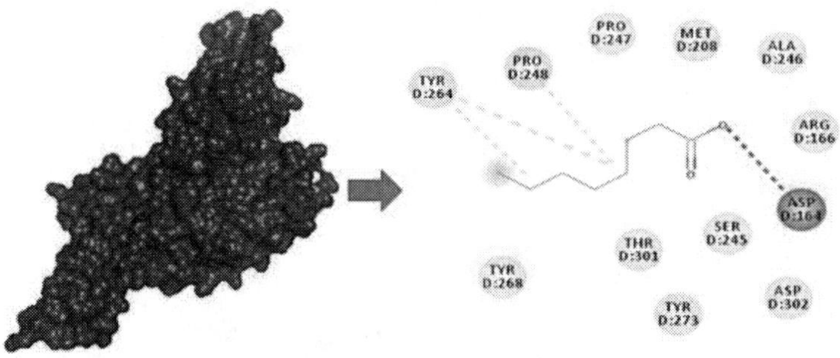

Figure 2. Caprylic acid's molecular docking position to the PL^{pro} receptor.

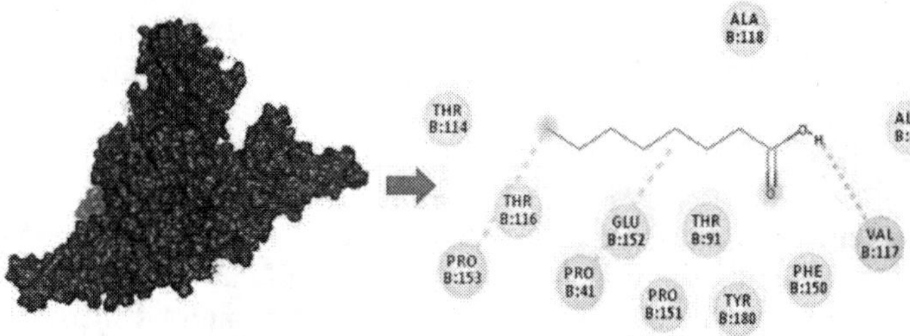

Figure 3. Caprylic acid's molecular docking position against the Receptor Binding Domain (RBD).

IOP Conf. Series: Materials Science and Engineering 1087 (2021) 012058 doi:10.1088/1757-899X/1087/1/012058

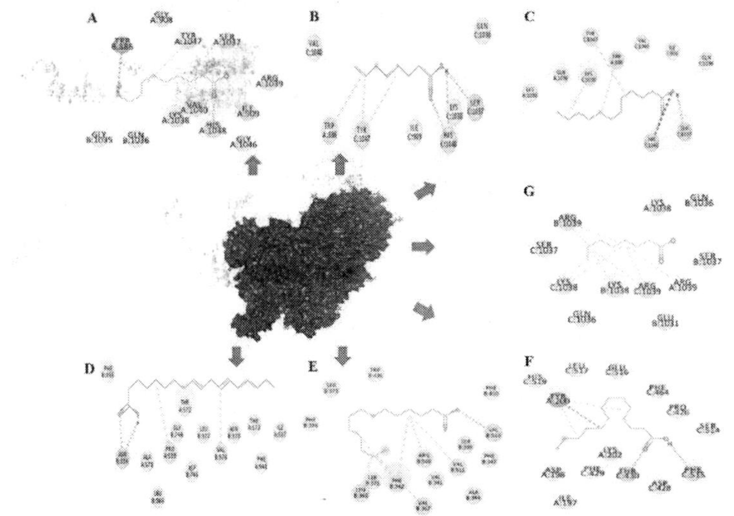

Figure 4. Molecular docking position: Capric acid (A), Caprylic acid (B), Lauric acid (C), Linoleic acid (D), Myristic acid (E), Oleic acid (F), Palmitic acid (G) to receptors spike ectodomain structure.

Table 2. Binding affinity between ligands (fatty acids from *Pliek U*) and some SARS-Cov-2 receptors.

Ligan C:D	PubChem CID	Binding affinity (Kcal/mol)								
		SARS-CoV-2 Receptor								
		M^{pro}	S Pro	PL^{pro}	RdRp	E Pro	Nsp 15	Nsp 7,8,12	SES	RBD
8:0 (FR)	379	-3.8	-3.8	-4.2	-3.9	-4.0	-3.8	-3.8	-4.5	-4.0
8:0 (NMD	379	-4.1	-4.7	-3.7	-4.2	-4.5	-3.7	-4.4	-4.5	-4.4
10:0 (FR)	2969	-4.1	-4.2	-4.3	-3.6	-4.4	-3.8	-4.1	-4.0	-4.9
10:0 (NMD)	2969	-3.8	-4.3	-3.3	-4.0	-4.4	-3.2	-3.7	-4.7	-3.9
12:0 (MUF)	3893	-4.4	-5.0	-4.2	-3.4	-4.6	-4.0	-3.8	-5.1	-4.1
12:0 (MD)	3893	-4.5	-4.4	-4.2	-3.7	-4.5	-4.1	-3.8	-5.0	-4.3
14:0 (SM)	11005	-4.3	-5.3	-4.3	-3.7	-4.9	-3.6	-3.5	-5.2	-4.0
14:0 (CPM)	11005	-3.9	-3.7	-3.5	-4.1	-4.7	-3.5	-4.1	-4.4	-3.9
16:0 (FR)	985	-4.5	-5.0	-4.2	-3.5	-4.4	-3.6	-3.7	-4.8	-4.1
16:0 (NMD)	985	-3.9	-4.5	-3.8	-3.8	-4.1	-3.8	-4.7	-4.4	-4.4
18:1 (SM)	445639	-4.5	-3.2	-3.7	-4.1	-4.9	-3.9	-4.9	-4.5	-4.2
18:1 (CPM)	445639	-4.6	-4.7	-4.0	-3.4	-4.9	-3.9	-3.6	-4.8	-4.1
18:2 (MD)	5280450	-4.5	-5.8	-4.4	-4.3	-5.3	-4.4	-4.2	-5.0	-4.8
18:2 (MUF)	5280450	-4.9	-5.4	-4.7	-3.8	-4.7	-4.6	-4.3	-5.5	-4.3

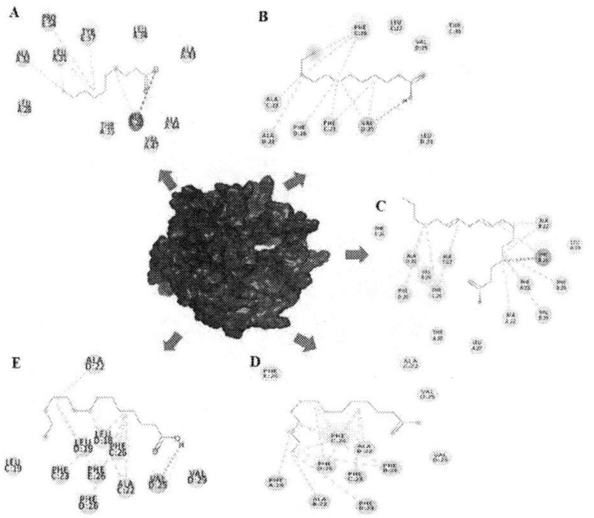

Figure 5. Molecular docking positions: Capric acid (A), Lauric acid (B), Linoleic acid (C), Myristic acid (D), Oleic acid (E) to protein E receptors.

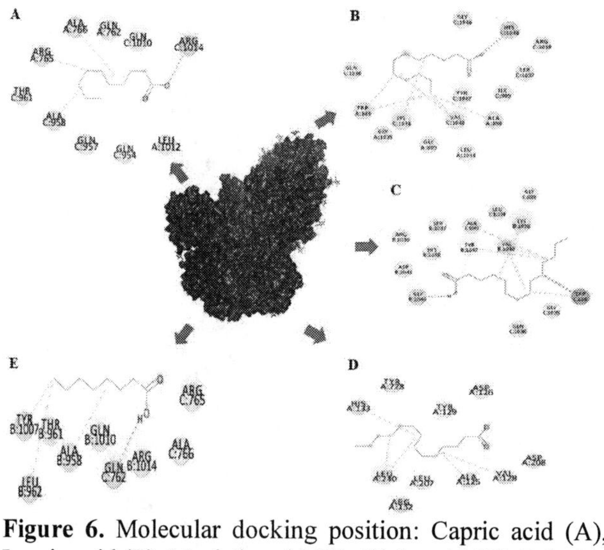

Figure 6. Molecular docking position: Capric acid (A), Lauric acid (B), Myristic acid (C), Oleic acid (D), Palmitic acid (E) to the Spike receptor.

4. Discussion

4.1. GC-MS analysis

Pliek U oil (PUO) and Pliek U oil ethanol extract (EPUOE) were analyzed using GC-MS. GC-MS is an instrument used for separation and identification. The results of GC were obtained in the form of chromatographic peaks, and MS will be obtained in the form of a mass spectrum. The compound results obtained from the analysis of Pliek U oil (PUO) and Pliek U oil ethanol extract (EPUOE) were caprylic acid, capric acid, lauric acid, myristic acid, palmitic acid, oleic acid, and linoleic acid [21].

4.2. Molecular docking

The binding affinity results show that linoleic acid ligands have the highest binding affinity compared to other ligands at the 3CLpro receptor (PDB code: 6LU7), Spike protein (PDB code: 6VXX), PLpro(PDB code: 6WX4), RdRp (PDB code: 6M71)), protein E (PDB code: 5X29), and Spike Ectodomain Structure (PDB code: 6VYB) with an energy binding affinity of -4,9, -5,8, -4,7, -4,3, - 5,3, and -5.5 kcal/mol. At the 6NUR receptor, the oleic acid ligand has the highest binding affinity value of -4.9 kcal/mol. At the 6YLA receptor, the capric acid ligand has the highest binding affinity value of -4.9 kcal/mol. Gibbs's energy theory states that the greater or more negative the energy produced from the bonds between the ligands and the receptors are, the more stable the bonds [21]. The higher or more negative ΔG value generated during the interaction of ligands and receptors, the stronger the ligand complex bonds with the receptors. This is due to compounds with receptors having stability and strength of non-covalent interactions, enabling easier entrance to cells and cause cells to die because they interfere with DNA replication or cellular metabolic processes [31].

Molecular docking results were visualized in three-dimensional (3D) and two-dimensional (2D) structures from the results of docking calculations carried out using the BIOVIA Discovery Studio Visualizer 2020. The interactions that occur were in the form of hydrogen bonds, hydrophobic interactions, electrostatic interactions, and unfavorable acceptors. Hydrogen bonds are interactions between hydrogen atoms (H), which are covalently bonded with atoms such as fluorine (F), nitrogen (N), and oxygen (O). The more hydrogen bonds with amino acid residues are formed, the stronger the bonds will be (Glowacki et al., 2013). Based on the 2D structure, hydrogen bonds are obtained in the interaction between capric acid and Spike Ectodomain Structure, capric acid and spike protein, caprylic acid and Spike Ectodomain Structure, caprylic acid and Receptor Binding Domain, lauric acid and Spike protein, lauric acid, and Spike Ectodomain Structure, lauric acid and protein E, linoleic acid and Spike Ectodomain Structure, myristic acid and Spike Ectodomain Structure, myristic acid and Spike protein, oleic acid, and Spike Ectodomain Structure, oleic acid and E protein, palmitic acid, and 3CLpro, as well as palmitic acid and spike protein. Most hydrogen bonds are found in palmitic acid fatty acids interacting with 3CLpro, namely the amino acid residues of Cys A: 145, Leu A: 141, and Ser A: 144. In addition, the most hydrogen bonds occur in the Spike Ectodomain Structure protein (PDB code: 6VYB).

Hydrophobic interactions are interactions that tend to avoid the liquid environment and gather in the globular protein structure [32]. Included in hydrophobic interactions are pi-Sigma and alkyl or pi-alkyl [33]. Hydrophobic interactions are amino acid residues that are nonpolar and tend to form groups in the interior of the protein so that they can prevent or block the protein from coming into contact with water [34]. Based on the 2D structure, hydrophobic interactions were obtained on the tethering between the Spike Ectodomain Structure receptor with capric acid ligands, caprylic acid, lauric acid, linoleic acid, myristic acid, oleic acid and palmitic acid (Figure 4), E protein with capric acid ligands, lauric acid, linoleic acid, myristic acid, and oleic acid (Figure 5), as well as spike proteins that bind to capric acid, lauric acid, myristic acid, oleic acid and palmitic acid ligands (Figure 6). Also, the bond between the PLpro receptor and the Receptor Binding The respective domains that each bind to caprylic acid ligands (Figure 2 and Figure 3), as well as bonds between linoleic acid and palmitic acid ligands to the 3CLpro receptor (Figure 1).

Electrostatic interactions are interactions that occur between atoms that are not covalent due to differences in polarity; thus, the interactions that occur are weak and loose but have the potential to inhibit the intended receptors because of the large number of bonds [35]. Included in electrostatic interactions are salt bridges and van der waals bonds [36]. Based on the 2D structure, electrostatic interactions were obtained in the interactions between linoleic acid and palmitic acid ligands against the 3CLpro receptor (Figure 1), the bond between the PLpro receptor and the Receptor Binding Domain, each of which binds to caprylic acid ligands (Figures 2 and 3), Spike Ectodomain receptors Structure with capric acid ligands, caprylic acid, lauric acid, linoleic acid, myristic acid, oleic acid and palmitic acid (Figure 4), protein E with capric acid ligands, lauric acid, linoleic acid, myristic acid, and oleic acid (Figure 5), as well as the spike protein which binds to the ligands of capric acid, lauric acid, myristic acid, oleic acid and palmitic acid (Figure 6).

Unfavorable Donor to Donor bond is a bond that shows the repulsive force between two molecules. The formation of this bond can reduce the stability of other types of bonds, affecting the stability of the ligands that will be used as drug candidates [37].

According to [38], the binding site area of the 3CLpro enzyme is located between the aspartate catalytic residue, which contains other supporting sub-units, the residue is located at Asp-A: 25, Thr-A: 26, Gly- A: 27 and His-A: 41. The linoleic acid ligand that binds to these receptors precisely binds to one of these residues. This proves that the ligand is predicted to be able to inhibit the performance of this receptor with maximum results.

Spike glycoprotein (S protein) receptors, both in the form of closed state (6VXX) and open state (6VYB) are receptors that are associated with the human ACE2 (hACE2) receptors; thus they do not have a target structure equipped with inhibitors in the Protein Data Bank (PDB). Inhibition occurs on the surface between the two receptors (S protein and hACE2) so that the binding site area is no longer at the spike glycoprotein receptors but between the two receptors [39,40]. Therefore, the blind docking method for molecular docking is used for the S protein receptor and as a potential receptor target for the discovery of new types of drugs [39].

5. Conclusion

GC-MS analysis showed that the PUO and EPUOE contained fatty acids (according to the order of Table 1) with oleic acid as a dominant substance. Based on the results of molecular docking, the poses with the highest affinity (most negative Gibbs free energy), are the linoleic acid fatty acids interacting with the 3CLpro receptor (PDB code: 6LU7), Spike protein (PDB code: 6VXX), PLpro (PDB code: 6WX4), RdRp (PDB code: 6M71), protein E (PDB code: 5X29), and Spike Ectodomain Structure (PDB code: 6VYB) with the affinity binding of -4,9, -5,8, -4,7, -4.3, -5.3, and -5.5 kcal/mol, respectively.

6. Reference

[1] Hui D S, I Azhar E, Madani T A, Ntoumi F, Kock R, Dar O, Ippolito G, Mchugh T D, Memish Z A, Drosten C, Zumla A, Petersen E 2020 *Int. J. Infect. Dis.,* **91** 264–266.
[2] Huang C, Wang Y, Li X, Ren L, Zhao J, Hu Y, Zhang L, Fan G, Xu J, Gu X, Cheng Z, Yu T, Xia J, Wei Y, Wu W, Xie X, Yin W, Li H, Liu M, Xiao Y, Gao H, Guo L, Xie J, Wang G, Jiang R, Gao Z, Jin Q, Wang J, Cao B 2020 *Lancet.* **395** 497–506.
[3] Liu X, Wang X J 2020 *bioRxiv* 2020 01.29.924100
[4] Estevam E, Griffin S, Nasim J, Zieliński D, Aszyk J, Osowicka M, Dawidowska N, Idroes R, Bartoszek A, Jacob C 2015 *Nat. Prod. Commun.,* **10** 1733–1738.
[5] Ningsih D S, Idroes R, Bachtiar B M, Khairan 2019 *IOP Conference Series: Mater. Sci. and Eng.,* **523** 12009.
[6] Nuraskin C A, Marlina, Idroes R, Soraya C, Djufri 2019 *IOP Conference Series: Mater. Sci. and Eng.,* **523.**
[7] Nuraskin C A, Marlina M, Idroes R, Soraya C, Djufri D 2020 *Open Access Maced. J. Med. Sci.,* **8** 181–184.
[8] Rahmad R, Earlia N, Nabila C, Inayati I, Amin M, Prakoeswa C R S, Khairan K, Idroes R 2019 *IOP Conference Series: Mater. Sci. and Eng.,* **523** 012011.
[9] Pratiwi S U T, Lagendijk E L, de Weert S, Idroes R, Hertiani T, den Hondel C Van 2015 *Int. J. Appl. Res. Nat. Prod.,* **8** 1–13.
[10] Suhartono E, Setiawan B, Santosa P B, Idroes R, Indrawan M S 2019 *J. Phys. Conf. Ser.,* **1374** 12057.
[11] Nasution R, Idroes R, Amna U, Al E 2020 *Open Access Maced. J. Med. Sci.*
[12] Mohamed S M, Hassan E M, Ibrahim N A 2010 *Nat. Prod. Res.,* **24** 1395–1402.
[13] Tallei T E, Pelealu J J, Pollo H N, Pollo G A V, Adam A A, Effendi Y, Karuniawan A, Rahimah S, Idroes R 2019 *Data in Brief.,* 104681.
[14] Nuraskin C, Marlina, Idroes R, Soraya C, Djufri 2020 *Rasayan J. Chem.,* **13** 18–23.
[15] Tallei T E, Linelejan Y T, Umboh S D, Adam A A, Muslem, Idroes R 2020 *IOP Conference*

Series: Mater. Sci. and Eng., **796** 012047.
[16] Zulfendi, Idroes R, Khairan 2019 *IOP Conference Series: Mater. Sci. and Eng.* **523** 12013.
[17] Paristiowati M, Moersilah M, Stephanie M M, Zulmanelis Z, Idroes R, Puspita R A 2019 *J. Phys. Conf. Ser.,* **1402** 55041.
[18] Rahimah S, Malinda W, Zaida, Sukri N, Salma J K, Tallei T E, Idroes R 2020 Sci. World J., **2020** 1–9.
[19] Salimon J, Mohd Noor D A, Nazrizawati A T, Mohd Firdaus M Y, Noraishah A 2010 *Sains Malays.,* **39** 761–764.
[20] Arpi N 2013 *Jurnal Teknologi Industri Pertanian Indonesia.* **5** 17–22.
[21] Earlia N, Muslem, Suhendra R, Amin M, Prakoeswa C R S, Khairan, Idroes R 2019 *Sci. World J.,* **2019** 1–7.
[22] Earlia N, Rahmad R, Amin M, Prakoeswa C, Khairan K, Idroes R 2019 *Sains Malays.* **48** 1019–1024.
[23] Zhao L, Chen Y, Wu K, Yan H, Hao X, Wu Y 2017 *Pestic. Biochem. Phys.,* **139** 87–91.
[24] Schoggins J W, Randall G 2013 *Cell Host Microbe.* **14** 379–385.
[25] Wu C, Liu Y, Yang Y, Zhang P, Zhong W, Wang Y, Wang Q, Xu Y, Li M, Li X, Zheng M, Chen L, Li H 2020 *Acta Pharm. Sin. B.*
[26] Suhartono E, Thalib I, Aflanie I, Noor Z, Idroes R 2018 *IOP Conference Series: Mater. Sci. and Eng.,* **350** 12008.
[27] Suhartono E, Noor Z, Edyson, Budianto W Y, Idroes R 2019 *AIP Conf. Proc.,* **2108** 020025.
[28] Morris G M, Lim-Wilby M 2006 *Encyclopedic Reference of Genomics and Proteomics in Molecular Medicine.,* **443** 1149–1153.
[29] O'Boyle N M, Banck M, James C A, Morley C, Vandermeersch T, Hutchison G R 2011 *J. Cheminformatics,* **3** 33.
[30] Trott O, Olson A J 2010 *J. Comput. Chem.,* **31** 455–461.
[31] Mardianingrum R, Herlina T, Supratman U 2015 *Chim. Nat. Acta* **3**. *(in Bahasa Indonesia).*
[32] Lins L, Brasseur R 1995 *FASEB Journal.,* **9** 535–540.
[33] Zubair M S, Maulana S, Mukaddas A 2020 *Jurnal Farmasi Galenika,* **6** 132–140. *(in Bahasa Indonesia).*
[34] Arwansyah A, Ambarsari L, Sumaryada T I 2014 *Curr. Biochemis. J.,* **1** 11–19.
[35] Sharp K A, Honig B 1990 *Annu. Rev. Biophys.,* **19** 301–332.
[36] Guan W, Ni Z, Hu Y, Liang W, Ou C, He J, Liu L, Shan H, Lei C, Hui D S C, Du B, Li L, Zeng G, Yuen K-Y, Chen R, Tang C, Wang T, Chen P, Xiang J, Li S, Wang J, Liang Z, Peng Y, Wei L, Liu Y, Hu Y, Peng P, Wang J, Liu J, Chen Z, Li G, Zheng Z, Qiu S, Luo J, Ye C, Zhu S, Zhong N 2020 N. *Engl. J. Med.*, **382** 1708–1720.
[37] Dhorajiwala T M, Halder S T, Samant L R 2019 *Biomed. Biotechnol. Res. J.,* **50** 553–558.
[38] Brik A, Wong C H 2003 *Org. Biomol. Chem.* **1** 5–14.
[39] Narkhede R R, Cheke R S, Ambhore J P, Shinde S D 2020 S*creening.* **5** 8.
[40] Walls A C, Park Y-J, Tortorici M A, Wall A, McGuire A T, Veesler D 2020 *Cell.* **181** 281-292.

Identification of Gayo arabic coffee beans and powder using the FTIR-PCA combination method

Z Akbar[1], R Idroes[1,2*] B Ginting[1], T Karma[1], S Rahimah[3], Z Helwani[4], M Yusuf[1],
[1]Department of Chemistry, Faculty of Mathematics and Natural Sciences, Universitas Syiah Kuala, Banda Aceh 23111, Indonesia
[2]Department of Pharmacy, Faculty of Mathematics and Natural Sciences, Universitas Syiah Kuala, Banda Aceh 23111, Indonesia
[3]Department of Food Industrial Technology, Faculty of Agroindustrial Technology Universitas Padjadjaran, Jatinangor 54363, Indonesia
[4]Department of Chemical Engineering, Faculty of Engineering, Universitas Riau, Kampus Bina WidyaPanam, Pekanbaru, Indonesia 28293

*E-mail: rinaldi.idroes@unsyiah.ac.id

Abstract. The identification of Gayo Arabica coffee beans and coffee powder has been done to classify coffee plantation from several different regions using FTIR Spectroscopy and Principal Component Analysis (PCA). Samples were taken by purposive sampling from the Arabica coffee plantations in Pegasing, Jagong, and Celala areas,Aceh Tengah Regency, and also in Bandar, Permata, and Wih Pesam areas, Bener Meriah Regency. FTIR analyzed both the dry Gayo Arabica coffee beans and powder in the wavelength range of 4000 cm^{-1}-400 cm^{-1}. The spectra resulted were analyzed using PCA. FTIR spectra show the absorption of typical functional groups of caffeine, namely O-H (3400 cm^{-1}), C-H aromatic (2900 cm^{-1}), C-H aliphatic (2800 cm^{-1}), C=O (1743 cm^{-1}), C=C (1640 cm^{-1}), C=C aromatic (1550 cm^{-1}), C-H alkanes (1450 cm^{-1}) and C-N (1240 cm^{-1}). There was a loss of absorption of C=O and C-N groups in coffee powder samples from the Pegasing, Jagong, and Permata areas caused by high-temperature heating during the coffee powder making process. The PCA showed that coffee bean samples could be distinguished properly based on the coffee origin location, namely Aceh Tengah and Bener Meriah Regencies. Meanwhile, coffee powder samples showed poor separation of PCA plot patterns between coffee powder from the two regions. It can be concluded that the PCA method can be used for the classification of both Gayo Arabica coffee beans and powder, where the classification of beans is better than powder.

1. Introduction

Coffee is an agricultural product with high economic value and is quite popular with many people worldwide. Many types of coffee can be found on the market, one of which is Gayo Arabica coffee. There are two central areas of Gayo Arabica coffee plantations in Aceh Province, namely Aceh Tengah and Bener Meriah Regencies. GayoArabica coffee types are suitable to grow in areas with an altitude range of 900-1700 MASL [1].

In general, coffee contains various types of volatile compounds such as aldehydes, furfural, ketones, alcohols, esters, formic acid, and acetic acid. The caffeine compound ($C_8H_{10}N_4O_2$) is an alkaloid compound that is very important in coffee. Although similar in type of content, each coffee's detailed chemical composition varies depending on the geographic factors, processing, and storage. This factor is also assumed to affect the taste of the coffee.

The identification of chemical composition in the organic sample can be made by several instrumental methods such as Uv-Vis [2–4], Chromatography-Mass Spectrometry [5–8], FTIR [9,10],laser[11,12], and even image processing[13].In identifying the chemical composition of coffee, the FTIR (Fourier Transform Infrared) spectroscopy method is the most commonly chosen, considered by the simple used and low cost. This method can identify the functional groups in coffee, enabling the chemical composition to be predicted [14].

However, FTIR spectra are so complex, complicating the differentiation between one coffee spectra and another. This makes it difficult for the coffee business to identify the possibility of knock-offs coffee. Therefore we need a statistical method that can identify complex data [15–18]. One of the popular statistical methods used to identify complex data is Principle Component Analysis (PCA)[19]. The combination of FTIR and PCA has been widely used in various fields, for example, to identify between Wheat Kernel and White Flour [20], falsification of Edible bird's nests (EBNs) as a medicinal ingredient[21], differences in animal hair as raw material for making brushes[22]distinguishing of cattle bones [23]and also in the forensic analysis [24], as well as differentiating between Arabica and Robusta coffee [25].

In this study, FTIR analysis was carried out from coffee bean and powder samples taken from several BenerMeriah locations and Central Aceh. The resulting FTIR spectra were then analyzed by PCA to classify the coffee samples based on location and treatment.

2. Research methodology

Samples of Gayo Arabica coffee beans and powder were taken from the plantation areas of Aceh Tengah (Pegasing, Jagong, and Celala areas) and Bener Meriah (Bandar, Permata, and Wih Pesam Areas). Sampling was done by purposive sampling. The pretreatment of the obtained coffee beans was done by exfoliating, washing, and drying. Coffee powder samples were pretreated by roasting the dry coffee beans until they were evenly black and grounded until they became coffee powder.

2.1. Tools and materials

The equipment used was the maceration toolset, rotary evaporator, FTIR Spectroscopy (Shimadzu 50-IR Thermo Fischer), and statistical applications for FTIR data processing; The Unscrambler X 10.4.

2.2. FTIR Spectroscopic analysis

FTIR analysis was carried out at a wavelength of 400 – 4000 cm^{-1}. Measurements were made in two repetitions to obtain a good spectrum. The spectrum data obtained were prepared for the chemometric purpose.

2.3. Preprocessing data and chemometric analysis

FTIR spectra were analyzed by The Unscrambler X 10.4 application to obtain the statistical information. Firstly, data preprocessing was needed to eliminate variations in data that were not related to analytical information [27]. The data preprocessing in this research was carried out by using The Unscrambler X 10.4 as the same as has been done by Van Horn et. Al. [28].

2.4. Principal component analysis (PCA)

FTIR spectra that have preprocessed were analyzed by The Unscrambler X 10.4 application to obtain the PCA plot.The plot was interpreted based on the grouping formed.

3. Results and discussion

3.1. FTIR spectrum

FTIR spectra of all samples can be seen in Figure 1 and Figure 2. In general, all the samples have similarities in functional groups that containedO-H (3400 cm^{-1}), C-H aromatic (2900 cm^{-1}), C-H aliphatic (2800 cm^{-1}), C=O (1743 cm^{-1}), C=C (1640 cm^{-1}), C=C aromatic (1550 cm^{-1}), C-H alkanes (1450 cm^{-1}) and C-N (1240 cm^{-1}). The differencesare only in the intensity and a little bit on the spectral patterns.This functional group's presence indicated the existence of a well-known compound found in coffee, namely caffeine compounds, with one characteristic of it being present in a C-N functional group. The other functional group detected was C-O at 1010 cm^{-1}that indicated the presence of alcohol, ether, ester, and carboxylic acid [29].

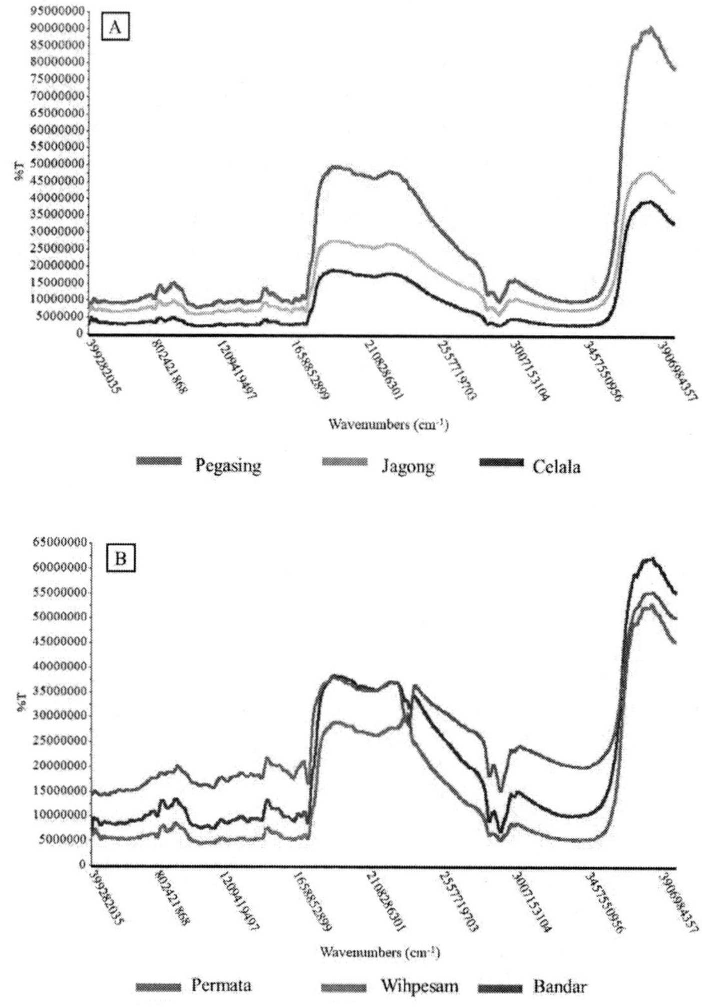

Figure 1. FTIR spectra of coffee powder from: (A) Aceh Tengah, and (B) Bener Meriah.

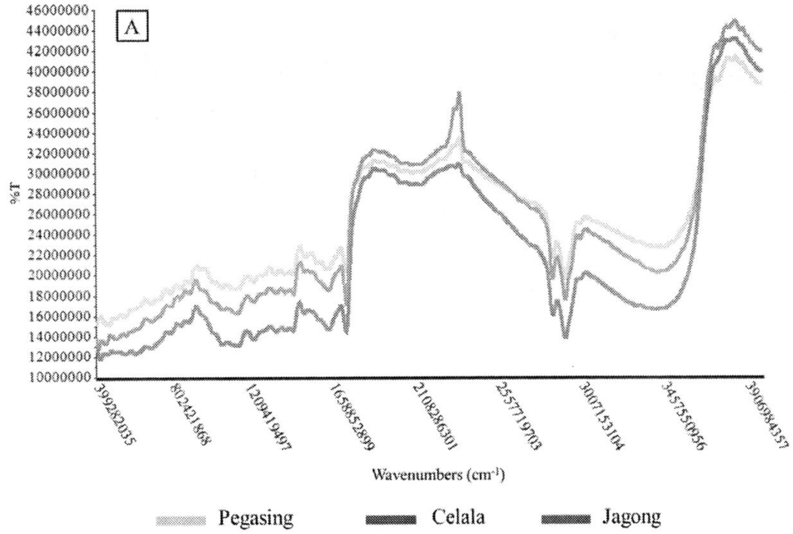

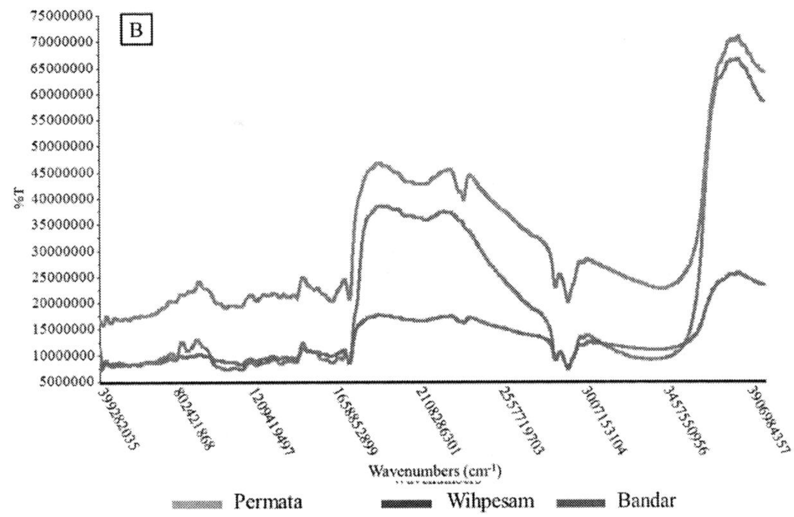

Figure 2. FTIR Spectra of coffee beans from: (A) Aceh Tengah, (B) Bener Meriah.

There was a loss of absorption of C=O and C-N groups in coffee powder samples from the Pegasing, Jagong, and Permata. This may due to the heating process in the sample during the coffee sample treatment process[30], resulting in the breakdown of the C-N and C-O spectra. Despite that, the C-N and C-O functional groups may still be presented in the coffee powder samples, but the concentration has decreased, and the absorbance peak looks like invisible.

3.2. Pretreatment/preprocessing of the FTIR spectrum data
FTIR spectrum data requires a pretreatment to reduce noise caused by external influences, multiplication effects (scattering), particle size, and multi-collinearity changes that can cause large variations in the

spectrum. Thus, it is impertinent to apply the pretreatment method to avoid noise effects [31]. According to Savitzky and Golay [27], the FTIR spectrum is corrected by several data pretreatment methods, namely Smoothing, Baseline Correction, Derivative, and Standard Normal Variant (SNV).

3.3. Principal component analysis (PCA)

The principal component analysis aims to classify samples.In this study, the classification was based on the geographical origin of the coffee. Figure 3 showed that coffee beans and powder samples could be distinguished based on the coffee's location of origin, namely Aceh Tengah and Bener Meriah. However, the coffee powder samples overlap in two regions (blue and yellow circles), so the classification results were not as good as the coffee bean samples (red and black circles). Figure 3 shows that identifying the origin of coffee is better to use coffee beans as samples. The use of coffee bean samples as analysis material certainly has many advantages, like not having to do pretreatment to many samples, thus saving time and money.

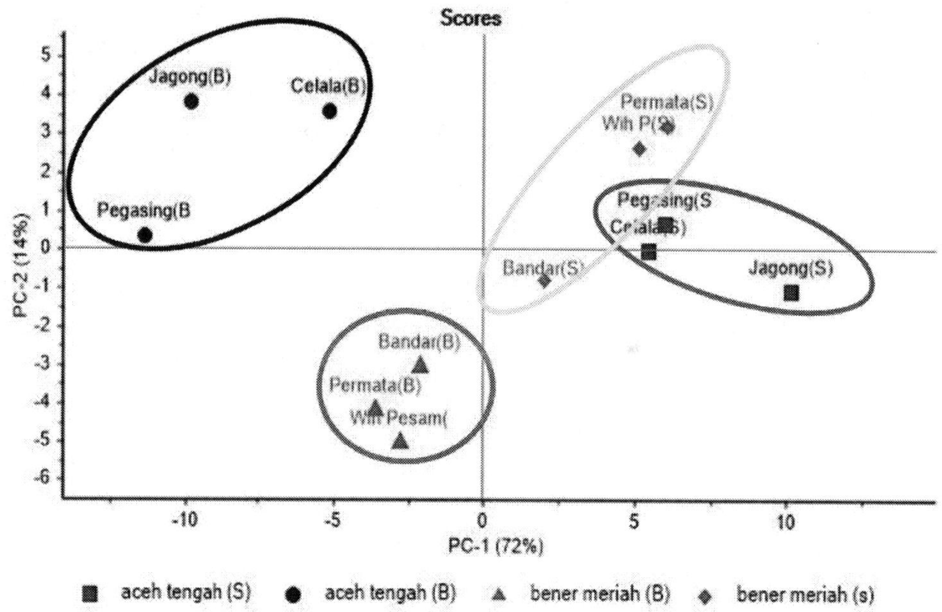

Figure 3. Classification of PCA plots by regencies.

4. Conclusions

Identification of both Gayo Arabica coffee beans and powder by FTIR showed the presence of an identical functional group of caffeine and several other compounds of alcohol, ether, ester, and carboxylic acid. There was a loss of absorption of C=O and C-N groups in coffee powder samples from the Pegasing, Jagong, and Permata areas caused by high-temperature heating during the coffee powder making process. The PCA showed that coffee bean samples could be distinguished adequately based on the coffee origin location, namely Aceh Tengah and Bener Meriah Regencies. Meanwhile, coffee powder samples showed poor separation of PCA plot patterns between coffee powder from the two regions. It can be concluded that the PCA method can be used for the classification of both Gayo Arabica coffee beans and powder, where the classification of beans is better than powder.

5. References

[1] Mawardo S, Hulupi R, Wibawa A, Wiryaputra S 2008 *Panduan Budidaya dan Pengolahan Kopi Arabika Gayo (Aceh Kopi Forum). (in Bahasa Indonesia).*
[2] Suhartono E, Thalib I, Aflanie I, Noor Z, Idroes R 2018 *IOP Conf. Ser. Mater. Sci. Eng.,* **350** 12008.
[3] Hasanah U, Sani N D M, Heng L Y, Idroes R, Safitri E 2019 *Biosensors,* **9** **(4)** 135.
[4] Hasanah U, Setyowati M, Efendi R, Muslem M, Md Sani N D, Safitri E, Yook Heng L, Idroes R 2019 *Biosensors,* **9** 60.
[5] Estevam E, Griffin S, Nasim J, Zieliński D, Aszyk J, Osowicka M, Dawidowska N, Idroes R, Bartoszek A, Jacob C 2015 *Nat. Prod. Commun.,* **10** 1733–1738.
[6] Pratiwi S U T, Lagendijk E L, de Weert S, Idroes R, Hertiani T, den Hondel C Van 2015 Int. J. *Appl. Res. Nat. Prod.,* **8** 1–13.
[7] Earlia N, Muslem, Suhendra R, Amin M, Prakoeswa C R S, Khairan, Idroes R 2019 *Sci. World J.,* **2019** 1–7.
[8] Earlia N, Rahmad R, Amin M, Prakoeswa C, Khairan K, Idroes R 2019 *Sains Malays.,* **48** 1019–1024.
[9] Suhartono E, Noor Z, Edyson, Budianto W Y, Idroes R 2019 *AIP Conf. Proc.,* **2108** 020025.
[10] Hasanah U, Setyowati M, Edwarsyah, Efendi R, Safitri E, Idroes R, Heng L Y, Sani N D 2019 *IOP Conf. Ser. Mater. Sci. Eng.,* **523** 12014.
[11] Hedwig R, Lahna K, Idroes R, Karnadi I, Tanra I, Iqbal J, Kwaria D, Kurniawan D P, Kurniawan K H, Tjia M O, Kagawa K 2019 *Microchem. J.,* **147** 356–364.
[12] Lahna K, Idroes R, Idris N, Abdulmadjid S N, Kurniawan K H, Tjia M O, Pardede M, Kagawa K 2016 *Appl. Opt.,* **55** 1731.
[13] Suhartono E, Setiawan B, Santosa P B, Idroes R, Indrawan M S 2019 *J. Phys. Conf. Ser,* **1374** 12057.
[14] Chatwal G R 2009 *Spectroscopy.* Bengaluru: Himalaya Publishing House.
[15] Idroes R, Husna I, Muslem, Mahmudi, Rusyana A, Helwani Z, Idroes G M, Suhendra R, Yandri E, Rahimah S, Sasmita N R 2019 IOP Conf. *Ser.:* Earth *Environ. Sci.,* **364** 12020.
[16] Husna I, Rusyana A, Muslem, Idroes G M, Suhendraand R, Idroes R 2020 *IOP Conf. Ser. Mater. Sci. Eng.,* **796** 012064.
[17] Idroes R, Japnur A F, Suhendra R, Rusyana A 2019 *IOP Conf. Ser. Mater. Sci. Eng.,* **523** 12007.
[18] Idroes R, Noviandy T R, Maulana A, Suhendra R, Sasmita N R, Muslem M, Idroes G M, Irvanizam I 2019 *Int. Rev. Modell. Simul.,* **12** 373.
[19] Wahidah S, Khairi, Lelifajri, Idroes R, Rahmadi, Lala A, Mahmudi, Muslem, Japnur A F 2019 *IOP Conf. Ser.: Earth Environ. Sci.,* **364** 12016.
[20] Wadood S A, Guo B, Zhang X, Wei Y 2019 *J. Food Sci. Technol.,* **54** *2045-54*
[21] Guo L, Wu Y, Liu M, Ge Y, Chen Y 2018 *J. Sci. Food. Agric.,* **98** 3057–65.
[22] Rafi M, Anggundari W C, Irawadi T T 2016 *Indones. J. Chem.,* **5** 229–234. *(in Bahasa Indonesia).*
[23] Nadia Y, Ramli M, Japnur A F, Rusyana A, Idroes G M, Suhendra R, Sasmita N R, Tallei T E, Idroes R 2019 IOP Conf. *Ser.:* Earth. *Environ. Sci.*, **364** 12040.
[24] Kumar R, Sharma V 2018 *Trends Anal. Chem.,* **105** 191–201.
[25] Sambudi S 2018 *Identifikasi keaslian kopi robusta dekafeinasi menggunakan teknologi uv-vis spectroscopy dan kemometrika.* Lampung: Universitas Lampung. *(in Bahasa Indonesia).*
[26] Departemen Pertanian 2009 *Pedoman Umum Pengembangan Usaha Agribisnis Perdesaan (PUAP).* Jakarta: Departemen Pertanian. *(in Bahasa Indonesia).*
[27] Savitzky A, Golay M J E 1964 *Anal. Chem.* **36** 1627–1639.
[28] Van Horn E, Kearney S P 2013 *Method of and system for reading visible and/or invisible code symbols in a user-transparent manner using visible/invisible illumination source switching during data capture and processing operations*. U.S. Patent 8, 408, 468.
[29] Mees C, Souard F, Delporte C, Deconinck E, Stoffelen P, Stévigny C, Kauffmann J-M, De Braekeleer K 2018 *Talanta.* **177** 4–11.
[30] Obeidat S M, Hammoudeh A Y, Alomary A A 2018 *J. Appl. Spectrosc.*, **84(6)** 1051- 55.

[31] Day R Al, Underwood A L 2002 *Analisis Kimia Kuantitatif Edisi Keenam.* Jakarta: Erlangga. *(in Bahasa Indonesia).*

The impact of solvent concentration on the characteristic of silica from rice husk ash using sol gel method

I Zuwanna[1], M Riza[1, 2,*], S Aprilia[1]
[1]Chemical Engineering Department, Syiah Kuala University, Banda Aceh, Indonesia, 23111

*E-mail: medyan_riza@unsyiah.ac.id

Abstract. Synthesis of silica particle was conducted using the sol gel method from Aceh Besar rice husk ash. Sol gel method is an environmental-friendly and easy process to extract silica particles (SiO_2) that consists of two stages of hydrolysis and polycondensation. The aim of this research was to analyse the influence of variations in the solvent concentration (1 N, 1.5 N, 2 N, 2.5 N and 3 N) using the sol gel method on the characteristics of silica particle. The rice husk ash was refluxed using solvent (sodium hydroxide) and the acidification process was using hydrochloric acid (HCl). The result of this study indicated that the highest yield in treatment was using 3 N NaOH, which was 39.66 percent. Based on XRD result, the silica was in the crystalline phase and had a particle size of 13,59-41,56 nm and functional groups of silanol (Si-OH) and siloxane (Si-O-Si). Based on the result of SEM characterization, uneven and agglomerated particle shapes were obtained.

1. Introduction

The nanotechnology expands continuously, and inspires researchers to put more effort in the nanotechnology field. The new technology is to use rice husk waste produced from rice milling. Rice husk ash is an agricultural residue that is abundant, cheap and able to produce silica and nano-silica when it was processed [1]. The properties of rice husks are more unique compared to other agricultural waste due to the high content of silica as well as the high porosity and high surface area which make them valuable for industrial use [2].

There are two methods in synthesizing rice husk ash silica, namely thermal treatment at a temperature of 500-1400 °C, which involves high temperatures and leaching using an acid or base solution, and being neutralized using an acid to make silica. The second method is an environmental-friendly method since it uses low energy and costs, CO_2 emissions from the Na_2CO_3 and SiO_2 reaction can be minimized more compared to the smelting method obtained during the process of making sodium silicate [3].

Due to their low price and simple process, nano-silica particles are used as raw materials in chemical processing, thickening fluids, improving surfactant efficiency, drug delivery systems, and fillers [4]. Nano-silica from palm kernel ash (PKSA) was extracted using the sol gel process [5].

Sol-gel technology has been used to manufacture biopolymer-silica hybrid materials with potential properties, stability of living matter and applicability in biomimetic processes. Preparation of this hybrid material is started by a regulated process of hydrolysis and condensation through the use of alkoxysilane such as tetraethoxysilane in alcohol and other polar solvents such as solution or colloid polymer suspension [6]. The sol-gel system consists of a silica hydrolysis process followed by a polycondensation reaction cycle [7]. Rice husk ash (RHA) pre-treatment with acids or alkali raises SiO_2 (>98 percent). Alkali extraction is an economical method that is capable of synthesizing nano-silica or high-purity silica and can be commonly used [8]. The aim of this research was to investigate the impact of variations in the alkaline solvents concentration using the sol gel process on the yield and structure of silica.

2. Materials and methods

2.1. Materials

Rice husk ash was obtained from rice mills in Aceh Besar and the residue of burning rice husk was used as raw material. Rice husk ash was crushed with mortar and sieved with 80 mesh sieves. The materials used are hydrochloric acid (HCl) p.a, sodium hydroxide (NaOH), distillate water, whattman filter paper and pH meter .

2.2. Extraction of silica

Twenty grams of rice husk ash was added to a three-cock flask and 200 mL of NaOH was added with the concentration of 1 N, 1.5 N, 2 N, 2.5 N and 3 N respectively. The solution was refluxed for 1 hour at 80 °C and was stirred continuously. Silica dissolves and produces sodium silicate solution (Na_2SiO_3) at this point. The solution was filtered using the filter paper Whatman No. 41 and the solution was cooled down at room temperature. The filtrate was then titrated using 3 N HCl with regular stirring up to pH 7. Silica gel starts to settle when the pH is < 10 and white silica gel is obtained. The silica gel shaped was set aside for 18 hours, then filtered and heated in the oven at a temperature of 120 °C for 12 hours to extract moisture and create xerogel. The xerogel obtained washed using 100 ml of distilled water and stirred for 15 minutes at temperature of 80 °C. The next step was to filter the mixture and heat the precipitate obtained in the oven for 4 hours at a temperature of 120 °C. Silica was crushed with a mortar and then placed in a jar for the Fourier Transform Infrared (FTIR) test to analyse the functional groups and the X-ray diffraction (XRD) to determine the structure of the silica. Figure 1 Shows the process of extracting silica using the Sol gel method.

Figure 1. Sol gel process.

3. Results and discussion

3.1. Impact of NaOH concentration on silica extraction

Rice husk ash has been dissolved using NaOH solvent with different concentrations of 1 N, 1.5 N, 2 N, 2.5 N and 3 N. A compound formed between silica ash and NaOH becomes a solution of sodium silicate. The sodium silicate solution is reacted with HCl to produce silica gel. Extraction of silica from rice husk ash using the sol gel method was achieved as the following reaction.

$$Na_2SiO_3 + 2\ HCl \rightarrow SiO_2 + 2\ NaCl + H_2O \quad (1)$$

$$SiO_2 + 2\ NaOH \rightarrow Na_2SiO_3 + H_2O \quad (2)$$

Reaction 1 shows that NaOH reacts with silica from rice husk ash and forms sodium silicate. Furthermore, sodium silicate forms silica acid when the pH becomes <10. The precipitate obtained is dried in an oven and the dry silica gel is crushed with a mortar. The yield of silica obtained from rice husk ash to silica powder following a chemical processes can be seen in Table 1. Treatment of variations in solvent concentration has been given in order to obtain the highest yield. It can be seen that there is an increase in yield at each degree of NaOH concentration. This is because more silica decays as the NaOH concentration increases [9].

Table 1. Silica yield of rice husk ash.

No.	Solvent (N)	Rice husk ash (gr)	Silica (gr)	Yield (%)
1	1	20	0.761	3.805
2	1.5	20	0.879	4.395
3	2	20	2.010	10.050
4	2.5	20	4.999	24.999
5	3	20	7.932	39.660

3.2. Characterisation of silica

3.2.1. Silica particle structure analysis using XRD (X-Ray Diffractometer). The x-ray diffractometer pattern results in Figure 2. illustrate the diffraction pattern of rice husk ash from Aceh Besar rice mills. The diffraction pattern formed at an angle indicated that the peaks of 2θ = 21-23° were indicated by the presence of silica compounds [10]. The diffraction pattern shows a broad, gently sloping hill. Sloping hills are silica with an amorphous structure and crystalline.

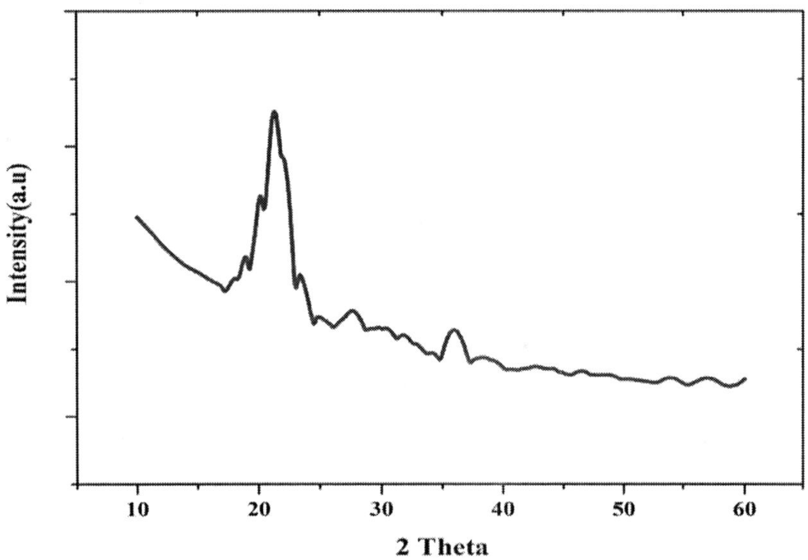

Figure 2. XRD diffraction pattern for rice husk ash.

XRD analysis was chosen to obtain a diffraction pattern that is useful for analysing the changes in crystal structure. Figure 3 shows the diffraction pattern of SiO_2 particle powder extracted using the sol gel process using different of NaOH concentrations.

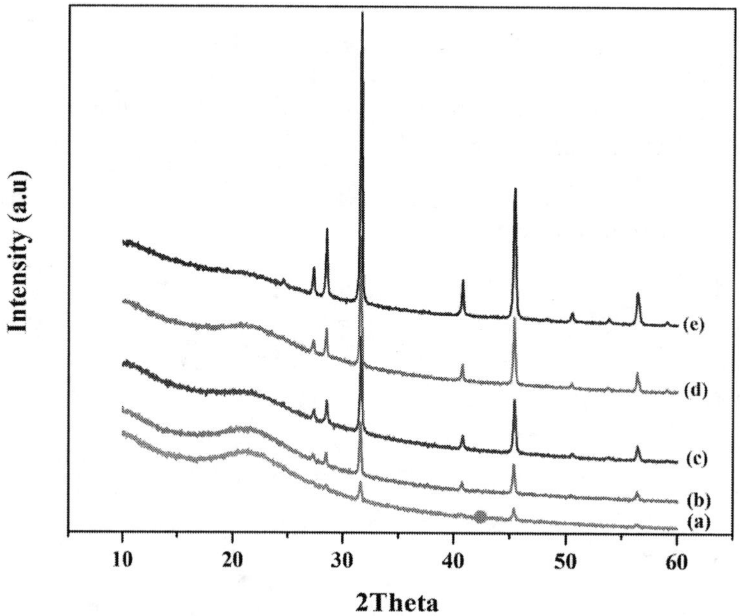

Figure 3. XRD silica diffraction patterns dependent on variations in solvent concentrations (a) 1 N, (b) 1.5 N (c) 2 N, (d) 2.5 N, (e) 3 N.

Figures 3a, b, c, d, and e display the diffraction pattern coordinates about 20.93 °, 21.93 °, 28.54 °, 31.62 °, and 45.41 ° respectively. By using the Debye equation, the silica particle size was between 13.59 nm and 41.56 nm. From this figure, it can be seen that the higher the concentration of NaOH, the higher the diffraction pattern, and silica contains the crystal phase. Peak in the XRD pattern was caused in the chemical pre-treatment process, which was involved in the extraction process and was deposited in the gelatine phase [11]. Due to the small SiO_2 particle size, the XRD peaks are large, suggesting that a high percentage of these particles are formless in all samples and are in the crystal phase according to XRD data [12]. Based on the XRD characterization, it can be inferred that the concentration variance significantly affects the diffraction peaks and size silica .

3.2.2. Silica functional group analysis by Fourier Transfer Infra Red (FTIR). Silica of rice husk ash was analyzed in the 500-4000 cm^{-1} wavelength range. The FTIR spectrum of rice husk ash is shown in Figure 4. The flexural vibration of the O–Si–O bond appeared at 474.49cm^{-1}. The wide band about 3618.46 cm^{-1} belongs to the vibrational stretching of O-H bonds in water molecules, bound by hydrogen bonds or O-H groups expressed in organic compounds [13].

Figure 5 shows the FTIR spectrum of industrial silica and rice husk ash added with the differences of NaOH concentration. The results show that chemical compositions of silica products obtained from various NaOH concentration was similar to each other, namely the silanol (Si-OH) and siloxane (Si-O-Si) groups. The FTIR spectrum in Figure 5 shows that the siloxane (Si-O-Si) group is present at wave numbers 457.29 cm^{-1} and 1074.76 cm^{-1}. The wave number 806.87 reflects the formation of the Si-O group, and the wave band 3650.08 cm^{-1} is the O-H silanol group due to the presence of water molecules adsorbed to silica, and part (a) indicates that industrial silica has the same functional group as silica particles from rice husk ash. The addition of NaOH concentration causes the range of O-H groups to be wider, where the higher the NaOH concentration the more OH groups found in the silica. The increase in O-H groups can affect the number silanol (Si-OH) groups in silica. As the silanol (Si-OH) increases the number of silica particles, it is easier to react with other substances [14].

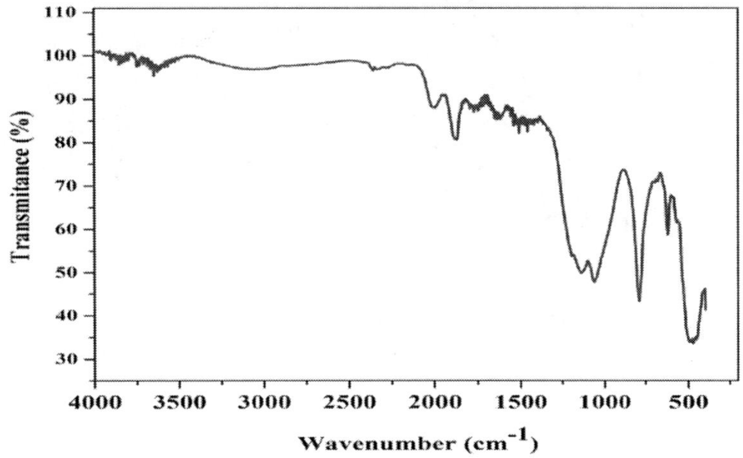

Figure 4. FTIR spectrum of rice husk ash.

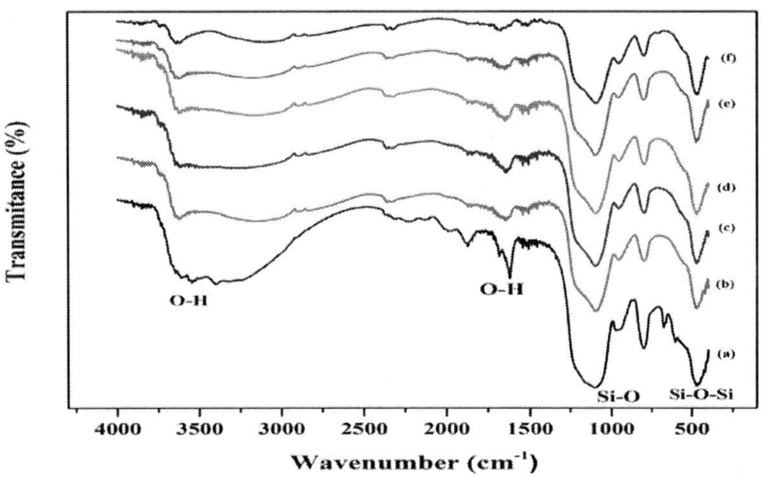

Figure 5. Silica wavenumber: (a) industrial silica, (b) 1 N, (c) 1.5 N, (d) 2 N, (d) 2.5 N, (e) 3 N.

3.2.3. Scanning electronic microcope analysis (SEM). Figure 6 shows the morphology of rice husk ash formed like hill. Rice husk ash, which had been burned at a temperature of 600 °C and had a structure that resembled hills without chemically impregnating [15]. The morphological characteristics of silica with different solvents are shown in Figure 7. Figure 7a-b reveals an irregular particle shape consisting of grouping and clustering patterns in Figure 7c-e to form clusters. The differences of solvent concentration did not affect the morphology of silica particles. Size of spherical silica nanoparticles varies between 10 nm and 60 nm, even after calcination and purification, the morphology of nano-silica remained unchanged [16]. This is due to the hygroscopic nature of silica (easy to absorb water) and the presence of a fairly diverse grain size with uneven surface distribution.

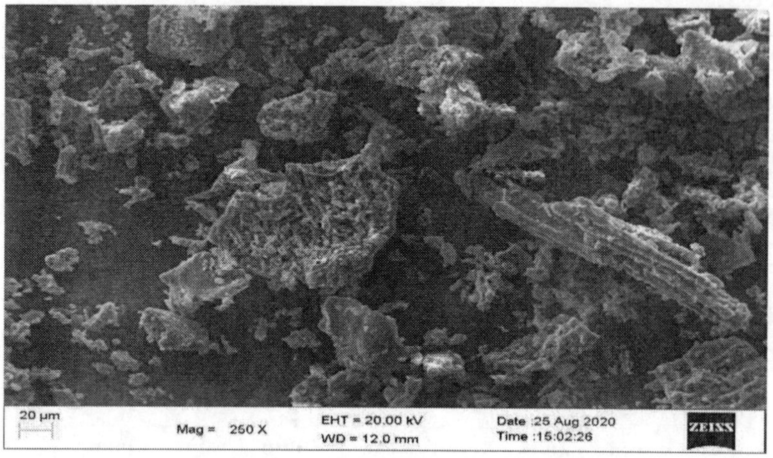

Figure 6. Morphology of rice husk ash.

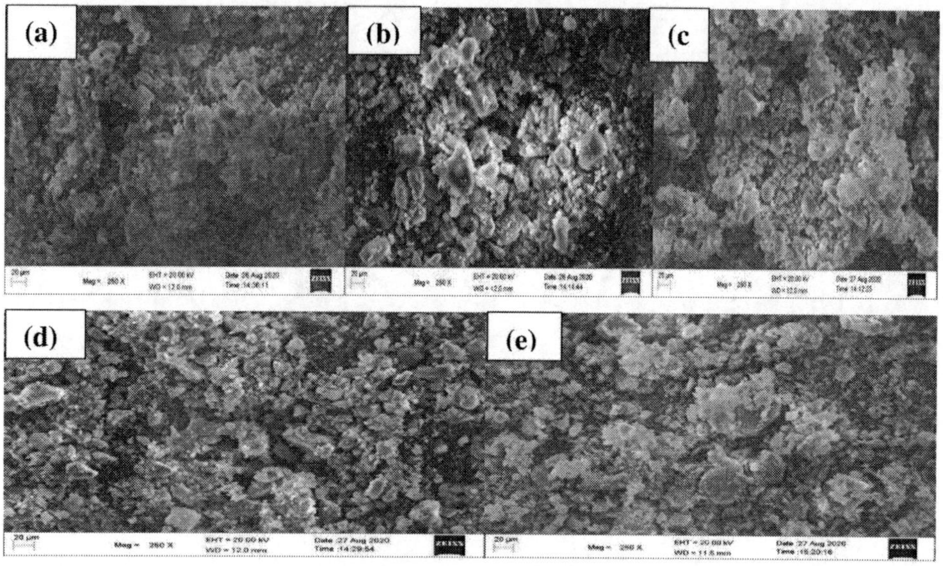

Figure 7. Silica morphology in different solvent: (a) 1 N, (b) 1.5 N, (c) 2 N, (d) 2.5 N, (e) 3 N.

4. Conclusion

The NaOH concentration impacts the yield of silica particles which can be synthesized from rice husk ash. The higher solvent concentration, the higher the yield produced. The highest yield for dissolving 3 N NaOH solution was 39.66 percent. Silica was in the crystalline phase and has a particle size of 13,59-41,56 nm. The result of the FTIR analysis showed that functional groups of silanol (Si-OH) and siloxane (Si-O-Si) continued to increase with the NaOH concentration, making it easier for the silica particles to react with other subtances. The general results of the particle shapes were uneven and consiststed of spherical or agglomerated and structural clusters.

5. Reference

[1] Ahmed A M, Ban F S 2017 *Am. J. Mater. Sci.,* **7** 223-231.
[2] Soltani N A, Bahrami, M I, Pech C, González L A 2014 *Chem. Eng. J.,* **264** 899-935.
[3] Selvakumar A K V, Umesh P, Ezhilkumar S, Gayatri P, Vinith V, Vignesh 2014 *Int. J. Chemtech. Res.,* **6(9)** 4455-4459.

[4] Liyang W, Jianchang X, Qian Y, Fusheng Z, Fangbai L, Lijuan Z 2020 *Chem. Eng. Sci.,* **216** 115538.
[5] Patrick E, Imoisili, Kingsley O, Ukoba, Tien C J 2020 *J. Mater. Res. Technol.*, **9** 307-313.
[6] Prerna D, Ramanand J 2012 *Pelagia Research Library, Der Chemica Sinica.,* **3** 589–601.
[7] Kelechi E, Afamefuna E O, Kenneth K A 2020 *J. King Saud Univ. Sci.,* **32** 550-554.
[8] Hossain S K S, Ram P, Roy P K 2020 *Ceram. Int.*
[9] Srivastava A K, Shringi N, Devra V, Rani 2013 *Int. J. Innov.,* **2** 2936–2942.
[10] Huadaij N, Nuntiya A 2008 *J. Nat. Sci.,* **7** 59-65.
[11] Ikram U H, Khalida A, Abdul M 2014 *J. Chem. Soc. Pak.,* **36** 3.
[12] Khorsand N H, Kiayee A H 2013 *Part. Sci. Technol.,* **31** 366–371.
[13] Azadeh T, Marzieh H, Vahid M 2012 *Mater. Res. Bull.,* **47** 2584–2589.
[14] Rahman I A, Padavettan V 2012 *J. Nanomater.,* 1-15
[15] Zarib N S M, Abdullah S 2018 *Int. J. Eng. Technol.,* **7** 190-195.
[16] Yadav V K, Fulekar M H 2019 *Mater. Today.,* 4351-4359.

The effect of KOH concentration and calcination temperature on the cement clinker catalyst activity in the transesterification of off-grade palm oil into biodiesel

Z Helwani[1], W S Negara[1], I Zahrina[1], S Z Amraini[1], G M Idroes[2], Muslem[3], R Idroes[4,5,*]

[1]Department of Chemical Engineering, Faculty of Engineering, Universitas Riau, Kampus Bina Widya Panam, Pekanbaru, Indonesia 28293

[2]Department of Chemical Engineering, Faculty of Engineering, Universitas Syiah Kuala, Banda Aceh 23111, Indonesia.

[3]Department of Chemistry, Faculty of Science and Technology, Universitas Islam Negeri Ar-Raniry, Banda Aceh 23111, Indonesia

[4]Department of Chemistry, Faculty of Mathematics and Natural Sciences, Universitas Syiah Kuala, Banda Aceh 23111, Indonesia

[5]Department of Pharmacy, Faculty of Mathematics and Natural Sciences, Universitas Syiah Kuala, Banda Aceh 23111, Indonesia

*E-mail: rinaldi.idroes@unsyiah.ac.id

Abstract. Cement clinker has been researched and has shown good performance as a heterogeneous catalyst in the biodiesel production process. The purpose of this study was to synthesize and characterize a cement clinker/KOH catalyst, to determine the effect of KOH concentration and calcination temperature on biodiesel yield. The catalysts were prepared by impregnation-calcination in the various KOH concentrations of 5, 10 and 15% and the calcination temperature of 500, 600 and 700 °C. The transesterification process was carried out for 3 hours with a mole ratio of oil: methanol of 4:1, reaction temperature of 65 °C, catalyst concentration of 6% wt oil, and stirring of 400 rpm. The results showed that the catalyst with the KOH concentration of 15% and the calcination temperature of 700 °C had the best performance with a biodiesel yield of 88.82%. The catalyst has a surface area of 9.622 m^2/g, basicity of $H\_ > 12.2$ and a crystalline phase of $3CaO.SiO_2$, K_2CO_3, $Ca(OH)_2$, and K_2O. From the research, it is known that calcination temperature and KOH concentration in the impregnation process are directly proportional to the biodiesel yield.

1. Introduction

The US Energy Information Administration (2019) predicted that world energy consumption will increase by 28% from 2015 to 2040, where 77% of energy source used is fossil fuels. This is a problem because this source is a non-renewable one that can experience scarcity. Therefore, the replacements, renewable fuels, are developed. There are some renewable energies developed by several country in the world, that are hydro, wind, solar [1], geothermal [2–8], and biodiesel [9,10] Biodiesel is a renewable fuel that made from vegetable oils that contain a lot of fatty acids [11,12]. To obtain biodiesel, the vegetable oils are transesterified to produce FAME (Fatty Acid Methyl Ester) or biodiesel and glycerol.

Riau is rich in diversity of natural resources for vegetable oil production. Based on data from the Directorate General of Plantation, Ministry of Agriculture of the Republic of Indonesia (2018)[13], this province has the largest oil palm plantations in Indonesia with CPO as the main product. Off-grade palm oil is the rest of the CPO mill's grading which tends to be unsuitable for processing into CPO. To improve the use and economic value, several researchers have tried to convert this product to biodiesel and succeed to obtain biodiesel yield from 70 % up to 90 % [10,14–16]. These studies prove that off grade palm oil is potential and suitable as a raw material for biodiesel production.

There are several parameters that affect the transesterification process in biodiesel production, namely temperature, reactant mole ratio, reaction time, stirring and catalyst. The catalyst is the most important one because it affects the reaction speed and yield. There are two types of catalysts in the transesterification, that are heterogeneous catalysts and homogeneous catalysts. The Heterogeneous catalysts have the advantages than homogeneous one because it is non-corrosive, environmentally friendly, simple waste handling, and easy to separate [17].

CaO are one of the popular heterogeneous catalyst used in transesterification because it have high activity, strong alkaline, light reaction conditions, long catalyst life, and low cost [17]. CaO is usually derived from limestone, calcium carbonate, and calcium hydroxide. Recent research even extracted CaO from eggshells waste[18]. Cement clinker is limestone that has undergone a pre-calcination process [19] have successfully used cement clinker impregnated with methanol as a catalyst to convert rubber seed oil into biodiesel. It reported to have better stability and recyclability than other solid base metal catalysts. [20] have succeed to support cement clinker with KOH in biodiesel production from soybean oil produced 84.25% yield [21] reported that cement clinker impregnated with KOH by wet-calcination impregnation method for transesterification of used cooking oil produced a very high biodiesel yield of 96.8%.

Recently, [22] have applied cement clinker catalyst without even impregnation to convert off grade palm oil into biodiesel. Even so, the biodiesel yield is relatively high up to 84.26%. In this study, the catalyst was developed from cement clinker impregnated with KOH and applied in biodiesel production from off grade palm oil. The performance of the catalyst also studied on different KOH concentrations and calcination temperatures.

2. Materials and methods

The materials used in this study were cement clinker obtained from PT Semen Padang, KOH, aquadest, off-grade palm fruits obtained from PTPN V, H_2SO_4, methanol, and hammet indicators obtained from PT. Bratachem. The tools used were spindle hydraulic press, condenser, three neck boiling flask, thermometer, magnetic stirrer and heating mantle.

2.1. Raw material preparation

The research began with washing the oil palm fruit, then steaming it for 120 minutes to soften the fruit. Furthermore, palm oil extraction is carried out using a spindle hydraulic press. The oil obtained was then analyzed for its FFA and water content.

2.2. Preparation of cement clinker/KOH catalyst

The cement clinker was dried at 105 °C for 2 hours. Then it was crushed and sieved with a 100 mesh sieve. The powder obtained was impregnated by KOH with a concentration of 5, 10 and 15% weight. The Impregnation is carried out by dropping the KOH solution into the cement clinker while stirring until it forms a paste. The mixture was then allowed to stand for 24 hours and dried at 110 °C. The catalysts product was calcined at the temperature variation of 500, 600, and 700 °C for 5 hours.

2.3. Study of the catalytic performance on transesterification process

150 grams of extracted off grade palm oil was put into the esterification reactor which was equipped with a stirrer and condenser. The reactor was heated until the temperature reached 60 °C. After that, 1% wt H_2SO_4 and methanol (methanol: oil molar ratio of 12: 1) were added. The reaction was carried out for 1 hour with a stirring speed of 400 rpm. The esterification product resulted was separated in a separating funnel. The lower layer was taken for FFA analysis purpose and for the transesterification stage [14].

The oil from esterification was heated at 65 °C and fed to the transesterification reactor which contains a mixture of methanol (methanol: oil molar ratio of 10: 1) and 6% weight -b oil cement clinker/KOH catalyst. The reaction was carried out for 3 hours with a stirring speed of 400 rpm. The product obtained was cooled and separated to obtain crude biodiesel (glycerol), remaining methanol and catalyst. Crude biodiesel is purified by washing with distilled water until the pH is neutral. Then the biodiesel is heated in the oven at 105 °C for 60 minutes to evaporate the residual methanol from the reaction and water [14].

2.4. Characterization of catalyst and biodiesel

The crystalline phase, basicity and the specific surface area of the catalyst were analyzed by X-Ray Diffraction (XRD), Hammet Indicators, and Brunaeur-Emmet-Teller (BET) methods, respectively. The characteristics of biodiesel also tested that included the acid number, flash point, cetane number, and viscosity (kinematic viscosity). The biodiesel composition was analyzed using GC-MS.

3. Result and discussion

3.1. Off-grade palm extraction

The extracted off-grade palm oil was characterized its properties of density, viscosity, water content and free fatty acid content tests. The characteristics of off-grade palm oil (Table 1) showed the high water content. The water can hydrolyze the oil to make FFA [23]. In addition, the catalytic activity on transesterification will decrease because the water will react with the catalyst [24] and can also cause the soap forming [25]. So that, water must be removed by heating. To reduce FFA levels on the extracted oil, the esterification reaction was carried out before transesterification one, so that the FFA was decrease from 9.28% to 1.38 [26].

Table 1. Physical properties and FFA of crude palm-oil off grade compared to palm oil.

Parameters	Off grade palm oil	CPO Standard SNI 01-2901-2006
Colour	Orange	Yellow
Density 40°C (kg/m^3)	892.11	-
Viskosity 40°C (mm^2/s)	29.47	-
Water content (%)	3.5	Max 0.5
FFA (%)	9.28	Max 0.5

3.2. Catalyst characterization used XRD analysis

To determine the effect of calcination temperature and KOH concentration on the crystalline phase of the catalyst, the XRD patterns were compared between those different treatment, that are 700 °C with 5%, 500 °C with 5%, 700 °C with 15%, and 500 °C with 15%. All XRD pattern showed the crystalline phases contained $3CaO.SiO_2$, K_2CO_3, $Ca(OH)_2$, and K_2O (Figure 1). The highest peak is $3CaO.SiO_2$ which is the main composition of cement. $Ca(OH)_2$ was formed through water absorption by CaO from the air during impregnation process or at the catalyst on the free space. The reaction of $Ca(OH)_2$ formation followed equation 1

$$CaO + H_2O \longrightarrow Ca(OH)_2 \quad (1)$$

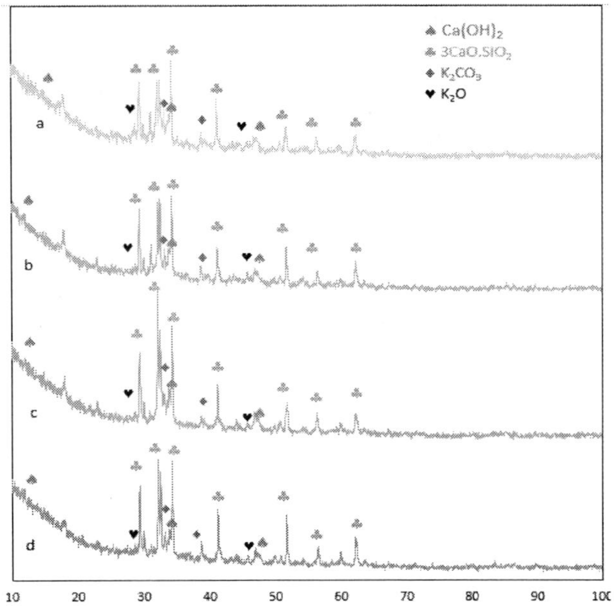

Figure 1. XRD diffractograms of cement clinker/KOH catalyst at different calcination temperature and KOH concentration of (a) 700 °C with 5%, (b) 500 °C with 5%, (c) 700 °C with 15%, and (d) 500 °C with 15%.

In addition, there is no KOH peak appeared on the diffractogram, but K_2CO_3 and K_2O peaks were observed. This proves that all KOH has been converted to K_2CO_3 and K_2O. According to [27], at 300 °C, KOH will convert into K_2O, H_2, O_2, and K_2CO_3. At 400 °C, K_2O will begin to change to K_2CO_3. At temperatures of ≥700 °C, K_2CO_3 is converted back into K_2O. K_2CO_3 has alkaline-lowering properties while K_2O is alkalizing. Therefore, K_2O is more expected than K_2CO_3[25]. The reaction for the formation of K_2O and K_2CO_3 can be seen in the following equation:

$$2KOH \longrightarrow K_2O + H_2O \quad (2)$$
$$2KOH + CO_2 \longrightarrow K_2CO_3 + H_2O \quad (3)$$
$$K_2O + CO_2 \longrightarrow K_2CO_3 \quad (4)$$

3.3. Brunaeur-Emmet-Teller (BET) and basic strenght

The catalyst that gave the highest biodiesel yield, Cement clinker/KOH catalyst under calcination temperature of 700 °C and KOH concentration of 15%, was analyzed by BET. The result showed that the surface area of the catalyst was 9.622 m^2/g. This value is even smaller than non-activated Cement Clinker and calcined Cement clinker catalysts that the surface area of 21.746 m^2/g and 25.497 m^2/g [22], respectively. The reduced surface area may happen during impregnation process that caused catalyst pore closure [25].

The cement clinker/KOH catalysts that were made in various KOH concentration (5%, 10% and 15%) and calcination temperature (500 °C, 600 °C, and 700 °C) were tested the basic strength. The results showed that the catalyst alkalinity ranged from $H\_ > 9.3$ - 12.2. The basicity of the catalyst increases with increasing KOH concentration. The Catalysts that have basicity of $H\_ > 9.3$ are the strong catalysts in the transesterification reaction [28].

3.4. Biodiesel yield and characterization

The effect of KOH concentration and calcination temperature on biodiesel yield can be seen in Figure 2. The highest yield of 88.82% produced by catalyst that prepared by the calcination temperature at 700 °C and a KOH concentration of 15%. The result showed that calcination temperature was directly proportional to biodiesel yield. The higher the calcination temperature, the more crystals formed [29]. In addition, the more K_2O is also formed. The more high temperatures, K_2CO_3 is converted to K_2O that increase the basicity, so that the catalytic activity was also increased [25].

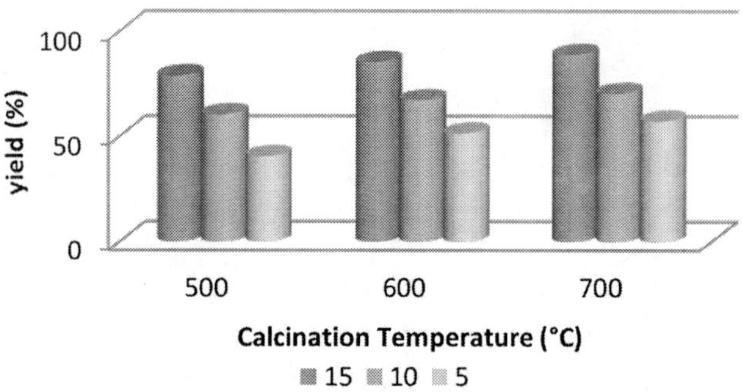

Figure 2. The effect of calcination temperature and KOH concentrations on biodiesel yield.

As the same as the calcination temperature, the KOH concentration was also directly proportional to the biodiesel yield. The higher KOH concentration, the more the alkalinity increased. So that, the catalyst activity also increased [29]. In contrast with the calcination temperature, too high a KOH concentration reduced the catalytic activity. At this condition, KOH can dissolve the cement clinker, thereby reducing the active site number of the catalyst [30]. In this study, the increase of the KOH concentration still increased the biodiesel yield. This showed that the KOH concentration has not passed the optimum concentration in the synthesis of cement clinker/KOH catalyst.

To determine the effect of activation and impregnation to the catalytic activity, the biodiesel yield produced by calcined-impregnated catalyst, calcined catalyst and non-activated catalyst were compared (Figure 3). The result showed that the use calcined-impregnated catalysts produced the highest biodiesel yield, followed by calcined catalyst and non-activated catalyst. The calcination caused the loss of moisture (water content) and CO_2 which is still absorbed in cement clinker [31]. The calcination also opened pores and formed the crystals phase which increased the surface area [29]. So that, the catalytic activity on transesterification reaction also increased.

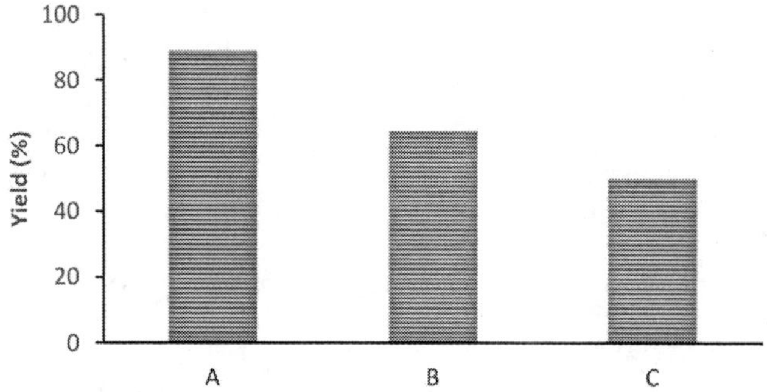

Figure 3. Biodiesel yield by using the various cement clinker catalyst treatments Calcination-Impregnation; (B) Calcination; (C) Without Activation.

Although the calcination increased the surface area, not all catalyst surfaces played a role in the transesterification process, but only the active side played one. Therefore, the basicity has a greater influence than the surface area. The KOH impregnation caused the increase in the catalyst active site numbers and catalyst alkalinity [32]. The KOH impregnation followed by the calcination caused the generation of active K_2O and K_2CO_3 phases and formed the crystals. This condition stabilized the catalyst and increased the surface area of the catalyst. Therefore, the impregnation-calcination catalyst showed a good catalyst performance in the transesterification process.

Table 2. Characteristic of the obtained biodiesel.

Characteristic	Result	SNI 7182:2015
Density (kg/m^3)	863.21	850-890
Kinematic viscosity (mm^2/s)	5.2	0.3-6.0
Acid number	0.29	Maks 0.5
Cetane number	52	Min 51
Flash point	125	Min 100

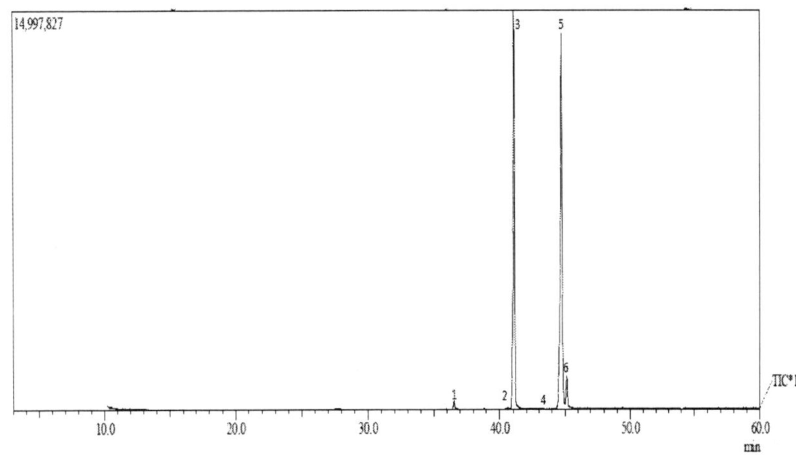

Figure 4. The GC chromatogram of the obtained biodiesel using cement clinker/KOH-700-15%.

Based on biodiesel characteristic (Table 2), the biodiesel made has met SNI standards. The Biodiesel density that is in range of standard indicates that the biodiesel will completely combust, so that it reduced the emissions and engine wear. Flash point values that are above the minimum standard indicates that biodiesel is safe for storage and transportation. The acid number below the maximum standard indicates that biodiesel is not corrosive. The viscosity that is in range of standard indicates that biodiesel has low effect to injection equipment on machines [33].

To determine the composition in biodiesel, a GC-MS test was performed. The results (Figure 4) showed that there were 2 highest peaks, namely methyl palmitate and methyl oleate with the level in the composition of 51.9% and 43.48%, respectively. This showed that the biodiesel has a high level of purity.

4. Conclusion

The best cement clinker/KOH catalyst is the catalyst that impregnated with 15% KOH and calcined at the temperature of 700 °C with the biodiesel yield of 88.82%. The catalyst has an alkaline value of ≥12.2, the surface area of 9.622 m^2/g and the crystalline phase of $3CaO.SiO_2$, K_2CO_3, K_2O, $Ca(OH)_2$. The biodiesel yield was directly proportional to the increase in calcination temperature and KOH concentration.

References

[1] Agarkar B, Barve S B 2011 *Int.* J. *Res. Eng. Technol.* **4** 188–191.
[2] Idroes R, Yusuf M, Alatas M, Subhan, Lala A, Saiful, Suhendra R, Idroes G M, Marwan 2018 *IOP Conf. Ser. Mater. Sci. Eng.,* **334** 012002.
[3] Idroes R, Yusuf M, Alatas M, Subhan, Lala A, Muhammad, Suhendra R, Idroes G M, Marwan 2019 *IOP Conf. Ser. Mater. Sci. Eng.,* **523** 012012.
[4] Idroes R, Yusuf M, Alatas M, Subhan, Lala A, Muslem, Suhendra R, Idroes G M, Suhendrayatna, Marwan, Riza M 2019 *IOP Conf. Ser. Mater. Sci. Eng.,* **523** 012010.
[5] Idroes R, Yusuf M, Saiful S, Alatas M, Subhan S, Lala A, Muslem M, Suhendra R, Idroes G M, Marwan M, Mahlia T M I 2019 *Energies.* **12** 4442.
[6] Marwan, Yanis M, Idroes R, Ismail N 2019 *Int. J. GEOMATE,* **17** 173–180.
[7] Marwan, Syukri M, Idroes R, Ismail N 2019 *Int. J. GEOMATE,* **16** 141–147.
[8] Putri D R, Nanda M, Rizal S, Idroes R, Ismail N 2019 *IOP Conf. Ser. Earth Environ. Sci.,* **364** 012003.
[9] Helwani Z, Ramli M, Rusyana A, Marlina M, Fatra W, Idroes G M, Suhendra R, Ashwie V, Mahlia T M I, Idroes R 2020 *Processes,* **8** 777.
[10] Helwani Z, Ramli M, Saputra E, Putra Y L, Simbolon D F, Othman M R, Idroes R 2020 *Catalysts,* **10** 724.
[11] Earlia N, Muslem, Suhendra R, Amin M, Prakoeswa C R S, Khairan, Idroes R 2019 *Sci. World J.,* **2019** 1–7.
[12] Earlia N, Rahmad R, Amin M, Prakoeswa C, Khairan K, Idroes R 2019 *Sains Malays,.* **48** 1019–1024.
[13] Kementerian Pertanian 2018 *Statistik Pertanian 2018.* Jakarta: Kementerian Pertanian. *(in Bahasa Indonesia).*
[14] Helwani Z, Fatra W, Saputra E, Maulana R 2018 *IOP Conf. Ser. Mater. Sci. Eng.,* **334** 012077.
[15] Putri F D, Helwani Z, Drastinawati 2015 *Jurnal Rekayasa Kimia dan Lingkungan.* **10** 99–105.
[16] Helwani Z, Ramli M, Saputra E, Bahruddin B, Yolanda D, Fatra W, Idroes G M, Muslem M, Mahlia T M I, Idroes R 2020 *Catalysts.* **10** 164.
[17] Liu X, He H, Wang Y, Zhu S, Piao X 2008 *Fuel.* **87** 216–221.
[18] Helwani Z, Fatra W, Saputra E, Herman S 2016 *Prosiding Seminar Nasional Industri Kimia dan Sumber Daya Alam.* 13–18. *(in Bahasa Indonesia).*
[19] Gimbun J, Ali S, Charan Kanwal C C S, Shah L A, Ghazali N H M, Cheng C K, Nurdin S 2012 *Adv. Mater. Phys. Chem.,* **2** 138–141.
[20] Soni S, Agarwal M 2013 *Int. J. Mod. Phys. Conf. Ser.* **22** 71–78.
[21] Muhamad Gh N H, Gimbun J, Nurdin S 2014 *Res. J. Appl. Sci.,* **14** 1425–1429.
[22] Putri K O, Miftahudin D N, Helwani Z, Bahri S 2019 *Jurnal Rekayasa Kimia Lingkungan.* **14** 147–154.
[23] Pahan I 2006 *Panduan Lengkap: Kelapa Sawit.* Jakarta: Penebar Swadaya. *(in Bahasa Indonesia).*
[24] Ulfayana S, Helwani Z 2015 *Abstract Book: Regional Conference on Chemical Engineering.* 7:22.
[25] Jin H, Kolar P, Peretti S W, Osborne J A, Cheng J J 2017 *Energies.* **10** 1920.
[26] Farag H A, El-Maghraby A, Taha N A 2013 *Int. J. Biol. Chem. Sci.,* **3** 1–8.
[27] Di Blasi C, Galgano A, Branca C 2009 *Energy & Fuels,* **23** 1045–1054.
[28] Ho W W S, Ng H K, Gan S, Tan S H 2014 *Energy Convers. Manag.,* **88** 1167–1178.
[29] Tang S, Wang L, Zhang Y, Li S, Tian S, Wang B 2012 *Fuel Process. Technol.,* **95** 84–89.
[30] Zhang J, Meng Q 2014 *World J. Eng. Res. Technol.,* **2** 184–191.
[31] Hussain A, Ali S, Ahmed I, Gimbun J, Albeirutty H M, Rehan A Z 2016 *Current Nanoscience.* **12** 576–585.
[32] Choi J W, Lee H, Lee J C, Lee S, Kim Y S, Yoon H J, Kim H C 2017 *Comput. Meth. Programs. Biomed.* **145** 35–43.

[33] Romero R, Luz S, Nativi R 2011 *Alternative Fuel* (InTech).

IOP Conf. Series: Materials Science and Engineering 1087 (2021) 012062 doi:10.1088/1757-899X/1087/1/012062

The utilization of activated carbon from *Jamblang* tree bark to adsorb lead heavy metal ion

Mariana[1,2,*], F Mulana[1], S Muchtar[1], A Fadhilah[3], C Y Febrina[3]
[1]Department of Chemical Engineering, Syiah Kuala University, Banda Aceh 23111, Indonesia
[2]Research Center for Environmental and Natural Resources, Universitas Syiah Kuala, Banda Aceh 23111, Indonesia
[3]Undegraduate School of Chemical Engineering, Universitas Syiah Kuala, Banda Aceh 23111, Indonesia

*E-mail: mariana_hasyim@yahoo.com

Abstract. This study aims to determine the characterization of activated carbon from *Jamblang* tree bark, the adsorption capacity and adsorption efficiency and the type of adsorption isotherm. The type of adsorbents used in this study are those without activation, with physical, chemical, and physical chemical activation. The preparation of adsorbent is initiated with carbonization of *Jamblang* tree bark at 400 °C for 3 hours, then physically activated by heating at 600 °C for 6 hours, chemically with 0.3 M HCl solution, and heating following with immersion with HCl solution. The adsorption process is performed atseveral adsorption time of 30, 60, 90, 120 and 150 minutes and initial adsorbate concentrations of 50, 75, 100, 125 and 150 ppm. The performance of prepared activated carbonsis tested on the adsorption of Pb metal ions. Several characterizations are carried out such as using Scanning Electron Microscopy (SEM) to evaluate the morphology of activated carbon, Fourier Transform Infra-Red Spectrophotometer (FTIR) determines functional groups, X-Ray Fluorescence (XRF) determines composition and Surface Area Analyzer (SAA) determines the surface area and pore size. The results show that the best adsorption capacity and efficiency are obtained at adsorption process usiang chemical-physically activated adsorbent with an optimum time of 120 minutes and an initial adsorbate concentration of 125 ppm. The achieved adsorption capacity and efficiency capacity are 57.28 mg/gr and 60.45%, respectively. The adsorption process follows the Langmuir isotherm model.

1. Introduction

Jamblang (*syzygiumcumini*) is a green tropical tree that is widely grown in Indonesia [1]. One of the treebarks can be used as a bioadsorbent to adsorb heavy metals [2]. The use of agricultural waste as a bioadsorbent has been widely used, however to our knowledge the use of *Jamblang* tree barkas an adsorbent has not been carried out [3], [4]. Therefore, we conducted research on the use of *Jamblang* tree barkas a bioadsorbent to adsorb lead metal ion in wastewater. *Jamblang* tree is one of the biomass from forest products that has wide leaves. Cellulosic biomass is formed from three main components namely cellulose, hemicellulose and lignin. Cellulose is the main component contained in plant cell walls and dominates up to 50% of plant dry weight [5]. Organics containing lignin, hemicellulose and cellulose can be used as raw material for producing activated carbon. Because it is very effective in adsorbing heavy metals in liquid waste. In addition, lignin and cellulose are mostly composed of the element carbon, which in general can be turned into carbon [6]. The produce adsorbent will be tested for its adsorption capacity and efficiency with lead solution.

2. Methodology

2.1. Materials and tools

The materials used in the study were *Jamblang* tree bark, hydrochloric acid solution, aquadest and lead solution. The equipment used was a ball mill, shaker, 80 mesh sieve, furnace, erlenmeyer, filter paper, funnel, magnetic stirrer, oven, measuring flask, spatula and scales.

2.2. Carbonization process of jamblang bark

Jamblang tree bark was crushed using a ball mill and sieved, then put into a furnace and baked at 400 °C for 3 hours.

2.3. Activation process

After carbonization, *Jamblang* tree barkwas physically activated by burning at 600 °**C** for 6 hours, chemically immersed in 0.3 M HCl solution for 24 hours and physical chemically by burning and immersing in HCl solution.

2.4. Adsorben characterization

The activated carbon that has been produced was characterized using Scanning Electron Microscopy (SEM) to see the morphology of activated carbon, Fourier Transform Infra-Red Spectrophotometer (FTIR) to determine the functional groups of activated carbon, X-Ray Fluorosity (XRF) to determine the composition of activated carbon, and Surface Area Analyzer (SAA) to determine the surface area and pore size of activated carbon.

2.4.1. Water content.

One gram of activated carbon was put into a weighed porcelain dish. The plate containing activated carbon was heated by placing it in an oven at 110 °C for 24 hours then cooled in a desiccator and weighed.

$$Water\ Content\ (\%) = \frac{a-b}{a}\ x\ 100\% \quad (1)$$

Where :
a = initial carbon weight (g)
b = dry carbon weight (g)

2.4.2. Ash content.

Two grams of activated carbon was put into a weighed porcelain dish. The plate containing activated carbon was put into the muffle furnace at 600 °C for 6 hours with air flowing into it then cooled in a desiccator and weighed.

$$\text{Ash Content}\ (\%) = \frac{a}{b}\ x\ 100\% \quad (2)$$

Where:
a = ash weight (g)
b = initial dry carbon weight (g)

2.4.3. Adsorption of lead metal ion.

100 ml of 1000 mg/L artificial $Pb(NO_3)_2$waste solution was contacted with 0.1 gram of adsorbent with variation of concentrate 50, 75, 100, 125 and 150 ppm and contact time 30, 60, 90, 120 and 150 minutes while stirring at speed of 70 rpm. After being contacted with adsorbents,the concentration of artificial $Pb(NO_3)_2$waste solution was measured using AAS. 100 ml of 1000 mg / L artificial $Pb(NO_3)_2$ waste solution was contacted with 0.1 gram of adsorbent with variation of concentrate 50, 75, 100, 125 and 150 ppm and contact time 30, 60, 90, 120 and 150 minutes while stirring at speed of 70 rpm. After being

contacted with adsorbents, the concentration of artificial $Pb(NO_3)_2$ waste solution was measured using AAS.

$$\text{adsorption (\%)} = \frac{(\text{initial concentration of adsorbate} - \text{final concentration of adsorbate})}{\text{initial concentration}} \, x \, 100\% \quad (3)$$

$$\text{adsorption capacity} = \frac{(\text{initial concentration of adsorbate} - \text{final concentration of adsorbate}}{\text{the amount of adsorbent}} \, x \text{ solution vol} \quad (4)$$

3. Results and discussion

3.1. Characterization of activated carbon

The water content of an activated carbon shows the hygroscopic properties of the activated carbon. The less water content contained in activated carbon would form wider pores. The wider pores increase the the surface area of the activated carbon that also increase the adsorption ability of activated carbon [7]. The results showed that the water content was 4-6%, as indicated by Table 1. In accordance with SII No. 0258-88, all prepared activated carbons have met the standard of water content.

Table 1. Quality of activated carbon.

No.	Parameter	Quality Standard SII No. 0258-88	Types of Adsorbent	Analysis Results (%)
1			Before Activation	5
2			Physical Activation	4
3	Water content	Max. 10%	Chemical Activation	6
4			Chemical-Physical Activation	5
5			Before Activation	14.5
6			Physical Activation	19
7	Ash content	Max.10%	Chemical Activation	7
8			Chemical-Physical Activation	9.5

Source: SII No. 0258-88

Ash content is the amount of metal oxide content which consists of minerals in a material that cannot evaporate during the ashing process. The ash content test aims to determine the metal oxide content of activated carbon. The ash content is expected to be as small as possible because a large ash content will reduce its adsorption capacity both in the form of gas and solution [7]. Samples that meet the standards were only activated carbon with chemical activation, namely 7% and physical and chemical activation, namely 9.5%. Both of these samples were activated using HCl solution, so that the HCl solution can dissolve the minerals from activated carbon by degrading cellulose in it [8]. The research results indicated that ash content of two samples did not meet the standards according to SII No. 0258-88. This happens because the higher of carbonization temperature resulted in the higher of the ash content in activated carbon [9]. Activated carbon with physical activation has a high ash content and exceeds the maximum limit of standard ash content because after carbonization at 400 ℃, it was reactivated by heating at 600 ℃.

3.2. Scanning electron microscopy (SEM) characterization

The Scanning Electron Microscope characterization was carried out to see the surface morphology of the adsorbent. The surface morphology of the *Jamblang* tree bark adsorbent was identified using SEM with an object magnification of 5000 times, the results of SEM analysis can be seen in Figure 1.

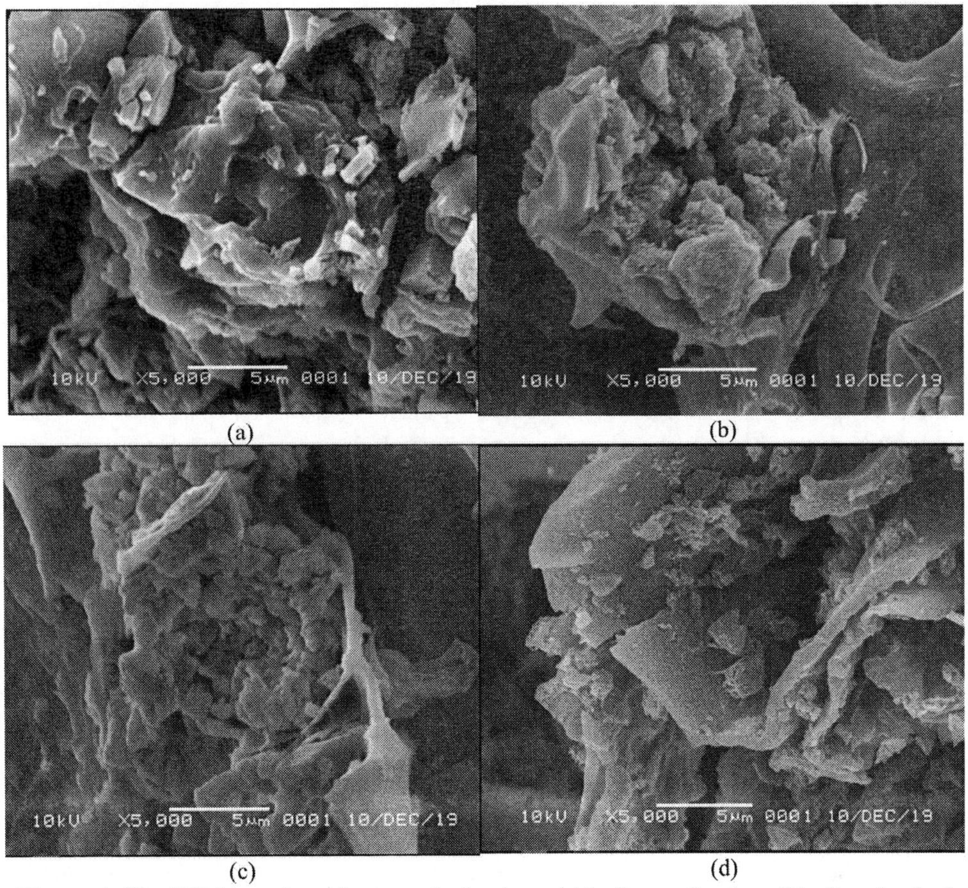

Figure 1. The SEM imaging of prepared adsorbent (a) before activation, (b) after physical activation, (c) after chemical activation, (d) after chemical physical activation with a magnification of 5000 times.

Figure 1 shows the differences surface morphology of activated carbon from *Jamblang* tree bark with different types of activation. The adsorbent before activation is still closed or not yet open because there are impurities or other components that fill in and cover the pores. The high ash content causes some of the pores on the surface of the adsorbent to remain closed with impurities. Meanwhile, the adsorbent using physical activation, chemical activation and physical chemical activation, the surface of the adsorbent pores is clean and not filled with impurities that cover the surface. Because the function of activation, both chemical and chemical activation, is to break hydrocarbon bonds on activated carbon or to oxidize surface molecules so that the pores of activated carbon will expand [10]. The results showed that the activated adsorbent had a clean surface compared to the adsorbent without activation. This is because the HCl used for activation can minimize tar formation and dissolve impurities, so that the pores are cleaner and the absorption process is maximized. Good pore formation can increase the ability of activated carbon to absorb heavy metals. The bigger the pores that are formed, the more the surface area of the activated carbon will increase, so it can increase the adsorption power of the activated carbon produced [11].

3.3. FTIR (Fourier Transform Infra-Red) characterization

This characterization was carried out to determine the functional groups or components of the *Jamblang*tree bark adsorbent. The chemical functional groups of *Jamblang* tree bark adsorbent with different types of activationwere identified using FTIR (Fourier Transform Infra Red) with a range of 500-4000 cm^{-1}. Table 2 shows the wavelengths and the main functional groups found at each peak in Figure 2.

Table 2. FTIR analysis standards.

No.	Types of compound	Wave numbers range (cm^{-1})	Fuctional groups
1.	H_2O	4000-3400	O−H
2.	CH_4	3000-2700	C−H
3.	CO_2	2400-2250	C=O
4.	CO	2250-2000	C−O
5.	Aldehide, Ketone, carboxyl acid	1900-1650	C=O
6.	Aromatic	1690-1450	C=C, benzene ring
7.	Alkane, Alcohol, Fenol, Eter, Fat	1475-1000	C−C, C−O, carbon bond
9.	Aromatic Substitution	500-850	Alifatic group
10.	CO_2	586-726	C=O

Jamblang bark is a plant with broad leaves, where plants with these characteristics have a chemical composition dominated by cellulose, namely 40-55% [5]. The *Jamblang* tree bark contains cellulose which will determine the adsorption capacity, where the cellulose contains carboxyl groups (-COO-) and hydroxyl groups (-OH) which can bind with metal ions [12]. The infrared spectra of the *Jamblang.* Uptake of aromatic –OH, C-H, C = O and C = C groups dominated the four adsorbent samples of *Jamblang* tree bark. The four carbon variations show that on the surface of the activated carbon of *Jamblang* tree bark there are a number of hydroxyl groups with active sites that are able to adsorb Pb (II) metal ion because the OH functional group has a negative charge, namely OH- ion which is very reactive to metals, so it can adsorb ions. metal. The C = O group is a unique group found in activated carbon and indicates that the bark of *Jamblang* forms an active carbon substance and this group can improve the performance of activated carbon in the adsorption process [13].

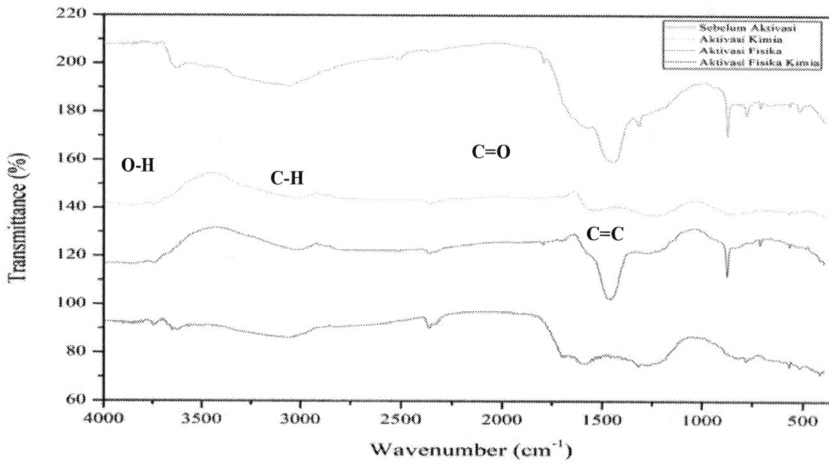

Figure 2. FTIR analysis results for adsorbent before activation, after physical activation, after chemical activation and after chemical physical activation.

Table 3. FTIR spectrum of several prepared adsorbents.

Functional Group	Bonds	Wavenumber (cm^{-1})				
		Reference	Before Activation	Physical Activation	Chemical Activation	Chemical Physical Activation
Hydroxyl	O–H	4000-3400	3614	3738	3745	3622
Aromatic alkyl	C–H	3000-2700	2800	2870	2820	2857
Aldehyde, ketones, carboxyl acids	C=O	1900-1650	1793	1793	1700	1693
Aromatic alkenyl	C=C	1690-1450	1452	1458	1558	1589

Based on Table 3, the FTIR spectrum of activated carbon from the activated *Jamblang* tree bark showed a shift in the wave number. This suggests that the activation process affects the adsorption intensity at this wavelength [14]. Activated carbon from *Jamblang* tree bark has several functional groups suitable for the structure of activated carbon in general which contains O-H, C-H, C = O, and C = C which are used as active groups to adsorb adsorbates.

3.4. Characterization of X-Ray Flourecence (XRF)

XRF is an analytical method used for elemental analysis in materials both qualitatively and quantitatively. The results of elemental analysis of activated carbon can be seen in Table 4.

Table 4. Composition of activated carbon of *Jamblang* tree bark.

Components/Weight (%)	Type			
	Before Activation	physical activation	chemical activation	chemical physical activation
P	0.87	1	-	0.69
S	1.3	0.3	1.95	1.3
K	3.46	4.26	2.2	1
Ca	87.7	89.4	40.7	38.1
Ti	0.1	-	0.26	-
Cr	0.10	-	0.22	0129
Mn	0.15	0.14	-	0.14
Fe	2.55	1.60	2.89	1.15
Ni	0.09	0.10	0.09	0.05
Cu	0.17	0.17	0.18	0.11
Zn	0.28	0.23	-	0.04
Sr	0.97	0.93	-	0.08
Mo	0.7	0.5	-	0.2
In	0.6	0.7	-	0.2
Ba	0.3	-	0.1	-
Yb	0.4	0.44	-	0.2
Re	0.3	0.2	0.36	0.2
Si	-	-	0.70	-
Cl	-	-	50.4	56.2

From Table 4 it can be seen that there are several elements in the activated carbon of *Jamblang* tree bark. Where the weight of elements Ca and K are higher than other elements. This element is a disruptor that can cover the pores of activated carbon. The addition of HCl in the activation process can reduce the levels of Ca and K. It can be seen that the activated carbon after chemical activation and physical chemical activation the content of Ca and K has decreased. By decreasing the levels of Ca and K, it will open more pores on the surface of the activated carbon so that the adsorption process will increase [15]. The similar results are also confirmed in previous SEM results.

Activated carbon after chemical activation and chemical physical activation has better adsorption ability due to the addition of an activator which is dominated by Cl which is an amorphous and porous solid which has a higher surface area so that it will increase its adsorption capacity. In addition, activated carbon after chemical activation is also dominated by Si where this element can adsorb [16].

3.5. Surface Area Analyzer (SAA)

Characteristics of the Surface Area Analyzer (SAA) using the Brunauer, Emmet, and Teller (BET) method aims to determine the surface area of the adsorbent. The results of surface area analysis can be seen in Table 5.

Table 5. Surface area and pore size of activated carbon from *Jamblang* tree bark.

Type of Adsorbent	Surface Area (cm^3/g)	Pore Volume (cm^3/g)	Pore Size (Å)	Ratio Surface Area to Pore Volume
Without activation	1378580	0.228	33.976	6046403.51
Physical activation	414610	0.088	46.769	4711477.27
Chemical activation	305030	0.056	43.985	5446964.29
Physical chemical activation	109600	0.030	53.778	365333333

From Table 5 it can be seen that the surface area of the adsorbent without activation is 1378580 cm^3/g with a pore volume of 0.228 cm^3/g. The surface area of the physical activation adsorbent was 414610 cm^3/g with a pore volume of 0.088 cm^3/g. The surface area of the chemical activation adsorbent was 305030 cm^2/g with a pore volume of 0.056 cm^3/g. The surface area of the chemically-activated adsorbent is 109600 cm^2/g with a pore volume of 0.030 cm^3/g. Adsorbent without activation has the largest surface area to pore volume ratio. The greater the value of the ratio of surface area to pore volume, the adsorbent will be more reactive [4].

The data from the BET test also showed that the average pore diameter for the adsorbent without activator was 33,976 Å, while the physical, chemical, and chemical physical activation adsorbents were 46,769 Å, 43,985 Å and 53,778 Å, respectively. Based on IUPAC, the diameter of the mesoporous lies in the range of 20-500 Å. This shows that the adsorbents obtained in this study were classified as mesoporous. It can be concluded that the pore size of physical and chemical activation is greater than the pore size without activation, physical activation and chemical activation so that the adsorption process on the adsorbent with physical and chemical activation is faster than other activations.

3.6. The effect of adsorbent types on adsorption efficiency and adsorption capacity of lead metal ion

In this study, the adsorption process of lead metal ion by prepared adsorbents was measured by using an AAS instrumental analysis and results are presented in Figure 3 and Figure 4. Figure 3 shows that the longer the contact time between the adsorbate and the adsorbent, the higher the adsorption efficiency until equilibrium occurs. The optimum time for the three types of activation occurs at contact time of 120 minutes. At the contact time of 150 minutes there is a decrease in adsorption efficiency because the Pb (II) ion in the adsorbent will saturate after longer contact time so that the process of desorption or release of Pb (II) ions which have been absorbed by the adsorbent so that Pb (II) is released back to form free Pb in solution [17].

Activation is one of the factors that influence adsorption efficiency. Based on the research, it shows that the adsorbent with physical and chemical activation has a faster adsorption efficiency compared to physical and chemical activation, this is because physical and chemical activation is a combination of physical and chemical treatments which in the adsorbent has the function of expanding and opening the pores of activated carbon and absorption of Pb content so that the adsorbent absorption is greater [18].

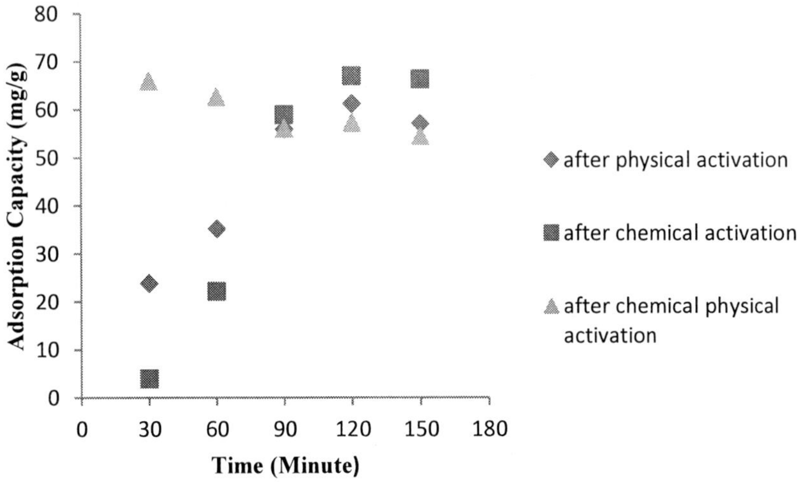

Figure 3. Relationship between contact time (minute), types of adsorbent and adsorption efficiency at adsorbate consentration of 94.76 ppm.

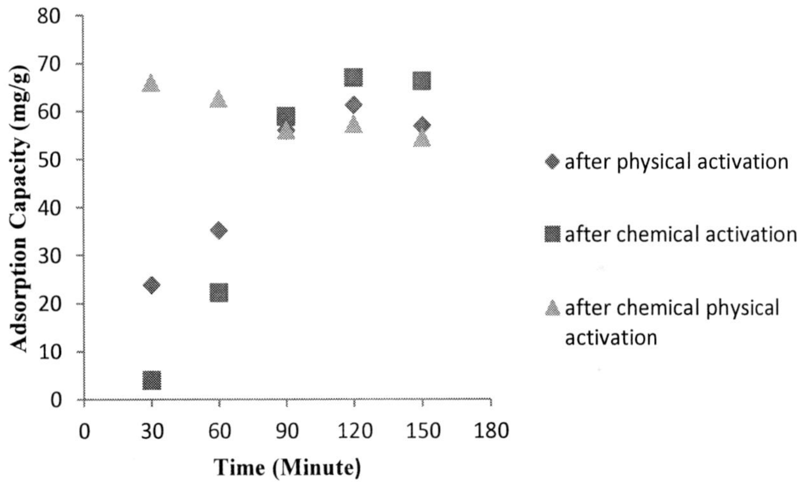

Figure 4. Relationships between contact time (minute) and types of activation toward adsorption capacity at initial adsorbate consentration of 94.76 ppm.

Figure 4 shows that the adsorption capacity increases with increasing contact time between the adsorbate and the adsorbent. The longer the contact time, the concentration of Pb (II) metal ion that is adsorbed will increase until equilibrium occurs. Adsorption capacity is directly proportional to the value of adsorption efficiency [17]. It was found that the greatest adsorption capacity occurred at the contact time of 120 minutes. Then it decreased at the contact time of 150 minutes this was due to the bond between the groups contained in the adsorbent with the metal getting weaker and finally released again [19]. The adsorption capacity of physical and chemical activation is slower than that of physical and chemical activation. In physical and chemical activation, the relatively short contact time can cause the adsorption process to be not optimal, this is because the adsorption rate is determined by the diffusion rate of the adsorbate species on the adsorbent solid [20].

3.7. Adsorption isotherms

The graphs of the Langmuir and Freundlich isotherm equations for the non-activated, physically activated, chemically activated, and chemical-physically activated carbons an adsorption time of 30 minutes can be seen in Figure 6 and Figure 5, respectively.

Figures 5 and 6 show the linearity of the two types of isotherms in adsorption with activated carbon, namely R2 = 0.6109 for Langmuir isotherm and R2 = 0.0336 for Freundlich isotherm. The adsorption isotherm pattern that is suitable for activated carbon of *Jambang*tree bark against Pb (II) ions can be seen by looking at the correlation coefficient (R2) which is close to the value 1.Based on the comparison of the two types of adsorption isotherms, the Langmuir type adsorption isotherm is closer to 1, namely 0.6109 compared to the Freundlich isotherm. Langmuir adsorption isotherm implies that the adsorption that occurs is chemical adsorption, this can occur because of the presence of a single layer (monolayer) of adsorbate on the adsorbent surface. All active sites on the adsorbent surface are homogeneous, where each active site can only adsorb one adsorbate molecule [21].

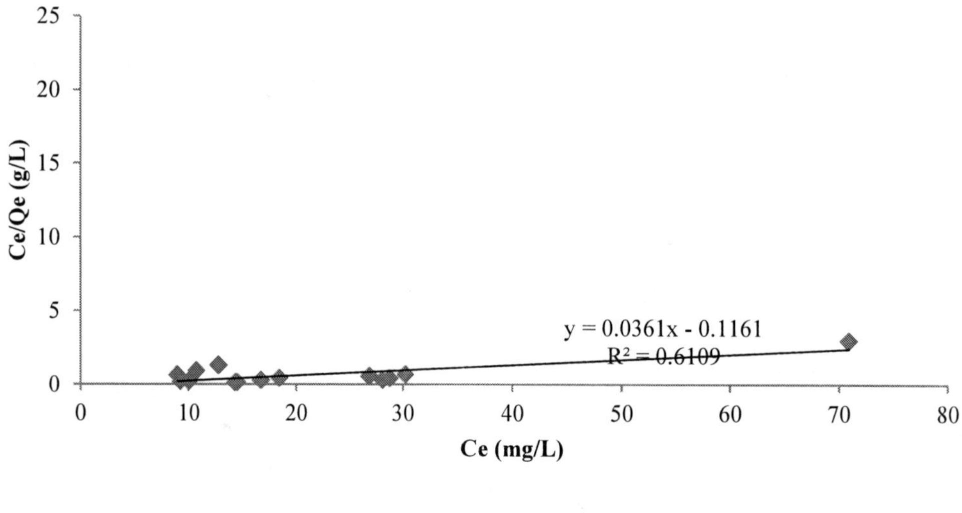

Figure 5. Langmuir isotherm model for Pb metal ion adsorption.

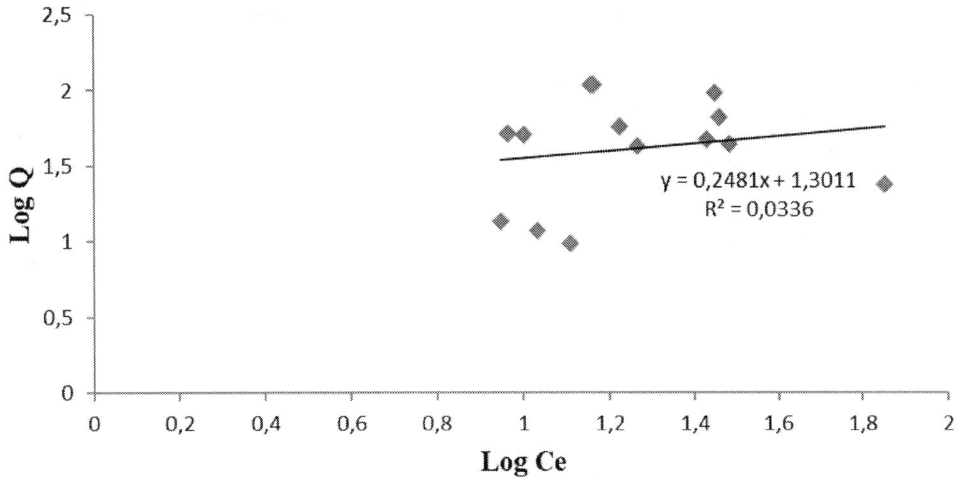

Figure 6. Freundlich isotherm model of Pb metal ion adsorption.

4. Conclusion

In this study, activated carbon from *Jamblang* tree bark has been prepared with different activation treatments. The results revealed that with physical and chemical activation, the adsorbent had cleaner and wider pores compared tothe other types of activation. These results were also confirmed by SAA characterization. The results of FTIR showed that the activated carbon of *Jamblang* tree bark is composed of functional groups of O-H, C-H, C = O and C = C. The water content of activated carbon of all types of activation complied with SII No. 0258-88 while for ash content, however, the adsorbent whichphysically activated did notmeet this standard. Activated carbon from *Jamblang* tree bark with physical and chemical activation had better and faster adsorption capacity with the optimum adsorption occurredat the adsorption time of 120 minutes. In addition, the adsorption isotherm of activated carbon from *Jamblang*tree bark was found to be Langmuir type.

5. Reference

[1] Sami F J, Syamsu N, Sukriani K, Sahibuddin A G, Trito R S. 2016. *Jf Fik Uinam,* **4** 130–138. *(in Bahasa Indonesia).*
[2] Eka C P, Florentina P. 2017. *Jurnal Integrasi Proses.* **6** 155–161. *(in Bahasa Indonesia).*
[3] Juniar L, Mariana, Mulyati S, Fathira D, Safitri R. 2020. *IOP Conf. Ser. Mater. Sci. Eng.,* **796** 012050.
[4] Mariana, Marwan, Mulana F, Yunardi, Ismail T A, Hafdiansyah M F. 2018. *IOP Conf. Ser. Mater. Sci. Eng.,* **334** 012029.
[5] Hermiati E, Djamali M, Titi C, Ono S, Bambang P. 2010. *Jurnal Litbang Pertanian.* **29** 121-130. *(in Bahasa Indonesia).*
[6] Pope J P. 2002. *Characterization and modeling of land subsidence due to groundwater withdrawals from the confined aquifers of the Virginia Coastal Plain.* Doctoral dissertation. Virginia: Virginia Tech.
[7] Laos L E, Arkilaus S. 2016. *Jurnal Ilmu Pendidikan Fisika.* **1** 32-36. *(in Bahasa Indonesia).*
[8] Setiyoningsih L A, Dwi I, Tri M. 2018. *Jurnal Kimia Riset.* **3** 13-19. *(in Bahasa Indonesia).*
[9] Arofah S, Muhammad N, Yasdi. 2019. *J. Eng.* **1** 38-51.
[10] Rahmawati, Aji W, Nafisah A, Kevin N D, Bagas R, Heriyanti. 2019. *Chempublish Journal.* **2** 11-20.
[11] Verayana, Paputungan M, Iyabu H. 2018. *Jurnal Entropi.* **13** 67-75. *(in Bahasa Indonesia).*
[12] Sari M F P, Loekitowati P, Mohadi R. 2017. *Jurnal Pengelolaan Sumberdaya Alam dan Lingkungan.* **7** 37–40. *(in Bahasa Indonesia).*

[13] Mentari V A, Handika G, Maulina S. 2018. *Jurnal Teknik Kimia.* **7** 16–20. *(in Bahasa Indonesia).*
[14] Hanum F, Gultom R J, Simanjuntak M. 2017. *Jurnal Teknik Kimia.* **6** 49-55. *(in Bahasa Indonesia).*
[15] Sari T I W, Muhsin, Hesti W. 2016. *Jurnal Konversi.* **5** 20-25. *(in Bahasa Indonesia).*
[16] Utari T. 1994. *Pembuatan Adsorben Alumina Dari Kaolin.* Doctoral dissertation. 27164. *(in Bahasa Indonesia).*
[17] Farah D R, Amalia V, Prabowo H E, Rohmatulloh Y. 2017. *Jurnal Al-Kimiya.* **4** 105-112. *(in Bahasa Indonesia).*
[18] Meisrilestari Y, Khomaini R, Wijayanti H. 2013. *Jurnal Konversi.* **2** 46-51. *(in Bahasa Indonesia).*
[19] Shamsuddin M S, Yusoff N R N, Sulaiman M A. 2016. *Procedia Chem.* **19** 558–565.
[20] Nurhasni, Frediyono F, Sya'ban Q. 2012. *Jurnal Valensi.* **2** 516-525. *(in Bahasa Indonesia).*
[21] Hamu G V, Imanuel G, Luther K. 2019. *Journal of Chemistry Notes.* **1** 12-23.

Selective esterification of glycerol diacetin and triacetin over rice husk biosilica catalyst with microwave heating

N Tasuna[1], K H Hidayatillah[1], Marwan[1,*], Zuhra[1]
[1]Chemical Engineering Department, Kopelma Darussalam, Syiah Kuala University, Banda Aceh, Indonesia 23111

*E-mail: marwan@unsyiah.ac.id

Abstract. Selectivity towards diacetin and triacetin was investigated in esterification of glycerol using rice husk biosilica catalyst. The reaction was carried out in a microwave-irradiated batch reactor using acetic acid to glycerol molar ratio of 1:3, catalyst amount of 3% and 5% of the glycerol volume, and reaction time of 20 and 30 minutes.The biosilica catalyst was prepared by treating rice husk with hydrochloric acid and followed by calcination at 600°C. SEM analysis of the resulted catalyst reveals significant change of rice husk morphology and its XRD analysis shows the characteristics of amorphous silica. Furthermore, FTIR analysis of the reaction sample reveals the presence of an ester group of triacetin at 1709.5 cm^{-1}. Product distribution of the glycerol esterification identified by GC-MS shows the highest total selectivity for diacetin and triacetin of 79.3% at the reaction condition of 30 minutes and 3% catalyst loading. However, individual selectivity for triacetin as the final product of this consecutive esterification steps was low, less than 5%.

1. Introduction

The rapid increase of biodiesel production has led to the wider availability of crude glycerol as a by-product in the chemical industry. Thus, the efficient utilization of excessive and cheap glycerol is becoming a new research topic at present, in which the etherification and esterification processes to fuel additive are considered as the promising route to the conversion of glycerol [1]. Esterification of glycerol with acetic acid using a catalyst to get mono, di and tri esters of glycerol acetates is one of the proposed solutions for the conversion of relatively low-priced glycerol to valuable products. Glycerol acetates, also known as acetinsconsisting ofmonoacetin, diacetin and triacetin have great industrial applications [2]. Monoacetin and diacetin exist in two different isomers. Monoacetin and diacetin have applications in the food industry, cryogenics and biodegradable polymers, while diacetin and triacetin may act as an additive in various fuels and triacetinis mainly used as a humectant [3]. Particularly, the diacetin and triacetin products have been established as oxygenated fuel additives that can enhance the cold flow properties of conventional diesel fuel, increase cetane number, and reduce noxious gas emission [4]. They could lead to possible attainment of 100% atom efficiency in diesel engines by blending triacetin with fatty acid methyl ester (FAME) since the physicochemical properties of triacetin blended FAME are similar to biodiesel fuel [5].

The glycerol esterification can be carried out with the help of an acid catalyst and the reaction is reversible, as illustrated by Figure 1. Without the presence of strong acids, the esterification reaction will proceed very slowly, but will reach equilibrium in a short time when strong acids and alcohols are employed [6]. The use of heterogeneous catalysts is an alternative in the glycerol esterification reaction. The advantages of solid catalysts consist of ease of product separation and catalyst recycling [7]. Silica catalysts can be used because they have acidic sites instead of acid groups in homogeneous catalysts. In addition, real silica is easily obtained in nature both in the earth's crust and from organisms. Rice husk is one of the potential sources of silica because rice husk has an abundant amount of silica which is 18.0-22.3% [8].Combustion of rice husk ash which is controlled at high temperatures (500-600°C) will produce silica ash which can be used for various chemical processes [9].

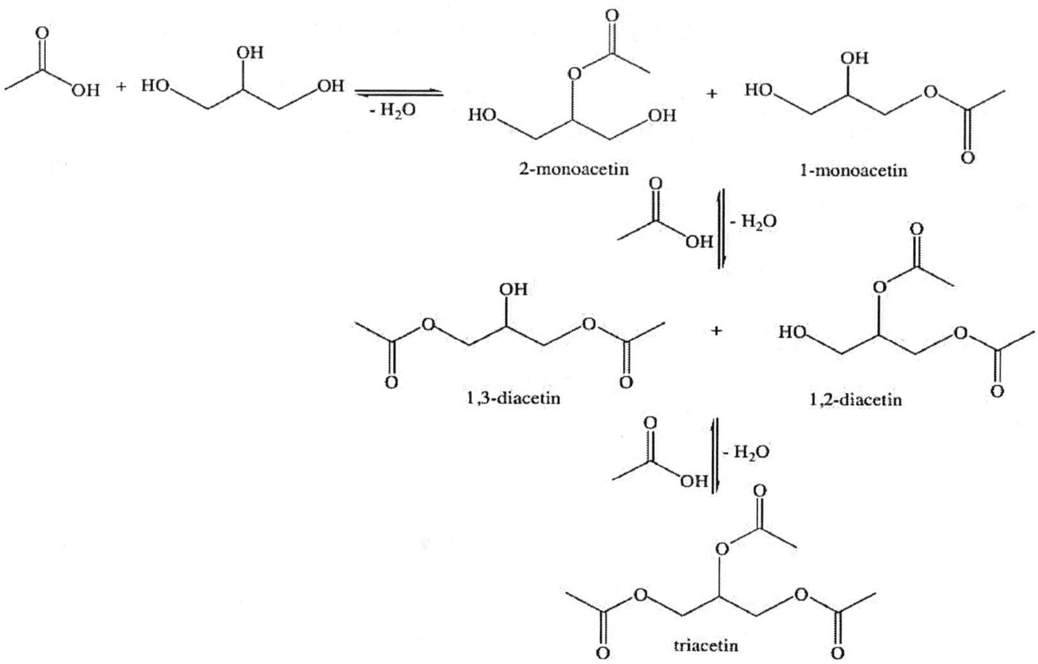

Figure 1. Three consecutive reversible steps of glycerol esterification with acetic acid [2].

The microwave heating method has different characteristics from conventional heating because heat is generated internally due to the vibrations of the molecules of the material that the microwaves want to heat [10]. As the result, microwave radiation can increase the speed of reactions in both batch and continuous processes. Since, thermal energy from microwaves is delivered directly to molecules that react through chemical reactions, heat transfer is more effective than conventional heating where heat is transferred from the environment [11].

2. Research methodology

2.1. Materials and instruments

The materials used in this study were rice husks from the Tungkopsubdistrict (Aceh Besar), filter paper (Whatman), technical grade glycerol (Merck), distilled water, acetic acid, and hydrochloric acid. All materials were purchased from commercial sources. Analytical instruments for catalyst and reaction characterizations consisted of X-Ray Diffraction (XRD) Shimadzu 6100, Fourier Transform Infra-Red Spectroscopy (FTIR)Shimadzu 8400, and Gas Chromatography-Mass Spectrometry (GC-MS) Shimadzu QP 2010 Plus.

2.2. Research procedure

2.2.1. Preparation of catalyst. Rice husk was washed with water and dried in the sun. 100 grams of dried rice husk was immersed in 600 mL of 3 N HCl solution at 100°C for 2.5 hours. Subsequently the wet material was dried and calcined at temperature of 600°C for 4 hours in the furnace. 25 grams of rice husk ash are washed with distilled water while filtering with Whatman filter paper to remove impurities. Finally, it was dried at 100 °C for 2 hours in an oven.

2.2.2. Esterification of glycerol. The reaction solution was prepared for glycerol to acetic acid molar ratio of 1: 3;20 mL glycerolwas charged into the reaction flask and followed by adding 43.2 mL acetic acid. This mixture of glycerol and acetic acid was preheated to 100 °C over a hot plate. A gradually increasing temperature was applied during preheating to avoid evaporation of reacting solution (acetic acid boiling point 117.9 °C). After the temperature reached 100°C, the biosilica catalystamounting 3% or 5% of the glycerol volume was charged into the reaction flask. The reactor was arranged as shown in Figure 2. Microwave unit used was a household type (Model NN-ST 342M, Panasonic) operating at frequency 2.45 GHz.The reaction was initiated by turningthe microwave on for irradiation time of 20 and 30 mins at stirring speed of 400 rpm.The microwave power output was set medium(520 watts). Solution reaction during reaction course observed using an infrared thermogun was in a range of 95-100 °C. After the irradiationperiod was completed, the reaction mixture was cooled down to room temperature and then filtered out to separate the catalyst from the glycerol acetate containing solution.

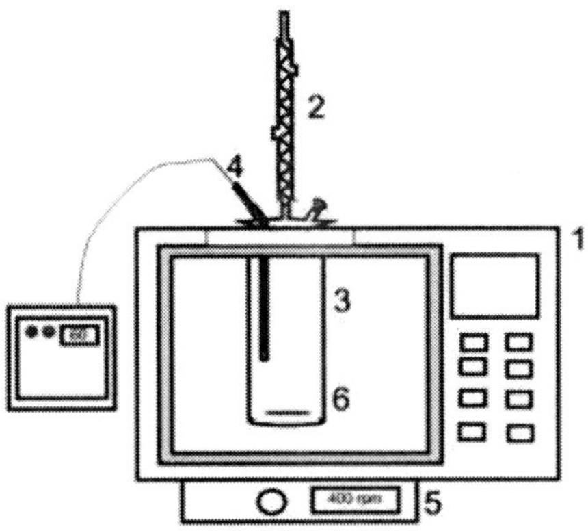

Figure 2. Experimental setup for the glycerol esterification (1. Microwave; 2. Reflux condenser; 3. Reaction flask; 4. Thermocouple; 5. Magnetic Stirrer; 6. Magnetic bar).

2.3. Characterization

Biosilica samples were accordingly characterized by SEM for surface morphology and XRD for silica content and crystallinity. XRD was equipped with software Match! 3 (Version 3.5.3.109). FTIR analysis was performed to determine functional groups in acetin products. The presence of certain functional groups was identified by comparing with standard spectra of diacetin and triacetin that already exist in the reference [12]. Later,the selectivity fordiacetinand triacetin was calculated from the product distribution resulting from GC-MS analysis.

3. Result and discussion

3.1. SEM analysis

Figure 3 shows the morphological differences between rice husk and rice husk ash. The morphology of rice husk looks more evenly and structured which shows the basic components of rice husk, while the morphology of rice husk ash is uneven, and pores of rice husk ash appear larger. Acid treatment of

IOP Conf. Series: Materials Science and Engineering 1087 (2021) 012063 doi:10.1088/1757-899X/1087/1/012063

lignocellulosic biomass such as rice husk leads tohydrolysis of hemicellulose and cellulose. As leaching of the celluloses took place, loss of rice husk structural form occurred [13].

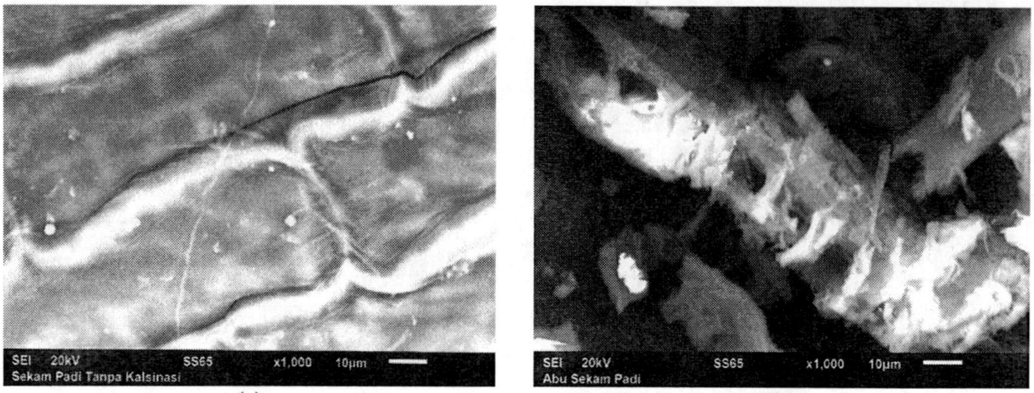

(a) (b)

Figure 3. SEM micrographs of (a) rice husk and (b) rice husk ash at 1000x magnification.

3.2. XRD analysis

XRD analysis was carried out on rice husk samples prior to and upon acid treatment and calcination. Figure 4 shows difference spectra profile between these two samples; raw rice husk shows presence of sharp peaks due to its crystalline property, while rice husk ash spectra reassembles amorphous phase. The acid treatment followed by calcination caused removal of crystalline components such as cellulose, hemicellulose and alkali metals, leaving amorphous silica as the main component in the treated rice husk [13]. It is revealed by the presence of the broad diffused peak at $2\theta = 22.07°$which corresponds to typical amorphous nature of silica.

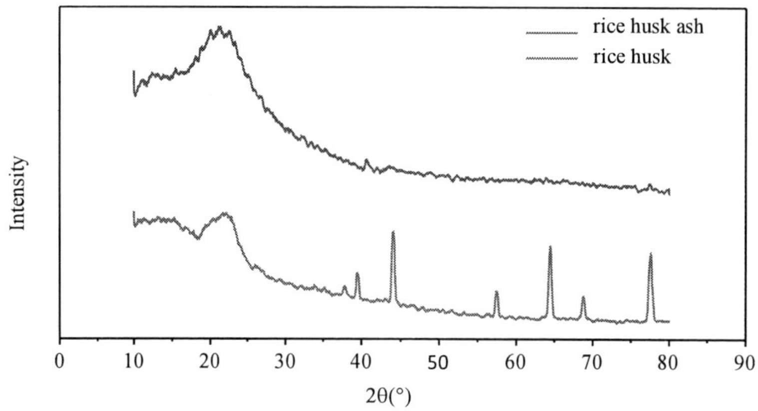

Figure 4. XRD spectra of rice husk and rice husk ash.

3.3. Product identification

Figure 1 shows that esterification of glycerol by acetic acid results in monoacetin, diacetin, triacetin and water. First, FTIR was used to identify the presence of functional groups related to products of the reaction. Figure 5 presents FTIR spectra of twodifferent reaction time samplesindicatingthe occurrence of O-H stretch of monoacetin, diacetin and remaining glycerolat 3100-3400 cm^{-1} and the ester group at 1690-1760 cm^{-1}.Later functional group corresponds to acetin and signal at 1709.5 cm^{-1} is related to a C=O stretch of the ester of triacetin confirming that the triacetin was obtained [12,14].

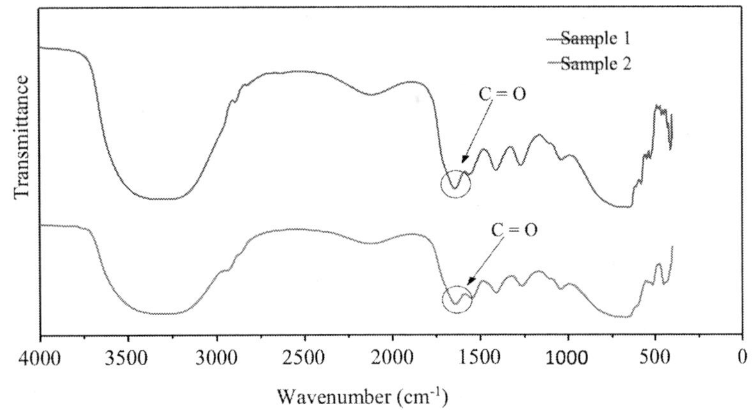

Figure 5. FTIR spectra ofreacting samples after reactions for 20 min (sample 1) and 30 min (sample 2) with glycerol to acetic acid molar ratio1:3 and catalyst content 5%.

3.4. Selectivity for diacetin and triacetin

Further chemical identification by GC-MS analysis was carried out to determine constituents in the reaction samples. Table 1 presents product composition for the reactions at different catalyst loading and irradiation time. It is foundthe reaction products consisting of 1-monoacetin, 2-monoacetin, 1,3-diacetin, 1,2-diacetin, and triacetin. It suggests the glycerol esterification by acetic acid taking place via consecutive route as given in Figure 1. Formation of monoacetin and diacetin isomers was also reported in our previous work [15,16].

Table 1. Product composition for the reaction with glycerol to acetic acid molar ratio of 1:3.

Catalyst	Time (minute)	Sample No.	Product Composition (%)				
			1-monoacetin	2-monoacetin	1,3-diacetin	1,2-diacetin	triacetin
3%	20	1	30.1	9.2	11.3	48.3	1.0
	30	2	11.2	9.5	9.4	68.0	1.9
5%	20	3	22.1	0.0	26.5	47.6	3.7
	30	4	24.5	15.4	9.6	47.3	3.1

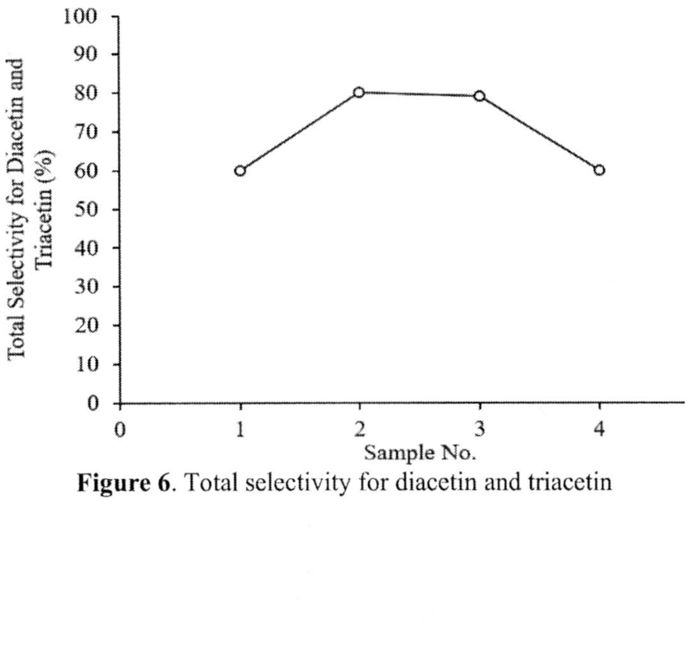

Figure 6. Total selectivity for diacetin and triacetin

Di and triacetins are commercially important chemicals used as fuel bioadditives. Therefore, itstotal selectivity for both acetins becomes interest of this work. Figure 6 shows the maximum total selectivity reached 79.3% from the reaction using catalyst loading of 3% and irradiation time of 30 min. Further increase in the catalyst loading did not improve the total selectivity. Whereas with conventional heating methods the total selectivity reaches more than 34.8% after 4 hours of reacting [15]. It reveals that microwave heating contributes significantly to reduction in reaction time and increase in the total selectivity for diacetin and triacetin.

4. Conclusions

Microwave as alternative heating was effective to enhance selectivity of glycerol esterification towards diacetin and triacetin; eco-friendlybioadditives for diesel fuel. Rice husk biosilica prepared by acid treatmentwas used to catalyze the reaction. FTIR and GC-MS analyses confirm that the reaction synthesizedmonoacetin, diacetin and triacetin. The total selectivity of diacetin and triacetin reached a maximum of 79.3% at reaction time of 30 min, a much shorter than time required by the reaction with conventional heating method. This microwave-assisted reaction system with solid catalyst is potential to replacetraditional homogenous esterification of glycerol.

5. References

[1] Sun J, Tong X, Yu L, Wan J 2016 *Catal. Today.* **264** 115-122.
[2] Ghaziaskar H S, Afsari S, Rezayat M, Rastegari H 2017 *J. Supercrit. Fluids.* **119** 52-57.
[3] Kale S, Umbarkar S B, Dongare M K, Eckelt R, Armbruster U, Martin A 2015 *Appl. Catal. A. Gen.,* **490** 10-16.
[4] Zhou L, Al-Zaini E, Adesina A A 2013 *Fuel.* **103** 617-625.
[5] Calero J *et al* 2015 *Renewable Sustainable Energy Rev.,* **42**1 437.
[6] Kong P S, Aroua M K, Daud W M, Lee H V, Cognet P, Pérès Y 2016 *RSC advances.,* **6 (73)** 68885-68905.
[7] Zięba A *et al* 2010 *Appl. Catal. A. Gen.* **387** 13.
[8] Bakar R A, Yahya R, Gan S N 2016 *Procedia Chem.,* **19** 189-195.
[9] Putro A L, Prasetyoko D 2007 *Akta Kimia Indonesia.* **3** 33. *(in Bahasa Indonesia).*
[10] Horikoshi S, Schiffmann R F, Fukushima J, Serpone N 2018 *Microwave Chemical and Material Processing.* Singapore: Springer Nature.
[11] de la Hoz A, Loupy A 2013 *Microwave in Organic Synthesis.* Weinheim: Wiley-VCH Verlag.
[12] Skoog D A, Holler F J, Crouch S R 1998 *Principles of Instrumental Analysis.* Boston: Cengage Learning.
[13] Lee J H, Kwon J H, Lee J W, Lee H S, Chang J H, Sang B I 2017 *J. Ind. Eng. Chem.* **50** 79-85.
[14] Lacerda C V, Carvalho M J, Ratton A R, Soares I P, Borges L E 2015 *J. Braz.,* **26(8)** 1625-1631.
[15] Marwan M, Indarti E, Darmadi D, Rinaldi W, Hamzah D, Rinaldi T 2019 *Bull. Chem. React. Eng.,* **14(3)**.
[16] Liao X, Zhu Y, Wang S G, Chen H, Li Y 2010 *Appl. Catal. B-Environ.,* **94(1-2)** 64-70.

The effect of milling time on formation of magnesium-activated carbon composites as hydrogen absorbing materials

F Mulana[1,*], Sofyana[1], Mariana[1], Azwar[1], Mustanir[2], Z Jalil[3]
[1]Department of Chemical Engineering, Syiah Kuala University, Banda Aceh 23111, Indonesia
[2]Department of Chemistry, Faculty of Mathematics and Natural Sciences, Universitas Syiah Kuala, Darussalam, Banda Aceh, Indonesia
[3]Department of Physics, Faculty of Mathematics and Natural Sciences, Universitas Syiah Kuala, Darussalam, Banda Aceh, Indonesia

*E-mail: faridmln@yahoo.com

Abstract. Hydrogen is a potential environmentally friendly fuel, but it has problems in terms of storage. Hydrogen storage using magnesium (Mg) and activated carbon (AC) is an effective and safe alternative because it can store hydrogen in a large enough capacity. This research was conducted to study the potential of Mg-AC material with addition of cobalt as catalyst in storing hydrogen by studying the effect of milling time and weight ratio of Mg-AC on the characteristics of the composites and the amount of hydrogen gas that can be adsorbed. Composites were formed using mechanical milling method using Fritsch 6 Planetary Ball Mill. The prepared composites were characterized using the X-ray diffraction (XRD), Fourier Transform Infra-Red (FTIR), Scanning Electron Microscopy (SEM) and Thermal Gravimetric Analysis (TGA). XRD analysis showed that MgH_2 compound was formed starting from 3 hr of milling time. The results of FTIR characterization showed that more C-H groups were formed with increasing milling time. The SEM graphs show that the longer milling time, the more evenly the distribution of particles in the Mg-AC composite and forming more nano particles. Meanwhile, the TGA analysis showed that with the addition of 5% AC, 10% and 15% then a weight reduction reached 9.6%, 10.6% and 14.2%, respectively.

1. Introduction

Hydrogen is an alternative energy source that can replace fossil fuels in the future [1]. Storage of hydrogen in the form of gas requires tanks that are expensive and unsafe especially for use in vehicles whereas liquefying hydrogen requires more energy so it is not economical solution. Storage alternatives in metal hydrides which have a high storage density of hydrogen (6.5 atoms/cm^3) compared to gaseous hydrogen (0.99 atoms/cm^3) or liquid hydrogen (4.2 atoms/cm^3) are promising [2]. Various materials such as transition metals and their alloys can absorb hydrogen to form interstitial bonds with hydrogen in the form of binary, tertiary and so on. However, not all transition metals and their alloys can be used as candidates for on board hydrogen storage [1].

Hydrogen storage materials research is currently developing rapidly and recently many researchers have focused their research on magnesium metal as the base material. Mg-base was chosen because magnesium is capable of absorbing quite a lot of hydrogen (containing 7.6% by weight of H_2), besides being light, easy to obtain and cheap [3,4]. Although magnesium has a large hydrogen absorption capacity, this metal also has disadvantages, namely that the kinetic process when hydriding and dehydriding is very slow and operates at relatively high temperatures around 350 - 400 °C [5]. In order for magnesium-based materials to be the top candidates as hydrogen storage materials, research on the manufacture of magnesium alloys with metals and other materials was carried out to reduce the hydriding/dehydriding temperature to around 80 - 150 °C without reducing the magnesium storage capacity hydrogen [4].

Some of the catalysts from the transition metal group that have been added to this magnesium-based material include partial replacement of magnesium with some metals. The replacement of magnesium with Al can reduce the absorption temperature of hydrogen and have better discharge capacity and cyclic stability [6]. The substituting Fe and Fe_2O_3 in magnesium did not significantly change the hydriding/dehydriding properties [7], the addition of Cr in magnesium hydride has succeeded in reducing the hydriding/dehydriding temperature and also accelerating the reaction [3]. The addition of Ti has an important role in accelerating the absorption process of hydrogen at temperatures lower than 300 °C [5].

Mohan et al [8] suggested that activated carbon can be used because it has good absorption capacity between 4.48-6.5% by weight, is easy to obtain and is relatively cheap. Some existing literature states that the hydrogen that can be absorbed by magnesium is far from its maximum capacity, which is around 3.82-5.15% by weight [2] so that in this study an attempt was made to modify magnesium-activated carbon composites by adding cobalt as a catalyst to see the adsorption/desorption process of magnesium to hydrogen.

2. Methodology

2.1. Materials and tools

The materials used in the study were commercial activated carbon with particle size of 325 mesh, magnesium metal (Sigma Aldrich), cobalt (Sigma Aldrich), ultra highpuriy hydrogen, 99.9% and nitrogen gas 95%.The used tools were mortar, Fritsch6ballmillplanetary, glove bag, oven with temperature range 25-400 °C (Gallenkamp), digital scales, 0-1000 gram (METLER Toledo) and desiccator.

2.2. Research procedure

The preparation of magnesium-activated carbon composites was carried out using a mechanical milling process using a Fritsch 6 planetary ballmill. This ballmill is also equipped with a connection valve to remove and enter gas. To avoid leakage between the vial and the lid, the vial was also equipped with a rubber o-ring and to avoid contact with air so all this preparation process was carried out in glove bag under nitrogen gas conditions.First, the amount of magnesium in the form of small chunks is crushed by a collision process to form granules using a mortar. Prior to mill, magnesium and activated carbon have been heated at 383 K for about 2 hours to remove moisture or other impurities contained therein. The research was performed by puttingvarious amounts of magnesium, activated carbon and cobalt into the ballmill vial. The vial ball mill was then closed. The valve on the lid of the vial is connected to a hydrogen cylinder and hydrogen gas with a pressure of about 0.5 MPa is fill in the vial ball mill (the process of entering and removing hydrogen gas is carried out 2-3 times).

After completing the process of filling in hydrogen gas, then the grinding process is carried out with Fritsch 6 at 300 rpm. To ensure sufficient hydrogen pressure in the vial, every 30 minutes the hydrogen in the vial will be added at a pressure of 0.5 MPa and at that time a sample will also be taken for the analysis process. When the grinding has finished, the grinding vial is opened to take the sample which is also carried out in a glove bag filled with nitrogen gas. To see further characterization the sample was put into a sealed bottle and stored in a desiccator.

2.3. Sample and data analysis procedure

X-ray diffraction (XRD) measurements were carried out using the Shimadzu XRD-7000 at 30 kV and 20 mA in the range of 15-50 °. This diffractometer uses Cu Kα radiation at a scanning rate of 2 deg/min. The functional groups of the sample composing compounds were tested using the FTIR tool in the infrared wavelength range of 400 to 4000 cm^{-1}. The thermal properties of magnesium-activated carbon (Mg-AC) composites were measured using the TG/DTA tool at a temperature of 30-475°C with a heating speed of 10°C/minute. Additional micro and morphological information were obtained from scanning electron microscopy (SEM Hitachi TM-3000) at 15 kV.

3. Results and discussion

3.1. X-Ray Diffraction (XRD) analysis

XRD is a tool used to characterize the crystal structure and the crystal size of a solid material. The peaks obtained from these measurement data are then matched against X-ray diffraction standards for most types of material.

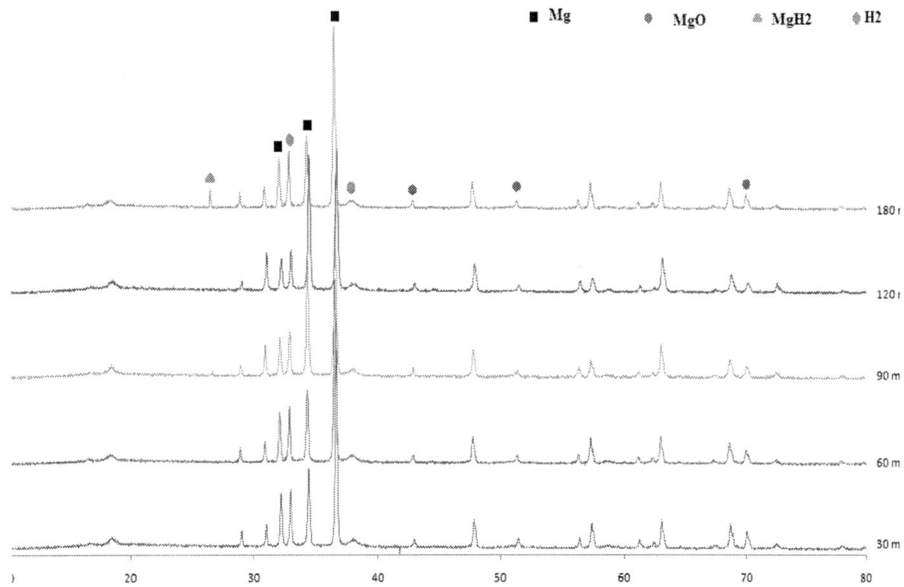

Figure 1. XRD analysis results with time variations on Mg-AC 5% composite.

Based on the XRD analysis results shown in Figure 1, it can be seen that at the milling time of 30 minutes to 120 minutes the MgH_2 compound has not yet formed. Yet, at milling time of 180 minutes MgH_2 had begun to form due to the prolonged milling time and also heatthat generated by the friction of the balls and walls against the milled composite accelerated the formation of magnesium hydride. The heat can be generated from the outer wall of the vial when the milling process is complete. This study is in accordance with research that performed by Yartys et al [9] and Shao et al [10] indicated that the absorption reaction of hydrogen to magnesium usually occurs at high temperatures around 300 K to 600 K and takes place for several hours. The composites also form an MgO compound which this oxide might occureduring sample preparationcausing an unwanted reaction between magnesium and oxide to occur.

XRD graph also showed that there are magnesium and free hydrogen compounds that have not reacted so that the longer milling time beyond 108 minutes would provide more optimal absorption of hydrogen by the composites. Milling time for composites with variations of Mg-AC 10% and Mg-AC 15% is also deemed necessary to be added to be 3 hr, 6 hr, 9 hr, 12 hr and 15 hr for further comparison of analysis results.

3.2. Fourier Transform Infra-Red (FTIR)

FTIR analysis was performed in order to analyze the functional groups contained in the prepared composites. The results of FTIR analysis can be seen in Figure 2.

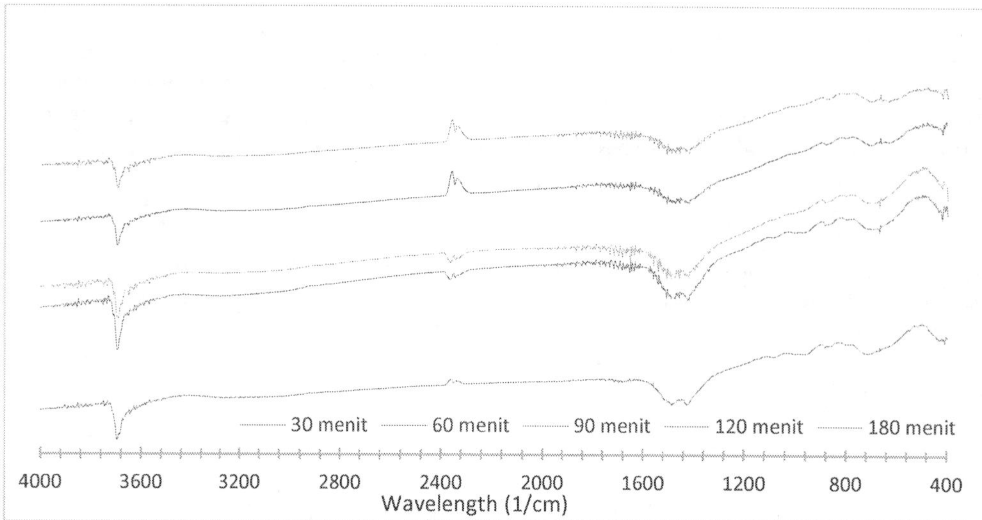

Figure 2. FTIR spectrum at various time variations for 5% Mg-AC sample.

Overall, the data results from the FTIR analysis for each variation of Mg-AC has the same fuctional groups. The groups contained in the composite sample include C-H, C = C and O-H. According to Mistry [11], the CH group is at a wavelength of 675 to 995 cm^{-1}, the C = C group is at a wavelength of 1500 to 1600 cm^{-1} and the OH group is at a wavelength of 3590 to 3650 cm^{-1}. The longer milling time performed, then the absorption peaks in the C = C bond weaken and moved to form a C-H bond as can be seen in the FTIR in Figure 2. This happens because the carbon continues to bind with hydrogen during the milling process to form a C = CH_2 bond. From the results of FTIR, it can also be seen that the formation of O-H bonds is also supported by the XRD results, namely that there are still exist oxides that could penetrate into samples during the sample preparation process so that the oxides react with magnesium and carbon during the milling process. Oxides are undesirable because they can react with composite compounds to form other compounds which cause the resulting compounds such as MgH_2 to not form optimally.

3.3. Structure surface analysis

The morphology of Mg-AC composites with milling time variations of 3 hours and 15 hours is shown in Figure 3.

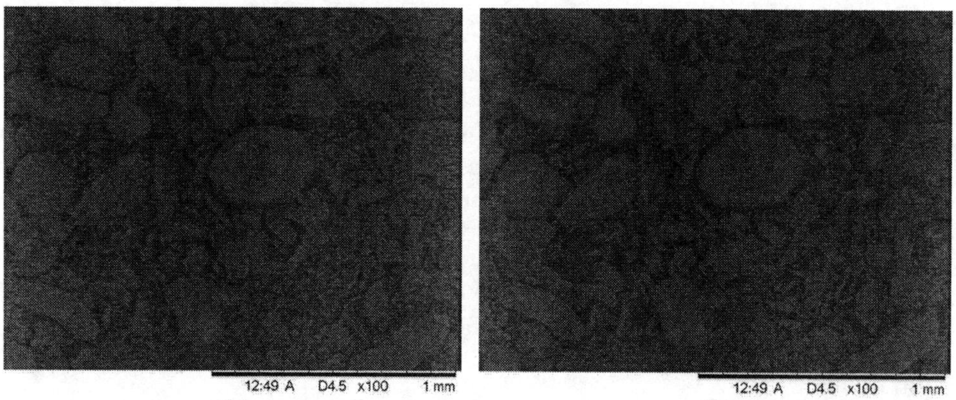

(a) (b)

Figure 3. SEM images at 100x magnification (a) Mg-AC 5% with a milling time of 3 hours; (b) Mg-AC 5% with a milling time of 15 hours.

Figure 3 shows that the longer the milling time produced the smaller the particle size and the more number of particles. At the 3 hr of milling time as shown in Figure 3 (a), it can be seen that the size and shape of the composites are not uniform. This appearance was also supported by the data from the XRD analysis that at the 3 hr of milling time the sample is still in amorphous shape, whereas in Figure 3 (b), the composite with a milling time of 15 hours had many particles that are in cubic shape and increasingly its uniform.

The milling time also affects the distribution of the Mg-AC composite sample, it can be seen from Figure 3 (a) that the sample with a milling time of 3 hr is still clearly visible, magnesium representated by the larger size of particles has not been mixed well with activated carbon and catalyst, while in samples with the longer milling time of 15 hr as can be seen in Figure 3 (b), it obviously appear that the magnesium which had not been mixed homogeneously with all activated carbon had less amount than that of the milling time of 3 hr. According to Crivello et al [12], carbon particles with sharp and hard edges can play a role as a grinding medium for magnesium. In addition, the hydrogen present in the atmosphere during the milling process can make magnesium particles wider and brittle by penetrating into magnesium crystals in the form of hydrogen atoms and this is very helpful in converting magnesium particles into nanoscale sizes [13].

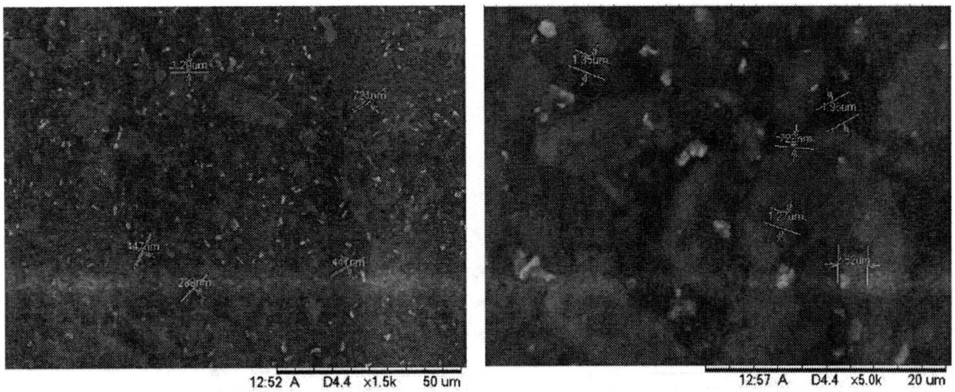

(a) (b)

Figure 4. SEM analysis of the 5% Mg-AC sample with a milling time of 3 hours (a) Magnification of 5000x; (b) Magnification 1500x.

It can be seen in Figure 4 (a) that the milling process for 3 hr is considered insufficient to produce a nano-sized sample, but that does not mean that no nanoparticles are formed. As can be seen in Figure 4, composite particles with nano size have been formed. In addition to affecting the particle size of the sample, milling time also affects the number of pores formed during the milling process, this is because the longer the milling process, the wider the surface of the sample, thereby increasing the rate of hydrogen absorption of the sample during the milling process.

3.4. Thermal analysis

Testing of the thermal gravimetry tool can be used to determine the heat stability of a material and can be used to determine the presence of decomposition reactions, oxidation reactions that cause weight loss (-TG) and weight gain (+ TG) of a material. Based on Figure 5, which is a function of temperature to sample weight of Mg-AC, it can be seen that the sample of 5% Mg-AC composite begins to decompose at a temperature of 315.88 °C to 447.83 °C with a weight reduction of 5.5%. Other TGA graph on samples of 10% Mg-AC composite and 15% Mg-AC composite, indicated that composites began to decompose at temperatures of 355.54 °C and 364.44 °C to 408.26 °C and 405.13 °C, respectively, with weight reduction of 4.7% and 3.8%. The figure also showed that at temperatures of 50 °C, there was also a slow reduction in sample weight, that might be because some of the hydrogen that binds to carbon decomposes easily.

According to Yartys [9], the amount of weight reduction up to 454 °C is caused by the release of the H_2 group from (-CH = CH-) n which is present in the decomposition reaction to (= C = C =) n as the final product. At a temperature of 466.84 °C, 422.99 °C and 451.66 °C also seen a second decomposition with a reduction in sample weight of 1.8%; 4.5%; and 5.5% so that the total decomposition of the composite samples of Mg-AC 5%, Mg-AC 10% and Mg-AC 15% was 9.6%; 10.6%; and 14.2%.

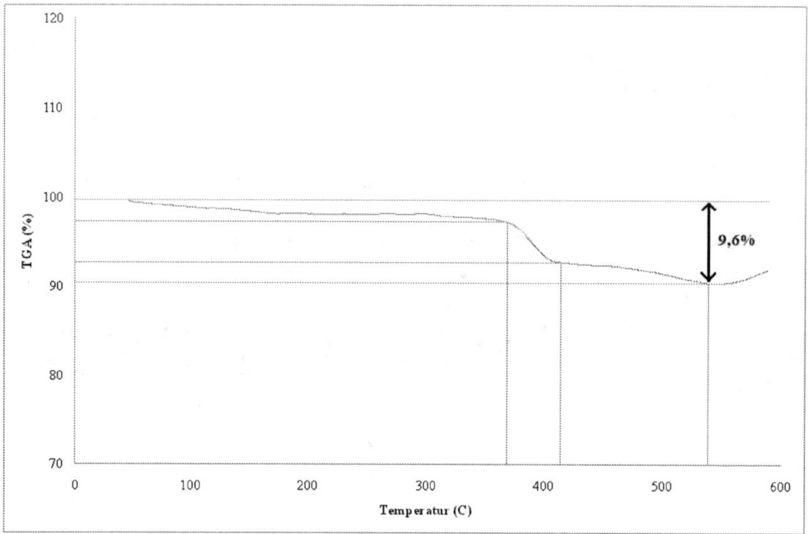

Figure 5. TGA analysis results on 5% Mg-AC composite.

The effect of temperature on hydrogen decomposition in Mg-AC composites can also be seen in the picture above which explains that the addition of activated carbon as an additive to magnesium greatly affects the amount of hydrogen adsorbed and the ability of magnesium to absorb hydrogen. When viewed from the point of view of hydrogen storage capacity in Mg-AC composites it is very promising due to the increase in the amount of hydrogen that can be adsorbed by the composite, however, when viewed from a decomposition temperature point of view, high temperatures are required to adsorb all of the hydrogen so that it requires high energy.

However, the analysis result from the TGA tool is not a result that shows the real number of hydrogen capacity stored in the composite material. These results can only be used for preliminary tests of composite materials such as seeing the decomposition temperature of the material being tested and comparing changes in the final weight and initial weight of the sample that occur due to changes in temperature.

4. Conclusion

Magnesium hydride was formed after 3 hr of mechanical milling under hydrogen atmosphere. Mainly three functional groups of C-H, C = C and O-H bond was found on Mg-AC composites. Three hours of milling time was insufficient to produce a nano-sized sample. Activated carbon in composites had a significant role to increase the adsorption capacity of Mg-AC composites. The thermal analysis indicated that the more added amount of activated carbon during composites formation, then the greater decomposition weight of Mg-AC composites. The thermal analysis of Mg-AC composite prepared by mechanical milling also indicated that composites had a good dehydriding in which the decomposition occurred in three phases at different temperature.

5. Acknowledgments
The authors would like to thank to Department of Chemical Engineering, Faculty of Engineering, Universitas Syiah Kuala for any support of the research. My thanks also to Cut Meutia ND and Ayu Mentari for their help and cooperation during conducting this research.

Reference
[1] Durbin D J, Jugroot C M. 2013, *Int. J. Hydrog. Energy,* **38(34)** 14595 14617.
[2] Darren P B. 2011. *Hydrogen Storage Material: The Characterisation of Their Storage Properties.* London: Springer.
[3] Grigorova E, Spassova M, Spasso T, Khristov M. 2014 *J. Mater., Sci.,* **49** 2647-2652.
[4] Jain I P, Lal C, Jain A. 2010 *Int. J. Hydrog.,* **35** 5133–5144.
[5] Song M Y, Kwak Y J, Lee S H, Park H R, Kim B G. 2013 *Met. Mater. Int.,* **19** 879-885.
[6] Guanglie L. 2001 *J. Alloys Compd.,* **321** 1-4.
[7] Baum L, Meyer M, Zélis Z L. 2007 *Physica B: Condens. Matter.,* **389** 189–192.
[8] Mohan M, SharmaV K, Kumar E A, Gayathri V. 2019 *Energy Storage.* **1** 35.
[9] Yartys V A, Lototskyy M V, Akiba E, Albert R. 2019 *Int. J. Hydrog. Energy,* **15** 7809–7859.
[10] Shao H, He L, Lin H, Li H W. 2018 *Energy Technol.,* **6** 445–458.
[11] Mistry B D. 2009. *A Handbook of Spectroscopic Data Chemistry* New Delhi: Oxford Book Company Press.
[12] Crivello J C, Dam B, Denys R V, Dornheim M. 2016 *Applied Physics A.* **122** 97.
[13] Zhang J, Li Z, Wu Y, Guo X. 2019. *RSC Advances* **9** 408–428.

Influence of EDTA on thermal properties of magnetic chitosan as an adsorbent of methylene blue

Rahmi[1,*], Lelifajri[1], C P Aulia[1], H Fathana[1]
[1]Chemistry Department, Universitas Syiah Kuala, Banda Aceh, 23111, Indonesia

*E-mail: rahmi@fmipa.unsyiah.ac.id

Abstract. Magnetic chitosan has been functionalized with EDTA in order to improve its adsorption capacity for the removal of methylene blue from water. The adsorbent was prepared using chitosan, PEGDE as a crosslinking agent, iron oxide prepared from local iron sand, and EDTA. The addition of EDTA on the adsorbent preparation was performed with different contents of EDTA. Adsorption experiments showed the EDTA addition improved the adsorption capacity of the adsorbent. The optimal content of EDTA in the adsorbent was 0.005 g. DSC and TGA analysis showed that the decomposition temperature of the adsorbent increased by increasing the EDTA content in the adsorbent.

1. Introduction

Water is a natural resource that is very important for living things, especially humans, for whom the most important need is drinking water. In order for water to be safely consumed, it must meet several criteria, one of which is that it must be free from contaminants. Many of these contaminant substances are produced from industries. The more rapid the development in the industrial sector, the more waste is generated. The resulting waste greatly affects environmental pollution, in both the air and water. Industrial waste, especially liquid waste, can contribute to water pollution [1]. Textile dye is a common cause of wastewater, a result of the global textile industry. In this industry, one of the main raw materials is textile dye and about 10-15% of the dyes that have been used are disposed of into water such as rivers. Textile dyes are generally made from compounds containing benzene groups such as methylene blue, rhodamine B, methanol yellow, tartrazine, malachite green, indigo carmine [2], basic yellow 28, and acid brown 75 [3].

Methylene blue is one of the most common dyes in liquid waste in the textile industry in Indonesia. Methylene blue is a dangerous dye which, if ingested, can cause respiratory tract irritation, and if it is touched can cause skin irritation [4-6]. The presence of methylene blue in a body of water, even at a concentration of only 1 ppm, can affect the photosynthetic process of plants in water because it inhibits the intake of sunlight [7]. Various attempts have been made to overcome the problem of dye pollution, including chemical treatment [8], biological methods, chemical physics methods [9], electrochemical processes, [10] and adsorption [11]. Adsorption is a separation method based on the adsorption or withdrawal of gas or liquid molecules from the adsorbent surface. Based on its performance, this method requires an adsorbent, which is a substance that acts as an adsorbent for molecules, whether liquid or gas. One of the natural polymers that can be used as dye adsorbent is chitosan [12].

Chitosan is a polysaccharide that can be easily obtained from natural resources such as shrimp, crab, shellfish and other animal waste. This compound is an alternative source that can be developed continuously due to its large number of renewable properties. Chitosan with its unique structure has a high affinity for dyes. Chitosan has been used to remove several dyes, such as Crystal violet [13], Congo Red & Direct Blue 1 and Remazol Brilliant Blue RN & Basic Blue 36 [14]. The choice of chitosan as an adsorbent is due to its non-toxic, biodegradable, and abundant availability. However, chitosan in pure form has disadvantages, such as high solubility at low pH and limited mechanical properties. Another thing that needs to be considered is the use of chitosan as an adsorbent, usually in the form of a powder. This fine size of chitosan makes it very difficult to separate chitosan from the solution after the adsorption process is complete. Therefore, it is necessary to modify chitosan.

Several attempts to modify chitosan to improve the mechanical properties and stability of chitosan in acidic conditions have been carried out in previous studies. One of them is the addition of crosslinking agents such as glutaraldehyde (GLA), epichlorohydrin, and polyethylene glycol diglycidyl ether (PEDGE). Modification of chitosan can also be done by adding iron sand. Iron sand aims to facilitate the separation of the adsorbent after application. Another compound that can be used to modify chitosan is ethylenediaminetetraacetic acid (EDTA). EDTA is a multidentate ligand capable of chelating with metal ions. Based on the research of Repo, 2013 [15], EDTA-modified chitosan was able to increase the adsorption capacity of chitosan.

Iron sand, which is magnetic, can convert chitosan-EDTA into magnetic chitosan-EDTA. The separation of adsorbent after adsorption process is easier, using a magnet separates out the magnetic chitosan-EDTA and leaves behind the solution. Previous studies used synthetic iron oxide particles in the preparation of magnetic-chitosan [16]. Meanwhile, we know that the availability of iron sand in Indonesia is quite abundant because Indonesia is an archipelago consisting of thousands of islands surrounded by beautiful beaches with abundant sand. The sand along this coastline contains valuable minerals such as iron, titanium, silica, and other elements. This iron sand containing iron oxide can later be used as a magnetic source in magnetic chitosan-EDTA preparation.

In this work, we focus on the discussion of the effects of EDTA addition on the methylene blue adsorption capacity of chitosan and the thermal properties of magnetic chitosan. Magnetic chitosan-EDTA adsorbents were prepared using iron oxide, PEDGE, and EDTA with different content. The adsorbent obtained was examined its adsorption capacity of methylene blue and characterized with DSC and TGA analysis.

2. Materials and methods

2.1. Preparation of iron oxide from iron sand

The iron sand was separated using a bar magnet which has paper around it, so that the iron sand containing Fe_3O_4 will stick to the bottom of the paper. This method was repeated many times until more iron sand containing Fe_3O_4 was obtained. The iron sand was washed with hot distilled water and then dried and repeated again the previous work steps. Iron sand containing 30 grams of Fe_3O_4 was mashed by ball milling for 30 hours.

2.2. Preparation of magnetic chitosan-EDTA

Chitosan (0.35 g) was dissolved with 20 mL 2% acetic acid and stirred for 2 hours with a magnetic stirrer. Then EDTA (0.005, 0.01 and 0.025 g) was added to chitosan solution and stirred for 2 hours. Next, PEDGE was added to the chitosan-EDTA solution and stirred for 2 hours. Then 0.5 grams of iron oxide was added and stirred again for 1 hour. After that, it was transferred using a dropper into a 3 N NaOH solution with a pH of 8 to form beads. The resulting beads were filtered and washed with distilled water until a neutral pH and then dried and stored [17]. The resulting adsorbent was then characterized using DSC and TGA.

2.3. Adsorption

The resulting magnetic chitosan-EDTA adsorbent was tested for its performance for methylene blue removal from water. Initially, the adsorbent weighed 0.1 grams and added to an Erlenmeyer flask, each of which was filled with 25 mL of 25 ppm methylene blue solution. The solution was stirred using a shaker at a constant speed of 150 rpm. Then the adsorbent was separated by pulling it with a magnet and the absorbance of the remaining solution was determined using a UV-Vis spectrophotometer.

3. Results and discussion

3.1. Magnetic chitosan-EDTA

The preparation of magnetic chitosan-EDTA was begun with dissolving chitosan using acetic acid and EDTA was added. After this, iron sand was added, creating a mixture of chitosan-EDTA and iron sand. Finally, this mixture was dropped into an NaOH solution to form magnetic chitosan-EDTA adsorbent. Magnetic chitosan-EDTA adsorbents that have been prepared in the form of beads were then used in the methylene blue adsorption process to determine the optimum composition of EDTA in order to achieve the best adsorption. The methylene blue adsorption process was performed for 25 minutes at a constant pH and methylene blue concentration. The adsorbed methylene blue was then separated from the magnetic chitosan-EDTA adsorbent using a permanent magnet and tested for its adsorption using a UV-Vis spectrophotometer. Furthermore, the adsorption capacity of each methylene blue that had been adsorbed using magnetic chitosan-EDTA was calculated with variations in the composition of EDTA and the resulting adsorption capacity was compared (Figure 1).

The results of the adsorption process show that the addition of EDTA can increase the adsorption capacity of the adsorbent. In the EDTA content of 0.005 grams, the adsorption capacity was greater than that of other EDTA content, namely 2.08 mg/g. Based on Repo, 2013, the increase in adsorption capacity was due to the increase in functional groups in chitosan [21]. EDTA reacts with chitosan through the bonds formed between NH_2 in chitosan and C = O from the carboxylate groups present in EDTA. However, when the EDTA content was increased, there was a decrease in the adsorption capacity. This was because EDTA reacts to the NH_2 groups in chitosan, which causes a decrease in the active side of chitosan, thus causing a decrease in adsorption capacity.

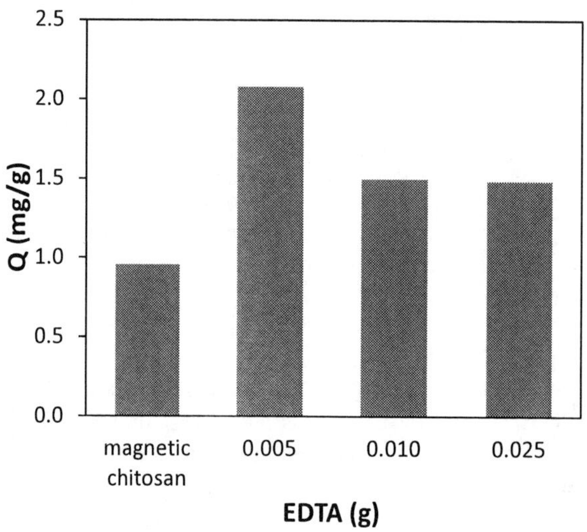

Figure 1. Adsorption capacity (Q) of magnetic chitosan-EDTA adsorbent at different content of EDTA (0.005, 0.01 and 0.025 g).

In order to study the thermal properties of the adsorbent, DSC and TGA analyses were performed. DSC is an instrument that measures the energy absorbed and the energy released by the sample when the sample is heated or cooled. Figure 2 is the result of the analysis of chitosan and magnetic chitosan-EDTA adsorbent with different content of EDTA in the adsorbent. The thermogram of chitosan obtained from DSC analysis in this study showed a wide endothermic peak at 170.28°C and an exothermic peak at 320.02°C. The endothermic peak that appears at 170.28°C is not the glass transition temperature (Tg) of chitosan, it is due to the presence of the sample moisture factor or the presence of residual solvent (acetic acid). This is supported by the TGA thermogram of chitosan (Figure 3) which shows the mass loss at a temperature of 170.28°C. The Tg value of chitosan is difficult to determine because of the semicrystalline nature of chitosan. However, based on research conducted by Sakurai (2000), the Tg value of chitosan was successfully obtained after several repetitions of the DSC thermal analysis test where the Tg value of chitosan obtained was 203°C. The exothermic peak at 320.02°C is an indication that the sample has begun to decompose.

DSC thermograms of magnetic chitosan-EDTA at different weight content of EDTA also had endothermic peaks. Similar to the pure chitosan thermogram, endothermic peak the magnetic chitosan-EDTA is due to the moisture and the presence of residual acetic acid or water in the sample. The exothermic peaks of magnetic chitosan-EDTA adsorbents with various content of EDTA appear at different temperatures. Based on the DSC thermograms, magnetic chitosan-EDTA adsorbents showed higher decomposition temperatures than chitosan. The higher the EDTA content in the adsorbent, the higher the decomposition temperature, which indicates an increase in thermal properties. Based on Figure 2, the lowest decomposition temperature is chitosan with a decomposition temperature of 320.02°C, followed by magnetic chitosan-EDTA containing 0.005 g EDTA with a decomposition temperature of 325.02°C, then magnetic chitosan-EDTA containing 0.01 g EDTA with a decomposition temperature of 327.85°C, and finally, magnetic chitosan-EDTA containing 0.025 g EDTA with a decomposition temperature of 329.34°C.

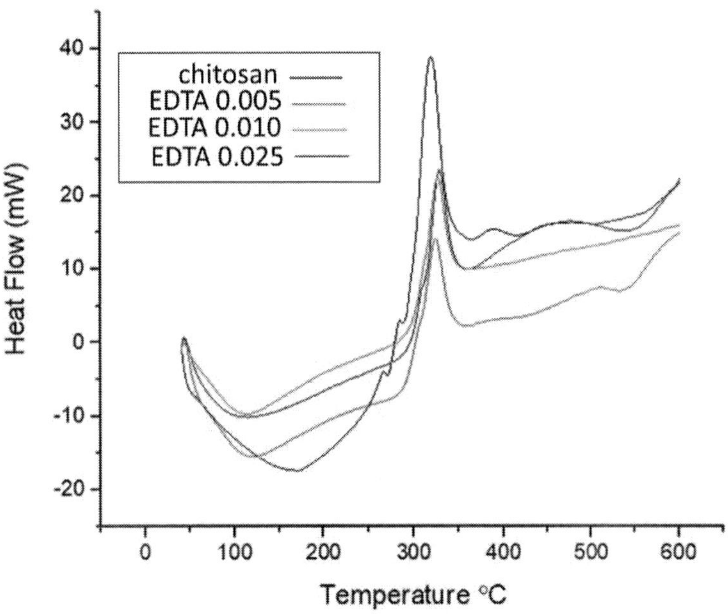

Figure 2. DSC analysis of magnetic chitosan-EDTA adsorbent at different content of EDTA (0.005, 0.01, and 0.025 g).

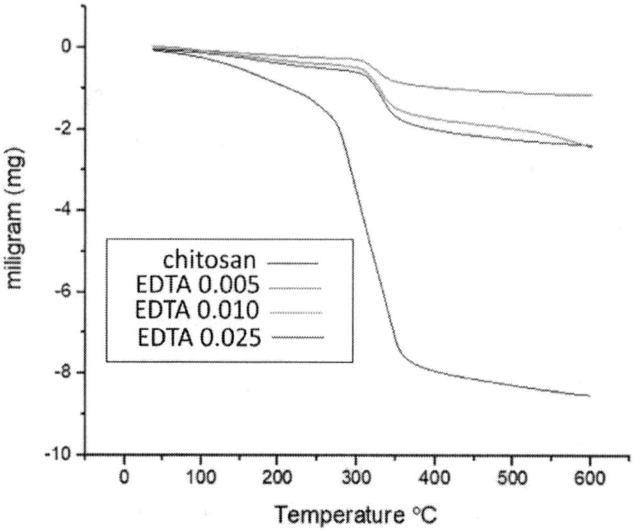

Figure 3. TGA analysis of magnetic chitosan-EDTA adsorbent at different content of EDTA (0.005, 0.01, and 0.025 g).

In this study, thermal analysis was also performed using TGA analysis. TGA is an instrument used to measure the change in weight of a sample when heated or cooled (temperature change). Figure 3 shows the curves of the four samples, namely chitosan, magnetic chitosan-EDTA adsorbents with different contents of EDTA (0.005, 0.01, 0.025 g). It can be seen that the four samples exhibited a decrease in weight at the beginning of the curve. The difference occurs at the time when the decomposition temperature is reached. In accordance with the results of the analysis by DSC, the mass reduction of chitosan occurred at a lower temperature compared to magnetic chitosan-EDTA adsorbents. The factors that influence the curve of the TGA analysis are the weight of the sample, the size of the particles, and the compactness of the sample. The decomposition temperatures of magnetic chitosan-EDTA adsorbents were higher than chitosan due to the presence of iron oxide which has higher decomposition temperature than chitosan. Moreover, the composite formation of chitosan, iron oxide, PEDGE, and EDTA will improve the stability of chitosan thereby increasing its decomposition temperature. Increasing in EDTA content in the adsorbent improved the decomposition temperature of the adsorbent. It was probably due to the reaction between EDTA and chitosan, which produced a more stable and compact compound. The compactness of the sample is directly proportional to decomposition temperature of the sample. Magnetic chitosan-EDTA adsorbents have less mass loss and chitosan has the most mass loss.

4. Conclusion

Magnetic chitosan-EDTA can be used as an adsorbent of methylene blue. The results of this adsorption study showed that the addition of EDTA could increase the adsorption capacity of chitosan for methylene blue adsorption and the optimal content of EDTA was 0.005 g. The thermal properties of magnetic chitosan-EDTA adsorbents are better than pure chitosan. The best thermal properties were obtained in magnetic chitosan-EDTA adsorbent containing 0.025 g EDTA.

5. References

[1] Kumari H J, Krishnamoorthy P, Arumugam T K, Radhakrishnan S, Vasudevan D. 2017 *Int. J. Biol. Macromol.,* **96** 324-333.

[2] Ramirez-Montoya L A, Hernandez-Montoya V, Montes-Moran M A. 2014 *J. Anal. Appl. Pyrolysis,* **109** 9-20.
[3] Chaari I, Fakhfakh E, Medhioub M, Jamoussi F. 2019 *J. Mol. Struct.,* **1179** 672-677.
[4] Sawafta R, Shahwan T. 2019 *J. Mol. Liq.,* **273** 274-281.
[5] Sabarinathan C, Karuppasamy P, Vijayakumar C T, Arumuganathan T. 2019. *Microchem. J.,* **146** 315-326.
[6] Rezakazemi M, Shirazian S. 2019 *J. Mol. Liq.,* **274** 778–791.
[7] Jauris I M, Fagan S B, Adebayo M A, Machado F M. 2016 *Comput Theor Chem.,* **1076** 42–50.
[8] Oller I, Malato S, S'anchez-P'erez J A. 2011 *Sci. Total Environ.,* **409** 4141-4166.
[9] Pavón-Silva T, Pacheco-Salazar V, Sánchez-Meza J C, Roa-Morales G, Colín-Cruz A. 2009 *J. Environ. Sci. Health A,.* **44** 180-215.
[10] Chen G. 2004 *Sep. Purif. Technol.,* **38** 11-41.
[11] Pavan F A, Mazzocato A C, Gushikem Y. 2008 *Bioresour. Technol.,* **99** 3162-3165.
[12] Hydari S, Sharififard H, Nabavinia M, Parvizi M R. 2012 *Chem. Eng. J.,* **193–194** 276–282.
[13] Ling S L Y, Yee C Y, Eng H S. 2011 *J Appl Sci.,* **11** 1445-1448.
[14] Kyzas G Z, Deliyanni E A. 2013 *Molecules,* **18** 6193-6214.
[15] Repo E, Koivula R, Harjula R, Sillanpää M. 2013 *Desalination,* **321** 93–102.
[16] Liu Z, Wang H, Liu C, Jiang Y, Yu G, Mu X, Wang X. 2012 *ChemComm.,* **48** 7350-7352.
[17] Gong J, Wang X, Shao X, Yuan S, Yang C, Hu X. 2012 *Talanta,* **101** 45-52.

Preparation and characterization of chitosan-starch Janeng membranes cross-linking with citric acid

Saiful[1,2,*], L Mardiana[1], Z Rahmah[1], M Ajrina[1], M R Imanda[1], Rahmi[1], Khairi[1], M Ramli[1], Marlina[1]
[1]Chemistry Department, Faculty of Mathematics and Natural Science, Syiah Kuala University Banda Aceh, Indonesia 23111
[2]Research Center for Environmental and Natural Resources, Syiah Kuala University, Jalan Tgk. Syech Abdul Rauf No. 5, Darussalam, Banda Aceh 23111, Indonesia

*E-mail: saiful@unsyiah.ac.id

Abstract. Neat chitosan membranes have shortcomings in their application owing to weak mechanical properties, thus, requiring modification. In this study, a chitosan membrane was modified utilizing starch and cross-linking agent (citric acid) addition. The characterization of chitosan-starch membranes using Universal Testing Machine, Scanning Electron Microscope (SEM), Fourier Transform Infra-Red (FTIR), Differential Scanning Calorimetry (DSC), and Thermogravimetric Analysis (TGA) had proven the cross-linking formation. The cross-linked membrane had better tensile strength (11.07 kgf/mm^2) than the uncross-linked membrane (66.07kgf/mm^2). The DSC thermogram of uncross-linked and cross-linked membranes showed endothermic and exothermic peaks at different temperatures. In addition to the membranes' thermal characteristics, the DTA thermogram showed the decomposition process on the uncross-linked membrane uncross-linking occurred at a temperature range of 277.74-363.27 °C, while the decomposition of the cross-linked membrane was observed at a range of 287.83-356.38 °C. The FT-IR spectra confirmed a decrease in the intensity of the absorption peak associated with cross-linking formation within the membrane.

1. Introduction

Chitosan-based membranes have been reported to have wide applicability, including in the desalination of seawater and brackish [1-4]. However, the application of neat chitosan membranes still has several weaknesses, such as brittleness, stiffness, and weak resistance against the acidic solution. The weaknesses can reduce the performance of the chitosan membrane in water purification. One alternative route to overcome the weaknesses is through physical and chemical modifications [5-7]. In this study, we explore the modification by means of polymer blending and cross-linking.

Chemical modification of the membrane surface through cross-linking is an effective technique to overcome the aforementioned weaknesses [8, 9]. Cross-linking reactions are used to link polymer chains, one to another, by forming permanent bonds through covalent/ionic bonds. It also provides a more flexible bond since it breaks the intramolecular bonds that contribute to the membrane stiffness. One of the polymers that can be used to blend with chitosan is starch [10]. Starch is a natural carbohydrate polymer that can be obtained from various bio-resources. *Dioscorea hispida* (the local name: janeng) tubers had been reported to be an alternative resource for starch due to their low price and high availability [11]. The use of starch in membrane manufacturing can increase the economic potential of janeng and environmentally friendliness. Starch possesses active sites, contributed by the presence of hydroxyl groups in their chemical structure that can form bonds with polymers or other compounds. The addition of a cross-linking agent assists the formation of the chitosan-starch cross-linking. The addition of a cross-linking agent in the polymer mixture can increase the stability (especially in acidic condition) and mechanical durability [10, 12]. One of the cross-linking agents that can be used is citric acid. Citric acid is chosen as a cross-linking agent stem from its low toxicity and non-carcinogenicity, where the product can be safe to use.

In this study, we explore the modification using polymer blending and cross-linking. A technique of blending chitosan and janeng starch was carried out to improve the mechanical properties, assited by cross-linking formation by citric acid. The success of this process was investigated by means of mechanical strength, thermal behavior, functional group and morphology analysis.

2. Materials and methods

2.1. Materials

Janeng was collected from Aceh Besar Regency (Aceh, Indonesia) and used for starch production. Medical grade chitosan with an acetylation degree above 85% was used in a membrane preparation. Glacial acetic acid was as a polymer-solvent, citric acid was used as a cross-linking agent, and sodium metabisulfite and Argentum nitrates were used in starch extraction. Glacial acetic acid, citric acid, sodium sulfite, and Argentum nitrates were purchased and supplied from Merck KgaA (Germany).

2.2. Starch isolation from janeng

The janeng-based starch (JS) was isolated using the method proposed by Saiful et al. [11]. Janeng tubers, as much as 5 kg, were firstly cleaned by soaking in distilled water. The procedure was followed by peeling and chopping the cleaned janeng tubers, then rinsed with distilled water. Next, sodium metabisulfite solution (1.12 g / l) was added to the janeng tuber pieces and processed using a crusher. The water and the pulp of the janeng are separated by filtration. The filtrate obtained is left to stand for 24 hours until the precipitate is produced. The precipitate obtained is JS, which is then washed repeatedly to remove the remaining HCN in the JS. The loss of HCN was indicated by a negative test result using 1 M $AgNO_3$ solution. Then, the JS was dried for 24 hours at 70 °C. After drying, the JS is mashed and sifted (100 mesh) to give JS powder.

2.3. Preparation of chitosan-starch membrane

The process of making the chitosan-starch cross-linking membrane using the citric acid cross-linking agent referred to the method developed by Lusiana et al. [14]. A 1.5% (w/v) pure chitosan solution was prepared by dissolving 2.5 gram chitosan in 200 ml of 1% (v/v) acetic acid solution. The starch used was priorly gelatinized, prepared by mixing 3 grams of starch in 200 ml of distilled water. The gelatinization process was carried out by heating the mixture at 70-80 °C for 5 minutes. Furthermore, the gelatinized starch was mixed in the chitosan solution, which was stirred using a magnetic stirrer. The chitosan-starch mixture was added with 0.01 mol citric acid (1.78 g) and the mixture was stirred at 70-80 °C for 24 hours to produce a casting solution. The thin membrane layer was printed on a granite plate with a thickness of 45 μm and oven-dried at 35 °C. The formed membrane was then washed with 1% NaOH to remove the solvent and followed by washing with distilled water repeatedly until the pH was neutral. Then, the membrane was allowed to dry at room temperature. These membranes were used for further characterization.

2.4. Characterization of chitosan-starch membrane

The citric acid uncross-linked and cross-linked chitosan-starch membranes were characterized for mechanical strength, Differential Scanning Calorimetry (DSC), Thermal Gravimetry Analysis (TGA), Scanning Electron Microscope (SEM) analysis, and Fourier Transform-Infrared (FT-IR) analysis.

2.5. *Tensile and elongation test*

The membrane tensile test referred to the ASTM D-368 method. This test was conducted at the Physics and Materials Laboratory of the Department of Physics, Faculty of Mathematics and Natural Sciences, Universitas Syiah Kuala. The membrane tensile test specimens are presented in Figure 1.

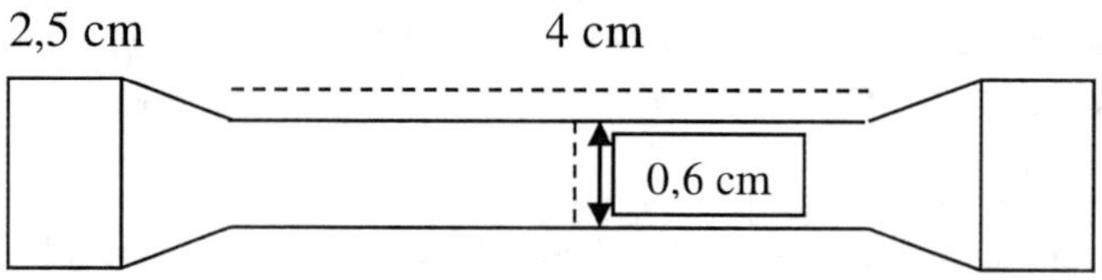

Figure 1. The shape of the tensile strength test specimen. Source: ASTM D-368.

In testing, the two ends of the specimen (the tip of the specimen) were clamped using a piece of mechanical test equipment; observations were made on the tensile strength (kgf/mm^2) and the elongation percentage.

2.6. *FT-IR analysis*

Membrane testing using FT-IR (Shimadzu FT-IR—Prestige 21 Series, Shimadzu, Japan) was carried out by cutting the membrane with a size of 5x5 cm and then analyzed at a wavenumber of 4000-500 cm^{-1} in transmittance mode. This test was carried out at the Instrumentation Laboratory, Department of Chemistry, Faculty of Mathematics and Natural Sciences, Universitas Syiah Kuala.

2.7. *Thermal analysis*

The membrane with a size of 1x1 cm was tested for thermal resistance using the Shimadzu DSC-60 differential scanning calorimeter and Shimadzu DTG-60 thermal gravimetric analyzer (Shimadzu, Kyoto, Japan) instruments. This test was carried out in the Department of Chemical Engineering, Lhokseumawe State Polytechnic.

2.8. *SEM analysis*

The membrane structure was analyzed using SEM JEOL JSM 6510 LA to see the membrane surface with several times the magnification. This test was carried out at the FMIPA Laboratory of the Bandung Institute of Technology. The membrane was cut to a size of 0.5 x 0.5 cm and tested using a magnification of 1500x

3. Results and discussion

3.1. *Janeng-based starch powder isolation*

JS powder has been isolated from the janeng tubers appeared to be white in visual color. The JS powder was free from cyanide toxins. Free cyanide was proven based on the negative results from the testing with 1 M AgNO3 reagent solution. The yield of was obtained to be 7% (w/w). These characteristics and yield percentage are consistent with the previous studies [11]. The starch obtained in this study was used as a base material for membrane preparation in this study. The SEM analysis on the JS powder in Figure 2 shows that the particles produced have a relatively uniform structure and shape with an average size of 2.5 μm and have a porous structure.

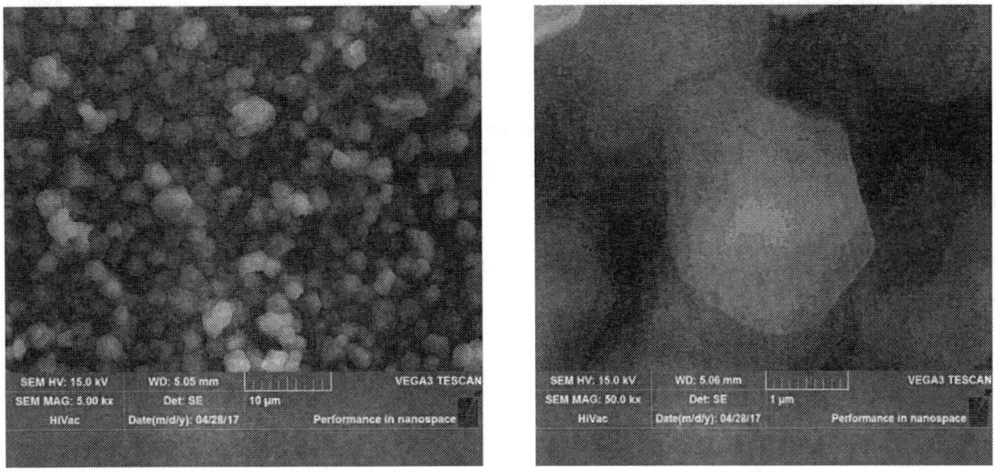

(a) (b)

Figure 2. SEM images of isolated starch particles from janeng tubers with magnification of (a) 5,000x and (b) 50,000x.

3.2. Membrane chitosan-starch

The citric acid uncross-linked and cross-linked chitosan-starch membranes were prepared by mixing a 1.5% (w/v) chitosan solution in 1% (v/v) acetic acid and a 1.5% (w/v) starch mixture in distilled water. The JS mixed into the casting solution had been priorly gelatinized. Gelatinization was needed to break the starch structure so that it could mix/absorb water by forming a gel through the formation of hydrogen bonds with amylose and amylopectin. Mixing the 1.5% (w/v) chitosan solution with gelatinized starch resulted in a homogeneous mixture. The visual images of citric acid uncross-linked and cross-linked chitosan-starch membranes can be seen in Figure 3a and b, respectively. The visual observation shows that the physical appearance of the membrane looks the same. However, based on direct observations on the manufacturing process, the uncross-linked chitosan-starch (CS0) membrane has a good physical appearance, elastic, shrink during the drying process, and easily removed from the mold. However, better physical appearance, elasticity, ease of being removed from the mold, and brittleness were observed on the citric acid cross-linked chitosan-starch (CSC) membrane.

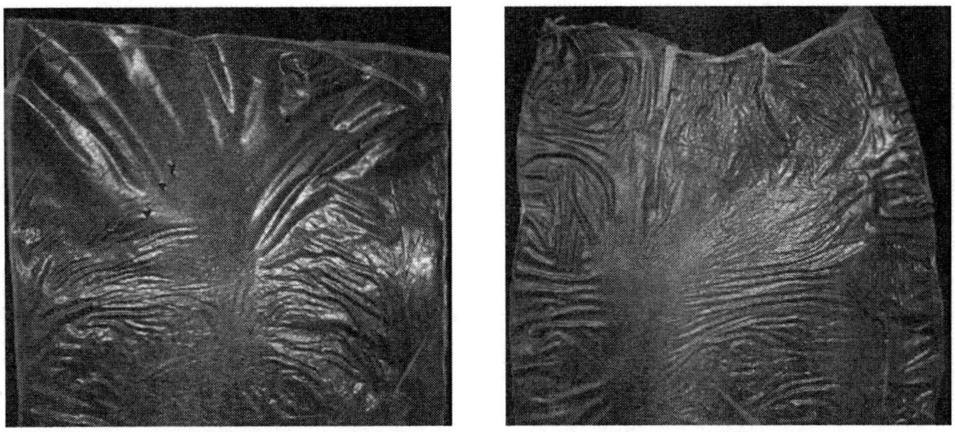

(a) (b)

Figure 3. A visual depiction of chitosan-starch membranes; (a). CS0 membrane; (b) CSC membrane.

3.3. Mechanical strength of the membrane

The mechanical properties of the membrane were analyzed to determine the tensile strength and elongation of the membrane. The membrane's mechanical properties are shown in Figure 4.

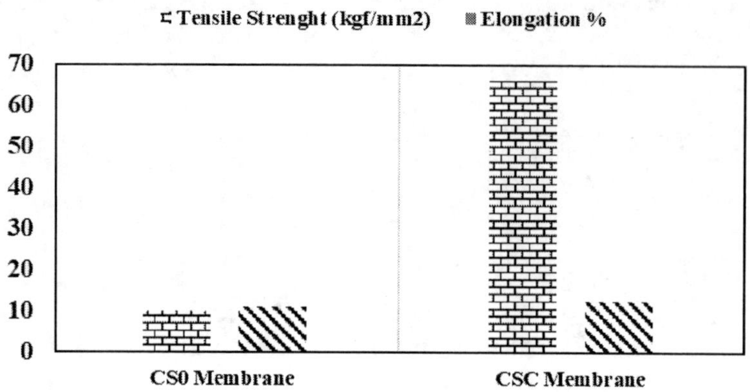

Figure 4. Tensile strength and elongation for CS0 and CSC membrane.

Figure 4 shows that the CSC membrane produced a much higher tensile strength than the CS0 membrane. The tensile strength value of the CS0 membrane was 10.07 Kgf/mm^2 and experiencing an increase after cross-linked up to 556% (66.07 Kgf/mm^2). Meanwhile, the elongation values for CS0 and CSC membranes were 11.27% and 12.7%, respectively. It reveals that the CSC membranes had higher elongation compared to CS0. The formation of cross-linking significantly affects the membrane tensile strength value. The cross-linking via citric acid addition had caused the membrane's structure to become denser, which eventually led to more robust mechanical properties. Lusiana et al. [13] reported that the cross-linking reaction between the amine groups in chitosan and the carboxyl groups in citric acid was assigned to increase membrane elasticity. Gohil et al. [14] also found that a cross-linking agent's addition caused the membrane structure to become denser.

3.4. DSC and TGA analysis the membrane

DSC thermal analysis is performed to measure the difference in heat or energy in a sample as a function of time or temperature. TGA thermal analysis is used to measure the change in weight of a compound as a temperature or time function. DSC and TGA thermal analysis were performed on CS0 and CSC membranes. The results of the DSC and TGA testing are in the form of a thermogram, which can be seen in Figure 5.

Figure 5a exhibits the thermogram of CS0 and CSC membranes, where the endothermic and exothermic peaks were observed at different temperatures. Endothermic peaks on the CS0 and CSC membranes were detected at 128.53 °C and 67.48 °C, respectively. The emergence of endothermic peaks is assigned to the evaporation of water in both membranes. Exothermic peaks on CS0 and CSC membranes were respectively detected at 318, 93 °C and 321.24 °C. The appearance of an exothermic peak represents a decomposition process. The TGA analysis supports this. The TGA thermogram results of the two membranes can be seen in Figure 4b, showing a single decomposition stage. The decomposition of the CS0 membrane was detected at temperatures ranged from 277.74 to 363.27 °C with a weight loss of 66.3%. In contrast, the CSC membrane decomposed at a temperature range of 287.83 – 356.38 °C, with a weight loss of 65.4%.

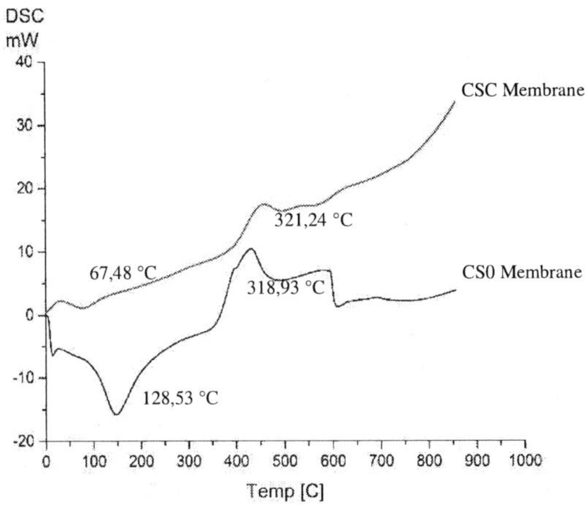

(a)

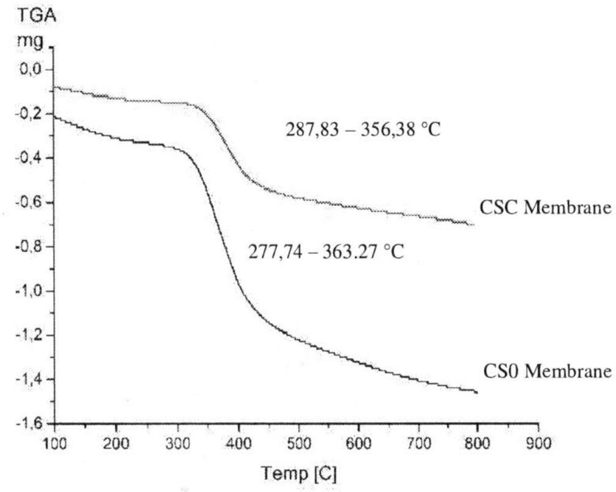

(b)

Figure 5. Thermogram results of (a) DSC and (b) TGA analysis for CS0 and CSC membranes.

3.5. FT-IR analysis

FT-IR analysis was performed to determine the functional group in the membrane formed between chitosan, starch, and citric acid. FT-IR testing was performed on the CS0 and CSC membranes, where the spectra can be seen in Figure 6.

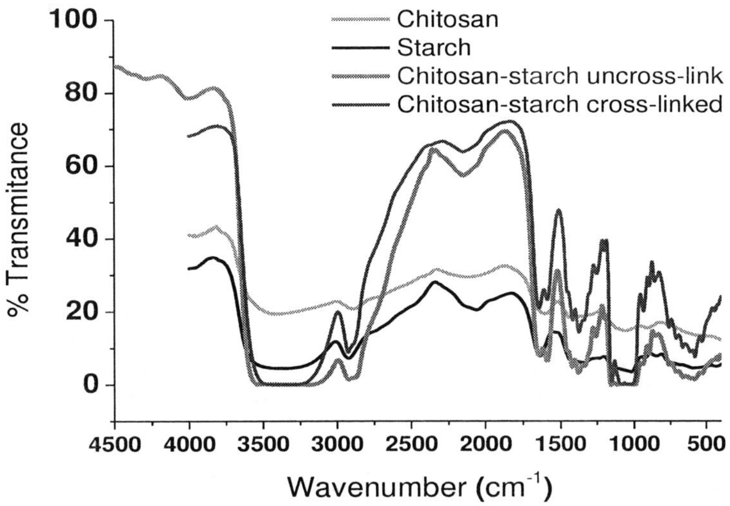

Figure 6. FTIR spectra of powder chitosan, powder starch, CS0 and CSC membrane.

Figure 6 shows a decrease in the intensity of the absorption peak obtained from the CSC membrane. Li et al. [5] stated that a decrease in peak intensity could be used to indicate the cross-linking formation. The –OH functional groups in CS0 and CSC membranes were detected at wavenumbers around 3250-3500 cm^{-1} with a strong intensity and broad peaks. The relatively stronger intensity and broader peaks indicated hydrogen bonds forming in the membrane between chitosan and starch. Furthermore, the -NH functional group from chitosan appears in the area 3000-3500 cm^{-1}, overlapping the -OH group spectral band that is widened due to the presence of hydrogen bonds. Stronger intensity is ascribed to the addition of -OH group from citric acid. It is in line with Lusiana et

IOP Conf. Series: Materials Science and Engineering 1087 (2021) 012066 doi:10.1088/1757-899X/1087/1/012066

al. [13], where they reported that the addition of starch and citric acid did not change the type of functional groups in chitosan compounds but only increased the percentage of –OH groups. Absorption peaks at 2922 and 2926 cm^{-1} indicate the presence of –CH groups on the CS0 and CSC membranes, respectively. The specific difference between the modified chitosan spectra can be observed at the sharp absorption at the wavenumber 1597cm^{-1}, which indicates a modification of the primary amine group to the secondary amine.

3.6. Scanning Electron Microscope (SEM) analysis

SEM analysis was performed to determine the surface morphology of the upper, lower, and cross-section parts of the membrane. The results of the SEM morphological analysis can be seen in Figure 7.

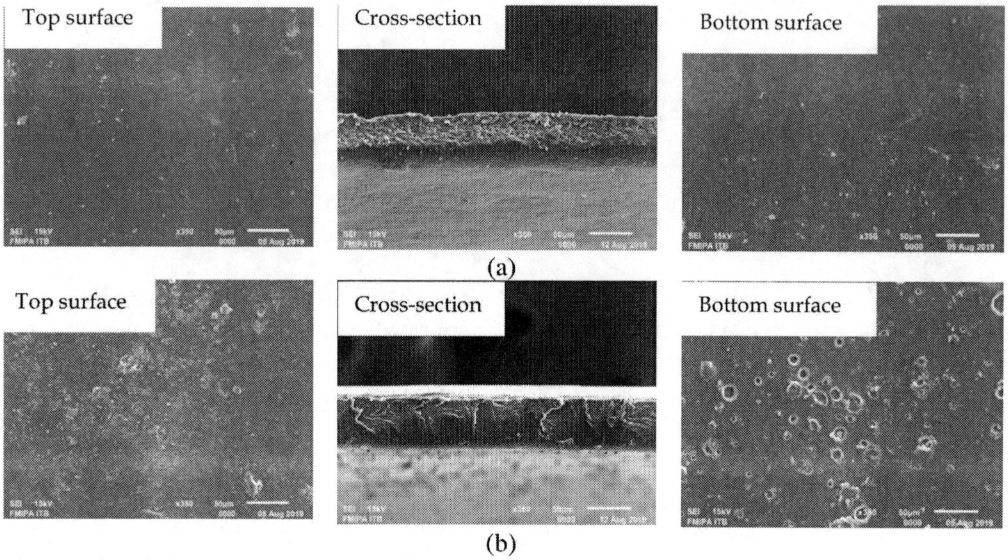

Figure 7. The SEM images of (a) CS0 and (b) CSC membranes on the top surface, cross-section and bottom surface.

Figure 7 shows that the upper surface of the CS0 membrane is smoother than the CSC membrane, which shows a coarser surface, as well as the lower surface of the membrane which shows the same conditions. The difference in the surface structure is due to the effect of adding citric acid to the membrane as a cross-linking agent. In the cross-section, the CS0 membrane looks coarser than the CSC membrane. Membranes with surfaces with different top and bottom surface structures fall under a category of asymmetric membranes. An asymmetric membrane is a membrane that has a structure and pore size that are not uniform.

4. Conclusion

The chitosan membrane was modified with the addition of starch, followed by a cross-linking reaction using citric acid. The uncross-linked membrane had a tensile strength of 11.07 kgf/mm^2, while the cross-linked membrane had better tensile strength (66.07 kgf/mm^2). The DSC results of the uncross-linked and cross-linked membrane showed endothermic and exothermic peaks at different temperatures. The membranes' DTA thermogram showed that the decomposition process on the uncross-linked membrane occurred at 277.74 – 363.27 °C, while in the cross-linked membrane, the decomposition point occurred at 287.83 – 356.38 °C. The FT-IR spectra confirmed a decrease in the intensity of the absorption peak associated with the formation of cross-links on the membrane.

5. Acknowledgement
This research was funded by Ministry of Education and Culture of the Republic Indonesia Competitive Research Grant.

6. References
[1] Saiful, Riana U, Marlina, Ramli M, Mahmud N 2019 *IOP Conf. Ser. Earth Environ. Sci.* **273** 012047.
[2] Shakeri A, Salehi H, Rastgar M 2017 *Carbohydr. Polym.* **174** 658-68.
[3] Saiful, Riana U, Marlina, Ramli M, Mahmud N 2018 *Aceh Int. J. Sci. Technol.* **7(8)**.
[4] Huang Y, Sun J, Wu D, Feng X 2018 *Sep. Purif. Technol.* **207** 142-150.
[5] Li H, Gao X, Wang Y, Zhang X, Tong Z 2013 *Int. J. Biol. Macromol.* **52** 275-279.
[6] Xie Y, Zhang B, Li M N, Chen H Q 2019 *Food Chem.* **289** 187-194.
[7] Chen Y, Qian J, Zhao C, Yang L, Ding J, Guo H 2019 *Eur. Polym. J.* **118** 17-26.
[8] Atangana E, Chiweshe T T, Roberts H 2019 *J. Polym. Environ.* **27** 979-95.
[9] Mukoma P, Jooste B R, Vosloo H C M 2004 *J. Power Sources.* **136** 16-23.
[10] Saiful, Rahmah Z, Ajrina M, Marlina, Rahmi 2020 *Rasayan J. Chem.* **13** 2062-73.
[11] Saiful, Helwati H, Saleha S, Iqbalsyah T M 2019 *IOP Conf. Ser. Mater. Sci. Eng.* **523** 012015.
[12] Prochaska K, Konował E, Sulej-Chojnacka J, Lewandowicz G 2009 *Colloids Surf. B Biointerfaces* **74** 238-243.
[13] Lusiana R A, Siswanta D, Mudasir M 2018 *Indonesian J. Chem.* **16** 144-150.
[14] Gohil J M, Bhattacharya A, Ray P 2006 *J. Polym. Res.* **13** 161-169.

Fourier Transformed Infrared (FTIR) spectroscopy analysis of patchouli essential oils based on different geographical area in Aceh

E Sufriadi[1,2,4], H Meilina[3,4], A A Munawar[4,5], R Idroes[2,6,*]
[1]Graduate School of Mathematics and Applied Sciences, Universitas Syiah Kuala, Banda Aceh 23111, Indonesia
[2]Department of Chemistry, Faculty of Mathematics and Natural Sciences, Universitas Syiah Kuala, Banda Aceh 23111, Indonesia
[3]Chemical Engineering Department, Faculty of Engineering, Universitas Syiah Kuala, Banda Aceh 23111, Indonesia
[4]RC PUI-PT Nilam Aceh, Universitas Syiah Kuala, Banda Aceh 23111, Indonesia
[5]Agricultural Technology Department, Universitas Syiah Kuala, Banda Aceh 23111, Indonesia
[6]Department of Pharmacy, Faculty of Mathematics and Natural Sciences, Universitas Syiah Kuala, Banda Aceh 23111, Indonesia

*E-mail: rinaldi.idroes@unsyiah.ac.id

Abstract. Studies on the quality difference between oils in various provinces have been conducted with varied results, especially between Acehnese Patchouli and Javanese Patchouli. However, studies on quality differences of Acehnese Patchouli based on geographic differences have never been carried out, thus, still leaving a big mystery in determining whether geographical differences such as coordinates and height from sea level produce different quality. The purpose of this study was to determine the effect of geographic location (coordinates and height of water MASL) of the patchouli cultivations on oil quality by conducting Principal Component Analysis (PCA) on the FTIR spectrum of patchouli oil from 3 different districts, namely South Aceh, Aceh Tamiang and South-West Aceh. The results indicate a water content difference as it is close to the O-H group wave number, which usually indicates the presence of Patchouli alcohol. The PCA analysis results show that there is a very significant difference between patchouli oil from South Aceh and Aceh Tamiang and South-West Aceh, caused by the height and patchouli variety difference. The PCA method validation is also quite reliable, which is proven by an explained variance plot.

1. Introduction

Patchouli plants have been known for thousands of years in China and India. Patchouli oil produced from the distillation of the patchouli plant has been used for hundreds of years in Traditional Chinese Medicine (TCM). Botanically, the two known species are *Pogostemon cablin* and *Pogostemon heynanus* [1]. Past studies on these two patchouli species have produced almost the same main chemical components but have different compositions of each compound [2]. PogostemonHeynanus has a lower patchouli alcohol content compared to Pogostemoncablin but has a higher alpha-caryophyllene content [3].

Studies on the chemical components formed between these two species have been widely carried out. Aceh is one of the provinces in Indonesia famous for Patchouli since the colonialism era in 1800[4]. The Dutch trade alliance has been trading patchouli leaves from Aceh to foreign countries ever since. Aceh is even known as a patchouli oil-producing area with excellent quality, as indicated with one of the main and most important contents of the oil, i.e., patchouli alcohol, which was found to have the highest content in Aceh [5].

For hundreds of years, the Patchouli planted in Aceh was the PogostemoncablinBenth species. This species has three varieties that are cultivated by farmers, namely Sidikalang, Lhokseumawe, and the Tapak Tuan varieties [2]. Past research has identified the differences between these three varieties, both in terms of yield and from aspects of chemical composition. In general, the chemical content of patchouli oil derived from PogostemoncablinBenth consists of sesquiterpenes, phenolic compounds, and a small portion of alkaloids, as shown in Table 1 below.

Table 1. Common chemical components found in patchouli oil.

No	Constituents	Concentration range (%)	Reference
1	α-Pinena	0.01-0.3	[6]
2	β-Pinena	0.02-1	[7]
3	Limonene	0.01-0.3	
4	δ-Elemene	0.01-1.9	
5	β-Pachoulene	0.03-12	[8]
6	β-Elemene	0.18-1.9	
7	Cycloseychellene	0.02-0.8	
8	(E)-β-Caryophyllene	0.75-6.8	[9,10]
9	α-Guaiene	2.9-23	[11,12]
10	Seychellene	2.3-13	[8,13]
11	α-Humulene	0.05-2	
12	α-Patchoulene	1.2-13	[14,15]
13	Germacrene D	0.0-0.2	[1,2,16]
14	Aciphyllene	0.7-4.2	[17]
15	α-Bulnesene	2.9-23	[13,15]
16	Norpatchoulenol	0.11-4.0	[12,18]
17	Caryophyllene oxide	0.0-4.6	[19,20]
18	Pogostol	0.2-6.2	[13,20]
19	Patchoulol	11-72	[11,13,14,18]
20	Pogostone		[10,12]

Aceh patchouli is known to have better quality than that of Java, Sulawesi, and other regions. Meanwhile, Aceh also has a variety of patchouli oil cultivation and refining locations under different conditions, such as rainfall and height from sea level.

Studies on the chemical composition of Aceh patchouli based on geographic differences have never been carried out. Thus it is crucial to see whether geographical differences such as coordinates and height from sea level give different qualities. Several instrumental methods have been widely used to identify the chemical composition on organic substances, such as Uv-Vis [21–23], laser [24,25], Mass Spectrometry [26–29], FTIR [30,31], and even image processing [32]. Functional group analysis using Fourier Transform Infrared Spectroscopy (FTIR) is a promising analysis technique to authenticate essential oils because it uses a fingerprint analysis approach. Essential oil fingerprint analysis was used for a qualitative approach, including authenticating plant species and assessing their quality [33]. One indicator in this study was to see the difference in the FTIR spectrum of patchouli oil from different geographic locations. In general, the main reference to identify differences in functional groups of FTIR absorption was based on some peaks seen in particular wavenumber with a specific intensity value. However, the complex spectra produced by FTIR analysis required the statistical method which can solve the complex data [34–37]. Principle Component Analysis (PCA) is the fit method for this purpose by classifying the data based on the similarities or differences [38,39]. This reference is expected to identify differences in chemical composition based on geographic differences in Patchouli oil [40].

2. Materials and methods

2.1. Materials and samples

This study used patchouli oil samples of *Pogostemon cablin* Benth species taken in 3 districts with varying geographic locations based on differences in coordinates and height from sea level. The samples came from the Regency of South Aceh, Aceh Tamiang, and South-West Aceh, with five sample points that were taken for each district. The distribution of sampling points in each district is shown in Table 2.

Table 2. Distribution of patchouli oil sampling locations in three districts.

Regency	Coordinate	Height (MASL)
South Aceh		
Point 1	3.150787, 97.387019	32.7
Point 2	3.246224, 97.298007	363
Point 3	3.105984, 97.403731	241
Point 4	3.057648, 97.328140	25.8
Point 5	3.216805, 97.381108	39.9
South-West Aceh		
Point 1	3.955137, 96.801492	15.6
Point 2	3.756932, 96.779986	9.0
Point 3	3.843926, 96.760195	18.9
Point 4	3.792979, 96.637504	23,1
Point 5	3.874802, 96.610691	84.6
Aceh Tamiang		
Point 1	4.325478, 97.852359	66.9
Point 2	4.329639, 97.831961	69.6
Point 3	4.331321, 97.953951	91.2
Point 4	4.290266, 97.977026	66.6
Point 5	4.337483, 97.887003	44.1

2.2. Measurement by FTIR and Principal Component Analysis (PCA)

The analytical equipment used was FTIR Shimadzu Model IR Prestige 21, 2012. The analysis was carried out at the Environmental Laboratory of the Chemical Engineering Department, Syiah Kuala University. The analysis was performed over a wavenumber range of 4,000-500 cm^{-1}. The measurement results were tabulated with Microsoft Excel version 365, and the FTIR spectrum for wavenumber (cm^{-1})versus transmittance was plotted. Principal component analysis (PCA) was performed using Unscrambler software version 10.4. The output obtained from the PCA analysis results is the Score Plot and Loading Plot, which can be used to see the effect of several variables on the chemical differences of patchouli oil from three different locations.

3. Results and discussion

3.1. Functional group identification using FTIR

Fourier transformed infrared spectroscopy is one of the most widely used techniques in the identification of functional groups in a mixture of chemical compounds. Figures 1a to 1c show the presence of an infrared spectrum and certain absorption bands observed in Patchouli Oil between 4000-500 cm. In general, 20-30 chemical components are often found in patchouli oil, although past studies show that the number of chemical components in patchouli oil can reach more than 50 compounds [16]. All of the chemical compounds in Patchouli Oil have various compound groups such as Sesquiterpene hydrocarbons (SQHCs), Aciphyllene, a-Copaene, Cycloseychellene, γ-Patchoulene, η-Selinene, Oxygenated sesquiterpenes, Norpatchoulenol, Nortetracyclopatchoulol, Pogostol, Selinane SQHCs, Sesquiterpene ketones, Pogostone, Eugenol, benzaldehyde and cinnamic aldehyde [11,17,41,42]. These various groups of the compounds produce tens of chemical components whose existence varies depending on the location of cultivation [43,44]. However, generally, the types of bonds that produce molecular vibrations in patchouli oil occur in the form of stretching OH at 3505 cm^{-1}, which comes from the alcohol group compound and displays an abundance of Patchouli alcohol (the essential component used in the perfume industry). The spectrum also showed vibratory stretching at 1635 cm^{-1}, confirming the presence of C = O carbonyl aldehyde. This proves that patchouli oil is rich in aldehyde compounds [45].

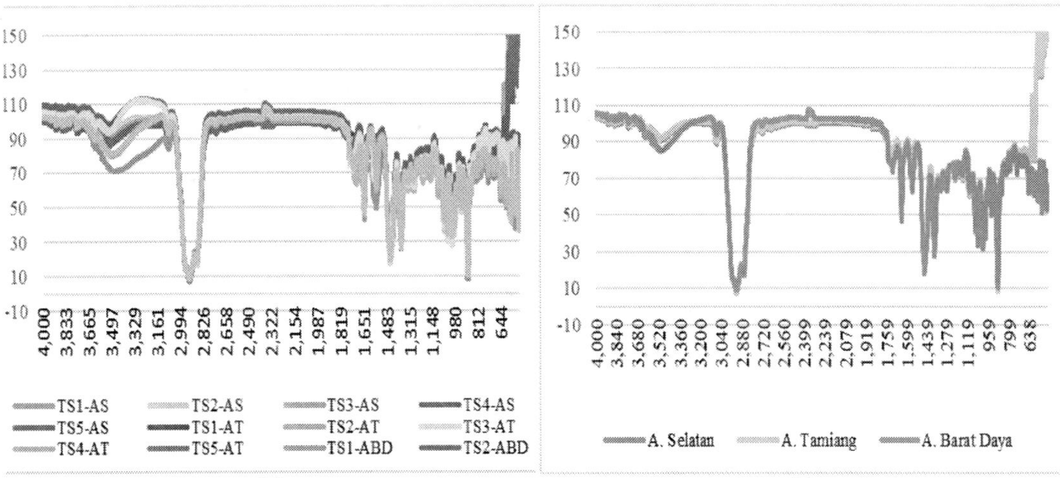

(a) (b)

Figure 1. FTIR spectrum for patchouli oil originating from South Aceh, South-West Aceh, and Aceh Tamiang Regency: (A) combined 15 points spectrum from three regencies; and (B) The spectrum of mean transmittance values for each regency. TS =sample point; AS = South of Aceh; ABD = Southwest Aceh; AT = Aceh Tamiang.

A peak at 1445 cm^{-1} indicates a stretching of C-H bonds in the alkane and a C = C stretching of the aromatic ring[46], and at the peak of 1373 cm-1 is characteristic of the O-H bond of carboxylic acids[47]. Of the three spectrums observed in Figure 1, there is a band widening in the spectrum of patchouli oil originating from South-West Aceh, especially at 3500-3040 cm^{-1}. This symptom indicates an unusual abundance of H_2O because it is close to the O-H group wave number, which usually provides information on the presence of Patchouli alcohol [48]. When viewed visually, of the five oil samples from South-West Aceh Regency, three of them look cloudy and dark in color, which indicates that the patchouli oil contains impurities. However, this assumption still needs to be proven by the Principal Component Analysis (PCA) analysis results.

3.2. Principal Component Analysis (PCA)

PCA analysis of a spectrum of 15 samples from three regencies with different coordinates and height (MASL) is shown in Figure 2. The analysis results show that there is a qualitatively significant difference between patchouli oil in South-West Aceh Regency and Aceh Tamiang and South Aceh Regency. The principal component analysis also shows that three PCs represent 96% of the variance of the sample.

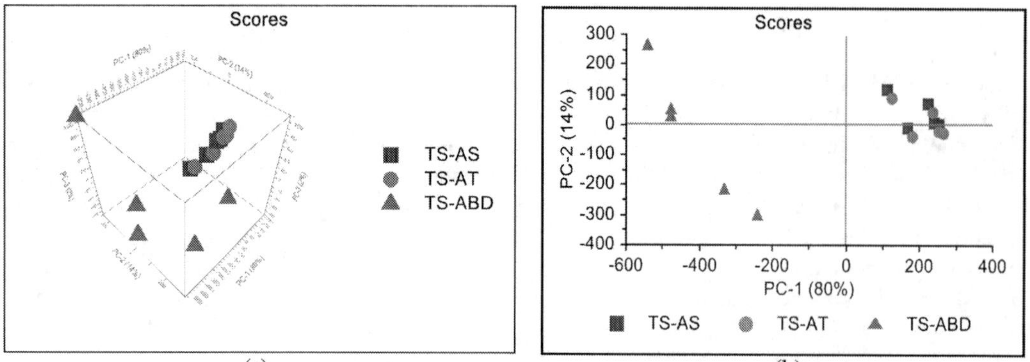

(a) (b)

Figure 2. PCA analysis results of patchouli oil FTIR spectrum originating from 3 districts in Aceh, namely South Aceh, Aceh Tamiang, and South-West Aceh, (A) Score plot PC1 vs PC2; and (B) 3D score plot of PC1, PC2, and PC3.

PC1-PC3 respectively gave variance values of 80, 14, and 2%. The data are clustered based on the patchouli oil regency origin. Although the number of samples is still relatively small, there is a good separation between the sample groups and the PC1 axis. Findings show a closer similarity between patchouli oil in South Aceh and Aceh Tamiang, and it looks significantly different from patchouli oil from South-West Aceh. It is suspected that the height of the cultivation location affects the similarity of patchouli oil properties, as indicated by the average height of the sample locations, namely: South Aceh has an average height of 140.48; South-West Aceh with 30.24; and Aceh Tamiang with 67.68 MASL. It appears that the height of the Aceh Strait is closer to Aceh Tamiang, although geographically, it is closer to South-West Aceh. Another assumption is the factor of differences in planted patchouli oil varieties. South-West Aceh may plant different varieties of Patchouli from South Aceh and Tamiang. However, these findings need to be tested further with other chemometric methods such as Discriminant Analysis [2,33].

The results of the loading plot (selected from PC-2) shown in Figure 3a show that in the range of 3600-2600 cm-1, there is a widening of the band, which indicates a difference in the water content of the sample3[49]. It also appears that there is much information that needs to be explored in the range between 1600-650 cm-1. The analysis's repetition is an important consideration when analyzing, by carrying out a specific analysis only in that range to obtain a better band.

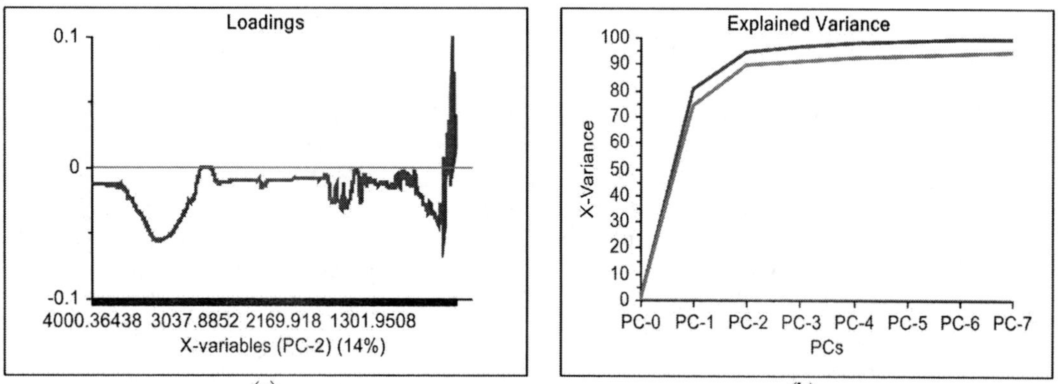

(a) (b)

Figure 3. Loading plot of PCA analysis results (a) and Explained Variance (b) of the Aceh patchouli oil. The curve explained variance shown in Figure 3a provides information that the PCA model used in this study is classified as robust, indicated by the red curve (validation), which is always in the same direction as the blue curve.

4. Conclusion

There are indications of water abundance in an unusual amount because it is close to the O-H group's wavenumber, which usually provides information on the presence of the Patchouli alcohol compound. The PCA analysis results show that there is a significant difference between patchouli oil from South Aceh Selatan and Aceh Tamiang and South-West Aceh, due to the difference in height or differences in the varieties of Patchouli. The loading plot strengthens the suspicion of contamination by water. The PCA method validation is also quite reliable, which is shown by the explained variance plot.

5. References

[1] Howarth J 2015 *Parfum Flavor* **40** 32–37.
[2] Diego M C R, Purwanto Y A, Sutrisno S, Budiastra I W 2018 *IOP Conf. Ser. Earth Environ. Sci.* **147** 012013.
[3] Tang Y, Zhong L, Wang X, Zheng H, Chen L 2019 *Bot. Stud.* **60** 1-11.
[4] Yu Z X, Wang L J, Zhao B, Shan C M, Zhang Y H, Chen D F, Chen X Y 2015 *Mol. Plant.* **8** 98–110.
[5] Sufriadi E, Aisyah Y, Harahap F, Fernando Y, Mardina V 2020 *IOP Conf. Ser. Mater. Sci. Eng.* **725** 012066.
[6] Murugan R, Mallavarapu G R, Padmashree K V, Rao R R, Livingstone C 2010 *Nat Prod Commun.***5** 1961–4.
[7] Yahya A, Yunus R M 2013 *Procedia Eng.* **53** 1–6.
[8] Swamy M K, Mohanty S K, Sinniah U R, Maniyam A 2015 *J. Essent. Oil-Bear. Plants* **18** 826–832.
[9] Adinew B 2014 *Int. J. Herb. Med.* **1** 22–31.
[10] Yuliani F, Riyanto S, Rohman A 2018 *Int. J. App. Pharm.* **10** 54–59.
[11] Blank A F, Sant'ana T C P, Santos P S, Arrigoni-Blank M F, Prata A P do N, Jesus H C R, Alves P B 2011 *Ind. Crops Prod.* **34** 831–837.
[12] Kusuma H S, Altway A, Mahfud M 2018 *J. Ind. Eng. Chem.* **58** 343–348.
[13] Donelian A, Carlson L H C, Lopes T J, Machado R A F 2009 *J. Supercrit. Fluids.* **48** 15–20.
[14] Dantas T N C, Cabral T J O, Dantas Neto A A, Moura M C P A 2020 *J. Ind. Eng. Chem.* **81** 219–227.
[15] Fan L, Jin R, Liu Y, An M, Chen S 2011 *J. Chromatogr. B Analyt. Technol. Biomed. Life Sci.* **879** 3653–7.
[16] van Beek T A, Joulain D 2018 *Flavour Fragr J.* **33** 6–51.
[17] Sundarajan S, Abdurahman N H 2019 *Int. J. Eng. Tech. Res.* **8** 1054–9.
[18] Nithyanand P, Beema Shafreen R M, Muthamil S, Murugan R, Karutha Pandian S 2015 *Ind. Crops Prod.* **69** 180–186.
[19] Dai O, Li X H, Zhou Q M, Peng C, Guo L, Xiong L 2018 *Phytochem. Lett.* **24** 56–59.
[20] Chen Y, Wu Y G, Xu Y, Zhang J F, Song X Q, Zhu G P, Hu X W 2014 *Bras. J. Farmacogn.* **24** 626–634.
[21] Suhartono E, Thalib I, Aflanie I, Noor Z, Idroes R 2018 *IOP Conf. Ser. Mater. Sci. Eng.* **350** 12008.
[22] Hasanah U, Sani N D M, Heng L Y, Idroes R, Safitri E 2019 *Biosensors.* **9** 135.
[23] Hasanah U, Setyowati M, Efendi R, Muslem M, Md Sani N D, Safitri E, Yook Heng L, Idroes R 2019 *Biosensors.* **9** 60.
[24] Hedwig R, Lahna K, Idroes R, Karnadi I, Tanra I, Iqbal J, Kwaria D, Kurniawan D P, Kurniawan K H, Tjia M O, Kagawa K 2019 *Microchem. J.* **147** 356–364.
[25] Lahna K, Idroes R, Idris N, Abdulmadjid S N, Kurniawan K H, Tjia M O, Pardede M, Kagawa K 2016 *Appl. Optics.* **55** 1731.
[26] Estevam E, Griffin S, Nasim J, Zieliński D, Aszyk J, Osowicka M, Dawidowska N, Idroes R, Bartoszek A, Jacob C 2015 *Nat. Prod. Commun.* **10** 1733–8.

[27] Pratiwi S U T, Lagendijk E L, de Weert S, Idroes R, Hertiani T, den Hondel C Van 2015 *Int. J. Appl. Res. Nat. Prod.* **8** 1–13.
[28] Earlia N, Muslem, Suhendra R, Amin M, Prakoeswa C R S, Khairan, Idroes R 2019 *Sci. World J.* **2019** 1–7.
[29] Earlia N, Rahmad R, Amin M, Prakoeswa C, Khairan K, Idroes R 2019 *Sains Malays.* **48** 1019–24.
[30] Suhartono E, Noor Z, Edyson, Budianto W Y, Idroes R 2019 *AIP Conf. Proc.* **2108** 020025.
[31] Hasanah U, Setyowati M, Edwarsyah, Efendi R, Safitri E, Idroes R, Heng L Y, Sani N D 2019 *IOP Conf. Ser. Mater. Sci. Eng.* **523** 12014.
[32] Suhartono E, Setiawan B, Santosa P B, Idroes R, Indrawan M S 2019 *J. Phys. Conf. Ser.* **1374** 12057.
[33] Fahmi Z, Mudasir, Rohman A 2020 *Indonesian J. Chem.* **20** 1–8.
[34] Idroes R, Husna I, Muslem, Mahmudi, Rusyana A, Helwani Z, Idroes G M, Suhendra R, Yandri E, Rahimah S, Sasmita N R 2019 *IOP Conference Series: Earth and Environmental Science.* **364** 12020.
[35] Husna I, Rusyana A, Muslem, Idroes G M, Suhendraand R, Idroes R 2020 *IOP Conf. Ser. Mater. Sci. Eng.* **796** 012064.
[36] Idroes R, Japnur A F, Suhendra R, Rusyana A 2019 *IOP Conf. Ser. Mater. Sci. Eng.* **523** 12007.
[37] Idroes R, Noviandy T R, Maulana A, Suhendra R, Sasmita N R, Muslem M, Idroes G M, Irvanizam I 2019 *Int. Rev. Model. Simul.* **12** 373.
[38] Wahidah S, Khairi, Lelifajri, Idroes R, Rahmadi, Lala A, Mahmudi, Muslem, Japnur A F 2019 *IOP Conf. Ser. Earth Environ. Sci.* **364** 12016.
[39] Nadia Y, Ramli M, Japnur A F, Rusyana A, Idroes G M, Suhendra R, Sasmita N R, Tallei T E, Idroes R 2019 *IOP Conf. Ser. Earth Environ. Sci.* **364** 12040.
[40] Smithson S C, Fakayode B D, Henderson S, Nguyen J, Fakayode S O 2018 *Foods* **7** 122.
[41] Das K 2016 *Patchouli (Pogostemon cablin Benth) oils*. Amsterdam: Elsevier Inc.
[42] Yang Y, Kong W, Feng H, Dou X, Zhao L, Xiao Q, Yang M 2016 *J. Pharm. Biomed. Anal.* **121** 84–90.
[43] Al Riza D F, Widodo S, Purwanto Y A, Kondo N 2019 *Flavour Fragr. J.* **34** 15–20.
[44] Mayerni R, Anwar A, Febriyetty L, Rukmana S, Sagita O, Sumatera W 2017 Identification of anatomy and quality of essential oil from patchouli plant (pogostemon cablin benth) in west pasaman.
[45] Li Y Q, Kong D X, Wu H 2013 *Ind. Crops Prod.* **41** 269–278.
[46] Niu B, Yan Z, Shao P, Kang J, Chen H 2018 *Nanomaterials* **8** 598.
[47] Hussin N, Mondello L, Costa R, Dugo P, Nik Yusoff N I, Yarmo M A, Ab. Wahabd A, Said M. 2012 *Nat. Prod. Commun.* **7** 927–930.
[48] Fahmi Z, Mada U G, Mudasir M, Mada U G 2017 *Indonesian J. Chem.* **20** 1-8.
[49] Bro R, Smilde A K 2014 *Anal. Methods.* **6** 2812–31.

IOP Conf. Series: Materials Science and Engineering 1087 (2021) 012068 doi:10.1088/1757-899X/1087/1/012068

The influence of temperatures, polarity, modifier and pressure to retention index in supercritical fluid chromatography: A review

R Idroes[1,2,*], Muslem[3], Mahmudi[4], Saiful[1], M Paristiowati[5], G M Idroes[6], R Suhendra[7], A Maulana[7], T R Noviandy[7], Irvanizam[7]

[1]Department of Chemistry, Faculty of Mathematics and Natural Sciences, Universitas Syiah Kuala, Kopelma Darussalam, Banda Aceh 23111, Indonesia
[2]Department of Pharmacy, Faculty of Mathematics and Natural Sciences, Universitas Syiah Kuala, Kopelma Darussalam, Banda Aceh 23111, Indonesia
[3]Department of Chemistry, Faculty of Science and Technology, Universitas Islam Negeri Ar-Raniry, Banda Aceh 23111, Indonesia
[4]Sekolah Tinggi Ilmu Kesehatan Assyifa Aceh, Banda Aceh 23242, Indonesia
[5]Department of Chemistry Education, Universitas Negeri Jakarta, Jakarta, Indonesia
[6]Department of Chemical Engineering, Faculty of Engineering, Universitas Syiah Kuala, Kopelma Darussalam, Banda Aceh 23111, Indonesia
[7]Department of Informatics, Faculty of Mathematics and Natural Sciences, Universitas Syiah Kuala, Kopelma Darussalam, Banda Aceh 23111, Indonesia

*E-mail: rinaldi.idroes@unsyiah.ac.id

Abstract. Changes in analysis parameters affect the retention index calculation on all chromatography systems. In Supercritical Fluid Chromatography, the analysis parameters are more than Gas Chromatography or Liquid Chromatography. In this study, the effect of all those parameters were presented based on the research that has developed to date. The data and the information were collected from the previous published articles. The changes in the analysis parameters which are temperature, polarity, modifier, and pressure on SFC affected the retention index calculation, but the significant levels were different. The pressure change has the least significant effect for both polar and non-polar analytes. The temperature and stationary phase polarity changes have insignificant effects for non-polar analyte, but significant for polar one. Meanwhile, the modifier change have a very significant effect, both for polar and non-polar analytes.

1. Introduction

Retention index studies on chromatography have shown great progress. The retention index calculation has been able to provide the accurate and reliable results. This success has changed the function of chromatography from being just a separation method to the chemical profile identification of the analytes [1].

In classical analysis, chromatography yields the separation information through the retention time. The number of retention time is proportional to the number of analytes contained in the sample. However, the retention time cannot provide the information of the chemical profile of the analytes. This is due to the dependence of the retention time to the analysis condition. If the analysis parameters such as temperature and column polarity are changed, the retention times also change [2]. Therefore, to determine the chemical profile of the analyte, it need to use either additional methods or instruments. If the analyte profiles is already known, identification is done by measuring the retention time of the standard compound [3]. However, if the analyte is a new or unknown compounds, a tandem instrument is carried out where each retention time is identified by the tandem, such as mass spectrometry [4–9], infrared spectroscopy [10], or nuclear magnetic resonance [11].

In modern analysis, the retention time can also be converted into a retention index by utilizing the dead time and the carbon number of the standard compound. The three retention parameters were applied to a mathematical algorithm introduced by Erwin Kovats in 1958 [12,13]. The retention index is robust against the changes of the analysis parameters. Although the retention of certain analyte varies, the retention index calculation always produce a relatively fixed value [14]. The retention index is related to the analyte carbon number. The carbon number is related to the analyte polarity. Therefore, if the retention index can be calculated, the carbon number and polarity of the analyte can be predicted [15].

The main challenges in determining the retention index are the accuracy of a dead time determination and a selectivity of the separation. The dead time is calculated either by measuring an inert gas in the chromatography column or by using a homologous series method [16]. With algorithms, the retention times of the homologous series are converted into the dead time [17,18]. As chromatography measurement products, both methods are always challenged by various analysis parameters [19]. As with the dead time, the selectivity of separation also plays an important role because it determines the accuracy of determining the retention time. In fact, this last one is only achieved at an optimum level through the determination of the suitable analysis parameters [20]. Therefore, the accuracy of the retention index is indirectly related to the analysis parameters.

The parameters which affect the retention index calculation in Gas Chromatography (GC) include polarity, temperature and column length [21]. In Liquid chromatography (LC) and supercritical fluid Chromatography (SFC), the number of these parameters increases, namely solvent composition, modifier, homologous series type and pressure [22,23]. The accuracy issue of the retention index calculation in GC system was mostly solved, where homologous series methods such as iteration, spreadsheets and statistics can produce the accurate dead time against the changes of analysis parameters [24]. In fact, this retention index has been applied in the identification of organic compounds [25–28].

In LC and SFC, the accuracy of the retention index calculation is weaker than in GC. This is because the analysis parameters is more complex. This problem has not been fully resolved neither by dead time algorithm nor by analysis method [29]. However, the study of retention index in LC is more developed than SFC. Calculations of both the retention index and the dead time in LC have several improvement and become closer to accurate [30,31]. In addition, the study in SFC itself has been slower to run. There were only a few related publications that discussed in this regard.

In this study, a discussion were carried out related to retention index studies on SFC. Various analysis parameters, which are temperature, polarity, modifier and pressure, affected the retention index calculation were described in detail. So that researchers can find out the latest developments in SFC. So that, it can facilitate the development of further research.

2. The effect of temperature and polarity on the retention index calculation

Temperature greatly determines the selectivity of the separation. It affects the strength of the interaction between the analyte and a stationary phase. This causes a shift in an elution ability to the marker measurement as well as the homologous series separation in the dead time determination [32]. The studies of 5 marker types representing 5 different polarity, as Table 1 illustrates, showed both in polar and non-polar columns fluoro cyclic marker provided the most constant dead times against the temperature changes. This marker has the lowest affinity compared to other ones in all column polarity types. The same thing is also shown in the separation of the n-alkanes homologous series under the same conditions. Both the fluoro-cyclic marker and the n-alkane homologous series have the same constant properties (in terms of separation density) [33]. Therefore, it can be stated that both compounds are assumed to represent the most suitable standard in determining the dead time for the retention index calculation purpose.

Tabel 1. The stationary phase of SFC column

Compound	Characteristics
Anthracene	Relatively polar
1,10-Decanediol	Polar
Methyl 3,5 dinitrobenzoate	Strongly polar
Methyl myristate	Relatively non polar
Fluoro cyclic	Non polar
n-Alkanes	Strongly non polar

In determining the retention index, the n-alkane homologous series still showed the lowest effect on the changes in temperature and column polarity compared to marker [34]. Among the five markers, methyl myristate dead time produced the most constant retention index compared to others. This retention index increased insignificantly with increasing the polarity. This corresponded to the non-polar character of the strong alkyl and the weak polar one which it owned. The temperature changes tested for all column polarity did not even have the effect to the retention index calculation from this compound [33].

The significant effect was shown in the retention index changes calculated by anthracene and methyl 3,3 dinitrobenzoate dead time. There is a significant change against the increase of temperature and column polarity. This was attributed to the polar character possessed by the phi bonds in aromatic rings [33].

This review concluded that the non-polar compounds for dead time calculation had the least significant effect to Retention index calculation against the changes of temperature and column polarity. Therefore, this compound type can be used as a reference for dead time and retention index studies in SFC.

3. The effect of modifier on the retention index calculation

In SFC, a modifier is added into the mobile phase to increase the analyte separation ability in the sample [35]. In some routine SFC analysis, several classes of compounds such as barbiturates, separated by a CO_2 mobile phase and a Polystriyrene-divinylbenzene (PS-DVB) column, showed very poor separation. Even with the use of the Octadecyl-silica (ODS) column, the barbitures were completely inseparable [36]. Therefore, a modifier such as methanol is applied to the mobile phase. In the suitable ratio of the mobile phase: modifier composition, the separation are better.

The addition of methanol and acetonitrile modifiers into the CO_2 mobile phase caused the change of the retention index. The retention index of the n-alkane and alkylarylketones homologous series, the changes was not very significant, but for the test substance is the otherwise [36].

The study of Barbiturates' retention index by using alkyl aryl ketones homologous series, CO_2 mobile phase, methanol modifier and PS-DVB column showed the significant changes in capacity factor and retention index. The presence of methanol in the mobile phase changes the interaction pattern between each analyte to both the mobile and stationary phases. The changes in modifiers and their composition ratios led to a significant difference in the retention of the relative unrelated compounds between the alkyl aryl ketones homologous series and barbiturates [36].

An alternative method to treat the significant effect of modifier, adjusting the pH on the mobile phase: modifier composition is necessary. Setting the pH of the composition appropriately minimized the retention index changes caused by modifier. However, the calculated retention index obtained wasn't categorized as constant yet. It mean that the effect of the modifier is still significant [36]. This condition indicates that in the laboratory analysis, the purposes of separation effectiveness and the retention index calculation cannot be done simultaneously.

4. The effect of pressure on the retention index calculation

Pressure is a unique parameter which only exists in SFC. This is related to the use of the supercritical mobile phase which is a function of temperature and pressure. Pressure is very difficult to determine accurately from one analysis to another. Even if the equipment is equipped with standard gauges, the average pressure in the column highly depend on both the column resistance to the mobile phase flow and the pressure exerted by the pump [23].

Retention index studies of several non-polar and polar compounds showed insignificant effect to the pressure changes. However, the changes between non-polar and polar analytes are very different. For non-polar analytes, the increase in pressure causes a decrease in the retention index, whereas for polar analytes, the increase in pressure causes the increase one [23].

5. Last reports

Recent research in the last two decades regarding the retention index in SFC is very difficult to find. Most of the research related to this one was published in the 80-90s and the numbers were minimal. In this section, several studies of retention index in SFC are presented in Table 2 to obtain an overview of the developments which have been achieved by the researchers. It is hoped that this will become a reference in continuing the research.

Table 2. The previous published articles of retention index in SFC

Years	Objectives	Findings	Ref
1988	The effect of changes in temperature, pressure and modifiers on the retention index of the n-alkanes, alkyl aryl ketones and alkyl benzene homologous series using the non-polar PS-DVB column	There was the insignificant effect in the retention index of the n-alkanes and alkyl aryl ketones homologous series to the changes in temperature, pressure and the addition of less than 15% organic modifier.	[34]
1989	The effect of methanol modifier on the retention index calculation of benzodiazepines using the alkyl aryl ketones homologous series	There was a significant decrease in the retention index of benzodiazepines with the increase in the percentage of modifiers	[35]
1989	The effect of methanol modifier on the retention index calculation of the Barbiturates using the alkyl aryl ketones homologous series	There was a significant decrease in the retention index of barbiturates with the increase in the percentage of modifier	[36]
1990	The effect of column polarity on the retention index calculation of the Poly-cyclic aromatic hydrocarbon (PAH)	There was an insignificant effect to retention index of PAH against the column polarity variations	[37]
1990	The effect of the pressure and modifier to retention index calculation of alkyl aryl ketones, alkyl benzenes and test substances using ODS column	The change in pressure caused insignificant changes in the retention index of both the homologous series and the test substances. There was only the retention index of alkyl aryl ketone which was not significantly affected by the addition of modifier, whereas others were otherwise.	[23]
1991	The comparison of the retention index calculation of aliphatic and cyclic compounds in SFC and GC	The retention index of cyclic compounds is more similar to reference than aliphatic ones	[38]

1992	The n-alkyl phenones homologous series were applied in the retention index calculation of secondary metabolite from Fusarium mycotoxins separated by capillary and packed column	The n-alkyl phenones homologous series were not sufficiently retained on non-polar packed column. It is required more polar homologous series for retention index on SFC packed column.	[39]
1992	The effect of temperature and pressure on the retention index calculation at various temperature and polarity	The retention index of non-polar compounds has the least significant effect on temperature and polarity changes	[33]
1993	The effect of temperature and pressure on the retention index calculation at various temperature and polarity	The retention index from the non-polar column has the least significant effect on the temperature change in the various test substance polarity	[40]
1999	The effect of the density change of the CO_2 mobile phase on the retention index calculation of cyclic and aliphatic compounds	The effect of density change is greater for cyclic compounds than for aliphatic one	[41]

6. Conclusion
The changes in the analysis parameters of temperature, polarity, modifier, and pressure have the effect to the retention index calculation in SFC, but differ in the level of significance. The pressure change has the least significant effect for both polar and non-polar analytes. The changes in temperature and polarity have insignificant effects for non-polar analytes, but significant for polar ones. Meanwhile, modifier changes have a very significant effect, both for polar and non-polar analytes.

7. Acknowledgement
We would like to thank Institute for Research and Community Services (LPPM) Universitas Syiah Kuala for supporting this research. This research was funded by Kementerian Pendidikan dan Kebudayaan through "Penelitian Dasar" scheme, grant number: 215/SP2H/LT/DPRM/2019

8. References
[1] Guardino X, Albaiges J, Firpo G, Rodriguez-Vinals R, Gassiot M 1976 *J. Chromatogr. A.* **118** 13–22.
[2] García Domínguez J A, Diez-Masa J C, Davankov V 2001 *Pure Appl. Chem.* **73** 969–992.
[3] Sinanoglou V J, Strati I F, Bratakos S M, Proestos C, Zoumpoulakis P, Miniadis-Meimaroglou S 2013 *ISRN Chromatogr.* **2013** 1–8.
[4] Earlia N, Rahmad R, Amin M, Prakoeswa C, Khairan K, Idroes R 2019 *Sains Malays.* **48** 1019–24.
[5] Earlia N, Muslem, Suhendra R, Amin M, Prakoeswa C R S, Khairan, Idroes R 2019 *Sci. World J.* **2019** 1–7.
[6] Nasution R, Idroes R, Amna U, Al E 2020 *Open Access Maced. J. Med. Sci.*
[7] Estevam E, Griffin S, Nasim J, Zieliński D, Aszyk J, Osowicka M, Dawidowska N, Idroes R, Bartoszek A, Jacob C 2015 *Nat. Prod. Commun.* **10** 1733–8.
[8] Helwani Z, Ramli M, Saputra E, Bahruddin B, Yolanda D, Fatra W, Idroes G M, Muslem M, Mahlia T M I, Idroes R 2020 *Catalysts* **10** 164.
[9] Helwani Z, Ramli M, Rusyana A, Marlina M, Fatra W, Idroes G M, Suhendra R, Ashwie V, Mahlia T M I, Idroes R 2020 *Processes* **8** 777.
[10] Wieboldt R C, Rosenthal R J 1988 *Mikrochim. Acta* **95** 203–207.
[11] Formisano C, Rigano D, Lopatriello A, Sirignano C, Ramaschi G, Arnoldi L, Riva A, Sardone

N, Taglialatela-Scafati O 2019 *J. Agric. Food Chem.* **67** 3159–67.
[12] Pang T, Zhu S, Lu X, Xu G 2007 *J. Sep. Sci.* **30** 868–874.
[13] Idroes R 2009 *Majalah Farmasi Indonesia.* **20** 77–83. *(in Bahasa Indonesia).*
[14] Idroes R 2010 *Indonesian J. Chem.* **10** 58–63.
[15] Idroes R 2005 *Malaysian J. Anal. Sci.* **9** 224–232.
[16] Quintanilla-López J E, Lebrón-Aguilar R, García-Domínguez J 1997 *J Chromatogr. A* **767** 127–136.
[17] Zhang X, Qi J, Zhang R, Liu M, Hu Z, Xue H, Fan B 2001 *J. Comput. Chem.* **25** 125–133.
[18] Idroes R, Muslem, Saiful, Mahmudi, Idroes G M, Suhendra R, Irvanizam, Zamzami M P 2019 *IOP Conf. Ser. Mater. Sci. Eng.* **364** 012033.
[19] Idroes R, Husna I, Muslem, Mahmudi, Rusyana A, Helwani Z, Idroes G M, Suhendra R, Yandri E, Rahimah S, Sasmita N R 2019 *IOP Conf. Ser. Earth Environ. Sci.* **364** 12020.
[20] Knox J H, Kaliszan R 1985 *J. Chromatogr. A* **349** 211–234.
[21] Suliman F 2002 *Talanta.* **56** 175–183.
[22] Bidlingmeyer B A, Warren Jr F V, Weston A, Nugent C, Froehlich P M 1991 *J. Chromatogr. Sci.* **29** 275–279.
[23] Smith R M, Sanagi M M 1990 *J. Chromatogr. A.* **505** 147–159.
[24] Idroes R, Muslem, Mahmudi, Saiful, Idroes G M, Suhendra R, Irvanizam 2020 *Heliyon.* **6** e03302–e03302.
[25] Pratiwi S U T, Lagendijk E L, de Weert S, Idroes R, Hertiani T, den Hondel C Van 2015 *Int. J. Appl. Res. Nat. Prod.* **8** 1–13.
[26] Husna I, Rusyana A, Muslem, Idroes G M, Suhendra R, Idroes R 2020 *IOP Conf. Ser. Mater. Sci. Eng.* **796** 012064.
[27] Idroes R, Japnur A F, Suhendra R, Rusyana A 2019 *IOP Conf. Ser. Mater. Sci. Eng.* **523** 12007.
[28] Idroes R, Noviandy T R, Maulana A, Suhendra R, Sasmita N R, Muslem M, Idroes G M, Irvanizam I 2019 *IREMOS* **12** 373.
[29] Pacáková V, Feltl L 1992 *Chromatographic retention indices: an aid to identification of organic compounds.* United Kingdom: Ellis Horwood Ltd.
[30] Ehrensberger C W 1996 *Beiträge zur Reproduzierbarkeit und zur Vorhersage von Retentionsdaten in der HPLC. (in Germany).*
[31] Idroes R 2009 *Indonesian J Pharm.* **20** 133–140. *(in Bahasa Indonesia).*
[32] McGinitie T M, Harynuk J J 2012 *J. Sep. Sci.* **35** 2228–32.
[33] Jones B A, Shaw T J, Clark J 1992 *J. Microcolumn Separations.* **4** 215–220.
[34] Smith R M, Sanagi M M 1988 *Chromatographia.* **26** 77–84.
[35] Smith R M, Sanagi M M 1989 *J. Chromatogr. A.* **483** 51–61.
[36] Smith R M, Sanagi M M 1989 *J. Chromatogr. A.* **481** 63–69.
[37] Raynor M W, Moulder R, Davies I L, Kithinji J P, Clifford A A, Bartle K D, Hall J E 1990 *J. High Resolut. Chromatogr.* **13** 22–26.
[38] Morishita F, Yaku K, Hasegawa H, Ichise M, Kojima T 1991 *Anal. Sci.* **7** 239–240.
[39] Christopher Young J, Games D E 1992 *J. Chromatogr. A.* **627** 247–254.
[40] Chester T L, Innis D P 1993 *J. Microcolumn Separations.* **5** 441–449.
[41] Yaku K, Hasegawa H, Morishita F, Okazaki S, Ichise M 1999 *Anal. Sci.* **15** 1065–1069.

The geological economic of sand quarry resources in Jantho, Aceh Besar

A Rais[1], M Irham[2,3,*], F Adrian[1], A Deli[4]
[1]Department of Engineering Geology, Universitas Syiah Kuala, Banda Aceh, Indonesia
[2]Faculty of Marine and Fisheries, Universitas Syiah Kuala, Banda Aceh, Indonesia
[3]Center for Environmental and Natural Resources Research (PPLH-SDA), Universitas Syiah Kuala, Banda Aceh, Indonesia
[4]Socio-Economic Agricultural Department, Faculty of Agriculture, Universitas Syiah Kuala, Banda Aceh, Indonesia

*E-mail: irham@unsyiah.ac.id

Abstract. Jantho, the Great Aceh district, is rich in its natural resources, which can benefit people. With the increasing need for humankind to increase minerals in both quantity and type, the knowledge about exploiting minerals into the material of high economic value also increases. Therefore, the research on sand quarry resources' economic potential needs to be carried out to estimate sand resources and their geological economy. This research was conducted in the area of Jantho with the scope of study covering characteristic lithology analysis, calculation of sand resources, and study of geological condition based on field data analysis and data derived from the literature. In estimating the resources, the contour method was used, and for the economic value, the sieve analysis method was employed. The result informed that sand resources are very economical when the sand is classified based on its type and size. The results also show that economically sand resources are four times more profitable when it is categorized based on type and size than when it is sold in the form of no sorting.

1. Introduction

Jantho City is a sub-district in Aceh Besar District, Aceh Province, Indonesia. Jantho City is also the capital and administrative center of Aceh Besar District. Geographically, Aceh Besar District is located at coordinates 5.2°-5.8° North and 95°-95.8° East. Aceh Besar District has an area of 2,969.00 Km^2 covering 23 Districts and 604 Villages [1]. The condition of the Aceh Besar Regency consists of mountainous areas, lowlands, and water areas.

Aceh Besar District is rich in natural resources and many potentials that have not yet been explored to be used for the benefit of the people. Along with the development of human needs for minerals, the knowledge on how to explore the mineral will also increase. This is due to technological advances and new discoveries in various industries that require mineral raw materials. One of the potential mineral rock quarries in the city of Jantho is C-quarry material in the form of sandstone [1].

Sandstone resources found in the city of Jantho are the object of research. Therefore, to find out the quantity and quality of these sandstone resources, it is necessary to assess and research resources. In estimating resources, the method used was the contour method [2]. This method is suitable for use in mineral deposits whose thickness changes [3], and for the economic value [4],[5],[6], a sieve analysis method was applied [7]. The calculation is focused on the quality and amount of deposits of the mineral deposits. The feasible mining activity was determined by the quality and volume of the extracted material [8],[5]. The results of the calculation and analysis of appropriate natural resources can determine production targets so that the management of natural resources is more beneficial and does not damage the environment. The problem is that there is no economic analysis on the source of a quarry of C-mine minerals in Jantho, which has traditionally been taken. Therefore, in this study, an estimate of the volume of sandstone resources in the study area will be carried out so that it can be estimated how much the economic value of the sandstone resources will be.

The expected benefit of this research is knowledge of sandstone resource reserves and geological economics. This research is expected to optimize mining, facilitate further activities, and maximize sandstone production in the study area. The results of this study can also be used as advice and evaluation in the context of sandstone management efforts so that it can be utilized by the government, the management, and the community.

2. Method

The research location was in Jantho, Aceh Besar district, with a research area of 140 m² and a mapping area of 25 km. Geographically the location of the study is at coordinates (5°19'40.82"N – 5°19'14.87"N and 95°35'14.00"E – 95°35'39.82"E) while the mapping location is at coordinates (5°18'34.35"N – 5°21'14.82"N and 95°33'54.01"E – 95°36'35.84"E) as shown in Figure 1. This research was done for about six months, starting from March 2018 to August 2018.

Before measurement measurements, methods of field mapping, tracking, sampling, and contour measurement using Theodolite were used [10],[11]. The field mapping phase is carried out by directly observing the geological conditions of the study area [12]. At each outcrop that is found, the outcrops are described, taking photos, sketching outcrops, taking stance and tilt, and taking the coordinates of the outcrop, then the coordinates are entered into the topographic map.

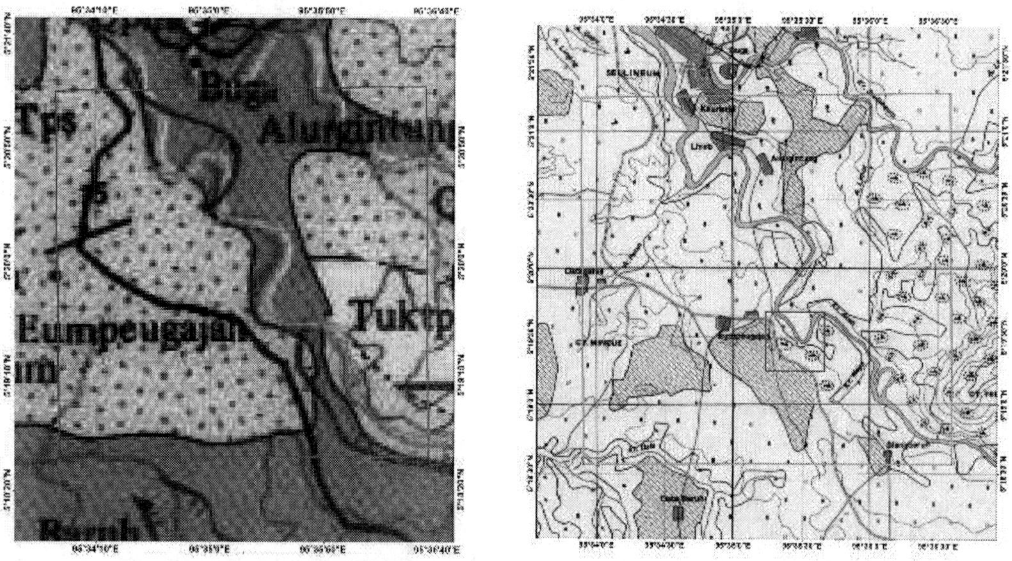

Figure 1. Study area located in Jantho, Aceh Besar district, Aceh Province [9].

Furthermore, to calculate resources, the contour method uses Theodolite. The processing data was based on the multilevel sieved analysis results, and the processing of contour data was generated from the Theodolite device. Theodolite data processing used AutoCad 2007 software, and then used mapcreatesurface tools to unify all the data that has been obtained in the field. The area calculation used tools area that is applied in the same contour elevation. The results of the filter analysis and contour analysis were then processed into the mining material economy analysis.

3. Results and discussion

3.1 Geology analysis

From the geological data and other supporting data obtained in the field during the study, geological information is presented in the form of a geological map that includes the distribution of lithology and other information as in Figure 2.

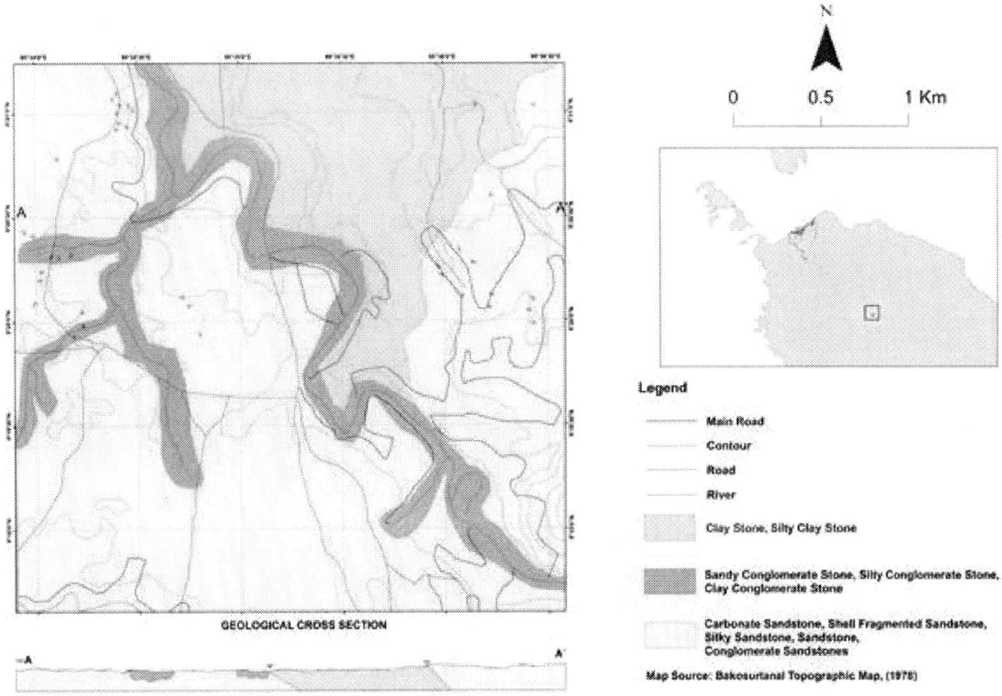

Figure 2. Geology map of the study area

The study area is in the Seulimum Formation, Indrapuri Member, Indrapuri Formation, and Alluvium. Based on geological maps [13][14]. The Seulimum Formation is composed of limestone sandstones, conglomerates, and a few mudstones. The Seulimum Formation has a Member Formation section in the form of a Lam Kabeue Limestone Member dominated by limestone coral rock [15]. Based on observations in the field, it is composed of several rock units, including Alluvium Unit, Carbonate Sandstone Unit, Batukonglomerate Unit, and Sandstone Unit.

The study area's geological order consists of carbonate sandstone units, carbonate claystone, and conglomerates, which are members of the Seulimum Formation. In addition, in the study area, the Alluvium Unit was the youngest unit consisting of gravel, sand, and mud found during the mapping. Carbonate sandstone is the rock that dominates in the Research area. The lithology of carbonated sandstone from every observation point is a type of bedding with medium sand to very fine sand size and brownish-gray with good porosity properties and a subrounded to rounded roundness. Shell Fragmented Carbonated Sandstone Unit also found in the study area. This type of rock unit has almost the same characteristics as carbonate sandstone, but its presence is only scattered at locations with the morphology of a 125-meter hill. The thickness of these rock formations ranges from ± 30 - 100 cm, with fine grain size and coarse sand, and has good grain uniformity and fine grain sorting. This rock also contains calcite components resulting from marine fossil ruins, and there are alternating clay layers with a layer thickness of 15-20 cm. Conglomerate stones found generally have lithology with a brownish-gray color and have grain sizes from cobble to pebble. The level of backwardness of conglomerate stones is subrounded to rounded and has good porosity.

The geological structure in the form of slope occurred in the outcrop of the Seulimeum Formation unit, where the rock lithology was in the form of carbonate sandstone. The sedimentary layers are deposited close to horizontal and basically parallel to the surface plane where the sedimentary layer is deposited. The composition of the layer, which is not a horizontal position, means that it has undergone another geological process after its deposition, for example, influenced by tectonic forces.

3.2 Sandstone resources

The results of the calculation of Sandstone resource volume in this research area can be seen in Table 1. It shows that the smaller the elevation, the larger the volume produced. This is because at a small elevation, the cross-sectional area is larger, resulting in a volume greater than the elevation volume above it.

Table 1. Interpretation of soil relative density based on cone resistance

No	Elevation (m)	Area (m^2)	Volume (m^3)
1	15.5	3,761.19	5,641.79
2	14.0	4,584.56	2,292.28
3	13.5	5,070.32	2,535.16
4	13.0	5,514.11	2,757.06
5	12.5	5,930.85	2,965.43
6	12.0	6,321.95	75,863.40
	Total Volume		92,055.11

The sieve analysis method is done by sifting the sample and then analyzing the distribution of fine and coarse aggregates. Figure 3 shows the cumulative aggregate filtering analyzed. The percentage of the largest cumulative pass is 98.84% found on a 2 mm sieve, and the smallest is 5.2% on a 0.075 mm sieve. This shows that the smaller the filter size, the smaller the cumulative percentage of passes.

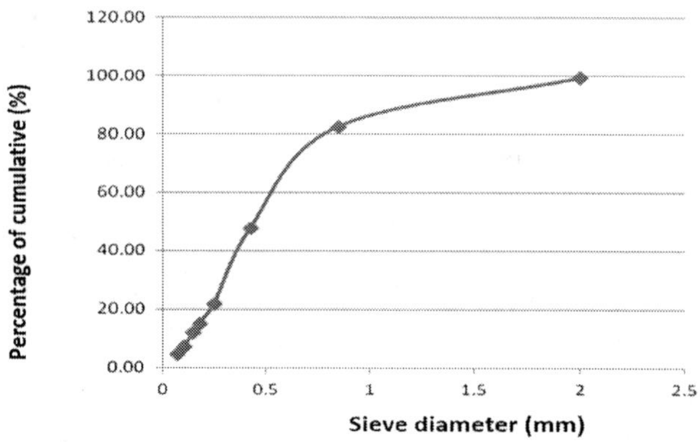

Figure 3. Graph of cumulative escaped percentages in a sample of 300 gr

Based on the Wentworth scale (1992) sandstone deposits in the study area can be grouped into 3 dominant sizes consisting of sand measuring 0.85 mm around 16.69%, sand measuring 0.425 mm around 34.73%, and sand measuring 0.25 mm around 25.96%. The predominant sandstone deposits are very dominant in the size of 0.425 mm (medium sand)

3.3 Geological economic analysis

Revenue obtained from the sale of sandstone as much as one truck is Rp. 650,000, therefore income is obtained per day if sold as many as 30 trucks per day will be obtained at Rp.19,500,000, -, so in a year the income earned is Rp. 7,020,000,000. In the calculation, the total volume of available resources is 92,055.11 m^3 with a truckload capacity of 4 m^3 per truck, so 23,013 trucks will be produced. The yield of minerals will run out in two and a half years if 30 trucks per day continuously operate.

Production costs and operational costs incurred in this analysis are excavator operating costs, labor costs, and business costs. Operational excavators are charged Rp. 150,000 per truck, then for 30 trucks, there is an expenditure of 4,500,000. While the amount of labor costs is Rp. 100,000 per truck that costs Rp. 3,000,000 per day. Business costs are estimated at Rp. 2,500,000 per day. Consequently, the total cost per day is Rp. 10,000,000, which means there is an Rp. 3,650,000,000 cost per year.

The net profit of a business is the income received is reduced by the number of costs incurred. In this case, the net profit obtained is equal to Rp. 3,370,000,000.

4. Conclusion

The estimation result of sand quarry resources is 75,863.40 m3 where sandstone deposits in the study area can be grouped into 3 granular sizes, namely sand of 0.85 mm about 16.69%; sand of 0.25 mm estimated as 25.96%; and sand of 0.425 mm as much as 34.73%. Based on the basic economic analysis, it concludes the research area is very economical.

5. Acknowledgments

We thank you very much to Syiah Kuala University through the facility support at Geological Engineering Laboratory during the research.

References

[1] Government of Aceh Besar 2018 Retrieved from http://www.acehbesarkab.go.id. *(in Bahasa Indonesia).*

[2] Agnerian H, Roscoe W E 2002 *CIM Bull.* **95** 100 – 107.

[3] Rauf A. 1999. *Perhitungan cadangan endapan mineral.* Yogyakarta: Jurusan Teknik Pertambangan, Fakultas Teknologi Mineral UPN "Veteran" Yogyakarta. *(in Bahasa Indonesia).*

[4] Jones O, Aspandiar M, Dugdale A, Leggo N, Glacken I, Smith B 2019 *The Business of Mining.* 121-179. doi:10.1201/9780429057540-5.

[5] Duczmal-Czernikiewicz A 2016 *Geologos* **22** 261-262. doi:10.1515/logos-2016-0026.

[6] Kesler S E, Simon A C 2015 *Geologos.* **22** 261 – 262.

[7] Irham M, Febriarista I, Sugianto S, Setiawan I 2017 *Jurnal Rekayasa Kimia & Lingkungan.* **12** 59 – 76. *(in Bahasa Indonesia).*

[8] Global Mineral Reserves and Resources (GMR&R) 2017 *Mineral Resources, Economics and the Environment.* 342-354. doi:10.1017/cbo9781139871426.015.

[9] Bakosurtanal Topographic Maps (BTM) 1998 Aceh topographic maps. *(in Bahasa Indonesia).*

[10] Werkheiser W 2017 Mapping the Future of America. United States Geological Survey.

[11] Griffiths J S 2019 *Engineering Geology: Geological Maps for Engineering Geology, Reference Module in Earth Systems and Environmental Sciences.*

[12] Donadio E, Spanò A 2015 *Proceedings of the 1st International Conference on Geographical Information Systems Theory, Applications and Management.*

[13] Bennett J D, Cameron N R, Bridge D McC, Djunuddin A, Ghazali S A 1981 *Geologic map of the Banda Aceh quadrangle, North Sumatra, Indonesia.* Geological Research and Development Centre.

[14] Wajzer M R, Barber A J, Hidayat S, Suhrasono 1991 *J. Southeast Asian Earth Sci.* **6** 159 – 476.

[15] Polom U, Arsyad I, Kümpel H J 2008 *Adv.Geosci.* **14** 135 – 140.

Bedrock identification using resistivity method at campus II of Universitas Syiah Kuala, Banda Aceh

M Syukri[1,2], A M Taib[2,*], Z Fadhli[2]
[1]Physics Department, Universitas Syiah Kuala, Banda Aceh, Indonesia
[2]Program Study of Geophysical Engineering, Universitas Syiah Kuala, Banda Aceh, Indonesia

*E-mail: amsir@unsyiah.ac.id

Abstract. Research of resistivity method at campus II Universitas Syiah Kuala has been conducted with the SuperSting 68 instrument. There were three lines, each 330 meters long with a spacing of 6 meters. The purpose of the research was to determine the depth of the bedrock at campus II of Universitas Syiah Kuala. The results showed that there was a conductive and resistive coating with a resistivity value of between 2–1250 Ωm. The rock types were composed of water-saturated rocks, Tuffs sandstone, and andesite. The supposed bedrock rock is an andesite rock with a resistivity of 170-1250 Ωm with good density. In conclusion, the separation between the sediment and the bedrock was visible at a depth of up 10 meters across the measuring track.

1. Introduction

Universitas Syiah Kuala (UNSYIAH) is currently developing campus II in Neuheun village, Aceh Besar, to support Aceh's higher educational activities further. The construction of Campus II UNSYIAH requires the construction plans to meet the development criteria and standards. These criteria help optimize buildings and physical infrastructures such as roads, pedestrian pavement, academic offices, lecturer buildings, laboratory, and garden. Thus, subsurface information is crucial. Subsurface information could be categorized as crucial information or data to optimize infrastructure construction, like Campus II development.

The condition of the rock structure is essential when planning the construction foundation. The type of rock and basement rock as supporting data help determine the characteristics and weight of the foundation that will be developed [1]. Bedrock information is vital subsurface data during the construction building. Furthermore, the foundation of a building is also crucial as the foundation continues the building load to the soil [2].

The resistivity method was applied in the western part of Campus II of UNSYIAH. The purpose of the research was to identify the subsurface based on the resistivity value. The result could help determine and describe the depth of bedrock and thickness of bedrock in Campus II of UNSYIAH. The resistivity method was historically used for these goals. It was able to determine and map the high resistivity value contrast between the conductive covering layer and the resistive bedrock [3], [4], [5], [6], [7]. This technique is non-invasive, rapid, and low-budget and can be efficiently executed in alluvial [8].

2. Methodology and Geology of the study area

The resistivity method is an efficient method in the geophysics survey to investigate the subsurface field based on the resistivity value of material in the subsurface [2]. Generally, the resistivity method was conducted by transmitting direct current (DC) into the ground surface as shown in Figure 2. Furthermore, the potential difference was measured at a different area on the ground surface [9]. The current can easily seep into a rock and soil, depending on the conductive properties of rock and soil.

The Schlumberger array was used as a type of electrode array in this research used four electrodes, including two current (C1 and C2) and two potential (P1 and P2) electrodes in a straight line. The Schlumberger array generally has better resolution, greater probing depth, and less time-consuming field deployment than another array such as Dipole-dipole, Pole-dipole, and Wenner array [10]. The Schlumberger array can be seen in Figure 1.

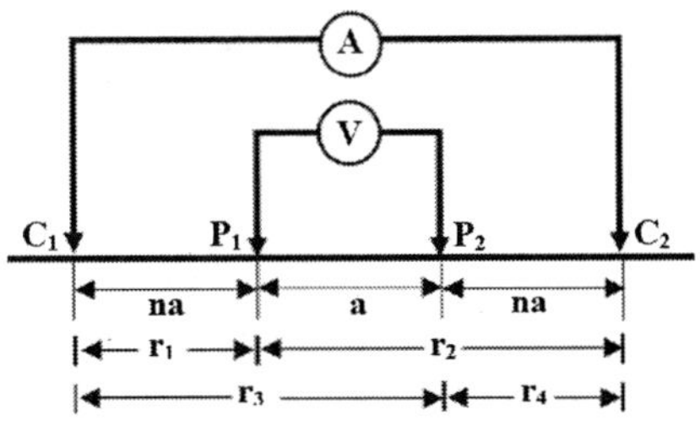

Figure 1. Schumberger array [10]

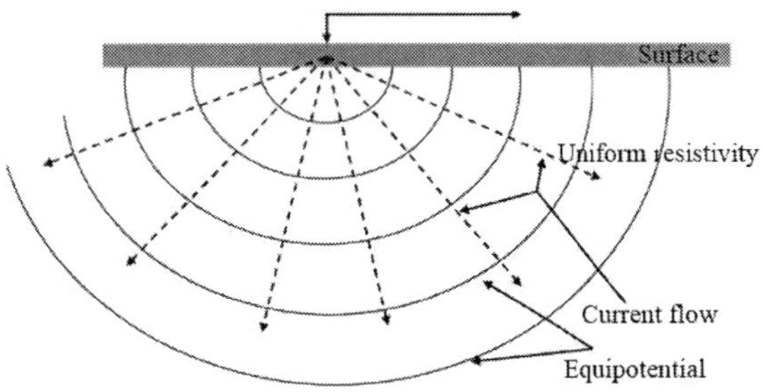

Figure 2. Current Flow Distribution [11]

The research area is a proposed location of Campus II of Universitas Syiah Kuala, Banda Aceh. The site is located in Aceh Besar District and situated between 5°36'43.32" N and 95°25'6.36" E and 5°36'51.04" N and 95°25'13.74" E in the southern part of the main UNSYIAH campus. The geology of campus II of Unsyiah located in Neuheun village, Aceh Besar has been well described [12] and consists mainly of alluvium, gravel, sand, mud, and Lamteuba volcano rock. The geology of the field is presented in Figure 3.

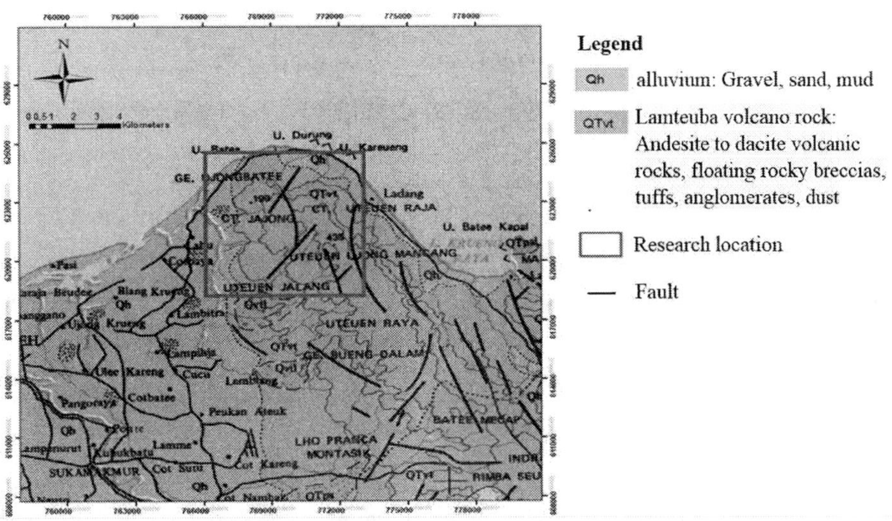

Figure 3. Geology map of Neuheun village, Aceh Besar [12].

The research field is presented in Figure 4. There are three measurement lines, using SuperSting R8 electrical resistivity with 56 electrodes and 6 meters spacing between electrodes. The total length for each measurement line is 330 meters. The resistivity measurement method conducted in the study is electrical resistivity imaging of Schlumberger array to identify bedrock zones; it will be interpreted into 2D cross-sections of the subsurface zones later on.

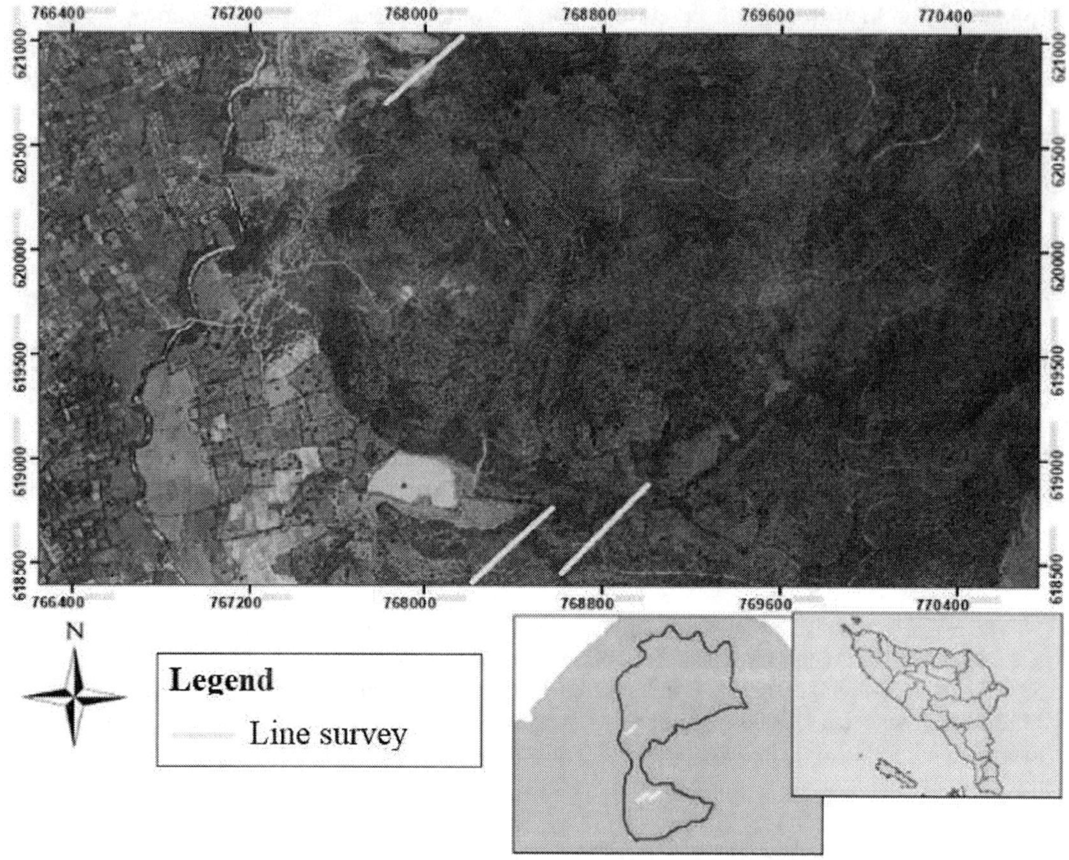

Figure 4. Measurement line of Campus II UNSYIAH

3. Results and discussion

The resistivity value for line 1 has a contrast between 2 - 1250 Ωm. The total length was 330 meters; the depth of investigation is about 63 meters (Figure 5). There are three different layers from the inversion resistivity value. The first layer contains sand showing a low resistivity value that ranges from 2 - 18 Ωm. The soil was located between andesite bumps at a depth of 20 - 40 meters with a thickness of approximately 4 - 20 meters. The resistivity value for the intermediate layer was 25 - 120 Ωm interpreted as the second layer and contains tuffaceous sandstone layers found in volcanic rock formations in Lam Teuba volcano with a thickness of approximately 0 - 60 meters. The third layer presented a high resistivity value (170 - 1250 Ωm) and was interpreted as an andesite rock. In this section, the bedrock appeared at a depth of 20 - 60 meters.

The hard/dense soil depth data can also be plotted in a base map, which is useful for knowing the position of each test point and the depth of stable soil at each location. Then an interpolation is made to determine the zoning of hard/dense depth, as shown in Figure 4. The zoning map shows that most of the tested locations have a stable soil depth of 10 meters below the existing ground surface. In certain places such as the area around Lambung, around Lamjame, around Neusu Aceh, around Lambaro Skep, around

Jeulingke, around Doi, and around Kopelma Darussalam exhibited stable depth from 5 to 10 meters. The shallowest hard/dense soil depth is only found around Lamjame of Jaya Baru District.

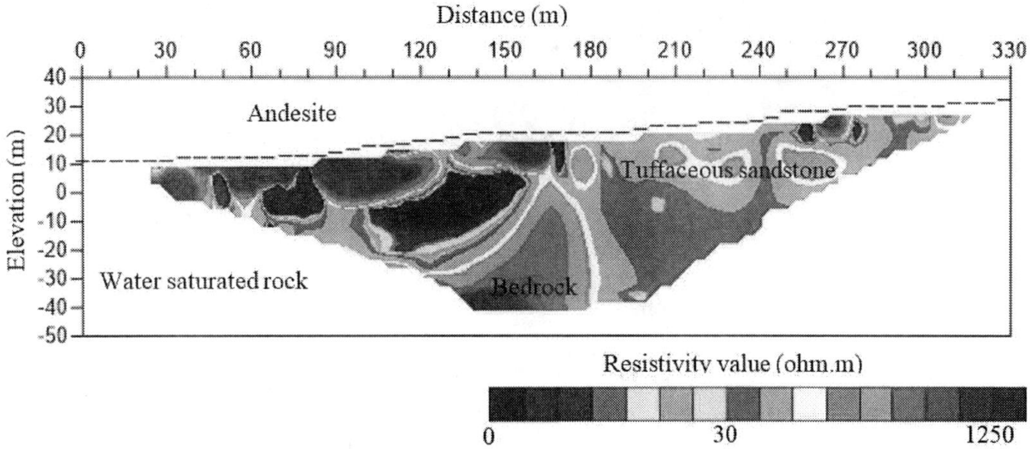

Figure 5. Inverted Resistivity value for line 1

The inversion result of the second line is presented in Figure 6. The length of this line is 330 meters, with a maximum depth of penetration of 67 meters. The lower resistivity with a value of 2 - 18 Ωm may be interpreted as water-saturated rocks at a depth of 0 - 30 meters in a distance of 70 - 160 meters. The second layers showed a low resistivity value of 25 - 120 Ωm and were presented as tuffaceous sandstone, which dominated this layer. The lowermost surface layer showed a relatively high resistivity value over 170 Ωm, which is bedrock at a depth of 10-60 meters in a distance of 50 - 200 meters.

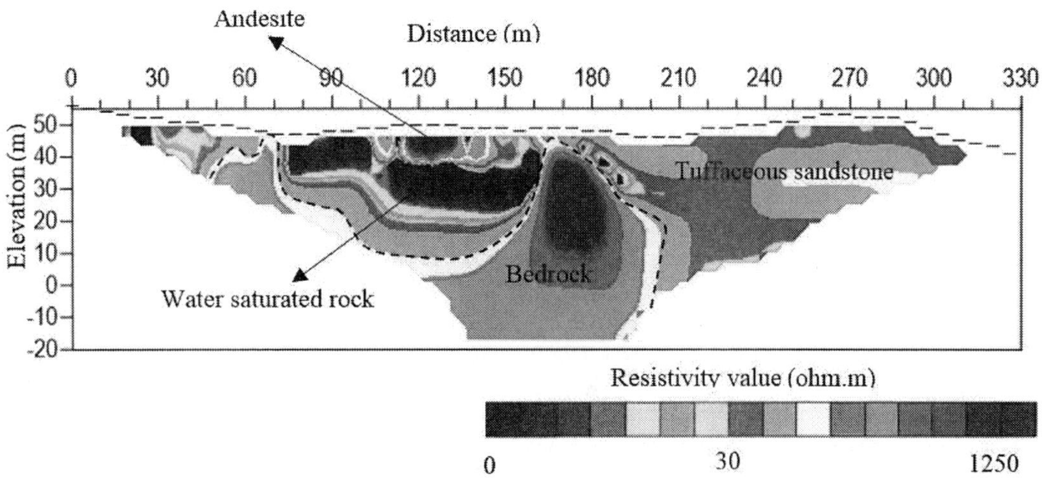

Figure 6. Inverted Resistivity value for line 2

The 2D inversion resistivity pseudosection of line 3 (Figure 7) trends from the Southwest to the northeast direction, the total length of the survey line was 330 meters. In comparison, the maximum depth of penetration was 67 meters, and the topography was approximately 70 meters high. In this section, the subsurface resistivity image along line 3 appears as an identical layer, where line 3 has three layers for low resistivity (2 - 18 Ωm) and is interpreted as a water-saturated rock with a value of 25 -

120 Ωm presented tuffaceous sandstone rock. The high resistivity value is shown in the third layer, which has a value up to 170 Ωm indicated as a bedrock at a depth of 10-45 meters.

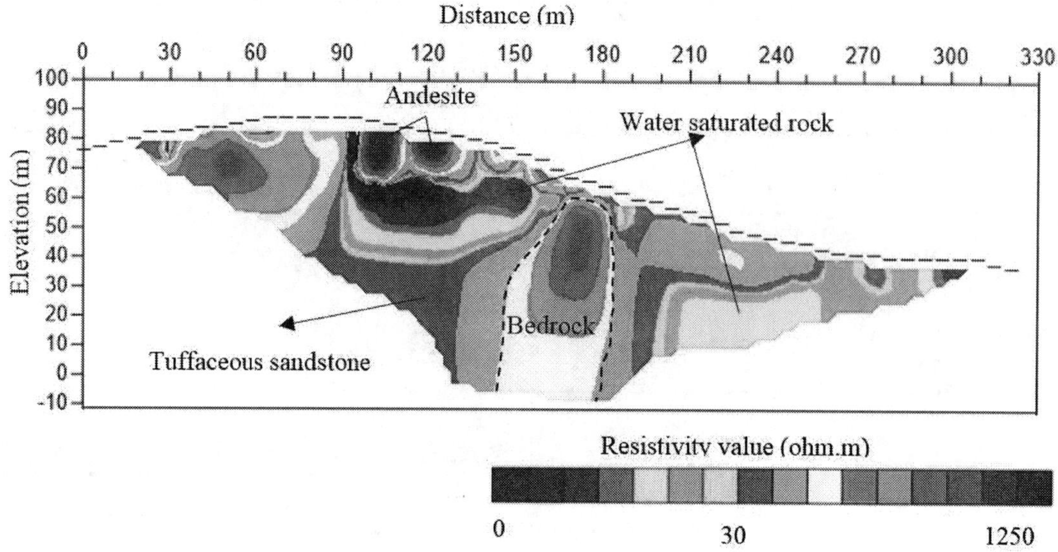

Figure 7. Inverted Resistivity value for line 3

4. Conclusion

The research results indicate a complex subsurface profile in the area. The measurement result of 3 lines found that the hard layer is located at a depth up to 10 meters in lines 2 and 3, while line 1 found the hard layer at a depth of up to 20 meters. Furthermore, based on the resistivity value of the geology map, the research location is composed of andesite rock as bedrock with a high resistivity value of up 170 Ωm.

5. Acknowledgments

The authors would like to thank the Geophysics teams, students, and staff of the Geophysical Engineering Program, Earth Sciences Engineering Department, Engineering Department, Universitas Syiah Kuala Banda Aceh (Indonesia). Special thanks are extended to the Ministry of Education and Culture, Indonesia, for financial support in the scheme of PLK-PNBP Unsyiah 2020 Grant.

References

[1] Pazha H, Nasro M, Agustina R D 2019 *J. Phys. Conf. Ser.* **1175** 012014.
[2] Rizka R, Satiawan S 2019 *JSAT* **2** 281430.
[3] Hamilton W, Bachman S B 1982 *Fundamental of Geological Surface.* Birmingham: University of Birmingham.
[4] Bavitra A, Amir H 2015 *Pillar of Physics* **6** 1-8. *(in Bahasa Indonesia).*
[5] Aziz N A, Abdulrazzaq Z T, Alwan H A Z 2015 *Iraqi J. Sci. Technol.* **6** 83-88.
[6] Syukri M, Anda S T, Safitri R, Fadhli Z, Saad R 2020 *Int. J.* **18** 123-129.
[7] Syukri M, Saad R, Anda S T, Fadhli Z 2019 *Int. J.* **17** 133-143.
[8] Cardarelli E, De Donno G 2017 *J. Appl. Geophys.* **141** 77-87.
[9] Samouëlian A, Cousin I, Tabbagh A, Bruand A, Richard G 2005 *Soil Tillage Res.* **83** 173-193.
[10] Loke M H. 2000. *A Practical Guide to 2d and 3d Surveys.* California: Stanford University.
[11] Telford W M, Geldart L P, Sheriff R E, Keys D A. 1990. *Applied Geophysics* 2nd ed. New York: Cambridge University Press.

[12] Bennet J D, Bridge D M, Cameron N R, Djunuddin A, Ghazali S A, Jeffry D H, Kartawa W, Keats W, Rock N M S, Thomson S J, Whandoyo, R 1981 *Peta Geologi Lembar Banda Aceh, Sumatera, Skala 1:250.000. (in Bahasa Indonesia).*

IOP Conf. Series: Materials Science and Engineering 1087 (2021) 012071 doi:10.1088/1757-899X/1087/1/012071

Identification of coal layer using gravity method in Lam Apeng Aceh Besar

D Darisma[1], **L Hakim**[1], **Marwan**[1, 2,*]
[1]Department of Geophysical Engineering, Faculty of Engineering, Universitas Syiah Kuala, Darussalam, Banda Aceh, 2311, Indonesia
[2]Department of Physics, Faculty of Mathematics and Natural Sciences, Universitas Syiah Kuala, Darussalam, Banda Aceh, 2311, Indonesia

*E-mail: marwan.geo@unsyiah.ac.id

Abstract. Aceh has a coal potential of around 450 million tonnes. One of the areas identified as having coal potential is Lam Apeng, Aceh Besar. In this study, the gravity method is used to model coal layers that are different from other layers based on rock density. The data used are 99 points with a space of 50 m and a coverage area of 450x450 m. From the results of the data corrected to become a complete anomaly Bouger (CBA), it was obtained a trend with a higher gravity value towards the northeast and the lower towards the southwest of the study area. CBA data are filtered using the moving average technique to obtain regional anomalies (associated with deep structures) and residual anomalies (associated with shallow structures). Regional anomalies and residual anomalies also show the same trend as CBA. Based on the 2D modeling results of the L-L 'section, there are three different rock layers, namely alluvium layer with density 2 gr/cm^3, sandy tuff layer with density 2.3 gr/cm^3, and coal layer. There are three coal layers with different density values, namely layer A with density 1.68 gr/cm^3, slope 125° and layer thickness 2-4 m; layer B with density 1.71 gr/cm^3, slope ± 120°, and layer thickness 7-8 m; and layer C with density 1.7 gr/cm^3, slope ± 120°, and thickness 4-6 m.

1. Introduction

The human population increases every year, causing the need for energy is also increase. These energy sources can be in the form of fossil fuels and coal. Indonesia has coal reserves of 161 billion tonnes, 53 percent of which are in Sumatra, and about 47 percent are in Kalimantan. There are several provinces in Sumatra that have large amounts of coal potential, such as South Sumatra, North Sumatra, and Aceh. Coal reserves in Aceh are predicted to be 450 million tonnes [1]. One area that has potential for coal reserves in Aceh is Lam Apeng, Aceh Besar District. However, the potential in this area has never been explored. For this reason, a geophysical survey is needed to map and model the coal layer in the area.

One of the geophysical methods that can be used to determine subsurface structures is the gravity method. The gravity method has been widely used to detect subsurface anomalies and geological structures [2–4]. The advantage of the gravity method is that it can describe the subsurface conditions of the earth based on differences in gravity acceleration. The differences in the gravity value can be related to fractures, rock porosity, crack density, and changes in fluid mass [5–7]. The gravity value collected in the field is then processed to obtain rock mass density, which can be interpreted to determine the subsurface structure related to the potential of coal in the research area. The coal layer is physically characterized by a black color and a lighter mass or lower density. Therefore, coal on a gravity map should be easily recognized [8–10].

The physiography of Aceh Besar and its surroundings is arranged as a wavy alluvium land area to the coastal lowlands with a northwest-southeast direction and a higher topography to the south. This area is known as Krueng Aceh Basin [11]. The western part of Aceh Besar is bordered by the Sumatran fault known as the Aceh Fault or including the Krueng Aceh Fault Segment. This area is located at the foot of the Bukit Barisan complex which is composed of Tertiary and Pra-Tersier rocks. In the eastern part, the Quaternary basin is bordered by hills composed of Plio-Pleistocene rocks originating from Seulawah Volcano. In this area, there is also a Seulimum fault that stretches in a northwest-southeast.

Lam Apeng is a research area located in Seulimeum Sub-district, Aceh Besar, Province of Aceh. Based on the regional geological map, the research location is covered by alluvium (Qh), which consists of mud, sand, gravel, tuff, with a wavy shape (Figure 1). In this area, there is also the Seulimeum Formation, which consists of tuffaceous sandstones, limestone, conglomerates, and mudstone [12].

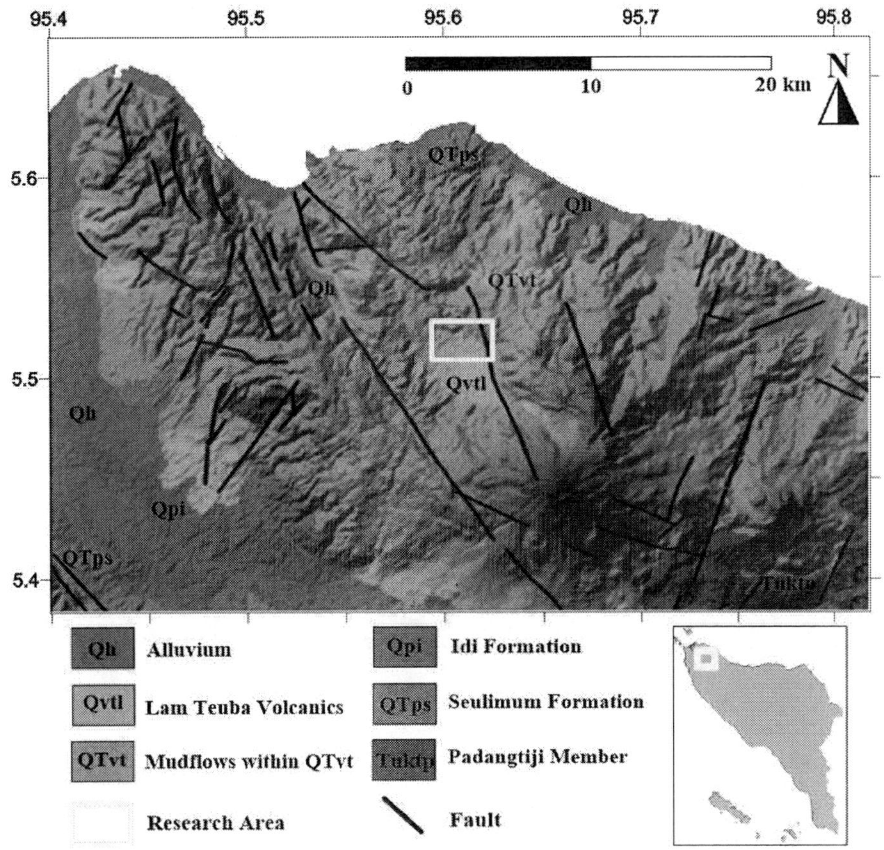

Figure 1. Geological map of Lam Apeng area (modified from [12]).

2. Material and methods

In this study, data measurement was carried out directly using the Scintrex CG-5 Gravimeter. The measurement points are 96 points with a distance between points of 50 m, covering an area of 450 x 450 meters according to field conditions (Figure 2). After the measurement, data correction is needed to remove noise and regional anomaly effects. The first correction is drift and tidal correction. Drift correction is performed to eliminate the effect of tool fatigue, while the tidal correction is to remove the effects of gravity from the Moon. Besides, latitude correction was also performed using the International Association of Geodesy 1967 equation.

The measured gravity data is generally not on the geoid, so it needs to be corrected for elevation, namely free air and Bouguer correction. Free air correction ignores the mass effect that may exist between the elevation at the measurement point and the sea level, whereas Bouger correction considers the mass. The last correction is terrain correction to remove topographic effects around the measurement area. All of the aforementioned corrections are done to obtain CBA data [14], [15].

Bouger anomaly is a combination of regional anomalies associated with deep structures and residual anomalies associated with shallow structures. The two anomalies must be separated to make it easier to interpret. Some of the regional anomaly separation methods were tested using the second polynomial, upward and downward continuation approach, but the results were almost the same as CBA. In this study, the anomaly separation technique used is moving average [16].

The separated residual anomaly data is then modeled using Grav2DC. The modeling results are in the form of a 2D subsurface rock density cross-sectional image, which shows the information on rock density, measurement line distance, and depth [17], [18]. The CBA value is interpreted with the geological conditions of the study area to conclude where the coal distribution is located in Lam Apeng area.

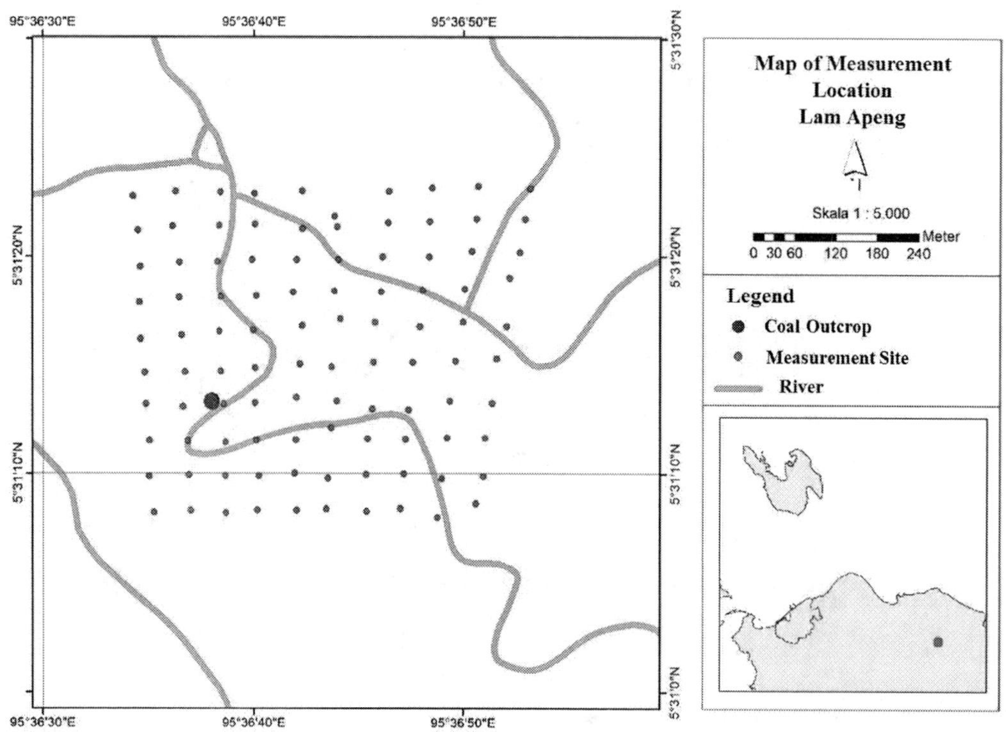

Figure 2. Map of gravity measurement sites in Lam Apeng.

3. Results and discussion

Field data that has been corrected into CBA data were then plotted onto the map as shown in Figure 3. The CBA values in this area range from 120-129 mGal. Areas with high anomaly values are located in the northwest of the study area, while areas with low anomalies are located in the southwest of the study area. Data with high anomalies are associated with high densities and vice versa. Because the CBA data is still influenced by regional and local components, the data was separated into regional and residual data to make it easier to interpret.

The residual data filtering process was carried out using the moving average technique. This technique separates data with a high frequency, which is considered to represent the gravity response of relatively shallow layers. The results of filtering CBA data into regional and residual anomalies can be seen in Figure 4 and Figure 5. In Figure 4, the highest regional anomaly is located in the northeast of the study area, and the lowest is in the southwest area. It is estimated that the rocks in the northeast area are more compact than the southwest area. The low and high anomaly trends are also almost the same when compared to the residual anomaly map (Figure 5). It is predicted that the rocks in the northeast area of the study area are more compact and older than the southwest area.

The difference in results between the distribution of regional anomaly density and the distribution of residual anomaly density shows a contrasting difference. One of these differences can be seen in the anomaly, which is located in the northwest. This anomaly does not exist in the regional anomaly, which indicates that the response is a local anomaly.

At the outcrop point in the regional anomaly map, the anomaly showing low density as a coal layer does not exist. This is because the coal is at a shallow depth, as shown in Figure 6. On the residual anomaly map, the location of the coal outcrop has a low density with a value of -1.5 to 1 mGal, which is located in the southwest of the study area. Based on the analysis of coal content, the variation in coal density has a value between 1.2 gr/cm^3 and 1.8 gr/cm^3 [19].

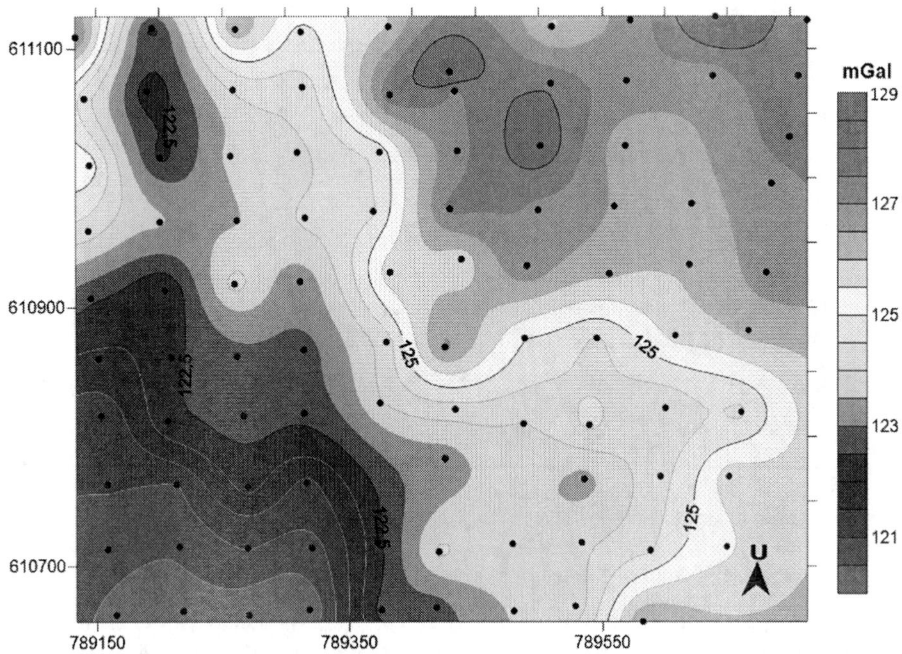

Figure 3. Map of Complete Bouguer Anomaly Lam Apeng.

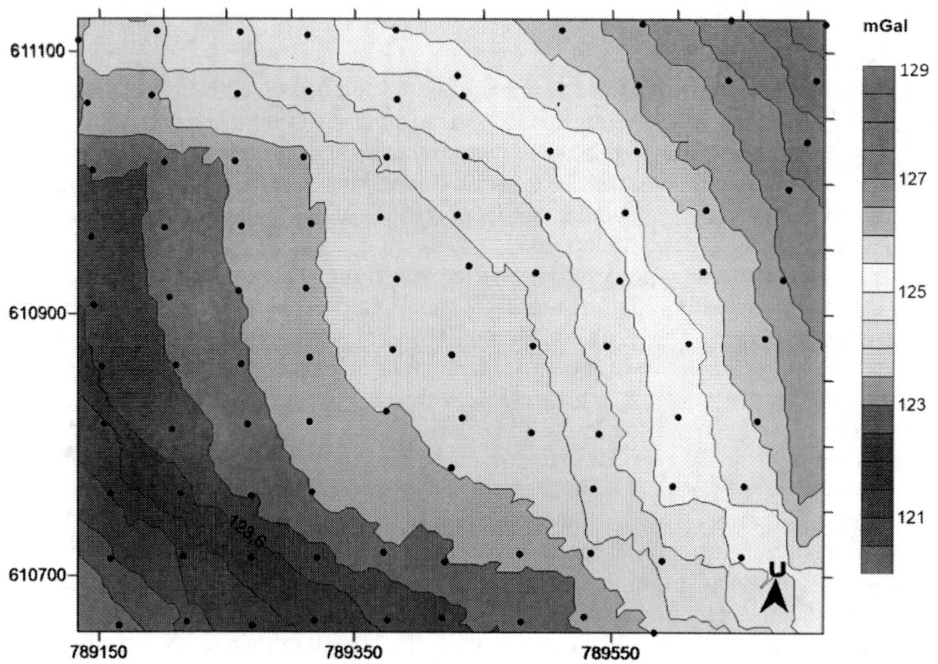

Figure 4. Map of regional anomaly Lam Apeng

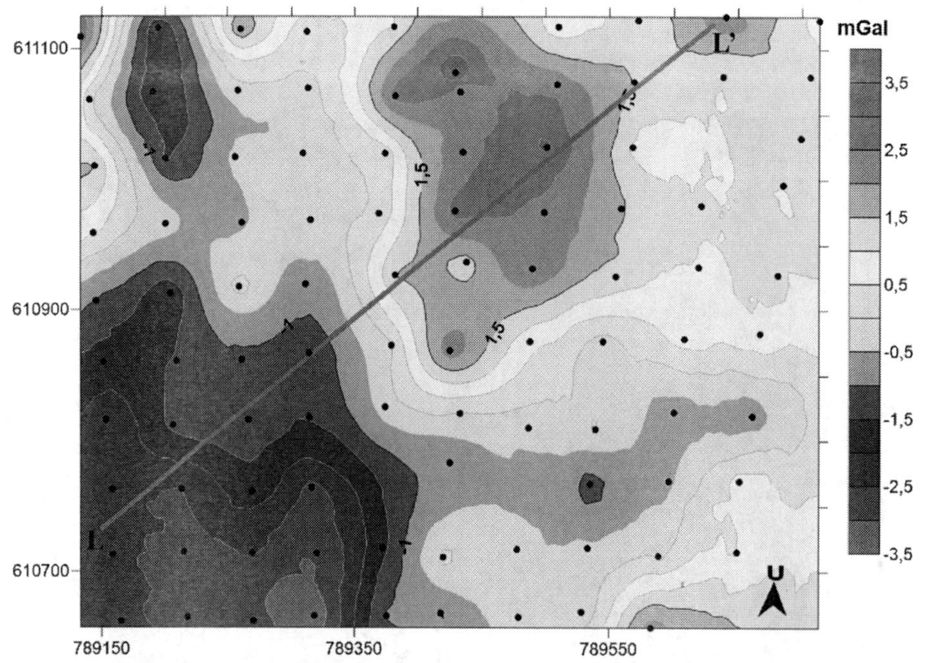

Figure 5. Map of the residual anomaly with cross-section line L-L'.

Furthermore, the 2D model line A-A of the study area was modeled with GRAV2DC (Figure 5). Forward modeling was performed to predict the density of each rock layer. Through this technique, field measurement profiles are matched against the model's gravity response profile. The process continues until it is deemed fit for the geological conditions, and the error is getting smaller. This modeling uses the Talwani principle, which creates a 2D model based on the polygon approach [20]. The model line determination was selected from the residual anomaly map with the aim that the resulting model is a local anomaly.

Figure 6 shows the results of forward modeling of the gravity anomaly, where the dotted line represents the residual anomaly, and the continuous black line represents the model. The misfit obtained was 3.08. The rock density value is determined based on the type of rock in the geological map of the study area. The 2D model in Figure 6 was then sketched out, as shown in Figure 7. There are three different layers, namely alluvium layer, coal layer, and sandy tufa layer. Along the surface layer is covered by an alluvium layer consisting of soil, gravel, and sand. This layer is estimated to have a thickness of 4 to 5 m. In some measurement areas, there is a layer of sandy tufa. This layer is estimated to have a density of 2.0 gr/cm^3.

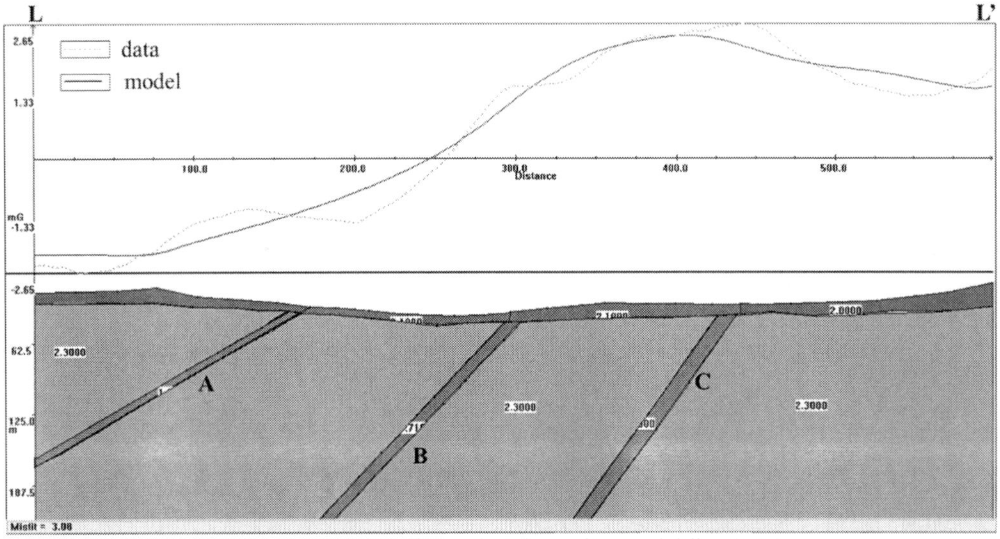

Figure 6. Two-dimensional gravity model line L-L'

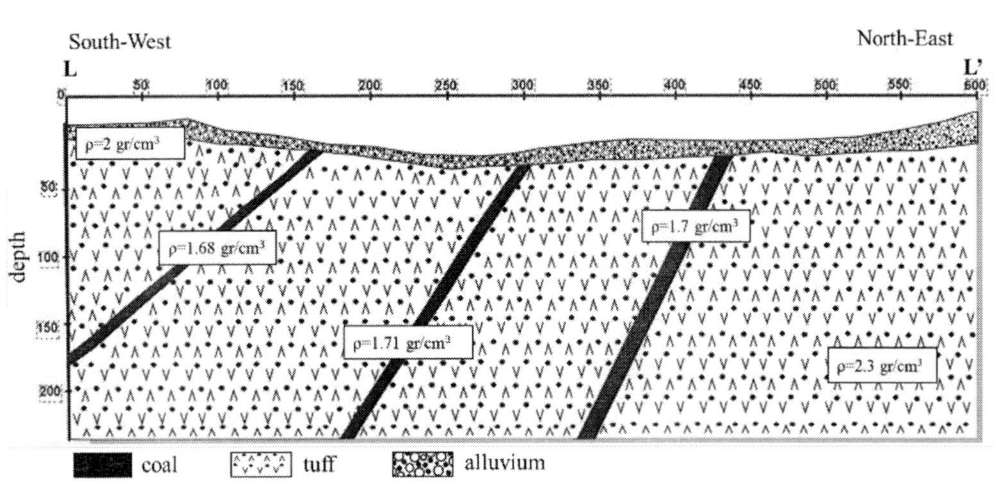

Figure 7. Sketch of the gravity model

Based on the results of the interpretation, there are three coal layers, namely coal layer A, coal layer B, and coal layer C (Figure 6). These three coal layers are classified as sub-bituminous coal.

Sub-bituminous is a sediment that has a moderate temperature, but this precipitate can be used for combustion. The characteristic of sub-bituminous deposits is that they contain wax and have a black color [21]. The coal layer A is exposed to the surface, which is at a distance of 150 m from the starting point of the data and at a depth of 4-5 m. This coal layer is estimated to have a slope of ± 125° with a layer thickness of 2-4 m and a density of 1.68 gr/m3. The coal layer B is estimated to be at a distance of 300 m from the starting point of the data with a slope of ± 120 ° and a layer thickness of 7-8 m and a density of 1.71 gr/m3. The C coal layer is estimated to be at a distance of 440 m from the starting point of the data with a slope of ± 120 ° and a layer thickness of 4-6 m and a density of 1.7 gr/m3.

The last layer is a sandy tuff layer, which is directly adjacent to coal and alluvium. This layer is estimated to be 5 to 220 m deep. This layer is arranged according to the shape of the surface topography, resulting in this layer sedimentation following its slope. The sandy tuff layer has a density of 2.3 gr/cm^3.

4. Conclusion

This study presents an analysis of gravity data to identify the coal layer in Lam Apeng area, Aceh Besar District. The gravity data in this area shows a trend that is getting higher to the northeast of the study area and lower to the southwest. At a location with a coal outcrop, the residual anomaly is low with a value around -1.5 mGal. The reason is that coal has a soft compacting rate compared to the surrounding environment. Based on 2D modeling of the L-L' section, there are three different layers in this area, namely alluvium layer (2 gr/cm^3), coal, and sandy tuff (2.3 gr/cm^3). There are 3 coal layers with different density values, namely layer A with a density 1.68 gr/cm^3, a slope ± 125° and a thickness 2-4 m; layer B with a density 1.71 gr/cm^3, a slope ± 120°, and a thickness 7-8 m; and layer C with a density 1.7 gr/cm^3, a slope of ± 120 °, and a thickness 4-6 m.

5. Acknowledgments

The author thanks students and technical staff of Geophysical Engineering Department Universitas Syiah Kuala for their help in collecting the data in the field.

References

[1] Gunara M 2017 *Seminar Nasional TEKNOKA ke -2.* **2**. *(in Bahasa Indonesia).*
[2] Abubakar M, Rusydy I, Setiya Nugraha G 2014 *J Nat.* **14** 1–6.
[3] Chisenga C, Dulanya Z, Jianguo Y 2019 *J. Afr. Earth Sci.* **149** 280–290.
[4] Nishijima J, Naritomi K 2017 *J. Hydrol. Reg. Stud.* **11** pp. 84–95.
[5] Pocasangre C, Fujimitsu Y, Nishijima J 2020 *Geothermics.* **83** 101735.
[6] Saibi H, Amrouche M, Fowler A-R 2019 *J. Asian Earth Sci.* **182** 103937.
[7] Yanis M, Marwan M, Ismail N 2019 *Indonesian Appl. Phys. Lett.* **9** 61-67.
[8] Bishop I, Styles P, Emsley S J 1997 *Geological Society, London, Engineering Geology Special Publications.* **12** 153 – 166.
[9] Hatherly P 2013 *Int. J. Coal Geol.* **114** 74–84.
[10] Siagian H P, Widijono B S 2013 *Jurnal Geologi dan Sumberdaya Mineral.* **14** 29–37. *(in Bahasa Indonesia).*
[11] Moechtar H, Subiyanto, Sugianto D 2009 *Jurnal Geologi dan Sumberdaya Mineral* **19** 272–283.
[12] Barber A J, Crow M J, Milsom J S. 2005. *Sumatra: Geology, Resources and Tectonic Evolution.* London: Geological Society of London.
[13] Bennet, J D, Bridge, D McC, Cameron, N R, Djunuddin, A, Ghazali, S A, Jeffery, D H, Kartawa, W, Keats, W, Rock, N M S, Thomson, S J, Whandoyo R. 1981. *Geologic Map of the Banda Aceh Quadrangle, Sumatra.*
[14] Darisma D, Marwan M, Ismail N 2019 *J. Aceh Phys. Soc.* **8** 1–5.
[15] Yanis M, Marwan 2019 *IOP Conf. Ser. Earth Environ. Sci.* **364** 12032.
[16] Abdelrahman E M, El-Araby T M 1996 *J. Appl. Geophys.* **36** 89–95.

[17] Oesanna M, Ismail N, Marwan 2018 *J. Aceh Phys. Soc.* **7** 56–60.
[18] Surya R A, Abdullah F, Darisma D 2019 *IOP Conf. Ser. Earth Environ. Sci.* **348** 12040.
[19] Speight J G. 2015. Physical Properties. in *Handbook of Coal Analysis.* Hoboken, NJ: John Wiley & Sons, Inc. pp. 170–197.
[20] Talwani M, Worzel J L Landisman M 1959 *J. Geophys. Res.* **64** 49–59.
[21] Falcon R M S, Snyman C P 1986 *Introduction to coal petrography: atlas of petrographic constituents in the bituminous coals of Southern Africa*. Johannesburg: Geological Society of South Africa.

Utilization of unmanned aerial vehicles in geothermal exploration: A review

Marwan[1], T R Noviandy[2], A Maulana[2], R Suhendra[2], M Yusuf [3], A Lala[3], G M Idroes[4], Muslem[5], Mahmudi[6], R Idroes[3,7,*]
[1]Department of Geophysical Engineering, Faculty of Engineering, Universitas Syiah Kuala, Kopelma Darussalam, Banda Aceh 23111, Indonesia
[2]Department of Informatics, Faculty of Mathematics and Natural Sciences, Universitas Syiah Kuala, Kopelma Darussalam, Banda Aceh 23111, Indonesia
[3]Department of Chemistry, Faculty of Mathematics and Natural Sciences, Universitas Syiah Kuala, Kopelma Darussalam, Banda Aceh 23111, Indonesia
[4]Department of Chemical Engineering, Faculty of Engineering, Universitas Syiah Kuala, Kopelma Darussalam, Banda Aceh 23111, Indonesia
[5]Department of Chemistry, Faculty of Science and Technology, Universitas Islam Negeri Ar-Raniry, Banda Aceh 23111, Indonesia
[6]Sekolah Tinggi Ilmu Kesehatan Assyifa Aceh, Banda Aceh 23242, Indonesia
[7]Department of Pharmacy, Faculty of Mathematics and Natural Sciences, Universitas Syiah Kuala, Kopelma Darussalam, Banda Aceh 23111, Indonesia

*E-mail: rinaldi.idroes@unsyiah.ac.id

Abstract. Geothermal is a source of heat energy from the earth and is a renewable energy source. Geothermal exploitation as an energy source is still ongoing. Recently there has been an increase in the use of an unmanned aerial vehicle (UAV) technology as an alternative option in aerial surveys and observations in geothermal areas. The use of UAVs in geothermal exploration is an alternative option. This is because the unmanned aerial vehicle has many advantages and features that can be utilized. Various studies that have been conducted have shown that the use of UAVs has become a method that brings changes in the exploration process of geothermal areas. UAVs are proven to assist in the geothermal exploration process and produce fast, reliable, and cost-effective solutions. UAVs also help explore geothermal areas that were previously difficult to study, namely areas with extreme environmental conditions and rugged access terrain. UAVs can also be used to search for potential sources as new manifestations in geothermal areas so that the points of manifestation in geothermal areas can be identified. This is very helpful in increasing the efficiency and less risk for the researcher.

1. Introduction

Geothermal is a source of heat energy from the earth and is a renewable energy source [1]. Indonesia is the largest geothermal producer in the world. This is indicated by 13% of the total number of volcanoes in the world in Indonesia, and 80% of which is estimated to have the potential for geothermal energy production. [2]. The potential of geothermal energy in Indonesia is estimated at 28,617, about 40% of the world's geothermal potential [3]. Geothermal can be said as an alternative energy resource for the world with high and renewable development prospects. This geothermal energy source can be a more environmentally friendly and prospective alternative to fossil energy sources [4].

Geothermal exploitation as an energy source is still being carried out. In Indonesia, geothermal exploitation is still centered on the island of Java, although seen from the geography and mapping of volcanic areas in Indonesia, almost half of Indonesia's geothermal potential is on the island of Sumatra. [5]. Aceh is a province located in Sumatra that has nearly 20 potential geothermal energy sources. The three areas that are very potential as geothermal energy are Mount Seulawah Agam, Jaboi, and Geuredong [6], [7]. Several studies have been carried out to explore geothermal energy at Seulawah Agam Volcano [8]. Geophysical research from the Seulawah Agam Volcano has reported deep and shallow data of the subsurface structures of the geothermal system [9]. In addition to this, research has been carried out to identify geochemistry in the southern zone, namely Ie Jue[10], Ie Brouk [11], Ie Seu'um [12], and Van Heutz Crater [7].

The resources in the geothermal film are highly varied. Not only the manifestations like hot spring and fumarole but also the geology sites [13] and the biodiversity [14],[15]. Therefore, an advanced technology that can process images [16] is needed to cover that all interests. Recently there has been increasing use of unmanned aerial vehicle (UAV) technology as an alternative option in aerial surveys and observations in geothermal areas [17]. UAV is a platform that can carry several different measuring devices and can be controlled manually by a trained person on the ground or controlled automatically using a programmed flight plan [18].

Figure 1. DJI drone modified with ICI thermal camera and UAV module [19]

The use of UAVs in geothermal exploration is an alternative option. Because the UAV has many features, namely equipped with a video recorder/camera, GPS, temperature sensor, and other components that can be used to support field research. The thermal sensors in the UAV can provide a choice for a high-resolution generation, as seen in Figure 1. This technology is promising to allow quick and safe surveys in geothermal areas with dangerous and difficult access to terrains [19].

In this paper, we will demonstrate the results of research conducted by various researchers related to the use of UAVs in the exploration process of geothermal areas. We will show the research results: the location, objectives, equipment used for exploration, and the output.

2. Recent Studies

Researchers have carried out various studies related to the use of UAVs in exploring geothermal areas. Research has been carried out in various geothermal manifestations worldwide, with various research objectives and different equipment uses. Table 1 presents the research results that have been carried out, covering the year, location, objectives, equipment, and the output.

Table 1. Recent studies of the use of UAVs in geothermal exploration

Year	Location	Objective	Equipment	Output	Ref.
2015	Tauhara thermal area, New Zealand	Demonstrating that drones have the potential to become a major tool in geothermal exploration	Using the DJI Phantom 2 Vision + quadcopter equipped with an ICI 640x480 uncooled thermal sensor with automated image capture (ICI UAV module)	RGB orthomosaic, thermal orthomosaic, DEM, and thermal 3D model	[20]
2015	Wairakei – Tauhara, New Zealand	Demonstrate the use of a small (<2kg) and cost-effective quadcopter to map the physical and biological characteristics of geothermal areas safely and accurately	Using the Blade 350 QX2 Quadcopter with a Spectrum DX5e DSMX 5-Channel transmitter equipped with a Sony HDR-AS100V, FLIR Tau 320 camera, and sensors for capturing infrared thermal video	RGB orthophoto and thermal orthophoto	[21]
2016	Karapiti thermal area, New Zealand	Shows the utility and economy of drones to produce accurate aerial photographs, DEM, and thermal imagery	Using the DJI Phantom 2 Vision + quadcopter equipped with an ICI 640x480 uncooled thermal sensor with automated image capture (ICI UAV module)	Thermal infrared orthophoto, RGB orthophoto, and DEM	[22]
2016	Waikite geothermal area, New Zealand	To assess the benefits of drone technology for finding hot tubs and other thermal features in difficult terrain such as wetlands and dense scrubs, and providing an estimate of the surface heat loss from thermal water in the survey area	Using the DJI Phantom 2 Vision + quadcopter equipped with an ICI 640x480 uncooled thermal sensor with automated image capture (ICI UAV module)	RGB orthophoto, thermal orthophoto, and DEM	[19]
2017	Hsiaoyukeng, Taiwan	Quadcopter design to collect thermal imagery that will be used for monitoring volcanic geothermal as a solution to dealing with difficult terrain, very	Using the AI-RIDER YJ-1000-QC Quadcopter that has been installed with GPS, compass, air pressure sensor, inertial measurement units	DSM and thermal orthoimages	[23]

Year	Location	Objective	Equipment	Output	Ref.
		varied topography, and extreme environmental conditions	(IMU), and XM6 thermal infrared camera		
2018	Geysir Geothermal Field	The use of UAVs to understand the relationship between deep and surface expressions in the Geysir geothermal field.	Using the DJI Matrice 100, which is equipped with a DJI Zenmuse X5R optical camera and a FLIR Tau 2 thermal camera	DEM, infrared mosaic, and thermal 3D model	[24]
2019	Various locations. Iceland	Proving that the use of an integrated drone system of optical and thermal cameras produces a fast, reliable, and cost-effective solution	Using the DJI Matrice 210 Drone, which is equipped with optical and thermal cameras	Thermal orthomosaic, DSM, and surface temperature map	[25]
2018	Khankala, Chechen Republic, Russia	Evaluating the use of UAVs for monitoring the geothermal plant environment	Using a Geoscan 201 UAV equipped with a Thermoframe-MX-TTX thermal imager and a Sony DSX-RX1 digital camera	RGB orthophoto and Thermal orthophoto	[26]
2020	Parco delle Biancane, Italy	Analyze surface temperature and distribution of thermal signatures in Tuscany's geothermal districts using data obtained through three separate surveys via satellite, the UAV	Using a FlyBit octocopter equipped with a FLIR VUE PRO R thermal imaging camera	RGB orthomosaic, thermal orthomosaic, and surface temperature maps	[27]

From Table 1, it can be seen that researchers have successfully utilized UAV technology for the exploration and monitoring of geothermal areas. In general, it can be seen that the main equipment needed to explore geothermal areas is a UAV equipped with optical and thermal cameras. UAV battery life must also be following the coverage area to be explored. Therefore it is advisable to bring additional chargers and batteries.

Before operating the UAV, it is very important to make a flight plan. One thing to note is the weather conditions in geothermal areas because unpredictable weather conditions can hinder the UAV operation. Temperature, fog, rain, and strong winds can all impact UAV performance [28]. Operating the UAV during strong winds and opposing wind direction also causes the UAV to consume more battery power, resulting in reduced flight time [21].

UAVs are used to collect RGB and thermal images of the geothermal area. Furthermore, the image that has been collected is analyzed to look for errors that may occur in the image retrieval process, such as images that are out of focus, blurry, and bouncy. These two-dimensional RGB and thermal images will later be converted to a three-dimensional point cloud using the software. The results obtained later can be a variety of map layouts. The layout maps that have been made by several researchers include the RGB orthophoto, thermal orthophoto, RGB orthomosaic, thermal orthomosaic, surface temperature map, digital elevation model (DEM), digital surface model (DSM), and thermal 3D model.

3. Conclusion

Various studies that have been conducted have shown that the use of UAVs has become a method that brings changes in the exploration process of geothermal areas. UAVs are proven to assist in the exploration process of geothermal areas and produce fast, reliable, and cost-effective solutions. UAVs also assist in exploring geothermal areas that were previously difficult to study, namely areas with extreme environmental conditions and difficult to access terrain. The utilization of UAVs can also be used to search for potential sources as new manifestations in geothermal areas so that the points of manifestation in geothermal areas can be identified. UAVs allow researchers to obtain images of the location of geothermal environments, such as cliffs, forests, and others, making it easier for researchers to reach points of manifestation. This allows mapping of areas using unmanned aerial vehicles in geothermal fields to be efficient for researchers' safety, especially for new areas where the situation is unknown.

References

[1] Mohammadzadeh Bina S, Jalilinasrabady S, Fujii H, Pambudi N A 2018 *Renew. Sust. Energ. Rev.* **93** 499–506.
[2] Maryanto S, Dewi C N, Syahra V, Rachmansyah A, Foster J, Nadhir A, Santoso D R 2017 *Geosciences* **7** 41.
[3] Nasruddin, Idrus Alhamid M, Daud Y, Surachman A, Sugiyono A, Aditya H B, Mahlia T M I 2016 *Renew. Sust. Energ. Rev.* **53** 733–740.
[4] PT Geo Dipa Energi 2012 *Tek. Pomits.* **1** pp. 1–6. *(in Bahasa Indonesia).*
[5] Bertani R 2016 *Geothermics* **60** 31–43.
[6] Putri D R, Nanda M, Rizal S, Idroes R, Ismail N 2019 *IOP Conf. Ser. Earth Environ. Sci.* **364** 012003.
[7] Idroes R, Yusuf M, Saiful S, Alatas M, Subhan S, Lala A, Muslem M, Suhendra R, Idroes G M, Marwan M, Mahlia T M I 2019 *Energies* **12** 4442.
[8] Marwan, Yanis M, Idroes R, Ismail N 2019 *Int. J. GEOMATE.* **17** 173–180.
[9] Marwan, Syukri M, Idroes R, Ismail N 2019 *Int. J. GEOMATE.* **16** 141–147.
[10] Idroes R, Yusuf M, Alatas M, Subhan, Lala A, Muhammad, Suhendra R, Idroes G M 2019 *IOP Conf. Ser. Mater. Sci. Eng.* **523** 012012.
[11] Idroes R, Yusuf M, Alatas M, Subhan, Lala A, Muslem, Suhendra R, Idroes G M, Suhendrayatna, Marwan, Riza M 2019 *IOP Conf. Ser. Mater. Sci. Eng.* **523** 012010.
[12] Idroes R, Yusuf M, Alatas M, Lala A, Suhendra R, Idroes G M 2018 *IOP Conf. Ser. Mater. Sci. Eng.* **334** 12002.
[13] Zhijie L, Guoying G 1986 *Geothermics* **15** 339–345.
[14] Nuraskin C, Marlina, Idroes R, Soraya C, Djufri 2020 *Rasayan J. Chem.* **13** 18–23.
[15] Nuraskin C A, Marlina, Idroes R, Soraya C, Djufri 2019 *Res. J. Pharm. Technol.* **12** 5247.
[16] Suhendra R, Arnia F, Idroes R, Earlia N, Suhartono E 2019 *2019 IEEE Int. Conf. Cybern. Comput Intell.* 35–39.
[17] Nishar A, Richards S, Breen D, Robertson J, Breen B 2016 *J. Unmanned Veh. Syst.* **4** 136–145.
[18] van Blyenburgh P 1999 *Air Sp. Eur.* **1** 43–47.
[19] Harvey M C, Rowland J V, Luketina K M 2016 *J. Volcanol. Geotherm. Res.* **325** 61–69.
[20] Harvey M, Luketina K 2015 *Proceedings of the 37th New Zealand Geothermal Workshop 2015.* 20.
[21] Nishar A, Richards S, Breen D, Robertson J, Breen B 2016 *Renew. Energ.* **86** 1256–64.
[22] Harvey M C 2016 *GRC Transactions.*
[23] Chio S H, Lin C H 2017 *Sensors* **17** 1649.
[24] Walter T R, Jousset P, Allahbakhshi M, Witt T, Gudmundsson M T, Hersir G P 2020 *J. Volcanol. Geotherm. Res.* **391** 106282.
[25] Bjornsson G, Grímsson G, Sigurdsson A, Laenen V 2019 *Thermal Mapping of Icelandic Geothermal Surface Manifestations with a Drone.*

[26] Cherkasov S V., Farkhutdinov A M, Rykovanov D P, Shaipov A A 2018 *J. Sustain. Dev. Energy Water Environ. Syst.* **6** 351–362.
[27] Silvestri M, Marotta E, Buongiorno M F, Avvisati G, Belviso P, Bellucci Sessa E, Caputo T, Longo V, De Leo V, Teggi S 2020 *Remote Sens.* **12** 2018.
[28] Sauter B. 2007. *Weather impacts on the aerostar unmanned aircraft system based on climatology over the US/Mexico border.* Army Research Lab White Sands Missile Range Nm Computational And Information.

Thermal conductivity of rice straw polypropylene particleboard

I Ismail[1,*], N Fitri[1], Mursal[1], S H S Md Fadzullah[2]
[1]Department of Physics, Faculty of Mathematics and Natural Sciences, Universitas Syiah Kuala, Banda Aceh 23111, Indonesia
[2]Centre for Advanced Research on Energy, Faculty of Mechanical Engineering, Universiti Teknikal Malaysia Melaka, Hang Tuah Jaya, 76100, Durian Tunggal, Melaka, Malaysia

*Email: ismailab@unsyiah.ac.id

Abstract. The availability of rice straw in Indonesia, including in some other Asian countries, is very abundant. However, it has not been used optimally. One possibility of using rice straw is for heat insulation in buildings. This study aims to analyze the thermal conductivity of rice straw polypropylene composite/particleboard. In this study, particulate composites from rice straw have been produced by using recycled plastic polypropylene (homopolymer) as adhesive. The particle size of rice straw used in this study was 20 mesh. The composition of rice straw and polypropylene were varied (the ratio of rice straw particle to polypropylene is 30/70 wt. %, 40/60 wt. %, 50/50 wt. %, and 60/40 wt. %). The composite samples were produced by using a hot press with 3 tons of load. The thermal conductivity of rice straw composite samples has been measured. We found that our composite thermal conductivity is 0.223 (±0.015) W/m °C for rice straw 30 wt.% and polypropylene 70 wt. %. For rice straw composition 60 wt. %, its thermal conductivity reduces to 0.191 (±0.002) W/m °C. The composite thickness swelling is found to be 0.76 (±0.26) % for rice straw 30 wt.% (after 24 hours immersing in water). The percentage of thickness swelling is 1.80 (±0.10) % for rice straw 60 wt. %. The results of our study revealed that the thermal conductivity of rice straw composite decreased as the thickness swelling increased. This behavior is related to the porosity of the composite.

1. Introduction

Indonesia is one of the agricultural countries in Asia that produces around 70 million tons of rice in 2018 [1]. Therefore, there is a lot of rice straw in the field of rice in Indonesia. If we assume the ratio of rice to straw is about 80:20, then there are about 14 million tons of rice straw produced in Indonesia in 2018. Almost all of these rice straws become waste in the rice fields; they have not been used optimally. Only a few are used for animal feed. Besides Indonesia, several other Asian countries such as China, Vietnam, and India also produce a lot of rice. The situation is the same as in Indonesia, where there is abundant rice straw waste in the rice fields. Meanwhile, a previous study found that there was about 36 - 39 % of cellulose contained in the rice straw [2].

Some previous studies have been conducted on rice straw composites. Yang et al. made a composite of rice straw particles mixed with wood particles using a urea-formaldehyde adhesive, which is potential for acoustic panel [3]. Akyildiz et al. showed that the particleboard could be made from rice straw and urea-formaldehyde matrix [4]. Zhongli Pan et al. prepared rice straw fiber particleboard by using polymeric methylene diphenyl diisocyanate (PMDI) mixed with rice bran as adhesive. The rice bran adhesive can be used up to 30 % to substitute PMDI to form particleboard [5]. Reasonable particleboard can also be formed by using rice straw and epoxy resin adhesive [6]. Besides urea-formaldehyde, PMDI, and epoxy resin, several researchers have used plastic as an adhesive in manufacturing composite. Yao et al. formed a rice straw fiber composite by using high-density polyethylene as a matrix [7]. Some other studies have used rice straw and polypropylene as a matrix in producing reasonably strong composites [8], [9], [10].

The application of composite material is extensive. One of them is for thermal insulation. Research on the thermal properties of composites has been conducted continuously [11], [12], [13], [14], [15]. Researchers continue to look for strong composite materials, low cost, and have low thermal conductivity (good thermal insulator). Jolanta Vejeliene et al. evaluated the thermal conductivity of several renewable resources. They found that the thermal conductivity of cattail, bentgrass, reed, and barley are in the range of 0.055 to 0.085 W/m °C [11]. Mounika et al. showed that the bamboo fiber polyester composite has a thermal conductivity of 0.185 W/m K. They found that bamboo fiber composite has better thermal insulator properties than glass fiber composite [12]. Tangjuank et al. found that the conductivity of papyrus fiberboard is 0.029 – 0.030 W/m °C [13]. Chandana et al. reported that the thermal conductivity of composite made of bamboo fiber using epoxy resin as a matrix is 0.173 – 0.197 W/m °C [14]. Raju and Kumarappa showed that the thermal conductivity of groundnut shell particle epoxy resin composite is 0.076 to 0.348 W/m °C [15].

A recent study proposed the particleboard made from rice straw particles (20 mesh) and polypropylene plastic waste as a matrix. It was found that the particleboard composite has excellent mechanical properties where its modulus of rupture and modulus of elasticity fulfill the Indonesia National Standard [10]. Nonetheless, the thermal conductivity of this particleboard composite is still unknown. On the other hand, thermal insulator panel having excellent mechanical properties is very important which can be used in building such wall and ceiling panels to prevent heat flow from outside building to reduce the consumption of electricity for air conditioner, especially in the tropical country. Based on the discussion above, we have investigated the thermal conductivity of rice straw polypropylene composite.

2. Material and Methods

Rice straw was obtained from the rice field in Aceh Besar district, which was close to the city of Banda Aceh. After cleaning, the rice straw was dried under the sun for several days. Furthermore, rice straw is chopped to a smaller size. In the next step, rice straw was dried by using an oven at 105 °C for 4 hours. Then, dried rice straw was milled then sieved to obtain a grain size of 20 mesh. Used polypropylene plastic (homopolymer) was collected from the city of Banda Aceh, which was cut by about 1 cm x 1 cm. The plastic waste was mixed with the coupling agent (methanol, xylene, anhydride maleic, and benzoyl peroxide) and heated at a temperature of 175 °C, which became a matrix. Then, the melted plastic was mixed with rice straw particles and poured into the sample mold to be pressed for 3 tons for 1 hour at a temperature of 175 °C to produce composite samples. The compositions of particleboard samples were varied: (a) 40 wt. % of rice straw and 60 wt. % of polypropylene matrix; (b) 50 wt. % of rice straw and 50 wt. % of matrix; (c) 60 wt. % of rice straw and 40 wt. % of the matrix; and (d) 70 wt. % of rice straw and 30 wt. % of the matrix.

The size of samples for thermal conductivity measurement was 10 cm x 10 cm (area) and x 1 cm (thickness). The thermal conductivity measurement was conducted by using a single-plate method instrument manufactured by Leybold. The thermal conductivity of the sample is determined by using equation (1) [16].

$$k = \frac{\Delta Q}{\Delta t} . \frac{d}{A.\Delta T} \quad (1)$$

Where d is the thickness of the sample, A is the area of the sample, ΔT is the temperature difference between the heated and the cooled side of the sample, ΔQ is heat flow through the sample during the time Δt. The system is entirely isolated. Therefore the amount of heat flow (ΔQ) is equal to the electrical energy (ΔW) for the heating element. Three repetitions of measurements were made to obtain each data point of thermal conductivity. The error-bar of thermal conductivity data was calculated from the repetition data by using the standard deviation equation.

The thickness swelling of samples (*TS*) was determined by using Equation (2) [17].

$$TS = \frac{T_W - T_A}{T_A} x100\% \quad (2)$$

Where T_A is the thickness of the sample in the air (before immersing in water), T_W is the sample thickness after immersion the sample in water for 24 hours. The size of the sample was 5 cm x 5 cm x 1 cm. Similar to thermal conductivity data, three repetitions of measurements of thickness swelling were made and averaged to have each data point. The error-bar of thickness swelling was also calculated by using the standard deviation equation.

3. Result and Discussion

Our results of thermal conductivity measurement of rice straw polypropylene composite are shown in Figure 1. For the rice straw composition 30 wt. % and polypropylene 70 wt. %, the thermal conductivity is 0.223 W/m °C. As the composition of rice straw increases to 40 wt. %, the conductivity of composite decreases to 0.219 W/m °C. For the composition of rice straw 50 wt. %, the thermal conductivity of composite is found to be 0.207 W/m °C. The value of thermal conductivity decreases as the composition of rice straw increases. The lowest value of conductivity of our composite is 0.191 W/m °C obtained at the rice straw composition 60 wt. %.

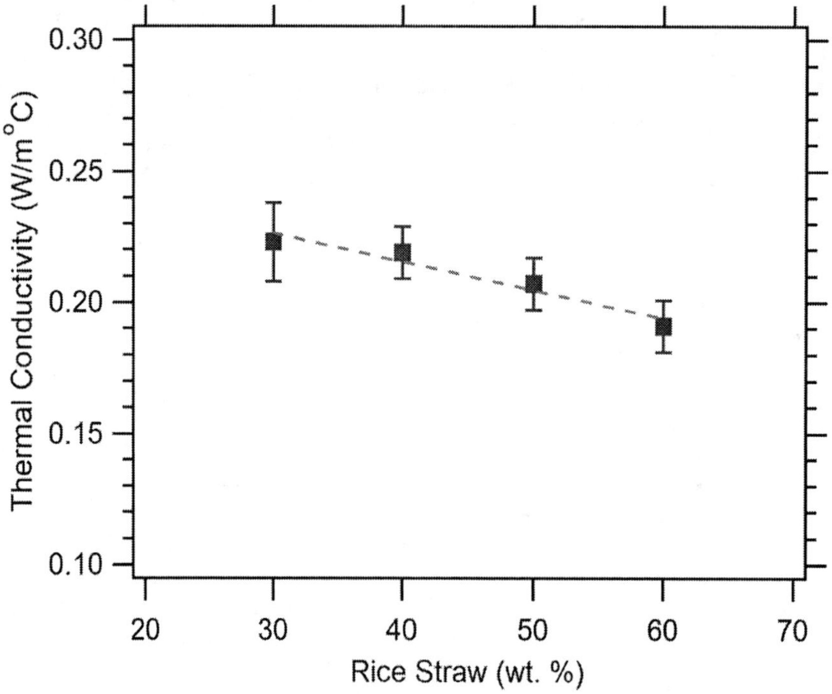

Figure 1. Thermal conductivity of the composite.

The comparison of thermal conductivity from the present study with other composites is displayed in Table 1. The value of thermal conductivity of rice straw polypropylene composites is almost the same value as the conductivity of the composite made from bamboo using a polyester matrix or epoxy matrix. The thermal conductivity of our composite is much larger than the thermal conductivity value of the papyrus fiber composite. However, the thermal conductivity of rice straw polypropylene composite is much smaller than the thermal conductivity of rice husk, bagasse, and corncob. Compared to concrete and glass, the rice straw polypropylene composite produced from this study has a better thermal insulator property. Thus, the rice straw polypropylene composite is a reasonable thermal insulator material.

Table 1. Thermal conductivity for selected composites.

Material/Composite	k (W/m °C)	Ref.
Bamboo fiber & polyester	0.185 – 0.211	[12]
Bamboo fiber & epoxy	0.173 – 0.197	[14]
Groundnut shell particle & epoxy	0.076 – 0.348	[15]
Rice husk, corncob, bagasse & starch	0.297 – 0.494	[18]
Papyrus fiber & rubber	0.029 – 0.030	[13]
Glass	0.78	[17]
Oak	0.17	[17]
Concrete	0.80	[17]
Rice straw particle & polypropylene	0.191 – 0.223	This study

The thermal conductivity of our composite sample is strongly dependent on the composition of rice straw particles, and it has a linear trend, as indicated by the dashed line in Figure 1. We speculate that this behavior is related to the porosity of the composite. The porosity of the composite must be related to the water absorption or thickness swelling of the composite. The standard for composite or particleboard requires the thickness swelling. For this reason, we have measured the thickness swelling of our composite sample after 24 hours of immersion in water for several compositions. The results can be seen in Figure 2.

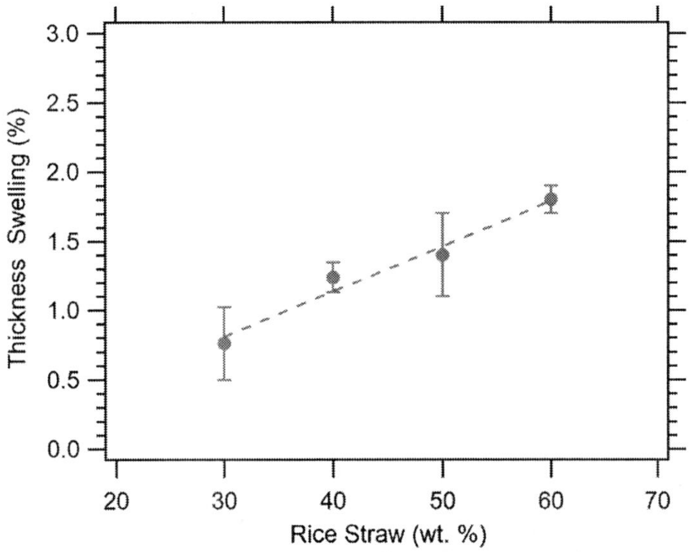

Figure 2. Thickness swelling of our composite.

For rice straw 30 wt. %, the swelling of thickness is 0.76 %. For rice straw 40 wt. %, the thickness swelling increases to 1.24 %. The swelling of thickness increases with the increasing of rice straw composition, as shown in Figure 2. The relationship between the thermal conductivity and the thickness swelling is shown in Figure 3. As the percentage of thickness swelling increases, the thermal conductivity decreases. Higher percentage thickness swelling means that this composite has a bigger percentage of porosity. So, the thermal conductivity is influenced by the porosity of the composite.

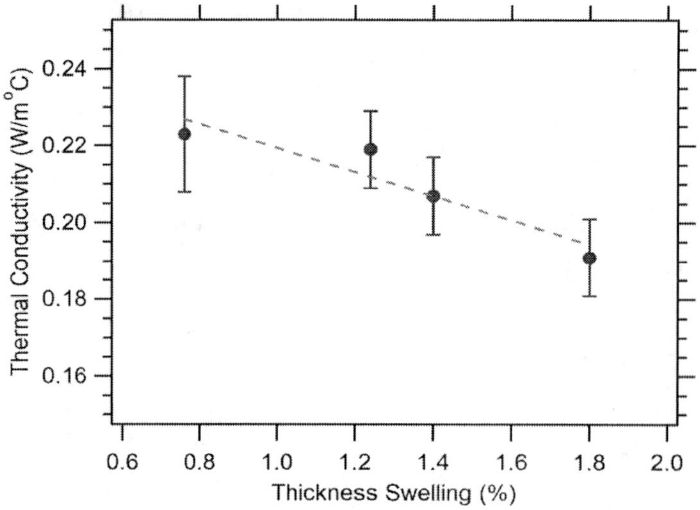

Figure 3. Thermal conductivity versus thickness swelling of rice straw polypropylene composite.

Since the thermal conductivity is influenced by the porosity as shown above, then we believe that the thermal conductivity should also be related to the density of the composite. The density of rice straw polypropylene composite for 20 mesh particle size has been measured and reported in the previous study [10]. Now let's plot the thermal conductivity against the density of our composite, as shown in Figure 4.

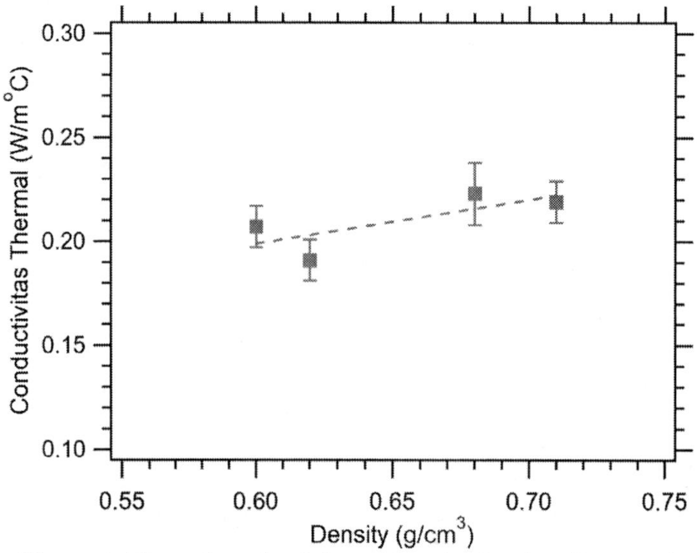

Figure 4. Thermal conductivity against the density of composite.

The dashed line is the linear fit to the data. The density data is from the previous work [10], while the thermal conductivity data is from this work. From Figure 4, we can see that the thermal conductivity is also influenced by the density of the composite. The higher density of rice straw polypropylene composite, the greater its thermal conductivity is. We believe this behavior applies to other composite materials.

4. Conclusion

The thermal conductivity of rice straw particle composite (20 mesh) using polypropylene matrix has been measured for various rice straw compositions. For 30 wt. % of rice straw composition, the thermal conductivity was found to be 0.223 W/m °C. The thermal conductivity of composite decreased as the composition of rice straw was increased. This behavior is related to the porosity of the composite. The lowest thermal conductivity was obtained at 60 wt. % composition of rice straw that was 0.191 W/m °C. Our finding indicates that the composite made of rice straw particles and polypropylene is potential for a thermal insulator.

5. Acknowledgment

This work was partially funded by the Research Grant (Hibah Laboratorium 2018) from the Universitas Syiah Kuala, Banda Aceh, Indonesia.

References

[1] Food and Agriculture Organization (FAO) 2018 Food Rice Market Monitor. 2018. **XXI**. Available at: http://www.fao.org/3/i9243en/I9243EN.pdf.
[2] Lu P, Hsieh Y L 2012 *Carbohydr. Polym.* **87** pp. 564-573.
[3] Yang H S, Kim D J, Kim H J 2003 *Bioresour. Technol.* **86** 117 – 121.
[4] Akyldiz M H, Kesik H I, Oncel M, Olgun C 2015 *PRO LIGNO* **11** 130 – 137.
[5] Pan Z, Cathcart A, Wang D 2006 *Ind. Crops Prod.* **23** 40–45.
[6] Ismail I, Aini Q, Zulfalina, Jalil Z, Md Fadzullah S H S 2018 *J. Phys. Conf. Ser.* **1120** 012014.
[7] Yao F, Wu Q, Lei Y, Xu Y 2008 *Ind. Crops Prod.* **28** 63–72.
[8] Grozdanov A, Buzarovska A, Bogoeva G G, Avella M, Errico M E, Gentille G 2006 *Agron. Sustain. Dev.* **26** 251–255.
[9] Mohammadi H, Mirmehdi S, Hugen L N 2016 *CERNE* **22** 449-456.
[10] Ismail I, Fitri N, Zulfalina, Md Fadzullah S H S 2018 *J. Phys. Conf. Ser.* **1120** 012015.
[11] Vėjelienė J, Gailius A, Vėjelis S, Vaitkus S, Balčiūnas G 2011 *Mater. Sci* **17** 208 – 212.
[12] Mounika M, Ramaniah K, Prasad R A V, Rao K M, Chandra Reddy K M 2012 *IOP Conf. Ser. Earth Environ. Sci.* **3** 1109 – 1116.
[13] Tangjuank S, Kumfu S 2011 *J. Appl. Sci.* **11** 26402645.
[14] Chandana E, Hussian S A 2013 *J. Mech. Civil Eng.* **9** 7-14.
[15] Raju G U, Kumarappa S 2012 *Polym. Renew. Resour.* **3** 117 – 138.
[16] Holman J P 2010 *Heat Transfer*. New York: Mc Graw Hill.
[17] Japanese Standards Association 2003 *Japanese Industrial Standard for particleboard JIS A 5908*. Tokyo: Japanese Standards Association.
[18] Kyauta E E, Dauda D M, Justin E 2014 *Am. J. Eng. Res.* **03** 34-40.

Spread of liquid waste in un-licited gold mining by geoelectrical method in Krueng Sabee sub-district Aceh

N Aflah[1,*], Syafrizal[2], Mulkal[1], H Harisman[1], M R Lubis[1,3], Muchlis[1]
[1]Mining Engineering Department, Universitas Syiah Kuala, Banda Aceh, Indonesia
[2]Physics Education Department, Universitas Malikul Saleh, North Aceh, Indonesia
[3]Chemical Engineering Department, Universitas Syiah Kuala, Banda Aceh, Indonesia

*E-mail: nurul_aflah@unsyiah.ac.id

Abstract. Illegal gold mining has increased significantly in Keude village of Krueng Sabee Aceh Jaya subdistrict. This causes environmental damage in the form of waste mercury pollution in the area. Environmental pollution caused by waste needs to be addressed. However, it needs to determine the appropriate method to overcome it by knowing the depth and the spread of mercury waste. This study aims to visualize subsurface conditions in the area of Krueng Sabee to provide information and data on the spread of mercury wastes. By using the method of geoelectric resistivity Schlumberger array, the contamination can be seen below the surface by the resistivity value of the sites. Results from these measurements after inverted using Earth Imager software of Res2Dinv can get a picture and visualize the location of subsurface conditions of the study. The results show that the resistivity value of waste mercury-contaminated soil at the study site ranged from 2–5 Ωm, and waste mercury contamination occurs vertically to a depth of 3 meters.

1. Introduction

The amount of illegally gold mining in Indonesia is increasing [1]. In early 2008, there are about 480 illegal gold mining in all over Indonesia, particularly in areas with potential gold deposits. Personal or a bunch of people, even by a private company, on private or public land without concession or the necessary paperwork, is doing [2] these illegal mining activities. In general, illegal gold miners use mercury for extracting precious metals [3]. Mercury is a hazardous metal that can harm organs [4] and the immune system. U.S. Food and Administration (FDA) determines 0.005 ppm as the threshold limit value for mercury in the environment.

In Aceh itself, one of the illegal gold mining takes place in Krueng Sabee area, Aceh Jaya district [5]. This area is located in southwest Aceh. It has tremendous gold deposits exploited illegally by the local communities. Because of their lack of knowledge about mining principles, such illegal activities have some bad impacts [6] on the nearby environment.

One of the impacts of the illegal mining activity in Krueng Sabee is the contamination of mercury in residential areas [7]. This happened because the machines used to extract the gold, called Gelendung, operates within the settlement. Right after using the mercury to amalgamate gold particles, the miners dump the mercury waste straight away [8].

The wasted mercury flows through the subsurface. It is hard to predict how far the waste will spread. An initial study needs to conduct to monitor the mercury proliferation since it may contaminate the groundwater, and the groundwater is the main water source for the communities living in that area. The use of polyurethane foam adsorbents to adsorb mercury can reduce mercury pollution [9]. This study uses the geoelectrical method to observe how far the mercury has contaminated the land (Figure 1). The geoelectrical method can distinguish between the contaminated area and the non-contaminated area based on the resistivity value [10]. The mercury waste has a higher resistivity value compared to the groundwater.

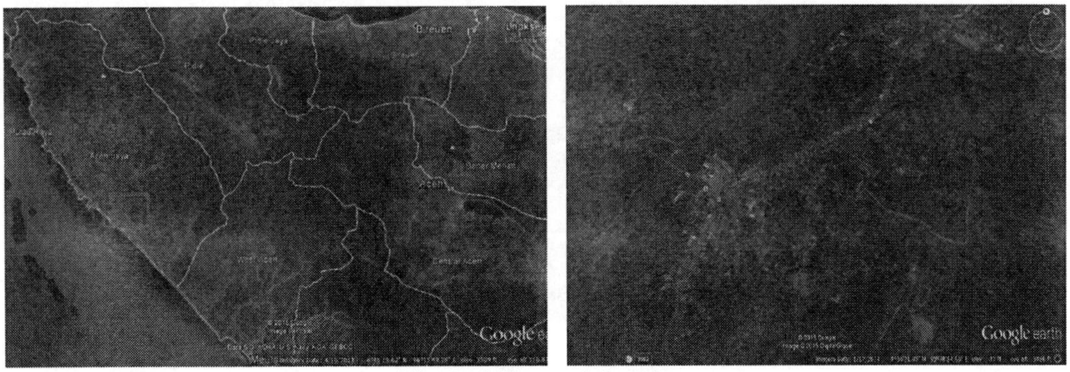

Figure 1. Research location

Keude Krueng Village, located in Krueng Sabee sub-district, Aceh Jaya district, has the potential for gold mining [11]. The local people extract the gold found with a machine called "gelundung". Gelundung uses mercury to separate gold from its gangue [12]. Mercury, also called quicksilver, is a chemical substance with the symbol of Hg. Hg stands for Hydrargyrum with liquid silver meaning. According to Rustagi and Singh [13], mercury is a heavy metal liquid at standard temperature and pressure conditions.

Mercury has silvery-white color and belongs to a fair conductor of electricity. However, it is a poor heat conductor with a freezing temperature of -38.9°C and a boiling temperature of 357°C. Another property of mercury is its tendency to dissolve other metals to form a metal alloy mixture, such as gold, known as amalgam [14]. The gold is dissolvable into the mercury. Therefore, mercury is useful to extract the gold [15] in the extraction process or amalgamation process.

Amalgamation is a process to bring free gold particles into contact with mercury to mix both substances to form a compound called amalgam [16]. In the final process, the collecting of mercury and gold amalgam takes place after separating the two substances. Taking the gold goes on, yet the mercury waste dumping occurs by just letting it flow out into the environment. If this disposal is not controlled, it can devastate the nearby environment such as the river, residential areas, and so on.

2. Methodology

The survey was conducted in a gold extraction area with the amalgamation process in Krueng Sabee [17]. Three sets of measurements along three different lines with each 55-meter length and 2.5 m electrode spacing were carried out to gather the data. These measurements used the geoelectrical method with Wenner-Schlumberger configuration in order to get more data points [18]. The more datum points measured horizontally, even vertically, the more accurate the model generated by Res2Dinv software [19]. The data acquisition process was done by using SuperSting resistivitimeter R8/IP [20] as indicated in Figures 2 to 4.

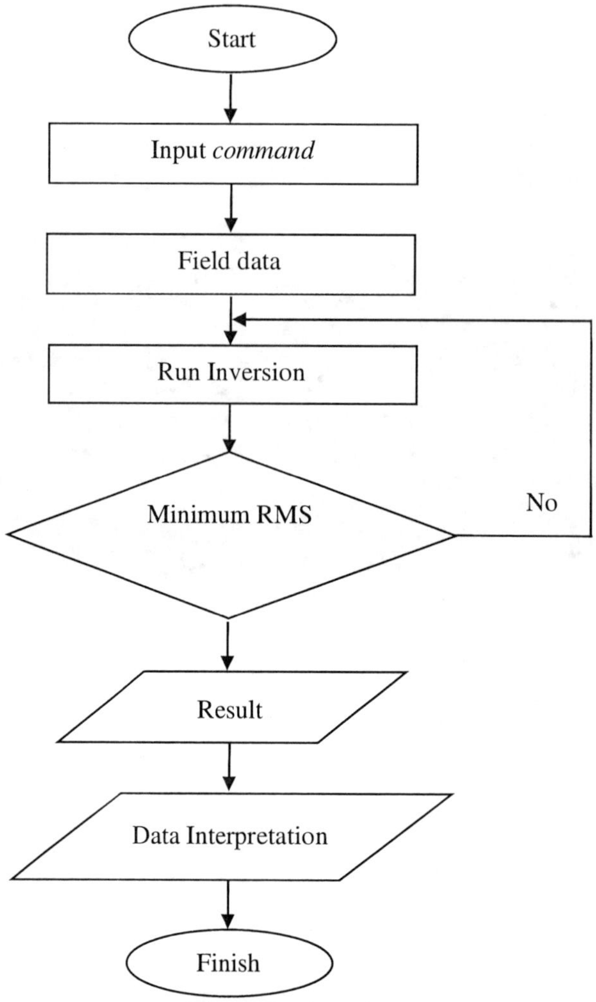

Figure 2. Flow chart

Figure 3. Multi-channel supersting

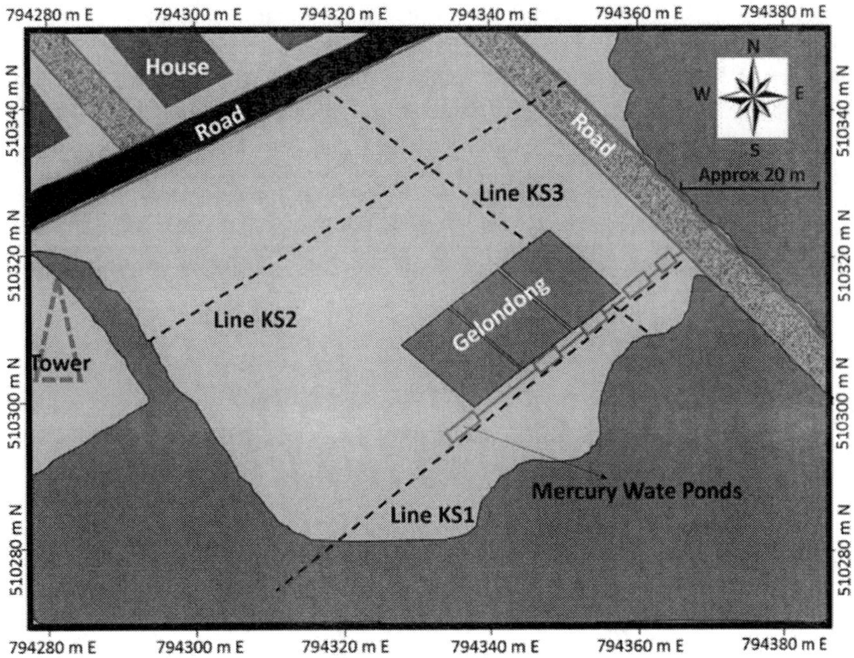

Figure 4. Survey line

3. Results and Discussions

The direct observation and the resistivity models from Res2Dinv software shows a clay layer of the measurement area (Figure 5). Its resistivity indicates it in the value range of 1–100 Ωm. Higher resistivity values between 100–250 Ωm (loose clay) were also obtained in three lines. At line KS1, the higher resistivity is in the range of 0–16 meters, while in KS3, it is in the range of 42–55 meters. At line KS2, it is also along the ditch in the range of 48–49 meters, and 49–55 meters along the dirt road.

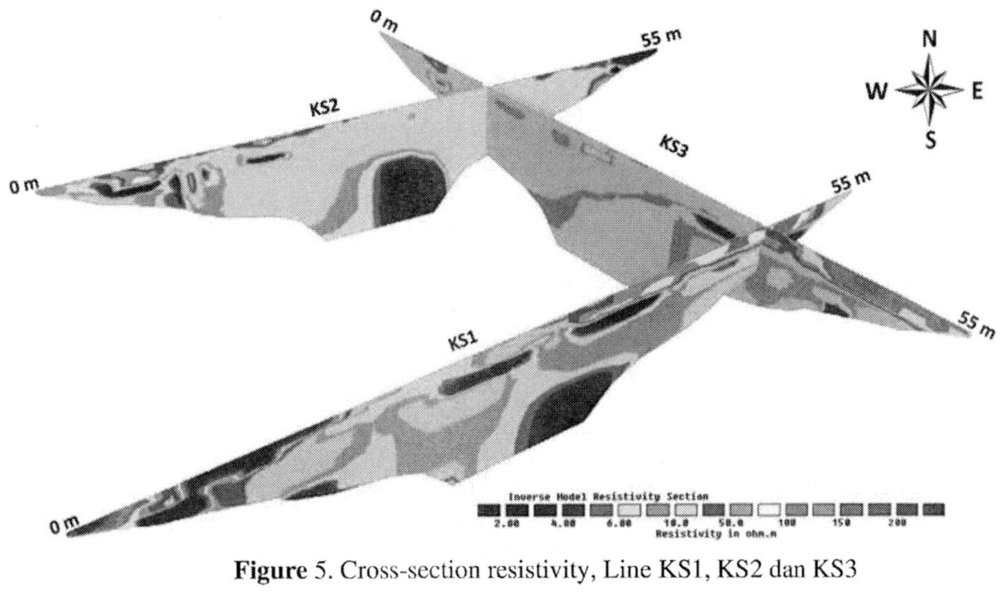

Figure 5. Cross-section resistivity, Line KS1, KS2 dan KS3

3.1. Line KS1

KS1 line was measured along the water disposal ponds contaminated by mercury. The ponds are sealed with concrete walls on all sides. They have a different size with 2 meter depth. There are five ponds along the KS1 line. The first pond distance is from 24–28 meters from the starting point of measurement. The others are from 36–39 meters, 42–45 meters, 40–43 meters, and 53–55 meters for the second, third, fourth, and fifth ponds, respectively.

At the line KS1, the lower resistivity (conductive) value is in the range of 2–5 Ωm. It was found at 22–28 meters, 31–38 meters, and 41–55 meters, with 2–3 m depth (showed by the dashed line in Figure 6). The blue colors inside the dashed line indicate the mercury-contaminated soil. The soil becomes contaminated as the mercury in the waste ponds comes through the concrete walls.

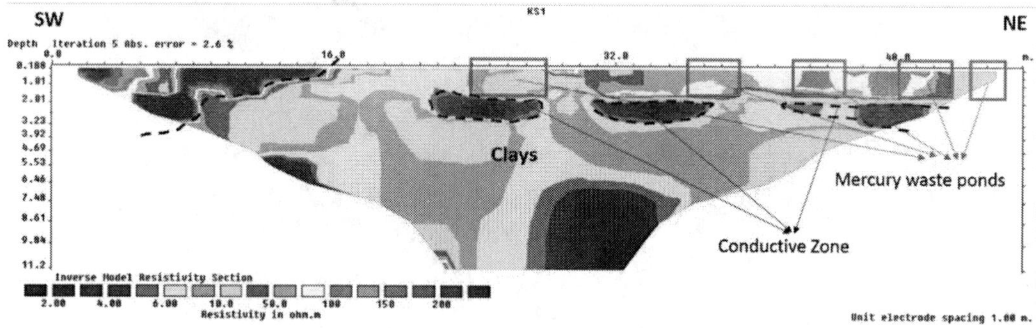

Figure 6. Cross-section resistivity, Line KS1

3.2. Line KS2

At line KS2, the lower resistivity values were in the range of 2–5 Ωm (Figure 7). It was found in the range of 0–21 meter distance from the starting point with 2–3 meter depth. However, the soil is not contaminated with mercury as there is no horizontal continuity with the contaminated soil detected at line KS1. Moreover, the lower resistivity zone is about 5 meter depth, whereas the mercury infiltrates the soil only up to 3 meters in KS1.

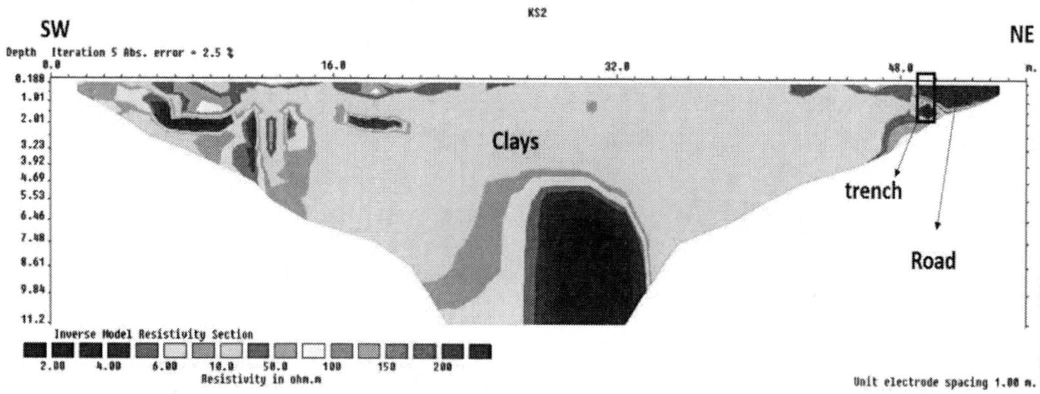

Figure 7. Cross-section resistivity, Line KS2

3.3. Line KS3

Line KS3 crosses line KS1 at 39 meter distance, while for line KS1, the cross point is at 46 meters from its starting point (Figure 8). The measurement at 41–45 meter distance in line KS3 shows almost the same depth in the conductive zone with a lower resistivity value (2–5 Ωm) as that in line KS1 with 34–39 m distance. The cross-section model clearly shows that there is no correlation between the contaminated zone (lower resistivity value) in KS1 and the lower resistivity zone in KS2.

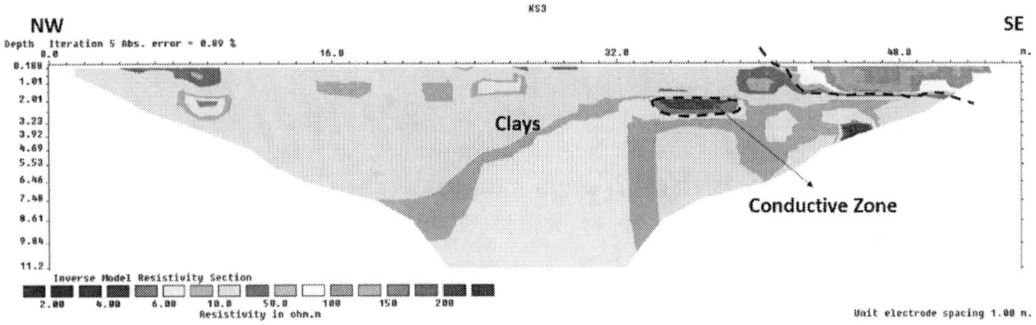

Figure 8. Crossection resistivity, Line KS3

4. Conclusion

Mercury waste has contaminated Keudeu village, sub-district of Krueng Sabee, up to 3 meter depth. The contamination does not spread horizontally. The resistivity value of the contaminated soil below the mercury waste ponds is in the range of 2–5 Ωm.

5. Acknowledgments

Authors wishing to thank all colleagues, especially technical staff and financial support from Ministry of Higher Education and Culture.

References

[1] Muslihuddin R O, Chirani N 2015 *Shape-mem. Polym. Biomed. Appl.* **267**.
[2] Carvalho F P 2017 *Food Energy Secur.* **6** 61.
[3] Bruno D E, Ruban D A, Tiess G, Pironne N, Perrota P, Mikhailenko A V, Ermolaev V A and Yashalova N N 2020 *Sci. Total Environ.* **715** 1.
[4] Engwa G A, Ferdinand P U, Nwalo F W, Unachukwu M N 2019 *Poisoning in the Moderm World – New Tricks for an Old Dog.* London: IntechOpen.
[5] Aminah 2018 *J. Ius Civ.* **112**.
[6] Yuka S, Elvince R, Tomonori K, Noriyasu N, Yuji N, Takanobu I, Osamu N, Takayuki M and Syamsidik 2013 *J. Ecotechnol. Res.* **17** 11.
[7] Suhendrayatna, Sofia, Ibrahim T, Elvitriana 2013 *AIWEST-DR* 23 December 2013 Banda Aceh.
[8] Reto C 2002 *Rural Remote Health* **1** 10.
[9] Darmadi, Lubis M R, Saputra R, Alfarabi M D, Sarah S, Adisalamun 2020 *IOP Conf. Ser. Mater. Sci. Eng.* **845** 012002.
[10] Shevnin V, Delgado-Rodiguez O, Fernandez-Linares L, Martinez H Z, Mousatov A, Ryjov A 2005 *Geofis. Int.* **44** 251.
[11] Sofia, Husodo A H, Sugiharto E 2016 *Int. J. Public Health.* **5** 257.
[12] Sumarjono E 2020 *Kurvatek.* **5** 1. *(in Bahasa Indonesia).*
[13] Rustagi N, Singh R 2010 *J. Occup. Environ. Med.* **14** 45.
[14] Nascimento R O, Chirani N 2015 *Shape-mem. Polym. Biomed. Appl.* **267**.
[15] Appel P W U, Andersen A, Na-Oy L D, Onos R 2015 *J. H&P.* **5** 12.
[16] Ucar Y, Brantley W 2017 *Biocompatibility of Dental Biomaterials.* Birmingham: Woodhead Publishing.
[17] Sofia, Husodo A H 2016 *Jurnal Manusia dan Lingkungan.* **23** 310. *(in Bahasa Indonesia).*
[18] Jamaluddin, Umar E P 2018 *IOP Conf. Ser. Earth Environ. Sci.* **118** 012006.
[19] Loke M H 1999 *A Practical Guide to 2D and 3D Survey.* Penang: Minden Heights.
[20] Reynold J M 2011 *An Introduction to Applied and Environmental Geophysics.* New York: John Wiley and Sons Ltd.

Graphite oxide tribo-layer formation under boundary lubrication of diesel fuel

Z Fuadi[1,*], R Kurniawan[1], K Kanda[2], K Adachi[2]
[1]Department of Mechanical and Industrial Engineering, Universitas Syiah Kuala, Banda Aceh, Indonesia
[2]Laboratory of Tribology and Nanointerface Engineering, Division of Mechanical Engineering, Tohoku University, Sendai, Japan

*E-mail: zahrul.fuadi@unsyiah.ac.id

Abstract. Friction and wear of mechanical components (such as engines) is controlled by some kind of boundary films (also referred to as tribo-layer). In some cases, such boundary film is formed on the contact interface due to a tribo-chemical process in the presence of liquid media. In this paper, we demonstrate the formation of graphite oxide tribo-layer on the contact interface of steel in the presence of diesel fuel liquid media. A Raman spectrometer was used to analyse the tribo-layer structure. Although the intensity is low, it is shown that the tribo-layer has a structure of graphite oxide, demonstrated by the presence of D band peak at approximately 1350 cm^{-1} and G band peak at approximately 1550 cm^{-1} in the Raman spectra. Such graphite solid films on the sliding surfaces have ability to control friction and wear. This finding demonstrates the possibility of in-situ formation of solid protective film of a tribo-pair components using the working fluid where the use of lubricants is restricted as they contaminate the working fluid.

1. Introduction

Due to environment and sustainability issues, the use of diesel fuels from renewable resources are continuously increasing. Biodiesel fuels are converted from vegetable oils or animal stocks by transesterification process. They can be used directly in diesel engines or mixed with petroleum diesel fuel at certain volume percentages. Palm methyl ester is one of the biodiesel fuel sources available in large amount due to high annual production of crude palm oil. The use of methyl ester as mixture in diesel fuel has several implications to the tribological behaviour of the mechanical systems such as fuel system and piston-ring component. Particularly, the oxygen contain in the methyl ester makes it corrosive in nature, compensating the material stability and integrity [1]. However, biodiesel also has several beneficial effects on the tribological properties of the materials, such as friction and wear reduction [2].

In order to control the friction and wear, additives are commonly used in lubricants as friction modifier. The wear protection is achieved by the formation of third body layer, or also called as tribo-layer/tribo-film on the contact interface during friction. Additives such as ZDDP, MoS_2, and PTFE protect the wear of material by forming a protective tribo-layer on the interacting surfaces, reducing direct material contact among bulk material asperities to prevent seizure and abrasion [3]–[6]. MoS_2, PTFE, and graphite belongs to the group of solid lubricants. They have lamellar structure interconnected by weak Van der Waals force, enabling they to withstand high load bearing capacity. The lamellar structures are deposited on the interacting surfaces during friction, acting as a protective layer.

Besides additives, solid coatings are also used to control friction and wear of mechanical components, such as diamond-like carbon DLC coatings [7]. Such solid coatings, as it is worn out during friction, produced a transfer layer of graphitic structure to the adjacent surface. The lamellar structure of graphite can easily slide among each other due to weak Van Der Waals force, resulting in low friction and wear of the bulk material [8]. Nowadays, carbon nanoflakes such as graphene are used as nano additives in lubricant because they produce similar effect to those achieved by solid carbon-based coatings.

The formation of protective tribo-layer on the contact interface that has carbon-based structure can also be achieved by in-situ tribo-chemical process [9]–[12] with the presence of hydrocarbon liquid media. The protective layer formation is assisted by high shear force on the contact interface during friction, which promotes decomposition of carbon from the material or liquid media, resulting in thin film of carbon structure on the contact interface.

In this study, we investigated the formation of tribo-layer on the contact interface by in-situ tribo-chemical process involving diesel fuels as the liquid media. The wear tests were conducted using a pin-on-disk tribometer at boundary lubrication regime. The objective of this study is to demonstrate possibility of forming a graphite oxide layer on the contact interface using in-situ tribo-chemical process using diesel fuel media. Such in-situ process can be used in tribological application in which lubricant cannot be used because it will contaminate the working fluid. In this study, the properties of tribo-layer was analysed using Raman spectroscopy method.

2. Methodology

A ball on disk tribometer was used in the investigation, as schematically shown in Figure 1. The tribometer consists of a ball and disk holder, which is rotated by a motor. The normal force is applied using a gravitational load lever, equipped with a load cell to measure the friction force. The contact interface was submerged in a media chamber filled with liquid. An 8 mm diameter ball specimen and a 5 mm thick with a 5 mm diameter disk were used as the specimens.

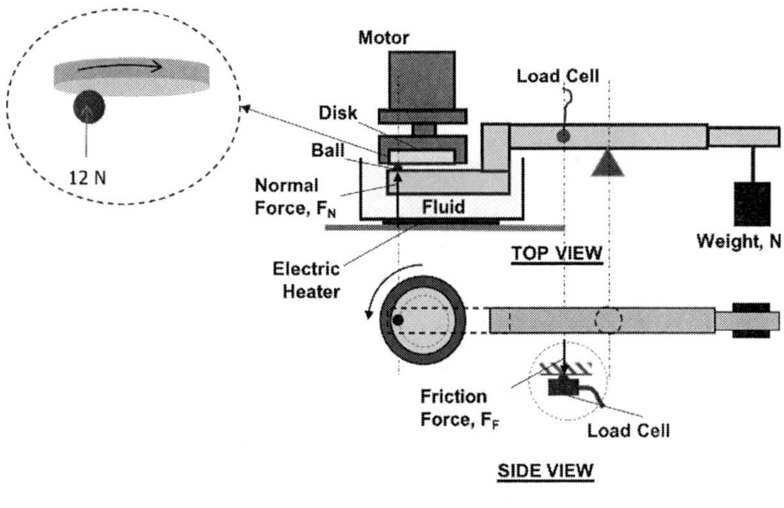

Figure 1. Schematic illustration of ball-on-disk tribometer

In order to simulate the severe contact condition between the material, the tests were conducted at boundary lubrication regime. The condition was achieved by applying a 12 N normal force, which resulted in approximately 1.2 GPa initial Hertzian contact pressure to the contact combination. The tests were conducted at ambient temperature of 27°C. The wear tests were conducted at sliding speed of 12 mm/s for a sliding distance approximately 750 meters.

In order to investigate the boundary film formation on the contact interface at various material and fluid media combinations, several materials and fluid media combinations were used. The material for the ball specimen is stainless steel SUS304 and the materials for the disk are stainless steel SUS 304 and grey cast iron. The materials are commonly used as materials for engines components. Particularly, SUS304 is a low carbon steel considered having high resistant to oxidative nature of palm methyl ester. Both the ball and disk specimen were polished to a roughness approximately Ra = 0.03 μm. For the fluid media, two compositions were prepared, i.e. petroleum diesel fuel and a mixture of 7.5% palm methyl ester in the petrol diesel fuel. The four material and fluid combinations are summarized in Table 1. The palm methyl ester was achieved from a transesterification process while the petrol diesel fuel was acquired from a commercial gas station. The mixture of the palm methyl ester and the petrol diesel fuel was prepared by direct mixture of the liquid using a physical mixing process.

Table 1. Material combinations for the wear tests

No	Material	Fluid Media	Temp., °C	Initial Contact Pressure, GPa
1	SUS 304 ball vs Cast Iron Disk	Petrol diesel fuel	27	1.2
2	SUS 304 ball vs Cast Iron Disk	7.5% palm methyl ester mixed petrol diesel fuel	27	1.2
3	SUS 304 ball vs SUS 304 disk	Petrol diesel fuel	27	1.2
4	SUS 304 ball vs SUS 304 disk	7.5% palm methyl ester mixed petrol diesel fuel	27	1.2

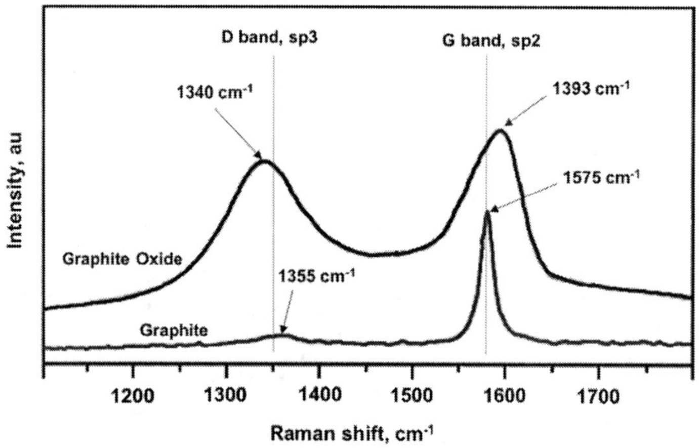

Figure 2. Raman shift reference of graphite and graphite oxide

The properties of boundary film on the contact interface were evaluated by using Raman spectroscopy method. The evaluation was conducted only to the ball's worn surface because the ball specimen was in static contact position whereas the disk was in constant rotation. Thus, the possibility of boundary film formation is higher on the ball's worn surface compared to that on the disk surface.

Figure 2 provides the reference Raman spectra for graphite (G) and graphite oxide (GO) [13]. The GO is characterized by the appearance of two peaks; D band and G band, also referred to as sp3 peak and sp2 peak, respectively, in the Raman spectra. D band peak appears at approximately 1355 cm^{-1} and G band peak appears at approximately 1593 cm^{-1}. Pure graphite has only one peak, which is Sp3, at approximately 1580 cm^{-1}. This reference was used for analysing the properties of boundary films in this study.

3. Results and discussion

In this paper, the main discussion is focused on the tribo-layer properties forming on the contact interface. Since the ball is in static position during the sliding test, there is a greater possibility that the tribo-layer is formed and accumulated on the ball's contact interface rather than that of the disk. Therefore, the investigation was conducted on the worn scar of the balls. Here, the worn track on the disk was not shown. Fig. 3 shows the results of wear test indicated by condition No. 1 and No. 2 in Table 1; condition No. 1 shown in (a), (b), and condition No. 2 shown in (c), (d). Figure 3(b) indicates the three-dimensional view of Figure 3(a), so as Figure 3(d) to Figure 3(c).

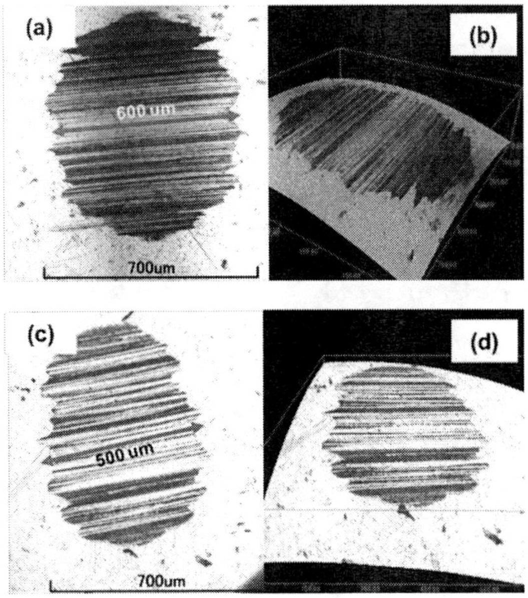

Figure 3. Worn scar of SUS304 ball sliding against cast iron disc; under petrol diesel fuel, (a) and (b), and under 7.5% palm methyl ester mixed petrol diesel fuel (c) and (d).

It can be observed in the Fig. 3 that the worn scar area of SUS304 ball is slightly larger in the case of condition No. 1, Fig. 3(c), than that in the case of condition No. 2, Fig. 3(a). In the case of condition No. 1, the width of the worn scar is 600 μm while in that of condition No. 2 it is 500 μm. This indicates some wear reduction of the material due to the presence of palm methyl ester components in the petrol diesel fuel, as reported in several previous works [10], [14], [15]. As given in Table 1, the fluid media in condition No. 2 contained 7.5% palm methyl ester while that of fluid media in condition No.1 has no contain of palm methyl ester. Despite the wear difference, it should be noted that there were no significant difference in the friction coefficient in the two cases. The average friction coefficient of the case in condition No. 1 was 0.3 while that in the case of condition No. 2 was 0.28.

In both case No. 1 and No. 2, the worn scar of the ball indicates a uniform wear mode in the form of worn grooves along the direction of sliding. Such regular grooves indicate abrasive wear which is commonly caused by counter surface having higher hardness. In this case, the cast iron is harder than SUS304 with a hardness of 294 (Brinell) [16] compared to 123 (Brinell) [17]. Thus, it can be concluded that the worn grooves on the SUS304 ball was abrasive wear. Some blackish mark along the sliding direction can be observed on the contact interface, indicated some tribo-layer had been formed on the contact interface during the friction process.

The worn scar area on SUS304 balls in the case of condition No. 3 and No. 4 are given in Fig. 4; Fig. 4(a) and (b) for condition No. 3 and Fig. 4(c) and (d) for condition No. 4. Figure 4(b) indicates the three-dimensional view of Fig. 4(a), so as Fig. 4(d) to Fig. 4(c). It can be observed from the Figure that the width of the wear track in both conditions are relatively similar, i.e. 500 μm. In both cases, the average value of coefficient of friction was relatively similar at 0.3.

Figure 4 illustrates that the wear behaviour of the SUS304 ball in the cases of condition No. 3 and No. 4 consists of two different wear modes. The first wear mode can be classified as abrasive wear and the second one as catastrophic wear due to the presence of pits and transferred material. Abrasive grooves is indicated by ploughing of the material sliding direction, similar to those occurred in condition No. 1 and No. 2. For condition No. 3 and No. 4, abrasive wear occurred for about 50% of the worn area. The second wear mode, catastrophic wear, involved the material transferred from the counter surface. In this case, it seems that the material was removed due to friction and reattached to adjacent location or to the counter surface. Such wear behaviour could be caused by material weakening due to some tribo-chemical reaction involving manganese and carbon [10]. In this case, the manganese came from the bulk material and rich carbon came from petrol diesel fuel fluid. The tribo-layer is formed on the worn surface of the SUS304 ball in both wear modes area; abrasive worn area and catastrophic worn area.

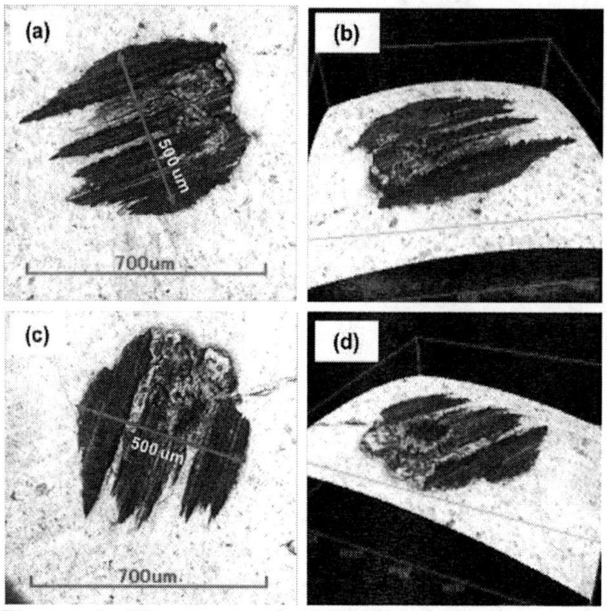

Figure 4. Worn scar of SUS304 ball sliding against SUS304; under petrol diesel fuel, (a) and (b), and under 7.5% palm methyl ester mixed petrol diesel fuel (c) and (d).

Figure 5 shows the Raman spectra taken on several positions on the contact interface of SUS 304 ball of condition No.1. In these cases, the SUS ball was sliding against cast iron disk under lubrication of petrol diesel fuel fluid. As shown in the Figure, position no. 1 is located in the middle of sliding track. A more detailed worn surface of position no. 1 reveals a visible tribo-layer, seen as blackish layer. The Raman spectra indicates two peaks of D band and G band at about 1350 cm^{-1} and 1550 cm^{-1}, respectively. These peaks are quite close to the sp3 and sp2 reference value, as indicated by the vertical lines. The appearance of D band and G band peaks can also be observed for Raman spectra taken at position no. 2, as indicated by the inset Figure, although the intensity is not as high at those taken at position no. 1. Similar to that in position no. 1, the tribo-layer in this position is also visible as blackish layer. Position no. 3 is outside the sliding track. It can be seen that in this position there is no trace of visible tribo-layer comparable to those in the other two positions. As expected, the Raman spectrogram on position no. 3 shows no peak, indicating that the surface contains no such tribo-layer of carbon structure.

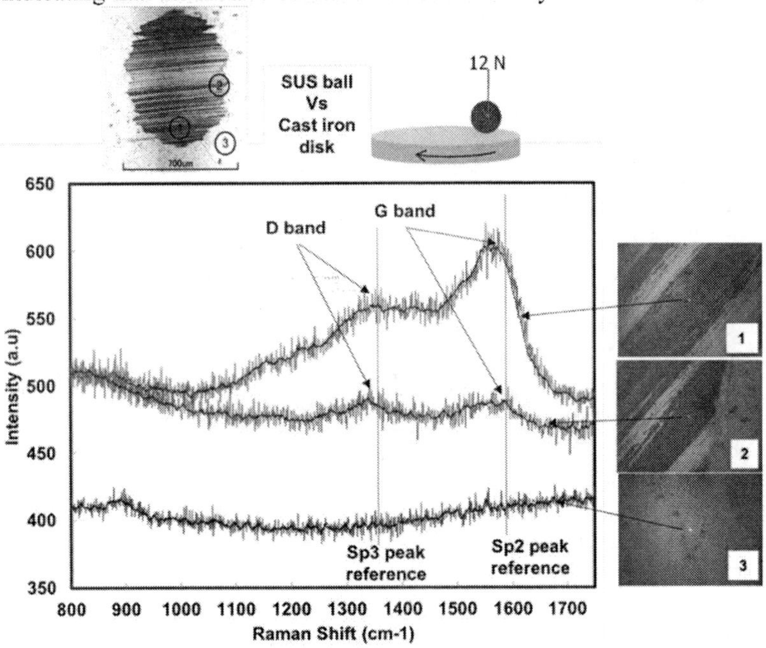

Figure 5. SUS ball vs Cast iron disc at 12 N normal load sliding under petrol diesel fuel

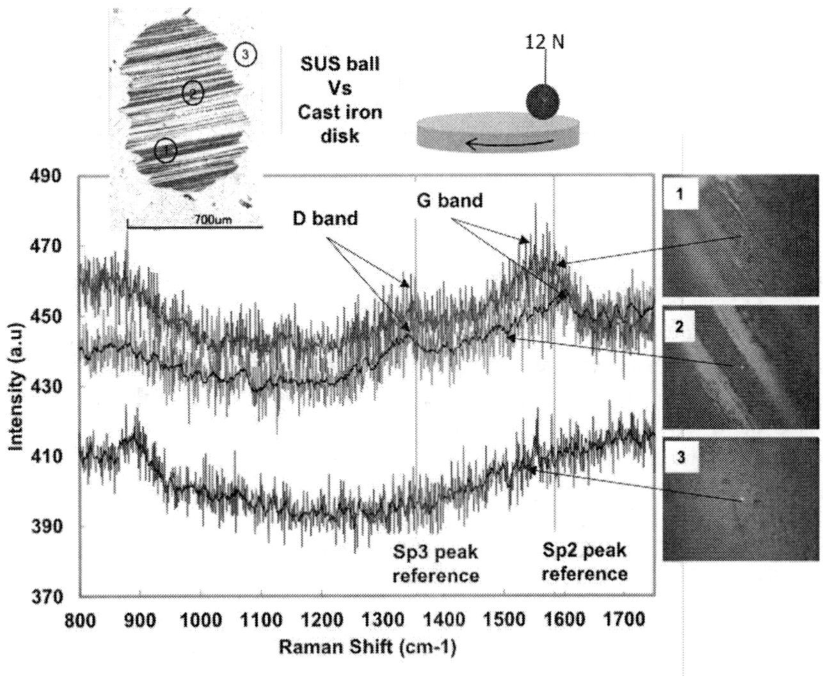

Figure 6. SUS ball vs Cast iron disc at 12 N normal sliding under 7.5% palm methyl ester mixed petrol diesel fuel

The Raman analysis results for condition No. 2 is given in Figure 6. In this case, the SUS304 ball was sliding against cast iron disk under petrol diesel fuel containing 7.5% palm methyl ester. As seen in the figure, the Raman analysis conducted to position no. 1 and no. 2 indicates the D band and G band peaks, although the intensity is quite low compared to those shown in Figure 5. For position no. 1, the D band peak occurred at about 1350 cm^{-1} and the G band peak occurred at about 1550 cm^{-1}. For position no. 2, the D band and G band peaks occurred at about 1345 cm^{-1} and 1600 cm^{-1}, respectively. As it is observed on the inset Figures for position no. 1 and no. 2, there are visible tribo-layer, which is blackish in colour, on the contact interfaces. Here, position no. 3 is outside the sliding track, provided for the purpose of comparison, thus no peak in reference to the sp3 and sp2.

The results of Raman analysis for test condition No. 3 is summarized in Figure 7. In this condition, the SUS304 ball was sliding against SUS304 disk under lubrication fluid of petrol diesel fuel. As mentioned previously, the worn scar of the ball in this condition consists of two wear modes. The Raman analysis were conducted on the part of worn surface of abrasive mode because the tribo-layer in this part is more visible than in the part of catastrophic wear mode. The blackish layer of tribo-film is visible at location no. 1 and no. 2, as given in the inset Figures of Figure 7. The Raman spectra of location no. 1 shows 2 peaks, occurred at about 1320 cm^{-1} for D band and about 1550 cm^{-1} for G band. For location no. 2, the resulted Raman spectra indicates the D band peak at about 1320 cm^{-1}. For G band, the peak is less visible, although still traceable at about 1580 cm^{-1}. Similar to other condition, location no 3 was taken outside the sliding track for comparison purpose.

Figure 8 shows the results of Raman analysis for test condition No. 4. In this case, the SUS304 ball was sliding against SUS304 disk under lubrication fluid of petrol diesel fuel mixed with 7.5% palm methyl ester. From the microscopic observation, the tribo-film on the contact interface is also visible as blackish layer, as given in inset Figure 1 and 2 of Figure 8. In location no. 1, the Raman spectra indicates the D band peak at about 1350 cm^{-1} and G band peak at about 1580 cm^{-1}. However, the Raman spectra taken at location no. 2 shows no traceable peaks of either D band or G band, despite visible trace of blackish layer of tribo-film on the contact interface. Among possible explanation is that the tribo-film in this specific location has a different structure, which requires further investigation. Location no. 3, as also mentioned in other test condition, was taken outside the sliding track.

Raman analysis conducted to all four condition reveals the existence of graphite oxide tribo-layer on the contact interface, evidenced from the occurrence of peaks of D band and G band comparable to those in Figure 1. The blue graph in Figure 1 refers to the Raman spectra of a highly ordered graphite, indicated by narrow G band (graphite lattice) at 1575 cm^{-1} and the weak disorder band caused by D band at approximately 1355 cm^{-1} [18]. The black graph refers to the Raman spectra of Graphite Oxide, indicated by the presence of D band peak at 1340 cm^{-1} and G band peak at 1593 cm^{-1}. The D band and G band experienced changes in the frequency due to amorphization of graphite as amorphous carbon, which contains a fraction of sp3 carbon [19]. A broader G band and D band is an indication of highly disordered graphite [13], which resulted in the shift of D band and G band frequency.

In reference to this, the D band's peak of the tribo-layer occurred on SUS304 balls in this analysis have frequency ranging from 1320 cm^{-1} (test condition No. 3) to 1350 cm^{-1} (test condition No. 1, No. 2, and No. 4) while that of G band's peak occurred at frequency ranging from 1550 cm^{-1} (test condition No. 1, No. 2, and No, 3) to 1600 cm^{-1} (test condition No. 4). Thus, it can be concluded that, since both D band and G band have a broad frequency band, the tribo-layer formed on the contact interface is graphite oxide with high disorder graphite. Here, the possibility of graphite oxide tribo-layer formation is higher in the case of SUS ball sliding against Cast iron disk rather than in the case of SUS ball sliding against SUS disk. This is evidenced from the presence of D band and G band peaks in all four locations selected on the case of SUS ball sliding against Cast iron. Whereas for the case of SUS ball sliding against SUS disk, only two of four selected locations show the occurrence of the D band and G band peaks.

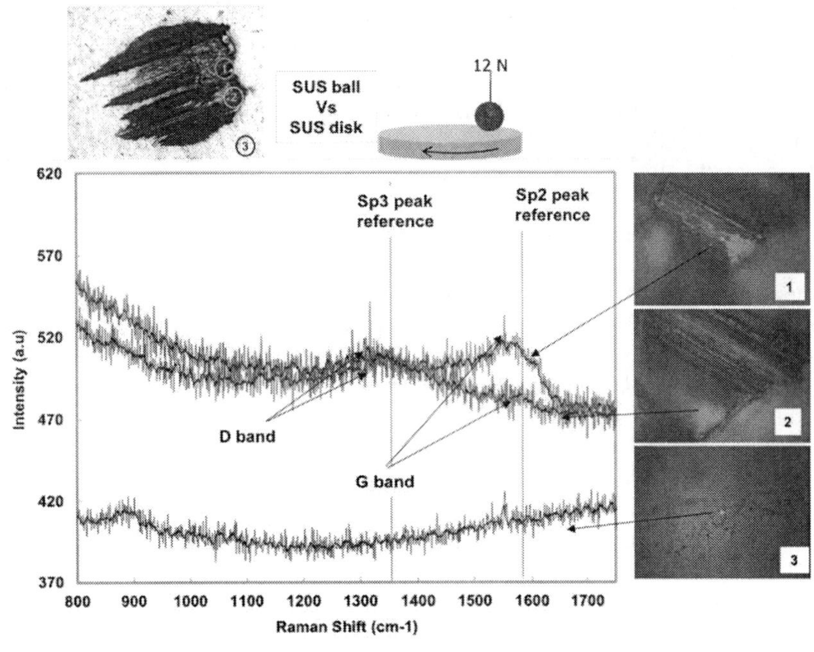

Figure 7. SUS ball vs SUS disk at 12 N normal Load 200 rpm under petrol diesel fuel

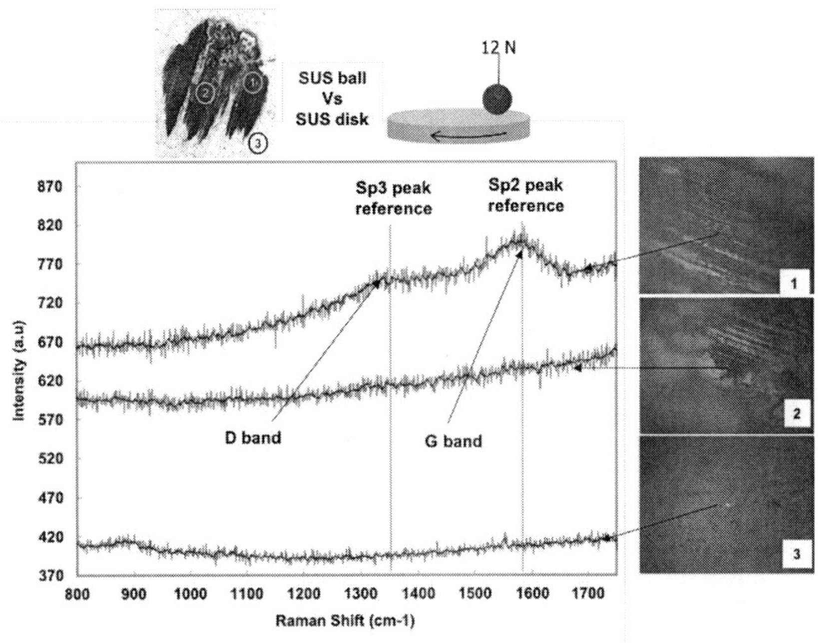

Figure 8. SUS ball vs SUS disk at 12 N normal Load sliding under 7.5% palm methyl ester mixed petrol diesel fuel

The higher possibility of graphite oxide formation in the case of SUS ball sliding against Cast iron disk is caused by the content of graphite flake in grey cast iron. Graphite flake is found easily deformed when shear force is involved, such as in tribological application [20]. Such formation of interlayered distorted graphite involves mechanical action and chemical reaction [21], [22], or also known as tribo-chemical reaction, with or without the presence of liquid lubricant media [9]. In the case of test condition No. 1 and No. 2, the graphite flake from cast iron disk could be transferred to the SUS ball during the tribo-chemical process, forming a layer of distorted graphite oxide.

Other source of carbon that was involved in the formation of graphite oxide is the lubricant fluid, which in this case is the diesel fuel. With high contact pressure and shear force at boundary lubrication condition, it is possible for the carbon to be decomposed during the tribo-chemical process leading to the formation of graphite oxide tribo-layer on the contact interface [10]. Therefore, in contact condition No. 3 and No. 4, the graphite oxide layer can still be found on the contact interface although the material does not contain graphite flake.

The mixture of palm methyl ester in the diesel fuel have not resulted in significant change in the property of tribo-layer in all four test conditions, at least for the volume percentage of 7.5%. However, the mixture has led to some reduction of material's wear, although not significant. Here, the purpose of adding palm methyl ester in petrol diesel fuel is to increase the renewable portion of diesel fuel. Therefore, this study has demonstrated that the graphite oxide tribo-layer can also be formed with the presence of palm methyl ester in the fuel.

Such tribo-chemical formation of graphite oxide tribo-layer provide a possibility for in-situ lubrication process of a mechanical system using the working fluid such as fuel system, hydraulic system, and combustion system. In such system, conventional lubricant fluid cannot be used because it will contaminate the working fluid. However, the formation of such carbon-based tribo-layer on the contact interface, which is beneficial for wear protection of the material, requires a suitable combination between the material and the working fluid. Therefore, further researches for understanding the optimal condition for such tribo-layer formation and its mechanism are required.

4. Conclusion

In this study we have demonstrated that tribo-layer having a structure similar to that of graphite oxide can be formed on the contact interface of metallic material under boundary lubrication using diesel fuel as the fluid lubricant. The structure is shown by the appearance of D band and G band in the Raman spectra at approximately 1350 cm^{-1} and 1550 cm^{-1} for all four test conditions conducted. The formation of the graphite oxide tribo-layer is assisted by a tribo-chemical process involving both the material and fluid media. Here, the role of palm methyl ester in the solid film formation is still not clear.

The formation of graphite-oxide tribo-layer on the contact interface during the friction process provide a possibility for an in-situ lubrication method using the working fluid. In order to achieve such lubrication process, it is necessary to conduct further investigation to find out the necessary condition as well as the formation mechanism.

5. Acknowledgment

The authors would like to thank all students for contributions in this research. Part of funding for this research was provided by LPPM Universitas Syiah Kuala, which is highly acknowledged.

References

[1] Zuleta E C, Baena L, Rios L A, Calderón J A 2012 *J. Braz. Chem. Soc.* **23** 2159-75.
[2] Peng D X 2015 *Adv. Mech. Eng.* **7** 1-10.
[3] Bagi S D, Aswath P B 2015 *Lubricants* **3** 687–711.
[4] Canan P, Spikes H A, Cameron A 1983 *ASLE Transactions* **26** 48-52.
[5] Holinski R, Gänsheimer J 1972 *Wear* **19** 329-342.
[6] Stolarski T 1976 *Tribol. Int.* **9** 161-163.
[7] Donnet C, Grill A 1997 *Surf. Coat. Technol.* **94-95** 456-462.
[8] Al Mahmud K A H, Kalam M A, Masjuki H H, Mobarak H M, Zulkifli N W M 2015 *Crit. Rev. Solid State Mater. Sci.* **40** 90-118.
[9] Erdemir A 2016 *Nature* **536** 67-71.

[10] Fuadi Z, Adachi K, Muhammad T 2018 *Tribol. Lett.* **66** 88.
[11] Ramirez G, Eryilmaz O L, Fatti G, Righi M C, Wen J, Erdemir A 2020 *CS Appl. Nano Mater.* **8** 8060-7
[12] Johnson B, Wu H, Desanker M, Pickens D, Chung Y W, Wang Q J 2018 *Tribol. Lett.* **66**.
[13] Kudin K N, Ozbas B, Schniepp H C, Prud'homme R K, Aksay I A, Car R 2008 *Nano Lett.* **8** 36-41.
[14] Fuadi Z, Faisal M, Dirhamsyah M, Tadjuddin M, Kurniawan R 2020 *IOP Conf. Ser. Mater. Sci. Eng.* **796** 12011.
[15] Fazal M A, Haseeb A S M A, Masjuki H H 2013 *Energy Convers. Manag.* **67** 251-256.
[16] Tuinstra F, Koenig J L 1970 *J. Chem. Phys.* **53** 1126.
[17] Ferrari A, Robertson J 2000 *Phys. Rev. B Condens. Matter Mater. Phys.* **61** 14095-107.
[18] Hinrichs R, Vasconcellos M A Z, Österle W, Prietzel C 2018 *Mater. Res.* **21** e20171000
[19] Österle W, Prietzel C, Dmitriev A I 2010 *Int. J. Mater. Res.* **101** 669-675.
[20] Österle W 2009 *Wear* **267** 781-788.

Simulation of synchronization photovoltaic system and low voltage grid

M K Akbar[1], S Syukriyadin[1,*], R H Siregar[1], S Syahrizal[1]
[1]Department of Electrical and Computer Engineering, Faculty of Engineering, Syiah Kuala University, Banda Aceh 23111, Indonesia

*E-mail: syukriyadin@unsyiah.ac.id

Abstract. In this research, the synchronization between the PV system and grid designed with synchronization parameters are voltage, frequency, and phase angle difference. As a result, by using a synchronization system, the PV system and grid could be parallelized and supplying power into the load at the same time. The synchronization method used is zero-crossing detection because it is simple than the other method. The synchronization system consists of several components; voltage detector, frequency detector, phase angle detector, and switch controller which all of them are controlled using the Arduino board. The switch controller will be functioned based on input from the detector to do synchronization or parallelize the PV system and grid. Testing is done by simulating running in Proteus. Based on the result of the simulation, the system is successful to do synchronization between the PV system and grid, it can be observed from the grid and PV inverter voltage graph and current which is measured from load, grid, and inverter.

1. Introduction

The Photovoltaic system (PV system) consists of a solar module to convert solar energy to electrical energy then saved into the battery. To serve the alternating current (AC) load, electrical energy saved in a battery (DC supply) is frequently converted into electrical energy using the inverter device. The PV system produces fluctuating electrical power influenced by solar radiation hit the solar module. When the intensity of solar radiation is high, the PV system produces more maximum electrical energy, meanwhile, when the intensity of radiation is low, so electrical energy produced is decreasing, moreover, in the dark, the PV system does not produce electrical energy at all.
The utilization of the PV system is expected to make loads such as lighting, water pump, and other electrical devices not only served by the electrical energy sources by the state electric company but also from the PV system so that the cost of electricity is cheaper. Therefore, a synchronization system to make the PV system work together with the state electrical company grid (hybrid) is needed. Hybrid power generation systems consist of two or more different electrical energy sources, whether connected to the grid or not, however when it connects into the grid, it will be more reliable because the grid can supply electrical energy into load when the PV system is disconnected from the system. As a result, the synchronization system is required so that electrical energy from the PV system and grid can supply and serve the load. Therefore, in this research, the synchronization system for the PV system and grid with synchronization parameters; voltage, frequency, and phase angle differences are designed.

Many previous kinds of research conduct synchronization between the PV system and grid by using different synchronization methods. In this research, synchronization done uses the Arduino controller and zero-crossing detection to detect phase angle differences between the PV system and the grid. In this research, advantages of synchronization system are more simple system because inverter used is a pure sine wave so that synchronization system is only handling switch controller

when voltage and frequency between the PV system and grid are same, and different of the voltage phase angle between both of them is not more than accepted maximum limit.
Testing of the system is conducted by doing the simulation of the synchronization process between the PV system and grid, by observing the condition of inverter voltage and grid voltage while synchronous or not synchronous. Moreover, it is also reviewed the current of each grid, inverter, and current at load. Simulation is run by using Proteus, an inverter is assumed presenting the PV on this research.

2. Grid synchronization

Synchronization means to minimize the differences in voltage, frequency, and phase angle between generator output and grid [1]. The necessity of synchronization and electric generator parallel is often based on the following things [2]:

1) The capacity of generator owned by existed system has been exceeded by new burden
2) The increasing of reliability (several generators vs single generator unit) must be considered
3) The efficiency of the generator

There are several electric generators that can be connected into a grid such as hydro, thermal, solar, etc, to supply power into the load. This generator unit can supply power if it has fulfilled rules. Intended rules are the variation or the change of voltage and frequency must on the limit of value, which have been determined, such as what has been explained in section 2.1 [2]. Before connecting the PV system to the grid, it must be synchronized with the parameter of the grid. Improper synchronization can influence power systems and produce electrical and mechanical transients that can ruin the prime mover, generator, transformer, and other system components [1].

The following is ideal synchronization [3]:

1) Able to detect the phase angles very well
2) Detect the varying frequencies efficiently
3) Reduce harmonic interference efficiently
4) Respond to the change on grid side well

2.1 Synchronization Parameters

The first problem on the synchronization process is equalizing the synchronization parameter, which the magnitude of voltage, frequency, phase, and phase angle between the grid and PV system must be matched. Table 1 showing several standards for interconnection between the PV system and the grid [4].

Table 1. Standards of PV system-grid connected

Voltage parameter	Standard
Magnitude	ANSI C84.1
Frequency	IEEE 929
Phase Angle	IEEE 929
Synchronization	ANSI 25

Table 2. Synchronization Parameters

Aggregate rating of DG units (kVA)	Frequency Difference (Δf, Hz)	Voltage difference (ΔV, %)	Phase angle difference ($\Delta \Phi$, °)
0-500	0,3	10	20
>500-1500	0,2	5	15
>1500-10000	0,1	3	10

Moreover, another standard is IEEE Std. 1547, IEEE Std. 1588-2008, IEC Std. 1727, IEEE Std. 929-2000 [3]. The synchronization parameter is voltage fluctuation, the magnitude of voltage, the sequences of phase, frequency, and phase [2]. Table 2 presents the synchronization parameter, however, the synchronization parameter allows little differences based on the standard used in every country [5].

2.2 Synchronization Schematic

Phase, frequency, and voltage are parameters and input on the synchronization system. Generally, the synchronization schematic is shown in Figure 1. After synchronization is done, power from the PV system will be connected to the grid on the PCC point [6].

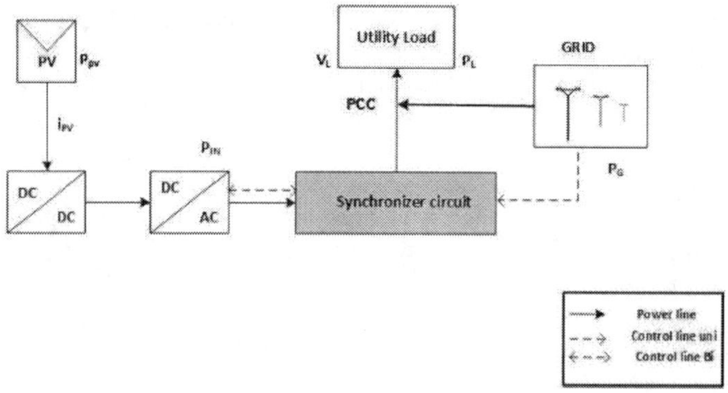

Figure 1. General Synchronization Schematic

3. Zero-crossing detection

The Zero-crossing detection method is measurement of voltage and detects zero points of voltage when voltage changes from negative value (-Ve) into positive value (+Ve) as shown in Figure 2. Because used for switching, then there must be such a thing as harmonic or noise [7]. Zero-crossing detection is also stated as a comparison, as shown in Figure 3, where zero-crossing detection uses a basic operational amplifier (Op-Amp) to compare two voltages simultaneously and change output according to the result of the comparison. Zero-crossing detection can be applied to the phase meter and time marker generator.

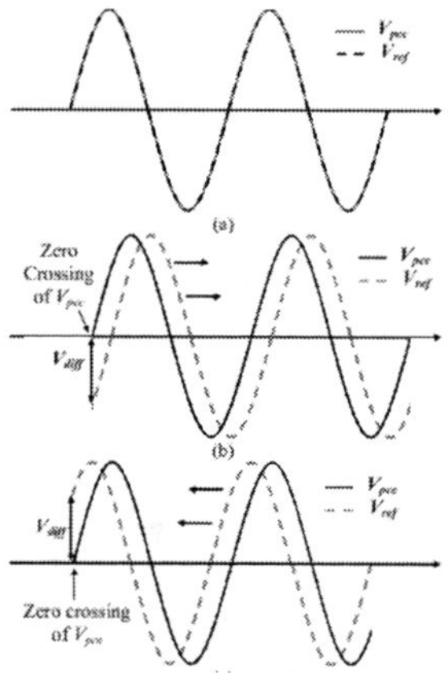

Figure 2. Zero crossing detection on AC voltage

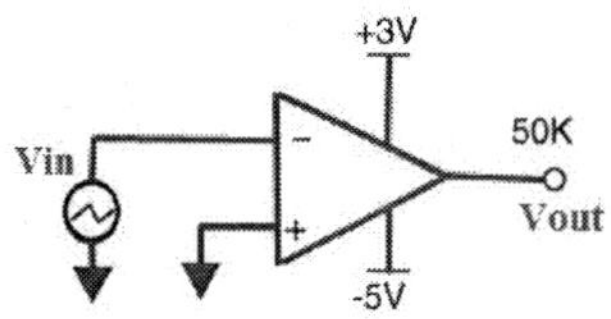

Figure 3. Basic circuit Op-amp

4. Configuration of system

The system proposed shown in Figure 4. This synchronization system integrated voltage detector, frequency detector, phase angle detector, and Arduino as the controller.

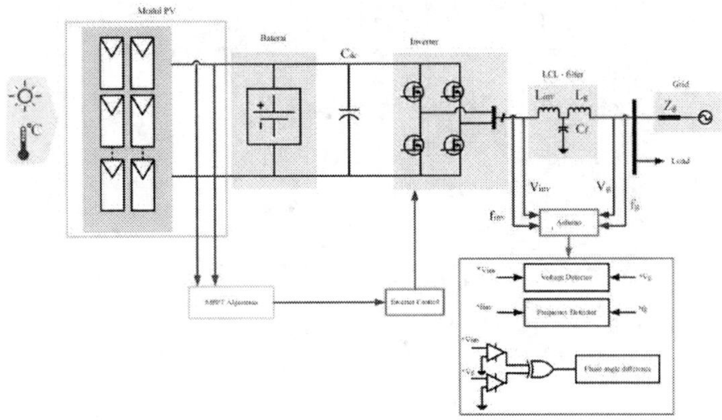

Figure 4. System Configuration and Control Structure

4.1 PV module configuration

The PV module used in this research is the *polycrystalline* solar cell with more complete specification as shown in Table 3.

Table 3. Solar Cell Specification

Model	BP350U
Serial	C1051213 4439776
Electrical Ratings	(At STC 1000 W/m^2, AM 1,5 spectrum, cell temperature 25°)
Peak Power (Pmax)	50 W
Warranted Minimum Pmax	45 W
Voltage (Vmp)	17,5 V
Current (Imp)	2,9 A
Open Circuit Voltage (Voc)	21,8 V
Short Circuit Current (Isc)	3,17 A
Minimum Bypass Diode	9 A
Maximum Series Fuse	20 A

4.2 Design of inverter

Inverter design has voltage output 220 V and frequency 50 Hz. The inverter circuit diagram used in this research is shown in Figure 5. Voltage waves produced by the inverter are pure sine waves as shown in Figure 6.

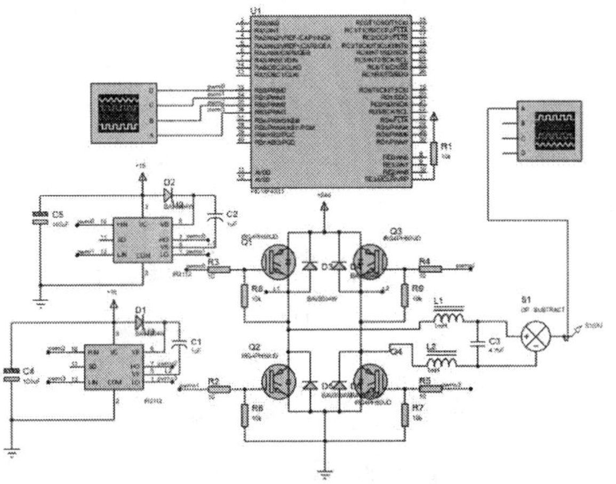

Figure 5. Inverter circuit diagram

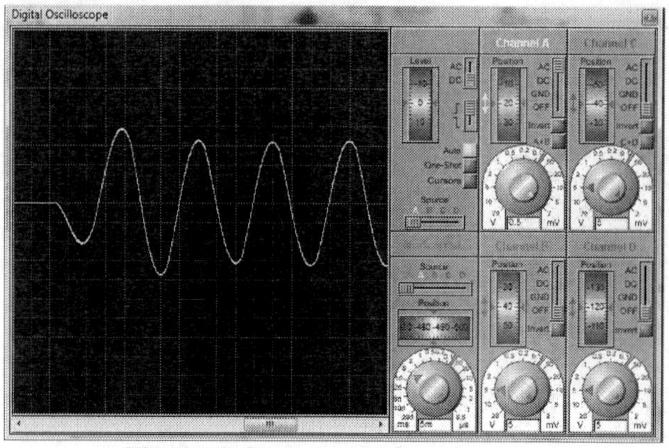

Figure 6. Output waveform inverter

4.3 Voltage detector

The voltage detector is used to read the voltage of the grid and inverter. The measurement of AC voltage is conducted by decreasing voltage because Arduino is only able to measure the voltage between 0V-5V and it is not able to read negative voltage value. Source voltage can be decreased with the transformer and voltage divider circuit as shown in Figure 7. The resistor on the voltage detector circuit diagram is a voltage divider series and a capacitor as a filter.

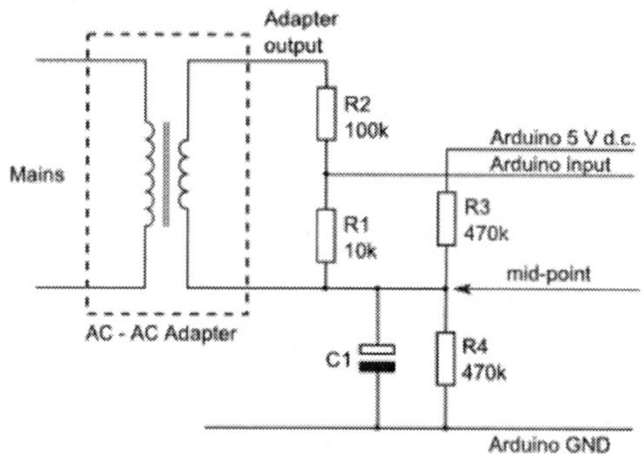

Figure 7. Voltage detector circuit diagram

The voltage detector design, which is designed on Proteus is shown in Figure 8. The result of voltage reading from the grid and inverter is illustrated on the LCD.

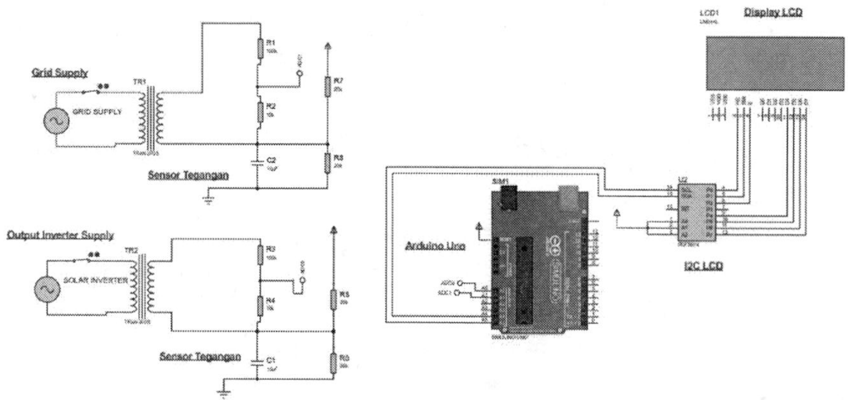

Figure 8. Design voltage detector

4.4 Frequency detector

Frequency on AC voltage can be calculated by measuring the time of the AC voltage wave. Time is the full-time wave, which means a half positive wave and a half of negative wave. To make it easier to detect and measure frequency, the diode rectifier is used as shown in Figure 9. Series to measure the AC frequency magnitude is shown in Figure 10, Optocoupler component is used as isolation for high voltage and low voltage from Arduino.

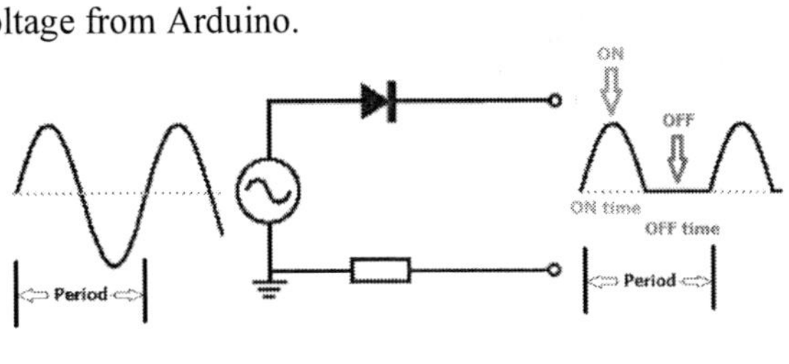

Figure 9. Period of sinewave

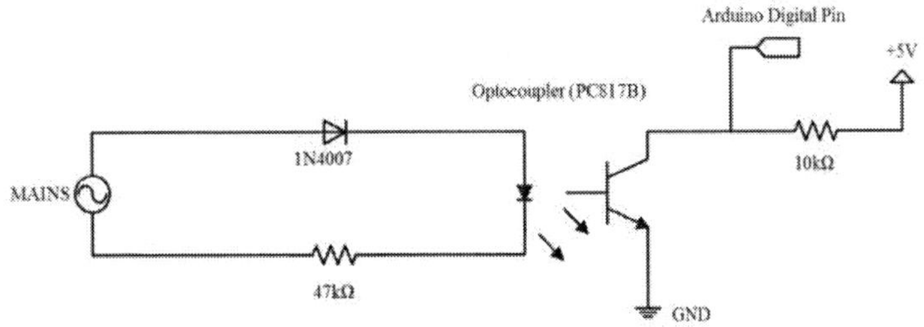

Figure 10. Frequency detector circuit diagram

4.5 Phase angle detector

Phase angle detector is a circuit diagram to know the differences of the phase angle. Measurement of the difference of the phase angle between grid and inverter is the same with the measurement of power factor, but the difference is only on implementation. To measure phase angle, first decreasing voltage from both sources using a transformer, then utilizing Op-Amp to detect zero points of grid voltage wave and inverter. Op-Amp convert sinusoidal wave signal becomes square wave with 4V amplitude, which is then linked into XOR gate input and processed by Arduino, shown in Figure 11. Moreover, the output of the XOR gate is shown in Figure 12. For example, the calculating of phase angle difference is shown in Figure 13. The simulation result of zero-crossing detection with several delay time values regulated is illustrated in Table 4.

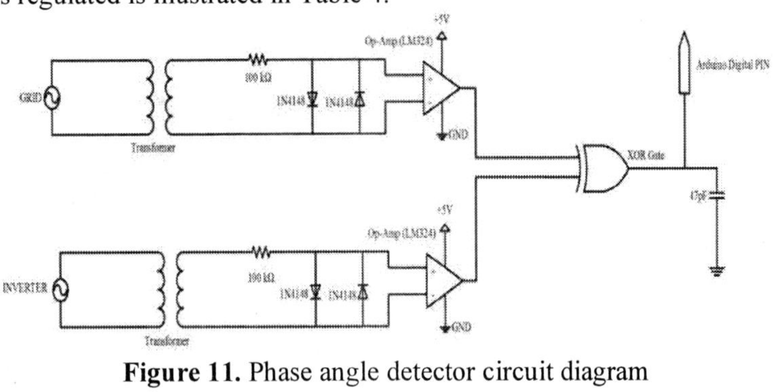

Figure 11. Phase angle detector circuit diagram

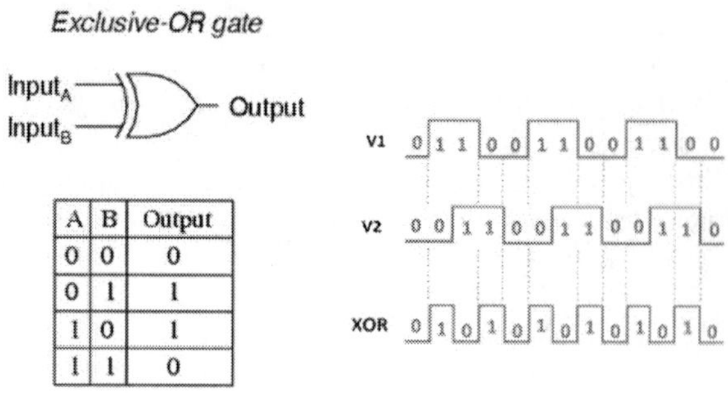

Figure 12. Gate XOR output

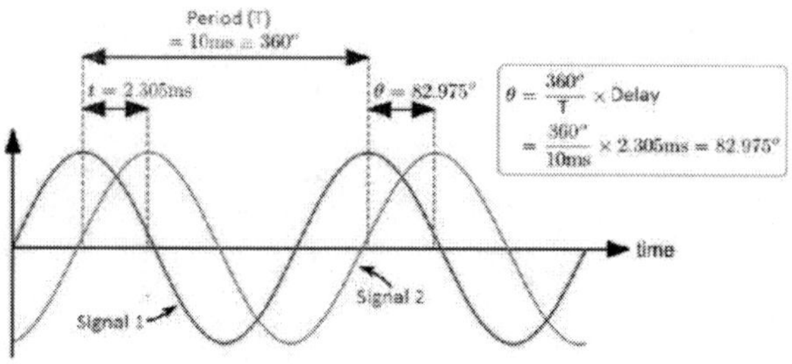

Figure 13. Example measure phase angle difference

Table 4. Phase angle simulation

No	Time Delay Setting between the grid and the inverter output	Zero-Crossing Detection Phase angle difference *(degree)*	Cos ϑ (p.f)
1	1 ms	18,15	0,95
2	2 ms	35,84	0,81
3	3 ms	54,96	0,58
4	4 ms	78,23	0,32
5	5 ms	90,67	0,01

5. Simulation and results

Voltage detector, frequency detector, and phase angle detector which have been designed before will be integrated becoming a system to do synchronization testing between the PV system and grid. Figure 14 is the display of LCD when synchronization between PV inverter and grid is successfully done.

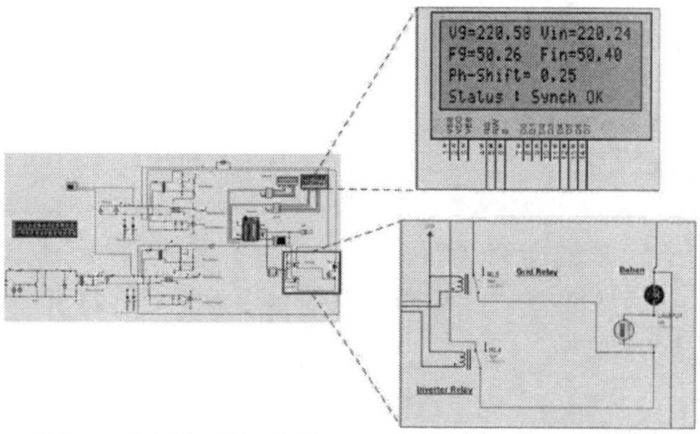

Figure 14. The Simulation system of synchronization

IOP Conf. Series: Materials Science and Engineering 1087 (2021) 012076 doi:10.1088/1757-899X/1087/1/012076

5.1 Result of simulation test

The test is not only to know the result of synchronization but also simulation done to know how the relay responds work properly, whether when all of the synchronization parameters are fulfilled or not. Table 5 informs the work of the relay which success to work is stimulated based on parameter value with 9 different possible conditions.

Table 5. Relay response with any different condition

No	Parameter					Status	Inverter Relay	Grid Relay
	Vgrid	Fgrid	Vinv	Finv	$\Delta\theta$			
1	1	1	1	1	1	Sinkron	1	1
2	1	1	0	1	1	Vinv out	0	1
3	1	1	1	0	1	Finv out	0	1
4	0	1	1	1	1	Vgrid out	0	0
5	1	0	1	1	1	Fgrid out	0	0
6	1	0	1	0	1	Fboth out	0	0
7	0	1	0	1	1	Vboth out	0	0
8	1	1	1	1	0	Phase out	0	0
9	0	0	0	0	0	Out of Sync	0	0

5.2 Phase synchronization

To be parallelized between the PV system and grid, both phases should be the same and not allowed to have phase angle different exceed the parameter limit determined. As a result, before synchronization of phase between PV inverter and grid must be conducted. The blue wave is grid voltage, red is inverter voltage, and green is PWM inverter wave. The yellow graph is the output of the zero-crossing detection series, which can be seen as yellow graph remarks zero points of grid voltage wave. Figure 15 shows the condition when the phase between grid voltage and inverter is not synchronous, meanwhile, Figure 16 shows the condition when the phase between grid voltage and inverter is synchronous.

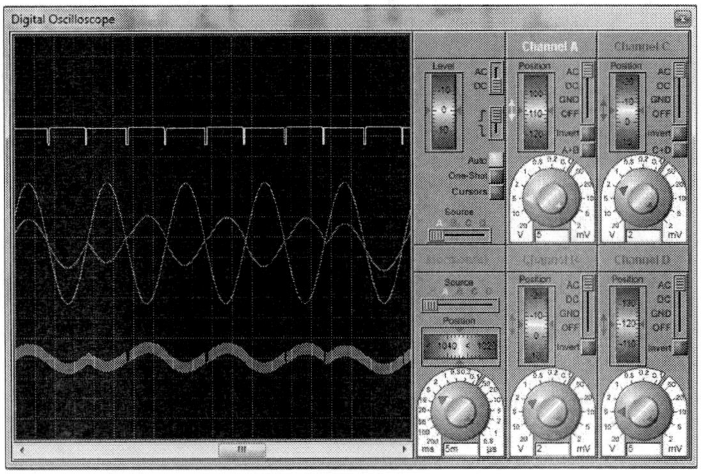

Figure 15. Vgrid and Vinv when asynchronous

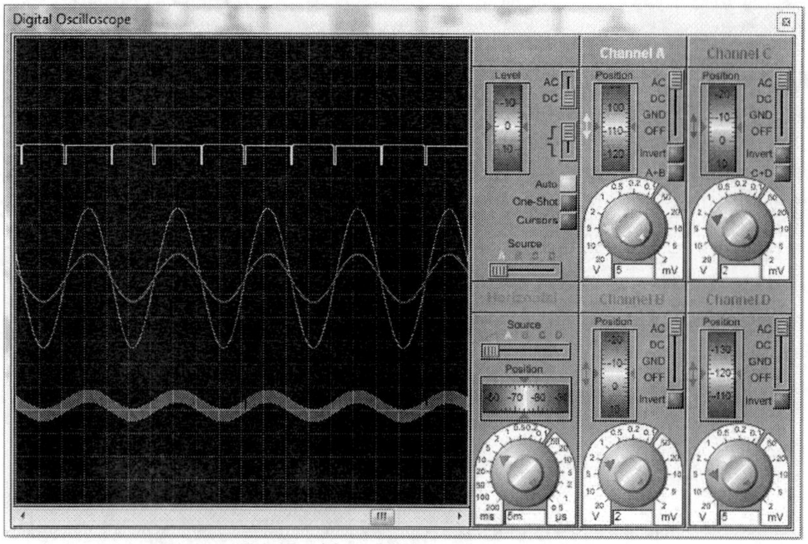

Figure 16. Vgrid and Vinv when synchronous

5.3 Grid and inverter voltage

When the simulation process is conducted, delaying happens several seconds until the voltage produced by the inverter has stabilized and steady-state so that it can be synchronized with the grid as shown in Figure 17. From those graphs, it can be seen that grid voltage and inverter are synchronous when it is 160ms, it can be seen voltage is synchronous although there are little differences in phase angle. The graph illustrated in Figure 17 is the inverter voltage and grid graph, which red graph is the grid voltage graph, meanwhile yellow is inverter voltage graph.

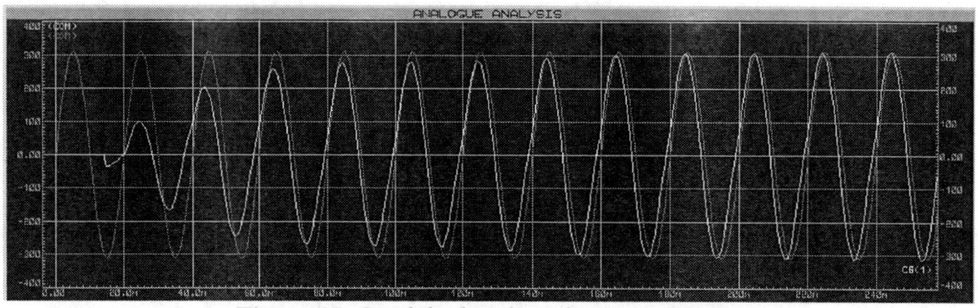

Figure 17. Graph both voltage inverter and grid

5.4 Grid and inverter current

To ensure the inverter helps to supply electrical energy to the load, so current measurement of the inverter, grid, and load is conducted. The result of current measurement when it is synchronous between inverter and grid is illustrated in Table 6. Based on Table 7, it remarks the current percentage supplied from the inverter is not constant, it caused inverter only supply current to support grid in serving the same load, or known as load sharing, the condition of two or more generators operates in parallel to supply the same load.

Table 6. Measurement current when synchronous

No.	Resistive load	Grid current	Inverter current	Load current
1	100 Ω	1,1 A	1,1 A	2,2 A
2	120 Ω	1,375 A	0,45 A	1,83 A
3	240 Ω	0,77 A	0,13 A	0,91 A
4	360 Ω	0,61 A	0,114 A	0,88 A

Table 7. Current supply percentage

No.	Load	Grid current	Inverter current	Load current (A)
1	100 Ω	50 %	50 %	2,2 A
2	120 Ω	75,72 %	24,27 %	1,83 A
3	240 Ω	85,71 %	14,28 %	0,91 A
4	300 Ω	84 %	15,90 %	0,73 A

6. Conclusion

The synchronization system designed in this research is successful to do synchronization between the PV system and the grid, simulated in Proteus. Based on the result of the simulation, the zero-crossing detection method can do synchronization with a very simple scheme, and the synchronization process can be done with Arduino control. The success of synchronization can be observed that PV inverter and grid can supply current to load simultaneously although the amount of current supplied from the inverter/PV system to load is not consistent or change depending on the load used. Arduino controls the synchronization process and switches the controller as the interconnection point between the PV inverter and grid. The system design also shows the value of voltage, frequency, and phase angle difference on the grid and inverter side in real-time that is illustrated in the LCD.

7. References

[1] Singh B, Dubey A N, Singh T, Tiwari N 2017 *IJFRCSCE* **3** 33.

[2] Thursila B V, Rahul K, Kumar K S, Varma V D S, Sarvani B N 1955 *Int. Res J. Eng. Technol.* **04**.

[3] Rahim N A, Bakar A H A, Tan C, Haidar A M A, Jaalam N 2016 *Renew. Sustain. Energy Rev..* **59** 1472.

[4] Ramaprabha R, Rajini V, Bansal K, Babu K N D 2016 *3rd International Conference on Electrical Energy Systems (ICEES).* 207.

[5] Berizzi A, Bovo C, Moshi G G 2013 *Renewable Energy for Unleashing Sustainable Development in* S Bologna and D Masera E Colombo, Ed. Milan, Italy: Springer, ch. **5**, p. 99.

[6] Anwar S, Asif M, Khalil A U R, Khalil F A 2018 *Professional Trends in Industrial and Systems Engineering (PTISE).* 629.

[7] Ahmed K H, Finney S J, Williams B W, Azrik M 2012 *IEEE International Symposium on Industrial Electronics.* Hangzhou 590.

IOP Conf. Series: Materials Science and Engineering 1087 (2021) 012077 doi:10.1088/1757-899X/1087/1/012077

The effect of bypass diode installation on partially covered solar panel output power

M Muliadi[1], I D Sara[1], S Suriadi[1,*]
[1]Department of Electrical & Computer Engineering, Faculty of Engineering, Syiah Kuala University, Banda Aceh, Indonesia

*E-mail: suriadimali@unsyiah.ac.id

Abstract. Shadow or partial shading is a bad phenomenon that affects the performance of photovoltaic (PV) systems in producing energy. Shadows can be caused by moving clouds, building shadows, tree shadows, bird droppings or dust covering the surface of solar panels. Therefore, this study was conducted to determine the effect of installing bypass diode on the output power of a partially closed solar panel based on three configurations of the solar module arrangement namely, a configuration without using bypass diode, two solar modules with one bypass diode, and one solar module with one bypass diode. Furthermore, the three configurations were tested with a shadow pattern scenario is 12.5% to 100% of solar panel area using Matlab software. The result shows the I-V and P-V characteristics when experiencing shading 12.5% - 100%. The peak power value or maximum power value of the one solar module configuration with one bypass diode is greater than the two solar module configuration with one bypass diode and the configuration without using a bypass diode. Different from shadow conditions 100% the effect of installing bypass diode on the configuration of two solar modules with one bypass diode and the configuration of one solar module with one bypass diode the resulting power is the same as the configuration without using a bypass diode, which is 179.2 W

1. Introduction

A unique renewable energy source such as the sun has an important role in electric power technology and can be used to meet the daily needs of electricity. Nowadays, solar energy is gaining popularity because of its readily available and inexhaustible nature. One of the devices used to convert solar energy into electrical energy is called a solar module [1][2]. The current and voltage characteristics of the solar module are non-linear and influenced by several factors such as the orientation of the solar module installation, solar radiation hitting the solar module due to changes in weather and shadow conditions that occur on the surface of the solar module [3]. Shadows are unusual occurrences related to the array of solar modules through passing clouds, building shadows, tree shadows, bird droppings, or dust covering the surface of the solar panels [4]. This condition can cause a significant decrease in output power. The more areas that hit the shadow, the less the output power of the solar panel [5][6]. However, several studies reveal that the output power of solar panels will be constant when the shadow or partial shading conditions that hit the solar panel area gets higher and reaches a certain critical point [5][7].

Therefore, this study aims to determine the effect of installing bypass diodes on output power solar panels of partially closed. Furthermore, testing is carried out based on the shadow scenario namely; the shadow condition is 12.5% to 100% of the total area of the solar panel covered by a shadow with the same image concentration of 400 W / m2; varying radiation intensity. In contrast to previous research on specifications, dimensions, and solar panel configurations, the effect of shadows on the output power of solar panels was tested using Matlab software based on three solar panel configurations, namely, a configuration without using a bypass diode, two solar modules with one bypass diode, and one solar module with one bypass diode.

2. Modelling of solar module

2.1 Characteristics of solar module

In photovoltaic (PV) system modelling, the general accuracy of cell characteristics can be predicted using a single diode model [8][9][10]. The equivalent diagram of solar cell with the one diode model is shown in Figure 1.

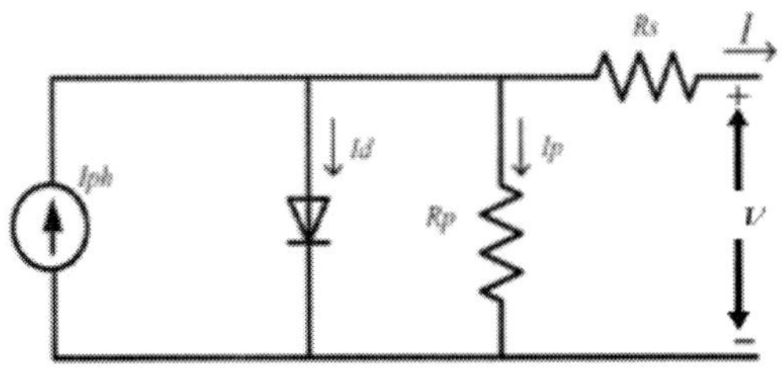

Figure 1. The Equivalent diagram of solar cell with one diode [8][9].

Figure 1 shows some of the parameters contained in a solar cell such as the photon current I_{ph}, the diode current I_d, the I_p current through the resistance R_p, R_s and R_p shows the series and parallel resistance. Based on these parameters the value of the output current I can be formulated as follows [9].

$$I = I_{ph} - I_d - I_p$$

$$I = I_{ph} - I_0 \left[e^{\frac{q*(V+I*R_S)}{n*k*T}} - 1 \right] - \frac{(V+IR_S)}{R_p}$$

The arrangement of solar cells in one large unit is also called solar module. The equivalent diagram for the solar module is shown in Figure 2. For solar module modelling, the equation is as follows [11]:

$$I_{ph} = [I_{sc} + K_i(T - 298)] * G/1000$$

$$I_{rs} = I_{sc}/[exp(qV_{oc}/N_s knT) - 1]$$

$$I_0 = I_{rs} \left[\frac{T}{T_r}\right]^3 exp \left[\frac{q*E_{go}}{nk}\left(\frac{1}{T} - \frac{1}{T_r}\right)\right]$$

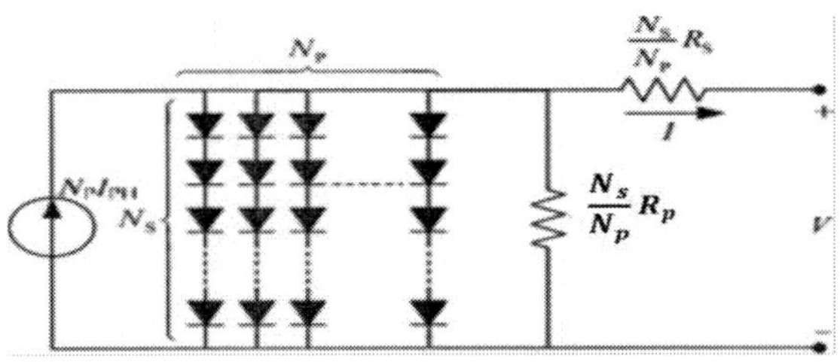

Figure 2. Equivalent diagram of solar modules [11].

So, output current of solar module is:

$$I = N_p\left\{I_{ph} - I_0\left[exp\left(\frac{\frac{V}{N_S}+I*\frac{R_S}{N_P}}{n*V_t}\right) - 1\right]\right\} - I_{sh}$$

with

$$V_T = \frac{k*T}{q}$$

and

$$I_{sh} = \frac{V*\frac{N_P}{N_S}+I*R_S}{R_P}$$

where:

I	=	Output current of solar cell/module (Amp)
I_{ph}	=	Photon current (Amp)
I_0	=	Saturation current of solar cell (Amp)
I_p	=	Current through resistance R_p (Amp)
V	=	Output voltage of solar cell (Volt)
I_{rs}	=	Reverse current of saturation diode (Amp)
I_{sc}	=	Short circuit current (Amp)
V	=	Output voltage of solar module (Volt)
V_{oc}	=	Open circuit voltage (Volt)
K_i	=	Temperature coefficient of short circuit current (0,0013 A/^{0}C)
T	=	Temperature of solar cell/module (^{0}K)
T_r	=	Temperatur referensi (^{0}K)
G	=	Solar iradiance (W/m^2)
N_s	=	Number of cell/module series
N_p	=	Number of cell/module parallel
n	=	Ideally factor of diode (1,2)
k	=	Boltzman's constant(1,38x10-23 J/^{0}K)
q	=	Charge of electron (1,6x10-19 C)
R_s	=	Series resistance (Ohm)
R_p	=	Shunt resistance (Ohm)

2.2 Photovoltaic components

The combination number of photovoltaic cells connected series and parallel to form a solar module. Furthermore, several solar modules are assembled into one arrangement called solar panels and their grouping series and parallel are called solar array as illustrated in Figure 3 below [1] [12].

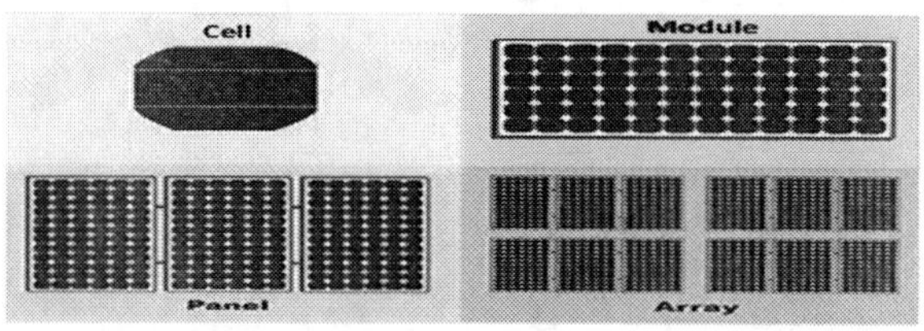

Figure 3. Solar cell, modules, panels, and arrays [1].

3. Effect of shadows and bypass diode on solar panels

The effect of shadows has a large impact on the efficiency of solar panels. Small shadows are usually caused by few thick clouds, shadows of buildings or trees, dirt, etc. This condition can cause the power

IOP Conf. Series: Materials Science and Engineering 1087 (2021) 012077 doi:10.1088/1757-899X/1087/1/012077

produced to decrease. Meanwhile, heavy shadows occur because sunlight is completely disturbed and does not reach the surface of the module. In this case, there is a possibility of partial shading or full shadow [4][10] [11][13] as illustrated in Figure 4.

Figure 4. The shape of shadow on solar panels [10][14].

The bypass diode is used to protect solar panels from effects of shadow [15]. Bypass diodes are usually installed to increase the output power of solar panels, although complications arise during partial shading [11][16]. Therefore, the bypass diode on the solar panel is useful for reducing losses by providing a parallel path for the panels experiencing shadow and for preventing hot-spots that can permanently damage solar panels [17].

4. Research methodology

In this study, the simulation was carried out using Matlab. The research flow is the selection of solar modules, then simulation of solar module configuration modelling, validation of solar module simulations in Standard Test Conditions (STC), solar module simulation scenarios based on the percentage of images on the module surface, data collection and discussion, and conclusions.

4.1 Selection of solar module

This study uses a polycrystalline type solar module with a specification of 55 Wp, dimensions of solar module with a length of 652 mm and a width of 639 mm consisting of 36 cells. Complete solar module specifications are listed in Table 1.

Table 1. Solar module specification of SPM055P-N

Parameter	**Nilai**
Number of Cell (cell)	36
Maximum Power (P_{max})	55 Wp
Maximum Voltage (V_{mp})	18.18 Volt
Maximum Current (I_{mp})	3.1 Amp
Open Circuit Voltage (V_{oc})	22.1 Volt
Short Circuit Current (I_{sc})	3.31 Amp
Standard Test Conditions (STC)	
Solar Irradiance	1000 W/m^2
Temperature	25 ^{0}C
A.M.	1.5

4.2 Configuration model of solar module

In this study used 8 (eight) solar modules so that the dimensions are long 2608 mm and wide 1278 mm. Furthermore, the solar modules are modelled based on the output current equation in equation (6). The solar module is simulated using Matlab to get the voltage and current characteristics. Furthermore, the results are validated based on the specifications of the solar module according to the manufacturer. The solar module configuration modelling is made as shown in Figure 5, namely the configuration of a solar module without using bypass diode, two solar modules with one bypass diode, and one solar module with one bypass diode.

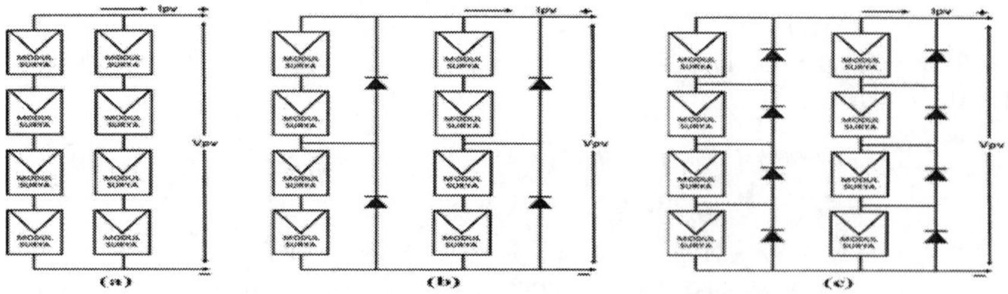

Figure 5. Configuration of solar module (a) without using bypass diode; (b) two solar modules with one bypass diode; (c) one solar module with one bypass diode.

The results of the validation, *I-V* and *P-V* characteristics at STC conditions (temperature of 25 ^{0}C and irradiance of 1000 W/m^2) were obtained with a total maximum power (P_{max}) of 430.5 Watt; V_{oc} is 88.27 Volt, and I_{sc} is 6.62 Ampere. For the *I-V* and *P-V* characteristic curves as shown in Figure 6.

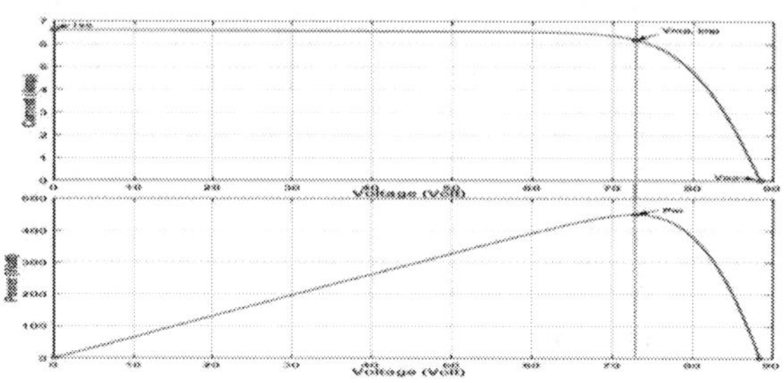

Figure 6. *I-V* dan *P-V* characteristics at STC conditions (irradiance 1000 W/m^2 and temperature 25 ^{0}C).

4.3 Simulation scenarios of solar module

The shadows are greatly influenced by how the sun circulates and they can cause a reduced amount of solar radiation absorbed by the solar module. In this study, the solar radiation absorbed by the solar module in the shadow less condition is assumed to be 800 W/m^2, while the sun shaded solar panel can absorb 400 W/m^2 of solar radiation. Therefore, the solar module simulation scenario is carried out based on the percentage of the solar panel area. It covered by shadows ranging from 12.5% to 100%. This scenario is made with the assumption that solar panels have first been installed on the rooftop of the household. Where the household are lined up or also called park housing, then next times the community has started to build terraced houses so that it due to solar panels installation is obstructed by the sun's rays. To avoid or put it in another place is also difficult because there is no more space and the roof is limited. The complete scenario is shown in Figure 7 below.

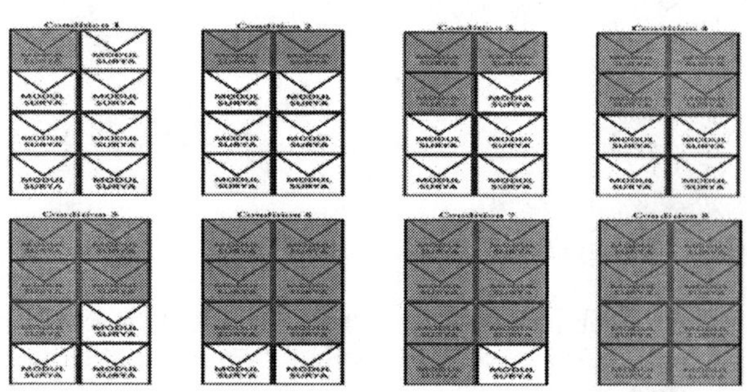

Figure 7. Shadow pattern scenario 12,5% - 100%.

IOP Conf. Series: Materials Science and Engineering 1087 (2021) 012077 doi:10.1088/1757-899X/1087/1/012077

4.4 *Modelling of solar panels using Matlab*

In this study, to determine the effect of the bypass diode installation on the output power of the solar panel, a test was performed using the Matlab simulation. The proposed model is shown in Figure 8.

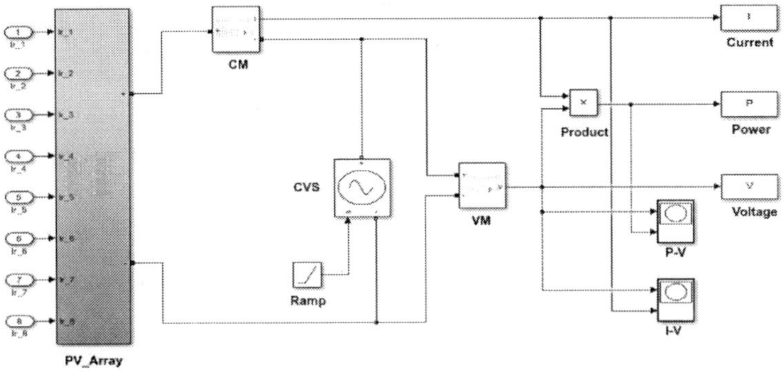

Figure 8. Modelling of solar panels using Matlab.

5. Result and discussion

The effect of installing a bypass diode on the output power of the solar panel at partial shading condition has been tested using Matlab.

5.1 *Testing of solar panel configuration without and with bypass diode installation on no shading condition*

At this stage, all configurations are tested under normal conditions or conditions without shading, assuming that the radiation that can be absorbed by each solar module is 800 W/m^2, temperature 25 ^{0}C. The simulation results show that maximum current (I_{mp}) is 4.964 A, maximum voltage (V_{mp}) is 72.8 V, and maximum power (P_{mp}) of all tested configurations is 361.4 Watts. The *I-V* and *P-V* characteristics are shown in Figures 9 and 10.

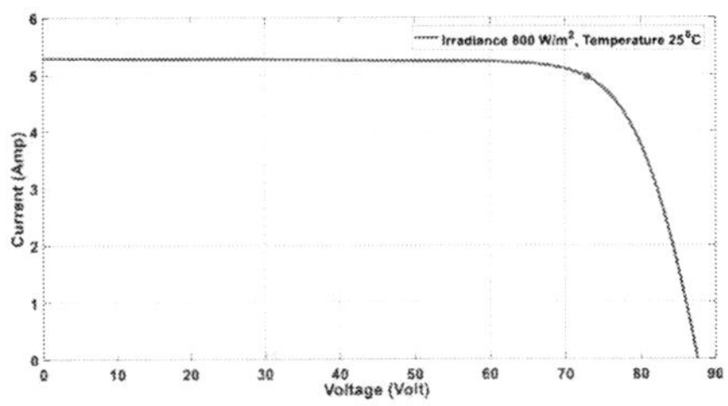

Figure 9. *I-V* curve with irradiance 800 W/m^2, temperature 25 ^{0}C.

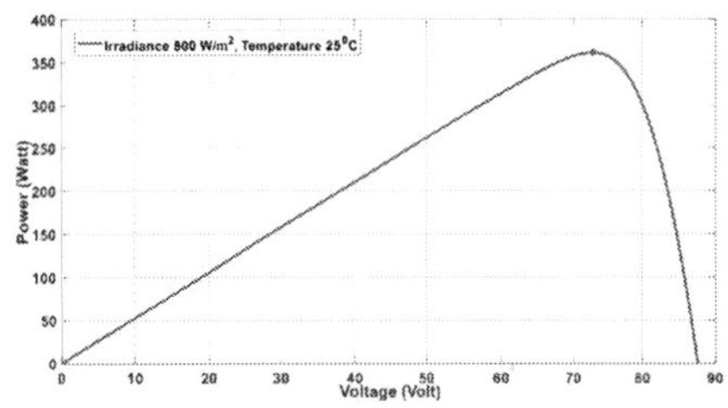

Figure **10.** *P-V* curve with irradiance 800 W/m^2, temperature 25 ^{0}C.

5.2 *Testing of solar panel configuration without and with bypass diode installation on partial shading conditions*

In this test, all solar panel configurations are tested with a shadow scenario of 12.5% to 100%. This test is conducted to determine the effect of installing the number of bypass diodes on the output power of the solar panel under partial shading conditions.

1) Testing on solar panel configuration without using bypass diode

In this test, based on 12.5% to 100% scenario, the resulting *I-V* and *P-V* curves in the solar panel configuration without using bypass diode are shown in Figure 11 and Figure 12 below.

IOP Conf. Series: Materials Science and Engineering 1087 (2021) 012077 doi:10.1088/1757-899X/1087/1/012077

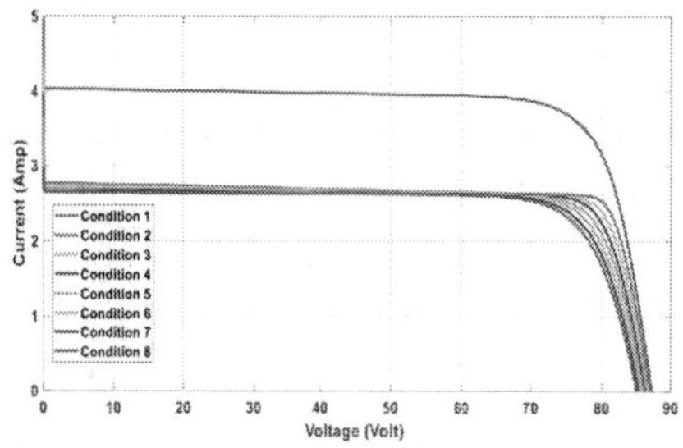

Figure 11. *I-V* curve on configuration without using bypass diode.

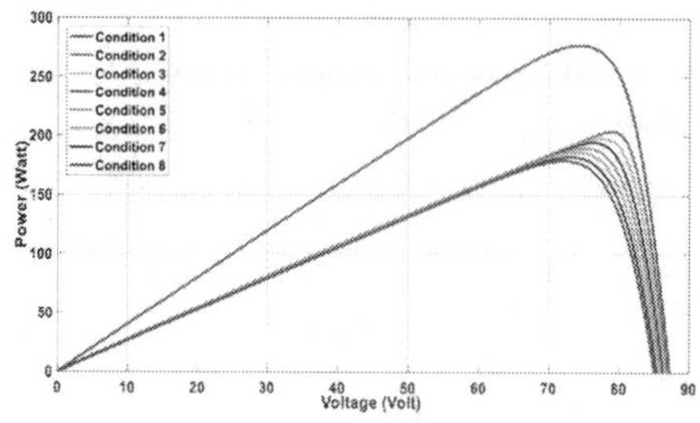

Figure 12. *P-V* curve on configuration without using bypass diode.

In Figure 11 and Figure 12, the condition of 12.5% shadowed solar panels shows the value of power and current is greater and the value of the voltage is smaller. This is due to the total number of solar modules working only partially so that the voltage is partially reduced. The *I-V* and *P-V* characteristics where their maximum point is observed results in the maximum power, current and voltage values which can be seen in Table 2 as follows.

Table 2. The point of maximum power, current, and voltage on the solar panel configuration without bypass diode.

The Condition of Solar Panel is Shaded	Maximum Peak Point On The Configuration Without Using Bypass Diode		
	P_{mp} (W)	I_{mp} (A)	V_{mp} (V)
Condition 12,5%	277.4	3.743	74.1
Condition 25%	204.3	2.586	79.0
Condition 37,5%	198.1	2.556	77.5
Condition 50%	194.7	2.545	76.5
Condition 62,5%	190.0	2.526	75.2
Condition 75%	186.5	2.521	74.0
Condition 87,5%	182.5	2.503	73.0
Condition 100%	179.2	2.489	72.0

2) Testing on configuration of two solar modules with one bypass diodes
With the same scenario, the test is carried out on a configuration of two solar modules with one bypass diode. The resulting *I-V* and *P-V* characteristic curves are as shown in Figure 13 and Figure 14.

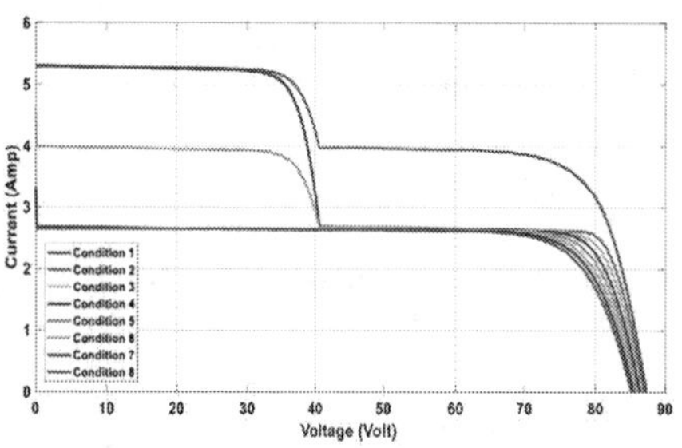

Figure 13. *I-V* curve on configuration of two solar modules with one bypass diode.

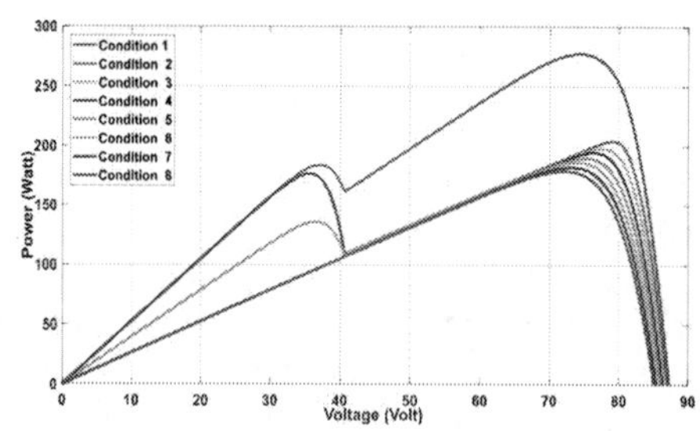

Figure 14. *P-V* curve on configuration of two solar modules with one bypass diode.

In Figure 13 and Figure 14, from the eight conditions tested, five conditions re resulting in two maximum peak points namely, on the shadow conditions 12.5%, 25%, 37.5%, 50%, and 62.5% and three conditions produce one the maximum peak point is on the shadow conditions (75%, 87.5%, and 100%). For the value generated from two maximum peak points in the configuration of two solar modules with one bypass diode on the scenario of 12.5% to 100% can be seen in Table 3 as follows.

Table 3. The point of maximum power, current, and voltage on the configuration of two solar modules with one bypass diode.

The Condition of Solar Panel is Shaded	Maximum Peak Point On The Configuration of Two Solar Modules With One Bypass Diode					
	Peak Point 1			*Peak Point 2*		
	P_{mp} *(W)*	I_{mp} *(A)*	V_{mp} *(V)*	P_{mp} *(W)*	I_{mp} *(A)*	V_{mp} *(V)*
Condition 12,5%	184.2	4.977	37.0	277.4	3.743	74.1
Condition 25%	176.7	4.950	35.7	204.3	2.586	79.0
Condition 37,5%	176.7	4.950	35.7	198.1	2.556	77.5
Condition 50%	176.7	4.950	35.7	194.7	2.545	76.5
Condition 62,5%	136.0	3.736	36.4	190.0	2.526	75.2
Condition 75%	186.5	2.521	74.0	-	-	-
Condition 87,5%	182.5	2.503	73.0	-	-	-
Condition 100%	179.2	2.489	72.0	-	-	-

In Table 3, when the condition of the solar panel is 12.5% shaded, the solar panel produces two maximum peak points. Peak point 1 has value P_{mp} 184.2 W, I_{mp} of 4,977 A, and V_{mp} 37.0 V while peak point 2 has a P_{mp} 277.4 W, I_{mp} 3,743 A, and V_{mp} 74.1 V so that at in this condition there is an increase in power of 184.2 W. For conditions of 25%, 37.5%, and 50%, the result is that peak 1 has value P_{mp} 176.7 W, I_{mp} 4.950 A, and V_{mp} 35. 7 V while the peak point 2 has the same P_{mp}, I_{mp} and V_{mp} values as the value generated in the configuration without using bypass diode. When the solar panel experiences shading disturbances, it will affect the performance of all modules in the PV, causing the condition of the solar panel to be covered in shadows, 12.5% to 62.5%, unable to work optimally in producing energy. Furthermore, for the conditions of 75%, 87.5%, and 100% only produce one maximum peak point, so that in this condition there is no effect given from the use of the bypass diode. So, configuration of two solar modules with one bypass diode can produce a maximum value that is greater when compared to a solar panel configuration without using a bypass diode.

3) Testing on the configuration of one solar module with one bypass diode
Based on the 12.5% - 100% image scenario, the test results carried out in the configuration of one solar module with one bypass diode are shown below.

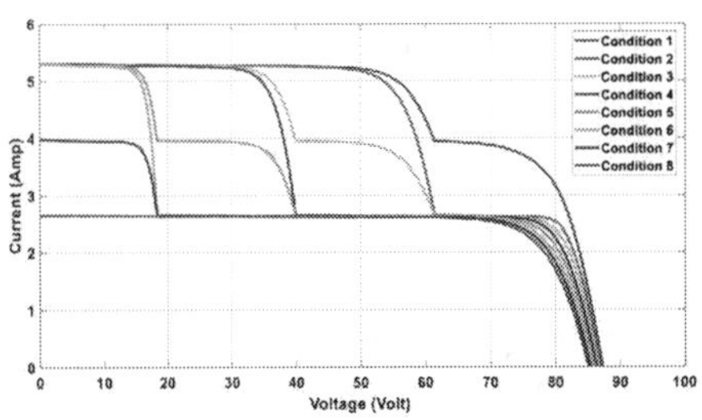

Figure 15. *I-V* curve on configuration of one solar module with one bypass diode.

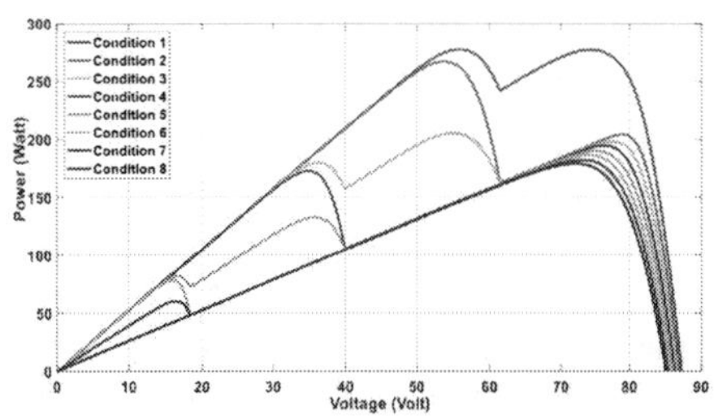

Figure 16. *P-V* curve on configuration of one solar module with one bypass diode.

In Figure 15 and Figure 16, when the solar panel experiences shading disturbances, where the *I-V* and *P-V* characteristics are observed, their maximum point results in maximum power, current and voltage which can be seen in Table 4 as follows.

Table 4. The point of maximum power, current, and voltage on the configuration of one solar module with one bypass diode

The Condition of Solar Panel is Shaded	Maximum Peak Point On The Configuration of One Solar Module With One Bypass Diode								
	Peak Point 1			*Peak Point 2*			*Peak Point 3*		
	P_{mp} (W)	I_{mp} (A)	V_{mp} (V)	P_{mp} (W)	I_{mp} (A)	V_{mp} (V)	P_{mp} (W)	I_{mp} (A)	V_{mp} (V)
Condition 12,5%	277.6	4.957	56.0	277.4	3.743	74.1	-	-	-
Condition 25%	267.1	4.965	53.8	204.3	2.586	79.0	-	-	-
Condition 37,5%	180.0	4.958	36.3	205.4	3.727	55.1	198.1	2.56	77.5
Condition 50%	172.8	4.950	34.9	194.7	2.545	76.5	-	-	-
Condition 62,5%	82.18	4.935	16.6	133.0	3.736	35.6	190.0	2.56	75.2
Condition 75%	78.48	4.937	15.9	186.5	2.521	74.0	-	-	-
Condition 87,5%	60.57	3.705	16.3	182.5	2.503	73.0	-	-	-
Condition 100%	179.2	2.489	72.0	-	-	-	-	-	-

Table 4 shows the results of modelling the configuration of one solar module with one bypass diode. When the solar panel experiences a shading disturbance of 12.5% and 25% the value of power, current and voltage at peak 1 are greater than the peak point value 2. However, when the solar panel experiences shading disturbances 50%, 75%, and 87.5% the value of power, current, and voltage at peak 2 is greater than the value of peak 1. Furthermore, when conditions 37.5% and 62.5% produce three peak points for power, current, and voltage. This result is also often referred to as multiple peak output. The maximum global peak (*GP*) value produced at 37.5% is at peak point 2 with P_{mp} 205.4 W, I_{mp} 3.727 A, and V_{mp} 55.1 V. However, the maximum global peak (*GP*) value The resulting 62.5% condition is at peak point 3 with P_{mp} 190.0 W, I_{mp} 2.526 A, and V_{mp} 75.2 V. the same as testing the configuration without using bypass diode, and the configuration of two solar modules with one bypass diode, namely P_{mp} 179.2 W, I_{mp} 2.489 A, and V_{mp} 72.0 V. In this condition all solar modules experience a shading condition so that solar radiation that can be received by the solar module is all 400 W/m^2.

Based on Table 2, Table 3, and Table 4, it can be seen that the maximum value of power and current resulting from the configuration of one solar module with one bypass diode produces a larger maximum value but the voltage varies depending on the number of global peak points and the number of peak points. (multiple peak output) generated. For larger power and larger current values, this is due to the total number of solar modules working only partially so that the voltage is reduced.

6. Conclusion

Based on the test results and data obtained, it can be concluded that the shading conditions experienced by solar panels can affect the characteristics of power, current and voltage generated. In the event of a shading disturbance, the configuration of one solar module with one bypass diode produces power, current and voltage characteristics with three curves, namely the curve of 1 peak point, 2 peak points, and 3 peak points. The configuration of two solar modules with one bypass diode produces power, current, and voltage characteristics with two curves, namely 1 peak and 2 peak points, while in a configuration without using bypass diode it only produces one peak point, the maximum value. If you compare the results from Table 2, Table 3 and Table 4, the peak power value or maximum power value of the configuration of one solar module is greater than the configuration of two solar modules with one bypass diode and the configuration without using a bypass diode when the shading disturbance occurs, namely in the condition 12.5% - 100%, it can be concluded that the bypass diode increases the maximum power value of the solar panel where one of the modules is shaded. For image conditions 100% the effect of installing a bypass diode in the configuration of two solar modules with one bypass diode and the configuration of one solar module with one bypass diode the resulting power is the same as the configuration without using a bypass diode, which is 179.2 W. The effect of installing a bypass diode is very high. important in improving the performance of the PV system against partial shading, except in the case of shadows covering the solar panels by 100%.

References

[1] Djalab A, Bessous N, Rezaoui M M, Merzouk I 2018. *Int. Conf. Commun. Electr.* 1–5.
[2] Rahmawati S, Iqbal M, Sara I D 2019 *IOP Conf. Ser. Mater. Sci. Eng.* **539** 012023.
[3] Diqing Z, Wu C, Li Z 2012 *Int. Comput. Sci. Conf.* 124–134.
[4] Khaing H H, Liang Y J, Nyein N, Htay M, Fan J 2014 *Int. J. Energy Power Eng.* **8** 1418–22.
[5] Patel H, Agarwal V 2008 *IEEE T. Energy Conver.* **23** 302–310.
[6] Bingöl O, Özkaya B 2018 *Sol. Energy.* **160** 336–343.
[7] Silvestre S, Boronat A, Chouder A 2009 *Appl. Energy* **86** 1632–40.
[8] Sera D, Baghzouz Y, Vegas L 2014 *WSEAS/IASME Int. Conf. Renew. Energy Sources.*
[9] Jazayeri M, Uysal S, Jazayeri K 2013 *HONET-CNS* 44–50.
[10] Chaudhary A, Gupta S, Pande D, Mahfooz F, Varshney G 2015 *Int. J. Eng. Res. Appl.* **5** 85-89.
[11] Nguyen X H 2015 *Environ. Syst. Res.* **4** 1–10.
[12] Mohanty P, Tyagi A 2012 *Introduction to Solar Photovoltaic Technology*. Amsterdam: Elsevier Inc.
[13] Santos P, Vicente E M, Ribeiro E R 2011 *XI Brazilian Power Electronics Conference, Praiamar*. pp. 676-681.
[14] Halim L 2017 *Perancangan dan Implementasi Solar Inverter dengan Pencatatan Daya pada Pembangkit Listrik Tenaga Surya Off-Grid.* Bandung: Lembaga Penelitian dan Pengabdian kepada Masyarakat Universitas Katolik Parahyangan.
[15] Veerapen S, Wen H 2016 *IEEE 8th International Power Electronics and Motion Control Conference (IPEMC-ECCE Asia).* pp. 3508-3513.
[16] Ishaque K, Salam Z 2013 *Renew Sustain. Energy Rev.* **19** 475–488.
[17] Rana A S, Nasir M, Khan H A 2018 *IET Renew. Power Gener.* **12** 143–148.

The over-height vehicle detection using the computer vision method

T M Iqbal[1,*], F Arnia[1], R Muharar[1]
[1]Department of Electrical and Computer Engineering, Univeritas Syiah Kuala, Aceh, Indonesia.
*E-mail: twk.m@mhs.unsyiah.ac.id

Abstract. Over-height vehicles are not only threats for the over-height vehicles themselves, but they also bring about a hazard to the tunnels they pass in, and to other drivers on the road way when a clash happens with a low clearance structure. Therefore, it is needed an over-height detection system which is affordable yet also reliable. Nowadays, there are the over-height detection system using laser and infrared. However, they are quite pricy. In this research a computer vision-based system is proposed to detect the height of vehicles and provide a warning for the over-height vehicles. The height determination was conducted using *Gaussian Mixture Model (GMM)* and *blob detection*. The *GMM* is used to detect the vehicle and the blob detection is applied to produce the vehicle coordinates which determines the direction of movement and the over-height vehicle. The over-height vehicles are detected by using 5 (five) variance total frames within 3 (three) conditions. The accuracy of the test proves that the method is reliable in determining the height of vehicles, achieving 100% accuracy of the detected vehicles. The significance of this work is the design of a vision-based method which can determine the height of the vehicles and is a low-cost alternative to the current costly laser and infrared detection systems.

1. Introduction

Statistics shows that the number of accidents occurred in Indonesia are relatively high. In 2019 there were 107.500 traffic accidents with the total victims of 23.530 [1]. The traffic accidents can be caused by several factors such as human errors, vehicle eligibility, and environmental factors. One of the examples of human errors is physical condition such as fatigue, while the vehicle eligibility factor is like the unoptimized condition of machine, brake, and wheels. Some of the environmental factors are climbing and downhill roads, as well tunnels and underpasses.

The accidents at the tunnel or underpass can be caused by the over-height vehicles. It causes vehicles striking the underpass. This condition brings about material losses, victims, underpass damage, and traffic lag/jam. To prevent those strikes, it is needed a system which can detect the over-height vehicle. The system can direct the vehicles passing the tunnels. Should the vehicles height is over the tunnels, the system will warn and alert drivers to stop or proceed to the exit way.

There are several pieces of previous research related to the over-height vehicles detection conducted by using various methods such as laser [2] [3] and optics [4] [5] [6] [7]. The study using laser method shows that the rate of error accuracy is 0.66 inch [6]. However, the laser or infrared methods are costly [5] [10]. Another study reveals that the infrared dual beam (Z-Pattern) method can detect the false over-height vehicles and alert the drivers [5]. This system is also relatively expensive as it can spend USD $150.000 – 200.000 (around Rp 2,9 billion) per installation and per direction. It includes equipment costs, excluding the maintenance fees. Meanwhile, the Sydney Harbor Tunnel using infrared method and pseudo-holographic system displaying "STOP" word at the water wall can spend AUD $150.000 – 200.000 (around Rp 2 billion) and maintenance fee AUS $25.000 – 35.000 (around Rp 352 million). The optic method (prototype) installed at the Fairlop Station Bridge, East London, UK costs £2.810 (around Rp 50,9 million); it counts only for the vehicle detection equipment (camera, computer, hard disk, router, mounting, cable, and outdoor box), excluding the warning system [8].

Among those methods, optics is one of the systems which needs the lowest costs in the over-height vehicle detection [4] [9] [10]. The previous research using optics to detect the over-height vehicle reveals the varied rate of accuracy using various methods. The blob detection method and Kanade Lucas Tomase (KLT) shows the accuracy of over height vehicle up to 92,2% [10]. Meanwhile, the blob detection method and Histogram of Oriented Gradients (HOG) marks the accuracy of vehicle detection at 95,83% and the over height vehicle detection at 95,77% [7]. The KLT detection method generates 68,7% for vehicle detection accuracy and 83,3% for over-height vehicle detection accuracy [5]. Lastly, KLT and SURF detection method marks 83,3% accuracy for vehicle detection and 96,6% accuracy for over-height vehicle detection [6].

Those researches show that the optics method can generate a more varied result of accuracy while the laser method needs costly expenses for equipment and maintenance. Therefore, it is required another optics method which can detect the over-height vehicle more accurately with affordable cost. This study proposes the use of Gaussian Mixture Models (GMM) feature. The GMM feature is applied to split between the foreground and background [11] [12]. The detected foreground object is assumed as a vehicle. The vehicle is then measured its height.

The previous research reveals that GMM method can generate high rate accuracies in detecting vehicles. The result of Nurhadiyatna et al. study shows that the foreground detection using GMM features combined with the Hole Filling (HF) algorithm marks the accuracy rate up to 97,9% [13]. Meanwhile, in Indrabayu et al., study, the accuracy rate generated by the Gaussian Mixture Models (GMM) in detecting the vehicle achieves 97,2% [14]. The results explain that the Gaussian Mixture Models (GMM) can yield a high rate of accuracy in vehicle detections. It, then, becomes the researcher's reason to choose the Gaussian Mixture Models (GMM) to be applied in detecting the over-height vehicles. It is expected that the system can detect the vehicles more accurately so that can warn and give alert precisely. Also, it is hoped that the system can cut the traffic accident and bridge strike numbers.

2. Literature Review

Gaussian Mixture Models (GMM) is the components of the Gaussian function forming a kind of density model from various threshold. This feature segregates between the foreground and background [11] [12] in the video. The segregation is produced by comparing an image from a frame with the next frame. If a density group moves, it will be classified as the foreground and the rest as background represented in a binary matrix. The detected foreground object is assumed as a vehicle. Later, this system will measure the height of the vehicle. Should the height of the vehicle is beyond the threshold, the system will trigger an alert in the form of voices, lights, and warning boards.

GMM is generated from the pixel color value based on the frame changes. The models is classified into two groups which are background and foreground [11]. Equation 1 is the equation for the Gaussian Mixture Model (GMM) [15]:

$$P(X_t) = \sum_{i=1}^{K} \omega_{i,t} * \eta(X_t, \mu_{i,t}, \Sigma_{i,t}) \quad (1)$$

With P(Xt) is the pixel probability of a frame. K is the sum of Gaussian Distribution, $\omega_{i,t}$ is the weight estimation of Gaussian to i in t frame. Then, $\eta(X_t, \mu_{i,t}, \Sigma_{i,t})$ is the probability of density function of Gaussian Mixture Model to i in t frame, represented in Equation 2.

$$\eta(X_t, \mu_{i,t}, \Sigma) = \frac{1}{(2\pi)^{\frac{n}{2}} |\Sigma|^{\frac{1}{2}}} e^{-\frac{1}{2}(X_t-\mu_t)^T \Sigma^{-1} (X_t-\mu_t)} \quad (2)$$

Where $\mu_{i,t}$ is the *mean* of *Gaussian* to i in t frame, and $\Sigma_{I,t}$ is the covariance matrix of *Gaussian* to i in t frame. The equation for covariance matrix is as Equation 3.

$$\Sigma_{i,t} = \sigma_k^2 I \quad (3)$$

Where σ_k^2 is the variance of *Gaussian* distribution, I is the identity matrix.

GMM feature is a statistic model which is effective to segregate between the foreground and the background with minimal movement. It is carried out by comparing an image of a frame with the next frame. Should there a density group moves, it will be classified as foreground and the rest as background represented in the form of binary matrix.

The steps in classifying the background is presented in Picture 1. First is matching every distribution until it is found the appropriate distribution. The matching is performed by determining ranges of a pixel with deviation standard value 2,5. It can effectively classify a distribution. Should a pixel does not match with all existing distributions, a new distribution is made and replace the low weight value, mean, and high variance value. The renewal of the weight value to the frame changes (time) presented in Equation 4.

$$\omega_{k,t} = (1 - \alpha)\omega_{k,t-1} + \alpha(M_{k,t}) \quad (4)$$

Where $\omega_{k,t}$ is the Gaussian distribution weight and α is learning rate of a model with value 1 if the model is matching and with 0 value if the model is not matching. After the weight value is updated, the weight value is normalized of all distribution to get value 1. The update of mean and variance score to frame (time) changes carried out with unmatched distribution presented in Equation 5 to 7.

$$\mu_{i,t} = (1 - \rho)\, \mu_{i,t-1} + \rho X_t \quad (5)$$

The variance value update to the frame (time) change is shown in the following equation:

$$\sigma_t^2 = (1 - \rho)\, \sigma_{t-1}^2 + \rho(X_t - \mu_t)^T(X_t - \mu_t) \quad (6)$$

Where :

$$\rho = \alpha\eta(X_t|\mu_k, \sigma_k) \quad (7)$$

And σ_t^2 is the Gaussian variance distribution value.

The next step is choosing the models of *background* distribution. The background determination based on the similarity value of the models in the distribution. Should the model has a large similarity value, the model is classified as *background*. In this step, the level of model similarity is sorted based on the weight value (ω) or deviation standard value (σ). The bigger the similarity level of Gaussian distribution, the bigger the weight value (ω) of Gaussian distribution and the smallest the deviation standard value (σ).

On the other hand, if the deviation standard value is smaller than the weight value, it is assumed as the moving object, or called as *foreground*. B distribution is determined as background model as presented in Equation 8.

$$B = \text{argmin}_b\left(\sum_{k=1}^{b} \omega_k > T\right) \quad (8)$$

Where T is the *threshold* of minimum data assumed as background. Then, foreground is determined from the smallest weight value. The feature is used widely, such as for modelling a dynamic environment in real-video and in the online-based fashion [16]. In transportation sector, this feature is used to detect vehicles [11].

3. Research methodology

This research was carried out by recording dataset video in two different locations, Simpang Surabaya, Banda Aceh and Pango bridge, Aceh Besar. The Simpang Surabaya location has traffic characteristics with a concrete wall background. Meanwhile, at the Pango bridge has traffic characteristics with a background of small plants, grasses, tress, and two-ways street. The vehicle detection was carried out by determining the length of time of object and movement direction in frame. This method aims to reduce errors in vehicle detections. In this study, there are five variance of frame i.e 3, 5, 7, 9, and 11. Meanwhile the over-height vehicle detection was performed in three conditions. First is without ROI and without vehicle movement direction detection. Second is with ROI and without vehicle movement direction detection. Lastly is with ROI and vehicle movement direction detection. The flowchart of the over-height vehicle detection is presented in Figure 1.

The flowchart begins with the video inputs. After that, the researcher determines the Region of Interest (ROI), as shown in Figure 2. The ROI is used to remove unnecessary objects. The ROI frame is used to detect foreground by using the GMM method. The next step is to remove noises. The noise removal is done by using the morphology filter. The filter cleans the noise by restructuring the morphological flat element of a pixel. The restructuring element is performed by using dilation and erosion operations. The form used in this filter is in square with 3 pixel width. The center element acts as the origin pixel determining the values of the surrounding pixels.

Next step is detecting the blob area. The output of the blob detection is the coordinate value of the object which is used as the basis in determining the movement direction and height. Then, the detection of the movement direction of the vehicle moving to the forward is determined by the coordinate y_now smaller from y_preceding. The reason is the coordinate O(0,0) is located at the left side of the top of the frame. The later step is counting the number of frames detecting the foreground object and the number of movement direction to ensure that the object is the vehicle. The counting of the consistency foreground_detected and direction_movement is bigger or equal than total_frames. The total_frames is the length time of the object in the frames and the variance of total_frames is 3, 5, 7, 9, and 11. The next phase is deciding the vehicle height by determining the coordinate object which is smaller than coordinate value of the threshold.

The threshold coordinate frame is determined from the result of the position point of camera calibration. The height of the camera position is at 2.5 meters and the direction is 90° and at 5 meters from the road side. On the other side of the road, the researcher places a 2.5 meters stick for calibrating the coordinate of real world to the pixel of frame. It is assumed that the threshold height is 2.5 meters. If x_foreground is smaller from x threshold then the object is determined as the over-height vehicle and the system will activate the trigger of warning system. Otherwise, should the height of the vehicle does not exceed the threshold, the system will ignore the vehicles. The camera calibration point and pixel coordinate is presented in Figure 3.

In Figure 3 the coordinate pixel x_threshold is 200 which is determined from the position of camera calibration point (P). The line OB-AC is the threshold line. The object A with the value coordinate x_object_A is 150 which is smaller than x_treshold. Then the object A is called as the over-height object. Meanwhile, the object B with the value coordinate x object_B is 230 which is bigger than x_threshold. The object B is then determined as a not over- height object. The position of camera calibration point is preferred to be placed at the center of the frame to decrease the effect of lens radial distortion [17].

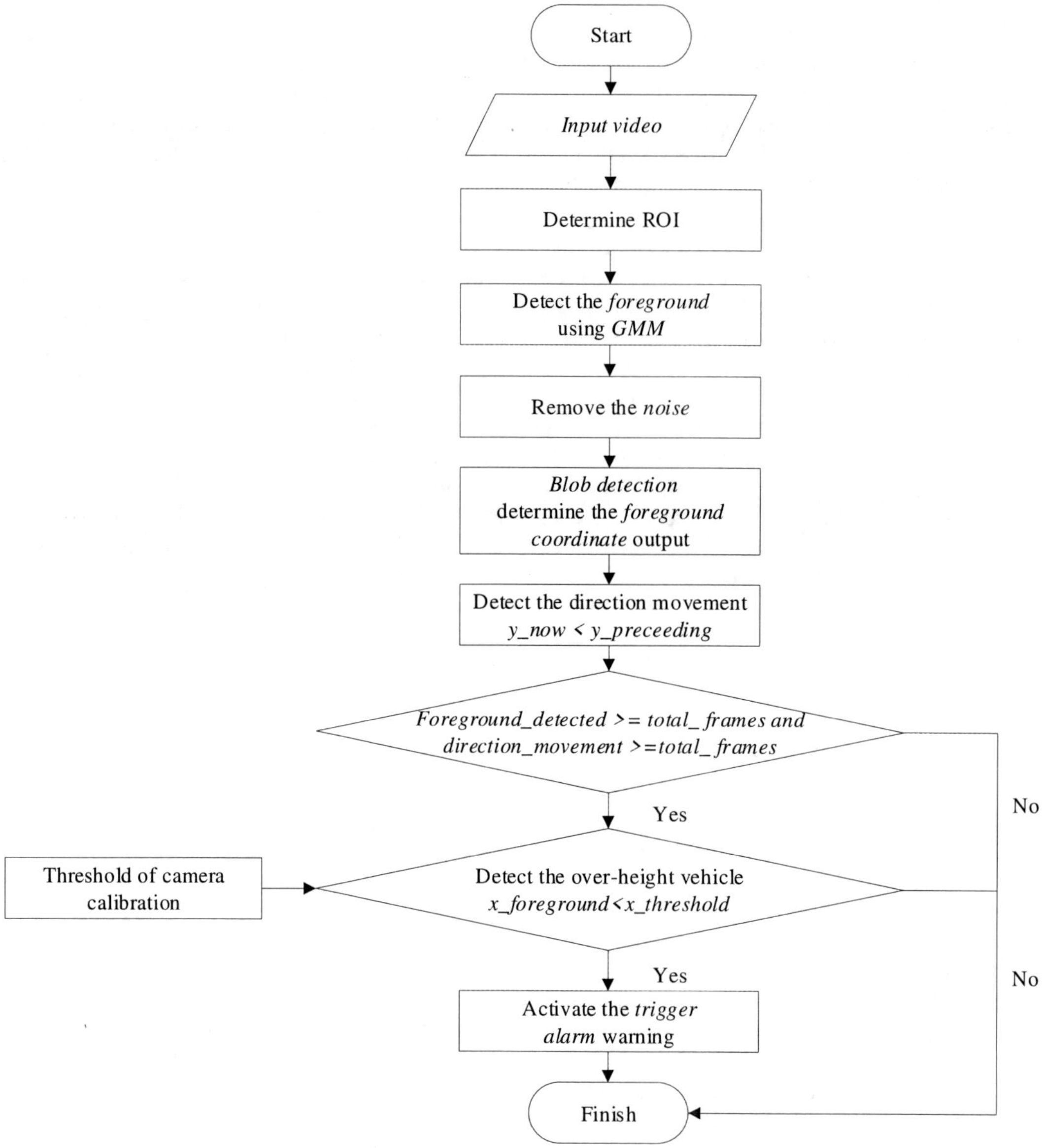

Figure 1. The system simulation flowchart of the over-height vehicle detection

Figure 2. The Region of Interest (ROI) area

The display of simulation system shows a passing vehicle height is over the threshold line. Therefore, the system activates the warning of over-height vehicle. The system also displays the counter of five variance total frames detection, i.e in 3, 5, 7, 9 and 11 frames. It also presents the conditions of over-height vehicle detection, i.e with the ROI, and direction detection, or not. The output display of the simulation system is shown on Figure 4.

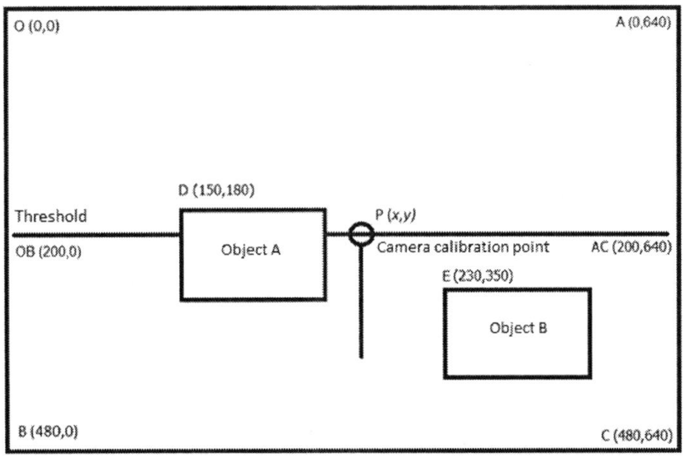

Figure 3. Camera calibration point and pixel coordinate

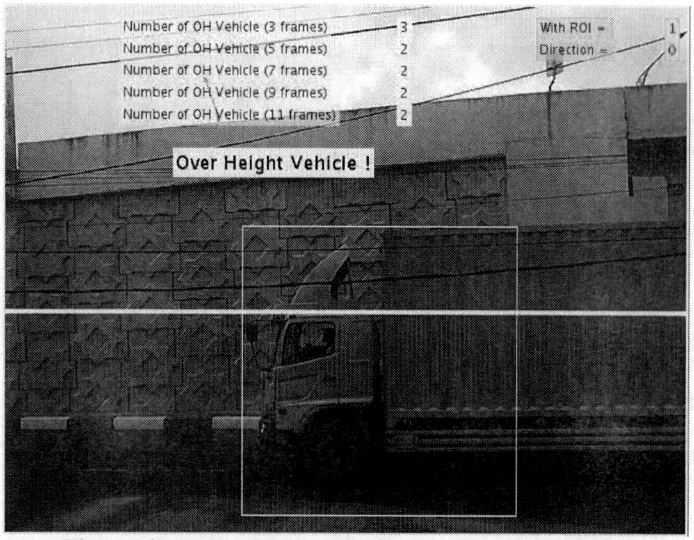

Figure 4. The output display of the simulation system

4. Result and discussion

This study used Gaussian Mixture Model and blob detection feature to detect the over height vehicles. The study was conducted at two locations, Simpang Surabaya and Pango bridge. The traffic documentation at the Simpang Surabaya was at the roadway from Simpang Surabaya to Lambaro with a concrete wall background. Meanwhile, the traffic documentation at the Pango bridge is at the roadway from Lambaro to Simpang Surabaya with a background of small plants, grasses, tress, and two ways street. The video was recorded for 7 minutes 54 second with 640x480 pixel resolution and frame rate 30 frames per second (fps) in cloudy conditions and light rains. The video testing data in both locations is represented in the Table 1 and Table 2

Table 1. Video testing data at Simpang Surabaya, Banda Aceh

Condition	*Frame*					*Actual*
	3	*5*	*7*	*9*	*11*	
Without ROI and without direction detection	12	9	8	8	6	
With ROI and without direction detection	5	4	4	4	4	2
With ROI and direction detection	2	2	2	2	2	

Table 2. Video testing data at Pango bridge, Aceh Besar

Condition	*Frame*					*Actual one way*	*Actual two ways*
	3	*5*	*7*	*9*	*11*		
Without ROI and without direction detection	34	23	19	19	16		
With ROI and without direction detection	15	10	8	8	7	2	6
With ROI and direction detection	4	2	2	1	1		

The result of the data shown that the there is an effect of frame number on the accuracy rate. The object consistency detected in a particular number of frames sequentially revealed the rate of accuracy [6]. The over-height vehicle detection using 3 frames shown the lowest accuracy rate. It is because the 3 frame cannot reduce the detection error. The over-height vehicle detection using 5 frames shown a rise in accuracy rate [6]. The climb is caused by the increase number of frames which can reduce the error in the over-height vehicle detection. The accuracy rise with the increase of frame number is presented in Figure 5. The number of frame which produced the smallest number of errors in the over-height vehicle detection is frames 11 and 7. However, the percentage of the accuracy average revealed that the optimal accuracy is at frames 5 and 7. Therefore, referring to the number of errors of the over-height vehicle detection and the accuracy rate, the optimal frame number is 7 frames. Figure 6 shows the traffic video accuracy enhancement graph at Simpang Surabaya while Figure 7 Presents the traffic video accuracy enhancement graph at Pango bridge.

The result revealed that the Gaussian Mixture Model (GMM) and blob detection can detect the over-height vehicle with the accuracy up to 100%. The background differences do not affect the system performance on conditions with ROI dan direction detection. It, thus, can be concluded that Gaussian Mixture Model (GMM) and blob detection can accurately detect the over-height vehicles. This study is further expected can be a reference for stakeholders in transportation sectors as an alternative to reduce the bridge strikes.

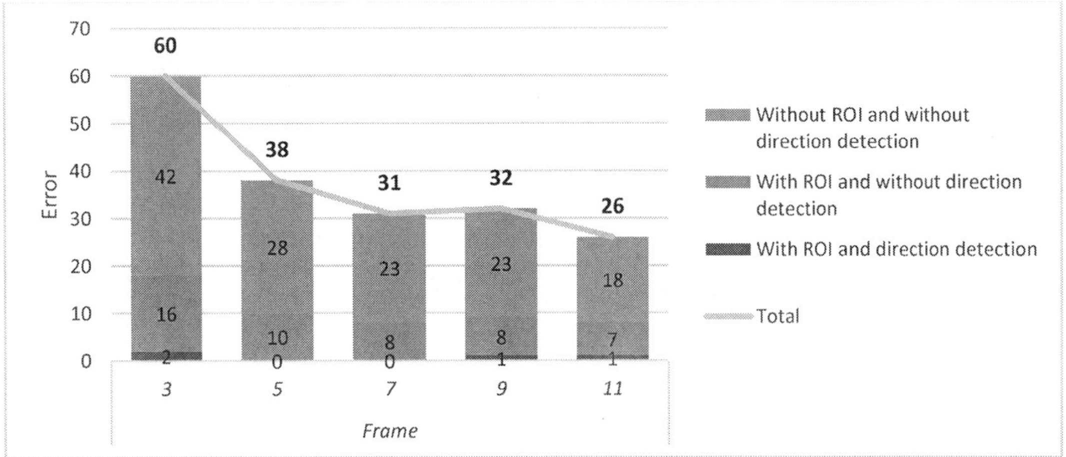

Figure 5. The effect of frame number in decreasing the error of the over height vehicle detection

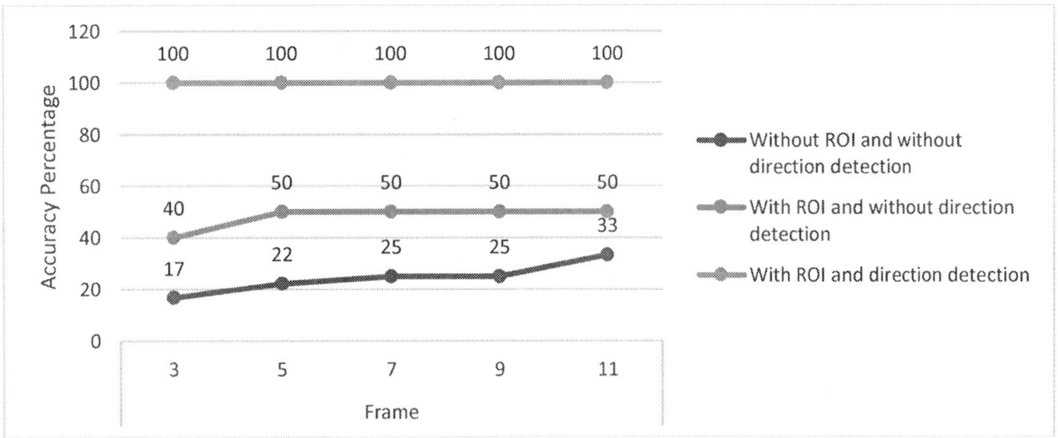

Figure 6. The traffic video accuracy enhancement graph at Simpang Surabaya

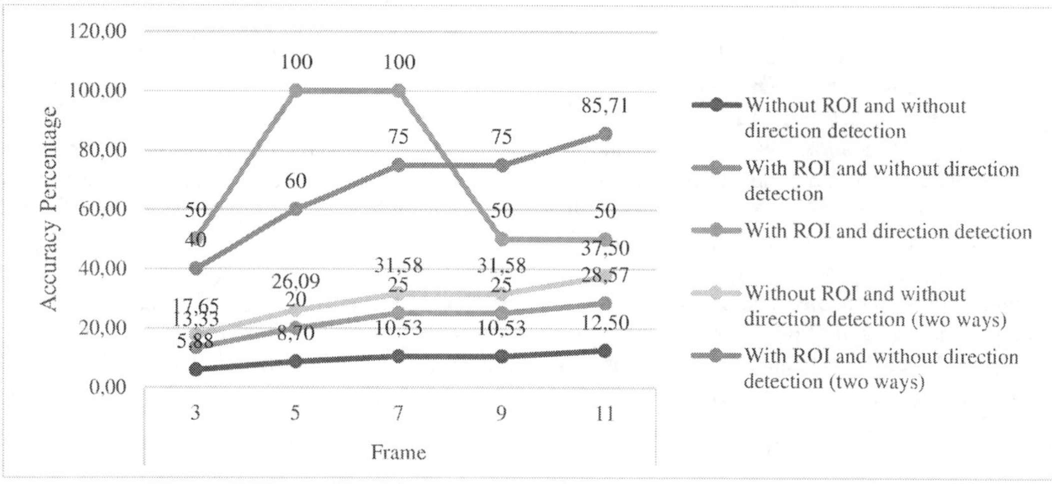

Figure 7. The traffic video accuracy enhancement graph at Pango bridge

5. Conclusion

The result of the study revealed that the GMM and blob detection can detect the over-height vehicles accurately up to 100%. The study was conducted within three conditions, without ROI and without direction detection, with ROI and without direction detection, and with ROI and direction detection. The accuracy rate significantly rises within condition with ROI and direction detection. This condition was applied at two locations with different backgrounds. One is with a concrete wall background and the other is with a background of grasses, small plants, tress, two ways street. The accuracy rate in both places achieves 100%.

The method to reduce the errors in detecting the over-height vehicles can be done by adding the number of frame. The best number of frame in detecting the over-height vehicles is 5 and 7 frame. Referring to the accuracy rate, frame 7 is marked as the most optimal number in detecting the over-height vehicles as it can obtain accuracy 100% with the lowest errors.

6. References

[1] Akbar N A, Amandac G 2019 Available: https://nasional.republika.co.id/berita/q37i3n423/jumlah-kecelakaan-lalu-lintas-naik-3-persen-pada-2019. *(in Bahasa Indonesia).*

[2] Singhal A, Kamga C, Agrawal A K 2018 *IET Intell. Transp. Syst.* **12** 689-695.

[3] Garcia F, Cerri P, Broggi A, Armingol J M, Escalera A D L 2009 *Computer Aided Systems Theory - EUROCAST 2009, EUROCAST 2009,* Las Palmas de Gran Canaria, Spain.

[4] Massoud M A 2013 *World Congress on Engineering,* London, U.K.

[5] Nguyen B, Brilakis I, Vela P A 2017 *J. Comput. Civ. Eng.* **31(5)**.

[6] Nguyen B, Brilakis I 2018 *Adv. Eng. Inform.* **38** 67-80.

[7] Dai F, Park M W, Brilakis M S D I 2015 *Automat. Constr.* **50** 29-39.

[8] Nguyen B 2017 Vision-Based Over-Height Vehicle Detection for Warning Drivers. Cambridge: University of Cambridge.

[9] Nguyen B, Brilakis I 2016 *Transp. Res. Procedia.* **14** 3915 – 24.

[10] Khorramshahi V, Behrad A, Kanhere N K 2008 *WASET, Open Science Index 15, Int. J. Transp. Veh.Eng.* **2** 681-685.

[11] Alfina 2016 *Perancangan dan Pengujian Piranti Pemantauan Visual untuk Menentukan Volume Lalu Lintas.* Banda Aceh: Universitas Syiah Kuala. *(in Bahasa Indonesia).*

[12] Goyal K, Singhai J 2018 *Artif. Intell. Rev.* **50** 241-259.

[13] Nurhadiyatna A, Jatmiko W, Hardjono B, Wibisono A, Sina I, Mursanto P 2013 *IEEE International Conference on Systems, Man, and Cybernetics,* Manchester, UK.

[14] Indrabayu, Bakti R Y, Areni I S, Prayogi A A 2016 *Int. Conf. Comput. Intell. Cybern*, Makassar, Indonesia.

[15] Stauffer C, Grimson W E L 1999 *CVPR*, Fort Colins, CO, USA.

[16] Najar F, Boururouis S, Al-Azawi R, Al-Badi A 2020 *Mixture Models and Applications.* Switzerland: Springer.

[17] Iqbal T M, Munadi K, Arnia F. 2011. *Seminar Nasional dan Expo Teknik Elektro 2011*, Banda Aceh. *(in Bahasa Indonesia).*

Provincial economic level analysis in Indonesia based on the geothermal energy potential and growth regional domestic products using cluster analysis

G M Idroes[1], S Syahnur[1,*], S A. Majid[1], N R Sasmita[2], R Idroes[3]
[1]Department of Economics, Faculty of Economics and Business, University of Syiah Kuala, Banda Aceh 23111, Indonesia
[2]Computational and Applied Statistics Research Group, Department of Statistics, Faculty of Mathematics and Natural Sciences, University of Syiah Kuala, Banda Aceh 23111, Indonesia
[3]Department of Chemistry, Department of Statistics, Faculty of Mathematics and Natural Sciences, University of Syiah Kuala, Banda Aceh 23111, Indonesia

*E-mail: kabari_sofyan@unsyiah.ac.id

Abstract. This study aims to determine the provincial economic level in Indonesia based on geothermal energy potential and growing regional domestic product (GRDP). The data used in this research is secondary data. The type of data is a cross-section from 34 Provinces in Indonesia in 2018. Geothermal Energy Potential Data is obtained from the Ministry of Energy and Mineral Resources (ESDM), and GRDP data for each province is obtained from the Indonesian Central Statistics Agency (BPS). This study divided data into 5 clusters using the k-means method. The results showed that there are 2 provinces in cluster 1, 1 province in cluster 2, 8 provinces in cluster 3, 22 provinces in cluster 4, and 1 province in cluster 5.

1. Introduction

Indonesia has the greatest potential for geothermal energy in the world with a total of 28.91 GW because it is located in a ring of fire in a volcanic route that is spread over 312 locations on several islands such as Sumatra, Java, Bali, Nusa Tenggara and Sulawesi [1]. Geothermal energy is environmentally friendly and quite economical [2] when compared to fossil fuels.

The resource of geothermal potential in Indonesia is predicted by about 40% of the total of the world's geothermal reserves of around 28,617 MW. Whereas, only about 4.5% is being exploited as electrical energy in the country. The power plant comes from a minimum of 10 locations: Darajat (260 MWe), Dieng (60 MWe), Kamojang (200 MWe), Gunung Salak (377 MWe), Sibayak (11 MWe), Lahendong (87 MWe), Wayang Windu (227 MWe), Ulu Belu – South Sumatra (110 MWe), Ulumbu – Flores (5 MWe) and Mataloko (2.5 MWe) [3].

Economic growth is a process of changing a country's economic conditions on an ongoing basis towards better conditions for a certain period. The economic growth of a region can be measured with the Gross Regional Domestic Product (GRDP). The existence of economic growth can be regarded as a success in heading for a better economic level.

This study aims to determine the economic level by clustering each province in Indonesia based on the potential of geothermal energy and the GRDP.

2. Literature Review

2.1 GRDP

GRDP (Gross Regional Domestic Product) is the total value added of goods and services resulting from all economic activities in all regions in a certain year period which is generally within one year.

2.2 Geothermal Potential

Geothermal potential in this study is divided into 5 potentials, namely speculative, hypothetic, possible, probable, and proven.

1) Speculative resources: the estimation of geothermal potential based on the existence of geothermal manifestations and heat flow.

2) Hypothetical resources: determined from regional geologic surveys and geochemistry of thermal features. Stored heat calculations and estimates used for resource sizing.

3) Possible reserves: the potential for geothermal energy in the earth is estimated using integrated surface geoscientific survey data.

4) Probable reserves: the estimation of potential geothermal subsurface by using integrated surface geoscientific survey data and the result of one discovery well.

5) Proven reserves: the potential geothermal energy below the surface is estimated using a geoscience work survey that includes one well and illustrates two production wells to obtain data on the subsurface of the system [4].

2.3 *Hierarchy Method*

The hierarchy method is a grouping method where the results are presented on several levels or in stages from groups n, (n-1), ..., 1 [5]. The commonly used distance function is the Euclid distance, which is defined as the distance between the i-th and k-th observations. Euclid's distance formula from the i-th object to the h-th object is formulated in equation (1):

$$d(i,h) = \sqrt{\sum_{k=1}^{p}(x_{ik} - x_{hk})^2} \; ; \; {}^{i=1,2,\dots,n}_{k=1,2,\dots,p\,;i \neq h} \tag{1}$$

Some kind of hierarchical method of merging (agglomerative) based on linkage includes single linkage, complete linkage, average environment and ward's method [3].

2.4. Non Hierarchy Method

In contrast to the hierarchical method, the non-hierarchical grouping procedure is not done in stages and the number of groups is also determined in advance. Methods that are classified as non-hierarchical grouping include k-means.

Grouping using the k-means method is based on the value of the membership function. The membership function is based on the minimum distance between objects and the center of the cluster (centroid). The k-means algorithm aims to minimize the objective function which is a quadratic error function. For example, there are n objects and p variables. The distance between the i-th object and the l-group is calculated using Euclid's distance squared formulated according to equation (2):

$$\boldsymbol{D}[\boldsymbol{i},\boldsymbol{l}] = \sum_{j=1}^{p} [\boldsymbol{X}(\boldsymbol{i},\boldsymbol{j}) - \bar{\boldsymbol{x}}(\boldsymbol{l},\boldsymbol{j})]^2 \; ; \; {}^{i=1,2,\dots,n}_{j=1,2,\dots,p} \tag{2}$$

With,

$\boldsymbol{X(i,j)}$: value between the i-th object with the j-th variable; $\boldsymbol{X(l,j)}$: the average of the j-th variable against the l-th group; $\boldsymbol{D[i,l(i)]}$: Euclid's distance between the i-th object and the cluster average of the object cluster (centroid) [5].

3. Research Method

3.1. Data Sources and Types
The data used in this research is secondary data. The entire data is in the form of cross-section data (cross-border data) from 34 Provinces in Indonesia in 2018. Geothermal Energy Potential Data is obtained from the Ministry of Energy and Mineral Resources (ESDM) and GRDP data from 34 Provinces in Indonesia is obtained from the Central Statistics Agency (BPS) Indonesia.

3.2. Data Collection Technique
The data of this research were collected in the same year in the form of tabulation by conducting the data cleaning process. Then, the data is analyzed.

3.3. Data Analysis Technique
Data analysis techniques used in this study are descriptive and cluster analysis, which is by describing the data that has been collected properly without and to provide conclusions that are generally accepted and group them according to the size of the similarity between objects using the K-Means method.

Descriptive analysis is in the form of a summary of the data used for each variable, such as the size of the central tendency and the size of the spread [6]. Cluster analysis used in this study follows the flowchart shown in Figure.

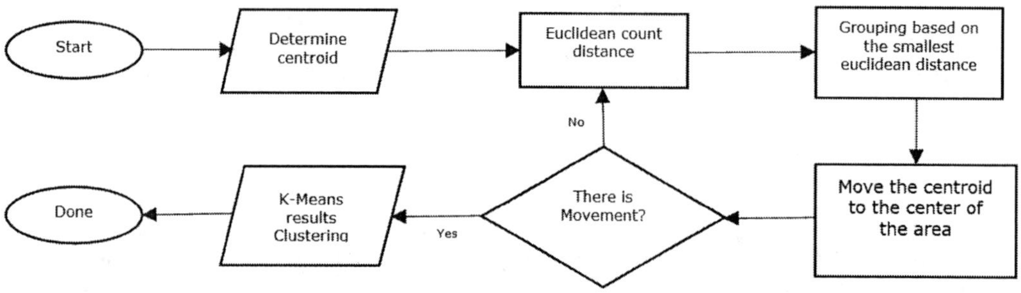

Figure 1. Flowchart Algoritma K-Means Clustering

3.4. Software
The software used in this study is Microsoft EXCEL® and IBM SPSS® Statistics version 22.

4. Results and Discussion

4.1. Descriptive Analysis
Statistically, a description of the economic level for each province in Indonesia can be determined based on the size of concentration and the size of its distribution. By using this measure of concentration, the average level of the economy in Indonesia will be known.

Table 1. Descriptive Statistics of 2017 Geothermal energy potential and GRDP 2017

Variable	Minimum	Maximum	Average	Standard Deviation
Speculative	0	1125	188.97	226.898
Hypothetic	0	734	118.74	187.247
Possible	0	1583	377.03	459.635
Probable	0	1283	72.26	223.582
Proven	0	1550	87.12	273.111
GRDP	-4.56	7.92	5.27	2.06

In Table 1 it can be seen that all geothermal potentials have a minimum value of 0 Mwe, while the maximum potential for geothermal energy is at an estimated potential of 1583 Mwe, after that the estimated potential of 377.03 Mwe achieves the highest average geothermal energy potential, then the standard deviation energy value geothermal potential is in the Suspected potential of 459.63 Mwe.

In Table 1 it can be seen that the minimum GRDP value is -4.56%, and the maximum GRDP value is 7.92%, while the average GRDP value is 5.27%, and the highest standard deviation GRDP value is 2.06%.

4.2. Cluster Analysis

Using the K-mean Method as a method in non-hierarchy to classify each province in Indonesia based on geothermal energy potential and GRDP variables for each province, it is necessary to standardize values. In detail, the results of the initial standardization are listed in Table 2.

Table 2. Initial grouping center on standardized values

Variable	Cluster				
	1	2	3	4	5
Zscore(Speculative)	-.27312	4.12533	.15879	-.48467	-.83284
Zscore(Hypothetic)	1.37394	3.28585	2.79986	-.55400	-.63411
Zscore(Possible)	2.36486	2.62376	1.15302	-.82028	-.82028
Zscore(Probable)	-.32321	5.41516	.25823	-.32321	-.32321
Zscore(Proven)	-.31898	5.35637	.45726	-.31898	-.31898
Zscore(GRDP)	2.82840	2.50349	-.02503	-.63797	3.21729

The standardization process carried out refers to the z-score with the following conditions: A negative value (-) means the data is below the total average and a positive value (+) means the data is above the total average.

Table 2 shows the initial data of the clustering process before iterating the cluster analysis. In the grouping process in this study, the number of iterations carried out to get the most optimal results requires an iteration of 5 times. Then, Table 3 shows the final grouping center on the standardized value.

Table 3. The final grouping center is the standardized value

Variable	Cluster				
	1	2	3	4	5
Zscore(Speculative)	-.34364	4.12533	.68271	-.36668	-.83284
Zscore(Hypothetic)	1.03748	3.28585	.85190	-.52463	-.63411
Zscore(Possible)	1.82095	2.62376	.82450	-.54734	-.82028
Zscore(Probable)	-.12194	5.41516	.15759	-.27767	-.32321
Zscore(Proven)	.24855	5.35637	.10758	-.29069	-.31898
Zscore(GRDP)	2.12649	2.50349	-.25074	-.36217	3.21729

Furthermore, Table 4 explains the differences in the variables formed in this cluster can be seen from the F value and the probability value for each variable. The greater the value of F and significantly smaller than 0.05, the greater the difference in variables in the cluster formed.

Table 4. ANOVA in cluster analysis

Variable	Cluster		Error		F Mean Square	Sig. df
	Mean Square	df	Mean Square	df		
Zscore(Speculative)	6.159	4	.288	29	21.351	.000
Zscore(Hypothetic)	6.303	4	.269	29	23.473	.000
Zscore(Possible)	6.554	4	.234	29	28.026	.000
Zscore(Probable)	7.838	4	.057	29	138.026	.000
Zscore(Proven)	7.717	4	.074	29	104.949	.000
Zscore(GRDP)	7.263	4	.136	29	53.335	.000

Thus the results of the clusters obtained in this study indicate that the variable probably most shows the differences between the provinces in the five clusters formed. This can be indicated by the value of F Maybe = 138,026 and significant = 0.00 for other variables interpreted with the same thing.

Table 5. Cluster Analysis Results Based on Geothermal Potential Indicators and GRDP

No	Cluster	Provinces
1	Cluster 1	(1) Jawa Tengah, (2) Jawa Timur
2	Cluster 2	(1) Jawa Barat
3	Cluster 3	(1) Aceh, (2) Sumatera Utara, (3) Sumatera Barat, (4) Jambi, (5) Sumatera Selatan, (5) Bengkulu, (6) Lampung, (7) Nusa tenggara Timur
4	Cluster 4	(1) Riau, (2) Kepulauan Bangka belitung, (3) Kepulauan Riau, (4) DI Yogyakarta, (5) Banten, (6) Bali, (7) Nusa tenggara barat, (8) Kalimantan barat, (9) Kalimantan tengah, (10) Kalimantan selatan, (11) Kalimantan timur, (12) Kalimantan utara, (13) Sulawesi utara, (14) Sulawesi tengah, (15) Sulawesi selatan, (16) Sulawesi tenggara, (17) Gorontalo, (18) Sulawesi barat, (19) Maluku, (20) Maluku utara, (21) Papua barat, (22) Papua
5	Cluster 5	(1) DKI Jakarta

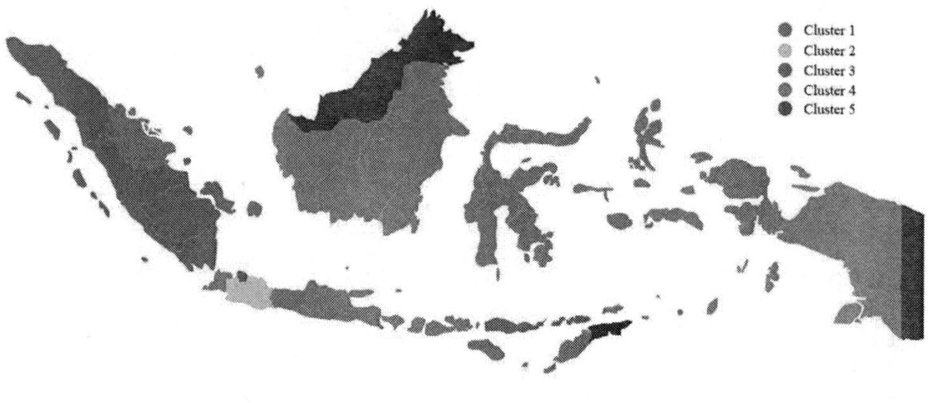

Figure 2. Visualization of the results of the cluster

Based on the results of cluster analysis, as shown in Table 5, it is found that there are 2 provinces in cluster 1, 1 province is in cluster 2, 8 provinces are in cluster 3, 22 provinces are in cluster 4 and 1 province is in cluster 5. In detail, cluster 1 is a province with potential low speculative and proven and high potential, hypothetical, probable, and high GRDP. In cluster 2 it contains provinces with high speculative potential, hypotheses, Possible, Probable, proven, and GRDP. In cluster 3 it contains provinces with high speculative, hypothetical, Possible, Probable, and proven potential and low GRDP. In cluster 4 it contains provinces with low speculative potential, hypotheses, possible, Probable, proven, and GRDP. In cluster 5 it contains provinces with low speculative, hypothetical, probable, and proven potential and high GRDP. Figure 2 illustrates the results.

5. Conclusions

The results of descriptive statistical analysis for geothermal potential variables indicate that the Estimated Potential is the potential with the lowest minimum value, the highest maximum value, the highest average value and the highest standard deviation value. Then, the minimum GRDP value is -4.56%, and the maximum GRDP value is 7.92%, while for the average GRDP value is 5.27%, and the highest standard deviation GRDP value is 2.06%.

Cluster analysis results from 34 provinces based on geothermal potential and GRDP can be grouped into 5 groups. The results of the grouping are classified into several regions; cluster 1 is a province with low speculative and proven potential and high hypothetical, Possible, Probable, and high GRDP potential, namely: Central Java and East Java. Cluster 2 is a province with high speculative, hypothetical, Possible, Probable, proven, and high GRDP potentials, is West Java.

Then, cluster 3 is a province with high speculative, hypothetical, Possible, probable and low proven provinces, namely Aceh, North Sumatra, West Sumatra, Jambi, South Sumatra, Bengkulu, Bengkulu, Lampung, and East Southeast Nusa Tenggara. Furthermore, cluster 4 is a province with a low speculative, hypothetical, probable, probable, and low GRDP potential, namely Riau, kep. Pacific Islands, Kep. Riau, in Yogyakarta, Banten, Bali, West Nusa Tenggara, West Kalimantan, Central Kalimantan, South Kalimantan, East Kalimantan, North Kalimantan, North Sulawesi, North Sulawesi, Central Sulawesi, South Sulawesi, Southeast Sulawesi, Gorontalo, West Sulawesi, Maluku, North Maluku, North Maluku West Papua, and Papua. Finally, cluster 5 is a province with low speculative, hypothetical, Possible, Probable, and proven potential, namely DKI Jakarta.

References

[1] Pambudi N A 2018 *Renew. Sustain. Energy Rev.* **81** 2893–901.
[2] Idroes R, Yusuf M, Alatas M, Subhan, Lala A, Muslem, Suhendra R, Idroes G M, Suhendrayatna, Marwan, Riza M 2019 *IOP Conf. Ser. Mater. Sci. Eng.* **523** 012010.
[3] Kuncoro M 2003 *Metode riset untuk bisnis dan ekonomi*. Jakarta: Erlangga. *(in Bahasa Indonesia).*
[4] Pertamina 2005 *Pertamina Geothermal Development: Resources & Utilization: Publication of Pertamina Geothermal Division.* Jakarta: Pertamina.
[5] Lazulfa I 2013 *Analisis Cluster Kabupaten/Kota di Jawa Timur Berdasarkan Tingkat Pencemaran Udara.* Doctoral Dissertation. Surabaya: Institut Teknologi Sepuluh Nopember. *(in Bahasa Indonesia).*
[6] Kristanto N H, Christopher A, Budi H. 2016. *Jurnal Teknik Informatika dan Sistem Informasi.* **2** pp. 9–15. *(in Bahasa Indonesia).*

The optimization of hybrid power generator system (PV-Wind turbine) using Homer software

S Suriadi[1,*], W I Daru[1], R Halid S[1], M Syukri[1], M Gapy[1]
[1]Electrical and Computer Department Engineering Faculty, University of Syiah Kuala, Darussalam Banda Aceh, Indonesia

*E-mail: suriadimali@unsyiah.ac.id

Abstract. There is a higher electricity demand due to increasing population and economic development. Conventional power companies struggle to meet these needs both in generation and distribution. However, the common use of electric and other power generators, driven mostly by fossils fuels, shows certain limitations including decreasing efficiency and energy resources. Therefore, the application of renewable energy sources is greatly significant. The purpose of this research was to analyse the potentials of hybrid power generator system (PV-Wind turbine) and the optimization of electric energy requirement using hybrid optimization model for electric renewable (HOMER) software. The results confirmed this study as useful and a valid reference for government in building power generation systems through renewable energy utilization.

1. Introduction

Indonesia population growth is similar to electric needs [1] known to increase every year as a result of having the highest world population [2]. In order to fulfil the incessant electricity demand, the country can no longer depend on conventional power generator highly driven by fossil fuels. Hence, it is necessary to introduce renewable energy, especially in remote communities [3]. These alternative sources are potentials for stand-alone applications with sufficient system reliability and are also readily available. One major practical approach in employing renewable energy maximally is by using hybrid system, which utilizes the primary energy for similar load [4]. Hybrid generators serve the purpose of providing assurance to primary energy, hence electricity is easily accessible. These systems are effective compared to the use of a single energy source. For instance, the application of wind energy possibly results to electric power instability. However, when two or more energy sources are employed, a more efficient and different situation is achieved. Therefore, one potential method is to utilize hybrid generator system in maximizing the energy from wind speed and solar radiation. This research uses simulation to utilize solar and wind as renewable energy. In order to achieve accurate calculation, HOMER software is employed.

2. Configuration System

A configuration system is required to arrange the generator capacity suitable to supply the load, based on the following:

2.1. Hybrid power generator system (HPGS)

Hybrid system is a combination of renewable and conventional energy resources [5]. This arrangement helps provide increased system efficiency and also balance the high energy demand, particularly targeted at remote communities. In addition, hybrid power generators commonly utilize renewable energy from solar and wind sources and are both operated and connected through the bus, known as a terminal for load distribution. The characteristics of both generator sources are influenced by natural conditions, where the availability of solar and wind energy appears intermittent. Furthermore, solar power generators function optimally during the day compared to night or cloudy atmosphere [6][7]. However, wind sources operate 24 hours every day, although the influence of wind speed fluctuates regularly. Figure 1 shown the configuration of hybrid generator by using solar and wind turbine.

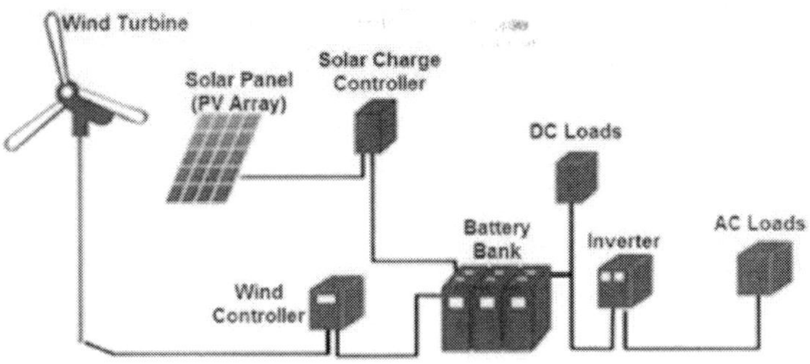

Figure 1. PV– Wind Turbine hybrid power generator system [7].

2.1.1. *PV Array*
PV Array is a solar module arranged from certain solar panels, and is combined in series or parallel. The electric power obtains similar direction based on solar radiation intensity which is accepted and then generates direct current (DC).

2.1.2. *Wind turbine*
Wind Turbine is an electric generator known to utilize kinetic energy from the wind and then converts to electricity. This mechanism is installed with long pole in order not to obstruct buildings or trees, hence maximum wind energy is generated.

2.1.3. *Hybrid controller*
Hybrid controllers are designed to integrate power from the wind turbine, PV array and back up batteries [8]. Subsequently, PV-wind charge controllers using buck boost converter are applied as the components of the hybrid controller. The buck boost converter functions to regulate the voltage stability fluctuates.

2.1.4. *Energy Storage*
Energy storage is very significant to hybrid power generator system (HPGS), and performs two primary functions. First, it is used to adapt to instability of electric load and energy from renewable resources. The second objective is to facilitate the control system and entire operations.

2.1.5. *Inverter*
The role of an inverter is to convert direct current (DC) to alternating current (AC). In addition, the quantity of inverter depends on the generator capacity and also the number of applied load.

2.2. *Homer software*
Energy micro-grid software is an international standard application used to optimize the micro-grid design in all fields, commencing from rural electrification, island utility and extends to campus connected with network. This idea was initially developed in Colorado National Renewable Energy Laboratory and was then perfected and distributed by HOMER company in the form of software. HOMER is an optimizing model for hybrid electric power targeted at all renewable energy sources in order to facilitate the generation and distribution of power through a stand-alone application. Therefore, optimal results are achieved in technical feasibility and economic terms [9][10]. HOMER software performs energy balance calculations installed in every configurations system. Then, proper configuration and optimization are also determined, including addressing the electric needs properly with predefined conditions and further predicts adequate installation and operating system. This model also evaluates expenses, including financial, replacement, and maintenance costs, as well as for fuels

and interest. Furthermore, the software models in three critical stages, termed simulation, optimization, and sensitivity processes [11][12].

2.3. *Analysis of Sensitivity*
In sensitivity analysis, the software tend to repeat the optimization process from every single sensitivity variable earlier determined. For instance, if the wind speed is determined as a sensitivity variable, the software simulates toward configuration system to conclude on several wind speed. The advantages of this application include ease of use, ability to perform simulation then optimizion based on a model and automatically determines the optimum configuration with the capacity to supply load at low expenses. Other advantages are the ability to adequately calculate economic parameters, e.g. net present cost (NPC) and cost of efficiency (COE) as the main output.

2.4. *PV power system*
Solar radiation is converted into DC electric power by PV module. The output power is determined using the equation.

$$P_{PV} = Y_{PV} f_{PV} \left(\frac{\bar{G}_T}{\bar{G}_{T,STC}}\right)\left[1 + \alpha_P\left(T_c - T_{c,STC}\right)\right]$$

Where:

- Y_{PV} = is measured the capacity from PV array, it means the output power is below the standard measurement (kW)
- f_{PV} is PV (%) derating factor
- G_T = amount of solar radiation which received by PV array adalah (kW/m^2)
- $G_{T,STC}$ = solar radiation when standard test.(1kW/m^2)
- α_p = load temperature coefficient (%/°C)
- T_c = recent PV temperature (°C)
- $T_{c,STC}$ = PV standard temperature determined (25°C)

2.5. *Wind turbine power system*
The output power from wind turbine is evaluated by the equation.

$$P_{WTG} = \left(\frac{\rho}{\rho_0}\right) P_{WTG\ STF}$$

Where :

- P_{WTG} = Output power from wind turbine (kW)
- $P_{WTG,STP}$ = overall wind turbine power by using standard temperature and pressure (kW)
- ρ = current air density (kg/m^3)
- ρ_o = air density on standard temperature and pressure (1.225/m^3)

2.6. *Renewable fraction*
Renewable fraction is described as the fraction energy sent to the load, and is determined using the equation:

$$f_{ren} = 1 - \frac{E_{nonren} - H_{nonren}}{E_{served} + H_{served}}$$

Where:

- E_{nonren} = electric energy production which is not come from renewable energy (kW/yr)
- $E_{grid,sales}$ = electric energy which sold to net (kWh/yr)(Including E_{served})
- H_{nonren} = thermal production which is not included into renewable energy (kWh/yr)

- E_{served} = total amount of load which is served (kWh/yr)
- H_{served} = total amount of thermal load which is served (kWh/yr)

3. Methodology

The method used to get the optimization result of hybrid power generator system (HPGS) is simulation using HOMER software. This study applied input data in the form of daily load information, wind speed, and solar radiation intensity. These parameters also covered for optimize hybrid renewable energy process.

3.1 The input of daily load data

The use of HOMER software provides comprehensive data profile on the monthly change of electric load. However, the load profile for tropical locations does not observe any significant change. Significantly due to the absence of climate change. The profile load which is located around Lhoknga Aceh Besar for 24 hours as shown in Figure 2.

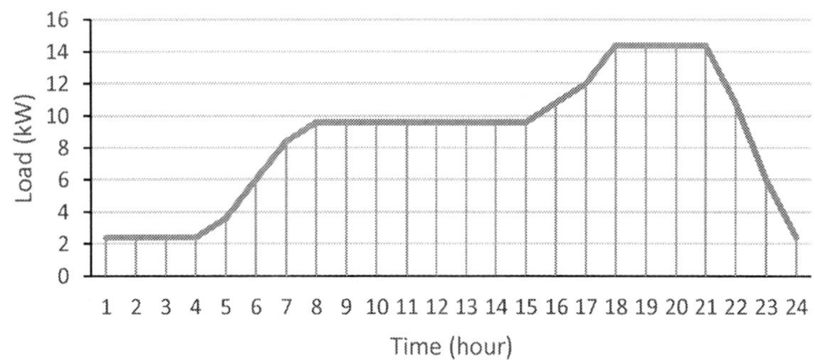

Figure 2. Load profile.

Figure 2 shows the daily load used to simulate HPGS for a one year period, and is assumed by using daily random variability of 10%. Subsequently, the average load is then applied at the consumer end in the case of daily generated load data. It is shown in Table 1.

Table 1. Daily load system

Electric load system	
Average (kWh/d)	165.29
Average (kW)	6.89
Peak (kW)	24.57
Load Factor	0.28

Based on Table 1, the average daily absorbed energy was estimated at 165.29 kWh, while the power average of 6.89 kW was used including the peak load valued at 24.57 kW for a one year estimation. Therefore, the load factor is specified at 0.28, and is gotten from the calculation of average and peak load values.

3.2. *The input data of wind speed*

The HOMER software is used to generate information on wind speed and result of average monthly wind speed based on time, and also serves as parameter in modelling the hybrid power generator system. Based on both results, it can be concluded that the ranges and the percentage of wind speed in Lhoknga Aceh Besar provides an overview of the characteristic of wind speed as shown in Figure 3

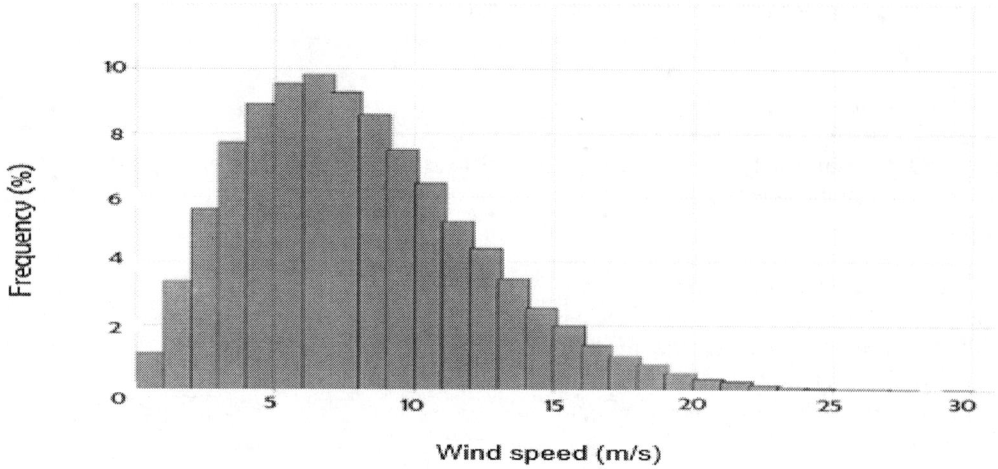

Figure 3. Wind speed characteristic.

3.3. *The input data of solar radiation*

The HOMER simulation requires latitude and longitude information in order to ascertain the potentials of solar radiation on certain places. This research area corresponds to the latitude and longitude values of 5°26' and 95°16', respectively, and is located within Lhoknga Aceh Besar. The obtained radiation data is as shown in Table 2.

Table 2. The average of solar radiation

Months	Clearness Index	Radiation (kWh/m2/hari)
January	0.57	5.39
February	0.577	5.75
March	0.557	5.79
April	0.542	5.64
Mei	0.496	5.01
June	0.506	4.99
July	0.493	4.90
August	0.47	4.81
September	0.459	4.74
October	0.465	4.67
November	0.489	4.66
December	0.528	4.88

3.4 The configuration of HPGS

The analysis of HPGS involves the use of HOMER software, where the configuration consists of certain basic components. Figure 4 shows the configuration of the HPGS design.

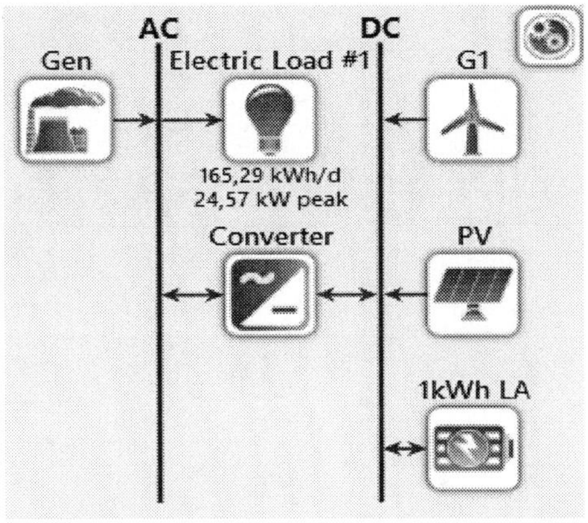

Figure 4. Hybrid power generator system configuration.

3.5. The strategy of HPGS optimization method
There are 2 (two) strategies involved in HPGS optimization in order to service the load, as explained below.

3.5.1. Load following (LF) method
The data above describes certain conditions using the load following (LF) method:
For the first condition, the output energy is sufficient to supply the load and also, the battery does not accept any type of power source. Therefore, the generator is considered non-functional. The second condition shows higher output power is more compared to the load demand, where the energy is then distributed to charge the battery. Therefore, the generator is also non-functional. Furthermore, the third condition provides very low energy, and is not possible to supply the load. Subsequently, two possible events tend to occur. First, provided $SOC=SOC_{minimum}$, then the generator tends to supply the load (which have been reduced with renewable energy). Diesel generators only achieve this purpose optimally without charging the batteries. Second, provided $SOC>SOC_{minimum}$, then there is a necessity to analyse in term of expenses incurred in using batteries compared to diesel generator. At higher expenses in operating the generator compared to batteries, the power from batteries (discharge) is applied, if not, the use of diesel generator is allowed.

3.5.2. Cycle charging (CC) method
The first condition of CC strategy shows the renewable energy is sufficient to supply the load and the batteries don't accept any energy due to the maximum state of charge (SOC). Under this condition, the generator is not operated. The second condition reveals the output of renewable energy is more compared to the load request, hence energy is distributed to charge the batteries. Under this condition, the generator is not also functional. The third condition occurs as the output energy appears very low, and is unable to supply the load demand. There are two possibilities to this occurrence. First, provided $SOC=SOC_{minimum}$, then the generator operates efficiently to supply the load (which has been decreased with the load of renewable energy). Diesel generator also functions at maximum to deliver the load and charge the batteries. Second, provided $SOC>SOC_{minimum}$, then there is need for analysis in terms of expenses incurred in using batteries compared to the cost of diesel generator. At higher expenses in maintaining the diesel generator compared to batteries, the power from batteries (discharge) is applied, if it not, the use of diesel generator is allowed.

4. Results and discussion
4.1. The result of simulation by using Load Following method
The simulation result was obtained from HOMER simulation are an optimization based on the components that have been entered into the system. The output of the optimization process is shown in Figure 5.

Sensitivity		Architecture						Cost				System	Gen			
Diesel Fuel Price ($/L)	Wind Scaled Average (m/s)	PV (kW)	G1	Gen (kW)	1kWh LA	Converter (kW)	Dispatch	COE ($)	NPC ($)	Operating cost ($)	Initial capital ($)	Ren Frac (%)	Hours	Production (kWh)	Fuel (L)	O&M Cost ($)
2,00	3,00	64,3			345	20,4	LF	$0,593	$439.558	$10.711	$302.633	100				
2,00	4,11	64,3			345	20,4	LF	$0,593	$439.558	$10.711	$302.633	100				
2,00	8,00	44,6	10		221	18,3	LF	$0,529	$393.732	$9.245	$275.554	100				
1,00	8,00	26,1	7	28,0	138	13,8	LF	$0,481	$371.049	$14.414	$186.783	75,3	1.586	14.928	5.606	1.332
1,00	3,00	35,0		28,0	151	12,8	LF	$0,505	$389.559	$17.326	$168.080	65,7	2.105	20.684	7.659	1.768
1,00	4,11	35,0		28,0	151	12,8	LF	$0,505	$389.559	$17.326	$168.080	65,7	2.105	20.684	7.659	1.768

Figure 5. The results of load following dispatch optimization.

Figure 5 shows the optimization result based on the load following dispatch strategy. The highlighted row signified the best optimization result, where the hybrid system uses photovoltaic (PV) in the amount of 26.1 kW, 1kW wind turbine with a total of 7 units, and 28 kW diesel generator. Under these circumstances, constant electric power was required to charge 128 unit of batteries at a storage capacity of 83.4 Ah specification, 12 volt DC with 100% SOC, and 40% minimum.

4.2. *The simulation result by using Cycle Charging method*

The simulation result using cycle charging method as shown in Figure 6

Architecture						Cost				System	Gen					PV	
PV (kW)	G1	Gen (kW)	1kWh LA	Converter (kW)	Dispatch	COE ($)	NPC ($)	Operating cost ($)	Initial capital ($)	Ren Frac (%)	Hours	Production (kWh)	Fuel (L)	O&M Cost ($)	Fuel Cost ($)	Capital Cost ($)	Production (kWh)
20,7		28,0	80	14,2	CC	$0,392	$302.141	$15.482	$104.226	36,5	2.470	38.290	12.509	2.075	6.255	61.956	30.875
20,7		28,0	80	14,2	CC	$0,392	$302.141	$15.482	$104.226	36,5	2.470	38.290	12.509	2.075	6.255	61.956	30.875
18,3	8	28,0	101	16,9	CC	$0,485	$374.163	$16.735	$160.237	58,4	1.245	25.126	7.766	1.044	7.766	54.862	27.341
24,4		28,0	122	17,6	CC	$0,502	$387.258	$20.191	$128.155	42,2	1.635	34.254	10.569	1.373	10.569	73.268	36.514
24,4		28,0	122	17,6	CC	$0,502	$387.258	$20.191	$128.155	42,2	1.635	34.254	10.569	1.373	10.569	73.268	36.514
44,6	10		221	18,3	CC	$0,529	$393.732	$9.245	$275.554	100						133.757	66.659
64,3			345	20,4	CC	$0,593	$439.558	$10.711	$302.633	100						193.000	96.183

Figure 6. The results of cycle charging dispatch optimization.

Figure 6 represents the optimization result based on cycle charge (CC) dispatch strategy. The marked portion specified the best optimization result, where the hybrid system uses photovoltaic (PV) in the amount of 18.3 kW, 8 unit of 1kW wind turbine and 28 kW diesel generator. Under these circumstances, continuity electric power was needed, so that it requires 101 unit of batteries with storage capacity similar to aforementioned specifications.

4.3. *HPGS technical comparing analysis based on both dispatch strategies*

In analyzing the optimal results from both strategies, proper attention on energy production and renewable fraction were observed. Subsequently, the economic section was defined based on the net present cost (NPC) value and its influence to the environment. Based on the three components stated, the best simulation approach based on dispatch strategy is the use of load following (LF). This tactic was able to utilize maximum renewable energy with renewable fraction of 75.3% compared to cycle charge (CC) of 58.4%. However, based on NPC, the LF showed lower expense due to the cost influence of generator fuels. Also, the role of diesel generator on CC delivered 34. 1% electricity from the total power and the annual fuel consumption was evaluated at 7,766 L. Meanwhile, for LF, the estimated yearly fuel consumption was quantified at 5,606 L, and the operation time (1,586 hours/year) appeared higher compared to CC (1,243 hours/year), as LF only produces sufficient power to distribute to load, while CC generator operates on maximum capacity to supply the load and charge the batteries.

4.4. *Sensitivity results*

Sensitivity results are represented in the form of an optimized graph type. Figure 7 shows the LF strategy consisting of four configurations.

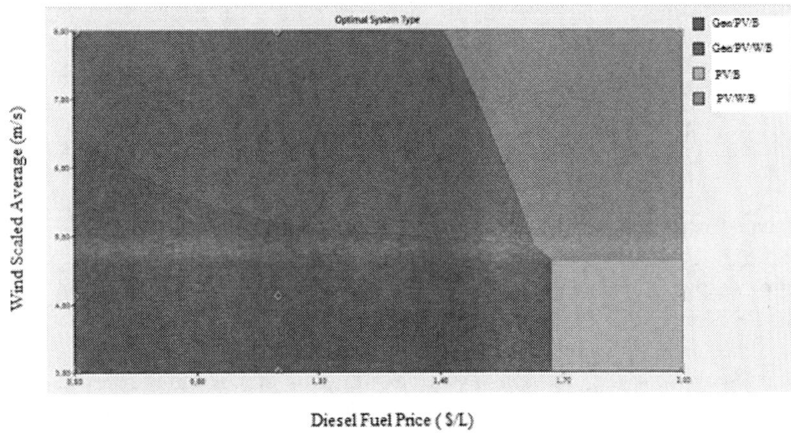

Figure 7. The results of sensitivity on LF strategy.

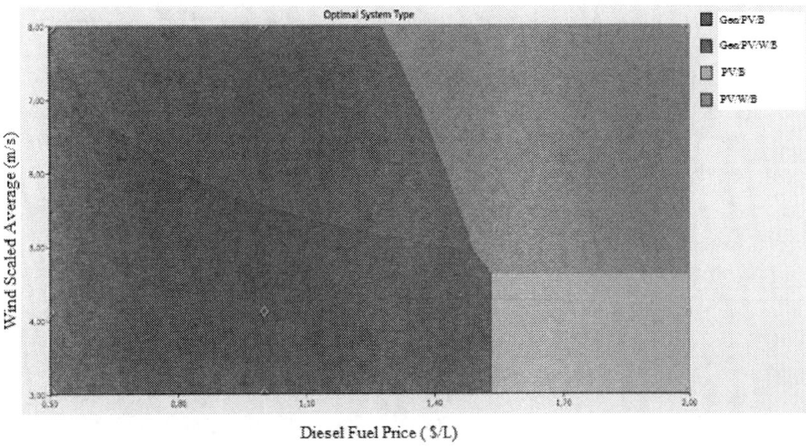

Figure 8. The results of sensitivity on CC strategy.

Figure 7 expands the analysis of sensitivity by using diesel cost and wind speed on LF. The type of generator -PV-wind- batteries are considered for use as the cost of diesel extended to 1.6 $/L with wind speed requirement at an average of 5 m/s. For generator-PV-battery with diesel price of 0.50 $/L up to 1.67 $/L, the average of wind speed appeared below 5 m/s, while PV-Wind battery was used for maximum wind speed and diesel cost at 7.99 m/s and 1.4 $/L, respectively. Based on Figure 8, generator-PV-wind- battery configuration obtained maximum energy under the conditions of diesel price at 1.5 $/L and wind speed of 5 m/s. The generator-PV- battery was used with diesel price from 0.50 $/L up to 1.5 $/L and wind speed under 5 m/s. Also, the PV-wind- battery was used as the wind speed approached 7.99 m/s with diesel price at 1.2 $/L.

5. Conclusion

Based on results and discussion, the simulation of HPGS (PV – wind turbine) using Homer software revealed the load following (LF) strategy as the best compared to cycle charge (CC). The load following dispatch optimization to supply a load of 165.29 kWh/ day at Lhoknga Aceh Besar was obtained at the combination of 26.1 kW PV, 7 kW wind turbine, 28 kW generator, 138 units of battery ,and 13.8 kW converter to produce NPC $ 371,049 and COE $0.481 per kWh. Obviously, the LF dispatch optimization is more economical with the smaller total NPC and the COE then CC dispatch optimization. Based on results simulation the NPC and COE of the CC dispatch optimization are respectively $ 374,163 and $0.485 per kWh. Furthermore, the use of HPGS based on both dispatch strategies tend to potentially produce the power to meet annual electricity demand, where LF acquired a total power of 72,535 kWh/year, while CC was estimated at 73,747 kWh/year. Based on the renewable fraction from both strategies, the highest value of 75.3% occurred with LF and involved less battery activities. This condition showed greater energy was produced by the LF system compared to the energy needed to supply load. Meanwhile, CC was estimated at 58.4 % renewable fraction.

References

[1] Perusahaan Listrik Negara (PLN) 2018 *Rencana Usaha Penyediaan Tenaga Listrik (RUPTL) 2018-2027*. Jakarta: Perusahaan Listrik Negara. *(in Bahasa Indonesia).*
[2] Daryanto Y 2007 *Kajian Potensi angin Untuk Pembangkit Listrik Tenaga Bayu*. Yogyakarta: Balai PPTAGG - UPT-LAGG. *(in Bahasa Indonesia).*
[3] Herlina 2009 *Analisis Dampak Lingkungan dan Biaya Pembangkit Listrik Tenaga Hibrida di Pulau Sebesi Lampung Selatan.* Tesis. Jakarta: Universitas Indonesia. *(in Bahasa Indonesia).*
[4] Quaschning V 2005 *Understanding Renewable Energy Systems*. London, VA: Earthscan.
[5] Aderemi B A, Chowdhury S P, Olwal T O, Abu-Mahfouz A M 2018 *Energies*. **11** 1572.
[6] Ren,Y, Suganthan P N, Srikanth N 2015 *Renew. Sustain. Energy Rev.* **50** 82-91.
[7] Sharif M S E, Khan M M Z, Moniruzzaman M, Bose A 2017 *Am. J. Mod. Energy* **3** 121-130.
[8] Aziz A S, Tajuddin M F N, Adzman M R, Ramli M A, Mekhilef S 2019 *Sustainability* **11** 683.
[9] Slameto A A 2015 *Data Manajemen dan Teknologi Informasi.* **16** 1-10. *(in Bahasa Indonesia).*
[10] Gilman P, Lambert T. 2005. *Homer the micro power optimization model software started guide.* National Renewable Energy Laboratory of United States Government.
[11] Suriadi, Taib S, Shawal J M 2013 *Proc. Aceh Dev. Int. Conf.* **3** 103-109.
[12] Suriadi, Aqsa A S H, Ramdhan H S, Syukri M, Gapy M 2019 *Prosiding Seminar Nasional dan Ekspo Teknik Elektro.* **7** pp. 42-46. *(in Bahasa Indonesia).*

IOP Conf. Series: Materials Science and Engineering 1087 (2021) 012081 doi:10.1088/1757-899X/1087/1/012081

Joint relay selection and energy harvesting to improve performance of cooperative communication systems

E Berutu[1], R Muharar[1], N Nasaruddin[1,*]
[1]Electrical and Computer Engineering Department, Universitas Syiah Kuala
Jl. Tgk. Syech Abdurrauf No.7 Darussalam, Banda Aceh 23111, Indonesia

*E-mail: nasaruddin@unsyiah.ac.id

Abstract - This paper focuses on a combining relay selection with energy harvesting (EH), known as joint relay selection with EH (JRSEH), to improve performance in wireless cooperative communication systems. Previous research was examined performance improvement with relay selection (RS) techniques without combined with EH (RS non-EH). This study aims to improve the performance by applying EH and the RS in the cooperative communication system. In the system, a source selects the best relay from several relays that produce the highest SNR value at the destination. The selected best relay performs EH based on time switching. Then, the best relay with EH obtains the power to forward information to the destination using the Amplify and Forward (AF) relay scheme. The channel from source to relay is Rician fading, and from the relay to a destination is Rayleigh fading. The research method used is system modeling and analysis of computer simulations. The system performance is calculated based on the bit error rate (BER) and the throughput in a multi-relay cooperative system with EH and RS. The results show that the performance of the proposed JRSEH higher than that of RS non-EH, multi-relay EH, and multi-relay non-EH.

1. Introduction

The wireless cooperative communication system has become the focus of the current research as a reliable strategy in overcoming fading problems and increasing channel capacity. The concept of a wireless cooperative communication system is a system that uses one or more relays to transmit information from source to destination [1]. Before the information is forwarded to the destination, the relay must process the information with a specific protocol. The commonly used protocols in relays are the amplify-and-forward (AF) and decode-and-forward (DF) protocols. A cooperative communication system that utilizes multiple relays (multi-relay) has advantages in throughput but requires more energy than a single relay system. One of the techniques to overcome energy shortages but maintain throughput is by performing energy harvesting (EH) on relays and reducing the number of relays that will forward information to a destination [2].

Several previous studies have examined the increase of energy efficiency in systems with relay selection techniques. Previous work in [3] used a relay selection strategy by applying an adaptive mode to the relay in the cooperative system. One of the techniques used is adaptive relay transmit power to determine the energy consumption and system throughput compared to non-adaptive relay techniques. Research in [4] has discussed increasing power efficiency by applying relay selection based on the cooperative region and comparing it with the technique without using a cooperative region. Several studies combine relay selection and energy harvesting (EH) to increase network throughput, such as the research in [5] examined the combination of relay selection with EH to improve the performance of the system. The channel model in the research is Rayleigh fading. Then, the EH protocol used in relays is time switching (TS) and forward signals by using the DF protocol. The best relay is selected based on the energy stored in the relay. The relay with the highest EH is considered to have the best channel between the source to the relay and is chosen to forward the information to the destination. Another research in [6] uses a relay selection strategy combined with EH with harvest-use-store power splitting techniques. EH on the relay is prioritized to charge the relay for further use in order to obtain a better network throughput.

This paper proposes joint energy harvesting (EH) and relay selection in a cooperative communication system. We consider the channel from source to relay is Rician fading, and from the relay to the destination is Rayleigh fading. The relay selection uses the opportunistic relay selection method as in [7]. The source selects the best relay that produces the highest SNR at the destination, and then the selected relay performs EH as the power to transmit information to the destination. The relay protocol is amplify and forward (AF), and the EH protocol is time switching. The performance parameters of the proposed system are bit error rate (BER) and throughput. Then, BER and throughput of the proposed method are compared with BER and throughput of relay selection systems without EH (RS non-EH), multi-relay EH, and multi-relay non-EH.

2. System Model

The system model of a multi-relay cooperative communication system of joint relay selection with the EH (JRSEH) can be seen in Figure 1, assuming the first relay is chosen as the best relay. The network component consists of several nodes, namely: source (S), relays ($R_1, R_2, \dots, R_M$), and destination (D). Each node is equipped with one antenna. The source uses several relays to transmit information due to environmental conditions that make it impossible for the source to communicate directly with the destination. Source and destination are assumed to have unlimited energy, but relay has limited energy. The energy that can only be harvested by the relay that comes from the source. The relay has two circuits, namely a circuit for conducting EH and a circuit for transmitting information to the destination.

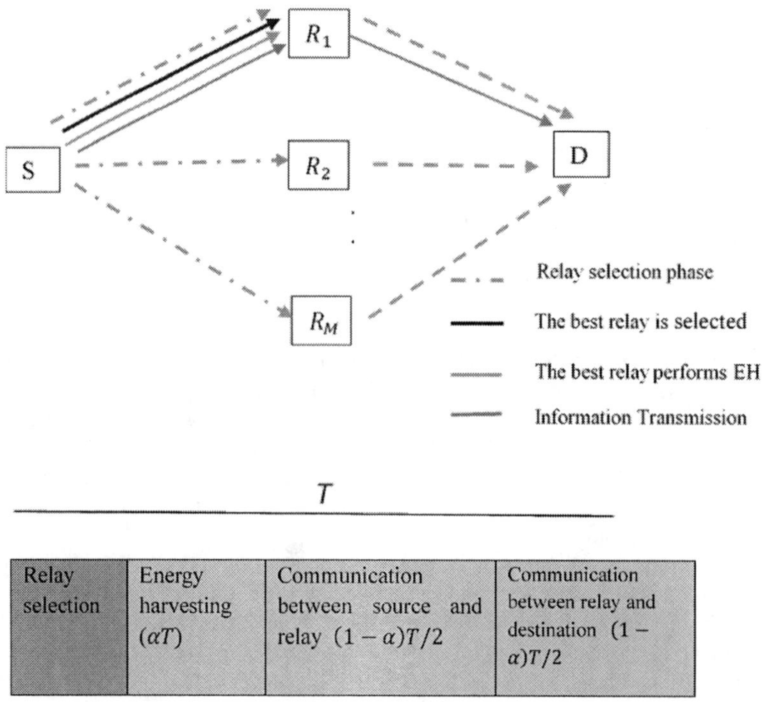

Figure 1. Joint relay selection with EH scheme and block time in cooperative communication systems.

In the model, the communication channel for the source-to-relay link is Rician fading with the fading coefficient is h. While the communication channel for the relay to the destination link is Rayleigh fading with the fading coefficient is g. Relay selection is carried out by opportunistic relay selection, where the source selects one relay by estimating the channel quality using channel state information (CSI) from source to relay and relay to the destination to produce the highest SNR at the destination [8], [9]. Perfect CSI is applied in channel estimation, which is the node that performs the estimation to know the channel condition. Even though in real conditions, this estimation is not possible to calculate directly. Therefore, the relay selection scheme combined with EH in this paper can be one of the solutions for the estimation, as shown in Figure 1, by assuming the selected relay is the first relay.

The process of the proposed joint relay selection with energy harvesting (JRSEH) on a wireless cooperative communication system is as follow.

- The source broadcasts the pilot bit to all relays, and then each relay forwards to the destination to estimate the channel quality from source to relay and relay to destination.
- The destination estimates SNR value based on the pilot bits, which obtained from each relay is to determine the appropriate relay to be selected as the best relay.
- The destination chooses the best relay that produces the highest SNR in the destination using the following equation:

$$R_b = \arg\max_{i \in M} (\gamma_{end_i}) \tag{1}$$

where γ_{end_i} is destination SNR through the relay i^{th}.

- The destination sends feedback to the source in the form of selected relay data.
- Based on the feedback obtained from the relay, the source sends the pilot bit to the selected relay to be active to the next stage, while the unselected relays are in an idle position (inactive).
- The source transmits energy with duration αT focused on the selected relay, and the selected relay performs EH during that duration. The path loss model used is the free space model [10], [11] is given by

$$P_{L_{xy}} = G_t . G_r \left(\frac{c}{4\pi . F . d_{xy}}\right)^2 \tag{2}$$

where G_t is antenna gain at the transmitter, G_r is antenna gain at receiver, $c = 3\,x10^8\ m/s$ is the speed of light in a vacuum, F is a carrier radio frequency, d_{xy} is distance x to y.

- The amount of energy saved by the selected relay at the time duration αT is calculated by [13]:

$$E_H = \mu P_s |h|^2 \alpha T\, P_{L_{SR}} \tag{3}$$

where P_s is the source power, h is the channel gain from source to relay, $P_{L_{SR}}$ is the free-space path loss model from source to relay, μ is the energy conversion efficiency value ($0 < \mu < 1$), α is EH time switching ratio, and T is the time duration in one transmission.

- The source then transmits the information to the selected relay at the time duration $(1-\alpha)\frac{T}{2}$ which is given by [12]:

$$y_{r(t)} = \sqrt{P_s\, P_{L_{SR}}}\ h\, x_s(t) + n_r(t) \tag{4}$$

where n_r is the noise in the relay, x_s is the information signal sent by the source

- The selected relay then amplifies the information signal is given by [13]:

$$\beta = \frac{P_r}{\sqrt{\frac{P_s|h|^2}{d_{SR}^m} + N_1}} \tag{5}$$

where d_{sr} is distance source to relay, where $N_0 = N_1 = 1$ is a variant noise.

- The selected relay uses all of the energy invested as the power to transmit information to the destination. The power of the relay is given by

$$P_{r_{EH}} = \frac{E_H}{\frac{(1-\alpha)T}{2}} = \frac{2\mu P_s|h|^2 \alpha T \, P_{L_{SR}}}{(1-\alpha)} \quad (6)$$

where E_H is energy harvested by a relay.

- The relay transmits the information signal to the destination at a time duration (1-α) T / 2.
- The signal received by the destination through the best relay is formulated by the following equation:

$$y_d(t) = \sqrt{P_{r_{EH}} P_{L_{RD}}}\, g\, \beta \sqrt{P_s P_{L_{SR}}} \cdot h.x_s(t) + \sqrt{P_{r_{EH}} P_{L_{RD}}}\, g \cdot \beta . n_r(t) + n_d(t) \quad (7)$$

where $P_{L_{RD}}$ is a free-space path loss model relay to the destination, n_d is the noise at the destination.

- The amount of SNR received at the destination through the selected relay is given by

$$\gamma_{EH_b} = \frac{P_s\, P_{r_{EH}}\, g^2 h^2 \beta^2\, P_{L_{SR}} P_{L_{RD}}}{P_{r_{EH}}\, P_{L_{RD}}\, g^2\, \beta^2 + N_0} \quad (8)$$

where N_0 is a variant noise.

- The throughput at the destination is calculated as [13]

$$\tau_{EH} = \frac{1}{2}(1-\alpha)(1 + \log_2 \gamma_{EH_b}). \quad (9)$$

where γ_{EH_b} is the SNR of the energy harvested at the best relay.

3. Setting and Simulation Parameters

This simulation is to calculate the bit error rate (BER) and throughput of joint relay selection and energy harvesting system. The relay selection method is opportunistic relay selection, which is then combined with energy harvesting. Joint relay selection with energy harvesting (JRSEH) system is expected to the high throughput compared with non-EH, multi-relay non-EH relay systems, or multi-relay EH. In this paper, the computer simulation uses several parameters as in Table 1. The number of sent data bits is 10^5 bits, with 20 iterations. The Channel model source to relay is Rician fading, and the relay to the destination is Rayleigh fading. The unit of throughput is bits / Hz / sec, meaning the number of bits received at the destination in 1 Hz within 1 second. Throughput of JRSEH, RS non-EH, Multirelay non-EH, and Multi-relay EH systems are calculated using the same parameters in Table1.

Table 1. Simulation Parameters.

Parameter	Symbol	Information
Number of bits sent	N_bits	10^5
Channel model	h, g	Fading Rician, Rayleigh
Resources	P_s	0.3 *watt*
Relay power non EH	P_r	0.3 *watt*
Rician value	k	15 dB [9]
Amount of relays	M	4
Energy conversion efficiency	η	0.9
Time switching ratio	α	0.2

Source to destination distance	d_{sd}	10 – 400 m
Source to relay distance	d_{sr}	0.2 d_{sd}; 0.4 d_{sd}; 0.6 d_{sd}; 0.8 d_{sd}
Relay to destination distance	d_{rd}	247 m; 0.6 d_{sd}; 0.4 d_{sd}; 0.2 d_{sd}
Antenna gain transmitter	G_t	23 dBi
Antenna gain receiver	G_r	23 dBi
Carrier frequency	F	450 MHz
Efficiency spectral	Δ	1 $bit/Hz/Sec$
Energy harvesting protocol	TS	Time Switching
Relay protocol	AF	Amplify and Forward
Variance noise	N_0	1

4. Results and Discussions

4. 1 Bit Error Rate (BER)

This section shows the bit error rate (BER) of a system with one relay, two relays, three relays, and four relays. Each relay performs EH during αT. The BER is calculated using the parameters in Table 1. Figure 2 shows the simulation results for the BER of the multi-relay cooperative communication system with EH.

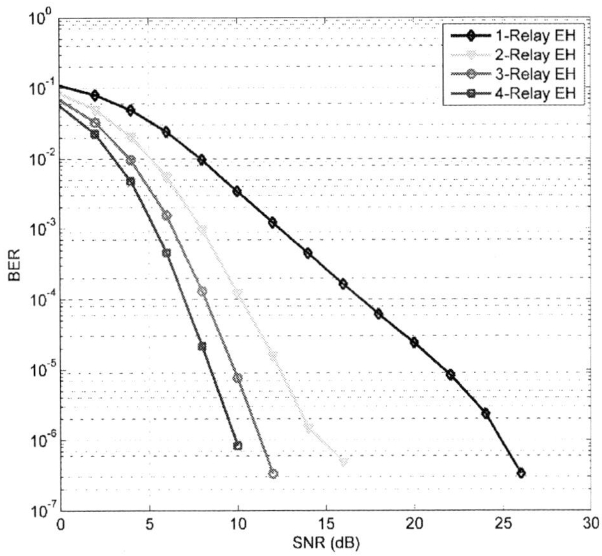

Figure 2. Bit Error Rate (BER) on a multi-relay system with EH.

Figure 2 shows that BER decreases when the number of relays increases from 1 relay to 4 relays, which means that the system performance is getting better if the more relays with EH are available in the system. Multi-relay systems have diversity gain when multiple relays forward information to a destination, where the destination combines all the information received by the relays. Thus, the more relays used in the system, the higher the diversity gain so that the performance is getting better.

The comparison of BER of the proposed and the other conventional system (RS non-EH, 4-relay EH, and 4-relay non-EH systems) are shown in Figure 3. The simulation results show that the cooperative system with RS non-EH has poor BER performance. Then, the BER performance of non-EH 4-relays is better than that of non-EH RS. However, the BER performance of the proposed and 4-relay with EH systems have the best performance compared to others. The BER performance of the proposed work is almost the same as the BER 4-relay with EH performance. Moreover, 4-relay with EH will consume higher energy because all relays are active. On the other hand, in the proposed system, only the best relay is active while the others are idle.

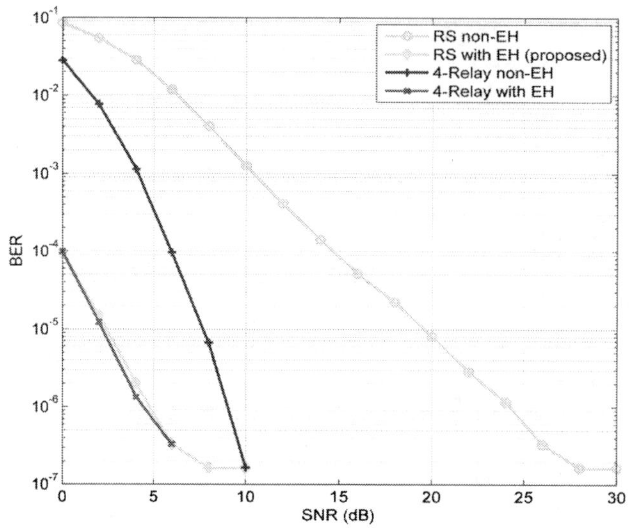

Figure 3. Comparison of Bit Error Rate (BER) of the proposed and other systems.

4.2. Throughput

The simulation of this section is to calculate the throughput of joint relay selection with EH (JRSEH) system, multi-relay non-EH system, multi-relay EH system, and relay selection non-EH (RS non-EH) system. The throughput is calculated using equation (9) and the parameters in Table 1. The throughput of the proposed system is expected to be increased or the same as the RS without EH because reducing the transmission power of the proposed work does not reduce the throughput performance of the system. Thus, the proposed system can save power compared to other conventional systems. The emission of power from the source to the relay in the EH multi-relay system is focused on one particular relay with an energy conversion efficiency (η) value is 0.9.

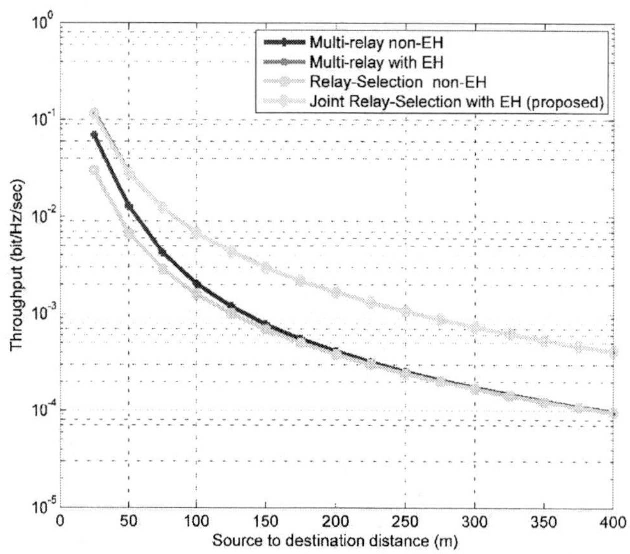

Figure 4. Throughput comparison of the proposed with other systems.

Figure 4 shows the simulation results of the throughput of multi-relay non-EH, multi-relay EH, RS non-EH, and JRSEH (proposed) cooperative systems. From the figure, we see that the throughput of the proposed system is almost close to the throughput of the multi-relay EH system but exceeds the throughput of multi-relay non-EH and RS non-EH systems. So, one of the objectives of this paper is achieved where the proposed work can reduce the power consumption of relays by maintaining throughput performance.

5. Conclusions

This paper has proposed a joint relay selection with EH (JRSEH) to improved the performance of wireless cooperative communication systems using amplify and forward (AF) schemes. The EH protocol used time switching. The computer simulation has been conducted to calculate the bit error rate (BER) and throughput of the proposed system, the relay selection non-EH (RS non-EH), multi-relay non-EH, and muti-relay with EH. The results found that the joint relay selection with the energy harvesting (JRSEH) system could provide the good BER and throughput compared with the previous system, which are RS non-EH, multi-relay non-EH system, and multi-relay EH systems. Furthermore, the proposed JRSEH can reduce the power consumption of relays by maintaining throughput performance.

6. Acknowledgment

This paper is supported by LPPM Universitas Syiah Kuala under grant number: 8/UN11.2.1/PT.01.03/PNBP/2020.

References

[1] Laneman J N, Tse D N C, Wornell G W 2004 *IEEE Trans. Inf. Theory.* **50** 3062–3080.
[2] Kansal M B, Hsu J, Zahedi S, Srivastava M B 2007 *ACM J.* **6** 1–35.
[3] Shah V, Mehta N B, Yim R 2010 *IEEE Trans. Wirel. Commun.* **9** 2505–15.
[4] Sheng Z, Ko B J, Swami A, Lee K, Leung K K 2010 *MILCOM 2010 Mil. Commun. Conf.* 543-548.
[5] Do N T, Bao V N Q, Beongku A 2015 *Int. Conf. Adv. Technol. Commun.* 243–247.
[6] Zhou Z, Peng M, Zhao Z, Wang W, Blum R S 2016 *IEEE J. Sel. Areas Commun.* **34** 969–982.
[7] Cui S, Haimovich A M, Somekh O, Poor H V 2009 *IEEE Trans. Inf. Theory.* **55** 5121–37.
[8] Soysa M, Suraweera H A, Tellambura C, Garg H K 2012 *IEEE Trans. Wirel. Commun.* **60** 840–850.
[9] Suraweera H, Karagiannidis G K, Smith P J 2009 *IEEE Trans. Wirel. Commun.* **8** 2783–8.
[10] Amin O, Lampe L 2012 *IEEE Commun. Lett.* **1** 412–415.
[11] Goldsmith A 2012 *Wireless communications.* Cambridge: Cambridge University Press.
[12] Eissa M M 2012 *Energy Efficiency-The innovative ways for smart energy, the future towards modern utilities (Intechopen).*
[13] Nasir A A, Zhou X, Durrani S, Kennedy R A 2013 *IEEE Trans. Wirel. Commun* **12** 3622–3636.

The Hamiltonian in the unidirectional surface wave propagation

D Fadhiliani[1], H Qausar[1], M Ramli[2,*], S Rizal[3], M Syafwan[4]
[1]Graduate School of Mathematics and Applied Sciences, Universitas Syiah Kuala, Banda Aceh, 23111, Indonesia
[2]Department of Mathematics, Universitas Syiah Kuala, Banda Aceh, 23111, Indonesia
[3]Department of Marine Science, Universitas Syiah Kuala, Banda Aceh, 23111, Indonesia
[4]Department of Mathematics, Universitas Andalas, Padang, 25163, Indonesia

*E-mail: marwan.math@unsyiah.ac.id

Abstract. Unidirectional surface water waves used in this study were obtained from the Benjamin Bona Mahony (BBM) equation. The BBM wave research is carried out by observing the envelope of BBM wave groups that involve following the Nonlinear Schrodinger (NLS) Equation. Soliton on Finite Background (SFB) is an excellent NLS solution for describing the dynamics of wave propagation. It is known that there are modulation instability and phase singularity phenomena which can significantly affect the amplitude of the wave in its propagation. Hamiltonian contains potential energy which can explain the singularity and amplitude dynamics of the phase. Hamiltonian equations using the displaced phase-amplitude variables have been obtained and are presented in this article.

1. Introduction

The Benjamin Bona Mahony (BBM) equation is a mathematical model that describes the unidirectional of surface water waves in its propagation. In this study, the BBM equation was studied in detail to support the activity of generating high, unbroken waves known as extreme waves [1,2]. The generation of extreme waves in the laboratory is carried out to test the resistance of ship models and offshore structures to extreme waves that can suddenly appear offshore [3,4].

Research on BBM waves was carried out by observing the wave itself and also using the BBM wave group envelope. It is known that the BBM wave group involves the Schrodinger Nonlinear (NLS) equation and the Soliton on Finite Background (SFB) is one of the excellent NLS solutions to describe the dynamics of wave propagation. Halfiani et al [5] and Fadhiliani et al [6] have carried out research on the envelope of the BBM wave group. Halfiani et al [5] state that there was a modulation instability in its propagation and the plot of the envelope shows a phase singularity phenomenon. Fadhiliani et al [6] in his research also stated the same thing by using argand diagrams. Argand diagrams are obtained by transforming SFB variables with displaced phase-amplitude variables introduced by Karjanto [7] to study the dynamics of another one-way wave, namely the Korteweg de Vries (Kdv) wave.

The phase singularity phenomenon can significantly affect the amplitude of the wave [7]. The singularity and the dynamics of the phase-amplitude in the NLS equation can be described using the Hamiltonian because it contains potential energy [8]. As we know, the Hamiltonian is the sum of the kinetic energy and potential energy of the system.

The use of Hamiltonian formulas in hydrodynamic problems for water waves has been widely used, including by Groves et al [9] and Wahen et al [10]. Sultana et al [11] in his research also used the Hamilton principle to investigate the evolution of water waves and derive Hamilton's canonical equation of motion. Craige et al [12] derives the Hamiltonian of the NLS equation and several other equations in the study of free surface water waves and continues by analyzing the stability of Benjamin-Feir [13]. Besides that, there is Kurnia et al [14] who applied the consistent Hamiltonian modeling method to obtain the high-order Hamiltonian equation with the Taylor expansion of the potential and vertical velocity around the water surface at rest based on the Hamiltonian formulation of water waves. Therefore, this study focuses on obtaining information about the distribution of energy by utilizing the Hamiltonian function and the dynamics of BBM waves in extreme waves generation in the laboratory.

2. The BBM Wave Group Envelope

The surface water wave propagation for normalized variables can be expressed in the form [15]:

$$\eta_t + \eta_x + \eta\eta_x - \eta_{xxt} = 0 \tag{1}$$

This equation is intended to model unidirectional wave propagation with large wavenumbers and small amplitudes. If the selected solution is in the wave group $\eta(x,t) = a(x,t)e^{i\theta} + c.c$ where $\theta = (kx - \omega t)$ dan k are the frequency and wavenumber, respectively. Halfiani shows that the envelope of the wave group propagates and involve the spatial NLS equation as follows:

$$A_\xi + i\beta A_{\tau\tau} + i\gamma|A|^2 A = 0 \tag{2}$$

where $\beta = -\left(2\omega p + \frac{\omega^2}{k}\right)/p^3$ and $\gamma = \omega\left(\frac{2}{p-1} - \frac{k}{8k^2\omega+2\omega-2k}\right)/p$ is a constant that depends on the frequency or wave number. Equation (2) is obtained by [5] for variable transformation to the space variable x and time t through the transformation $\xi = \varepsilon^2 x$, $\tau = \varepsilon\left(t - \frac{x}{p}\right)$, $a(x,t) = \varepsilon A(\xi,\tau)$. Parameter $p = (1 + 2k\omega)/(1 + k^2)$ are elements that appear in the variable transformation obtained to eliminate the resonance term in the process of determining the solution through the multiple scale method [5]. For problems involving the wave signal as the initial signal in the wavemaker to describe the propagation in space can using the NLS equation. The independent variables ξ and τ have meaning depending on the problem described. In the dispersive wave problem, ξ and τ represents the spatial and the time variable, respectively.

The solution of the spatial NLS equation which can describe the occurrence of extreme waves in a wave pool is SFB. SFB is a non linear extinction of a monochromatic signal with an amplitude of r_0, that is $A_0(\xi) = r_0 e^{-i\gamma r_0^2 \xi}$ which is interfered with by a modulating wave with small wavelength intervals resulting in an exponential increase in instability. These SFB waves reach their maximum amplitude at $\left(0, \frac{2n\pi}{\nu}\right)$, $n \in \mathbb{Z}$ [16]. The spatial SFB solution can be written as:

$$A(\xi,\tau) = A_{SFB}(\xi,\tau)A_0(\xi) \tag{3}$$

In the propagation of the BBM wave group, modulation stability and the phenomenon of phase singularity appears [5]. The phase singularity occurs when the real value of the amplitude $|A_{SFB}|$ disappeared [16]. Phase singularity, here, is indicated by the presence of wave dislocation which is indicated by the merging of two waves (wave merging) or wave splitting [17].

3. Variational Formulation (Lagrangian and Hamiltonian Structures)

The variational formulation of the NLS equation to be used is in the form of a Lagrangian and Hamiltonian structure. As stated by Groesen et al [18] that the equation for the wave-ship interaction

is based on the Lagrangian principle of variational, which leads to the formulation of the combined system as the Hamiltonian system. Hamiltonian functions to regulate the dynamics of phase-amplitude because it contains potential energy [8].
Let A be the solution to the NLS equation,

$$A_\xi + i\beta A_{\tau\tau} + i\gamma|A|^2 A = 0, \tag{4}$$

Then the Lagrangian $\mathcal{L}$ associated with equation (4) is :

$$\mathcal{L}(A) = \frac{1}{4} i\left(A^* \partial_\xi A - A\partial_\xi A^*\right) + \frac{1}{2}\beta|\partial_\tau A|^2 - \frac{1}{4}\gamma|A|^4, \tag{5}$$

with Euler-Lagrangian :

$$\frac{\partial \mathcal{L}}{\partial A} = \frac{\partial}{\partial \tau}\left(\frac{\partial \mathcal{L}}{\partial A_\tau}\right) + \frac{\partial}{\partial \xi}\left(\frac{\partial \mathcal{L}}{\partial A_\xi}\right). \tag{6}$$

The Hamiltonian H structure is derived based on the Lagrangian. The Hamiltonian density $\mathcal{H}$ is expressed as follows:

$$\begin{aligned}\mathcal{H}(A) &= \frac{1}{4} i\left(A^* \partial_\xi A - A\partial_\xi A^*\right) - \mathcal{L}(A) \\ &= -\frac{1}{2}\beta|\partial_\tau A|^2 + \frac{1}{4}\gamma|A|^4,\end{aligned} \tag{7}$$

and Hamiltonian H expressed as follows:

$$H(A) = \int_{-\infty}^{\infty} \mathcal{H}(\mathcal{A})\, d\tau. \tag{8}$$

The Hamiltonian function to prove the singularity for a finite time in the NLS equation. The NLS equation in the Lagrangian form is a dynamic system. It is integrable and as a consequence it has an infinite number of conserved quantities. Hamiltonian is one of the quantities in question, besides that there is also a form:

$$C_1 = \int_{-\infty}^{\infty} |A|^2\, d\tau. \tag{9}$$

It is a simple physical meaning for he first three conservation integrals C_1. For the lowest order conserved quantity, i.e the wave energy', 'mass', 'wave action', 'plasmon number', or 'wave power' in optics. For the second-order conserved quantity C_2 called the '(linear) momentum' and can be written as

$$C_2 = \int_{-\infty}^{\infty} i(A^* \partial_\tau A - A\partial_\tau A^*)\, d\tau. \tag{10}$$

4. Displaced phase-amplitude variables

Transformation to displaced phase-amplitude variables is carried out to represent the phase-amplitude of the SFB wave with a finite background as a solution to the NLS equation. The form of a solution to the NLS equation using displaced phase-amplitude variables is given:

$$A(\xi,\tau) = A_0(\xi)F(\xi,\tau), \tag{11}$$

where $A_0(\xi) = r_0 e^{-i\gamma r_0^2 \xi}$ and F will be determined later. The variables used are $\phi(\xi,\tau)$ as the displaced phase, $G(\xi,\tau)$ as the displaced amplitude parameter and for special cases F is used as follows:

$$F(\xi,\tau) = G(\xi,\tau)e^{i\phi(\xi,\tau)} - 1, \qquad G \text{ dan } \phi \in \mathbb{R}, \tag{12}$$

so that the NLS equation solution will be observed with the following form:

$$A(\xi,\tau) = \left(G(\xi,\tau)e^{i\phi(\xi,\tau)} - 1\right) r_0 e^{-i\gamma r_0^2 \xi}. \tag{13}$$

Furthermore, equation (13) will be substituted to the Hamiltonian to obtain the transformed Hamiltonian (depending on G and ϕ) [19].

5. Results and Discussion

From equations (7) and (8), the form of the Hamiltonian equation H is as follows:

$$H(A) = \int_{-\infty}^{\infty} \left(-\frac{1}{2}\beta|\partial_\tau A|^2 + \frac{1}{4}\gamma|A|^4\right) d\tau. \tag{14}$$

Next, the solution of SFB, $A(\xi,\tau)$ will be substituted which contains the displaced phase-amplitude variables in equation (14). The first derivative A (ξ, τ) with respect to τ will be obtained:

$$\partial_\tau A = (\partial_\tau G + iG\partial_\tau \phi) r_0 e^{i(\phi - \gamma r_0^2 \xi)}, \tag{15}$$

then the first part of equation (14) will be obtained:

$$\begin{aligned} \frac{1}{2}\beta|\partial_\tau A|^2 &= \frac{1}{2}\beta\left|(\partial_\tau G + iG\partial_\tau \phi) r_0 e^{i(\phi - \gamma r_0^2 \xi)}\right|^2 \\ &= \frac{1}{2}\beta {r_0}^2[(\partial_\tau G)^2 + G^2(\partial_\tau \phi)^2]. \end{aligned} \tag{16}$$

The second part is related to potential energy, as follows:

$$\begin{aligned} \frac{1}{4}\gamma|A|^4 &= \frac{1}{4}\gamma\left|\left(G(\xi,\tau)e^{i\phi(\xi,\tau)} - 1\right) r_0 e^{-i\gamma r_0^2 \xi}\right|^4 \\ &= \frac{1}{4}\gamma {r_0}^4 (G(G - 2\cos\phi) + 1)^2, \end{aligned} \tag{17}$$

so that the result of the Hamiltonian $H(G,\phi)$ transformation is obtained as follows:

$$H(A) = \int_{-T/2}^{T/2} \left(-\frac{1}{2}\beta {r_0}^2[(\partial_\tau G)^2 + G^2(\partial_\tau \phi)^2] + \frac{1}{4}\gamma {r_0}^4 (G(G - 2\cos\phi) + 1)^2\right) d\tau. \tag{18}$$

Equation (18) resulted from the Hamiltonian transformation using the displaced phase-amplitude variables which will be studied in future studies. meanwhile, the form $\phi(\xi,\tau)$ as the displaced phase and $G(\xi,\tau)$ as the displaced amplitude parameter has been previously described (see Fadhiliani et al [20]).

6. Conclusion

In the propagation of BBM waves, modulation instability occurs and the phenomenon of phase singularity appears, this is known from the analysis of the BBM wave envelope that involves following the NLS equation. The NLS solution, known as SFB, can describe the dynamics of the

BBM wave and to be presented, a transformation of the SFB variable is carried out with displaced phase-amplitude variables. The singularity and amplitude dynamics of the phase can be shown using the Hamiltonian because they contain potential energy. The Hamiltonian equation using the displaced phase-amplitude variables has been derived and is presented in this article.

7. Acknowledgments

Penelitian Disertasi Doktor, DRPM with contract number 68/UN11.2.1/PT.01.03/DPRM/2020 and Program Riset Unggulan Unsyiah Percepatan Doktor, Universitas Syiah Kuala with contract number 320/UN11.2.1/PT.01.03/PNBP/2020 which is funded this research.

References

[1] Dean R G 1990 *Freak Waves: A Possible Explanation, Water Wave Kinetics*. Amsterdam: Kluwer.
[2] Onorato M, Osborne A R, Serio M, Bertone S 2001 *Phys. Rev. Lett.* **86** 5831–4.
[3] Toffoli E M, Gregersen B 2011 *Open Ocean Eng. J.* **4** 24-33.
[4] Dysthe K, Krogstad H E, Muller P 2009 *Rogue Waves.* Amsterdam: Elsevier.
[5] Halfiani V, Salmawaty, Ramli M 2017 *Far East J. Math. Sci.* **102** 1077–90.
[6] Fadhiliani D, Halfiani V, Ikhwan M, Qausar H, Munzir S, Rizal S, Syafwan M, Ramli M 2020 *Heliyon.* **6** e04004.
[7] Karjanto N 2006 *Mathematical Aspects of extreme water waves.* Enschede: University of Twente.
[8] Sulem C, Sulem P L 1999 *The Nonlinier Schrödinger Equation: Self Focusing and Wave Collapse*. New York: Springer-Verlag.
[9] Groves M D, Stylianou A 2014 *Comptes Rendus de l'Académie des Sciences - Series I.* **352** pp. 205-211.
[10] Wahlen E 2008 *Phys. Letters A.* **372** 2597-602.
[11] Sultana S, Rahman Z 2013 *Fluida Dynamics* **3** 75-81.
[12] Craig W, Guyenne P, Sulem C 2010 *Wave Motion* **47** 552-563.
[13] Craig W, Guyenne P, Sulem C 2012 *Eur. J. Mech. B Fluids* **32** 22-31.
[14] Kurnia R, van Groesen E 2014 *Coast. Eng.* **93** 55-70.
[15] Benjamin T B, Bona J L, Mahony J J 1972 *Philos. Trans. Royal Soc. A.* **272** 47–57.
[16] Huijsmans R H M, Klopman G, Karjanto N, Andonawati. 2005. *Proc. Rougue Waves 2004.*
[17] Karjanto V, van Groesen E 2007 *Phys. Lett. A.* **371** 173-179.
[18] Groesen E van, Andonowati 2017 *Appl. Math. Model.* **42** 133-144.
[19] Groesen E van, Andonowati, Karjanto N 2006 *Phys. Lett. A.* **354** 312–319.
[20] Fadhiliani D, Ramli M, Rizal S, Syafwan M 2019 *IOP Conf. Ser. Earth Environ. Sci.* **364** 012009.

IOP Conf. Series: Materials Science and Engineering 1087 (2021) 012083 doi:10.1088/1757-899X/1087/1/012083

Soliton solution of stationary discrete nonlinear Schrödinger equation with the cubic-quintic nonlinearity

H Qausar[1], M Ramli[2,*], S Munzir[2], M Syafwan[3], D Fadhiliani[1]
[1]Graduate School of Mathematics and Applied Sciences, Universitas Syiah Kuala, 23111, Indonesia
[2]Department of Mathematics, Universitas Syiah Kuala,Banda Aceh, 23111, Indonesia
[3]Department of Mathematics, Universitas Andalas, Padang, 25163, Indonesia

*E-mail: marwan.math@unsyiah.ac.id

Abstract. This research discusses stationary discrete nonlinear Schrödinger equation with cubic-quintic nonlinearity. This equation is interesting to study because it has a unique solution known as a soliton. This solution has a fixed profile and speed when propagating and in the context of applications in the optical field, soliton can also be engineered as a carrier of information that can propagate on media with very long distances without experiencing significant interference. This paper only focuses on on-site type soliton (soliton that peak in the middle on one site). The method of determining solution on stationary discrete nonlinear Schrödinger equation with cubic-quintic nonlinearity is divided into two cases. The first case for the value of parameter C is zero and the soliton solution is determined analytically. In this case the soliton solution can be stated explicitly, therefore the soliton solution will be displayed and also the boundaries on the parameters that make the solution in the form of on-site soliton. The second case for the value of parameter C is not zero and the soliton solution is determined using a numerical approach namely Trust Region Dogleg Method. In this case the soliton solution cannot be stated explicitly, therefore only boundaries of the parameters that make the solution in the form of on-site soliton will be displayed.

1. Introduction

Discrete nonlinear schrödinger's equation (DNLS) is one of the basic equations in the nonlinear lattice dynamics model [1]. This equation has a wide range of applications including electrical circuits [2], denaturation and phase transformation associated with double strands (DNA) [3], biomolecular chain [4], propagation of the optical beam on a nonlinear waveguide [5], and material form (Bose-Einstein condensation on an optical lattice) [6]. The Bose-Einstein condensation (KBE) is the fifth form of matter in a particle when it approaches 0 Kelvin, which was theoretically predicted by S.N Bose and A. Einstein in 1925 [7]. The first successful experiments to prove the existence of KBE were carried out by Wolfgang Ketterle, Eric Cornell and Carl Wienmann. For this achievement, they were awarded the Nobel Prize in physics in 2001.

The DNLS equation is interesting to study because it has a special solution known as a soliton. This solution has a fixed profile and speed when propagating [8]. In the context of applications in the field of optics, solitons can also be engineered as information carriers that can propagate on media over long distances without experiencing significant disturbances. This is considered very important in the development of information and communication technology in the future [5]. This paper focuses only on on-site solitons (solitons peaking in the middle at one site). This is due to the results of research conducted by Eilbeck and Johansson [9] which states that the on-site type solitons tend to be stable.

The general form of the DNLS equation can be written as follows

$$i\dot{\psi}_n = -C(\psi_{n+1} - 2\psi_n + \psi_{n-1}) + F(\psi_{n+1}, \psi_n, \psi_{n-1}) \tag{1}$$

with $\psi_n \equiv \psi_n(t) \in \mathbb{C}$ is a wave function with respect to time $t \in \mathbb{R}^+$ dan $n \in \mathbb{Z}$, $\dot{\psi}_n$ denotes the derivative of the function ψ_n respect to t, C denotes the coupling parameter between two adjacent sites and F is a nonlinear term. The nonlinear term in the (1) equation has several forms, namely [9]

1. Cubic type nonlinearity : $|\psi_n|^2\psi_n$
2. Quintic type nonlinearity : $|\psi_n|^4\psi_n$
3. Ablowitz-Ladik type nonlinearity : $\frac{1}{2}|\psi_n|^2(\psi_{n+1} + \psi_{n-1})$.

Based on the three nonlinear terms, the nonlinear terms of the cubic and quintic types were selected, so that the (1) equation could be written

$$i\dot{\psi}_n = -C(\psi_{n+1} - 2\psi_n + \psi_{n-1}) - B|\psi_n|^2\psi_n + Q|\psi_n|^4\psi_n \tag{2}$$

with B and Q respectively denote the cubic and quintic nonlinear coefficients. This selection is based on the research of Chong and Pelinovsky [10] which states that the nonlinear effect on some waveguide materials is better modeled by using the cubic-quintic type nonlinearity, precisely when $B > 0$ and $Q > 0$.

Soliton solutions in the Equation (2) can be determined semi-analytically by various methods, some of which are the variational approximation method, anti-continuity principle, center manifold reduction and the Nehari manifold approach [11]. Often the soliton solution calculated by the semi-analytic method is confirmed using the soliton solution of the stationary equation that is determined by numerical methods, as done by Carretero et al [12] using the Newton method. In addition, setting limits on parameter values is also important to finding soliton solutions for stationary DNLS equations [13]. In a previous research, Qausar et al [14] determined nontrivial onsite soliton produced by cubic-quintic discrete nonlinear Schrödinger using the Trust Region-Dogleg method and taking the parameter values $0.1 \leq w \leq 0.41$ and $0.16 \leq C \leq 9.3$. In this paper, we will determine soliton solutions of stationary DNLS equations using different parameter values and also comparisons of solitons based on differences in selecting the coefficient of cubic nonlinearity B and quintic Q.

2. Materials and Methods

2.1. Stationary DNLS Equation with Cubic-Quintic Nonlinearity

Stationary DNLS form with Cubic-Quintic Nonlinearity can be obtained by performing the substitution $\psi_n(t) = u_n e^{-iwt}$ to the equation (2) with w as the propagation parameter is obtained.

$$wu_n + C(u_{n+1} - 2u_n + u_{n-1}) + B|u_n|^2u_n - Q|u_n|^4u_n = 0. \tag{3}$$

In general, the solution u_n is a complex value, however the soliton solution must satisfy $u_n \to 0$ when $n \to \pm\infty$, therefore the equation (3) can be simplified without loss of generality by considering only real value solutions. So that the equation becomes:

$$wu_n + C(u_{n+1} - 2u_n + u_{n-1}) + Bu_n^3 - Qu_n^5 = 0 \tag{4}$$

with B is the cubic constant, Q is the quintic constant [9]. A soliton is said to be on-site when it peaks in the middle and meets $u_n = u_{-n}$ [5].

2.2. Trust Region Method

A nonlinear system of equations is given which can be written as follows

$$\begin{aligned} F(\mathbf{u}) &= \mathbf{0} \\ \mathbf{u} &= (u_1, u_2, \ldots, u_m) \in \mathbb{R}^m \end{aligned} \tag{5}$$

with $F : \mathbb{R}^m \to \mathbb{R}^m$ is a mapping in the form of $F(\mathbf{u}) = (F_1(\mathbf{u}), F_2(\mathbf{u}), \ldots, F_m(\mathbf{u}))^T$ and for every $n \in \{1, 2, \ldots, m\}$ applies $F_n : \mathbb{R}^m \to \mathbb{R}$ continuous and differentiable. In the context of the stationary DNLS equation with cubic-quintic nonlinearity, define

$$F_n(\mathbf{u}) = wu_n + C(u_{n+1} - 2u_n + u_{n-1}) + Bu_n^3 - Qu_n^5. \tag{6}$$

The vector $\mathbf{u}^*$, which satisfies (5) is called the solution or root of the equation (5). Finding a solution to (5) can be done by finding the root of the nonlinear least-squares problem with no constraints as follows.

$$\min_{\mathbf{u} \in R^m} f(\mathbf{u}) = \frac{1}{2}\|F(\mathbf{u})\|^2, \tag{7}$$

with $\|\cdot\|$ is the Euclidean norm. In order to solve the problem (7) the Trust Region Method is used. The first thing the Trust Region Method does in determining a solution is to construct a model function that can approach F around point $\mathbf{u}_k \in \mathbb{R}^m$ as follows

$$M_k(\mathbf{u}) = F(\mathbf{u}_k) + J(\mathbf{u}_k)(\mathbf{u} - \mathbf{u}_k) \tag{8}$$

with $J(\mathbf{u}_k)$ is the Jacobian matrix of F at point $\mathbf{u}_k$. The problem of determining a solution to the model function $M_k(\mathbf{u})$ is done by applying (7) so that the problem becomes

$$\begin{aligned} \min_{\mathbf{u} \in R^m} m_k(\mathbf{u}) &= \frac{1}{2}\|M_k(\mathbf{u})\|^2 \\ &= \frac{1}{2}\left\|F(\mathbf{u}_k) + J(\mathbf{u}_k)(\mathbf{u} - \mathbf{u}_k)\right\|^2 \\ &= \frac{1}{2}F(\mathbf{u}_k)^T F(\mathbf{u}_k) + (\mathbf{u} - \mathbf{u}_k)^T J(\mathbf{u}_k)^T F(\mathbf{u}_k) + \frac{1}{2}(\mathbf{u} - \mathbf{u}_k)^T J(\mathbf{u}_k)^T J(\mathbf{u}_k)(\mathbf{u} - \mathbf{u}_k). \end{aligned} \tag{9}$$

with $J(\mathbf{u}_k)^T F(\mathbf{u}_k)$ is a gradient of F at point $\mathbf{u}_k$.
Define the *trust region* R_k as follows

$$R_k = \{\mathbf{u} \in \mathbb{R}^m : \|\mathbf{u} - \mathbf{u}_k\| \leq \Delta_k\}. \tag{10}$$

with $\Delta_k > 0$ is called the *trust region radius*. After creating the model function (9), then determine $\mathbf{u} \in R_k$ such that $m_k(\mathbf{u})$ is minimum, mathematically denoted as

$$\mathbf{v}_{k+1} = \arg\min_{\mathbf{u} \in R_k} m_k(\mathbf{u}) \tag{11}$$

equation (11) often referred as *Trust Region Subproblem* [15].

2.2.1. New Point Selection Provisions

Suppose point $\mathbf{v}_{k+1}$ satisfies the *Trust Region Subproblem* (11). Basically, this is the point that should be selected as the next iteration, namely $\mathbf{u}_{k+1}$. However, because $\mathbf{v}_{k+1}$ is calculated based on the model function m_k, this causes the objective function value f in $\mathbf{v}_{k+1}$ to have two possibilities, increasing or decreasing. Therefore, a provision is needed in the selection of $\mathbf{u}_{k+1}$ and in this paper the provisions for selection $\mathbf{u}_{k+1}$ that have been defined by Nocedal and Wright [16] are used, as follows

$$\mathbf{u}_{k+1} = \begin{cases} \mathbf{v}_{k+1} \text{ jika } \frac{f(\mathbf{u}_k)-f(\mathbf{v}_{k+1})}{m_k(\mathbf{u}_k)-m_k(\mathbf{v}_{k+1})} > \eta, \\ \mathbf{u}_k \text{ lainnya,} \end{cases} \tag{12}$$

with $\eta \in (0, 1/4)$ fixed. Note that refusing to select a new point does not stop the algorithm, because by decreasing the trust region radius, $\mathbf{v}_{k+2} \neq \mathbf{v}_{k+1}$ will be obtained.

2.2.2. Updating the Trust Region

The selection of the new trust region R_{k+1} around $\mathbf{u}_{k+1}$ also uses the provisions defined by Nocedal and Wright [16], as follows

$$\Delta_{k+1} = \begin{cases} \frac{\Delta_k}{4} \text{ jika } \frac{f(\mathbf{u}_k)-f(\mathbf{v}_{k+1})}{m_k(\mathbf{u}_k)-m_k(\mathbf{v}_{k+1})} < \frac{1}{4}, \\ \min(2\Delta_k, \Delta_{\max}) \text{ jika } \frac{f(\mathbf{u}_k)-f(\mathbf{v}_{k+1})}{m_k(\mathbf{u}_k)-m_k(\mathbf{v}_{k+1})} > \frac{3}{4}, \\ \Delta_k \text{ lainnya.} \end{cases} \tag{13}$$

This provision is made so that Δ_k does not pass through $\Delta_{\max}$, and also calculate the ratio between the decrease value of the objective function $f(\mathbf{u}_k) - f(\mathbf{v}_{k+1})$ and the model function $m_k(\mathbf{u}_k) - m_k(\mathbf{v}_{k+1})$.

Trust Region Algorithm

- (Initialization) Select $k = 0$, $\Delta_{\max} > 0$, $\Delta_k \in (0, \Delta_{\max})$, $\eta \in (0, 1/4)$, $\mathbf{u}_k \in \mathbb{R}^m$, $\epsilon > 0$.
- While $\|\nabla f(\mathbf{u}_k)\| \geq \epsilon$ repeat
  - Calculate $\mathbf{v}_{k+1}$ as the minimum point on (11).
  - Determine $\mathbf{u}_{k+1}$ using (12).
  - Compute Δ_{k+1} using (13).
  - Establish new functions model $m_{k+1}(\mathbf{u})$.
  - $k \leftarrow k + 1$.
- Done
- Take value $\mathbf{u}_k$.

2.3. Dogleg Method

The dogleg method is a method used to find solutions in the *Trust Region Subproblem*. This method combines the *steepest descent* method through the Cauchy point and the Quasi-Newton method through the Quasi-Newton point in a point selection that makes (9) a minimum. Another interesting point is that this method works with two lines reminding people of the dog's paw, and the knee is at its minimal point in the *steepest descent* direction which is $\mathbf{v}_k^u = \mathbf{u}_k - \alpha_k^u J(\mathbf{u}_k)^T F(\mathbf{u}_k)$, through $\mathbf{v}_k^u$ the dogleg trajectory continues to the Quasi-Newton point $\mathbf{v}_k^{qn}$ [17]. Therefore the *dogleg* path can be described as follows

$$\mathbf{v}(\tau) = \begin{cases} \mathbf{u}_k + \tau(\mathbf{v}_k^u - \mathbf{u}_k) & \text{for } \tau \in [0, 1], \\ \mathbf{v}_k^u + (\tau - 1)(\mathbf{v}_k^{qn} - \mathbf{v}_k^u) & \text{for } \tau \in [1, 2]. \end{cases} \tag{14}$$

Dogleg Algorithm

- Calculate $\mathbf{v}_k^u$

- If $\|\mathbf{v}_k^u - \mathbf{u}_k\| \geq \Delta_k$ stop and take value $\mathbf{v}_{k+1} = \mathbf{u}_k + \frac{\Delta_k}{\|\mathbf{v}_k^u - \mathbf{u}_k\|}(\mathbf{v}_k^u - \mathbf{u}_k)$
- Calculate $\mathbf{v}_k^{qn}$
- If $\|\mathbf{v}_k^{qn} - \mathbf{u}_k\| \leq \Delta_k$ stop and take value $\mathbf{v}_{k+1} = \mathbf{v}_k^{qn}$
- Others
  - Determine τ^* s.t. $\|\mathbf{v}_k^u + \tau^*(\mathbf{v}_k^{qn} - \mathbf{v}_k^u) - \mathbf{u}_k\| = \Delta_k$
  - Take value $\mathbf{v}_{k+1} = \mathbf{v}_k^u + \tau^*(\mathbf{v}_k^{qn} - \mathbf{v}_k^u)$
- Done

3. Results and Discussion

3.1. Selecting B and Q

This research will determining the limits of the parameter values that produce the on-site soliton in cases $B = 1$ and $Q = 0.5$, in addition to determining the limits of the parameters for cases $B = 2$ and $Q = 1$ as a comparison.

3.2. Creating the Model Function

It should be noted that the stationary DNLS equation with cubic-quintic nonlinearity (4) is computed in the domain $n \in \{-N, -N+1, \ldots, N\}$, with $N \in \mathbb{Z}$ and $u_{-N-1} = u_{N+1} = 0$. Therefore $F(\mathbf{u}) = (F_{-N}(\mathbf{u}), F_{-N+1}(\mathbf{u}), \ldots, F_N(\mathbf{u}))^T$ with $\mathbf{u} = (u_{-N}, u_{-N+1}, \ldots, u_N) \in \mathbb{R}^{2N+1}$ and make (5) and (6) into a nonlinear system of equations with $2N + 1$ equations which can be written as follows

$$F(\mathbf{u}) = \begin{bmatrix} F_{-N}(\mathbf{u}) \\ F_{-N+1}(\mathbf{u}) \\ \vdots \\ F_{-1}(\mathbf{u}) \\ F_0(\mathbf{u}) \\ F_1(\mathbf{u}) \\ \vdots \\ F_{N-1}(\mathbf{u}) \\ F_N(\mathbf{u}) \end{bmatrix} = \begin{bmatrix} wu_{-N} + C(u_{-N+1} - 2u_{-N} + 0) + Bu_{-N}^3 - Qu_{-N}^5 \\ wu_{-N+1} + C(u_{-N+2} - 2u_{-N+1} + u_{-N}) + Bu_{-N+1}^3 - Qu_{-N+1}^5 \\ \vdots \\ wu_{-1} + C(u_0 - 2u_{-1} + u_{-2}) + Bu_{-1}^3 - Qu_{-1}^5 \\ wu_0 + C(u_1 - 2u_0 + u_{-1}) + Bu_0^3 - Qu_0^5 \\ wu_1 + C(u_2 - 2u_1 + u_0) + Bu_1^3 - Qu_1^5 \\ \vdots \\ wu_{N-1} + C(u_N - 2u_{N-1} + u_{N-2}) + Bu_{N-1}^3 - Qu_{N-1}^5 \\ wu_N + C(0 - 2u_N + u_{N-1}) + Bu_N^3 - Qu_N^5 \end{bmatrix} = \mathbf{0} \tag{15}$$

and the Jacobian matrix at point $\mathbf{u}_k$ can be written as

$$J(\mathbf{u}_k) = \begin{bmatrix} \frac{\partial F_{-N}(\mathbf{u}_k)}{\partial u_{-N}} & \frac{\partial F_{-N}(\mathbf{u}_k)}{\partial u_{-N+1}} & \cdots & \frac{\partial F_{-N}(\mathbf{u}_k)}{\partial u_N} \\ \frac{\partial F_{-N+1}(\mathbf{u}_k)}{\partial u_{-N}} & \frac{\partial F_{-N+1}(\mathbf{u}_k)}{\partial u_{-N+1}} & \cdots & \frac{\partial F_{-N+1}(\mathbf{u}_k)}{\partial u_N} \\ \vdots & & \ddots & \vdots \\ \frac{\partial F_N(\mathbf{u}_k)}{\partial u_{-N}} & \frac{\partial F_N(\mathbf{u}_k)}{\partial u_{-N+1}} & \cdots & \frac{\partial F_N(\mathbf{u}_k)}{\partial u_N} \end{bmatrix} \tag{16}$$

and for $i, j \in [-N, N]$ applies

$$\frac{\partial F_i}{\partial u_j} = \begin{cases} w - 2C + 3Bu_i^2 - 5Qu_i^4 & \text{if } i = j, \\ C & \text{if } i = j + 1 \text{ or } i = j - 1, \\ 0 & \text{others.} \end{cases} \tag{17}$$

Function model (8) and (9) can be created from (15) and (16).

3.3. Exact On-site Soliton Solution for $C = 0$

Note that if the value is $C = 0$ then the equation (15) becomes:

$$F(\mathbf{u}) = \begin{bmatrix} F_{-N}(\mathbf{u}) \\ F_{-N+1}(\mathbf{u}) \\ \vdots \\ F_{N-1}(\mathbf{u}) \\ F_N(\mathbf{u}) \end{bmatrix} = \begin{bmatrix} wu_{-N} + Bu_{-N}^3 - Qu_{-N}^5 \\ wu_{-N+1} + Bu_{-N+1}^3 - Qu_{-N+1}^5 \\ \vdots \\ wu_{N-1} + Bu_{N-1}^3 - Qu_{N-1}^5 \\ wu_N + Bu_N^3 - Qu_N^5 \end{bmatrix} = \mathbf{0} \quad (18)$$

Solving (18) can be done by finding a solution for each F_n because for every $i, j \in [-N, N]$ and $i \neq j$, the functions F_i and F_j do not have interrelated variables. Rewrite F_n as follows

$$wu_n + Bu_n^3 - Qu_n^5 = 0 \quad (19)$$

- For $B = 2$ and $Q = 1$

$$u_n \in \left\{0, \pm\sqrt{1 \pm \sqrt{1+w}}\right\} \quad \text{with} \quad -1 < w < 0$$

so that the following on-site soliton solutions can be formed

$$u_n = \begin{cases} u_0 = \sqrt{1 + \sqrt{1+w}}, \\ u_{-1} = u_1 = \sqrt{1 - \sqrt{1+w}}, \\ u_i = 0, \quad i = \pm 2, \pm 3, \ldots, \pm N. \end{cases}$$

- For $B = 1$ and $Q = 0.5$

$$u_n \in \left\{0, \pm\sqrt{1 \pm \sqrt{1+2w}}\right\} \quad \text{with} \quad -0.5 < w < 0$$

so that the following on-site soliton solutions can be formed

$$u_n = \begin{cases} u_0 = \sqrt{1 + \sqrt{1+2w}}, \\ u_{-1} = u_1 = \sqrt{1 - \sqrt{1+2w}}, \\ u_i = 0, \quad i = \pm 2, \pm 3, \ldots, \pm N. \end{cases}$$

3.4. On-site Soliton Solution Comparison

Based on research conducted by Carretero et al [12], Chong et al [10] and Kusdiantara [13], soliton solutions in the stationary DNLS equation with cubic-quintic nonlinearity can be obtained in

$$C > 0 \quad \text{and} \quad -1 < w < 0. \quad (20)$$

However, the resulting solitons are not necessarily on-site. Therefore, to find a soliton solution of the on-site type and its constraints, we will examine C and w that meet the (20) with the following steps

1. Selecting pair (B, Q)
   In this research, the value was selected $(B, Q) = (1, 0.5)$ or $(B, Q) = (2, 1)$.

2. Selects C and w values on the constraint (20)
   A point is selected along C and w with a difference of 0.1 so that it can be written $C = 0.1, 0.2, \ldots$ and $w = -0.9, -0.8, \ldots, -0.1$.
3. Enter value B, Q, C, w to F
4. Determine $\mathbf{u}^*$ on F using *Trust Region Dogleg* Method
   The calculation is done by using the MATLAB software program and using the *fsolve subroutine* and selecting $\mathbf{u}_0 = (\text{sech}(-N), \text{sech}(-N+1), \ldots, \text{sech}(N))$ as the initial value.
5. The $\mathbf{u}^*$ value is obtained and then the on-site soliton criterion is checked.
6. The values of C and w are taken as parameter limits if $\mathbf{u}^*$ is an on-site soliton.

Carry out the steps as above, which means that at first the values of $(B, Q) = (1, 0.5)$, $C = 0.1$ are selected and the solution is sought at all values of $w = -0.9, -0.8, \ldots, -0.1$ so that the solution graph is obtained as follows

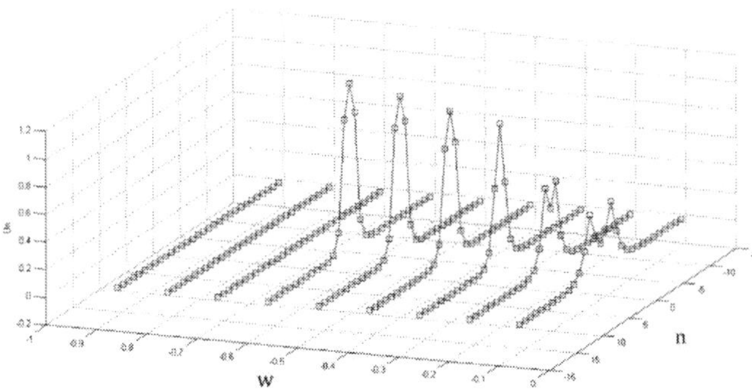

Figure 1. Solutions for value $(B, Q) = (1, 0.5)$, $C = 0.1$ and $w = -0.9, -0.8, \ldots, -0.1$

In Figure 1 it can be seen that the solution that forms the on-site soliton is only at the value of $w = -0.6, -0.5, -0.4, -0.3$,, therefore the values of C and w are taken and entered into Table 1. Furthermore, still with the value of $(B, Q) = (1, 0.5)$, but the value of C is increased to be $C = 0.2$ and the solution is determined at all values of $w = -0.9, -0.8, \ldots, -0.1$ in order to obtain the following solution graph.

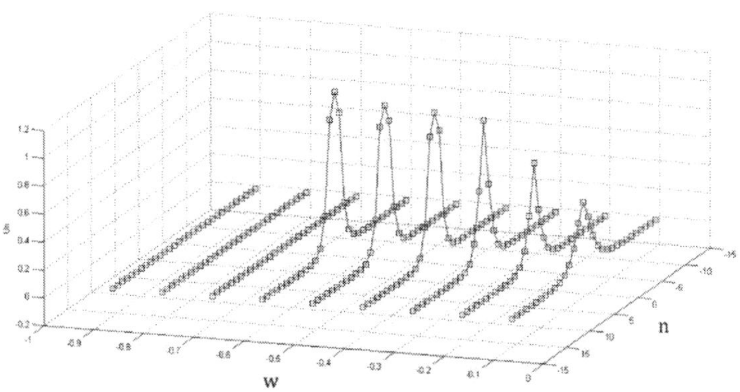

Figure 2. Solutions for value $(B, Q) = (1, 0.5)$, $C = 0.2$ and $w = -0.9, -0.8, \ldots, -0.1$

In Figure 2 it can be seen that the solution that forms the on-site soliton is only at the value of $w = -0.6, -0.5, \ldots, -0.1$,, therefore the values of C and w are taken and entered into Table 1. These steps continue until no C and w are found to form the on-site soliton. The same is done for the value of $(B, Q) = (2, 1)$.. After the simulation with these steps, the on-site soliton limits are obtained as follows.

Table 1. Limitation of on-site soliton parameters for $B = 1$ and $Q = 0.5$

C		w	
min	max	min	max
0.1	0.1	-0.6	-0.3
0.2	0.2	-0.6	-0.1
0.3	0.6	-0.5	-0.1
0.7	1.3	-0.4	-0.1
1.4	2.5	-0.3	-0.1
2.6	3	-0.2	-0.1
3.1	3.7	-0.1	-0.1
3.8	5.1	-0.2	-0.1
5.2	12	-0.1	-0.1

3.5. On-site Soliton Solution Comparison

The following is a graph of the on-site soliton solutions searched using the *Trust Region Dogleg* Method and will be compared based on the selection of constants B, Q and parameter w, C. For convenience, solutions with $B = 1, Q = 0.5$ are labeled with DNLS1 and solutions with $B = 2, Q = 1$ are labelled with DNLS2.

Table 2. Limitation of on-site soliton parameters for $B = 2$ and $Q = 1$

C		w	
min	max	min	max
0.1	0.1	-0.9	-0.8
0.2	0.2	-0.9	-0.6
0.3	0.5	-0.9	-0.3
0.6	1.8	-0.9	-0.1
1.9	2.6	-0.8	-0.1
2.7	3.6	-0.7	-0.1
3.7	5.1	-0.6	-0.1
5.2	7.2	-0.5	-0.1
7.3	7.4	-0.3	-0.1
7.5	10.3	-0.4	-0.1
10.4	11.2	-0.3	-0.1
11.3	13.9	-0.2	-0.1
14	15.7	-0.3	-0.1
15.8	18.7	-0.2	-0.1
18.8	27.6	-0.1	-0.1

4. Conclusions

Based on the results of the research and discussion that has been described, it can be concluded that:

1. The on-site soliton solution for $C = 0$ can be constructed with the following conditions:
   - DNLS2 ($B = 2, Q = 1$) with parameter value constraints $-1 < w < 0$

$$u_n = \begin{cases} u_0 = \sqrt{1 + \sqrt{1 + w}}, \\ u_{-1} = u_1 = \sqrt{1 - \sqrt{1 + w}}, \\ u_i = 0, \quad i = \pm 2, \pm 3, \ldots, \pm N. \end{cases}$$

- DNLS1 ($B = 1, Q = 0.5$) with parameter value constraints $-0.5 < w < 0$

$$u_n = \begin{cases} u_0 = \sqrt{1 + \sqrt{1+2w}}, \\ u_{-1} = u_1 = \sqrt{1 - \sqrt{1+2w}}, \\ u_i = 0, \quad i = \pm 2, \pm 3, \ldots, \pm N. \end{cases}$$

2. The on-site soliton solution for $C \neq 0$ can be found using the parameter constraints listed in Table 2 for DNLS2 and using the parameter constraints shown in Table 1 for DNLS1.
3. In case $C \neq 0$, the maximum limit value of C on DNLS2 which can construct on-site solitons is up to $\approx 2x$ greater than that of DNLS1.

5. Acknowledgements

This research is funded by Pendidikan Magister menuju Doktor untuk Sarjana Unggul, DRPM, with contract number 106/UN11.2.1/PT.01.03/DRPM/2020 and Penelitian Dasar, DRPM, with contract number 80/UN11.2.1/PT.01.03/DRPM/2020.

References

[1] Kevrekidis P G 2009 *The Discrete Nonlinear Schrödinger Equation: Mathematical Analysis, Numerical Computations and Physical Perspectives*. New York City: Springer.
[2] Sato M, Yasui S, Kimura M, Hikihara T, Sievers J 2007 *Lett. J. Explor. Front. Phys* **80** 30002.
[3] Mabrouk S M 2018 *Int. J. Curr. Eng. Technol.* **9** 655-660.
[4] Gninzanlong C L, Ndjomatchoua F T, Tchawoua C 2018 *Chaos: Interdiscip. J. Nonlinear Sci.* **28**043105.
[5] Asfa R, Azadi A F, Netris Z P, Syafwan M 2018 *Journal Mathematics and Its Applications.* **15(2)** pp. 113-126. *(in Bahasa Indonesia).*
[6] Kruse J, Fleischmann R 2017 *J. Phys. B At. Mol. Opt. Phys.* **50** 055002.
[7] Mendonca J T, Tercas H 2013 *Physics of Ultra-Cold Matter: Atomic Clouds, Bose-Einstein Condensates and Rydberg Plasmas*. New York: Springer.
[8] Mejia C, Vicencio R A, Malomed B A 2013 *Phys. Rev. E.* **88** 052901.
[9] Eilbeck J C, Johansson M 2003 *Proceedings of the Third Conference on Localization and Energy Transfer in Nonlinear Systems. Madrid, Spain.* Pp. 44-67.
[10] Chong C, Pelinovsky D E 2011 *Discrete Contin. Dyn. Syst. Ser. S.* **4** 1019–1031.
[11] Chen G, Ma S 2014 *Appl. Math. Comput.* **232** 787–798.
[12] Carretero G, Talley J, Chong C, Malomed B A 2006 *Physica D.* **216** pp. 77-89.
[13] Kusdiantara R 2018 *Homoclinic snaking in discrete systems*. Doctoral Dissertation. University of Essex.
[14] Qausar H, Ramli M, Said M, Mahdhivan S, Hadi S, Vera H 2020 *IAENG Int. J. Appl. Math.* **50** 379-383.
[15] Keyvan A, Kimiaei M, Shiker M 2016 *4OR-Q J Oper Res.* **14** 133-152.
[16] Nocedal J, Wright S J 2006 *Numerical Optimization*. New York City: Springer.
[17] Abaye Y M 2018 *Trust Region Newton with Dogleg Method.* Doctoral Dissertation. Addis Ababa University.

Performance analysis of mAlexnet by training option and activation function tuning on parking images

S Rahman[1], M Ramli[2*], F Arnia[3], R Muharar[3], A Sembiring[4]

[1]Doctoral Program, School of Engineering, Universitas Syiah Kuala, Banda Aceh
[2]Department of Mathematics, Universitas Syiah Kuala, Banda Aceh
[3]Department of Electrical and Computer Engineering, Universitas Syiah Kuala, Banda Aceh
[4]Department of Informatics Technology, Universitas Harapan Medan

*E-mail: marwan.math@unsyiah.ac.id

Abstract. Information on the availability of parking spaces is one of the main needs in urban areas. This information can minimize the impact of vehicle growth, including exhaust gas emissions, traffic jams, and fuel use. In general, the detection of parking space availability can be done in two ways, namely the sensor system and computer vision. Computer vision monitoring of parking spaces is more promising for use in the future. A single-camera can monitor multiple parking spaces by making use of computer vision. CNN is a technique for parking space classification. CNN's pre-trained dedicated to parking space classification is mAlexnet. mAlexnet can classify parking spaces well, but not perfect. in this paper, we try to observe and improve the performance of mAlexnet. We tried training options and activation function tuning. From the results of testing the combination of SGDM training option and the LeakyReLu activation function, the performance of mAlexnet improves.

1. Introduction

In recent years, information on the availability of parking spaces has been urgently needed [1]. The high growth in the number of cars raises several problems that must be resolved immediately. With the availability of information on parking spaces, it can overcome traffic congestion problems caused by drivers looking for the nearest parking location. Information on the availability of parking spaces can also save fuel consumption by vehicles. Fuel is a world problem that must be saved. in addition, by not moving around the car looking for a parking space will minimize the impact of vehicle exhaust emissions [2,3].

The space availability information is the result of real-time information extraction from parking spaces. Various ways were developed to convey this information, including by using sensor systems and smart cameras. The use of ultrasonic sensors can solve this problem, but in its implementation, it requires expensive installation and maintenance costs [4]. Using a smart camera or computer vision is a way to hope for the future. Smart cameras are able to recognize parking spaces simultaneously with a single camera. However, in the implementation of the smart camera, there are more challenges such as changes in light in parking spaces, the presence of obstacles such as trees and poles, and the use of a large amount of computing.

Extraction of information on the availability of parking spaces using cameras has been developed, in general, it is divided into two, namely by marking the parking slot and detecting directly without a marker. Recognize the mark given on the parking slot [5,6], if there is, the slot is still free and if the mark is not detected then it is busy. This method is the same as using a sensor, it needs installation and maintenance so that the mark remains visible on the camera. The most promising step is how to recognize parking spaces without markers. Convolutional neural network (CNN) is a technique in machine learning that can extract features and classify data [4,7]. Parking space classification is very suitable using CNN to distinguish between busy and free spaces. CNNs specifically developed for parking space classification are mLenet and mAlexnet which are Lenet and Alexnet mini networks [4,7]. Testing of parking space images recommends mAlexnet because it is more accurate in parking space classification. In this study, we analyzed the performance of mAlexnet if the training options and their activation functions were changed. Changing training options and activation functions is expected to improve the performance of mAlexnet.

2. Convolutional Neural Network (CNN)

CNN is part of Deep Learning. CNN is a multilayer perceptron which is designed to process image data. CNN is often used to extract features and classify objects in an image. The way CNN works is to extract the features on the raw image at the convolutional network layer and the activation function followed by pooling. The features of this layer are sent to the next layer and so on. The last layer is assigned to classify features, features are classified as fully connected [8]. An example of the CNN architecture is shown in Figure 1.

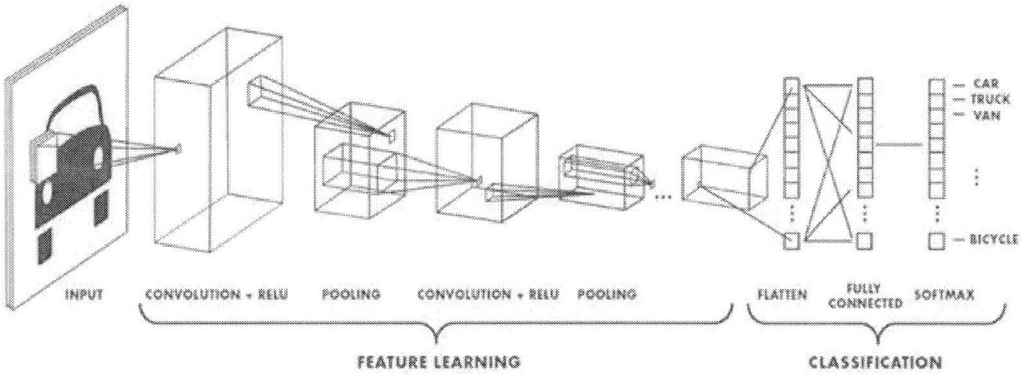

Figure 1. CNN Architecture.

2.1 Training option CNN

Training options that are often used in deep learning are sgdm (stochastic gradient descent with momentum), adam (adaptive moment estimation), and RMSProp (root mean square propagation). Each training option both SGDM, Adam and RMSProp has its advantages [9,10]. In this study, a comparison of the performance of each training option was carried out. Training options are best used to improve the performance of mAlexnet. After the training options are determined, the next step is to compare the use of activation functions on the CNN architecture.

2.2 Activation function CNN

Artificial neural networks generally have an activation function. The activation function is useful for directing the input to the desired output. CNN is part of deep learning which has layer end activation function. Activation functions that can be used include ReLU, ELU, LeakyReLU and tanh [11–13]. After we found the best training option, we tried each activation function. The combination of training options and activation functions is expected to improve the performance of mAlexnet. The following activation function graphs are shown in Figure 2.

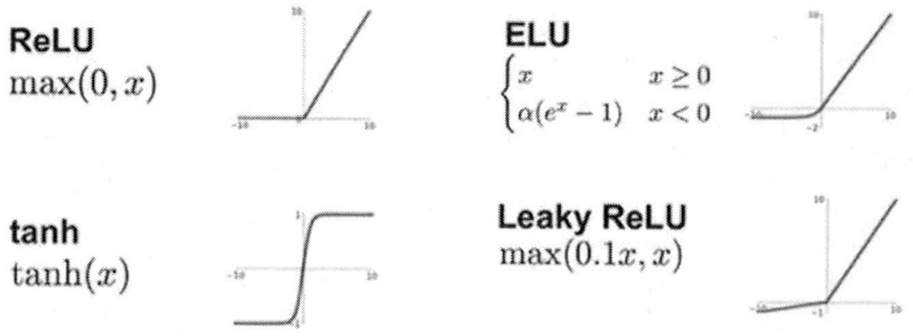

Figure 2. Activation function graphs.

2.3 mAlexnet

mAlexnet is CNN pre-trained specifically for the needs of parking space classification. MAlexnet architecture has three convolutional layers namely conv-1, conv-2 and conv-3. Each layer followed by ReLu, LRN and maxpolling which are useful for feature extraction. In the feature classification section, there are two fully connected ones given ReLu between them [7]. mAlexnet architecture is shown in Figure 3.

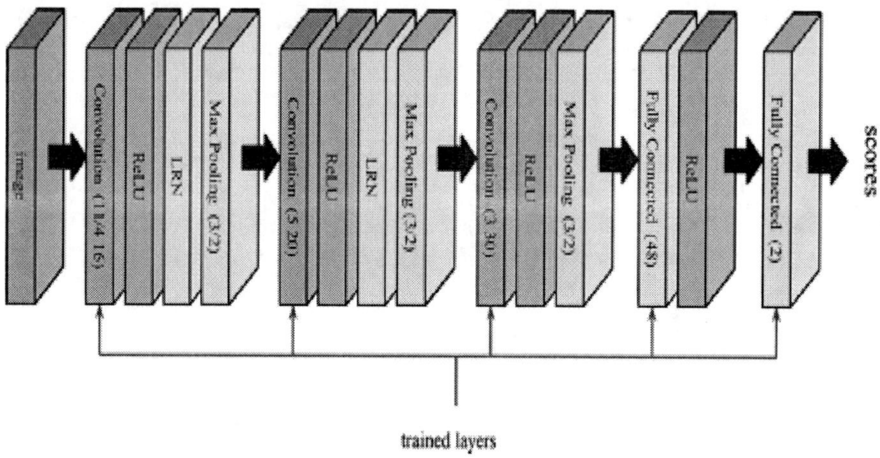

Figure 3. mAlexnet architecture.

3. Result and discussion

Comparative testing of the use of training options and activation functions using the CNR-Park dataset which was taken randomly as many as 4000 images measuring 150 x 150 pixels. The dataset is labeled busy and free, each with 2000 images. The busy label is for images of parking spaces filled with cars and free for parking spaces that are not filled with cars. Dataset with a size of 150 x 150 pixels is shown in Figure 4.

Figure 4. Image dataset.

3.1 Training options tuning

The training model in neural networks is one of the determining factors in the success of a neural network architecture in object classification. In CNN deep learning there are several training options that are often used, namely SGDM, Adam, and RMSProp. We tried to set training options to measure the performance of mAlexnet, which is a CNN pre-trained specifically for parking space classification. In this experiment, we did not change the architecture of mAlexnet. The experimental results show that the accuracy rate of validation using SGDM is 97.77%, ADAM is 97.45%, and RMSProp is 96.73%. From these results, SGDM is better for mAlexnet training for parking space classification. Figure 5 shows the validation graph in each iteration.

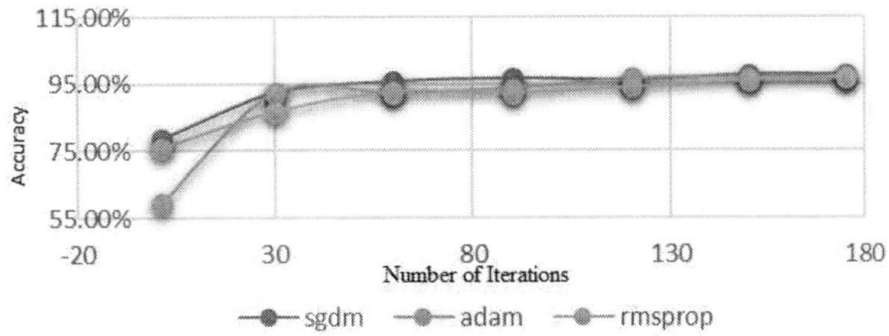

Figure 5. Performance of the training options

3.2 Activation functions tuning

From the previous test, we used SGDM in the mAlexnet training process. We experimented using several activation functions. The activation functions used are ReLU, ELU, LeakyReLU, and tanh. From the experimental results, we note that the performance of ReLU is 97.77%, ELU is 97.30%, LeakyReLU is 98.34%, and tanh is 93.04%. Overall, the results of each experiment using the activation function are shown in Table 1.

Table 1. The performance of the activation function.

Activation Function	Iteration							
	1	30	60	90	120	150	175	Final
ReLU	78.79%	92.36%	95.58%	96.57%	96.00%	97.56%	97.51%	97.77%
ELU	78.85%	91.37%	94.44%	96.31%	97.40%	97.30%	97.09%	97.30%
Leaky ReLU	76.72%	94.85%	95.95%	97.25%	97.19%	97.92%	98.13%	98.34%
Tanh	74.17%	86.85%	90.75%	92.67%	94.13%	92.52%	92.83%	93.04%

The use of the LeakyReLu activation function gets better results than other activation functions as shown in Table 1. LeakyReLU is the only one that gets stable training results and increases with each iteration. The following is a graph of the performance of the activation function shown in Figure 6.

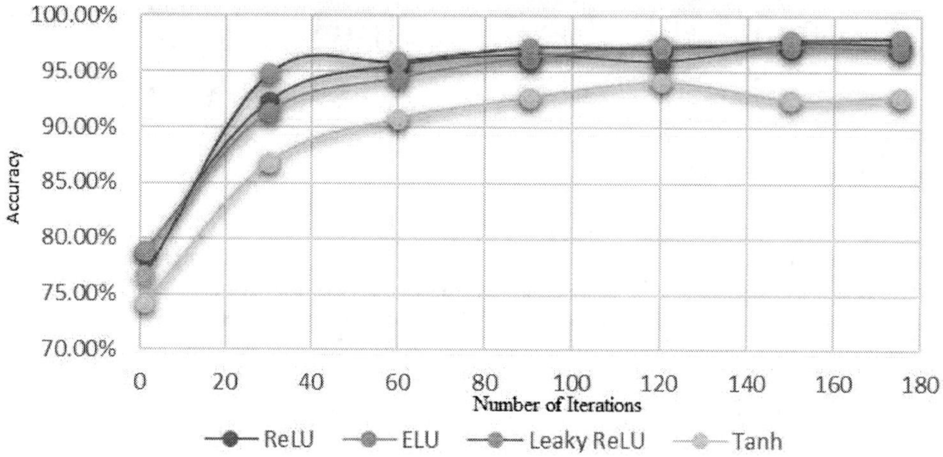

Figure 6. The performance of the activation function in improving validation accuracy.

4. Conclusion

Experiments were carried out using the CNRPark dataset which was taken 4000 images randomly. The test was conducted to review the performance of training options and activation functions. From the observation that the use of SGDM training options is better than the others. Then we tried to change the activation function trained with sgdm. the LeakyReLU activation function has better performance than other activation functions. It can be concluded that the use of the LeakyReLU activation function and training with SGDM can improve the performance of mAlexnet.

5. Acknowledgments

This research is funded by Professor Research Grant, Universitas Syiah Kuala 2020, with contract number 3/UN.11.2.1/PT 01.03/PNBP/2020.

References

[1] Lin T, Rivano H and Le Mouël F 2017 *IEEE Transactions on Intelligent Transport. Syst.,* **18** 3229–3253.

[2] Sukarno I, Matsumoto H and Susanti L 2016 *Future Cities Environ.,* **2** 1-11.

[3] Kakouei A, Vatani A and Idris A K B 2012 *J. Environ. Health Sci. Eng.,* **9** 1-5.
[4] Amato G, Carrara F, Falchi F, Gennaro C, Meghini C and Vairo C 2017 *Expert Syst. Applicat.,* **72** 327–334.
[5] Delibaltov D, Wu W, Loce R P and Bernal E A 2013 *16th International IEEE Conference on Intelligent Transport. Syst. (ITSC 2013),* 2387–92.
[6] Rahman S, Ramli M, Arnia F, Muharar R, Luthfi M and Sundari S 2020 *J. Physics: Conf. Ser.,* **1566** 12092.
[7] Amato G, Carrara F, Falchi F, Gennaro C, and Vairo C 2016 *2016 IEEE Symp. Comput. Communic. (ISCC),* 1212–17.
[8] Yamashita R, Nishio M, Do R K G, and Togashi K 2018 *Insights Imag.,* **9** 611–629.
[9] Dogo E M, Afolabi O J, Nwulu N I, Twala B and Aigbavboa C O 2018 *Int. Conf. on Comput. Tech. Electr. Mech. Syst. (CTEMS),* 92–99.
[10] Jiang X, Hu B, Satapathy C S, Wang S H and Zhang Y D 2020 *Sci. Program.,* **2020** 1-36.
[11] Dewa C K 2018 *Procedia Computer Sci.,* **144** 124–132.
[12] Zhang Y, Hua Q, Xu D, Li H, Bu Y and Zhao P 2019 *IGARSS 2019-2019 IEEE Int. Geosci. Remote Sensing Symp.,* 10023–26.
[13] Dubey A K and Jain V 2019 *Appl. Comput. Auto. Wireless Syst. Electric. Eng.,* **553** 873–880

The comparison of k-modes clustering and ROCK clustering to the poverty indicator in Samadua Subdistrict, South Aceh

H Sofyan[1], M Iqbal[1,*], M Marzuki[1], M Muhammad[2]
[1]Statistics Department, Syiah Kuala University, Indonesia
[2]Regional Planning and Development Agency of Aceh Province (Bappeda), Indonesia

*E-mail: m.iqbal@mhs.unsyiah.ac.id

Abstract. The cluster methods including k-modes clustering and ROCK clustering are rarely used for analyzing categorical data. Much of the research data, however, are categorical, such as the poverty indicator dataset. Many studies have examined each of these two methods, but there is still limited study for comparing them. This study compared between k-modes clustering and ROCK clustering on a dataset of poverty indicators at a category scale. There were 1449 households with the lowest 40% income in Samadua Sub-district, South Aceh Regency were used as data in this study. The result of this research was that the optimal number of clusters for k-modes clustering was 2. Similarly, the optimal number of clusters for ROCK clustering was 2 with a threshold value (θ) of 0.29. The SW/SB and R-Squared values for the k-modes clustering were 0.0149 and 0.7558, respectively, compared to 0.0682 and 0.5732 for ROCK clustering. From these two values, it can be concluded that the k-modes clustering is better than ROCK clustering. The significant difference between the two clusters based on deciles of households with the lowest 40% income of the k-modes clustering is ownership of defecation facilities, ownership of television and motorbikes.

1. Introduction

Grouping data (observation) into a group / cluster based on the same characteristics which is similar, is the method of statistics which commonly used nowadays. This grouping method was conducted by the basis of similarity from each observation. This grouping will be more convenient when the relationship between each characteristic has been analyzed.

The Cluster Method is based on the type of dataset that will be used. If the dataset is numeric, therefore the cluster method that is used such as hierarchical clustering, non-hierarchical clustering (k-means, k-median), and fuzzy clustering (fuzzy c-means). Whereas if the dataset is category type, therefore the cluster method that can be used are cluster k-modes and Robust Clustering using Links (ROCK clustering).

The k-modes is one of the clustering methods that is used for the category dataset. k-modes is modification method from K-means method, which in k-means method the data type should be numeric ratio or interval. Dataset of category type is commonly based on the certain criteria in conducting the clustering. Those criteria will be later formed the categories in each variable of indicator [1].

Besides the k-modes, there is also ROCK clustering method that is commonly known for the category type of data. It is compatible for the extreme data (outlier). This method is common for applying the link concept as the similarity measurement in the cluster forming. The link value is determined by the number of threshold ("θ") where it range between 0 until 1 [2, 3].

One of the highest poverty rate provinces in Indonesia is Aceh. The poverty in Aceh in 2018, according to The Central Bureau of Statistic (BPS) of Aceh, carry through 15.97% (839 thousand inhabitants). This number has decreased from 2017. Moreover, this number is the lowest poverty rate since 2002. However, the poverty number in Aceh remains distant above the national number of povertys, which is 9.82% [4].

South Aceh is one of the 23 Districts/Cities in Aceh Province. This district has poverty rates as much as 14.07% (32.51 inhabitants) on 2017. The number remains in higher position than the national poverty percentage, yet under the Aceh Province, namely 15.97%. South Aceh become the 5 lowest districts according to the poverty percentage rate of Aceh Province [5].

Samadua is one of the subdistricts in South Aceh which located close to the district capital, Tapaktuan. Samadua is the seventh most populated subdistrict in 2017, the population reached to 15.537 inhabitant. The growth of the population in Samadua is the third least, it was 0.54 in 2016 until 2017 [4]. The number of households that earn income that least than 40% in Samadua reach 1449 habitants, with the total population of 5608 inhabitant (Integrated Database, 2015).

Another poverty study was conducted by Ulfah and 'Uyun [6]. It concerned on the group determination of impoverished household (RTM) in Special Region of Yogyakarta by using the k-means and fuzzy c-means (FCM) methods. There were 15 poverty indicator which were examined which was separated into 10 aspect, namely the income, food, clothing, residence, clean water, health, education, property, lighting, and inhabitant number. The obtained result from the study is that k-means method was better compared to the FCM method since the obtained accuracy value for k-means was 83.33% (FCM accuracy value was 50.00%). This conclusion is similar to the conducted research by Hidayat et al. [7]. The k-means method is more proper to be used for grouping in poverty in Central Java.

This research was conducted in two part, namely descriptive analysis and group determination. The descriptive analysis was conducted to analyze the figure of poverty in Samadua Subdistrict, South Aceh District. The group determination was conducted by determining the optimum group and the best group by using k-modes and ROCK clustering for the grouping of poverty household numbers. This research was expected to provide the statistical information for the poverty household numbers grouping in Samadua Subdistrict, South Aceh District.

2. Materials and method

The data that had been used for this research was collected from Integrated Database (BDT in Indonesia) of The Regional Planning and Development Agency (BAPPEDA) of Aceh Province in the year of 2015. Those are 1449 household data which had 40% lowest income in Samadua Subdistrict, South Aceh District. The data consist of poverty indicators which were assigned by BPS. The poverty indicators that were used for this research is only 12 indicators, namely the floor type on shelter (X_1), the type of wall of the shelter (X_2), the disposal facility usage (X_3), the main lighting source (X_4), the main source of drinking water (X_5), the main fuel for cooking (X_6), the ownership of KIS/BPJS Health Insurance/Jamkesmas (X_7), the income source of the household head (X_8), the education background of the household head (X_9), the ownership of refrigerator (X_{10}), the ownership of television (X_{11}), and the ownership of motorcycle bike (X_{12}).

The data analysis steps which were conducted started from the data coding for each variable, descriptive analysis, optimal clusters determining from k-modes clustering and ROCK clustering based on the SW/SB ratio value and R-squared, conducting the k-modes clustering and ROCK clustering comparison method based on the SW/SB ratio value and R-squared [8], and conducting the further analysis according to the optimal method obtained.

3. Discussion analysis

3.1. Descriptive analysis
This research was using 12 predictor variables from the poverty indicators, village variables, and welfare rate variables. Entire predictor variables were analyzed descriptively by observing the mode from the 40% lowest income households in Samadua Subdistrict, South Aceh District, as shown in Table 1. The 40% lowest income households generally had the unplastered floor/red brick floor with the percentage of 88.06% (1276 households). For the type of the shelter wall, it is dominated by the regular wall as much as 83.85% (1215 households).

Table 1. Modes for predictor variables

Variables	Mode	Numbers	Persentage (n=1449)
Floor type of the shelter	Unplastered floor / red bricks	1276	88.06%
Wall type of the shelter	Regular wall	1215	83.85%
Disposal facility usage	Private use	687	47.41%
The ain lighting source	PLN Electricity	1418	97.86%
The main drinking water source	Protected water well	510	35.20%
The main cooking fuel	Firewood	912	62.94%
Ownership of KIS/ BJS Health Insurance / Jamkesmas	No	869	59.97%
The main head of household income source	Daily construction worker	650	44.86%
Head of household education background	Elementary / Extraordinary disable elementary school	839	57.90%
Ownership of refrigerator	No	1075	74.19%
Ownership of television	Yes	950	65.56%
Ownership of motorcycle bike	No	808	55.76%

The 40% lowest income households generally had their own facility of disposal (47.41%), they also used PLN electricity as the main source of lighting (97.86%), used the protected water well to as the source of drinking water (35.20%), used firewood as the main cooking fuel (62.94%). The 40% lowest income households also did not have any access to KIS/ BPJS Health Insurance/ Jamkesmas (59.97%), did not own any refrigerator (74.19%), did not own any motorcycle bike (55.76%), and did not own any television (65.56%). Whereas for the household main income source variable, approximately they worked as daily constructional workers (44.86% household), and their highest education background only reached Elementary/ Extraordinary elementary school for disable person (57.90%).

3.2. Cluster analysis

3.2.1. K-modes clustering
The clusters numbers determination on k-modes clustering was conducted by observing the ratio value from standard deviation on the group (SW) and the standard deviation between group (SB). The cluster numbers in this research range between 2 until 10. Then, the comparison of ratio value from each cluster was conducted. The clusters numbers with the least ratio is the optimal cluster numbers. Ratio between SW and SB from each cluster number was shown on Table 2.

Table 2. Ratio value of SW/SB and R-squared based on the cluster numbers on k-modes clustering

No	Cluster numbers (k)	Ratio (SW/SB)	R-squared
1	**2**	**0.0149**	**0.7558**
2	3	0.0304	0.5999
3	4	0.0355	0.6217
4	5	0.0498	0.5274
5	6	0.0417	0.6656
6	7	0.0928	0.3259
7	8	0.0872	0.3901
8	9	0.0960	0.3762
9	10	0.1032	0.3699

The determination of optimum cluster number is also regarded to the R-squared value. R-squared is the ratio of squares between groups numbers (SSB) with the total numbers of squares (SST). If the R-squared value resulted is close to 1, the better formed clusters. R-squared which was formed from each cluster is shown in Table 2.

The optimal cluster numbers are 2 clusters (k=2), it has the least SW/SB ratio, namely 0.1049. This is also supported by the R-squared value which is high, namely 0.7558 or 75.58%. It can be concluded that if the cluster numbers are 2 clusters (k=2), it is the optimal clusters numbers that are formed.

3.2.2. ROCK clustering

The determination of optimal cluster numbers on ROCK clustering is similiar to the determination of optimal cluster numbers on k-modes clustering, namely by determining the SW/SB ratio with the R-squared. On ROCK clustering, first of all, we need to determine the maximum number of clusters which available to be formed. The optimum cluster numbers will be formed after the threshold or theta (θ) value is given. In this research, the maximum numbers of clusters formed are 10 clusters. Whereas, for the threshold are 0.01, 0.05, 0.10, 0.11, 0.12, 0.13, 0.14, 0.15, 0.16, 0.17, 0.20, 0.21, 0.22, 0.23, 0.24, 0.25, 0.26, 0.27, 0.29, and 0.30.

The threshold values of 0.31, 0.32, 0.33, 0.34, and 0.35 were not analyzed due to the similar result as the 0.30. The threshold 0.29 has the least SW and SB ratio, namely 0.0682 as Table 3 summarizes. The R-squared value which was formed is 0.5732 or 57.32%. This R-squared value was the highest R-squared which was approached to 1. By the threshold value of 0.29, hence the optimal cluster numbers formed are 2 clusters.

Table 3. SW/SB ratio and R-squared value based on the cluster numbers in the ROCK clustering

No.	*Threshold* (θ)	Number of formed *clusters* (*k*)	SW/SB	*R-squared*
1	0.01	2	0.3518	0.0478
2	0.05	2	0.3225	0.0567
3	0.10	8	0.0813	0.4861
4	0.11	8	0.1583	0.1997
5	0.12	8	0.1086	0.3464
6	0.13	8	0.1836	0.1564
7	0.14	6	0.1021	0.3750
8	0.15	2	0.0807	0.4900
9	0.16	2	0.0896	0.4320
10	0.17	2	0.1243	0.2880
11	0.18	2	0.0721	0.5458

12	0.19	2	0.3014	0.0644
13	0.20	2	0.1790	0.1633
14	0.21	5	0.0992	0.3887
15	0.22	5	0.0892	0.4367
16	0.23	5	0.0830	0.4759
17	0.24	5	0.0793	0.4985
18	0.25	4	0.1040	0.3666
19	0.26	5	0.0779	0.5074
20	0.27	3	0.0713	0.5516
21	0.28	4	0.0797	0.4961
22	0.29	2	0.0682	0.5732
23	0.30	1	0.4086	0.0361

3.2.3. Determination method of optimum clusters

The optimum determination method was conducted to compare k-modes clustering with the optimal cluster numbers and ROCK clustering with optimal cluster numbers. There were 2 cluster for the cluster (k) on k-modes clustering method, and so were for the ROCK clustering method, comparison of both methods used the measurement instrument of SW/SB ratio and R-squared. The comparison for the determination of optimum method in this research is shown in Table 4. The least SW/SB ratio value which was obtained by k-modes clustering was 0.0149. The R-squared value from k-modes clustering is also the highest compared to the ROCK clustering, namely 0.7558 or 75.58%. Whereas, the value of SW/SB and R-squared from ROCK clustering consecutively were 0.0682 and 0.5732 or 57.32%.

Table 4. The best method determination

Measurement Instruments	Cluster Methods	
	K-modes clustering (k = 2)	ROCK clustering (k = 2)
SW/SB	**0.0149**	0.0682
R-squared	**0.7558**	0.5732

3.3. Further analysis of k-modes clustering

The grouping of the 40% lowest income households resulting 2 optimal clusters by using k-modes clustering. The number of whole grouped household were 1449 households. The obtained grouping result are in cluster 1, there were 790 households and there were 659 households in cluster 2. These are the mode from each variable in each cluster.

Table 5. Modes of each cluster on k-modes clustering

No.	Variables	Cluster 1	Cluster 2
1	Floor type of the shelter	unplastered floor/ red bricks	unplastered floor/red bricks
2	Wall type of the shelter	Regular wall	Regular wall
3	Disposal facility usage	No disposal facility	Private use
4	The main lighting source	PLN Electricity	PLN Electricity
5	The main drinking water source	Pumped well	Pumped well
6	The main cooking fuel	Firewood	Firewood
7	Ownership of KIS/ BJS Health Insurance/Jamkesmas	Yes	Yes
8	The main head of household income source	Daily construction workers	Daily construction workers

9	Head of household education background	Elementary / Extraordinary disable elementary school	Elementary / Extraordinary disable elementary school
10	Ownership of refrigerator	No	No
11	Ownership of television	No	Yes
12	Ownership of motorcycle bike	No	Yes

Table 5 shows the modes for each variable on each cluster. The floor type, the wall type, the main source of lighting, the main source of drinking water, the main cooking fuel, the ownership of KIS/BPJS Health Insurance/Jamkesmas, the household head main income, the Head of household education background, and the ownership of refrigerator have the similarity modes both in cluster 1 and cluster 2. Households in both of these clusters generally own the unplastered floor/red bricks floor type, regular wall type, the main lighting source was from PLN electricity, the main drinking water source was from pumped well. The main cooking fuel was firewood, they owned KIS/ BJS Health Insurance/Jamkesmas, the main income of household head was from daily constructions workers, the education background was Elementary / Extraordinary disable elementary school, and approximately they did not own any refrigerator at their houses. However, for the variable of disposal facility, television ownership, and motorcycle bike ownership have different modes.

Households on cluster 1 generally did not have disposal facility. These cluster is also did not own television and did not own any motorcycle bike. Whereas, on cluster 2, generally, the households have their own disposal facility privately. Cluster 2 also had the television and motorcycle bike. The percentage of househoulds in each cluster is shown in Table 6.

The village of Payonan Gadang, Lubuk Layu, Suaq Hulu, Luar, Ujung Kampung, Air Sialang Hulu, Subarang, Gunung Ketek, Kota Baru, Madat, Dalam, Tengah, Baru, Ladang Kasik Putih, Ladang Panton Luas, Alur Simerah, Gunung Cut, dan Batee Tunggai had the households that generally classified on cluster 1. The characteristic on this cluster 1 were the non-existence of disposal facilty, no ownership of television, and no ownership of motorcycle bike. Whereas, in Ujung Tanah, Air Sialang Tengah, Air Sialang Hilir, Balai, Gadang, Alur Pinang, and Kuta Balang Villages, they are conculed to cluster 2. The cluster 2 characteristics were generally the existence of private disposal facility, ownership of television and motorcycle bike. However, Tampang, Jilatang, and Arafah Villages did have similarity on numbers of households between cluster 1 and cluster 2.

Cluster 1 had 790 households (Table 7). The number of households consecutively by welfare rates from the highest to the least are decile 3, decile 2, decile 4, and decile 1. Decile 2 and decile 4 have nearly similar number of households. The difference between decile 1 and decile 3 reached 18.73%. The number of households on decile 1, decile 2, decile 3, and decile 4 consecutively were 109 households (13.80%), 219 households (27.72%), 257 households (32.53%), dan 205 households (25.95%).

Table 6. The percentage of each cluster households

No.	Villages	Cluster 1	Cluster 2	Cluster 1 Percentage	Cluster 2 Percentage
1	Ujung Tanah	16	18	47%	53%
2	Payonan Gadang	57	42	58%	42%
3	Lubuk Layu	33	30	52%	48%
4	Suaq Hulu	41	33	55%	45%
5	Luar	38	20	66%	34%
6	Tampang	13	13	50%	50%
7	Jilatang	36	36	50%	50%
8	Ujung Kampung	9	8	53%	47%

9	Air Sialang Hulu	30	22	58%	42%
10	Air Sialang Tengah	27	32	46%	54%
11	Air Sialang Hilir	30	37	45%	55%
12	Subarang	34	16	68%	32%
13	Gunung Ketek	24	21	53%	47%
14	Kota Baru	29	19	60%	40%
15	Madat	21	12	64%	36%
16	Dalam	14	8	64%	36%
17	Tengah	39	17	70%	30%
18	Balai	9	15	38%	63%
19	Baru	30	29	51%	49%
20	Gadang	8	9	47%	53%
21	Ladang Kasik Putih	14	11	56%	44%
22	Ladang Panton Luas	30	11	73%	27%
23	Alur Simerah	33	19	63%	37%
24	Alur Pinang	39	51	43%	57%
25	Gunung Cut	32	22	59%	41%
26	Kuta Blang	52	65	44%	56%
27	Batee Tunggai	48	39	55%	45%
28	Arafah	4	4	50%	50%

Cluster 2 had 659 households (Table 7). The number of households consecutively by welfare rates from the highest to the least are decile 4, decile 3, decile 2, and decile 1. Decile 2 and decile 4 have nearly similar number of households. The difference between decile 1 and decile 4 reached 54.63%. The number of households on decile 1, decile 2, decile 3, and decile 4 consecutively were 13 households (1.97%), 78 households (11.84%), 195 households (29.59%), and 373 households (56.60%).

Cluster 1 had higher proportion number of household percentage on decile 1 than the cluster 2. So, did on the decile 2 and decile 3. The comparison of proportion number of household percentage between cluster 1 and cluster 2, for decile 1, decile 2, and decile 3 consecutively were 13.80%:1.97%, 27.72%:11.84%, dan 32.53%:29.59%. Whereas, for decile 4, cluster 2 had higher proportion number of household percentage than cluster 1. The difference of the proportion number of household percentage on decile 4 reached 30.65% with the proportion number of household percentage on cluster 1 and cluster 2 consecutively were 25.95% and 56.60%.

Table 7. Comparison of welfare rates on each cluster

Welfare Status	Cluster			
	1		2	
	Number	Percentage	Number	Percentage
Decile 1	109	13.80%	13	1.97%
Decile 2	219	27.72%	78	11.84%
Decile 3	257	32.53%	195	29.59%
Decile 4	205	25.95%	373	56.60%
Total	790	100.00%	659	100.00%

4. Conclusions

The comparison between two methods of grouping namely k-modes clustering and ROCK clustering were conducted for 12 poverty indicators in Samadua Subdistrict, South Aceh District. The result of the analysis provided several following conclusions:

- By using grouping method of k-modes clustering, 2 optimum clusters (k=2) were obtained. Whereas by using grouping method of ROCK clustering, 2 optimum clusters (k=2) were also obtained at the threshold value of 0.29 (θ=0.29).
- By the comparison between k-modes clustering with k=2 and ROCK with k=2 clustering, obtaining k-modes clustering is the best method since k-modes clustering had SW/SB ratio value lower than the ROCK clustering, namely 0.0149 (k-modes clustering) and 0.0682 (ROCK clustering). The R-squared value of k-modes clustering was higher than the ROCK clustering's, namely 0.7558 (k-modes clustering) and 0.5732 (ROCK clustering).
- Cluster 1 and cluster 2 has the similar mode on the variables of the floor type of shelter, the wall type of shelter, the main source of lighting, the main source of drinking water, the main cooking fuel, the ownership of KIS/BPJS Health Insurance/Jamkesmas, the main income source of household heads, the education background of the household heads, and the ownership of refrigerator with the each variable category consecutively were unplastered floor/ red bricks, regular wall, PLN electricity, pumped well, firewood, owning the access for KIS/BPJS Health Insurance/Jamkesmas, daily construction workers, Elementary / Extraordinary Elementary School, and the non-ownership of refrigerator. Whereas for the variable of the use of disposal facility, households on cluster 1 generally did not own the disposal facility and cluster 2 did have the disposal facility privately. Households in cluster 1 also generally did not own any television and motorcycle bike. Otherwise, households on cluster 2 own the television and motorcycle bike. This could be concluded that it is necessary to renovate the disposal facility in Samadua Subdistrict since the 40% lowest income households which did not own disposal facility privately reached 52.59%, particularly for households on cluster 1.

References

[1] Huang J Z 2009 *Encyclopedia of Data Warehousing and Mining, Second Edition.* Pennsylvania: IGI Global.

[2] Guha S, Rastogi R, Shim K 1999 *Proceed. the 15th Int. Conf. on Data Eng.*, 512–521.

[3] Tyagi A, Sharma S 2012 *Int. J. Comp. Sci. Eng. (IJCSE),* **4** **(5)** 809–815.

[4] Badan Pusat Statistik Kabupaten Aceh Selatan 2018 *Kabupaten Aceh Selatan dalam Angka 2018.* Tapaktuan: Badan Pusat Statistik Kabupaten Aceh Selatan. *(in Bahasa Indonesia).*

[5] Badan Pusat Statistik Provinsi Aceh 2017 *Aceh dalam Angka 2017*. Banda Aceh: BPS Aceh. *(in Bahasa Indonesia).*

[6] Ulfah A N, 'Uyun S 2015 *Jurnal Teknik Informatika dan Sistem Informasi,* **1** **(2)** 139–148. *(in Bahasa Indonesia).*

[7] Brock G, Pihur V, Datta S, Datta S 2008 *Solid State Comm.,* **25** **(4)** 371–372.

[8] Dewi A, Sutikno, Kuswanto H 2012 *Metode Cluster untuk Pengelompokan Desa Pedesaan di Provinsi Riau.* **4** 1–16.

AUTHOR INDEX